ENVIRONMENTAL INVESTIGATION AND REMEDIATION

1,4-DIOXANE AND OTHER SOLVENT STABILIZERS

ENVIRONMENTAL INVESTIGATION AND REMEDIATION

1,4-DIOXANE AND OTHER SOLVENT STABILIZERS

Thomas K.G. Mohr

with chapters by
Julie A. Stickney and William H. DiGuiseppi

CRC Press
Taylor & Francis Group
Boca Raton   London   New York

CRC Press is an imprint of the
Taylor & Francis Group, an **informa** business

CRC Press
Taylor & Francis Group
6000 Broken Sound Parkway NW, Suite 300
Boca Raton, FL 33487-2742

© 2010 by Taylor and Francis Group, LLC
CRC Press is an imprint of Taylor & Francis Group, an Informa business

No claim to original U.S. Government works

Printed in the United States of America on acid-free paper
10 9 8 7 6 5 4 3 2 1

International Standard Book Number: 978-1-56670-662-9 (Hardback)

Library of Congress Cataloging-in-Publication Data

Mohr, Thomas, 1959-
    Environmental investigation and remediation : 1,4-dioxane and other solvent stabilizers / authors, Thomas Mohr, Julie Stickney, Bill DiGuiseppi.
        p. cm.
    "A CRC title."
    Includes bibliographical references and index.
    ISBN 978-1-56670-662-9 (hardcover : alk. paper)
    1. Dioxane--Environmental aspects. 2. Stabilizing agents--Environmental aspects. 3. Soil remediation. 4. Groundwater--Pollution. 5. Solvents. I. Stickney, Julie. II. DiGuiseppi, Bill. III. Title.

TD427.D547M64 2010
628.1'68--dc22

2009049461

Visit the Taylor & Francis Web site at
http://www.taylorandfrancis.com

and the CRC Press Web site at
http://www.crcpress.com

In memory of my father, Eric Mohr, who always encouraged me to give back to my profession.

In honor of my wife, Maria Kamau, without whose assistance, advice, and support this book would not be possible.

For my son, Kimathi, may his future be bright, clean, and healthy.

Contents

Acknowledgments

This book has its genesis in the network of groundwater professionals who are my colleagues at the Santa Clara Valley Water District, my peers at California's Water Boards and the Department of Toxic Substances Control, USEPA, members of the Groundwater Resources Association of California, and many others, all of whom shared their knowledge and experiences with 1,4-dioxane. But before I could tap into this rich network and knowledge base, an incentive to pursue information about 1,4-dioxane was required. I thank Jim Crowley, my former supervisor at the Santa Clara Valley Water District, for granting my request for time to conduct my initial research on the possible presence of 1,4-dioxane at solvent release sites in Silicon Valley, the effort that culminated in the *Solvent Stabilizers White Paper*. The talented assistance of Santa Clara Valley Water District's librarian, Bob Teeter, was instrumental in compiling the information in the *White Paper* and is gratefully acknowledged.

The *2001 White Paper* marked the end of Santa Clara Valley Water District's support for 1,4-dioxane research, and the beginning of my personal quest to assemble and distribute all the information I could find on this most interesting contaminant. I corresponded or spoke with a long list of groundwater professionals and scientists whose work at some time included 1,4-dioxane, and who were generous with their time and knowledge. While it would be easy to forgo listing this crowd of contributors and supporters, their help was instrumental to my research, and for that, I am grateful. I have attempted to list as many as I can recall:

Santa Clara Valley Water District staff, including Celia Norman, Cris Tulloch, George Cook, Donna Drogos (now with Alameda County); San Francisco Bay Regional Water Control Board staff, including Vince Christian, Keith Roberson, Greg Bartow, and Stephen Hill; Central Valley Regional Water Quality Control Board's Alex McDonald; State Water Resources Control Board's John Russell and James Giannopoulos; California Department of Toxic Substances staff, including Calvin Willhite (Berkeley), Brian Lewis (Sacramento), Laura Rainey (Glendale), Karen Baker (Glendale), and Bart Simmons (retired hazardous materials laboratory director); California Department of Public Health staff, including Bill Draper (retired), Heather Collins, Alan Sorsher, and Steve Dizeo; USEPA's Wayne Praskins, Charles Berry, Christian Daughton, Marti Otto, Jean Munch, Kendra Morrison, Herb Levine, Bruce Macler, Kathy Davies, Linda Fiedler, Bernie Zavala, Dick Willey, Andy Crossland, Kay Wischkaemper, and Chau Vu; ATSDR's Carole Hossom (GA); Arizona Department of Environmental Quality's Craig Kafura; Florida Department of Environmental Protection's Bill Linn; Michigan Department of Environmental Quality's Mitch Adelman; Oregon Department of Environmental Quality's Greg Aitken, Max Rosenberg, and Deborah Bailey; US Navy's Carl Bonura (CA); Alameda County Water District's Ted Trenholme and Steven Inn (CA); Orange County Water District's Adam Hutchinson, Roy Herndon, Mike Wehner, Virginia Greber, Jenny Glasser, and Lee Yoo; Water Replenishment District of Southern California's Ted Johnson; Stanford Linear Accelerator Center's Susan Witebsky, Trevor Dumitru, and Dellilah Sabba; Calgon's Charles Drewery; Applied Process Technologies' Reed Bowman, Chuck Borg, Gary Johnson, Ray Milne, and Terry Applebury; ProHydro's Sandy Britt (NY); Accelerated Remediation Technologies' Marco Odah (KS); Basin Water's Rob Haney (CA); Microvi BioTech's Dr. Fatemeh Shirazi and Dr. Joe Salanitro (KS); Columbia Analytical Services' Dee O'Neill (WA); Entech Laboratories' Craig Thom (CA); Kiff Analytical Laboratories' Rich Premzic and Joel Kiff (CA); Lancaster Labs' Don Wyand (PA); Supelco's Bob Shirey (PA); Zymax Laboratories' Jesper Nielsen (CA); Pall Life Sciences—Farsad Fotouhi (MI); Aerojet's Rodney Fricke and Gerry Swanick (CA); UTC's Tim Marker and Tim Rumbolz (CA); Dow Chemical's Jim Mertens (MI); Dupont's

Dave Ellis (OH); Shaw Environmental's Paul Hatzinger and Rob Steffan (NJ), and Greg Nicoll (CA); Geosyntec's Evan Cox (Ontario, Canada); Locus Technologies' Elie Haddad (now with Haley and Aldrich), Alan Lui, and Jim Boarer; Komex's Jon Rohrer (now with Aquiver) and Anthony Brown (CA); DPRA's Bob Morrison and Emily Vavricka; O'Brien & Gere's Chris Voci (now with Geosyntec) (PA); Disposal Safety Inc.'s Steven Amter (Washington, DC); Groundwater Services Inc.'s Tom McHugh (TX); North Carolina State University's Bob Borden; Oregon State University graduate student Carl Isaacson (now with US EPA's Athens GA Lab) and Professor Jennifer Field; North Carolina State University Professor Michael Hyman; UC Berkeley Professor Lisa Alvarez-Cohen and graduate student Shailey Mahendra (now a professor at UCLA); California State University Chico Professor Mitchell Johns; Johns Hopkins University Professor Bill Ball; EAWAG's Christian Niederer (Swiss Federal Institute of Aquatic Science and Technology); California State University East Bay Professor Jean Moran (formerly a researcher with Lawrence Livermore National Laboratories); AECOM (formerly Earth Tech) staff, including Bill DiGuiseppi (CO), Ed Meyers (FL), Dora Chiang (GA), Caroline Whitesides (CO), Stephanie Price (CO), Charlene Peoples (CO), Barbara Lemos (TX), and Linda Villareal (CA); Engineering & Consulting Resources, Inc.'s Richard Doherty (MA); INTERA's Richard Jackson (CO); Tetra Tech's Mike Berman (VA); Camp Dresser McKee's Patrick Evans (WA); Malcolm Pirnie's Elizabeth Hawley, Rula Deeb, Jim Strandberg, and Mike Kavanaugh (CA); MACTEC Engineering and Infrastructure's Vittal Hosengedi and Sarah Raker (CA); Arcadis' Fred Payne (MI), Mark Fenner (CO), Steven Feldman (NY), Ruddy Clarkson (NY), Julie Stickney (NY) (now with Syracuse Research Corporation), Alison Jones (CA), John Horst (PA), and Eric Nichols (NH); AMEC-Geomatrix's Bettina Longino, Peter Bennett, and Murray Einarson (CA); Blasland Bouck and Lee's John C. Alonso (FL), Jim Linton (FL), Michael Gefell (NY), Julie Sueker (CO), Patrick Keller (CO), and Steven Huntley (CA) (BBL is now Arcadis); MWH Laboratories' Andy Eaton (CA) and MWH Global's Ken Quinn; MicroSeeps' Bob Pirkle (PA); Regenesis' Stephen Koenigsberg and Anna Willett (now with ENVIRON and ITRC, respectively); Daniel B. Stephens' Steve Cullen (CA), Nicole Sweetland (CA), Neil Blanford (NM), and Jenny Sterling (CA); Solutions IES' Matt Zenker (NC); Brian Haughton, Barg Coffin Lewis & Trapp, LLP (CA); Michael L. Caldwell, Zausmer, Kaufman, August & Caldwell, PC (MI); Steven L. Hoch and Brownstein Hyatt Farber Schreck, LLP (CA); Todd Engineers' Sally McCraven, Bill Motzer, and David Abbott (CA); R3 Environmental Technology Ltd's Dr. Paul Bardos (England); and Haley and Aldrich's Lorne Everett (CA). Valuable and essential support was provided by Richard Levin, Michael Mora, Jan Hutchins, Joe Dinis, Rusty Lutz, Brian Wagner, Marv Cohen, Bill Walters, Roy Abbott, Jim Fitzgerald, Brian Korek, Brandon Cunningham, Rafael Camarota, Mahesh Grossman, and several others who went out of their way to encourage me.

I greatly appreciate the skilled editing of this book by Mary Eberle (WordRite, Boulder, CO), who patiently deciphered my rough drafts and helped me to find simple and clear ways of presenting my information. I am also grateful to the editorial team at Taylor & Francis as well as the project team including Joseph Clements, Andrea Dale, Marsha Pronin, and Pat Roberson. I am thankful for the contributions of my coauthors, Bill DiGuiseppi and Julie Stickney, and the chapter reviewers, who are acknowledged in each chapter.

I am most grateful to my wife, Maria Kamau, not only for tolerating the seemingly endless phases of research and writing, but also for lending her database design and project management skills to research and create the patent database, and for helping me focus on completing the project.

Authors

Thomas K.G. Mohr works with the Santa Clara Valley Water District as senior hydrogeologist where he oversees regional groundwater monitoring and provides oversight for remediation of solvent release sites in Silicon Valley. He has published on the application of stable isotope data for forensic investigations of perchlorate and nitrate contamination, bioremediation of gasoline, a method for assessing and ranking the potential for legacy contamination from past dry cleaners, and on a new refinement to methods for the laboratory analysis of 1,4-dioxane. Mohr previously managed and monitored environmental impacts from landfills for the County of Yolo and the City of Sunnyvale, and worked as a consultant conducting site characterization, remediation of leaking fuel and solvent tank sites, bioremediation of gasoline, and a variety of related projects.

Mohr was the 2006/2007 President of the Groundwater Resources Association of California and currently serves on GRA's Board of Directors. He has chaired six major symposia for GRA on perchlorate, dry cleaners, and 1,4-dioxane, and has been an invited speaker or session chair at symposia hosted by the International Society of Environmental Forensics, the National Groundwater Association, Battelle International, the California State Water Resources Control Board, and USEPA. He also enjoys mentoring students and has lectured at his alma mater, the University of California at Davis, as well as at Stanford University, San Jose State University, and California State University East Bay. Mohr is continuing his pursuit of chlorinated solvent forensics and related topics through research, writing, and collaborations.

William H. DiGuiseppi, PG, is a principal hydrogeologist with AECOM, a Fortune 500 provider of environmental, water, transportation, and infrastructure engineering services to clients in over 100 countries. In his 18 years with AECOM, his career has focused on soil and groundwater investigation and remediation programs at industrial sites and federal facilities. As the leader of the Environmental Technical Practices Network at AECOM, he has played a pivotal role in global technology transfer, ensuring that project teams leverage best practices to provide innovative and cost-effective solutions to complex remedial and compliance challenges. DiGuiseppi holds a BS in geology from George Mason University and an MS in geology from the University of Utah. Since entering the environmental field in the 1980s, he has addressed wide-ranging contamination issues, including PCBs, heavy metals, petroleum products, and solvents, in a variety of media, including soil, groundwater, soil vapor, indoor air, surface water, and sediments. DiGuiseppi developed and successfully implemented innovative approaches to characterizing and remediating chlorinated solvent DNAPL source areas and has successfully closed perchloroethene and trichloroethene DNAPL sites for manufacturers and the USAF with single-site VOC mass removal as great as 100,000 pounds.

DiGuiseppi has extensive knowledge of solvent stabilizer chemistry and behavior as well as on regulatory and remedial issues, especially for 1,4-dioxane, having provided remedial solutions for multiple sites since early 2002. DiGuiseppi's experience with 1,4-dioxane includes evaluation, bench-scale and pilot testing, and full-scale implementation of multiple remedial technologies, including molecular filtration, zeolite and carbon sorption, *in situ* chemical oxidation, *in situ* bioremediation, *ex situ* advanced oxidation, and monitored natural attenuation. DiGuiseppi is the lead AECOM hydrogeologist for remediation of one of the most extensive 1,4-dioxane plumes in the United States, where the primary drinking water aquifer is impacted to a depth of over 250 feet, a width of 1 mile, and a length of almost 6 miles, with a footprint of nearly 4000 acres.

DiGuiseppi has presented, authored, and coauthored papers and cochaired sessions on 1,4-dioxane at national and international conferences, including "1,4-Dioxane: Biodegradation and Other Treatment Approaches," Session Co-Chair, Battelle, Baltimore, Maryland, 2007; "Sorption of 1,4-Dioxane from Ground Water onto Multiple Sorption Media," American Chemical Society/American Institute of Chemical Engineers (ACS/AIChE), Denver, Colorado, 2007; "Treatment Options for Remediation of 1,4-Dioxane in Groundwater," *Environmental Engineer*, Spring 2007; "Evaluation of Natural Attenuation at a 1,4-Dioxane Contaminated Site," *Remediation*, Winter 2008; "1,4-Dioxane Treatment using Advanced Oxidation," Groundwater Resources Association of California (GRAC), San Jose, California, 2008; "Advanced Oxidation for *Ex Situ* Treatment of 1,4-Dioxane and Chlorinated Solvents in a Large-Scale Groundwater Extraction System," US Air Force Center for Engineering and the Environment (AFCEE), San Antonio, Texas, 2008; "Emerging Contaminants," Session Co-Chair, Battelle, Monterey, California, 2008; "Using Mass Flux Estimation to Assess the Effectiveness and Sustainability of 1,4-Dioxane Remedial Strategies," National Defense Industrial Association (NDIA), Denver, Colorado, 2009; "Aggressive Source Control to Support a Large-Scale Groundwater Remediation Program," NGWA, Tucson, Arizona, 2009; and "Innovative Assessment of TCE and 1,4-Dioxane Natural Attenuation in Aerobic Aquifer," Battelle, Monterey, California, 2010 (in preparation).

Julie A. Stickney has 19 years of experience in the design and management of chemical toxicity evaluations and human health and ecological risk assessments. She has expertise in the critical review of mechanistic toxicology studies and the evaluation of chemical-specific modes of action. Dr. Stickney received BS and PhD degrees in toxicology from Northeastern University in 1986 and 1990, respectively. She was certified as a diplomate of the American Board of Toxicology in 2004. Dr. Stickney is currently employed by SRC, a not-for-profit environmental research company. She is also an adjunct scientist and member of the Maine Center for Toxicology and Environmental Health at the University of Southern Maine.

Technical Editor: Mary C. Eberle, MS, is the owner of Wordrite Editorial Services. She was managing editor of *American Mineralogist* and has done technical editing for the journals *Geology, Geological Society of America Bulletin*, and *Geophysics*, as well as for U.S. Geological Survey publications, Colorado, Louisiana, and Oklahoma Geological Survey reports and symposium volumes, and numerous science and natural history books and chemistry and geology textbooks.

Acronyms

AAV	Acid acceptance value
AES	Alcohol ethoxy sulfate
ALP	Alkaline phosphatase
ALT	Alanine aminotransferase
AMO	Alkane monooxygenase
AOP	Advanced oxidation process
AOT	Advanced oxidation technology
AST	Aspartate aminotransferase
ASTM	American Society for Testing and Materials
ATSDR	Agency for Toxic Substances and Disease Registry
BCEE	*bis*-2-Chloroethylether
BHC	Benzene hexachloride (also Lindane)
BOD	Biochemical oxygen demand
BSM	Basal salt medium
BTU	British thermal units
CASRN	Chemical Abstracts Service Registry Number
CCL	Contaminant candidate list
CCl_4	Carbon tetrachloride
CDPH	California Department of Public Health
CDPHE	Colorado Department of Public Health and Environment
CFR	Code of Federal Regulations
CHO	Chinese hamster ovary (cells)
CLP	Contract Laboratory Program
CNS	Central nervous system
COC	Constituents of concern (or chemicals of concern)
COD	Chemical oxygen demand
CPK	Creatinine phosphokinase
CSF	Cancer slope factor
CSIA	Compound-specific isotope analysis
CT DPH	Connecticut Department of Public Health
CYP	Cytochrome P450
d, h, m, s	days, hours, minutes, seconds
DCE	1,1,-Dichloroethylene
DCM	Dichloromethane (also methylene chloride)
DEG	Diethylene glycol
DMBA	Dimethylbenzanthracene
DMT	Dimethyl terephthalate
DNAPL	Dense nonaqueous-phase liquid
DPE	Dual-phase extraction
DTSC	Department of Toxic Substances Control
DX	1,4-Dioxane
EC_{50}	Median concentration causing an effect in 50% of the test organisms
ECD	Electron capture detector
EDF	Ethylene glycol diformate (also ethylene-1,2-diformate)

EDTA	Ethylenediaminetetraacetic acid
eV	Electron-volt
FBR	Fluidized-bed reactor
FDA	U.S. Food and Drug Administration
FID	Flame ionization detector
f_{oc}	Fraction of organic matter
Freon 11	Dichlorodifluoromethane
Freon 113	1,1,2-Trichloro-1,2,2-trifluoroethane (also CFC-113)
Freon 12	Trichlorofluoromethane
FRI	Focused remedial investigation
GAC	Granular activated carbon
GAMA	Groundwater Ambient Monitoring and Assessment Program
GC	Gas chromatography
GC-IRMS	Gas chromatography-isotope ratio mass spectrometry
GCM	Group contribution model
GC-MS	Gas chromatography-mass spectrometry
GGT	γ-Glutamyl transpeptidase
HEAA	Hydroxyethoxyacetic acid
HPV	High production volume
i.p.	Intraperitoneal
i.v.	Intravenous
IRIS	Integrated risk information system
ISCO	*In situ* chemical oxidation
IUPAC	International Union of Pure and Applied Chemistry
KBV	Kauri-butanol value
K_D	Distribution coefficient
KIE	Kinetic isotope effect
K_{OC}	Organic carbon partition coefficient
KOW	Octanol/water partition coefficient
LC	Liquid chromatography
LC_{50}	Median concentration lethal to 50% of the test organisms
LD_{50}	Median lethal dose
LDH	Lactate dehydrogenase
LLE	Liquid–liquid extraction
LMS	Linear multi-stage (model)
LOAEL	Lowest observed adverse effects level
LOQ	Limit of quantitation
M	Molar
m/z	Mass-to-charge ratio
MC	Methyl chloroform (also 1,1,1-trichloroethane)
MCI	Molecular connectivity index
MCL	Maximum contaminant level
MCLG	Maximum contaminant level goal
MDEQ	Michigan Department of Environmental Quality
MDL	Method detection limit
mM	Milli-moles
MMA	Michigan Manufacturers Association
MNA	Monitored natural attenuation
MPP	Macro-porous polymer
MRE	Minimum relative entropy
MRL	Minimal risk level

MS	Mass spectrometry
MS/MSD	Matrix spike/matrix spike duplicate
MSDS	Material safety data sheet
MTBE	Methyl *tert*-butyl ether
MTD	Maximum tolerated dose
MW	Molecular weight
n-BGE	*n*-Butyl glycidyl ether
ND	Not detected (also nondetect)
NESHAP	National Emission Standards for Hazardous Air Pollutants
NL	Notification level
nm	Nanometer
NOAEL	No observed adverse effect level
NPDES	National Pollution Discharge Elimination System
OCWD	Orange County Water District
ODC	Ozone depleting compound
ODC	Ornithine decarboxylase (activity; toxicology context)
OEHHA	(California) Office of Environmental Health Hazard Assessment
OVIs	Organic volatile impurities
Pa	Pascals (SI unit for pressure, 1 N/m^2)
PBPK	Physiologically based pharmaco-kinetic (model)
PCA	Potentially contaminating activity
PCE	Perchloroethylene (also perchloroethene or tetrachloroethylene)
PCEMI	Perchloroethylene as a manufacturing impurity
PCR	Polymerase chain reaction
PDB	Passive diffusion bag
PDX	1,4-Dioxane-2-one
PEG	Polyethylene glycol
PET	Polyethylene terephthalate
PHG	Public health goal
PID	Photo-ionizing detector
PNEC	Predicted no effect concentration
POEA	Polyoxyethyleneamine
POTW	Publicly owned treatment works (wastewater treatment plants)
ppb	Parts per billion
PPG	Pittsburg plate and glass
ppm	Parts per million
ppt	Parts per trillion
PRG	Preliminary remediation goal
PRP	Potentially responsible party
PSMS	Polysulfone membrane sampler
PT	Purge-and-trap
PZ	Prohibition zone (Re: Ann Arbor, MI)
QSARs	Quantitative structure–activity relationships
RAP	Remedial action plan
RBC	Rotating biological contactor
RCRA	Resource Conservation and Recovery Act
REACH	Registration, Evaluation & Authorization of Chemicals
R_f	Retardation factor
RfC	Reference concentration
RfD	Reference dose
RoD	Record of decision

RPPS	Rigid porous polyethylene samplers
RSC	Relative source contribution
RSD	Relative standard deviation
RVD	Relative vapor density
SCE	Sister chromatid exchange
SDWA	Safe Drinking Water Act
SIM	Selective ion monitoring (less commonly: selected-ion mode)
SMILES	Simplified Molecular Input Line Entry System
SMZ	Surfactant-modified zeolite
SPDE	Solid phase dynamic extraction
SPE	Solid phase extraction
SPME	Solid phase micro-extraction
SVE	Soil vapor extraction
SWAP	Source Water Assessment Program
TBA	*tert*-Butyl alcohol
TCA	1,1,1-Trichloroethane (also methyl chloroform)
TCD	Thermal conductivity detector
TCE	Trichloroethylene (also trichloroethene)
TDI	Tolerable daily intake
TEG	Triethylene glycol
THF	Tetrahydrofuran
THP	Tetrahydropyran
TICs	Tentatively identified compounds
TOC	Total organic carbon
TPA	Tetradecanoylphorbol-13-acetate
TRI	Toxics release inventory
TRV	Toxicity reference value
TWA	Time-weighted average
UCMR	Unregulated contaminant monitoring requirement
UF	Uncertainty factor
USAF	United States Air Force
USP	*U.S. Pharmacopeia*
UV	Ultraviolet
VOCs	Volatile organic compounds
WHO	World Health Organization
wt	Weight
ZVI	Zero-valent iron

Introduction

WHAT ABOUT 1,4-DIOXANE?

Dave Matthews, a colleague of mine at the Santa Clara Valley Water District, asked me this question in November 2000, shortly after I began working there to provide stakeholder oversight on solvent release cases. He had read somewhere that it is an additive to the widely used degreasing solvent methyl chloroform (1,1,1-trichloroethane) and noted that he was not aware of any testing for 1,4-dioxane at a large aerospace facility with multiple solvent release sites in San Jose. I had never heard of 1,4-dioxane, so I requested "a few hours" from my supervisor, Jim Crowley, in order to do some research. My supervisor was quick to embrace the idea as he too was curious about the potential for solvent stabilizers to degrade groundwater, a critical resource for Silicon Valley's water supply. Silicon Valley has an abundance of electronics firms—printed circuit board manufacturers, "wafer fabs" (semiconductor production plants), memory chip producers, and much more—all of which use chlorinated solvents, including many that use methyl chloroform. Thus began the pursuit that led to the *2001 Solvent Stabilizers White Paper* in which I assembled a wide range of information on 1,4-dioxane that proved to be useful to regulators and remedial project managers dealing with solvent release sites. I thought this pursuit would end with the White Paper, but that was only the beginning.

After I presented the White Paper at a 2001 symposium hosted by the Groundwater Resources Association of California (GRAC), I posted it online.* Eventually, groundwater professionals from around the country and several from overseas downloaded the paper and followed up with e-mail and telephone inquiries. These e-mails provided a wealth of additional information on how 1,4-dioxane was showing up in different types of industrial groundwater contamination sites, how it was migrating through groundwater in different hydrogeologic settings, and how different states were regulating 1,4-dioxane.

The White Paper also led to three key surveys for the presence of 1,4-dioxane at solvent release sites. First, the San Francisco Bay Water Board (SFBWB) asked for 1,4-dioxane analyses at 15 Silicon Valley sites known to have used and released methyl chloroform. 1,4-Dioxane was found at 12 sites, mostly at low concentrations; however, concentrations at some sites exceeded 1000 micrograms per liter (μg/L). The aerospace facility for which the 1,4-dioxane question was first asked proved to have widespread 1,4-dioxane contamination: one monitoring well had a concentration of 11,000 μg/L. Second, the chief geologist of the Department of Toxic Substances Control (DTSC), Brian Lewis, arranged for his agency to conduct a sampling survey for 1,4-dioxane at about three dozen sites; it was found at more than half. Brian Lewis also arranged the third effort, in which I was invited to participate in a series of monthly teleconferences of the U.S. Environmental Protection Agency (USEPA) Superfund Groundwater Forum, which culminated in an informal survey of existing 1,4-dioxane data in Superfund case files nationwide. More than 50 Superfund sites had 1,4-dioxane detections, some at very high concentrations; however, groundwater at most of the more than 2000 Superfund cases had not been analyzed for 1,4-dioxane.

The Superfund Groundwater Forum discussions of 1,4-dioxane led USEPA Region 3 to recommend testing for 1,4-dioxane in 2003 at Bally, Pennsylvania, where water from a solvent-contaminated

---

* http://www.valleywater.org/Water/Water_Quality/Protecting_your_water/_Solvents/index.shtm

municipal well was being treated using air stripping, and the treated effluent was distributed to supply the town's drinking water. The municipal well was located 1000 ft from a facility that produced urethane panels for refrigeration insulation and had discharged spent solvent waste to unlined lagoons between 1960 and 1965. The well test showed that 1,4-dioxane was not removed by air stripping. The initial detections in the air stripper effluent that was pumped into the water system were as high as 60 µg/L.

While the White Paper was a catalyst that led to several discoveries of 1,4-dioxane at sites across the United States, some of the information in the White Paper was acquired by networking with California regulatory agency professionals at the SFBWB, DTSC, Department of Health Services, and Office of Environmental Health Hazard Assessment, all of whom preceded me in researching the nature and occurrence of 1,4-dioxane. Some dismissed 1,4-dioxane because it is not regulated, while others were intrigued and required testing and remediation at sites they oversaw, yet all were overwhelmed with their daily case loads and few had the opportunity to research this contaminant in depth. The White Paper clearly filled an information void, and after it was downloaded more than 1000 times, I decided that this book must be written.

The book is organized around the uses of chlorinated solvents to which solvent stabilizers, including 1,4-dioxane, were added and focuses on how different uses changed the composition of solvent wastes that were released to groundwater through spills, leaks, dumping, unlined lagoons, unlined landfills, leaking sewer lines, and other avenues. This focus on what I have termed "contaminant archeology," that is, studying the industrial fate of the contaminant before it was discharged, affords some unique material that may provide insights for forensic investigations into the origins of solvents in contaminated groundwater. In the course of researching the contaminant archeology aspects of solvent stabilizers, I encountered more than 300 chemicals patented for use as solvent stabilizers for the four major chlorinated solvents in nearly 100 patents. The documented use of several dozen stabilizers for the major chlorinated solvents is detailed in Chapter 1.

The environmental fate and transport properties of confirmed solvent stabilizers are profiled in Chapter 3; however, the focus of the book is 1,4-dioxane. Accordingly, Chapter 2 focuses on the chemistry and the many industrial uses and manifestations of 1,4-dioxane. For example, I was surprised to learn that nearly every landfill serving universities and other institutions with life science research laboratories in the 1960s–1980s was found to harbor 1,4-dioxane contamination from discarded liquid scintillation cocktail waste. Chapter 4 examines the challenges of laboratory analysis of 1,4-dioxane—coaxing this hydrophilic ether compound out of the water and reliably separating and quantifying it at low concentrations—and new methods for 1,4-dioxane analysis, such as USEPA Method 522, which uses solid-phase extraction coupled with tandem mass spectrometry.

The toxicology of 1,4-dioxane is featured in Chapter 5, written by Julie Stickney of the Syracuse Research Corporation, a leading researcher of 1,4-dioxane toxicity. Chapter 6 profiles the regulation and risk assessment of 1,4-dioxane and the wide range of state guidance on 1,4-dioxane cleanup levels and drinking water levels. William DiGuiseppi of AECOM Environment (formerly Earth Tech) has acquired a wealth of hands-on experience managing 1,4-dioxane treatment projects, including the large and complex but successful cleanup of Air Force Plant 44 in Tucson, Arizona. Consequently, he is ideally suited to write about treatment technologies for 1,4-dioxane in groundwater (Chapter 7) and has provided several examples of solutions employed at cleanup sites across the United States.

The seven case studies in Chapter 8 describe the nature and extent of 1,4-dioxane releases, their regulation, and their remediation in a variety of geologic settings including limestone, sandstone, glacial outwash plains, and a variety of alluvial environments. In these case studies, the amounts released range from small to massive, resulting in plumes spanning a few hundred feet to several miles. The consequences of the 1,4-dioxane spills summarized in the case studies range from inconsequential impacts to groundwater locked up in bedrock and used by no one to widespread contamination of private and municipal wells and long-term consumption of 1,4-dioxane in drinking water. The corresponding regulatory responses range from treatment train modification to purchasing and condemning private wells, and even an outright ban on using groundwater for any purpose—the establishment of a "groundwater exclusion zone"—to clear the way for the inexorable march of a massive and unstoppable 1,4-dioxane plume beneath Ann Arbor, Michigan, on its way to the Huron River where it will be discharged, diluted, and dispersed.

Opportunities to use solvent stabilizers as a forensic tool to solve contamination problems such as deconvoluting commingled plumes of a common solvent or determining the age of a release are examined in Chapter 9. In Chapter 10, the regulatory policy implications of the discovery of 1,4-dioxane at solvent release sites after cleanup has begun are discussed. 1,4-Dioxane is emblematic of "emerging contaminants," a cast of chemicals that includes *N*-nitrosodimethylamine, 1,2,3-trichloropropane, perfluorooctanoic acids, perfluorooctane sulfate compounds, and a number of others. The manner in which these compounds are addressed holds profound implications for the indirect potable reuse of recycled water—a critically important resource in the arid southwest—and the discharge of treated wastewater to freshwater bodies whose downstream uses include drinking water. The regulatory policy toward 1,4-dioxane, whether declared in written guidance or inferred from the practices of caseworkers on individual cases, is profiled in Chapter 10 as well. Gaps in the regulatory framework that facilitated the situation in which we now find ourselves—blindsided once again—are also examined in Chapter 10. Regulatory solutions via "green chemistry" and the European Union's REACH (Registration, Evaluation and Authorization of Chemicals) program are discussed.

WHY SHOULD YOU CARE?

This book embodies a great deal of information on 1,4-dioxane, but how can this be of use to you? Let us make it tangible with some real-world examples that should convince you that what you do not yet know about 1,4-dioxane could indeed harm the interests of your client, employer, site owners, or investors. The mobility, persistence, and treatment challenges combine to make 1,4-dioxane a particularly vexing contaminant. It is more mobile than any other contaminant you are likely to find at solvent release sites. If you worked at fuel-leak sites in the 1990s, you could think of 1,4-dioxane as "MTBE on steroids" to get a better sense of what this book is about. The mobility, persistence, and treatment of 1,4-dioxane in groundwater present formidable challenges to site characterization and remediation.

Because analytical methods for 1,4-dioxane were not widely available before the late 1990s, many solvent contamination site investigations did not include testing for 1,4-dioxane. Now that the analytical methods are available and more regulators have taken an interest in 1,4-dioxane, it is being discovered long after the extent of solvent plumes has been delineated, the treatment technology selected, the capture zone established, and the health risk assessment completed. The Record of Decision (RoD) or the equivalent regulatory cleanup decision has been established at many solvent release sites without consideration or examination of 1,4-dioxane. When 1,4-dioxane is later discovered, the adopted and implemented cleanup plan that was working just fine and well on the way to site closure can be turned upside down. In short, 1,4-dioxane has every potential to be, and has been, an "RoD reopener."

Consider the experience of a consultant managing a San Jose, California, solvent recycling facility cleanup site. In 1998, the contracted laboratory inadvertently analyzed for an expanded list of analytes for a few site monitoring well samples and reported 1,4-dioxane at concentrations up to 56,000 µg/L. Analysis for 1,4-dioxane was not a site monitoring requirement; however, the regional Water Board requested that all results of tests performed at the site on an elective basis be reported, which the consultant conscientiously did. Follow-up sampling and analysis for 1,4-dioxane led to the discovery that the 1,4-dioxane plume had migrated much farther than the chlorinated solvent plume that had been the subject of the remedial investigation and cleanup, and that the maximum 1,4-dioxane concentration encountered was 340,000 µg/L. California's Notification Level for 1,4-dioxane is 3 µg/L.

The wider occurrence of 1,4-dioxane meant first that additional monitoring wells were needed to delineate the downgradient extent of 1,4-dioxane, and also that the extraction trenches and extraction wells installed for solvent removal were no longer sufficient to capture all of the groundwater contamination caused by the solvent recycling operation. Moreover, the treatment technologies employed for solvent removal were not sufficient to remove 1,4-dioxane, which is relatively immune to removal by the conventional treatment technologies used for chlorinated solvents, for example, air stripping and granular activated carbon. Finally, the health risk assessment had addressed only the chlorinated and ketone solvents, but not 1,4-dioxane, which has been classified as a probable human carcinogen; hence this too had to be done at considerable expense to the responsible party. In other cases, because private wells had tested "nondetect" for solvents, regulators assured off-site domestic well owners that they could consume their well water free of any concern of contamination. The news, a few years later, that they had in fact been drinking 1,4-dioxane all along was met with incredulity and outrage.

At numerous sites, including some of those profiled in Chapter 8, once discovered, 1,4-dioxane becomes the main driver for site cleanup. Yet considerable uncertainty remains in the toxicology of 1,4-dioxane, and some toxicologists advocate setting regulatory threshold orders of magnitude higher than the current guidance values used by states, which range from 3 µg/L in California to 85 µg/L in Michigan. Nevertheless, arguments for a higher threshold of toxicity and questions about the reliability of carcinogenicity assays were not persuasive to the Colorado Department of Public Health and Environment, which established the United States' only legally binding standard for 1,4-dioxane (a Colorado Maximum Contaminant Level of 6.1 µg/L in groundwater).

Shortly after I completed the manuscript for this book, USEPA released a draft toxicity review of 1,4-dioxane. The draft review applied a linear low-dose extrapolation to obtain an oral cancer slope factor for 1,4-dioxane, which suggests that the drinking water equivalent level may need to be lowered 17-fold. The controversy over 1,4-dioxane toxicity is far from over—stay tuned.

Many readers of this book will be seeking strategies to avoid engaging their clients in expensive cleanups for 1,4-dioxane, particularly where no exposure through drinking or showering in 1,4-dioxane-contaminated groundwater can be shown. There will certainly be cases where active remediation may not be necessary (i.e., where passive remediation will suffice), and I hope you will find this book useful in helping to guide you toward your objective. But a more important goal for this book is to increase the awareness of regulators, water utilities, and consultants alike to identify and eliminate additional instances of ongoing exposure to 1,4-dioxane.

At some as yet undiscovered locations in the United States and in all the nations engaged in industrial production that use methyl chloroform, it is very likely that people are unwittingly and unnecessarily drinking well water contaminated with 1,4-dioxane—*right now*. This has happened with significant frequency, and it will happen again. Is it happening at a well downgradient of a site you are investigating and managing, perhaps much further downgradient than you first expected was relevant? Examine the contents of these pages and apply that knowledge to do the right thing by eliminating ongoing exposure to 1,4-dioxane where it is found or by confirming its absence.

Thomas K.G. Mohr
Santa Clara, CA, March, 2009

Disclaimer

The support of the Santa Clara Valley Water District (SCVWD) for my research and compilation of the *2001 Solvent Stabilizers White Paper* is gratefully acknowledged and played a seminal role in the early genesis of this book; however, this book was researched, written, edited, and rewritten without funding or support from SCVWD. The views presented here are those of the chapter authors only and are not the official policies, positions, or opinions of their employers, the SCVWD, AECOM, or the Syracuse Research Corporation. Similarly, while the lead author is a Director and Past President of the Groundwater Resources Association of California (GRAC), the contents of this book have not been reviewed or endorsed by GRAC's Board of Directors and do not represent the official viewpoints of GRAC. Finally, the reviewers who generously volunteered their time and knowledge to read and comment on draft chapters bear no responsibility for any errors, omissions, or other possible problems in the final manuscript.

Contact the Authors

Your comments, questions, corrections, and criticisms are welcome. Please contact the authors by e-mail or telephone at the following addresses and phone numbers:

Thomas K.G. Mohr
    mohrhydrogeo@aol.com   (408) 265-2600
Bill DiGuiseppi
    Bill.Diguiseppi@aecom.com   (303) 804-2356
Dr. Julie Stickney
    stickney@syrres.com   (207) 883-9824

Errata

The authors have acquired a great deal of data, information, and knowledge through their years of focused attention on 1,4-dioxane, which we believe has culminated in an authoritative and reliable reference. Nevertheless, if you spot an error, please send us an e-mail and we will post it to the errata page on the book's Web site, www.The14DioxaneBook.com. The Web site also hosts current information on 1,4-dioxane regulation, 1,4-dioxane news, abstracts of new papers on 1,4-dioxane, additional case studies, reader-contributed materials, and profiles of service providers who can assist you with your 1,4-dioxane challenges.

1 Historical Use of Chlorinated Solvents and Their Stabilizing Compounds

Thomas K.G. Mohr

This book deals with four the major chlorinated solvents: methyl chloroform, trichloroethylene, perchloroethylene, and dichloromethane. Carbon tetrachloride is also an important chlorinated solvent, but it was more commonly used between 1900 and 1960. Its primary use since then has been as an intermediate in the production of chlorofluorocarbons and other chemicals (Doherty, 2000a; Shepherd, 1962). The more common names for these four major solvents include 1,1,1-trichloroethane or TCA (for methyl chloroform), trichloroethene or TCE (for trichloroethylene), tetrachloroethene or PCE (for perchloroethylene), and methylene chloride (for dichloromethane). Tables 1.1 and 1.2 present identifying information and chemical structures for the four major chlorinated solvents, as well as lists of synonyms and trade names.

Solvent stabilizers have been added to a wide variety of solvents, including chlorinated and other halogenated solvents, petroleum solvents, ketone solvents, ether solvents, alcohol solvents, and others. This book focuses on the stabilizers for the chlorinated solvents, which were by far the most frequently used solvents for degreasing, dry cleaning, cold cleaning, and dozens of other industrial applications. This book also provides brief discussion of stabilizers for the following additional solvents: carbon tetrachloride, Freon 11, Freon 12, and Freon 113 (dichlorodifluoromethane, trichlorofluoromethane, and 1,1,2-trichloro-1,2,2-trifluoroethane, respectively; Freon is a trademark name for chlorofluorocarbons registered by E. I. Du Pont de Nemours and Company Corporation of Delaware) and new replacement solvents limonene, 1-bromopropane, siloxanes, terpenes.

1.1 HISTORY OF CHLORINATED SOLVENTS

Chlorinated solvents have played an integral role in the world's industrial societies for more than 70 years. The majority of manufactured consumer products made of metals or incorporating electronic components have undergone some form of cleaning process using solvents to remove residuals from fabrication. Chlorinated solvents are nonflammable, easy to recover from the vapor phase, and easy to recycle when laden with oils and other debris by degreasing and dry-cleaning processes. The relatively low boiling points of these chemicals make them economical for use in vapor degreasing and dry cleaning, and their high solvency makes them extremely effective in a wide variety of challenging cleaning applications, for example, cold cleaning and ultrasonic cleaning.

From the 1940s until recently, chlorinated solvents played a part in the production of most manufactured products. Today, many manufacturing processes continue to leverage the cleaning power and effectiveness of halogenated solvents. However, in compliance with regulations that prevent uncontrolled releases to the environment, these processes use equipment that has been engineered to eliminate or minimize emissions. Chlorinated solvents provided an enormous benefit to the industry and consumers alike; however, they pose a threat to water quality where they were released to soil and groundwater. The unintended adverse consequences of past uncontrolled or undercontrolled solvent uses have taught us valuable lessons, which have guided major improvements in

TABLE 1.1
Major Chlorinated Solvents: Formulas and Structures

	Solvent				
	Carbon Tetrachloride	**Methyl Chloroform**	**Dichloromethane**	**Trichloroethylene**	**Perchloroethylene**
CASRN[a]	56-23-5	71-55-6	75-09-2	79-01-6	127-18-4
Formula	CCl_4	$C_2H_3Cl_3$	CH_2Cl_2	C_2HCl_3	C_2Cl_4

Structure

<pre>
       Cl   Cl H   Cl   Cl   H   Cl   Cl
       | | | | \ /   \   /
  Cl— C— Cl   Cl— C — C— H   H— C— H   C=C   C=C
       | | | | / \   /   \
       Cl   Cl H   Cl   Cl   Cl   Cl   Cl
</pre>

[a] CASRN = Chemical Abstracts Service Registry Number.

TABLE 1.2
Major Chlorinated Solvents: Identification

Carbon Tetrachloride	Methyl Chloroform	Dichloromethane	Trichloroethylene	Perchloroethylene
Tetrachloromethane	1,1,1-Trichloroethane,	Methylene chloride,	TCE, trichloroethene,	PCE,
Perchloromethane	TCA, TCE	methane dichloride	1,1,2-	tetrachloroethylene,
Methane tetrachloride	Trichloro-1,1,1-ethane	Methylene dichloride	trichloroethylene	tetrachloroethene,
Carbon tet	Methyltrichloromethane	Freon 30	1,1-Dichloro-2-	ethylene tetrachloride
Freon 10	Trielene	Aerothene MM™	chloroethylene	Perc, Perclene™
Halon 104	Aerothene TT™	DCM	Ethylene trichloride	(Diamond Shamrock/
Tetraform	Chlorothane NU™		Acetylene trichloride	DuPont)
Tetrasol	Chlorothene SM™		TRI	Persec (Vulcan
Tetrachlorocarbon	Chlorothene VG™		Tri-clene	Materials)
Carbona	Chlorothene NU™		Tri-plus	DOWPER™ (Dow
	Solvent 111		Triclene™	Chemical)
	"TRIC"		Trilene™	Perchloroethylene
	Inhibisol™		Trielin	SVG (Dow
	Genklene™		Narcogen	Chemical)
	1,1,1-TCE		Dow-Tri	Perchlor, Percosolv,
			Neu-Tri™	Perklone™, Perstabil
			Hi-Tri™	(Solvay et Cie)
			Solvent O-T-634C	Soltene™ (Solvay
				et Cie)

Sources: National Library of Medicine, 2006, Hazardous Substances Data Bank (HSDB). National Institutes of Health. http://toxnet.nlm.nih.gov/ (accessed March 15, 2006); Doherty, R.E., 2000a, *Journal of Environmental Forensics* 1: 69–81; Doherty, R.E., 2000b, *Journal of Environmental Forensics* 1: 83–93; Solvay SA, 2002a, Chlorinated solvents stabilisation. GBR-2900-0002-W-EN Issue 1—19.06.2002, Rue du Prince Albert, 33 1050 Brussels, Belgium. http://www.solvaychemicals.com (accessed November 9, 2003); and MSDSs cited in Chapter 1 reference list.

Notes: Most commercial trade names in this list are registered trademarks; these are indicated by capitalization; the symbol ™ indicates that the solvent name was once registered with the U.S. Patent and Trademark office, but the name may no longer be registered or has since been assigned to another product.

equipment design, operator training, monitoring, regulation, and enforcement for continuing solvent use. Substantial progress is being made toward environmentally safe uses of chlorinated solvents under controlled conditions. Concurrently, new solvents that pose a somewhat lower threat to drinking water, the atmosphere, and worker health and safety have been developed and introduced.

The enormous volume of chlorinated solvents used in the United States alone has left a monumental legacy that will keep contaminant hydrogeologists and environmental engineers occupied for decades. A compelling aspect of the historical use of chlorinated solvents is that most were ultimately released to the environment, usually to the atmosphere, but also to soil and groundwater. Solvents that were not emitted or discharged were incinerated or placed in a landfill. When solvents were added to products such as paints, inks, or cleaning agents, the fate of the solvent was often evaporation. Degreasing and dry-cleaning solvents were discarded when they were no longer usable. In earlier decades, waste solvents were often dumped to the ground or disposed in landfills, which in the mid-twentieth century were usually unlined or clay-lined landfills. Since the enactment of the Resource Conservation and Recovery Act in 1980, waste solvent has been subjected to "cradle-to-grave" management and hazardous waste manifesting. Waste minimization has emerged as an economical and regulatory compliance-driven solution to the challenge of safe solvent handling and disposal. Chlorinated solvent consumption in 2006 was about 10% of the consumption in 1976. The reduction was due to regulatory bans on ozone-depleting compounds (ODCs), more efficient degreasing and dry-cleaning machinery, and replacement of chlorinated solvents by new "designer solvents" and other alternatives.

The history of chlorinated solvent production, use, and regulation has been thoroughly addressed in Richard E. Doherty's two landmark articles in the *Journal of Environmental Forensics* (Doherty, 2000a, 2000b). Doherty's articles profile which solvents were favored for different applications and how those preferences changed in response to the recognition of worker health and safety issues and new regulatory restrictions adopted to protect the environment. The sequence of regulations controlling solvent use, described further in Chapter 6, played a key role in the evolution of which solvents were put to use for various purposes. Economic and other forces also helped in determining specific solvent usage. For example, during World War II, chlorinated solvent markets were controlled by the government: the military received the first priority followed by essential uses such as grain fumigation and refrigeration; dry cleaning received the lowest priority (Doherty, 2000a). Figure 1.1 charts

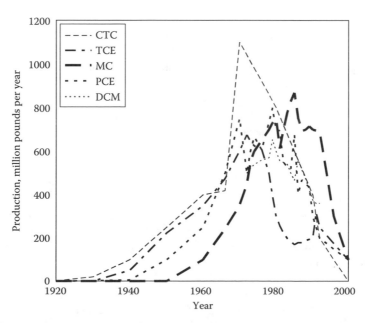

FIGURE 1.1 History of U.S. production of the major chlorinated solvents. (After Doherty, R.E., 2000a. *Journal of Environmental Forensics* 1: 69–81. With permission.)

the U.S. production history of the major chlorinated solvents. An approximate U.S. production of chlorinated solvents in the early 1970s is represented by the following 1972 figures (Considine, 1974):

Methyl chloroform	447 million pounds
Trichloroethylene	675 million pounds
Dichloromethane	500 million pounds
Perchloroethylene	500 million pounds

Methyl chloroform was substituted for trichloroethylene (TCE) beginning in the mid-1960s. The change was largely due to the enactment of regulations that restricted the use of TCE. Methyl chloroform production peaked in the mid-to-late 1980s. Reduced production was mandated by the Montreal Protocol on substances that deplete the ozone layer, an agreement among the world's nations to curtail the use of ozone-degrading compounds, including methyl chloroform. The Montreal Protocol went into effect on January 1, 1989; it called for 100% reduction in the 1989 methyl chloroform production levels by January 1, 1996 (UNEP, 1999).

1.1.1 MAJOR APPLICATIONS OF CHLORINATED SOLVENTS

The major chlorinated solvents were previously put to the following primary uses:

- Degreasing
- Extraction of oils or other food products
- Textile cleaning
- Wool processing
- Preparation of inks, paints, varnish, and lacquer
- Production of pharmaceuticals
- Cleaning of electronics, semiconductors, and precision instruments
- Cleaning of aircraft and other military hardware
- Minor uses in veterinary and medical applications
- A carrier for insecticides
- Applying paints to automobiles and other metal surfaces

Solvent boiling point, solvent power (solvency), stability, availability, and cost were factors that determined which solvents were used for particular applications. For example, perchloroethylene boils at 121°C, whereas dichloromethane boils at 39.7°C. Therefore, dichloromethane may have been preferred for cleaning electronics or other parts having plastics that are susceptible to melting, and perchloroethylene may have been preferred for removing waxes and resins whose melting points exceed the boiling point of other solvents.

1.1.1.1 Chlorinated Solvents in Historical or Current Use

1.1.1.1.1 Carbon Tetrachloride

Carbon tetrachloride is no longer produced for commercial use as an end product; it is still produced as an intermediate for the production of other chemicals. It was phased out because of its toxicity and its role as an ODC in the upper atmosphere. Carbon tetrachloride was the first of the major chlorinated solvents to be used in North America, beginning in 1898; its production in Europe began in the 1890s (Doherty, 2000a). It was used in fire extinguishers, in dry cleaning, as a grain fumigant, as a solvent for the extraction of oils from seeds and animal products, in rubber production, in metal degreasing, in rubber cement, in floor wax, as a refrigerant, as a household cleaning agent, in insecticides, and in numerous other applications (Irwin et al., 1997; Doherty, 2000a; European Chemicals Bureau, 2000b; ATSDR, 2005). Carbon tetrachloride was used to make Freon compounds that were widely used as aerosol propellants in the 1950s and 1960s (ATSDR, 2005; R.E. Doherty, personal communication, 2007). The majority of these uses of carbon tetrachloride was discontinued before 1970; some pesticide formulations included carbon tetrachloride until 1986. The regulation of ozone-depleting chemicals through the Montreal Protocol and the Clean Air Act Amendments of 1990, as well as bans by the Food and Drug Administration and other regulators on the basis of carcinogenicity, has led to the complete phaseout of carbon tetrachloride use (ATSDR, 2005; Morrison et al., 2006).

1.1.1.1.2   Trichloroethylene

TCE has been used in a multitude of applications (Shepherd, 1962; Irwin et al., 1997; Doherty, 2000a; European Chemicals Bureau, 2000b; HSIA, 2001; Mohr, 2001; ATSDR, 2005; Morrison et al., 2006):

- Dry cleaning, leather cleaning, and scouring wool
- Vapor degreasing of metal furniture, fabricated metal products, electric and electronic equipment, and military and transport equipment
- Cleaning liquid oxygen and hydrogen tanks in aerospace and other operations
- A carrier solvent for the active ingredients of insecticides
- A solvent in the rubber industry
- Adhesive formulations
- A solvent in printing inks, paints, lacquers, varnishes, adhesives, and paint strippers
- A solvent for dissolving and applying resin and rust-preventive coatings that dry as solvents vaporize
- A solvent in the pulp and paper industry
- A solvent in the textile industry for spotting fluids and as a solvent in waterless dyeing and finishing
- A solvent base for metal phosphatizing systems used to prepare metal surfaces for rust protection and painting
- A solvent in specialty glass production
- A component of household cleaners and consumer automotive products
- A component of drain cleaners and septic tank cleaners
- A component of metal polishes
- An extraction solvent for natural fats and oils such as palm, coconut, and soybean oils
- An extraction solvent for spices and hops and for removing caffeine from coffee
- A vapor-phase sterilizing agent for canned foods
- A solvent for the preparation of cosmetics
- A solvent for the preparation of pharmaceuticals
- An insecticidal grain fumigant
- An anesthetic in veterinary and medical procedures

The use of TCE as a skin, wound, and surgical disinfectant and as an inhalation anesthetic for childbirth and other short-duration medical procedures was banned in the regulations issued by the U.S. Food and Drug Administration in 1977 (ATSDR, 2005). In some developing nations, TCE continues to be used as an anesthetic even today. Because pharmaceutical grade TCE is no longer produced, industrial and reagent grades are used instead to substitute for more expensive anesthetic agents (Totonidis, 2005).

Metal degreasing consumed up to 80% of all TCE in European markets in the 1980s; similar consumption patterns were also present in North America in the 1970s (Doherty, 2000b; European Chemicals Bureau, 2004). In the 1950s and 1960s, TCE was used for aluminum degreasing in the United States, where chemists, for the major solvent producers, had worked out successful stabilization formulae. In Europe, however, reports of "spontaneous decomposition" of TCE, when used for vapor degreasing of aluminum parts, reflect a disparity in the art of stabilization of TCE between European and American solvent producers in that timeframe (Shepherd, 1962).

1.1.1.1.3 Perchloroethylene

The primary use of perchloroethylene is dry cleaning; about 60% of the 1991 perchloroethylene production was used for dry cleaning, whereas only 15% of that year's production was used for vapor degreasing (USEPA, 1994b). A minor use of perchloroethylene was as a dielectric fluid. Improvements in dry-cleaning machinery to minimize solvent losses, in response to increasing the regulatory oversight of perchloroethylene use in dry-cleaning operations, have greatly curtailed perchloroethylene consumption. Compared to other solvents, the solvent losses from perchloroethylene vapor degreasers were more easily controlled because of the higher vapor density of perchloroethylene (Dow Chemical Company, 2006d). By 1998, the total annual U.S. production of perchloroethylene had declined to 341 million pounds; the total consumption by dry cleaners had fallen to 40% of the total perchloroethylene production, and only 10% was used for metal degreasing. The rest was used as a feedstock for the production of fluorocarbons (chlorofluorocarbons and hydrofluorocarbon) for refrigerant, blowing agent, and fluoropolymer applications.

Perchloroethylene was preferred for cleaning alkali metals such as magnesium and aluminum, particularly before stabilized methyl chloroform became available, because perchloroethylene is more stable against reactions with metals than the other chlorinated solvents (Dow Chemical Company, 2001; Solvay SA, 2002a; Morrison et al., 2006). Perchloroethylene has seen uses similar to those of TCE, including vapor degreasing of metals and electronics. Perchloroethylene was preferred for removing wax and resin residues with high melting points, such as those used in the plating industry, because of its high boiling point. Its high boiling point results in a high-temperature vapor that completely dries wet work[*] because it vaporizes the moisture trapped in porous metals, deeply recessed parts, and blind holes (Dow Chemical Company, 2000). The high-temperature perchloroethylene vapor, compared with solvents that boil at lower temperatures, also provides a longer contact time because the room-temperature work—considered "cold" relative to the high-temperature solvent vapor—takes longer to warm to perchloroethylene's high boiling point (Morrison et al., 2006; Dow Chemical Company, 2006d). Perchloroethylene was also used in ultrasonic cleaning.

Perchloroethylene is used as an extractant in the production of pharmaceuticals and in veterinary formulations for deworming pets (NEMA worm capsules). It has also reportedly been used in the past as an anthelminthic[†] in the treatment of hookworm and some trematode[‡] infestations in humans on the Indian subcontinent and other subtropical regions (USEPA, 2001).

In addition to dry cleaning, perchloroethylene was used for cleaning leather; as a scouring, sizing, and desizing agent for processing textiles; as a carrier solvent for fabric dyes and finishes; and as a water repellant in textile manufacture. It was used in mixtures with grain protectants and some liquid grain fumigants. Perchloroethylene with a surfactant additive is preferred for vapor drying of glass products and some metal products requiring a high-quality surface finish. Aerosol formulations for the automotive aftermarket used perchloroethylene, particularly for brake cleaning (HSIA, 1999). Other applications were as an insulating fluid and cooling gas in electrical transformers; in paints, lacquers, and varnishes; in oil refineries for the regeneration of catalysts; in paint stripping; in cleaning printing equipment; in aerosol formulations; in polishes and lubricants; and as a solvent in various consumer products for household use (Doherty, 2000a; European Chemicals Bureau, 2004, 2005; Morrison et al., 2006).

1.1.1.1.4 Methyl Chloroform

Before being phased out in the mid-1990s owing to its listing as a stratospheric ODC, methyl chloroform was most commonly used for metal cleaning by cold cleaning, vapor degreasing, and

---

[*] In this book, the adjective "wet" is used in reference to water and not other liquids; the verb "wetting" takes on another meaning, as described in Chapter 2.

[†] Drugs that expel parasitic worms (helminthes) from the body.

[‡] Parasitic flatworms, also called flukes. Some species chemically castrate their hosts: a capsule of perchloroethylene was most likely seen as a benign and preferred alternative!

ultrasonic cleaning baths. It reportedly first became widely used in the late 1950s as a cold-cleaning solvent for metal-cleaning applications that previously used carbon tetrachloride, because methyl chloroform was significantly less toxic (HSIA, 1994; Doherty, 2000b). Because of the inherent chemical instability of methyl chloroform vapors in the presence of aluminum and other alkali metals, it was not widely used for vapor degreasing until a suitable stabilized vapor-phase formulation became available. As discussed further in Section 1.2.6.6, the first stabilizer that enabled methyl chloroform use in vapor degreasing of aluminum and other metals was 1,4-dioxane. The addition of 1,4-dioxane to methyl chloroform was first patented in 1957 (Bachtel, 1957).

Methyl chloroform was commonly used as a solvent in aerosol formulations and in aerosol propellant formulations for various household products sold in spray cans. As an aerosol solvent, methyl chloroform was used to reduce the flammability of hydrocarbon propellants (HSIA, 1994). A number of consumer spray aerosol products contained methyl chloroform, including hair sprays, cosmetics, oven cleaners, spot removers, furniture polishes, automotive lubricants, automotive choke cleaners, water repellents, and adhesives (Aviado et al., 1976).

Methyl chloroform was promoted by Dow Chemical as a dry-cleaning solvent in the early 1980s. It was used as a primary dry-cleaning solvent (particularly in leather cleaning) by a very small number of dry-cleaning operations—only 50 plants in the United States. The instability of methyl chloroform led to problems with machine and equipment corrosion; therefore, it is no longer used as a dry-cleaning solvent, but is still found in some spot-removal solutions at dry cleaners that use perchloroethylene, hydrocarbon, or other solvents (Linn et al., 2004). Methyl chloroform has also been used in spray products to condition suede and for glossing and weatherproofing leather products.

In addition to its major application as a metal cleaner, methyl chloroform was also used in the following products and processes, among others (World Health Organization, 1990; USEPA, 1994a; Doherty, 2000a):

- Automotive products
- Coatings and finishes
- Consumer and industrial adhesives
- Various paints, coatings, and inks
- A textile-scouring agent
- A cleaner for machinery and tools used in the textile industry
- A dye carrier
- An electronics cleaner to remove flux and dry-film photoresist developer from printed circuit boards
- A grease cutter in drain cleaners and septic tank cleaners
- A cleaner for plastic molds
- A pharmaceutical extractant
- A plastic film cleaner in the film and television industry
- A solvent for various insecticides
- A degreening product for citrus fruits
- A postharvest fumigant for strawberries
- A solvent for natural and synthetic resins, oils, waxes, tars, and alkaloids
- A coolant in cutting oils
- An asphalt-extracting agent for paving aggregate
- A solvent for the application of water and oil repellents to paper and textiles

Methyl chloroform was banned from use as the active ingredient in pesticides in 1998. It was considered excellent for cleaning plastics, polymers, and resins because it is less likely to degrade these materials than perchloroethylene or TCE (Tarrer et al., 1989). Tables 1.3 through 1.5 show the end-use pattern for TCE, perchloroethylene, and methyl chloroform, respectively, for the United States.

TABLE 1.3
Estimated U.S. End-Use Pattern for Trichloroethylene

Year	Metal Degreasing (%)	Chemical Intermediate (%)	Export (%)	Other (%)
ca. 1972	87	—	8	3
1985	80	5	10	5
1983	66	5	22	7
1997	65	35	—	—
1999	42	54	—	4

Sources: National Library of Medicine, 2006, Hazardous Substances Data Bank (HSDB). National Institutes of Health. http://toxnet.nlm.nih.gov/ (accessed March 15, 2006); Halogenated Solvents Industry Alliance (HSIA), 2001, Trichlorethylene white paper. http://www.hsia.org/white_papers/paper.html (accessed April 2002); U.S. Environmental Protection Agency (USEPA), 1989a, Locating and estimating air emissions from trichloroethylene and perchloroethylene. U.S. Environmental Protection Agency, Office of Air Quality Planning and Standards, Research Triangle Park, NC. EPA/450/2-89/013; and Doherty, R.E., 2000a, *Journal of Environmental Forensics* 1: 69–81.

TABLE 1.4
Estimated U.S. End-Use Pattern for Perchloroethylene

Year	Dry Cleaning (%)	Metal Degreasing (%)	Chemical Intermediate (%)	Export (%)	Other (%)
1962	90	—	—	—	—
1967	88	—	—	—	—
1974	59	21	6	11	3
1980	53	21	12	—	12
1989	50	15	28	—	~10
1998	25	10	50	—	15
2000	21	18	50	—	11

Sources: National Library of Medicine, 2006, Hazardous Substances Data Bank (HSDB). National Institutes of Health. http://toxnet.nlm.nih.gov/ (accessed March 15, 2006); Halogenated Solvents Industry Alliance (HSIA), 1999, Perchlorethylene white paper. http://www.hsia.org/white_papers/paper.html (accessed April 2002); U.S. Environmental Protection Agency (USEPA), 1989a, Locating and estimating air emissions from trichloroethylene and perchloroethylene. U.S. Environmental Protection Agency, Office of Air Quality Planning and Standards, Research Triangle Park, NC. EPA/450/2-89/013; and Doherty, R.E., 2000a, *Journal of Environmental Forensics* 1: 69–81.

1.1.1.1.5   Dichloromethane

Dichloromethane became an attractive replacement solvent as regulations were promulgated to protect worker health and safety in the 1970s and the ozone layer in the 1980s and 1990s. It has no flash point under conditions of normal use and therefore reduces the flammability of other solvents when mixed as an azeotrope. Among the benefits promoted for the use of dichloromethane in the solvents industry are its atmospheric characteristics. Compared to other chlorinated solvents, dichloromethane is not a major contributor to atmospheric pollution through the formation of smog, to the depletion of the stratospheric ozone layer, or to global warming (HSIA, 2003).

Dichloromethane provides a higher solvency than TCE, perchloroethylene, or methyl chloroform. Its aggressive solvency makes it an ideal paint remover, and it does not harm wood in the removal process (Dow Chemical Company, 2002; HSIA, 2003). Paint-stripping formulations make up the largest end use of dichloromethane; it has a unique ability to penetrate, blister, and lift a wide

TABLE 1.5
Estimated U.S. End-Use Pattern for Methyl Chloroform

Year	Vapor Degreasing (%)	Cold Cleaning (%)	Aerosols (%)	Adhesives (%)	Coatings and Inks (%)	Electronics (%)
1980		70	5	—	—	3
1984	22	40	10	12	1	6
1985	22	41	7	10	2	6
1989	34	12	10	8	5	4
1991		50	11	11	8	4
1995		25	—	5	3	5

Sources: U.S. Environmental Protection Agency (USEPA), 1994a, Locating and estimating air emissions from methyl chloroform. U.S. Environmental Protection Agency, Office of Air Quality Planning and Standards. EPA/454/R-93/045 and Kavaler, A.R. (Ed.), 1995, *Chemical Marketing Reporter* 247 (February 27): 9, 37.

variety of paint coatings. Dichloromethane is found in industrial and commercial furniture strippers and in home paint removers, and it is used extensively in both flow-over and dip tanks in furniture-refinishing operations. It is also used in aircraft-maintenance formulations to inspect surfaces for damage (ATSDR, 2000).

Dichloromethane is also used extensively as a foam-blowing agent for the production of softer grades of flexible polyurethane foams for furniture and bedding industries. Evaporation of the solvent during production causes expansion of the cells in urethane polymer foam, which reduces the urethane density while maintaining its flexibility. Dichloromethane controls the temperature of the reaction, which may otherwise become sufficiently high to burn or scorch the foam interior (HSIA, 2003).

Dichloromethane use as a propellant in aerosols such as insecticides, hair sprays, and spray paints increased significantly after the regulation of chlorofluorocarbons. Dichloromethane was a replacement for fluorocarbons in aerosols (World Health Organization, 1984b). Dichloromethane is used in aerosols as a strong solvent, a flammability suppressant, a vapor-pressure depressant, and a viscosity thinner. However, labeling requirements and health and environmental concerns led to a decline in the use of dichloromethane in consumer aerosol products in the late 1980s. Dichloromethane was used in hair sprays before it was banned by the Food and Drug Administration in 1989. It continues to be used in specialized spray paints and lubricants (ATSDR, 2000).

Table 1.6 provides the approximate breakdown of the major uses of dichloromethane from the 1970s through the 1990s; usage figures vary widely by information source and year.

Dichloromethane is used as a reaction and recrystallization solvent in the extraction of several pharmaceutical compounds and in the production of many antibiotics and vitamins. It has been used as a carrier to coat tablets. Dichloromethane is eliminated from the coating of the finished tablet in order to comply with Food and Drug Administration regulatory standards.

A wide variety of nonpharmaceutical applications have been found for dichloromethane (World Health Organization, 1984b; USEPA, 1994c; ATSDR, 2000; European Chemicals Bureau, 2000d; HSIA, 2003; National Library of Medicine, 2006). It has been used in cold and vapor degreasing of metals. It is involved in the manufacture of synthetic fibers and polycarbonate resin used for the production of thermoplastics. Dichloromethane is a component in fire-extinguishing products and serves as a coolant and refrigerant. It is used as a solvent in the production of cellulose triacetate as a base for photographic film, in the solvent welding of plastic parts, as a metal-finishing solvent, and as an extraction solvent for naturally occurring, heat-sensitive products including edible fats, cocoa butter, caffeine, and beer flavoring in hops. Dichloromethane serves as a releasing agent to prevent the manufactured part from permanently bonding to the mold and for cleaning electronic

TABLE 1.6
Estimated U.S. End-Use Pattern for Dichloromethane

Year	Paint Removers (%)	Metal Cleaning[a] (%)	Foam Blowing (%)	Aerosols (%)	Pharmaceuticals (%)	Chemical Intermediates (%)	Other Uses (%)
1978	30	22	5	17	5	—	21
1981	30	11	6	20	6	11	16
1985	30	5	15	30	—	—	20
1989	28	8	9	18	—	11	26
1995	40	13	6	—	6	10	22

Source: Compiled in National Library of Medicine, 2006, Hazardous Substances Data Bank (HSDB). National Institutes of
   Health. http://toxnet.nlm.nih.gov/ (accessed March 15, 2006).
[a] Includes vapor degreasing and cold cleaning combined.

components. In the agricultural sector, the chemical has applications as an insecticidal postharvest fumigant for grains, as a fumigant for strawberries, and as a ripening or "degreening" agent for citrus fruits. The dental profession makes use of dichloromethane in pour molding of dental material; a 50:50 mixture of dichloromethane and methyl methacrylate forms a cold-curing monomer to treat acrylic teeth used to improve bonding.

Using dichloromethane to make decaffeinated coffee met with Food and Drug Administration approval; however, most coffee processors no longer use it for this purpose because of concerns over the consumer perception that a minor amount of dichloromethane may remain in the coffee as a residue (ATSDR, 2000). Nevertheless, improvements to making decaffeinated coffee with dichloromethane now produce coffee that meets the requirements of the Food and Drug Administration for the elimination of residues (HSIA, 2003).

1.1.1.2   Cold Cleaning

The manufacture of parts and products using metals and other materials often leaves them soiled with residues of cutting oils, machine oils, buffing compounds, soldering flux, rosins, lubricating grease, finely divided metal grindings, drill turnings, and residual powders. These soils must be removed before coatings such as rust preventives, primers, paints, waterproofing agents, varnishes, polishes, or other finishes can be applied. Many of these soils are insoluble in water. The two main methods for cleaning parts are alkali detergent washes and solvents in both liquid and vapor phases. The general categories of solvents used in washing manufactured parts include petroleum solvents, alcohols, ketones, and chlorinated aliphatic hydrocarbons.* Chlorinated solvents have been preferred for degreasing parts because they have high solvency for both cold cleaning and vapor degreasing. Their higher vapor density makes it easier to control the vapor level in a vapor degreaser, and their lower heats of vaporization make them more energy and cost effective than other solvents with higher solvency such as acetone and methanol. Chlorinated solvents also have low flammability compared to acetone and methanol (Archer and Stevens, 1977).

"Cold cleaning" refers to several common parts-washing technologies that use tanks for dipping soiled parts in a solvent at room temperature. Cold cleaning also refers to wiping or spraying parts or large equipment such as aircraft wings and fuselages, locomotive engines, and automotive engines. Cold-cleaning systems most often consist of simple tanks fitted with a lid and filled with a

---

* Chlorinated aliphatic hydrocarbons are straight-chained hydrocarbon compounds in which chlorine atoms are
   substituted for hydrogen atoms—for example, carbon tetrachloride, methyl chloroform, dichloromethane, TCE, and
   perchloroethylene.

solvent. Cold cleaning in open tanks usually employed a layer of water atop the solvent to prevent solvent losses from evaporation (Gregersen and Hansen, 1986). Baskets filled with parts are immersed in the solvent tank and allowed to soak long enough to loosen and dissolve oils, grease, rosins, or flux. In processing printed circuit boards, a photoresist stripper employs cold cleaning by using an in-line design to dip parts on a conveyor rack (USEPA, 1989b).

Physical agitation of parts in a solvent bath aids cleaning. Operators use brushes or spray wands fitted with nozzles to clean parts, or they may use mechanical agitators or flow provided by circulation pumps. Circulation pumps produce solvent flow at the desired velocity. Agitation is achieved with oscillators that move parts back and forth through the cleaning solution or with ultrasonic units that use sound waves to induce high-frequency vibrations in and around the part (AFCEE, 1999; PRO-ACT, 1999; Dow Chemical Company, 1999b).

In ultrasonic cleaning, the cleaning action of the solvent is supplemented by sonically induced cavitation, which produces intense physical cleaning. Electricity at approximately 20 Hz powers a crystal or magnetic transducer, which expands and contracts with each cycle to generate shock waves in the solvent. Shock waves cause the rapid formation and collapse of low-pressure bubbles throughout the solvent, which creates a vigorous scrubbing action. Ultrasonic cleaning is often utilized in the manufacture of electronic components and printed circuits as part of a cold-cleaning or vapor-degreasing operation. Methyl chloroform is a common ultrasonic cleaning agent (Considine, 1974). Ultrasonic cleaning is especially useful for cleaning metal chips from blind holes or removing small insoluble materials from fine machined surfaces, such as particulate residue from honing operations and some buffing compounds. Ultrasonic cleaning is most efficient in nonagitated liquids at temperatures 25–40°C below the solvent's boiling point (Dow Chemical Company, 1999b).

Methyl chloroform was preferred for cold cleaning because its worker-exposure threshold was significantly higher than that of other chlorinated solvents. Cold cleaning applications for methyl chloroform included maintenance cleaning of electronic components and precision instruments, small parts, aircraft parts, and other parts that require a high degree of cleanliness (USEPA, 1989b). TCE, perchloroethylene, and dichloromethane, while being very effective as cold-cleaning solvents, have lower worker-exposure thresholds and are regulated as hazardous air pollutants and as volatile organic compounds (USEPA, 1994a). In 1987, 70 million pounds of chlorinated solvents were used by cold cleaners in the United States. However, the amount of chlorinated solvents used for cold cleaning decreased dramatically as methyl chloroform was phased out following the Montreal Protocol in the 1990s. Methyl chloroform was replaced primarily with dichloromethane, but also with 1-bromopropane, limonene, terpenes, and other "designer" solvents (Petroferm, Inc., 1997; Dow Chemical Company, 1999b). Solvents used in immersion tanks were usually recycled on-site by filtering or distilling the solvent (AFCEE, 1999; PRO-ACT, 1999).

1.1.1.3 Vapor Degreasing

Degreasers have played a vital role in industrial manufacturing, particularly in the automobile, heavy equipment, aircraft, appliance, railroad, and electronics industries. Approximately 25,000 vapor degreasers operated in the United States in 1979 (Tarrer et al., 1989). Degreasing is favored for the removal of oils imparted to the item being cleaned ("the work") from cutting, molds, dies, welding, soldering, stamping, and processing, as well as for the removal of soil carried in cutting oils, such as metal fines and fluxes. Parts that have been die cast, stamped, machined, welded, soldered, or molded carry residues that can be cleaned by vapor degreasing. Vapor degreasers introduce solvents to small inaccessible crevices or "blind holes," allowing quick and efficient cleaning and drying (Mertens, 2000a). Vapor degreasing is used for removing mineral oil contamination prior to electrocleaning and subsequent electroplating.

Vapor degreasing produces very clean, dry parts, suitable for subsequent finishing work such as the application of rust protection, primer, or paint. Clean dry parts are needed to facilitate quality control inspection and for subsequent assembly of machines. Parts requiring further metal work, such as welding, heat treatment, or machining, also benefit from vapor degreasing between steps to remove all traces of processing oils. Welding oily metal parts forms carbonaceous deposits because of pyrolysis of organic matter at high temperatures; such deposits are extremely difficult to remove. Vapor degreasing oily metal parts before welding greatly reduces the subsequent cleaning requirements. Electron-beam welding requires that parts be completely free of oily wastes and is therefore preceded by high-quality vapor degreasing (Reuter, 2002). Vapor degreasing is also a final step before packing to deliver clean parts and before shipping copper, aluminum, and plated chromium, zinc, and silver products in a clean, bright, and shiny condition [American Society of Testing and Materials (ASTM), 1962].

1.1.1.3.1 Types of Vapor Degreasers

Three types of vapor degreasers are used for solvent cleaning: batch vapor, in-line machines, and non-air-interface machines. Most of the degreasing machines operating since about 1960 fall into the batch vapor category. Conveyor systems such as monorail degreasers are in-line machines. Non-air-interface machines—vacuum degreasers that use subatmospheric pressure in the cleaning chamber—were introduced in the late 1990s to eliminate emissions and worker exposure. Energy costs in vacuum degreasers are reduced because parts can be dried at temperatures lower than the boiling point of the solvent. With a solvent such as perchloroethylene that has a high boiling point, it is possible to clean heat-sensitive parts at lower temperatures and still get them completely dry under vacuum, but the cycle time is longer (Dow Chemical Company, 1999b).

1.1.1.3.2 Vapor Degreaser Operations

Vapor degreasing systems typically consist of an open tank that is partially filled with solvent. A heater in the bottom of the tank heats the solvent to its boiling point, which generates vapors. Solvent vapors fill the tank to a level controlled by the elevation of the degreaser's cooling or condenser coils, positioned around the top of the tank's perimeter. These coils circulate cold water or a refrigerant, creating a layer of cold air that confines vapors to the tank and minimizes vapor escape. The condenser coils chill the air in the freeboard zone above the desired solvent vapor elevation. National Emission Standards for Hazardous Air Pollutants (NESHAP) require that refrigeration must create a cool air zone that is 30% or less of the solvent's boiling point. Table 1.7 lists the minimum temperatures needed for the major chlorinated solvents (Thomas, 1995).

The work requiring cleaning enters the degreaser at room temperature. When the work enters the heated vapor zone inside the tank, cool surfaces cause condensation of the solvent on the soiled metal surface. The solvent then dissolves the oil, grease, flux, or rosin and loosens metal chips and other debris, which drip off into the bottom of the tank. Solvent vapor condensed by cooling coils is returned to a clean solvent reservoir, which overflows back into the boiling solvent compartment (Petering and Aitchison, 1945).

TABLE 1.7
NESHAP Temperature Requirements for Freeboard Refrigeration Devices in Vapor Degreasers

Solvent	Boiling Temperature [°C (°F)]	Required Air Blanket Temperature [°C (°F)]
Dichloromethane	40 (104)	12 (31)
Methyl chloroform	74 (165)	22 (50)
Trichloroethylene	87 (189)	26 (57)
Perchloroethylene	121 (250)	36 (75)

Sources: Thomas, T., 1995, Clean Air Act compliance for solvent degreasers—regulatory strategies for manufacturers affected by Clean Air Act amendments NESHAP for halogenated solvent cleaners. University of Tennessee Center for Industrial Services, Tennessee Department of Environment and Conservation.

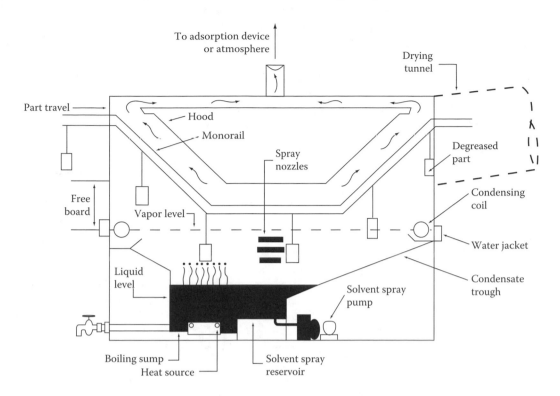

FIGURE 1.2 Configuration of typical offset vapor degreaser. (Adapted from University of Tennessee Center for Industrial Services, 1995.)

The cleaning action of a vapor degreasing system is dependent on the temperature differential between the vapor and the work. The efficiency of the process decreases as the work becomes warmer. Some degreaser operations use cool liquid solvent to presoak the work before bringing it into contact with solvent vapor; thereby extending the cleaning time, and promoting good solvent penetration from vapor condensation on the work (Petering and Aitchison, 1945). Some degreasing operations use multiple dips, where the work is moved from a dirty tank to subsequently cleaner tanks, followed by a final cleaning in clean solvent vapor (Dow Chemical Company, 1999b). The higher boiling point of perchloroethylene prolongs the condensation effect in a continuous degreaser because it takes more time to bring the cold work to perchloroethylene's higher vapor temperature. Compared to the solvents that have lower boiling points, perchloroethylene is therefore considered to have better cleaning power per unit of time that the work spends in a continuous vapor degreaser (Johnson, 1973).

Some degreaser systems are equipped with a wand that can be used to spray warm solvent onto the work to complete the cleaning process. The physical action of the spray pressure knocks off the strongly bound soils (AFCEE, 1999; Pro-ACT, 1999; Dow Chemical Company, 1999b). However, high-pressure sprays can disrupt the vapor blanket, leading to solvent vapor losses, unless care is taken to spray the work beneath the vapor blanket or inside an enclosed compartment of a vapor degreaser. A diagram portraying the features of a typical vapor degreaser is shown in Figure 1.2.

Solvent vapors usually remain below the level of the condensing coils; nevertheless, open-top degreasers routinely incurred substantial solvent losses. Solvent vapors diffuse from the dense solvent vapor blanket into the overhead air mass at a rate dependent on solvent vapor density and boiling point and the temperature differential with the ambient air. Tests have shown that vapor loss from an idling open-top vapor degreaser can reach 0.20 (lb ft^2)/h for TCE, 0.26 (lb ft^2)/h for dichloromethane, and 0.29 (lb ft^2)/h for perchloroethylene (Dow Chemical Company, 1999b).[*]

1.1.1.3.3 Solvent Losses

Understanding the rate of solvent losses from degreasers gives an appreciation of the frequency at which solvents were replenished. The rate of solvent handling provides information about the volume of frequent minor losses from spills, drips, container waste, and other factors contributing to solvent releases associated with topping off or changing out solvents.

Solvent loss occurred in open-top vapor degreasers when the vapor blanket was disrupted by drafts, movement of the work through the degreaser on a conveyor system, drag-out by the work as it exits the vapor blanket, incomplete drying of the work, and frequent opening of closed degreasers during operation. Other causes of solvent losses included diffusion, discarded solvent in degreaser and still sludges, and leaks in solvent pipes and valves in the degreaser. Operator guidance since the 1960s has advised that the work should move through the solvent vapor blanket no faster than 11 ft/min (ASTM, 1962; Thomas, 1995). Where problematic solvent levels are measured in the operator zone around the degreaser, operator guidance calls for slowing the conveyor rate to 3 ft/min (Howell and Tarrer, 1994). Solvent may also be lost when the condensed solvent fails to drain because of "cupping" attributes in the shape of the work, leading to liquid solvent carryout. Estimates for solvent losses due to these combined factors are as high as 70% of the solvent added to a degreaser over the course of a year (Mertens, 2000a). A description of degreasing practices from the early 1960s detailed an approach to solvent vapor management in which no hood is needed; in this approach, a large portable blower and roof ventilator are sufficient to vacate solvent vapors (Stoddard and Wells, 1962).

Solvent consumption rates due to emissions have been estimated at 1–2 gallons solvent per ton of small parts cleaned when a covered, in-line conveyor system is used, whereas small parts cleaned at a similar rate in manually operated open-top degreasers consumed 20% more (ASTM, 1962). The solvent and manufacturing industries substantially decreased solvent emissions from degreasing equipment over the decades during which chlorinated solvents were used for vapor degreasing. Improvements were motivated first by cost savings and later by regulation and enforcement of emissions standards. Today, equipment is designed with vapor locks, vapor-detection devices, and other controls, so that solvent emissions are substantially reduced. Even so, vapor degreasers equipped with emission-control devices can still lose 5–25% of the solvent through vapor escape. In enclosed (non-air-interface) degreasers, almost all of the solvent is captured and condensed for reuse. Solvent losses in enclosed degreasers are less than 5% (Reuter, 2002).

Table 1.8 summarizes annual solvent loss to emissions and annual generation of waste solvent in degreasers studied at several Air Force bases in the southeastern states. The study shows that methyl chloroform degreasing operations generally produced more liquid waste than perchloroethylene degreasing operations. Differences in emissions and waste generation were also the result of differences in the quantity of parts cleaned and in the types of cleaning operation conducted in each vapor degreaser.

Table 1.9 provides a manufacturer's estimate of solvent losses at a major printed circuit board operation where Freon 113 was the primary solvent used. Although evaporative losses would be higher for Freon 113 than for chlorinated solvents, the values provide a general estimate of quantities released because of spills and leaks.

1.1.1.3.4 Water Separators

A smaller but significant source of solvent losses and possible releases to soil and groundwater involves the management of condensate water produced in vapor degreasers. Water enters degreasers when atmospheric water vapor condenses on the degreaser cooling coils, from moisture present on

---

[*] Based on tests in an idling open-top degreaser having a 24 × 58 in. (~10 ft^2) opening; units are pounds of solvent per square feet of exposed solvent per hour of idling.

TABLE 1.8

Annual Solvent Waste Generation and Emission Losses in Vapor Degreasers at Air Force Bases in Southeastern States in the Late 1980s and Early 1990s

Solvent	Reservoir Volume (gallons)	Batch Changes	Annual Usage (gallons)	Annual Waste (gallons)	Emissions Loss (%)
Perchloroethylene	35	6	1780	480	73
Perchloroethylene	50	17	1824	420	77
Perchloroethylene	110	4	2740	968	65
Methyl chloroform	50	10	654	275	58
Methyl chloroform	110	6	3850	990	74
Methyl chloroform	110	8	4180	1320	65

Source: Howell, S.G. and Tarrer, A.R., 1994, Minimizing pollution in cleaning and degreasing operations. EPA/600/SR-93/191.

the work being cleaned, from leaks in the cooling water coils or steam heating lines in the solvent boiler, and from water-soluble cutting oils and buffing compounds (ASTM, 1962; Dow Chemical Company, 1999b). As discussed further in Section 1.2, the presence of water in the degreaser may lead to hydrolysis of the solvent, removal of water-soluble solvent stabilizers, and corrosion of the metal surfaces in the degreaser and on the work. Presence of water in the solvent is indicated by spots or even rust on the metal parts after cleaning. Another indication of excess water in the solvent is the formation of a dense white fog in the vapor zone, called "ghosting."

Vapor degreasers are equipped with water separators. Condensed water and solvent drop into a trough below the condenser coils and drain to the separator. Because water floats atop the solvent and does not dissolve significantly in the solvent, the water can be siphoned off. A relatively moisture-free solvent passes through the solvent return line at the bottom of the water separator and is then discharged to the degreaser or the storage tank (ASTM, 1962; Dow Chemical Company, 1999b). Operator guidance called for flushing with fresh water and keeping the water separator full by periodically adding solvent and water. Flushing prevented the buildup of corrosive acids from hydrolysis of solvents and corrosion of water separator pipes and tanks. The guidance specified that the pH of

TABLE 1.9

Typical Solvent Losses in a Major Printed Circuit Board Manufacturing Facility

Solvent Handling Process	Cause of Loss	Solvent Lost (%)
Storage in holding tanks	Evaporative losses	1
	Leaks	1
Transport within facility	Spills	8
Vapor degreasing	Evaporative losses, leaky seals, etc.	12
	Drag-out	40
	Spills and leaks	8
On-site recovery by distillation	Evaporative losses, leaky seals	2
Hand cleaning	Evaporative losses	15
Off-site for recycling of waste solvent	—	18

Source: U.S. Environmental Protection Agency (USEPA), 1991, Conservation and recycling practices for CFC-113 and methyl chloroform. U.S. Environmental Protection Agency, Office of Air and Radiation. EPA/400/1-91/017.

Note: The primary solvent used at the facility was Freon-113; loss rates for other solvents may have lower evaporative losses.

water in the separator should be monitored. If operators found the water layer to be acidic, it was replaced or buffered with an addition of a small amount of sodium bicarbonate to maintain the pH above 7. Some operations flushed the water layer weekly (Dow Chemical Company, 1999b). Operators were also directed to open the valve that drained the water off the top of the water separator frequently to allow accumulated water to drain out (ASTM, 1962). Wastewater from degreasers can be expected to contain solvents at concentrations equal to the solvent's solubility in warm water.

The volume of water condensed in a vapor degreaser is largely a function of the humidity at the degreaser location. The fate of water removed from degreasers varied with regulations in effect while the degreaser was operated and with the degree of operator compliance, as well as regulatory outreach and enforcement. Prior to regulation, degreaser wastewater was discharged to sewers, dry wells, storm drains, evaporation ponds, or the ground, or placed in drums and sent to unregulated landfills. Under actively enforced regulation, degreaser wastewater is managed through pretreatment using carbon filtration or on-site distillation, and waste discharges are subjected to inspection and sampling by local agency industrial wastewater pretreatment inspectors. After the mid-1980s, solvent-contaminated wastewater was managed as a RCRA (Resource Conservation and Recovery Act) waste, and discharge to sewer lines was prohibited. However, a common release mechanism for solvents from degreasing operations before and after RCRA implementation was through spills onto cracked concrete on shop floors or from poorly maintained sumps beneath the degreasers.

1.1.1.3.5 Heating Solvents

Vapor degreasers heat solvents to their boiling points with electric, steam, or gas heaters. Accumulation of sludge and metal fines on the heating elements may cause "hot spots" that result in solvent decomposition and burnout of the heating elements (Campbell, 1962). Dow Chemical advised operators to remove degreasing sludge when the solvent reached boiling temperatures listed in Table 1.10. To remove sludge, operators boiled solvent down to within 2 inches of the heating coils while diverting the solvent return line to a clean drum. The solvent sludge mixture was then drained and placed in drums for further distillation or disposal. Solvent sludge mixtures approaching 60–70% oil are not recoverable by further distillation (Dow Chemical Company, 1999b).

In addition to avoiding hot-spot formation in the boiling solvent, accumulation of oil presented problems for critical cleaning applications. Most cutting oils have low vapor pressures; nonetheless, parts per million concentrations have been found in the vapor phase. The range of vapor pressures for oily wastes is small; solvent boiling temperature controls the quantity of oil in the vapor phase. Only a few parts per million oil could be found in vapors of dichloromethane, which has a low boiling point, but up to 800 ppm oil could be found in vapors of perchloroethylene, which has a high boiling point (Dow Chemical Company, 1999b).

1.1.1.3.6 On-Site Solvent Recovery in Degreasing Operations

The primary costs for operating vapor degreasers include solvent, energy for heating, and labor. New designs of degreaser covers and fume hoods to recover solvents from vapors were motivated by

TABLE 1.10
Solvent Boiling Temperature for Oil-Laden Solvent

Solvent	Normal Boiling Temperature [°C (°F)]	Boiling Temperature at 25% Oil [°C (°F)]
Dichloromethane	39.7 (104)	42.2 (108)
Methyl chloroform	73.8 (165)	78.8 (174)
Trichloroethylene	87 (189)	89.4 (194)
Perchloroethylene	121 (250)	125 (257)

Sources: ASTM (1962), Johnson and Wedmore (1983), and Dow Chemical Company (1999b).

rising solvent costs and by solvent losses of up to 70% in open-top vapor degreasers. Another incentive for solvent recovery from degreaser vapors was the introduction of standards for permissible worker exposure and restrictions on air emissions. The most common method for recovering solvent from vapors is carbon adsorption. Carbon cartridges designed to allow high air flows adsorb most of the solvent vapor. Periodically, steam is passed through the carbon bed to carry the vapors to a condenser. A water separator removes the water from the solvent, which is then recovered and returned to storage tanks. As discussed in Section 1.2.7.5, carbon adsorption also removes some solvent stabilizers from the solvent, while introducing water contamination to the solvent. Carbon bed desorption is also accomplished by using hot air flow on a daily 6–8-h cycle; solvent vapors are condensed and returned to the solvent reservoir in the degreaser, and the carbon is regenerated for further use (Reuter, 2002). Cryogenic solvent-recovery techniques are also used (Douthitt, 1990).

1.1.1.4   Dry Cleaning

Dry cleaning is the term of art for a number of textile-cleaning processes applied to clothing, fabric, rugs, draperies, and other textile articles, which uses petroleum solvents and halogenated solvents. The term "dry" refers to the absence of water as a primary means for laundering clothes. Clothes obviously need to be cleaned regularly; laundering with water and detergent is insufficient for some fabrics and some soils and does not produce the same degree of cleanliness as can be obtained with chlorinated solvents. The details of how perchloroethylene was used, filtered, recovered, discharged, and vented from dry-cleaning operations relate to the nature and volume of perchloroethylene releases and to the partitioning and release of solvent stabilizers, as discussed in Section 1.2.7.

Perchloroethylene became the solvent of choice for the dry-cleaning industry after surpassing petroleum solvents approximately in 1960. Carbon tetrachloride, TCE, methyl chloroform, and chlorofluorocarbons have also been used in the dry-cleaning industry. Today, more than 80% of dry cleaners use perchloroethylene; some still use petroleum solvents, and a growing number of operators are switching over to machines that use carbon dioxide or other solvents or cleaning agents deemed to be "green," with fewer regulatory restrictions and environmental consequences.

The dry-cleaning industry is composed of two main sectors: the commercial sector and the industrial sector. Previously, the coin-operated dry-cleaning sector was economically important, but machine inefficiency and regulation led to its elimination. Coin-operated dry-cleaning machines were banned in December 1994. In the early 1990s, the U.S. dry-cleaning industry included about 30,500 commercial dry cleaners, 1400 industrial cleaners, and 3000 coin-operated dry cleaners (USEPA, 1995). Industrial dry cleaners are large operations that clean uniforms, aprons, overalls, shop towels, or other items. Approximately 50% of the industrial dry cleaners use petroleum solvents. A typical industrial cleaner of clothing may have a 500-pound capacity washer-extractor and 3–6 dryers, each with 100-pound capacity. The annual solvent usage ranges from 200 to 1300 gallons per machine (USEPA, 1995).

1.1.1.4.1   Dry-Cleaning Process

The descriptions in this section apply primarily to dry-cleaning operations that used earlier-generation machines from the 1950s through the early 1990s. Operations from this timeframe contributed substantially more perchloroethylene to the subsurface than current or recent operations. Since 1986, there have been substantial improvements in solvent stewardship and dry-cleaner operator training, adoption and enforcement of regulations, and improvements in equipment (Mohr et al., 2007).

The perchloroethylene dry-cleaning process consists of washing clothes in liquid solvent, extracting solvent from the clothes, and drying the clothes. Two general categories of dry-cleaning machines have been used: transfer machines and dry-to-dry machines. Transfer machines perform the washing and extraction steps; clothes are then manually transferred to a separate dryer. Dry-to-dry machines perform all the three steps in the same machine (USEPA, 1980).

Dry cleaning involves two wash cycles, each followed by drying. In the first wash, soiled garments are loaded into a rotating, perforated, stainless-steel cylindrical basket with a capacity varying from 20 to 100 pounds of clothes or fabrics. Clothes are completely soaked and tumbled in detergent-charged solvent for 8–12 min at a temperature ranging from 70–85°F. A typical machine may pump perchloroethylene into the rotating basket filled with soiled clothing at a rate of 1500 gallons per hour; for an 8-min cleaning cycle, soiled clothes would be soaked in 200 gallons of solvent (Cantin, 1992). The detergent-charged solvent dissolves greases and oily stains and is continually filtered during the wash cycle. Soil-laden solvent is first pumped through a button trap to remove pins and buttons from the solvent to prevent damage to the pump. Charged solvent is then pumped through either a disposable cartridge filter, a spin-disk filter, or a regenerative or flex-tube filter to remove soil, lint, and other insoluble materials (CARB, 1996).

A typical dry-cleaning machine has one or two perchloroethylene solvent tanks: one for the pure solvent and the other for the "charged" solvent, made up of distilled solvent with a small amount of detergent and water to help clean water-soluble stains. Solvent is recycled through continuous distillation. Fresh solvent is added to the solvent tanks to replace the solvent lost to emissions, filters, and water separation.

To extract perchloroethylene from washed clothes, the solvent bath is drained; then clothes are spun at high speed to wring out excess solvent. Both transfer and dry-to-dry machines combine the washer-extractor functions in a single unit. In transfer machines, an operator moves clothes from the washer-extractor to the dryer-reclaimer. In the solvent-reclaiming dryer, heated air is blown over the clothes in a closed, recirculating loop to vaporize the solvent, which is recovered in a water-cooled condenser. After a timed drying step lasting 15–20 min, fresh air is blown over the clothes for approximately 5 min to complete the drying cycle and aerate the clothes. The air stream and perchloroethylene vapors are vented to a carbon adsorption system (USEPA, 1980).

In the second wash, clothing is rinsed in fresh, pure solvent for 4–5 min to remove detergent residue. Like the charged solvent, the pure solvent is filtered continuously by one of the three filtration systems described in Section 1.1.1.4.9. Rinsed clothing is rapidly spun to extract the solvent. The spent solvent is drained into the charged solvent tank and is used for the first wash for the next load of soiled clothing.

Freshly cleaned clothing is heated to about 160°F for approximately 40 min. The remaining perchloroethylene fumes and solvent are vaporized by warm air and then condensed over cooling coils to recover the solvent. The residual amount of perchloroethylene removed from clothing following the second wash ranges from 30 to 60 pounds for 100 pounds of clothing (Cantin, 1992). Most of the perchloroethylene is removed from the clothing via the last two extraction and drying steps; however, trace levels of perchloroethylene may remain in dry-cleaned clothing.

1.1.1.4.2 First-Generation Dry-Cleaning Machines

Transfer machines first appeared on the market in the 1930s and were used exclusively until the 1960s (Schmidt et al., 2001). In 1995, 34% of the dry-cleaning machines in the United States were transfer machines. Transfer machines were banned in California on October 1, 1998 (Izzo, 1992). Transfer machines usually included a washer-extraction unit and a separate dryer unit. Clothing and solvent are agitated in the drum, which is then spun at high speed to extract the solvent. An attendant must manually transfer the clothing to a dryer. The clothing is tumbled in recirculating warm air to vaporize the residual solvent. After the drying cycle, cool air is circulated through the clothing to reduce wrinkling. Some transfer machines comprised a washer unit, a separate spin-dry extraction unit, and a tumbler dryer or reclaimer to vaporize the remaining solvent and fumes.

Transfer machines allowed increased production because a new load of clothing can be washed while the previous load is drying. Compared with dry-to-dry machines, maintenance of transfer machines was easier because of their basic construction with less automation. There was also less fabric damage because the drum remains cool after the previous load is removed (USEPA, 1995). The disadvantages of transfer machines included additional labor needed to manually transfer the solvent-soaked clothing from the washer-extractor to the dryer, worker exposure to solvent vapors during transfer of clothing, and loss of solvent during transfer.

To minimize worker exposure to perchloroethylene vapors during transfer of clothing from the washer to the dryer, a hamper enclosure made of impervious plastic was used to enclose the clothing when the clothing was transferred from the washer to the dryer. In a second approach, a room enclosure, composed of a metal frame covered with an impervious plastic, surrounded both the washer and dryer units and drew solvent vapors through a device to control vapor emissions. First-generation machines consumed and emitted more solvent than later-generation machines. Perchloroethylene consumption in first-generation machines was typically in the range of 78–100 kg per 1000 kg of clothes cleaned, depending on the primary control technology as well as on operator and maintenance practices (CARB, 1996).

1.1.1.4.3   Second-Generation Vented Dry-to-Dry Machines

Second-generation dry-cleaning machines were the vented dry-to-dry machines introduced in the late 1960s. Washing, extracting, and drying were done in one unit, thereby avoiding the transfer of solvent-soaked clothing from the washer to the dryer. Second-generation machines were originally designed to release residual vapors to the atmosphere, but they could be retrofitted with a vapor-control device such as carbon absorbers or refrigerated condensers. In 1995, of U.S. dry cleaners, 66% had vapor-control devices consisting of refrigeration units, and 32% had vented units without vapor-control devices (USEPA, 1995).

Carbon adsorbers were retrofitted to both transfer and vented dry-to-dry machines to improve solvent recovery. Instead of venting solvent vapors directly to the atmosphere, retrofitted machines recover solvent by directing the solvent-laden vapors through an activated carbon bed that adsorbs the solvent. The adsorbed solvent is recovered by passing low-pressure steam or hot air through the carbon bed; then the mixed steam and solvent vapor are passed through a water-cooled condenser and collected in a phase separator. The desorbed activated carbon bed is dried and reused, and the recovered solvent is returned to the solvent tank. The removal efficiency is about 85% compared to noncontrolled vented dry-to-dry machines (USEPA, 1995).

1.1.1.4.4   Third-Generation Closed-Loop (Nonvented) Dry-to-Dry Machines

Third-generation or closed-loop dry-to-dry machines with built-in refrigerated condensers came on the market in the late 1970s. A third-generation machine is essentially a second-generation machine retrofitted with a refrigerated condenser. Closed-loop machines do not vent solvent vapors to the atmosphere; instead they recycle it continually throughout the dry-cleaning cycle. In 1995, 34% of dry-cleaning facilities used third-generation machines.

1.1.1.4.5   Fourth-Generation Closed-Loop (Nonvented) Dry-to-Dry Machines

In the early 1990s, fourth-generation machines came on the market. They feature a nonvented closed-loop dry-to-dry approach and have an additional internal vapor-recovery device, usually a built-in carbon adsorber installed in series with the refrigerated condenser. Another common feature of fourth-generation machines is secondary containment below the machine to capture accidental releases of perchloroethylene. Increased vapor controls in fourth-generation machines provided nominally higher solvent mileage than third-generation machines (CARB, 1996).

1.1.1.4.6   Fifth-Generation Closed-Loop (Nonvented) Dry-to-Dry Machines

Fifth-generation machines were introduced in the late 1990s. These machines are fitted with a single-beam infrared photometer to monitor the perchloroethylene concentration in the machine cylinder. An interlock on the machine door prevents the attendant from opening the door until perchloroethylene concentration in the cylinder falls below 290 ppm.

While fourth- or fifth-generation machines can achieve a factory solvent-mileage design rating of 10 kg of solvent per 1000 kg of clothes, operator and maintenance practices are critically important for achieving the design performance. During sustained commercial operation, however, an average solvent mileage of 20 kg of perchloroethylene per 1000 kg of clothing cleaned is a reasonable

TABLE 1.11
Solvent Mileage by Machine Type

Machine Type	Perchloroethylene Consumption per 1000 kg of Clothes Cleaned
First-generation transfer	78–100 kg (8–10%)
Second-generation dry-to-dry	77–94 kg (8–10%)
Third-generation closed loop	20–40 kg (2–4%)
Fourth-generation closed loop with refrigeration	10–20 kg (1–2%)
Fifth-generation closed loop with vapor locks	10–20 kg (1–2%)

Source: Mohr, T.K., et al., 2007, *Study of Potential for Groundwater Contamination from Past Dry Cleaner Operations in Santa Clara County.* Santa Clara Valley Water District, San José, CA.

target for the dry-cleaning industry as a whole. Therefore, fifth-generation machines provide a 5–10-fold improvement in solvent mileage compared to first-generation machines (USEPA, 1995).

1.1.1.4.7   Self-Service Coin-Operated Dry-Cleaning Machines

Self-service coin-operated dry cleaners were introduced to the market in the early 1960s. Coin-operated machines were usually located in self-service laundries and were operated by consumers. The coin-operated machines are small dry-to-dry machines with 8 to 10 pounds clothing capacity that used perchloroethylene. These machines were inefficient, and maintenance issues invariably led to spills and emissions violations. Coin-operated dry-cleaning machines were prohibited beginning in 1994.

Table 1.11 summarizes the perchloroethylene mileage for five generations of dry-cleaning machines.

1.1.1.4.8   Chlorinated Solvents Used for Dry Cleaning

By 1915, carbon tetrachloride provided a nonflammable alternative to the hydrocarbon solvents used for dry cleaning at the time. It was used by the majority of dry cleaners in the United States in the years prior to World War II. Carbon tetrachloride was replaced by perchloroethylene and TCE as dry-cleaning and degreasing solvents because they are less toxic, easier to recover, and easier to ship. By 1950, three times as many dry cleaners were using perchloroethylene compared to those using carbon tetrachloride. Carbon tetrachloride is no longer used in the dry-cleaning industry.

TCE was first used as a dry-cleaning solvent in Germany in the 1920s and was introduced in the United States for this purpose in about 1930. TCE was more easily reused and cleaned faster and more safely than carbon tetrachloride. However, TCE damaged certain types of fibers and synthetic substances. Because of human toxicity and photochemical reactivity concerns, TCE is no longer used in the dry-cleaning industry (Johnson, 1973).

Perchloroethylene was introduced to the U.S. dry-cleaning industry in the late 1930s. Perchloroethylene became the solvent of choice for most small dry cleaners because it is not flammable, it safely cleans virtually all types of fabrics without damaging the material or bleeding dyes, and it is less toxic than carbon tetrachloride. As of 1952, the dry-cleaning industry used 80% of the perchloroethylene produced. However, from that point forward, such usage decreased; from 1972 to 1975, perchloroethylene usage for dry cleaning decreased from 75% to 63% of the total production (Doherty, 2000a). The decline in perchloroethylene consumption corresponded to improvements in dry-cleaning equipment and vapor-recovery systems, as well as to trends toward wash-and-wear clothing. In 1975, one 55-gallon drum of perchloroethylene cleaned 8000 pounds of clothing, whereas in 1993, the same amount cleaned 16,000 pounds of clothing. By 1990, however, only 50% of the perchloroethylene produced was used for dry cleaning. Improved efficiencies in handling and recovery have caused dry-cleaner perchloroethylene consumption to substantially decrease although perchloroethylene still dominates the U.S. dry-cleaning industry.

In 1964, E. I. du Pont de Nemours and Company, introduced 1,1,2-trichloro-1,2,2-trifluoroethane, also known as Freon 113, under the trade name Valclene™ (Johnson, 1973). Valclene™ is less aggressive than perchloroethylene and evaporates from fabrics at lower temperatures and was thus useful for cleaning delicate fabrics. Valclene was preferred over perchloroethylene for rubber, painted fabrics, polystyrene, vinyl-coated fabrics, polyvinyl chloride fibers, some pigment colors, furs, and leathers (Johnson, 1973). Although Freon 113 had favorable attributes for delicate clothing, it was not widely used. It was banned in 1996 in accordance with the Montreal Protocol because of its potential to deplete ozone and contribute to global warming (Johnson, 1973; CARB, 1996).

In the early 1980s, Dow Chemical introduced methyl chloroform as a dry-cleaning solvent (Linn et al., 2004). It was not widely used in the dry-cleaning industry as a primary dry-cleaning solvent because it damages delicate fabrics. Methyl chloroform is also somewhat unstable when mixed with water, where it breaks down and forms acids that corrode dry-cleaning machines (CARB, 1996). Methyl chloroform was also restricted from most uses following the passage of the Montreal Protocol and Clean Air Act Amendments (Johnson, 1973; CARB, 1996).

1.1.1.4.9 Filtering Perchloroethylene

Three types of disposable cartridge filters are used by 90% of dry cleaners: carbon filters, paper filters, or combination paper and carbon filters. Disposable cartridges are changed after cleaning 450–1000 pounds of clothing. Operators observe the solvent line pressure to determine when filters need to be changed, as indicated by increased pressure in the solvent line behind the filter (Cantin, 1992; CARB, 1996). Operators typically drain used or spent cartridges for 24 h for paper filters or 48 h for carbon cartridges prior to reuse or disposal at a facility that conducts hazardous waste treatment. Spent cartridges can retain as much as 1 gallon of perchloroethylene before draining, and from 0.13 to 0.25 gallon perchloroethylene remains in the cartridge after draining (CARB, 1996). The filter cartridge became prominent in the dry-cleaning industry in about 1961.

A smaller number of dry cleaners use spin-disk filters. Spin-disk filters are made up of multiple high-surface-area polyester disks mounted in series to filter the dirty solvent. Spin-disk filtration systems may incorporate powdered activated carbon and diatomaceous earth. Filters are regenerated by spinning the disks and directing the solvent and the "caked" impurities through a drain valve into a distillation still. Spin-disk filters last longer and produce less waste than the cartridge systems, but skilled labor and frequent maintenance are required for proper operation (Hygnstrom, 1995).

A third type of filter is the regenerative or flex-tube filters, made of braided tubes coated with a combination of diatomaceous earth and activated carbon. The filter is cleaned by resuspending the powder and dirt in solution by a countercurrent of air and backwashing; the resuspension creates a mixture that is then redeposited back on the filter to form a highly porous "cake" and uniform filter media for the proper removal of contaminants. The regeneration action is repeated with each cleaning cycle. Regenerative filter systems are more expensive than spin-disk filters and require skilled labor to operate properly (CARB, 1996).

1.1.1.4.10 On-Site Distillation for Perchloroethylene Recovery

Approximately 90% of dry cleaners use distillation as a companion process to purify and recover used solvent (Cantin, 1992). Distillation stills are built into most modern dry-cleaning machines; stand-alone stills, called "muck cookers," were used for earlier-generation machines. Steam is used to heat the perchloroethylene to its boiling point (121°C/250°F). Solvent and water vaporize and rise to the top of the still, where vapors are cooled back into a liquid that then goes to a water separator. Sludge or "still bottoms"—composed of nonvolatile residues such as detergents, waxes, dyestuffs, sizing, oils, and grease—are removed from the sump where they accumulate and are disposed of as hazardous waste. The sludge or still bottoms can contain as much as 50–70% perchloroethylene. A "muck cooker" is used to recover most of the perchloroethylene. A cooker or still can recover as much as 95% of the solvent from the spent sludge (Morrison, 2002b).

1.1.1.4.11   Water Separator

Water separators have been used in solvent distillation stills for dry cleaners and vapor degreasers since the early 1900s. The distilled solvent and water mixture sits in a gravity separator unit. Heavier perchloroethylene drains from the bottom of the unit to the solvent tank, and excess water is drained off the top of the separator. In earlier decades, separated water was drained to floor drains or onto the ground; later it was routed to the sanitary sewer system (Morrison, 2002a). In the past, it was legal to discharge wastewater to the sanitary sewer system, and high levels of perchloroethylene in the separator water were routinely so discharged. For example, levels of perchloroethylene up to 1120 mg/L, with an average of 152 mg/L, were detected in samples of separator water; as much as 30% of the samples were pure solvent (Cohen and Izzo, 1992). Among older dry-cleaning machines (ca. 1986), discharges from water separators commonly caused soil and groundwater contamination (Hoenig, 2002). Leaking sewer lines also facilitated perchloroethylene release to the subsurface (Cohen and Izzo, 1992).

1.1.1.5   Off-Site Solvent Recycling

Industrial plant managers and industrial engineers found various ways of dealing with the costs of solvent consumption and disposal of waste solvent. Improvements in degreasing and dry-cleaning equipment significantly reduced consumption. As solvent costs rose, operators switched to more efficient equipment. When regulations prohibited land disposal of waste solvent, the cost of permitted disposal was a further incentive for improving the efficiency of solvent use in various operations and to use on-site solvent-recovery operations. Table 1.12 summarizes the relative ranks of solvents recycled in the United States in 1981.

Prohibitions of land disposal of waste solvent redirected most of this waste stream to the solvent-recycling industry, which collected still bottoms from degreasing operations for distillation and solvent reclamation. Solvent recycling was already a viable business before the adoption of land disposal prohibitions because it offered attractive economic incentives for the generators of waste solvent. Recovered solvents were sold, and distillation residues were discarded or treated as hazardous waste. Prior to the adoption and enforcement of the Resource Conservation and Recovery Act in 1985, still bottoms and other waste solvent from recycling operations were generally sent off-site; the waste went into landfills and deep injection wells or was burned as fuel for blast furnaces in the steel industry or as fuel for cement kilns. In 1985, because steel mills abruptly stopped taking waste solvent to avoid the burden of obtaining permits as hazardous waste storage facilities, cement kilns became a primary destination for waste solvent (Douthitt, 1990). The oil residue in waste solvent provided enough heat energy (8000–9000 Btu) to make the waste attractive as fuel for kilns and boilers, provided the metal and chlo-

TABLE 1.12

Ranks of Chlorinated Solvents by Relative Quantity Recycled in 1981

Solvent	Solvent Recycling Rank
Methyl chloroform	2
Dichloromethane	5
Trichloroethylene	7
Perchloroethylene	8

Source: Douthitt, C.A., 1990, Resource recovery through solvent reclamation. Presentation to the U.S. Department of Energy Industrial Solvent Recycling Conference October 15, 1990, Charlotte, NC. NASR (National Association of Solvent Recyclers), www.p2pays.org/ref/26/25692.pdf (accessed February, 2006).

Note: Other recycled solvent ranks: methyl ethyl ketone (#1), acetone (#3), xylene (#4), and toluene (#6).

rine contents were sufficiently low. Cement kilns burn waste at temperatures above 2600°F, assuring destruction efficiencies comparable to those of an incinerator (NCDHS, 1983; Dawson, 1989).

Recovered solvents were considered to be of good quality, meeting the same technical specifications required for virgin solvents, although solvent producers cautioned customers about the impurities in and instability of recycled solvents (NCDHS, 1983). Recycling companies restabilized reclaimed solvents and sold "restabilizers" to companies operating on-site solvent distillation units (NCDHS, 1983; see Section 1.2.7.7 for more information on replenishing stabilizers in reclaimed solvents).

Commercial solvent-recycling operations used two pricing models: custom toll recycling, which served generators of large quantities of waste solvent, and open-market recycling. In custom toll recycling, the generator's spent solvents are kept segregated, batch processed separately to the generator's specification, and then returned for reuse. The minimum batch size for toll recycling, determined by processing equipment capacity, was generally 1000–2000 gallons (Dawson, 1989). The toll recycling approach provided assurance to the generator that clean solvent returned for reuse would not contain any contaminants foreign to the generator's operation. The reclaimed solvent is returned to the generator at a price similar to or slightly lower than virgin solvent purchase, which was a cheaper alternative to the off-site disposal of the waste solvent (Dawson, 1989).

In open-market recycling, spent solvents are commingled with similar wastes from many generators and processed for resale as refined solvents. Because the recycler derives revenue from the resale of recycled solvents, the generator may pay a price lower than the alternative disposal cost, while obtaining assurance that the waste is managed in an environmentally acceptable manner.

The prevailing method for solvent reclamation is distillation or vaporization. Many solvents can be processed into a reusable state by simple vaporization in either a pot-still or a thin-film evaporator. In a pot-still evaporator, the spent solvent is heated to boiling by steam or hot oil heat exchangers. Solvent vapors exit the system to a water-cooled heat exchanger and are then condensed to provide clean solvent. The nonvolatile residues, or still bottoms, are pumped to storage where they await final disposal or transport to a cement kiln or boiler for burning.

In thin-film evaporators, waste solvent is spread in a thin layer on an internal surface by rotating wiper blades. The internal surface is heated by a steam or hot oil jacket, causing organic solvents to undergo flash evaporation. Vapors are condensed to produce clean solvent liquid. Nonvolatile residues continuously exit the bottom of the evaporator and are handled in the same manner as still bottoms. Thin-film evaporators are fed from an external storage tank and can therefore operate continuously. In thin-film evaporators, nonvolatile residues are removed without prolonged heating: they thereby avoid chemical breakdown that can cause odor or color problems in the reclaimed solvent (Dawson, 1989).

Both pot-still and thin-film evaporators can be operated under vacuum to reduce the boiling point of the solvent and the amount of energy required. Neither system is capable of separating different chlorinated solvents. If the waste solvent is properly segregated from the waste solvent of a different composition, the recovered solvent will be relatively pure. If volatile organic impurities are present in the waste, they will also be present in the reclaimed solvent. Some solvent recyclers install fractionating columns—including packed columns, bubble-cap columns, sieve-plate columns, and valve-tray columns—to separate solvents and ensure good quality control. Waste solvents are heated, and vapors enter the fractionating column. Components having lower boiling points migrate to the top of the column, whereas those with higher boiling points are condensed internally and subsequently migrate to the lower part of the column. As vapors with lower boiling points exit the top of the column, they are condensed to a liquid. Fractionating columns perform better when the boiling-point difference between the mixed solvents is large, generally at least 20–30°C (Dawson, 1989).

Solvent recyclers produce refined products used in almost all phases of industry, including pharmaceutical applications to manufacture drugs. Some solvent grades require further processing following reclamation, including extraction, filtering, restabilization, and drying (water removal). The greatest challenge faced by solvent recyclers is the removal of excessive moisture. The reuse of solvent with moisture contamination creates corrosion problems in degreasing equipment, coating problems in paint solvents, and hydrolytic breakdown problems in chlorinated solvents. Solvent recyclers remove water by using fractional distillation, if available, or physical or mechanical removal methods, such as filtering the refined solvent through a desiccant. Anhydrous calcium chloride is very effective in the removal of trace water contamination. Refined solvents contaminated by less than 0.5% moisture can be dried by calcium chloride down to 200 ppm moisture, but chloride contamination of the refined solvent may occur, which can cause corrosion problems and solvent breakdown. Ion-exchange resins are also used for the removal of trace water contamination. In solvents with less than 0.5% moisture, water molecules attach to the resin, and dry solvent passes through, reducing moisture levels to less than 100 ppm in the best possible case. Ion-exchange resins used to remove water from the solvent can be regenerated. The best available technology for moisture removal is the molecular sieve bed, which traps water molecules in the interstices of the molecular sieve, allowing the dry solvent to pass through. Molecular sieves can dry solvents contaminated with 5% water to less than 1% and can dry solvents with trace moisture from 0.5% down to less than 200 ppm. Molecular sieves can also be regenerated for continued use (Dawson, 1989).

1.2   ROLE OF SOLVENT STABILIZERS

Chlorinated solvents can become unstable when subjected to environmental stresses in the broad variety of industrial applications for which they are used. Solvents are exposed to a wide range of physical conditions and a great diversity of materials and substances that place unusual demands on solvent performance. Chlorinated solvents must perform equally well in both liquid and vapor phases and must remain inert to impurities. Solvents must remain stable under challenging physical and chemical conditions, including high temperatures, ultraviolet light, moisture and water vapor, exposure to alkali and acidic substances, and exposure to reactive metals.

As this section demonstrates, solvents often need to be fortified against the severe physical and chemical conditions of the various environments in which they are used. "Solvent stabilizers" is the term of art chosen for the group of chemicals added to chlorinated solvents to ensure that they will not break down during their intended industrial applications. Unstabilized solvents will deteriorate—some very quickly with dramatic and dangerous reactions, others slowly—over repeated use cycles and in storage. In the most demanding applications, operators must regularly replenish stabilizers to replace those consumed or lost during use. Stabilizer replenishment is commonly accomplished by adding fresh, stabilized solvent to make up for solvent losses during operation (USEPA, 1989b). The need to replenish spent stabilizers was documented as early as 1937 (Dinley, 1937). As detailed in Section 1.1, solvents are lost from the operation when vapors escape, when solvent dissolves into condensed water, and when solvent remains on parts or in textiles. Improvements to machines that use solvents, such as vapor degreasers and dry-cleaning machines, have minimized or completely eliminated solvent losses, as required by increasingly restrictive emissions regulations. Stabilizers are nevertheless consumed in newer equipment. Stabilizer replenishment is now accomplished with specialty stabilizer packages sold by the solvent manufacturer.

1.2.1   TYPES OF SOLVENT STABILIZERS AND SOLVENT BREAKDOWN

Solvent stabilizers, also called inhibitors (and in the early literature, "anticatalysts"), are selected for their ability to prevent the three main types of reactions that produce decomposition of chlorinated solvents, which leads to acid formation: (1) hydrolysis; (2) oxidation, initiated by exposure to air, by thermal breakdown, or by ultraviolet light; and (3) condensation reactions with alkali metal salts.[*] The main categories of solvent stabilizers are acid acceptors, antioxidants, and metal inhibitors. Ultimately, all stabilizers serve to mitigate the formation of acids, because each type of reaction

---

[*] A condensation reaction joins two molecules to create a complex molecule and a much simpler molecule as a by-product of the reaction. The simple molecule eliminated from the two reactants can be water, ammonia, or alcohol. Each reactant contributes to the eliminated molecule (Carey, 1987). The two combining molecules each contribute a single moiety (a portion of a molecule having a characteristic chemical property) to the eliminated molecule.

produces acids. Antioxidants prevent reactions that form acids. Acid acceptors act to neutralize acids once they form in the solvent or are introduced into the operation from the work. Metal inhibitors prevent reactions between the solvent and alkali metals and their salts.

Solvents stored in drums may eventually deteriorate and form enough acid to corrode the drums. This happens when drums were not stored with airtight bungs or seals. Water vapor entering a drum will condense and form liquid water, which floats atop the solvent. Hydrolysis of impurities in the solvent carried over from production will form acids that then progressively deteriorate the solvent. A ring of pinholes at the solvent–water line near the top of the drum is a telltale sign of solvent deterioration. Various stabilizing formulations were developed to mitigate acid formation during solvent storage. Early solutions that stabilized TCE, perchloroethylene, and dichloromethane against acid formation during storage included addition of gasoline; however, large quantities were required, and the benefit of using a relatively nonflammable cleaning agent was negated (Pitman, 1933). Amylene, also known as 1-pentene, was added to TCE as an acid inhibitor more than 75 years ago (Harris, 1933). A 1959 patent notes that metal tanks corroded and failed within months because of the formation of acid that corroded tanks (Beckers, 1959). Such corrosion leads to leaks and contamination of underlying soil and groundwater.

Perchloroethylene is susceptible to auto-oxidation when exposed to ultraviolet light in the presence of moisture. And in the absence of a metal stabilizer, methyl chloroform exposed to aluminum salts will undergo a violent reaction that quickly reduces the solvent to a tarry mass and evolves phosgene gas with potentially harmful or even fatal consequences to the equipment operator.

Carbon tetrachloride corrodes metals and brass in particular. Early fire extinguishers used carbon tetrachloride in brass containers. Valves on these carbon tetrachloride fire extinguishers corroded after lengthy storage and prevented them from discharging in the moment of need (Snelling, 1914). Solvent stabilizers were proposed to remedy the problem in a 1913 patent application filed in Pittsburgh, Pennsylvania.

To understand the challenges of solvent stabilization, some familiarity with the mechanics of industrial applications using chlorinated solvents is necessary. Section 1.1.1 provides an overview of the major industrial applications employing chlorinated solvents, with an emphasis on how these industrial processes may have contributed to the composition of waste solvents and the contamination patterns found in the investigation of contaminated sites.

1.2.2 CONSEQUENCES OF SOLVENT BREAKDOWN

Degreasing operations require periodic shutdowns for the cleaning of degreasing equipment to remove accumulated metal fines, to recycle solvent through distillation, and to maintain pumps, seals, valves, nozzles, and other parts. When solvents become unstable, the degreasing equipment itself may be damaged or require intensive cleaning, which results in additional labor, production downtime, and loss of solvent. In early open-top vapor degreasers, operators faced potential hazards from exposure to solvents during routine operations, even when the solvent was performing as intended. The hazards increase considerably when solvents break down owing to runaway acid formation. Degreaser maintenance practices have a bearing on the nature and composition of the waste solvent that was stored in tanks, leaked, spilled, sent to landfills, or sent off to facilities for solvent recycling. Accordingly, the degreaser operating practices prescribed in guidance from degreasing equipment manufacturers such as Baron-Blakeslee and from solvent providers such as Dow Chemical may provide clues to the expected composition of waste solvent.

1.2.2.1 Operator Safety

Reactions of acids and alkali metals with chlorinated solvents can produce several hazardous by-products. Foremost among these is phosgene gas, also known as carbonyl chloride, or $COCl_2$. Phosgene forms during thermal decomposition of chlorinated solvents when they are overheated in a vapor degreaser. Subjecting chlorinated solvent vapors to intense ultraviolet light from arc welding will also form phosgene gas. Health and safety guidance for arc welding calls for no welding within 200 ft of a degreaser using chlorinated solvents (AFSCME, 2006).

Several studies have been performed for measuring the by-products that form when TCE is photocatalytically decomposed because of arc welding in proximity to a vapor degreaser. The major products are phosgene and carbon dioxide; small quantities of dichloroacetic acid and trichloro-acetaldehyde remain at the end of the reaction (Blake et al., 1993; Rice and Raftery, 1999). Perchloroethylene is known to break down to form phosgene, chloroacetic acid, and trichloroacetic acid (Skeeters, 1960a). Of the chlorinated solvents, perchloroethylene gives off the most phosgene when heated by flame or when in contact with a very hot surface or when subjected to intense ultraviolet light (Smallwood, 1993).

Dichloromethane and carbon tetrachloride can produce phosgene; methyl chloroform also forms phosgene as a photocatalytic breakdown product, but to a lesser degree than TCE (Dreisbach, 1987; European Chemicals Bureau, 2000a). Phosgene is a particularly dangerous toxin. It was used as a chemical warfare agent in World War I. There are few symptoms immediately following exposure. Symptoms may appear within 24 h of exposure, but can show up as long as 72 h afterward. Phosgene is a gas at room temperature and combines with water inside the respiratory tract to form carbon dioxide and hydrochloric acid, which dissolves the membranes in the lungs; fluid then fills the lungs, and death results from a combination of blood loss, shock, and respiratory failure. Even a very small amount of phosgene may be deadly, although early symptoms of exposure such as dizziness, chills, and cough usually take 5 or 6 h to manifest (AFSCME, 2006). Phosgene can react with DNA and with enzymes (polymerases) that are responsible for the replication of DNA in cells. Phosgene is considered carcinogenic, even at low exposure levels. The lethal concentration of phosgene gas for which half the exposed people would survive (human LC_{50}) is 500 ppm (Lim et al., 1996).

Illnesses and fatalities have been attributed to degreasing operations in which destabilized solvents have evolved phosgene gas. In one incident, TCE was overheated when it was mistakenly used in a perchloroethylene degreaser. The thermal breakdown of TCE caused the death of an operator (Spolyar et al., 1951). The literature holds many other examples of similar fatalities, such as the chain-smoking proprietor of a dry-cleaning establishment who smoked while operating dry-cleaning machinery in an atmosphere laden with perchloroethylene vapors. They formed phosgene gas as he drew the vapors through his burning cigarette (Derrick and Johnson, 1943).

Several of the metal inhibitors used for the normally nonflammable methyl chloroform are highly flammable and can change the flash point associated with this solvent. Under severe conditions, stabilizers in methyl chloroform solvent can render it flammable, a fact that proved fatal to a welder working on the solid rocket boosters for the space shuttle. On August 10, 1983, an accident killed a welder who was working in a vapor degreaser pit filled with methyl chloroform vapors at a Utah facility that cleaned and inspected the used cases for the solid rocket boosters used on the space shuttle. The welder had been told that methyl chloroform would not burn. While in a basket suspended from an overhead crane inside the pit, he was attempting to weld a bracket into place when a ball of flame developed by his chest and then expanded; he jumped or fell into the pit. A fireball erupted from the pit, and the welder was burned to death or asphyxiated by the dense methyl chloroform vapors. Vapor combustibility tests showed that the liquid solvent would not burn, but the vapors would burn after the liquid was gone. Testing showed that degreasing grades of methyl chloroform were easier to ignite than vapors of pure or technical grade methyl chloroform. A high proportion of solvent stabilizers remaining in the solvent after the accident made it easier to ignite a sample (de Nevers, 1986).

Several stabilizers of methyl chloroform, such as nitromethane and 1,4-dioxane, may impart flammability to methyl chloroform if concentrated; these two compounds may have caused the Utah welding accident. The properties of stabilizers most commonly added to the major solvents and the industrial operations that lead to their partitioning are discussed in Chapter 3. Operator safety for industrial processes using chlorinated solvents is the subject of a large body of regulation.

Occupational safety and health regulations for worker exposure to solvents and toxic stabilizer compounds are discussed in Chapter 6. Proper operation of degreasers and periodic testing of solvent stability prevents solvent-breakdown reactions and toxic reaction products such as phosgene gas.

1.2.2.2 Off-Spec Work

A solvent that has become acidic may stain or damage the work. Parts from which grease, soldering flux, cutting oils, buffing oils, and other oily wastes and debris must be removed can become discolored or receive spots and insoluble deposits from a solvent that has deteriorated. A contaminated solvent will discolor or stain metals, particularly brass and aluminum. Sensitive instruments, printed circuit boards, thermal switches, or other parts may become damaged if the solvent has accumulated too much oil that consequently vaporizes at a higher temperature. Temperature-sensitive parts may melt when the solvent boils too hot. In addition to spots and stains, the work may become corroded from the acid evolved during solvent-breakdown reactions.

A solvent that has deteriorated because of loss or imbalance of the stabilizer package may also become malodorous, colored, or too acidic for dry-cleaning clothing and textiles. Fabrics cleaned with decomposing solvent appear "dull and lifeless," especially fabrics in pastel colors, and the odor of the residual solvent in the fabric and cleaning plant becomes increasingly unpleasant. Perchloroethylene that has become too acidic from decomposition reactions can damage the clothing by leaching dyes and fabric colors. The leached dye discolors the perchloroethylene and makes it unsuitable for further use; its recovery is also difficult (Skeeters, 1960b).

In order to clean dirt and grime from uniforms, clothing, shop overalls, and other items, dry-cleaning solvent is often infused with detergent and water to remove both water-soluble and water-insoluble fraction of the dirt (Skeeters, 1960a). However, some stabilized compositions are not fully compatible with soaps and detergents, which leads to solvent failure in dry-cleaning applications.

Management of waste solvents has been the subject of a series of increasingly stringent regulations. Problems with spotting, staining, discolored textiles, and malodorous solvents often led operators to distill or discard the offensive solvent. Disposal or recycling of waste solvents was usually done in the most cost-effective and expedient manner available or was not actively regulated. For example, until 1986, dry cleaners could legally discharge condensate water to sanitary sewers. Today, the connection of dry-cleaning machinery to sewers is prohibited; all waste must be inventoried, manifested, and sent to a licensed facility for recycling. Chapter 6 provides a full discussion on the evolution of waste-handling regulations and their effect on the nature and chemical composition of waste discharges.

1.2.2.3 Production Line Shutdown

Chlorinated solvents have played an integral role in supporting production lines for a wide variety of products such as automotive parts, armaments, aircraft, electronics, precision instruments, metal parts, textiles, semiconductors, and printed circuit boards. The value of goods produced from these assembly lines can be very high. The cost of a degreaser shutdown due to solvent that has lost its stability and has become acidic can be enormous: the loss of production and the cost of repairing the damaged degreaser far outweigh the cost of the solvent. The solvent must be drained from the degreaser, and all the surfaces must be cleaned with an alkaline detergent. The tarry material may have deposited in inaccessible components and surfaces, requiring the replacement of pipes and other equipment.

In addition to damage to products cleaned in degreasers from staining or spotting, the runaway reaction of an out-of-balance solvent can damage the degreasing equipment itself. To prevent this, operators test solvents frequently for indicators of stability versus impending breakdown. (Testing protocols for determining solvent stability are described in Section 1.2.7.) Preventing loss of production and damage to degreasing equipment was presumably a strong motivation to discard and replace solvents. Prior to regulation and wider understanding of the environmental and/or economic consequences of releasing solvent to the ground, operators faced with the issue of discarding waste solvent may have opted to dump it on-site or to contract with haulers to remove the waste to landfills, recyclers, or unknown destinations.

Solvents can rapidly deteriorate to the point where the work and the degreasing equipment are damaged. For example, methyl chloroform and methylene chloride readily undergo a decomposition reaction in the presence of aluminum chloride; to a lesser extent, TCE and perchloroethylene may also react with aluminum chloride. Solvent degradation is accompanied by a rapid rise in temperature and a discoloration of the solvent. The reaction advances until a black tarry mass forms and deposits on the work and the degreasing equipment (Starks and Kenmore, 1960a; Petering and Aitchison, 1945). Unstabilized methyl chloroform exposed to aluminum chloride or other alkali metal salt catalysts completely decomposes to tar in 20 min (Hardies, 1966). In addition to the tarry by-product of solvent decomposition, acids formed in the reaction may corrode the metal surfaces in a degreaser or dry cleaner. The first sign of acidity in a degreaser is often corrosion at the vapor–atmosphere interface, where rust appears on the walls of the unit around the cooling coils and condensate trough (Dow Chemical Company, 1999b).

The overall consequences of solvent decomposition—worker exposure to harmful or fatal gases, damage to the products being cleaned, and damage to the degreasing equipment with corresponding loss in production—provide a strong motivator to keep solvents in good condition. Operators dealing with solvent that has become acidic owing to loss of stabilizers or other factors seek either to recover the solvent through on-site distillation or to replace the solvent. Solvent replacement could involve on-site recycling or collection and hauling to off-site solvent recyclers or to a hazardous waste incinerator or landfill; more expedient on-site disposal would have been in burn trenches, by drum burial, or directly disposing to the ground, before such practices were prohibited by actively enforced regulation.

The consequences of solvent breakdown were also strong motivators for the solvent producers to develop stabilizer packages capable of withstanding the numerous adverse conditions encountered in the industrial uses of chlorinated solvents. Equipment manufacturers for degreasing and dry cleaning were also challenged to design solutions to ensure that solvents were not subjected to extremes of temperature and stabilizer loss. Examples include increasing the freeboard in the degreaser to minimize vapor loss and the resulting imbalance of stabilizers in the solvent, as well as improvements to water separators and carbon adsorption systems for vapor recovery.

1.2.3 Stability of the Major Solvents

Among the four major chlorinated solvents, perchloroethylene is often listed as the most stable, whereas methyl chloroform is considered the least stable. Mixtures of chlorinated solvents, such as perchloroethylene and methyl chloroform, tend to decompose at an accelerated rate, compared to perchloroethylene alone (Goodner et al., 1977). Table 1.13 summarizes the relative stability of the major chlorinated solvents to aggressive agents encountered in operating environments.

TABLE 1.13
Solvent Stability in Harsh Operating Environments

Solvent	Oxidation	Hydrolysis	Alkali Metals	Pyrolysis	UV Light
Dichloromethane	Stable	Stable	Slightly unstable in vapor phase	Slightly unstable	Stable
Methyl chloroform	Slightly unstable	Slightly unstable	*Unstable*	*Unstable*	Stable
Perchloroethylene	Slightly unstable	Stable	Stable	Stable	*Unstable*
Trichloroethylene	*Unstable*	Stable	Slightly unstable	Slightly unstable	Slightly unstable

Sources: Solvay, S.A., 2002a, Chlorinated solvents stabilisation. GBR-2900-0002-W-EN Issue 1—19.06.2002, Rue du Prince Albert, 33 1050 Brussels, Belgium. http://www.solvaychemicals.com (accessed November 9, 2003) and Irani, M.R., 1977, United States Patent 4,032,584: Stabilized methylene chloride. Assignee: Stauffer Chemical Company, Westport, CT.

TABLE 1.14
Purity of Solvents Sold for Industrial Uses

Solvent	Grade	Purity (%)	References
Methyl chloroform	Degreasing	95	Ashland Chemical Company (1995)
	Technical	96.5	Dow Chemical Company (1990)
	Technical	94–98	Unocal (1989)
Dichloromethane	Technical	98	Vulcan Chemicals (1987)
	Technical	99.5	Dow Chemical Company (1995)
	Technical	99.9	Dow Chemical Company (1999a)
	Aerothene MM™	99.4	Dow Chemical Company (2006a)
Trichloroethylene	Neu-Tri™	99.4	Dow Chemical Company (1981, 2006a)
	Technical, O-T-634	> 95	Dow Chemical Company (1985a)
Perchloroethylene	Perchloroethylene-SVG™	99.5	Dow Chemical Company (1986)
	DOWPER CS solvent™, 1985	97	Dow Chemical Company (1985b)
	DOWPER solvent™, 2006	99[a]	Dow Chemical Company (2006a)
	Perclene™ dry-cleaning grade	99–100	Occidental Chemical Corporation (1991)
	SVG double-stabilized	99[b]	Dow Chemical Company (2006a)

[a] 1% stabilizer.
[b] 0.1% stabilizer.

The relative stability of the major solvents can be inferred from the quantity of stabilizers required. Material Safety Data Sheets (MSDSs) and other documents indicate the purity of solvents sold for industrial applications in recent years, as presented in Table 1.14.

Solvent purity listed in Table 1.14 reflects both the addition of stabilizers and impurities from manufacturing. Technical grade TCE (also called "industrial grade") has been known to include about 0.15% carbon tetrachloride as the major chlorinated hydrocarbon impurity (Willis and Christian, 1957). The compound 1,1,2-trichloroethane can be an impurity in the manufacture of methyl chloroform and is more reactive than methyl chloroform (World Health Organization, 1990).

Solvent impurities may also be introduced from cross-contamination in unclean tank cars, tank trucks, drums, piping, and hoses. Mixtures of chlorinated solvents, even if stabilized, may have substantially reduced stability when in contact with aluminum, compared to the stability of the individual solvents. For example, stabilized TCE can become unstable in the presence of as little as 2% of stabilized dichloromethane. When replacing one chlorinated solvent with another, it is therefore necessary to remove all traces of the first solvent before using the new solvent (EuroChlor, 2003). Most facilities would not frequently change solvents. However, solvent-recycling facilities that use the same distillation equipment for different solvents must clean out drum-filling pipes between batches. Several anecdotes describe drum-filling pipes that were flushed to the ground or storm drain sumps between batches of solvents at a solvent recycler in Silicon Valley, California.

1.2.3.1 Performance Criteria for Solvent Stabilizers in Various Applications

The industrial chemists who worked to solve the solvent-breakdown problem overcame a major challenge of solvent stabilization through much experimentation. Additives were needed that would inhibit a variety of reactions with the aggressive substances found in industrial processes, especially where physically destructive conditions are present. The chemists' pursuit was to find stabilizers that would meet most of the following criteria (Levine and Cass, 1939; Klabunde, 1949; Kauder,

1960; Richtzenhain and Stephan, 1975; Irani, 1977; Blum, 1984; Smallwood, 1993). The ideal stabilizer must

- Adequately inhibit deleterious reactions to prevent solvent breakdown caused by the action of light, air, or temperature on the solvent
- Protect the solvent from the action of small amounts of strong acid or from metals and their salts
- Correct any incipient localized breakdown of the solvent
- Not impart increased toxicity to the solvent or endanger worker health and safety
- Remain stable through use in the expected industrial process
- Stand up to repeated phase transitions from liquid to vapor and back to liquid
- Remain stable at high temperatures
- Partition to the vapor phase to provide sufficient stability in degreasing
- Be sufficiently volatile to exist in both the liquid and the vapor phases of the solvent and to be recoverable from distillation residues
- Boil at temperatures not more than 15°C higher or lower than the solvent boiling point
- Not be consumed by reactions
- Prevent the solvent from taking on the character of a mixture of solvents, not require addition in large quantities
- Not stain the metal part, textile, or clothing being cleaned
- Not leave deposits on the parts being cleaned
- Not promote the formation of corrosion products or insoluble sludge
- Be fully soluble in the solvent
- Be at least more soluble in the solvent than in water so that water would not extract the stabilizer
- Be inexpensive
- Not react with other stabilizers
- Be compatible with the metals used in the construction of the degreaser and with the many and varied contaminants encountered in the cleaning of metal parts from modern fabricating operations
- Not weaken textile fibers
- Not shrink textile fibers
- Not bleed dyes
- Not be malodorous or leave residual odor
- Be compatible with soaps and detergents if used in dry-cleaning applications
- Not react with free fatty acids
- Not be so strongly alkaline that its volatility is reduced in the presence of acids
- Be easily measurable to confirm its concentration during use

1.2.3.2 Selection of Solvent Stabilizers

Selection of stabilizers was determined by the conditions of the particular application. The many permutations of the various requirements placed on solvents led to decades of competitive inventions of solvent stabilizer packages that would improve solvent performance in specific applications. For example, operations that cut and polish metals may use buffing compounds and drawing oils that must be removed from the metal parts by a vapor degreaser. Buffing compounds and drawing oils are sources of fatty acids. Buffing compounds contain 50% stearic acid, whereas drawing oils can contain 5–10% oleic acid (Klabunde, 1949). The stabilizer package must therefore not react with fatty acids.

As detailed in the first half of this chapter, perchloroethylene has been used in a great variety of applications, for example, in dry cleaning, in electrical transformers as a heat-transfer medium, in ultrasonic cold cleaning, in vapor degreasing, and as a paint thinner. Perchloroethylene with a minor complement of stabilizers may remain stable for multiple uses when degreasing large pieces of stainless steel. The same formulation may become unstable when cleaning a more reactive metal or the same metal in which fine turnings, grindings, and other metal fines create a substantially increased surface area over which reactions may occur or which may cause hot spots near heating coils and lead to pyrolysis of the solvent. Determination of the preferred stabilizer formulation therefore depends on the severity of the environment to which the solvent is subjected.

Among the most demanding applications requiring stabilization of perchloroethylene is its use as a dielectric fluid. A dielectric fluid must remain stable for up to 30 years without breaking down to form electrically conductive or corrosive materials. Breakdown of perchloroethylene by dehydro-chlorination forms hydrogen chloride, which is conductive and therefore deleterious to the dielectric fluid and the electrical device used (EPO, 1989).

Another application in which perchloroethylene and TCE are subjected to severe stress is in phosphatizing baths. Nonaqueous phosphatizing baths are used to apply phosphate coatings to metallic surfaces, for example, automobile exteriors, to reduce corrosion and improve paint adhesion. Besides orthophosphoric acid and an agent to solubilize the acid, the phosphatizing bath is mostly TCE or perchloroethylene. These two solvents can undergo rapid decomposition by a mechanism different from the type of decomposition for which most stabilizers were developed; thus, an application-specific group of stabilizers is required (Fullhart and Swalheim, 1962).

1.2.4   CAUSES OF SOLVENT BREAKDOWN

The primary causes of solvent breakdown—oxidation, hydrolysis, pyrolysis, ultraviolet light, and reaction with alkali metals—have been introduced. This section examines the chemical reactions that cause solvent deterioration. The structures of the chlorinated solvents (Table 1.1) play a decisive role in the nature of their decomposition.

1.2.4.1 Oxidation

TCE and perchloroethylene are susceptible to oxidative attack on the ethylene carbon–carbon double bond. Oxidative attack is a problem for the ethylene compounds perchloroethylene and TCE, but is not a problem for the alkane compounds methyl chloroform, dichloromethane, and carbon tetrachloride, which have single carbon–carbon bonds or only a single carbon atom (Levine and Cass, 1939). The initial deterioration of the ethylene bond in TCE is due to the oxidation initiated by light (photolysis, in which light energy is absorbed by a molecule). The reaction is strongly accelerated by ultraviolet rays from natural light, fluorescent lighting, and arc welding, by heat, and by the presence of catalysts, including metals such as iron and aluminum, and metallic salts such as ferric chloride and aluminum chloride. TCE is relatively immune to photolysis, except in the presence of oxygen (Shepherd, 1962). The products resulting from the oxidation of TCE also serve to promote further degradation of TCE. Humidity does not appear to affect the rate of oxidation of TCE (Solvay SA, 2002b). Air, moisture, and acid favor further deterioration once oxidation has been initiated. Because acid is evolved, the deterioration of TCE is autocatalytic (Pitman, 1933). As the breakdown of TCE progresses, the reaction takes the form of a self-accelerating Friedel–Crafts reaction. Once started, the reaction will proceed rapidly and, in some cases, explosively with the evolution of heat and large quantities of hydrogen chloride.

The acceleration of TCE degradation by ultraviolet light when oxygen is present suggests a free radical mechanism. A variety of products are formed, including phosgene gas, carbon monoxide, dichloroacetic acid, formic acid, glyoxylic acid, hydrochloric acid, chlorine gas, and a variety of polymer compounds (Shepherd, 1962).

Most halogenated solvents are resistant to attack by oxygen until a chlorine atom is removed. Removing a chlorine atom and an adjacent hydrogen atom leaves a double bond, which is more susceptible to oxidation, particularly in the presence of metals such as aluminum, copper, and vanadium (Howell and Tarrer, 1994).

The mechanics of solvent breakdown are profiled in a study by Arthur Tarrer of Auburn University, with researchers from Tyndall Air Force Base and from the U.S. Army Construction Engineering Research Laboratory (Tarrer et al., 1989; Howell and Tarrer, 1994). An initiator, such as heat or ultraviolet light, removes hydrogen from an unsaturated molecule, RH, forming a free radical, R. Oxygen then combines with the free radical to form a peroxide radical, ROO·, and that then removes hydrogen from a new unsaturated molecule, R′H, thereby propagating the chain reaction (Tarrer et al., 1989).

The susceptibility of TCE to oxidation alone, without an ultraviolet light initiator and free radical mechanism, was summarized by Shepherd, citing work in the German literature by Erdmann in 1911, as follows (Erdmann, 1911; Shepherd, 1962):

Trichloroethylene is oxidized in a decomposition mechanism involving formation of a highly reactive epoxide which is readily isomerized to acetyl chloride.

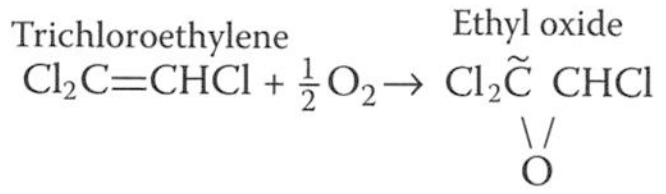

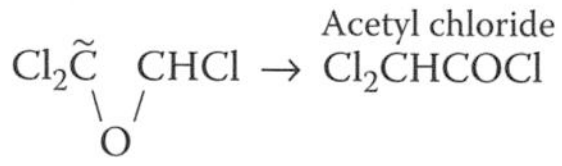

$$\overset{\text{Trichloroethylene}}{Cl_2C{=}CHCl} + \tfrac{1}{2}O_2 \rightarrow \overset{\text{Ethyl oxide}}{Cl_2\tilde{C}\underset{\underset{O}{\diagdown\diagup}}{}CHCl}$$

$$Cl_2\tilde{C}\underset{\underset{O}{\diagdown\diagup}}{}CHCl \rightarrow \overset{\text{Acetyl chloride}}{Cl_2CHCOCl}$$

In the presence of moisture, the acetyl chloride is hydrolyzed to dichloroacetic acid and hydrogen chloride:

$$\overset{\text{Acetyl chloride}}{Cl_2CHCOCl} + H_2O \rightarrow \overset{\text{Dichloroacetic acid}}{Cl_2CHCOOOH} + HCl$$

The epoxide has been verified experimentally by McKinney et al. (1955). Perchloroethylene also forms an epoxide (Shepherd, 1962).

Commercial grades of perchloroethylene have been subject to free radical attack initiated by ultraviolet light; it is the least stable solvent in the presence of ultraviolet light. High-purity perchloroethylene is very inert to the action of air, light, heat, moisture, and the metal surfaces it contacts during commercial storage and use. However, most perchloroethylene is not pure. Perchloroethylene obtained from the chlorination or chlorinolysis of lower aliphatic hydrocarbons may contain small amounts of saturated and unsaturated chlorinated hydrocarbons such as dichloroethylene, TCE, methyl chloroform, and unsymmetrical tetrachloroethane (Skeeters, 1960a). These compounds are more susceptible to free radical attack than perchloroethylene and may initiate and catalyze the breakdown of perchloroethylene. The presence of the lower-molecular-weight chlorinated hydrocarbons in perchloroethylene leads to the formation of oxidation products that corrode metal surfaces. These lower-molecular-weight chlorinated hydrocarbons are thought to be the principal initial source of chloro-oxygen-containing impurities such as phosgene, chloroacetic acid, and trichloroacetic acid. It is these impurities that catalyze the decomposition of perchloroethylene (Skeeters, 1960b). Methyl chloroform was widely used as a spotting agent to remove spots from clothing prior to dry cleaning. Its introduction to the perchloroethylene vapor environment in a dry cleaner could also contribute to acid formation in perchloroethylene (Linn et al., 2004). Experimental data show that heat accelerates the deleterious effects of light and oxygen on the decomposition of solvents (Starks and Kenmore, 1960b).

1.2.4.2 Ultraviolet Light Breakdown

Perchloroethylene shows poor light stability. Fluorescent light and ultraviolet light cause perchloroethylene to decompose, producing a measurably increased acidity and a corresponding change in pH (Stevens, 1955). Tests for stability of solvents often involved a multiday or multiweek reflux test.

To show the instability of unstabilized perchloroethylene to light, a typical test might proceed for 72 h, during which perchloroethylene with water and copper strips is heated to boiling and the vapor is exposed to fluorescent light. A change in pH of 1 unit or more is taken as an indicator of light instability. Without a light stabilizer, perchloroethylene can change from pH 6.8 to less than pH 2 in 72 h in the presence of light (Stevens, 1955). The decomposition of perchloroethylene is more pronounced when it is in contact with iron or copper. Products of perchloroethylene decomposition cause metal corrosion. Perchloroethylene will also degrade in the presence of light from a welder's arc (Dow Chemical Company, 1999b).

1.2.4.3 Thermal Breakdown

The major chlorinated solvents are stable and resistant to thermal breakdown at their normal boiling temperatures, but are vulnerable to thermal breakdown and acid formation if operating temperatures exceed their stability range (Petering and Aitchison, 1945; Archer and Stevens, 1977). The order of increasing thermal stability does not directly correspond to the order of the solvents' boiling points. The order, from least to most thermally stable, is methyl chloroform (b.p. = 74°C) < dichloromethane (b.p. = 41°C) < TCE (b.p. = 86.7°C) < perchloroethylene (b.p. = 121.4°C) (Archer and Stevens, 1977). TCE is stable up to 130°C; at higher temperatures, it is subject to pyrolysis. Perchloroethylene is stable up to 150°C in the presence of air and moisture (Solvay SA, 2002b). In the absence of any catalysts, such as in the sealed environment of an electrical transformer, perchloroethylene may be stable up to 500°C (World Health Organization, 1984a). Decomposition in the presence of air and moisture begins above 125°C for TCE and above 150°C for perchloroethylene. For safety reasons, equipment operators are advised to avoid exceeding temperatures of 110°C and 140°C for TCE and

TABLE 1.15
Pyrolysis Test Results for the Major Chlorinated Solvents

Solvent	Metal	Vapor Condition	Temperature at Which Slow Increase in Pyrolysis Begins (°F)	Temperature at Which Rapid Increase in Pyrolysis Begins (°F)
Dichloromethane	Black iron	Vapor only	525	700
	Black iron	With dry air	<400	600
	Aluminum	Vapor only	376	525
	Copper	Vapor only	475	525
Methyl chloroform	Black iron	Vapor only	525	625
	Black iron	With dry air	—	325
	Aluminum	Vapor only	—	675
	Copper	Vapor only	~350	675
Trichloroethylene	Black iron	Vapor only	—	825
	Black iron	With dry air	—	550
	Aluminum	Vapor only	~550	725
	Copper	Vapor only	—	575
Perchloroethylene	Black iron	Vapor only	—	>900
	Black iron	With dry air	—	>900
	Aluminum	Vapor only	—	>750
	Copper	Vapor only	—	>850

Source: Archer, W.L. and Stevens, V.L., 1977, *Industrial & Engineering Chemistry, Product Research and Development* 16(4): 319–326.

Note: Pyrolysis test consisted of passing chlorinated solvent vapor (0.2 mol/h), with or without air, through a heated metal pipe whose internal surface area was 200 in.^{2} at several elevated temperatures. Acidity of heated vapors was then determined by collecting 40 mL of condensed solvent from a cold trap; acid generated was determined by titrating with caustic soda.

perchloroethylene, respectively (Solvay SA, 2002b). Table 1.15 summarizes the susceptibility of the major solvents to pyrolysis in the presence of metal catalysts and air.

When chlorinated solvents are overheated, a chloride ion gets split off, leading to the formation of hydrochloric acid. Prolonged contact of solvents with oils, grease, and other soils at elevated temperatures also causes solvent instability problems (Archer, 1984). For example, thermal breakdown of TCE under conditions of normal use and in the absence of stabilizers probably proceeds according to the following reactions (Shepherd, 1962):

$$2ClCH{=}CCl_2 \longrightarrow (ClCH{=}CCl_2)_2, \tag{1.1}$$

$$(ClCH{=}CCl_2)_2 \longrightarrow HCl + Cl_2C{=}CHCC{=}CCl_2. \tag{1.2}$$

1.2.4.4 Acid Breakdown

All halogenated solvents in contact with free-phase moisture and air will slowly react with water in a process called hydrolysis. Rates of hydrolysis for the chlorinated solvents vary widely. Methyl chloroform has a hydrolysis half-life of a little more than two years, whereas TCE and perchloroethylene have immeasurably slow hydrolysis half-lives. Hydrolysis produces hydrogen chloride in chlorinated solvents. If exposed to water, hydrogen chloride produces hydrochloric acid. If not eliminated when formed, hydrochloric acid will catalyze and increase the rate of hydrolysis and produce acid at a faster rate. Hydrochloric acid also removes the protective oxide coating on metal surfaces, exposing fresh metal to the solvent and enabling metal-catalyzed solvent deterioration. In the earliest study of problems with hydrolysis of chlorinated solvents, Levine and Cass (1939) described carbon tetrachloride as decomposing as a result of hydrolysis, "splitting of acid and reacting with water."

Acid can also be introduced into the degreaser solvent from oils, greases, soldering flux, and other material present on the work. The oxidation and decomposition of cutting oils dissolved from the work form acid, usually hydrochloric acid or acetic acid, which then reacts with the solvent (Starks and Kenmore, 1960b). Acids formed by oxidation of cutting oils or hydrolysis of solvents can corrode the work and the cleaning equipment itself. In dry-cleaning applications, acids that form in perchloroethylene can leach dyes and fabric colors; this leaching damages the clothing and makes solvent recovery difficult (Cormany, 1977). A source of acid in dry cleaning is sebaceous oils and other bodily soil on the clothing. These soils contain short-chain, free fatty acids such as butyric acid and valeric acid that acidify moisture in the cleaning process, which leads to machine corrosion, odors on the clothing, and undesirable textile effects such as swales. Acidity cannot be tolerated in dry cleaning, where TCE was used to a limited extent in the 1930s through the 1950s. Free acidity was also undesirable for TCE when it was used to make decaffeinated coffee (Pitman, 1943).

TCE and perchloroethylene are relatively immune to hydrolysis, however, compared to carbon tetrachloride or methyl chloroform. Dichloromethane hydrolyzes very slowly (Solvay SA, 2002b). Table 1.16 provides commonly cited values for hydrolysis half-lives of the major solvents under ambient conditions in soil and groundwater.

TABLE 1.16
Generalized Abiotic Hydrolysis Half-Lives of the Major Chlorinated Solvents

Solvent	Abiotic Hydrolysis Half-Life (years)	References
Perchloroethylene	1.3×10^6	Jeffers et al. (1989)
Trichloroethylene	1.3×10^6	Jeffers et al. (1989)
Dichloromethane	704	Mabey and Mill (1978)
Carbon tetrachloride	41	Jeffers et al. (1989)
Methyl chloroform	2.5	Vogel and McCarty (1987)

Source: Data and citations from Pankow, J.F. and Cherry, J.A., 1996, *Dense Chlorinated Solvents and Other DNAPLs in Groundwater*. Waterloo, Ontario, Canada: Waterloo Press.

TABLE 1.17
Acidity Generation by Boiling Trichloroethylene and Perchloroethylene

Test Condition	Trichloroethylene	Perchloroethylene
Dry solvent under nitrogen	10.2[a]	10.5[a]
Wet solvent under nitrogen	7.7	12.5
H_2O-saturated solvent exposed to soft steel under nitrogen	11.0	11.6
Wet solvent exposed to soft steel under nitrogen	5.6	—
Dry solvent in oxygen	820	782

Note: Reported as cubic centimeters of 0.01 N hydrochloric acid per 25 mL solvent (from Shepherd, 1962).
[a] Air leakage introduced oxygen.

In the absence of air and metals, solvents will undergo hydrolysis at slow rates and form only minor quantities of acids. TCE and perchloroethylene both hydrolyze at a rate of 0.01 mg/L per day in the absence of air and metals (6.8×10^{-7} mol% and 6.2×10^{-7} mol%), respectively (Carlisle and Levine, 1932). Citations of Carlisle and Levine's work record the rates of solvent decomposition listed in Table 1.17 when boiling TCE and perchloroethylene are subjected to the specific conditions given.

Through hydrolysis, methyl chloroform reacts abiotically to form 1,1-dichloroethylene:

$$
\begin{array}{ccccccc}
\text{Cl}\ \text{H} & & & & & & \text{H}\quad\text{Cl}\\
|\quad| & & & & & & \backslash\quad/\\
\text{Cl--C--C--H} & + & \text{H}^+ & \longrightarrow & \text{HCl} & + & \text{C}=\text{C}\\
|\quad| & & & & & & /\quad\backslash\\
\text{Cl}\ \text{H} & & & & & & \text{H}\quad\text{Cl}
\end{array}
\qquad (1.3)
$$

Methyl   Hydrogen Hydrogen 1,1-dichloroethylene
chloroform   chloride

Abiotic degradation of methyl chloroform has also been described as follows (Cornell Dubilier, 2003):

$$
\text{CH}_3\text{CCl}_3 \longrightarrow \text{CH}_2\text{CCl}_2 + \text{CH}_3\text{COOH} \longrightarrow \text{CO}_2 + \text{H}_2\text{O} + \text{Cl}^-
$$

Methyl   1,1-dichloroethylene Acetic acid Mineralization   (1.4)
chloroform

1.2.4.5 Solvent Deterioration Catalyzed by Metals and Metal Salts

Methyl chloroform is most vulnerable to reaction with alkali metals, a problem also faced by dichloromethane and carbon tetrachloride when those two compounds are in the vapor phase. TCE and perchloroethylene are comparatively less vulnerable to reaction with alkali metals (Smallwood, 1993). The absence of hydrogen atoms in perchloroethylene makes it less sensitive to the action of bases and light metals (Solvay SA, 2002a). Chlorinated organic products in contact with a light or alkali metal such as aluminum or magnesium produce some hydrochloric acid and a Lewis acid catalyst, for example, anhydrous aluminum trichloride ($AlCl_3$). $AlCl_3$ will catalyze condensation of chlorinated solvents onto themselves or onto other organic substances, especially chloride ions or other nucleophiles.[*] The products of this reaction include a brownish-black, tarry condensate and more hydrochloric acid, which in turn reacts further with aluminum. Such a reaction increases the amount of $AlCl_3$ present; hence, the reaction is perpetuated. In some cases, the reaction can be

---

[*] A nucleophile, or *nucleus-seeking compound*, is typically negatively charged with unshared electron pairs that can be used to form a covalent bond with a carbo-cation (Carey, 1987).

violent and produce a considerable amount of hydrochloric acid and heat. Heat accelerates the reaction of condensation and evaporates some of the remaining solvent. $AlCl_3$ has a catalytic effect on TCE, dichloromethane, and especially methyl chloroform (Kauder, 1960). $AlCl_3$ is generally considered to have no catalytic effect on pure perchloroethylene (Solvay SA, 2002a). However, in earlier decades, impurities in perchloroethylene may have been carried over from production and reacted with alkali metals.

Methyl chloroform reacts with aluminum ($AlCl_3$), splitting off hydrochloric acid and leaving 1,1-dichloroethylene, shown here in a generalized reaction:

$$CH_3CCl_3 \xrightarrow[\text{Dehydrohalogenation}]{AlCl_3} CH_2CCl_2 \quad + \quad HCl$$

Methyl chloroform   1,1-dichloroethane   Hydrochloric acid   (1.5)

Aluminum surfaces react with oxygen to form a protective aluminum oxide coating. If a machining, abrasive, or cleaning process removes this coating, the "virgin" metal is available to react with a solvent. The aluminum reaction with methyl chloroform is most easily observed in the "aluminum scratch test," as described in Section 1.2.5.2. Hydrochloric acid is formed during the scratch test and dissociates easily. The chlorine ions react with aluminum in the following sequence:

Hydration of oxide film:

$$2AlO_3 + 3H_2O \longrightarrow 2Al(OH)_3 + \tfrac{3}{2}O_2. \tag{1.6}$$

Reaction of hydrated oxide film and chlorine (dissolution of film):

$$Al(OH)_3 + 3HCl \longrightarrow AlCl_3 + 3H_2O. \tag{1.7}$$

Reaction of aluminum and hydrochloric acid (dissolution of aluminum):

$$Al + 3HCl \longrightarrow AlCl_3 + \tfrac{3}{2}H_2. \tag{1.8}$$

Precipitation of aluminum hydroxide:

$$AlCl_3 + 3H_2O \longrightarrow Al(OH)_3 + 3HCl. \tag{1.9}$$

The compounds produced by the reactions are aluminum hydroxide and hydrochloric acid; hydrochloric acid is not consumed and acts as a catalyst (Cornell Dubilier, 2003).

TCE, dichloromethane, and methyl chloroform react with aluminum to give aluminum chloride and oligomers of the solvent (a few solvent molecules bonded together). For TCE, the aluminum chloride goes on to react with the solvent to form hydrogen chloride, which acts as a catalyst and is not itself consumed by the reaction. Once initiated, this self-accelerating, "autocatalytic" dehydrochlorination (loss of hydrogen chloride) reaction produces a vigorous subsequent reaction between TCE and aluminum. Aluminum chloride does not have a similar reaction with dichloromethane (EuroChlor, 2003).

Aluminum chloride promotes a dimerization reaction with TCE to produce hydrochloric acid:

$$Cl_2C = CHCl \xrightarrow{AlCl_3} Cl_2CHCCl_2CH = CCl_2$$

Trichloroethylene   (1.10)

$$Cl_2CHCCl_2CH = CCl_2 \xrightarrow[\text{Dimerization}]{} CCl_2 = CClCH = CCl_2 + HCl \tag{1.11}$$

TABLE 1.18
Average Metal Corrosion Rates of Chlorinated Solvents

Solvent	Metal Inhibitor	Corrosion Rate[a] (mils/year)[b]		
		2024 Aluminum	1010 Iron	Zinc
Dichloromethane, technical grade	No	<1	<1	<0.1
Methyl chloroform	Yes	<3	<1	<1
Trichloroethylene with antioxidant	No	<0.1	<0.1	<0.1
Perchloroethylene with antioxidant	No	<0.1	<0.1	<1

Source: Archer, W.L. and Stevens, V.L., 1977, *Industrial & Engineering Chemistry, Product Research and Development* 16(4): 319–326.

[a] Penetration per year, determined in seven-day reflux tests.

[b] 1 mil = 0.001 in. = 0.0254 mm.

The direct reaction between a chlorinated solvent, such as methyl chloroform, and a metal, such as aluminum, iron, or zinc, produces the metal chloride and the saturated dimer of the chlorinated reactant (Archer, 1984):

$$2Al + 6CH_3CCl_3 \xrightarrow[\text{Dimerization}]{} 3CH_3CCl_2CCl_2CH_3 + 2AlCl_3$$

Methyl chloroform

(1.12)

The saturated dimer can in turn react with the metal to give the unsaturated *cis* and *trans* dimers, $CH_3CCl{=}CClCH_3$.

Reactions of stabilized and unstabilized solvents with various metal alloys have been the subject of extensive study. Numerous patents describe the rapid devolution of methyl chloroform refluxed with aluminum, which proceeds to a black tarry mass in as little as 20 min. The presence of a separate water phase floating atop the solvent accelerates the rate of metal corrosion for all solvents (Archer, 1984). Results of additional reflux (boiling and condensing solvent) and high-temperature exposure corrosion tests are listed in Tables 1.18 and 1.19.

TABLE 1.19
Metals Corroded by Trichloroethylene and Dichloromethane at More Than 50 mils Penetration per Year

Solvent	Metal	Temperature (°C)
Trichloroethylene	Austenitic Cr–Ni stainless steel (18-8; 304/304L/347)	80–107
	Lead	−4–24
	Mo superalloy (Ni–Cr–Fe–9Mo; 625/725)	107–135
	Mo superalloy (Ni–16Cr–16Mo; C276)	107–135
	Mo superalloy (Ni–20Cr–16Mo–4W; 686)	107–135
	Mo superalloy (Ni–22Cr–16Mo; C22/59)	107–135
	Mo superalloy (Ni–23Cr–16Mo–1Cu; C20000)	107–135
	Nickel (200)	107–135
	Nickel–copper (400)	107–135
	Steels, carbon/low alloy	80–107
Dichloromethane	Austenitic Cr–Ni stainless steel (18-8; 304/304L/347)	88–139

Source: NACE International, The Corrosion Society, 2002, Corrosion survey database (COR·SUR). http://www.knovel.com/knovel2/Toc.jsp?BookID=532&VerticalID=0 (accessed 2006).

Note: 1 mil = 0.001 in. = 0.0254 mm.

1.2.5 Testing Solvent Stability and Performance

The balancing act of maximizing solvent utility while controlling and preventing solvent deterioration requires early detection of changes in solvent characteristics. Visual inspection of the solvent is not a reliable means for determining its condition. A solvent that is perfectly clear can be depleted of its stabilizers, whereas one that is badly discolored and cloudy can still be well stabilized. The most reliable method for confirming the presence and quantity of stabilizers is laboratory analysis by gas chromatography–mass spectrometry; however, the associated delay and expense are too impractical for most operations, so surrogate tests were developed (Howell and Tarrer, 1994). Operators of equipment using chlorinated solvents—such as degreasers, phosphatizing lines, and dry cleaners—routinely test solvents for the presence of acid. A variety of testing approaches have been developed since 1970.

The main approaches for testing solvents check for acids by titrating a base [the acid acceptance value (AAV) test] and test for reaction with alkali metals (the aluminum scratch test). Tests with more specific endpoints—such as reaction with copper (ASTM D 3316), zinc, or brass, and dye-bleed tests—have also been developed for specific applications. Light transmittance has been used in dry-cleaning operations as an indicator of when a solvent is due for replacement. Solvents usually transmit light in the range from 450 to 600 nm; a 50% reduction in transmittance in solvent that has been filtered is often used as the rule of thumb for when the solvent must be replaced (Tarrer et al., 1989). Transmittance is measured by using a visible light spectrometer or a colorimeter. Color, odor, presence of dirt and grease, and boiling temperature have also been used by dry-cleaner operators to gauge when the solvent is due for filtration, distillation, or replacement. All the solvent performance tests are temperature sensitive; a water bath is usually used to hold solvent temperatures constant to produce comparable results.

Solvent power is a measure of the solvent's effectiveness at dissolving oils, soldering flux, and buffing compounds. The most common test for judging changes to solvent power yields the Kauri-butanol value (KBV) (ASTM D 1133). The test uses Kauri gum,[*] which is very soluble in butanol but less soluble when butanol is diluted with a solvent that does not dissolve the resin. The test uses a burette, a flask, and a precision balance and provides a relative ranking of solvency. The solvent tested is added in small amounts until the solution becomes cloudy because of the precipitation of Kauri gum from the butanol solution. The more the solvent added before the solution becomes cloudy, the greater the solvent power and the higher the corresponding KBV. When tested as new solvent, uncontaminated by cleaning waste, water, or other impurities, the major chlorinated solvents have the following lower to higher order of solvent power and KBVs: perchloroethylene (KBV = 93) < methyl chloroform (KBV = 124) < TCE (KBV = 130) < dichloromethane (KBV = 178)[†] (Tarrer et al., 1989; Dow Chemical Company, 2002, 2006c; Solvay SA, 2002b). The solvent power of a solvent decreases with continued use and an increasing fraction of oily waste, grease, or soil.

Another test used to determine solvent stability involves the Acid Number. The Acid Number establishes the amount of fatty acid in a solvent, determined as the number of milligrams of potassium hydroxide to neutralize 1.28 mL of the solvent, by using a burette, a pipette, and a flask with potassium hydroxide, methanol, and phenolphthalein (Tarrer et al., 1989). Fatty acids such as stearic acid and oleic acid are introduced to the solvent when the work is coated with buffing compounds and drawing oils (Klabunde, 1949).

---

[*] Kauri gum is formed when resin exudes from a crack in the bark of the kauri tree (*Agathis australis*), found in New Zealand, and hardens on exposure to air. It also exists in fossil form. Its uses range from chewing gum to industrial applications.

[†] Dichloromethane's Kauri Butanol Value (KBV) is listed as 87 for unstabilized methylene chloride and 178 for Solvaclene, metal-stabilized dichloromethane, both sold by Solvay SA. Dow lists a KBV for Dow Methylene Chloride (dichloromethane) as 136 expressed as cubic centimeters of solvent per 20 g of Kauri-Butanol solution. Dow's KBV value for perchloroethylene is 90, the value for methyl chloroform is 124, and that for NEU-TRI trichloroethylene is 129 (Dow Chemical Company, 2002, 2006c).

Additional measures of solvent stability and solvent oily waste content include viscosity (measured in a viscometer), electrical conductivity, and specific gravity. A hydrometer is often used for measuring differences in specific gravity for nonaqueous materials whose density is markedly higher or lower than water, such as fuels, or dense nonaqueous phase liquids including the major chlorinated solvents. However, a hydrometer is not sensitive enough to detect the small variations in specific gravity that occur when a solvent becomes contaminated with oily waste. The mixtures of lighter oils and heavy solvents approach the specific gravity of water and may become lighter than water as increasing amounts of oil are dissolved into the solvent. Perchloroethylene, TCE, and dichloromethane have initial specific gravities of 1.62, 1.46, and 1.36, respectively; as the oil content approaches 10% by volume, the specific gravities of all three solvents converge on 0.92 (Solvay SA, 2002b; see Chapter 3 for further discussion). To determine the specific gravity of used solvents, a pycnometer or an electronic specific gravity meter is used. A pycnometer is a bottle that holds a specific volume of liquid and is weighed on a balance (Tarrer et al., 1989).

Solvent performance tests were generally run on samples obtained from the condensate tank on a vapor degreaser, because the condensate solvent was expected to hold the lowest level of stabilizers for dichloromethane, TCE, and perchloroethylene. Maintaining the stabilizers at recommended levels in the condensate ensures proper stabilizer levels throughout the vapor degreaser environment (Dow Chemical Company, 1999b).

1.2.5.1 Acid Acceptance Value

The total AAV test has been the most common method for determining whether acid in a solvent was sufficiently inhibited with acid-acceptor compounds. The standard for this test was ASTM D 2942, "Test Method for Total Acid Acceptance of Halogenated Organic Solvents—Nonreflux Method."

The AAV test gauges solvent condition by monitoring the concentration of neutral acid acceptors in the solvent. An aliquot of acid is added, and the solvent is titrated with sodium hydroxide to determine the amount of base required to neutralize the acid. Various degreasing equipment manuals advise operators to maintain AAV above 0.03% sodium hydroxide. For methyl chloroform, solvent with an AAV of less than 0.08% sodium hydroxide was considered to have a "borderline" condition, and solvent having an AAV of less than 0.04% was considered to be "unacceptable." Guidelines such as those mentioned in Table 1.20 informed operators when to distill, remove for recycling, or discard waste solvent.

Dow Chemical currently provides its own acid acceptance test kit to monitor stabilizer concentration for Dow Methylene Chloride Vapor Degreasing Grade, TCE Neu-Tri solvent, and perchloroethylene SVG (Dow Chemical Company, 1999b). Dow's test kits incorporate alkalinity testing as well as acid acceptance monitoring. In some cases, where the acid acceptance was maintained at a satisfactory level, corrosion could still be found in the degreaser equipment when there had been a drop in the pH of the solvent before the loss of acid acceptor. Solvent pH is normally greater than 8.5, primarily because of the reserve alkalinity from the antioxidants in the stabilizer system.

TABLE 1.20
European Guidance for Acid Acceptance Values Requiring Solvent Management Action

Solvent	Fresh	Add Fresh Solvent	Replace or Distill Solvent
Tovoxene (trichloroethylene)	>0.16	0.05	0.02
Soltene™ (perchloroethylene)	>0.10	0.03	0.02
Solvaclene™ (dichloromethane)	>0.30	0.10	0.08

Source: Solvay, S.A., 2002a, Chlorinated solvents stabilisation. GBR-2900-0002-W-EN Issue 1—19.06.2002, Rue du Prince Albert, 33 1050 Brussels, Belgium. http://www.solvaychemicals.com (accessed November 9, 2003).

Note: AAVs are expressed in wt% NaOH.

Consequently, testing is needed for both reserve alkalinity and acid acceptance (Dow Chemical Company, 1999c).

Other acid tests and related solvent tests include the following:

- ASTM D 21-6-78—an amine acid acceptance method that measures the concentration of an amine (basic) inhibitor by titration with standard acid.
- ASTM D 2942-74—a test that determines the total AAV and measures the total concentration of an amine and neutral-type (α-epoxide) inhibitors in a solvent.
- ASTM D 1364-90—a test that measures the water content of volatile solvents. The presence of water promotes corrosion of metals and hydrolysis of the chloride–carbon bond. The hydrolysis releases hydrogen chloride, which immediately combines with water to form hydrochloric acid. This test uses an analysis of Karl–Fischer type, which involves placing a sample into a drying oven at a predetermined temperature for a predetermined period of time. The water in the sample is vaporized and carried into a reaction vessel with methanol. The methanol traps the water, which is titrated to an endpoint with a Karl–Fischer reagent to determine the amount present.

1.2.5.2 Aluminum Scratch Test

The aluminum scratch test, ASTM D 2943-76, determines whether sufficient metal inhibitor (a compound that inhibits the catalyzing action of metal surfaces) is present to prevent reaction between aluminum and methyl chloroform. The test involves holding a clean coupon (a coupon is the metallurgical term for a thin strip of metal used for testing metal properties) of aluminum beneath the surface of the liquid solvent and scratching away the protective oxide coating. If the metal stabilizers in methyl chloroform (most commonly 1,4-dioxane) are depleted, a reaction will ensue that produces a blood-red color emitted from the metal surface. The color and bubbling are attributed to the complex formed between aluminum chloride and methyl chloroform, following dehydrochlorination and the production of hydrogen chloride gas. The degree of solvent discoloration, the amount of dark residue material formed, and the presence or absence of hydrogen chloride bubbling are used to gauge the inhibitor strength (Archer, 1984; Tarrer et al., 1989).

Other tests for inhibition of solvent breakdown by metals are as follows:

- The National Institute of Cleaning and Dying Standard 3–50 tests for an 18 mg or smaller loss in three uniformly sized copper strips (Skeeters, 1960b)
- Federal Specification OT-634A, "Trichloroethylene, Technical, 4/17/56" tests quantitatively for the extent to which metal-catalyzed oxidation decomposition of TCE has progressed

1.2.6   IDENTIFYING SOLVENT STABILIZERS USED IN CHLORINATED SOLVENTS

Determining the right mix of additives that will successfully stabilize a chlorinated solvent for a particular application was the subject of intensive research by industrial chemists working for the major solvent producers. Obtaining a successful formulation was a definite advantage in a competitive market. Proprietary formulations were therefore held as trade secrets, making difficult the task of establishing which stabilizers were used in commercial formulations of the major chlorinated solvents. Nevertheless, there are several ways of determining the identity and composition of stabilizers in commercial solvent formulations. The presence of specific stabilizer compounds in commercial solvent formulations may be documented in the following:

- MSDSs
- Chemical Safety Data Sheets from the Manufacturing Chemists Association (predating MSDSs)

- Marketing literature by the major solvent producers
- Patents
- Industrial literature from technical journals and trade magazines
- Toxicology studies that analyzed and identified the complete range of toxic substances present in solvents

MSDSs may not disclose all additives, indicating that formulations are proprietary, that quantities are less than 1%, or that components are either nontoxic or nonvolatile. Disclosure rules varied through time, and different states required disclosure of different information. For example, California's Proposition 65 rule requires listing reproductive toxins. Worker health and safety rules and transportation regulations require disclosure of some regulated compounds in the solvent formulation.

Product fact sheets and consumer guidance from the major solvent producers provide a rich resource for understanding the nature of solvent use, demands for stabilization, and the operator's challenge for maintaining solvent stability. The literature from Dow Chemical, available online, is particularly informative; however, with a few exceptions, the solvent marketing literature does not disclose the ingredients of solvent-stabilizer formulations. The approximate proportion of a given solvent formulation including its stabilizers can be inferred by contrasting the density of virgin, unstabilized solvent with that of a stabilized formulation, because stabilizers are in general substantially less dense than chlorinated solvents and most are less dense than water.

Hundreds of patents have been issued for stabilizing chlorinated solvents in the major industrial countries. The first patent for stabilizing a solvent was issued in 1914 to Walter Snelling of Pittsburgh, Pennsylvania, for stabilizing carbon tetrachloride fire extinguishers against corrosion of brass containers. Snelling used aliphatic hydrocarbons such as ethylene and acetylene (ethyne) compounds added at 0.5–3% to combine with "free chlorine" to inhibit corrosion.

Patents for stabilizing chlorinated solvents and new "designer solvents" continue to be issued today. The listing of a stabilizer formulation in a patent does not prove that it was deployed for commercial use; many patented formulations were never brought to market. In demonstrating how a particular new formulation presents an improvement over the prior art, some patents present the composition of degreasing grades of solvents in use at the time, determined by laboratory analysis. This information can be used to piece together the history of typical solvent formulations by the classes of compounds used to achieve stabilization of solvents against attack by acids, metals, and ultraviolet light. In some instances, patents list the specific compounds and the quantities of stabilizers of the leading formulations at the time the patent was issued. Patents also reveal the relative proportions of stabilizer compounds that work together to prevent reactions with the solvent. Some stabilizers work synergistically with other stabilizer compounds, and the patent literature reveals the stoichiometric relationships between them.

The industrial literature—including that from institutions promulgating specifications and performance standards for solvent use in military and industrial applications—is another source of information on identity and quantities of stabilizers in chlorinated solvents. Promulgation of air-quality regulations pursuant to the Clean Air Act led to an industry-wide effort to minimize solvent losses in vapor degreasing operations. Similarly, the land ban on disposal of liquid waste solvent led to increased attention to maximizing solvent recovery through recycling operations. The literature on solvent recycling includes information on replenishment of solvent stabilizers lost during distillation or other solvent-recovery processes and thereby reveals typical stabilizer formulations in the 1980s.

Studies on the toxicity of the major chlorinated solvents distinguished between the toxicity of pure solvents and that of technical grade solvents. A few solvent stabilizers are considerably more toxic than the solvents they stabilize. A number of articles on toxicology reveal the composition of the solvents assayed; some endeavor to separate the toxic effects of epoxide stabilizers from those of the chlorinated aliphatic solvents.

Tables 1.21 through 1.24 present a compilation of the documented presence of stabilizers in different solvent formulations obtained from cited sources, as identified at the beginning of this section. The

TABLE 1.21
Some Examples of General Citations of Stabilizers for the Major Solvents, by Year

Methyl Chloroform

Archer et al. (1977)

1,3-Dioxolane; nitromethane; 1,2-butylene oxide; isobutyl alcohol; and toluene

1,2-Butylene oxide; acetonitrile; 1,3,5-trioxane; and nitromethane

tert-Butyl alcohol; nitromethane; and methyl butynol

1,2-Butylene oxide; *tert*-amyl alcohol; methyl ethyl ketone; and nitromethane

Nitromethane; acetonitrile; 1,2-butylene oxide; and isopropyl nitrate

Goodner et al. (1977)

DOW Chlorothene VG™: nitromethane; 1,4-dioxane; and acetonitrile

IARC (1979)

Nitromethane; *n*-methylpyrrole; 1,2-butylene oxide; 1,3-dioxolane; and secondary butyl alcohols

Dow Chemical Company (1980)

2-Methyl-3-butyn-2-ol; nitromethane; nitroethane; *t*-amyl alcohol; and 1,2-butylene oxide

Archer (1984)

USA: 0.5–0.8% 1,2-butylene oxide; 0.4–0.7% nitromethane; 2.0–3.5% 1,4-dioxane; 1.0–2.0% *sec*-butanol; and 1.0% 1,3-dioxolane

Europe: 0.6–1.0% 1,2-butylene oxide; 0.4–1.0% nitromethane; 3.5% 1,4-dioxane; 2.0–6.5% *tert*-butynol; 2–3% methyl butynol; 2% isopropylnitrate; and 3% acetonitrile

Japan: 0.1–0.6% 1,2-butylene oxide; 0.1–0.7% nitromethane; and 3.5% 1,4-dioxane

Gregersen and Hansen (1986)

Epoxides; 1,4-dioxane; nitro compounds; carbonyl compounds; and *tert*-alcohols

Quast et al. (1988)

Methyl chloroform has a purity of 94%, with 5% stabilizers, including 1,2-butylene oxide, *tert*-amyl alcohol, methyl butynol, nitroethane, nitromethane, and <1% minor impurities

Dow Chemical Company (1992)

Dow Aerosol Grade: 1.5% *tert*-butyl alcohol; 0.5% 1,2-butylene oxide; and 1.9% dimethoxymethane

Meike (1993)

A typical cutting/tapping fluid based on a chlorinated hydrocarbon solvent was composed of 76% methyl chloroform, 16.3% 1,4-dioxane, 4.2% dichloromethane, and 3.5% 1,2-epoxyethane

Michigan ORR (2000)

tert-Butyl alcohol, *sec*-butyl alcohol, methylal (dimethoxymethane), and 1,2-butylene oxide

INEL (2006)

Dowclene EC: 1.9% 1,4-dioxane

Trichloroethylene

Stauffer (1956)

Triethylamine, *p-t*-amyl phenol, and diisobutylene

1976 and 1978 compositions cited in European Chemicals Bureau (2000c)

1976: 0.19% 1,2-butylene oxide; 0.04% ethyl acetate; 0.02% *n*-methylpyrrole; 0.03% diisobutylene; and 0.09% epichlorohydrin

1978: 0.023% diisobutylene; 0.024% 1,2-butylene oxide; 0.052% ethyl acetate; 0.008% *n*-methylpyrrole; and 0.148% epichlorohydrin

TABLE 1.21   (continued)
Some Examples of General Citations of Stabilizers for the Major Solvents, by Year

Gregersen and Hansen (1986)

0.001–2% of bis(2-propyl)amine, 2-isopropyl-5-methyl-phenol, butylene oxide, and ethyl acetate

Arena and Drew (1986)

Pentanol-2-triethanolamine; 2,2,4-trimethylpentene-1; and isobutanol

Dichloromethane

Gregersen and Hansen (1986)

Up to 1% phenol compounds, amines, epoxides, 1,4-dioxane, nitromethane, and methanol

20 ppm amylene, 50 ppm amylene, 0.2 wt% ethanol, or 150 ppm cyclohexane

Huntingtons Chemicals (2006)

Commercial methylene chloride is normally inhibited with small quantities of stabilizers (typically 0.005–0.2% by weight) to prevent acidification and corrosion

Typical stabilizers are methanol, ethanol, amylene (2-methyl-but-2-ene), cyclohexane, or tertiary butylamine

Perchloroethylene

Gregersen and Hansen (1986)

0.1%, phenols, or mixtures of epoxides and esters

Morrison (2003)

Morpholine derivatives and mixtures of epoxides and esters

MDEQ (2004)

Dry-cleaning grade: 4-methylmorpholine, diallylamine, tripropylene, cyclohexene oxide, benzotriazole, and β-ethoxypropylnitrile

Linn (2002)

DOWPER™ Solvent: 4-methylmorpholine

DOWPER™ CS Solvent ("charged solvent"): nonylphenol—4-mol ethylene oxide adduct (this was a solvent charged with detergent; it is no longer on the market)

Vulcan Materials' PerSec: diallylamine and tripropylene compounds

PPG's dry-cleaning grade PCE: cyclohexene oxide and β-ethoxyproprinitrile

PPG Perchlor Type 236 Stabilizer Concentrate MSDS: cyclohexene oxide, β-ethoxyproprinitrile, *n*-methyl morpholine, and 4-methoxyphenol

n-Propyl Bromide[a]

Sclar (1999)

<0.5% 1,2-butylene oxide, <2.5% 1,3-dioxolane, and <0.25% nitromethane

Cheap Solvents (2002)

20–30% dimethoxymethane, 15–20% 2-methyl-2-propanol, 5–8% 1,2-butylene oxide (Stabilizer Solvent is a concentrate to refortify *n*PB for depleted stabilizers)

Petroferm, Inc. (2005)

10–15% 1,2-butylene oxide; 1–3% 2-methyl-2-propanol; and 1–3% dimethoxymethane

Swanson et al. (2002)

1,3-Dioxolane and 1,2-butylene oxide in *n*PB adhesive; contains 3% stabilizer

[a]   nPB is n-propyl bromide, a recently introduced replacement solvent for methyl chloroform.

TABLE 1.22
Antioxidant Compounds Used in Chlorinated Solvents (Partial List)

Solvent and List of Antioxidant Compounds	Quantity (vol%)[a]	References
Methylene chloride		
2-Methyl-2-butene	0.5	Irani (1977)
Phenol	0.05–0.5	Daras (1960), U.N. Environment Programme (UNEP) (1984)
Thymol	0.001–0.5	Pray and Chisholm (1960)
Cresol	Present in vapors	Ramos (1974)
Stabilized trichloroethylene		
Thymol	0.001–0.5, 0.28	Shepherd (1962), World Health Organization (1985), Pray and Chisholm (1960)
p-t-Amyl phenol	1–5	Shepherd (1962), Fullhart and Swalheim (1962), Watson and Rapp (1959)
Phenol	0.001–0.5 (0.05–0.5 g/L)	Pray and Chisholm (1960)
Pyrrole		Smallwood (1993)
n-Methyl pyrrole	0.022–0.028	Copelin (1959), Shepherd (1962), Tarrer et al. (1989), Starks (1957), OEHHA (1999)
1-Ethoxy-2-imino-ethane	0.01–0.02	Smallwood (1993)
Perchloroethylene		
2,6-di-*tert*-butyl-*p*-cresol[b]	30 ppm, 80 ppm	Dempf et al. (1977), Howell and Tarrer (1994)
Thymol	0.01% wt/wt	Reynolds and Prasad (1982)
Methyl chloroform		
Resorcinol	0.0001–1	World Health Organization (1990)

[a] Except as noted.
[b] Also known as Ionol™.

patent literature is cited where a particular patent discloses stabilizer compositions in use at the time the patent was issued, but this practice is not intended to imply that the stabilizers claimed in the patent are present in the solvent. Patent *claims* on stabilizer packages are nevertheless useful and informative as an indirect means of understanding the empirical art of stabilizing solvents. A comprehensive compilation of stabilizers claimed in United States, British, and Canadian patents issued to United States, British, Canadian, German, Belgian, French, and Italian companies is presented in Appendix 1.

1.2.6.1 Solvent Stabilizer Chemicals

The major solvents have been stabilized with groups of chemicals identified as having an inhibiting effect on the reactions or catalysts that break down solvents. Each chemical added serves a distinct purpose, such as neutralizing acids, supplying an antioxidant, and inhibiting reactions with metal chloride salts. Finding the right chemicals to perform these tasks was the subject of much research and experimentation. For many years, the nature of the reactions that cause solvent breakdown was not identified or well understood by the industrial chemists whose task it was to prevent these reactions. In 1943, an inventor wrote in his patent, "The nature of deteriorative changes being unknown, the operative mechanism is also unknown. Stabilizing action must therefore be determined empirically" (Pitman, 1943).

The patent literature shows that industrial chemists understood which classes of compounds had the capacity to inhibit reactions. These chemists experimented with many candidate compounds

TABLE 1.23
Acid-Acceptor Compounds Used in Chlorinated Solvents

Solvent and Trade Name and List of Acid-Acceptor Compounds	Quantity (vol%)	References
Methylene chloride		
Amylene	<1.0	Solvay SA (2002a)
Methylene chloride, Technical Grade		
Propylene oxide	0.5	Dow Chemical Company (1995)
Dichloromethane, Technical MIL-D-6998		
Propylene oxide	2.0	Vulcan Chemicals (1987)
Methylene chloride, Technical		
Cyclohexane	<0.10	Dow Chemical Company (1999a)
Neu-tri solvent: Trichloroethylene #56530		
1,2-Butylene oxide	0.60	Dow Chemical Company (1981)
Stabilized trichloroethylene [a]		
1,2-Butylene oxide	0.053	Watson and Rapp (1959)
Trimethylpentene (diisobutylene)	0.30	Watson and Rapp (1959), Pray and Chisholm (1960), Fullhart and Swalheim (1962)
	0.03	
Stabilized trichloroethylene		
Isopropyl alcohol	0.25	Monroe and Rapp (1959)
Pyridine	0.001–0.1	Ferri and Patron (1959)
Triethylamine	0.05–0.20	Shepherd (1962), McKinney et al. (1955), Tsuruta and Fukuda (1983), Graham (1967)
Epichlorohydrin	0.31	Tsuruta and Fukuda (1983), Tarrer et al. (1989), Henschler
	0.09–0.8	et al. (1984), World Health Organization (1985)
Perchloroethylene		
Cyclohexene oxide	0.002	Blum (1984), Joshi et al. (1989), Dempf et al. (1980), Pray
	0.25	and Chisholm (1960)
	0.01–0.5	
Butoxymethyl oxirane (*n*-butyl glycidyl ether, BGE)	0.004	Joshi et al. (1989), Dow Chemical Company (1987)
	0.5	
4-Methylmorpholine	0.003	Blum (1984)
	0.005	
	0.008	
n-Methylmorpholine	0.0053	53 ppm in DOWPER: NIH (1986)
Epichlorohydrin	0.10	Van Waters and Rogers, Inc. (1990), Beckers (1974)
	0.25	
Methyl chloroform		
1,2-Butylene oxide	0.47	Dow Chemical Company (1990), Occidental Chemical
	0.3–0.6	Corporation (1989)
n-Methyl pyrrole	Present	Smallwood (1993)
Nitromethane	0.34	Dow Chemical Company (1990) and Occidental Chemical
	0.2–0.5	Corporation (1989)

[a] Patent citation of a typical formulation of TCE in 1959 was 300 ppm 1,2-butylene oxide, 2500 ppm isopropylacetate, and 1500 ppm trimethylpentene (also called diisobutylene).

TABLE 1.24
Common Metal-Inhibitor Compounds Used in Chlorinated Solvents

Solvent and Trade Name and List of Metal-Inhibitor Compounds	Quantity (vol%)	References
Methylene chloride		
1,4-Dioxane	<1	World Health Organization (1984b),
	0.8	Beckers (1973)
Cyclohexane	Present	Rowe and Cawley (1977)
1,3-Dioxolane	~ 2	Beckers (1973)
Stabilized trichloroethylene		
Isoamyl alcohol		Petering and Aitchison (1945)
Perchloroethylene		
Cyclohexane	<1	Stevens (1955)
Ethyl acetate	0.1–0.5	Stevens (1955), Copelin (1959)
Methyl chloroform		
1,4-Dioxane	2–2.7	Occidental Chemical Corporation (1989)
1,4-Dioxane	3	Ashland Chemical Company (1995)
1,4-Dioxane	0–4	Unocal (1989)
1,3-Dioxolane	2	Great Western Chemical Company (1990)
1,3-Dioxolane	1–5	Ashland Chemical Company (1996)
Formaldehyde dimethyl hydrazone	Present	Tarrer et al. (1989)
2-Methyl-3-butyn-2-ol	Present	Ishibe and Metcalf (1982)

before arriving at the right chemical to reliably stop solvent breakdown. Far from trial and error, industrial chemists applied sophisticated methods to select from thousands of candidate stabilizers. A number of patents claim groups of three or more stabilizer chemicals that work synergistically, such that the absence of one compound precludes the stabilizing action of the others. The extensive list of more than 500 stabilizer compounds named in the patent literature and presented in Appendix 1 testifies to the enormous effort carried out by industrial chemists since 1914 to enable use of the major chlorinated solvents in demanding applications.

In the early decades of solvent stabilization, some inventors made claims for single additives to solve several stabilization problems. As the art of solvent stabilization evolved, industrial chemists developed solvent stabilizer packages commonly consisting of four or more chemicals to address the multiple reactions contributing to solvent breakdown. In recent decades, solvents were sold with stabilizer packages customized for very specific applications particular to each solvent, such as vapor degreasing parts with iron and zinc but not aluminum or magnesium.

As with any endeavor involving organic chemistry, compiling lists of chemicals can be difficult because of the many acceptable synonyms for organic compounds as well as the larger list of trade names for some compounds. Appendix 1 lists some of the stabilizers claimed in the selected patents cited.

A number of articles, product fact sheets, and other sources provide numerous nonspecific lists of stabilizer compounds that name the class of compounds used or provide a longer list of chemicals that might have been used. Some listings of stabilizers for a given solvent include all the stabilizer formulations used by the leading solvent producers, giving the incorrect impression that one manufacturer's grade of solvent might contain all the listed compounds. These listings are useful to obtain a sense of which classes of compounds were required to neutralize acid, inhibit acid formation, and impede solvent reactions with alkali metals. The examples in Table 1.21 are ordered by chemical species and year cited.

1.2.6.2 Antioxidants

Antioxidants reduce the solvent's potential to form oxidation products (Archer, 1984). The degree to which a solvent can undergo oxidation depends on its chemical structure, its exposure to ultraviolet light, and, if used in a vapor degreaser, its boiling point. Solvents with lower boiling points—dichloromethane (40°C) and methyl chloroform (74°C)—are less susceptible to oxidation than solvents with higher boiling points, such as perchloroethylene (121°C). The ethylene bond in TCE and perchloroethylene increase their vulnerability to oxidative attack.

Antioxidants usually fall into three chemical groups: phenols, amines, and amino-phenols, all of which contain an unsaturated benzene ring with either an amine group or a phenol group. Antioxidants suppress the free radical chain reaction that decomposes unsaturated solvents by forming stable resonance hybrids after losing a hydrogen atom to an oxidation-free radical and slowing the propagation step of auto-oxidation (Joshi et al., 1989).

Table 1.22 provides a list of commonly cited antioxidant compounds added to the major chlorinated solvents.

Although some sources list antioxidants in methyl chloroform, they are largely unnecessary. Methyl chloroform is not prone to oxidation reactions because it does not have a double bond (Archer, 1984). The use of phenolic compounds as antioxidants for TCE has caused problems for extracting essential oils from flavorings and for making decaffeinated coffee. For example, *p-t*-amyl phenol is unsuitable because of toxicity reasons.

A 1957 patent mentions that *n*-methyl pyrrole is a widely used antioxidant for TCE and is effective against normal air, light, and heat decomposition (Starks, 1957). However, contemporaneous patents note that pyrroles used as antioxidants have a tendency to form a dark purple sludge and that TCE stabilized with pyrrole becomes discolored over time when stored for several months in drums (Kauder, 1960; Willis and Christian, 1957). One proposed remedy to prevent the decomposition of pyrrole-stabilized TCE is the addition of diisopropylamine (Ferri and Patron, 1959). Furthermore, *n*-methyl pyrrole itself was shown to have stability problems, and an organometallic chelate compound was required to stabilize it (Starks, 1957).

Motorola's industrial chemists capitalized on the propensity for ethylene compounds to undergo oxidative attack by proposing that these compounds be used as stabilizers for methyl chloroform. Motorola's patent notes that Chlorothene VG, a widely used grade of methyl chloroform sold by Dow Chemical, will withstand semiconductor degreasing conditions for only one or two hours. For methyl chloroform, Motorola's chemists proposed a stabilizer package that uses TCE, perchloroethylene, and fluorine as free radical scavengers to meet the stringent demands of degreasing semiconductor devices while also satisfying EPA regulations. Free radical scavengers combine with trichloromethyl radicals as follows (Goodner et al., 1977):

$$R_3CH + Cl_3C \longrightarrow R_3C + Cl_3CH. \tag{1.13}$$

Use of TCE, perchloroethylene, and fluorine as free radical scavengers in methyl chloroform will trap trichloromethyl radicals as intermediates before the acidic decomposition of these radicals takes place.

1.2.6.3 Light Inhibitors

A variety of chemicals have been patented to stabilize perchloroethylene against the oxidative attack of ultraviolet light. "Light inhibitors" claimed in patents have included *n*-methyl morpholine, isoeugenol (4-hydroxy-3-methoxy-1-propenylbenzene), alkyl cyanide compounds such as alkylaminoalkylcyanide, and nitrile compounds such as β-ethoxyacetonitrile and β-methoxyacetonitrile (Stevens, 1955; Strain and De Witt, 1956; Dial, 1957; Starks, 1957; Skeeters, 1959, 1960b). The mechanism for inhibiting the breakdown of perchloroethylene from ultraviolet light is not discussed in these patents. It is likely that these inhibitor compounds interrupt the free radical chain reaction, but do not physically impede the action of the ultraviolet light catalyst on the carbon–carbon double bond.

1.2.6.4 Thermal Stabilizers

A few compounds are described as providing thermal stability to perchloroethylene and TCE. Such stabilizers include diisobutylene, cyclohexene, amylene, pyrrole, and 1,2-epoxide (Stauffer, 1956; Shepherd, 1962; Smallwood, 1993).

1.2.6.5 Acid Acceptors

Acid acceptors, also referred to as antacids in the older patent literature, react with and chemically neutralize trace amounts of hydrochloric acid or acetic acid formed by hydrolysis or introduced during degreasing operations. The hydrolysis reaction cannot be prevented; hence, acid acceptors—which react with the acid to produce a benign alcohol—are added to solvents (DeGroot, 1998). Acid-acceptor compounds are either neutral (e.g., epoxide compounds) or slightly basic (e.g., amine compounds); they react with acid in the solvent and form an alcohol in the process (Archer, 1984; Tarrer et al., 1989). If left unneutralized, hydrochloric acid and acetic acid can cause progressive solvent degradation. Compounds commonly used to neutralize acids in chlorinated solvents are summarized in Table 1.23.

Equation 1.14 (from Tarrer et al., 1989) demonstrates a typical acid acceptance reaction affected by an acid-accepting epoxide stabilizer:

$$\underset{\text{Ethylene oxide}}{\overset{\overset{\displaystyle O}{\diagup\ \diagdown}}{CH_2 \!-\!-\! CH_2}} \;+\; HCl \;\longrightarrow\; \underset{\text{1-Chloroethanol}}{\overset{\overset{\displaystyle O\,H\quad\; Cl}{|\qquad\; |}}{CH_2 \!-\!-\! CH_2}} \tag{1.14}$$

Amines were among the more commonly used acid-acceptor compounds until about 1955, when DuPont introduced a nonalkaline formulation based on a pyrrole and Westvaco introduced a neutral formulation. These essentially replaced amines by 1961 (Doherty, 2000a). Used alone, amines can be disadvantageous because they are dissipated by reaction with acids. Amines have the further disadvantage that amine salts (1) accelerate the corrosion of metals such as zinc in galvanized degreasers and (2) can complex with copper and copper compounds (Stauffer, 1956; Copelin, 1959).

Certain combinations of acid-acceptor compounds can work together to regenerate the stabilizer by returning it to its original state when the reaction terminates. An example of this is the synergistic combination of an amine and an organic epoxide, which together are many times more effective than either additive used alone. The regeneration of an amine acid inhibitor (such as triethylamine) or the organic nitrogen-ring amine compound groups (pyridines and picolines) with an epoxide (such as 1,2-butylene oxide, propylene oxide, or cyclohexene oxide) may proceed as follows (Copelin, 1957):

$$\text{Amine} + HCl \;\longrightarrow\; \text{Amine} \bullet HCl, \tag{1.15}$$

$$\text{Amine} \bullet HCl + \text{Epoxide} \;\longrightarrow\; \text{Chlorohydrin} + \text{Amine}. \tag{1.16}$$

Because the amine compound is regenerated, less is needed; the mole ratio of epoxide to amine can be 150:1. Eventually, epoxides will need to be replenished because they are consumed during the acid acceptance reaction.

Early efforts to find good acid-acceptor compounds included additions of the following:

- Caffeine, quinine, and limestone (calcium carbonate) to TCE (Dinley, 1937)
- Resins such as gum mastic, sandarac, and rosin to protect carbon tetrachloride and perchloroethylene for drum storage in contact with iron and copper (Stewart and DePree, 1933)

- Gasoline (Pitman, 1933)
- Camphor and turpentine (Dinley, 1937)

1.2.6.6 Metal Inhibitors

Metal inhibitors deactivate the catalytic properties of metal surfaces and complex any metal salts that might form. Metals and their salts promote solvent breakdown through catalysis. Metal inhibitors may also terminate free radicals through hydrogen donation (Archer, 1984; Howell and Tarrer, 1994). Metal inhibitors are Lewis bases that inhibit solvent-degradation reactions in the presence of a metal and its chloride (e.g., aluminum and aluminum chloride).[*] 1,4-Dioxane is a Lewis base because the oxygen molecules in 1,4-dioxane have electrons available for sharing (a base is a proton acceptor; a Lewis base is an electron-pair donor). Other Lewis bases used as metal inhibitors include 1,3-dioxolane, *tert*-amyl alcohol, methyl ethyl ketone, isopropyl nitrate, and nitromethane (van Gemert, 1982).

Metal inhibitors compete with the solvent for the aluminum chloride produced at microcorrosion sites on the aluminum surface. They either react with the active aluminum site, forming an insoluble deposit, or complex with aluminum chloride, preventing degradation of the solvent. Successful inhibition involves complexing of the chemisorbed aluminum chloride product with electronegative chemical groups in the molecular structure of the stabilizer (Archer, 1982). The solubility of the complex formed by the stabilizer and the metal chloride in the solvent determines the effectiveness of the stabilizer. Highly soluble stabilizer–metal chloride complexes are undesirable, because instead of forming a protective coating over the active reaction sites on the metal surface, the complexes are removed by dissolution (Archer, 1982). Excess water promotes corrosion at the solvent-water interface because the metal chloride reaction product dissolves easily from the metal surface into the water phase (Archer, 1982).

The main function of a metal stabilizer is to compete with the solvent for electron-deficient sites on aluminum chloride adsorbed to the metal or metal oxide surface. The Lewis base stabilizer converts aluminum chloride into an insoluble coating on the metal surface, following the series of steps presented by Archer (1982):

$$Al(OH)_3 \leftrightarrow Al(OH)_2^+ + OH^-, (1.17)$$

$$Al(OH)_2^+ + Cl^- \longrightarrow Al(OH)_2Cl \text{ (soluble aluminum hydroxychloride salt)}, \qquad (1.18)$$

$$2[Al^{3+}] \cdot Cl_{adsorbed} + 3CH_3Cl_3 \longrightarrow 2[Al^{3+}] + 4Cl^- + 3(CH_3Cl_2C\bullet), \qquad (1.19)$$

$$Al_{surface} + 3(CH_3Cl_2C\bullet) \longrightarrow \tfrac{3}{2}(CH_3CCl_2CCl_2CH_3). \qquad (1.20)$$

Methyl chloroform   2,2,3,3-Tetrachlorobutane

In addition to the radical dimer product 2,2,3,3-tetrachlorobutane, an equimolar amount of 1,1-dichloroethane and a minor amount of chloroethane have been identified as products of the aluminum + methyl chloroform reaction, which suggests a reductive reaction pathway confirmed by an absence of vinyl chloride (van Gemert, 1982).

---

[*] A Lewis base is any molecule or ion that can form a new coordinate covalent bond, by donating a pair of electrons. Lewis bases are preferred as metal inhibitors because they have available electrons capable of complexing as electron donors with an electron-deficient atom such aluminum. In this context, aluminum chloride is a Lewis acid.

Aluminum is a highly reactive metal; its standard potential is −1.66 V, that is, it is strongly reducing. The insoluble oxide coating on aluminum will by itself prevent reaction with methyl chloroform; clean but oxide-coated aluminum dipped into methyl chloroform produces no reaction. However, any scratches in the aluminum oxide surface produce sites for microcorrosion to begin, which leads to the degradation of the solvent and the consumption of aluminum in the catastrophic decomposition reaction.

Methyl chloroform is relatively immune to the free radical mechanism for oxidation reactions; however, methyl chloroform can react violently with exposed aluminum or aluminum chloride salts. Methyl chloroform will also react with iron or tin in a similar fashion, but the corrosion rates are much slower than the reaction with aluminum (Archer and Stevens, 1977). Dichloromethane and, to some degree, TCE are also susceptible to degradation in the presence of aluminum chloride salts or aluminum swarf (aluminum fines), particularly in the vapor phase. Perchloroethylene does not appear to have any significant reactivity toward aluminum (EuroChlor, 2003). A variety of metal stabilizers that are Lewis bases are added to chlorinated solvents to prevent or inhibit these reactions.

For inhibition to occur, there should be one or preferably two electronegative functional groups in the molecular structure of the selected metal inhibitor. Nitrogen functional groups have greater activity than sulfide groups, which in turn have greater activity than ether compounds. The activity of the functional group in this hierarchy is in reference to the ability to complex with Lewis acids such as aluminum chloride. Cyclic structures such as 1,4-dioxane and 1,3-dioxolane are generally more active than their straight-chain analogues. Compounds commonly used as inhibitors incorporate ether linkages, sulfide linkages, carbonyl groups, nitriles, amines, and alcohols (Archer, 1982). Because 1,4-dioxane has two ether linkages, it is considered "difunctional" and exhibits greater activity than a compound with a single ether linkage such as tetrahydropyran.

Compounds found to be effective metal inhibitors for methyl chloroform are not necessarily effective for other chlorinated solvents. For example, the hierarchy of functional groups in a mixture of 90% dichloromethane and 10% toluene (a common aerosol spray formulation) is opposite the hierarchy for methyl chloroform metal inhibitors (Archer, 1982). Other solvents require larger quantities of the same stabilizers that are effective for methyl chloroform. For example, 1,1,2-trichloroethane is easily stabilized by furfuryl alcohol and dimethyl oxalate, whereas 1,1-dichloroethane can be stabilized by furfuryl alcohol, dimethyl oxalate, pyrazine, glycidol, or 1,4-dioxane. Stabilizers preferred for the commonly used dichloromethane-toluene aerosol aluminum paint formulation contain an oxygen functional group, such as dimethoxymethane, dimethyl carbonate, or 1,4-dioxane (Archer, 1979).

The ability of solvent stabilizers selected to inhibit the aluminum + methyl chloroform reaction to protect the solvent from breakdown, ordered from best to worst, is 1,4-dioxane > nitromethane > *tert*-amyl alcohol > 1,2-butylene oxide[*] (van Gemert, 1982). However, when aluminum powder is added to a commercial formulation of methyl chloroform containing all these stabilizers, the first stabilizer to react is 1,2-butylene oxide, followed by *tert*-amyl alcohol. The reaction of aluminum with 1,4-dioxane and nitromethane is slow in comparison, but all of these stabilizers are more reactive toward aluminum than toward methyl chloroform. An explanation for this difference notes that epoxides[†] and alcohols form reaction products with aluminum chloride and are consumed, whereas 1,4-dioxane and nitromethane form complexes and are preserved. The 1,4-dioxane–aluminum chloride complex is highly insoluble; however, 1,2-butylene oxide will dissolve it and release the

---

[*] 1,2-Butylene oxide, the most widely used acid acceptor for methyl chloroform, can also be considered a Lewis base (van Gemert, 1982).

[†] 1,2-Butylene oxide is an epoxide.

1,4-dioxane in its original form and a soluble aluminum-epoxide reaction product. This reaction can be conceptually represented as follows (van Gemert, 1982):

$$\text{AlCl}_3 \cdot \text{O} \overbrace{}^{\text{Very insoluble}} \text{O} + 3 \triangle\!\!\!\backslash\!\!\!\text{O} \rightarrow \text{Al} \left(\text{O} \overset{\text{Soluble}}{} \underset{\text{Cl}}{} \right)_3 + \text{O} \overbrace{} \text{O} \tag{1.21}$$

Aluminum chloride and 1,4-dioxane	3(1,2-butylene oxide)	Aluminum epoxide product	1,4-dioxane

1,4-Dioxane has always been used in combination with 1,2-butylene oxide to stabilize vapor degreasing formulations. The electronegativity and molecular size of the complex formed by the stabilizer and the metal determine its solubility. In order to form an effective physical barrier at the reaction site, the complex should have only limited solubility in the solvent system (Archer, 1979).

Carbon tetrachloride also corrodes aluminum, after an initial induction period. The corrosion occurs by direct chemical attack and not by an electrochemical corrosion cell (Stern and Uhlig, 1952, 1953). The reaction of carbon tetrachloride with aluminum produces hexachloroethane and aluminum trichloride as follows (Minford et al., 1959):

$$2\text{Al} + 6\text{CCl}_4 \longrightarrow 3\text{C}_2\text{Cl}_6 + 2\text{AlCl}_3. \tag{1.22}$$

Perchloroethylene can react with fresh zinc surfaces, found in galvanized metals, to form toxic and flammable dichloroacetylene. The zinc surface eventually becomes passivated, and the reaction ceases, but this process can take up to one month. Since this reaction was discovered in the late 1970s, dry-cleaning machine manufacturers have stopped making distillation units from galvanized metal and use stainless steel instead. Reactions of perchloroethylene with galvanized metals are also a potential problem for degreasing equipment and distillation equipment for solvent recycling (EuroChlor, 2003).

Inventors of stabilizer packages found that TCE can be stabilized against metals by adding 1 mol of an oxygen-containing compound per 99 mol of chlorinated solvent (Petering and Aitchison, 1945). Alcohol compounds are among several means of providing oxygen to stabilize solvents. The more alcohol or other oxygen compound present, the more effective the composition in restraining the metal-induced decomposition of solvents. For cleaning zinc, magnesium, and their alloys, as well as iron, steel, and copper, other metal stabilizers for TCE included oximes (Petering and Aitchison, 1945) and cyclohexene, which boils with TCE, allowing recovery on distillation. Cyclohexene is 17 times more effective as a metal stabilizer for TCE than cyclohexane (Larchar, 1950).

Among the earliest citations of solvent stabilizers to inhibit reactions with metals, corrosion of brass by carbon tetrachloride was inhibited by unsaturated hydrocarbons according to a 1914 patent, and the stabilization of carbon tetrachloride with a thiourea compound dissolved in methyl ethyl ketone was claimed in 1936 (Snelling, 1914; Missbach, 1936).

1,4-Dioxane provides two oxygen atoms to methyl chloroform. At the typical formulations listed below, the mole ratio of oxygen provided by 1,4-dioxane to methyl chloroform was 2:51.5. Oxygen-containing stabilizer compounds named in the patent literature include epoxides, oximes, ethers, and alcohols. Table 1.24 provides a listing of commonly utilized metal inhibitors for the major chlorinated solvents.

Stabilizing methyl chloroform against reaction with aluminum was a challenge that significantly delayed its commercial use in vapor degreasing and other metal-cleaning applications. These problems were memorialized in a 1957 patent to Dow Chemical, which tells us "No inhibitor heretofore used commercially has been capable of really stabilizing methyl chloroform in contact with aluminum" (Bachtel, 1957). Although the addition of 1,4-dioxane solved most aspects of the problem, it created others. Methyl chloroform stabilized with 1,4-dioxane, nitromethane, and *sec*-butyl alcohol does not react with aluminum, but its boiling vapor reacts with zinc. The compound 1,4-dioxane catalyzes the reaction with zinc, leading to damaged degreasing equipment and solvent distillation equipment made of galvanized steel sheeting. When 1,4-dioxane becomes concentrated in the degreaser boiling sump, the attack on zinc in galvanized metals can be severe (Spencer and Archer, 1981). Adding a vicinal* monoepoxide such as 1,2-butylene oxide prevents the reaction between methyl chloroform and zinc in the vapor phase. The most common formulations of methyl chloroform contain 1,2-butylene oxide for this reason (Brown, 1962).

Pure 1,4-dioxane is susceptible to forming peroxides over time when stored. The compound 1,3-dioxolane will also combine with oxygen from the air to form explosive peroxides. Methyl chloroform stabilized with 1,4-dioxane or 1,3-dioxolane can develop peroxides or acidity during storage; however, 1,4-dioxane develops peroxides to a lesser extent than 1,3-dioxolane (Manner, 1977).

Efforts by Dow's competitors to stabilize methyl chloroform against the aluminum reaction involved other oxygen-containing compounds. Adding *tert*-butyl alcohol with 1,2-butylene oxide and nitromethane slows but does not completely inhibit the reaction, as shown by the formation of an amber precipitate. Pittsburgh Plate and Glass (later PPG), one of the producers of methyl chloroform since the 1960s, claimed that the addition of 1,2-dimethoxymethane prevents decomposition in the presence of aluminum and its salts (Hardies, 1966).

WHEN WAS 1,4-DIOXANE FIRST USED TO STABILIZE METHYL CHLOROFORM?

1,4-Dioxane was first identified in 1863 (Lourenço, 1863; Wurtz, 1863). It first became available for commercial use in the late 1920s (Reid and Hofman, 1929). Carbide and Carbon Chemicals Corporation was supplying 1,4-dioxane in the early 1930s for use as a solvent for cellulose acetate and plastics manufacturing (Jones et al., 1933). Reid and Hofman (1929) listed other uses motivating commercial production for 1,4-dioxane as a solvent:

- Degreasing
- Wool scouring
- Producing dyes
- Printing and staining
- Varnishes
- Paint and varnish removers
- Cosmetics
- Glues
- Shoe creams
- A preservative
- A fumigant
- A deodorant

The earliest mention of 1,4-dioxane as a solvent stabilizer was in 1945 in a patent for stabilizing solvents, in particular, TCE (Petering and Aitchison, 1945). However, the patent does not clearly indicate that 1,4-dioxane was preferred for use with TCE (see Box: Was 1,4-Dioxane a Stabilizer for Trichloroethylene?).

continued

---

* Vicinal means any two functional groups bonded to two adjacent carbon atoms; an epoxide is a cyclic ether with only three ring atoms (one oxygen and two carbons) in the shape of an equilateral triangle.

The first patent to claim 1,4-dioxane as a stabilizer for chlorinated solvents was filed in 1954 by Dow Chemical (Bachtel, 1957). Dow claimed 1,4-dioxane specifically to inhibit the reaction between methyl chloroform and aluminum and presented experimental data to show its effectiveness. Aluminum alloy coupons were scratched with a stylus to strip away the aluminum oxide surface coating. These were then immersed in unstabilized methyl chloroform, which led to immediate solvent breakdown and evolution of a carbonaceous, tarry mass. Methyl chloroform stabilized with 3.5% 1,4-dioxane remained water white with no visible reaction when in contact with scratched aluminum alloy coupons. Similar tests showed the effectiveness of 1,4-dioxane to inhibit the methyl chloroform reaction with aluminum in a boiling reflux chamber for 30 days and for storage in black iron drums for 2 months. Inclusion of *sec*-butyl alcohol or *tert*-amyl alcohol, in addition to 1,4-dioxane, prevented discoloration of the solvent.

Dow Chemical incorporated 1,4-dioxane into its chlorothene line of methyl chloroform products with the May 1960 introduction of Chlorothene NU.

WAS 1,4-DIOXANE A STABILIZER FOR TRICHLOROETHYLENE?

Petering and Aitchison's (1945) patent claims a variety of oxygen-containing compounds as stabilizers for chlorinated solvents, and TCE in particular. The patent (U.S. Patent 2,371,645) clearly lists 1,4-dioxane and even shows its molecular structure as an example of an "inner ether." However, the patent recommends propyl ether as a stabilizer for TCE, because it boils with the solvent, and emphasizes that the selection of the ether or oxide to stabilize TCE should be predicated upon its ability to distill with the solvent without loss of the stabilizer. In the claims section, the class of inner ethers is named, but the only compound specifically claimed by name is propyl ether. The patent also claims stabilizers that are "volatile with the chlorinated solvent" and stabilizers that are "simple aliphatic ethers" (1,4-dioxane is a cyclic ether).

The boiling-point difference between TCE and 1,4-dioxane is 14°C, a smaller difference than with methyl chloroform and 1,4-dioxane (which have a 26.9°C difference). The large boiling-point difference and tendency for 1,4-dioxane to partition and remain in the boiling sump did not prevent its widespread use for methyl chloroform; therefore, the smaller boiling-point difference would not by itself preclude the use of 1,4-dioxane to stabilize TCE.

In a 1960 patent filed in 1954 and issued to the Diamond Alkali Company of Cleveland, Ohio, an encyclopedic list of more than 100 compounds for use as stabilizers is provided, including "dioxan" (Skeeters, 1960b). This patent cites five other patents, none of which mention 1,4-dioxane. This patent is for stabilizing perchloroethylene, but the claims are for other compounds and do not specifically mention 1,4-dioxane.

A sampling survey of products made by processes that use organic solvents was completed in Japan in the 1980s. A large number of solvent samples (1179) were collected and analyzed by gas chromatography. A small amount of 1,4-dioxane was detected as an additive in every sample that contained methyl chloroform; however, the study does not mention any detections of 1,4-dioxane in products produced with TCE (Inoue et al., 1983).

At a number of sites with monitoring wells, 1,4-dioxane is found together with TCE where no methyl chloroform is detected. Collocation of 1,4-dioxane with TCE does not in itself prove that 1,4-dioxane was a stabilizer for TCE. At many facilities that operated over long periods of time, several different solvents were used in response to changing environmental and worker health and safety regulations. For example, in the late 1960s and early 1970s,

continued

TCE was replaced with methyl chloroform when it became apparent that TCE was contributing to smog in the Los Angeles basin and possibly to worker-exposure-related illness. In the mid-1980s, methyl chloroform fell out of favor when it was recognized as an ozone-depleting chemical. Many facilities switched from methyl chloroform to dichloromethane, perchloroethylene, or TCE in conjunction with adopting upgraded degreasers and ultrasonic cleaning equipment. A facility may have an old or recent release of TCE, followed or preceded by a 1,4-dioxane-containing methyl chloroform release. The abiotic half-life of methyl chloroform, approximately two years, is considerably shorter than either the abiotic or biodegradation half-life of TCE. Consequently, most of the methyl chloroform from an old spill will have converted to 1,1-dichloroethylene or 1,2-dichloroethane, while TCE will persist. This situation can result in the appearance that 1,4-dioxane was present in TCE. At some sites, the 1,4-dioxane could have been used separately for another purpose.

Of greatest significance is the fact that TCE is inherently more stable toward reactions with aluminum and other alkali metals than methyl chloroform. Smaller quantities of stabilizers were required for TCE, and TCE was more easily stabilized against metal reactions by using other chemicals. Tests of TCE stabilization using 0.1% 1,4-dioxane did not measurably improve TCE stability against reactions with iron, zinc, copper, and aluminum (Willis and Christian, 1957). The smaller boiling-point difference between 1,4-dioxane and methyl chloroform would probably not have produced partitioning in as pronounced a manner as occurs for 1,4-dioxane in the boiling sump of a methyl chloroform degreaser.

Officials at DOW Chemical assertively emphasize that 1,4-dioxane was not a constituent of DOW's TCE grades (J.A. Mertens of Midland, Michigan, Dow Chemical Corporation, personal communication, 2000). Finally, none of the extensive material collected for this book and used to confirm the presence of specific stabilizer compounds in chlorinated solvents assertively states that 1,4-dioxane was used to stabilize TCE.

1.2.7 FATE OF STABILIZERS IN INDUSTRIAL APPLICATIONS AND SOLVENT RECYCLING

The premise of "contaminant archeology" (as detailed in Chapter 9) is that understanding the nature of the industrial process in which the contaminant was used will inform the environmental scientist of the likely composition and properties of the waste industrial effluent that was released to soil and groundwater. Understanding the demands on solvents for various uses will provide clues regarding the types of stabilizers that were needed for a particular process. Contrasting the physicochemical properties of stabilizers with those of their host solvents reveals which stabilizers would more likely end up in the vapor degreaser still bottoms and stored in waste tanks that ultimately leaked. Contaminant archeology focuses on the clues that suggest the whole composition of the waste and how that composition may drive the subsurface behavior of the waste mixture, as well as the constituents of concern that should be considered for investigating a release of the waste solvent mixture. The properties that have the greatest bearing on the separation, concentration, or elimination of stabilizers in chlorinated solvents include boiling point, solubility in solvent and water, vapor pressure, organic carbon partition coefficient, thermal stability, and consumption in stabilizing reactions. These stabilizer and solvent properties and their consequences are discussed in more detail in this section.

1.2.7.1 Boiling-Point Differences

As described in Section 1.2.3.1, a stabilizer should not partition differentially: the stabilizer should have the same concentration in the vapor phase as it has in the liquid phase. Two or more organic solvent or stabilizer compounds, when brought to a boil, will distribute themselves differently between the two phases. The compound with a higher vapor pressure, or lower boiling point, will concentrate in the vapor phase. The compound with the lowest vapor pressure, or highest boiling point, will concentrate in the liquid phase. Stabilizers that concentrate in the vapor phase will eventually become depleted in open-top vapor degreasers (which were commonly used in earlier decades) because a substantial quantity of solvent vapor is lost to the atmosphere with evaporation. If a batch of stabilized solvent is evaporated under these conditions, the stabilizer will leave before the solvent is evaporated, resulting in lower stabilizer concentrations in the liquid solvent (DeGroot, 1998).

Stabilizers whose boiling points are higher or lower than the solvent to which they are added will partition differentially when the solvent is boiled for vapor degreasing or other applications.[*] For example, 1,4-dioxane boils at 101°C, whereas methyl chloroform boils at 78°C. Thus, 1,4-dioxane is a "high-boiler" relative to methyl chloroform. Although enough 1,4-dioxane will vaporize to stabilize methyl chloroform in the vapor phase, a larger proportion preferentially remains in the liquid phase in the vapor degreaser boiling sump. The proportion has been determined: 27% of the 1,4-dioxane will partition to the vapor phase, while 73% will remain in the liquid phase (Spencer and Archer, 1981). Through continued use, a vapor degreaser iteratively partitions 1,4-dioxane such that the proportion of 1,4-dioxane in the sump will increase over time. In several weeks of daily use, the liquid solvent can take on a composition that is as much as 10–20% 1,4-dioxane. Because 1,4-dioxane is flammable, this may pose a fire hazard (Archer et al., 1977).

The iterative degreasing cycle that partitions 1,4-dioxane in still bottoms presents a substantially greater mass strength for sources of 1,4-dioxane groundwater contamination—far greater than what might be inferred from the initial composition of stabilized methyl chloroform. As investigations of dozens of 1,4-dioxane plumes mapped since 1995 have shown, the mass of 1,4-dioxane released in methyl chloroform degreasing waste or solvent reclamation waste was indeed enough to sustain large plumes.

Laboratory analysis of new and spent methyl chloroform[†] sampled from a vapor degreaser at Hayes International Corporation showed a 68% increase in 1,4-dioxane concentration (Tarrer et al., 1989). Another vapor degreaser was operated under controlled conditions in a laboratory to measure changes in stabilizer composition over time. The starting composition of the stabilized methyl chloroform was 2.8% 1,4-dioxane. After operating the degreaser for 24 days, the 1,4-dioxane content increased to 7.5%. Other stabilizers of methyl chloroform were also partitioned in this experiment. *tert*-Amyl alcohol increased from 1.5% to 3.2%; nitromethane decreased from 0.5% to 0.25%; and 1,2-butylene oxide decreased from 0.75% to 0.5% (Spencer and Archer, 1981).

The decrease in nitromethane is due to its greater tendency to partition into the vapor phase. At 74°C, the boiling point of methyl chloroform, nitromethane partitions to 62% in the vapor phase and 38% in the liquid phase, even though its boiling point is 101°C (Spencer and Archer, 1981; Budavari, 1996). At 25°C, nitromethane has a vapor pressure of 35.8 mm Hg, whereas methyl chloroform has a vapor pressure of 127 mm Hg (Daubert and Danner, 1989; Sunshine, 1969). The relative loss of nitromethane to the vapor phase is compensated by including another nitroalkane as a stabilizer of methyl chloroform. Nitroethane has a vapor pressure of 20.8 mm Hg and so is more favorably

---

[*] Partitioning stabilizers from boiling solvent may also depend on vapor pressure differences, vapor density contrasts, vapor removal, and other factors such as "drag-out." Vapor removal or drag-out is caused by retention of solvent on the parts exiting the vapor degreaser and by moving parts or drafts that stir the vapors and cause them to escape the cooling blanket. Although not accounting for uncontrolled factors such as drag-out, the tendency of a stabilizer compound to partition out of a mixture into the vapor phase can be determined by combining the Clausius–Clapeyron equation, Antoine's equation, and Raoult's law. This approach becomes difficult when consideration is given to the waste oil component of the mixture. Nevertheless, boiling-point difference is a good surrogate indicator of the potential for a stabilizer compound to become concentrated in the liquid phase or to become depleted.

[†] "Spent" solvent is a relative term; degreasing operator guidance calls for removing and concentrating the still bottoms oily waste sludge when the oil content reaches 25–30% (Dow Chemical Company, 1999b). Temperature is a surrogate for measuring oil content: Dow recommends that still bottoms be removed when the boiling temperature reaches 42°C for dichloromethane (a 1°C increase from the normal boiling point for dichloromethane), 89°C for trichloroethylene (DOW NEU-TRI Solvent; a 2.7°C increase), and 125°C for perchloroethylene (a 3.6°C increase) (Dow Chemical Company, 1999b). Still bottoms for methyl chloroform should be removed when the boiling temperature reaches 77°C (a 3°C increase) (NZDL, 1981).

TABLE 1.25
Partitioning of Trichloroethylene Stabilizers from New and Spent Solvent, Robbins Air Force Base

Solvent	1,2-Butylene Oxide[a]	Epichlorohydrin	Ethyl Acetate	Methyl Pyrrole
New trichloroethylene	1.64	1.66	34.6	15.9
Spent trichloroethylene	0.685	1.69	28.5	21.8
Percent change[b]	−58	+2	−18	+37

Source: Tarrer, A.R., Donahue, B.A., Dhamavaram, S., and Joshi, S.B., 1989, *Reclamation and Reprocessing of Spent Solvents.* Park Ridge, NJ: Noyes Data Corporation.

[a] Concentrations expressed as wt% × 103.

[b] Decreases are due to loss of stabilizer.

retained in the liquid phase; adding nitroethane therefore inversely complements the behavior of nitromethane (Spencer and Archer, 1981).

In TCE, similar apportioning of stabilizer compounds has been measured. Samples of new and spent TCE from Robins Air Force Base outside Macon, Georgia, were analyzed by Tarrer et al. (1989); results are summarized in Table 1.25. Epichlorohydrin remained essentially unchanged, 1,2-butylene oxide decreased by 58%, ethyl acetate decreased by 18%, and methyl pyrrole increased by 37%.

Stabilizers used in perchloroethylene are also subject to partitioning because of boiling-point differences. Samples of new and spent perchloroethylene from a degreasing operation at Kelly Air Force Base near San Antonio, Texas, were analyzed and reported in Alfred Tarrer's book, *Reclamation and Reprocessing of Spent Solvents* (Tarrer et al., 1989). Cyclohexene oxide is negligibly diminished by about 6%, whereas *n*-BGE increases by about 75%. Table 1.26 presents data on the partitioning of perchloroethylene stabilizers from new and spent solvents.

Aniline and morpholine, both claimed in patents as stabilizers for perchloroethylene, boil at higher temperatures than perchloroethylene. Amines such as aniline react to form anilidine, an undesired impurity that is retained in the distillation residue (Skeeters, 1960b). Aniline was one of the earliest stabilizers for chlorinated solvents (Ellis, 1925). It was also one of the earliest stabilizers for which toxicity and worker health and safety issues were raised (Dinley, 1937).

Boiling point and vapor pressure are integrally linked parameters. Both should be considered when weighing whether partitioning of a stabilizer from its solvent will likely occur.

Other impurities in solvents can also be partitioned because of boiling-point differences. For example, one of the methods of producing TCE also produces perchloroethylene (Shepherd, 1962). If perchloroethylene is carried over as a minor impurity into TCE, the perchloroethylene is

TABLE 1.26
Partitioning of Perchloroethylene Stabilizers from New and Spent Solvent, Robbins Air Force Base

Solvent	Cyclohexene Oxide	Butoxymethyl Oxirane[a]
New perchloroethylene	1.06	4.26
Spent perchloroethylene	0.998	7.45
Percent change (%)	−6	+75

Source: Tarrer, A.R., Donahue, B.A., Dhamavaram, S., and Joshi, S.B., 1989, *Reclamation and Reprocessing of Spent Solvents.* Park Ridge, NJ: Noyes Data Corporation.

Note: Concentrations expressed as wt% × 10³.

[a] Also called *n*-BGE; 1-butoxy-2,3-epoxypropane; 1-butoxy-2,3-epoxypropane; and 2,3-epoxypropyl butyl ether.

concentrated in the vapor degreaser still bottoms owing to its much higher boiling point (121.4°C versus 86.7°C for TCE). However, perchloroethylene is generally produced with the least impurities and requires the smallest quantities of stabilizers, as suggested by Table 1.14.

Some investigators have concluded that the starting concentration of perchloroethylene as an impurity of TCE is probably too small to manifest as a detectable constituent at TCE waste release sites, even after iterative concentration in a vapor degreaser (Lane and Smith, 2006). Other investigators have drawn the opposite conclusion: perchloroethylene—if initially present at 1% in TCE—would reach a volume fraction of nearly 8% after multiple iterative cycles of boiling and condensation in a vapor degreaser, as calculated from the Clausius–Clapeyron equation. The operation of this process therefore allows enough perchloroethylene to be present in groundwater contaminated by TCE waste solvent to give the appearance that perchloroethylene was used for degreasing (Morrison et al., 2006). Moreover, direct chemical analysis of TCE of reagent grade, technical grade, and pharmaceutical grade has reportedly detected the presence of perchloroethylene in all three. The analysis was done to check for more toxic impurities in commercial grades of TCE, which are currently used to anesthetize patients for surgery as a cost-saving measure in hospitals that cannot afford more expensive agents of anesthesia (Totonidis, 2005). TCE produced in Europe had up to 0.03% perchloroethylene as an impurity (European Chemicals Bureau, 2004). Typical industrial grade TCE has been known to include about 0.15% carbon tetrachloride as the major chlorinated hydrocarbon impurity (Willis and Christian, 1957).

Since the 1980s, the major producers of chlorinated solvents have sold increasingly reliable grades of TCE. In earlier decades, however, quality control may not have been as stringent, and carryover of solvents from production led to impurities in the finished product sold for vapor degreasing. The patent literature from the mid-1950s notes that it was not then feasible to produce high-purity perchloroethylene for commercial use. The presence of lower aliphatic hydrocarbons introduces instability to perchloroethylene. Perchloroethylene produced through chlorination and chlorinolysis* may contain trace impurities including dichloroethylene, TCE, trichloroethane, and other compounds. Some patents from this era claim methods for restoring perchloroethylene to purity (Skeeters, 1960b). Analysis has shown that perchloroethylene with a purity of less than 50 ppm of chlorinated ethanes is still susceptible to decomposition, leading to a maximum acidity of 47 ppm (EPO, 1989).

The converse may also occur in which TCE is present in perchloroethylene as a degreasing blend designed to extend the life of the perchloroethylene. Solvent producers typically recommend replacing perchloroethylene in degreasing applications when the oil contamination reaches 25–30% or when the specific gravity drops to 1.4 from perchloroethylene's normal value of 1.6 (Mertens, 2000b). Producer guidance also advises that perchloroethylene should be replaced in degreasing operations when the AAV drops into the range 0.02–0.06%. TCE is sometimes added to perchloroethylene to allow continued use at higher acidity and higher oil content. The addition of up to 19% TCE to perchloroethylene extended its operational life. By lowering the boiling point from 121°C for perchloroethylene alone to 115°C for the perchloroethylene–TCE mixture, energy savings were also achieved (Roehl, 1981).

Partitioning of stabilizers from their host solvents due to boiling-point differences may increase as the boiling point of the solvent increases with continued use. A pure solvent has a defined boiling point. When oil, grease, or other compounds with higher boiling points are added to solvents during the vapor degreasing process, the boiling point of the mixture increases according to the molar concentration and the vapor pressure of each component. Solvent boiling temperature increases several degrees as the percentage of waste oil builds in the solvent sump. Changing out the solvent is recommended when the waste oil content in the degreaser approaches 25%; boiling temperature

---

* In the chlorinolysis method for producing perchloroethylene, hydrocarbons are chlorinated at or near pyrolytic conditions to produce a mixture of carbon tetrachloride and perchloroethylene.

of the liquid solvent is used as a surrogate for determining the volumetric fraction of waste oil (Solvay SA, 2002b).

Solvent vapor losses from the open-top vapor degreasers used in earlier decades were substantial. The average emission of TCE from an open-top vapor degreaser was 2.6 g/min; about 10% of the total degreaser emissions escaped into the workplace. Solvents were frequently replenished by topping off the boiling sump with new solvent, sometimes daily (Wadden et al., 1989).

Tables 1.27 through 1.30 contrast the boiling points of stabilizers with the boiling points of their host solvents to estimate the tendency for stabilizers to be either concentrated in the still bottoms or driven off as vapor losses. Those stabilizers that tend to partition to the vapor phase will be retained in the waste solvent, but at a lower weight fraction than what was originally present. Stabilizers that are progressively depleted owing to vapor loss, consumption by reaction, or removal in the water may be replaced by periodically adding new stabilized solvent or by adding a stabilizer concentrate, as discussed in Section 1.2.7.7, Replenishing Solvent Stabilizers.

1.2.7.2 Solubility Differences

Most solvent stabilizer compounds are soluble in water; many are highly soluble or fully miscible in all proportions. When water condenses in dry-cleaning equipment and vapor degreasers, those stabilizers that are more soluble in water than the solvent may be extracted from the solvent by the layer of water floating atop the solvent in the sump. Water may also disperse within the liquid

TABLE 1.27
Boiling-Point Contrasts for Stabilizers of Methyl Chloroform

Stabilizer	Maximum Stabilizer (wt%)	Boiling Point (°C)	Boiling-Point Difference[a] (°C)	Rank
Concentrated in Still Bottoms				
Resorcinol	—	280	205.9	1
n-Methyl pyrrole	—	115	40.9	2
1,3,5-Trioxane	1	114.5	40.4	3
Nitroethane	0.8	114	39.9	4
Toluene	—	110.6	36.5	5
Isobutyl alcohol	—	108	33.9	6
2-Methyl-3-butyn-2-ol	7	103.5	29.4	7
t-Amyl alcohol	3	102.4	28.3	8
Nitromethane	2	101.2	27.1	9
1,4-Dioxane	8	101.1	27	10
Isopropyl nitrate	—	101	26.9	11
sec-Butyl alcohol	1.22	99.5	25.4	12
Lost to Vapor				
tert-Butyl alcohol	—	82.4	8.3	13
Acetonitrile	—	81.6	7.5	14
Methyl ethyl ketone	—	80	5.9	15
1,3-Dioxolane	—	78	3.9	16
1,2-Butylene oxide	—	63.3	−10.8	17
Dimethoxymethane	—	44	−30.1	18

Source: See Appendix 2 for a tabulation of physical and chemical properties of solvent-stabilizer compounds with full citations.

[a] Methyl chloroform's boiling point is 74.1°C.

TABLE 1.28
Boiling-Point Contrasts for Stabilizers with Dichloromethane

Stabilizer	Maximum Stabilizer (wt%)	Boiling Point (°C)	Boiling-Point Difference[a] (°C)	Rank
Concentrated in Still Bottoms				
Thymol	0.5	233	193.25	1
Phenol	0.05	181.75	142	2
Ethanol	0.2	135	95.25	3
4-Cresol	0.5	129.5	89.75	5
Nitromethane	0.43	101.2	61.45	6
1,4-Dioxane	8	101.1	61.35	7
Cyclohexane	0.01	80.7	40.95	8
1,3-Dioxolane	1	78	38.25	9
Methanol	0.05	64.7	24.95	10
Lost to Vapor				
tert-Butyl amine	–	44.04	4.29	11
2-Methyl-2-butene	2.5	39	−0.75	12
Propylene oxide	–	34.23	−5.52	13
Methanamine	1	−6	−45.75	14

Source: See Appendix 2 for a tabulation of physical and chemical properties of solvent-stabilizer compounds with full citations.

[a] Dichloromethane's boiling point is 39.75°C.

solvent as a mixture of fine free-phase water droplets. Dispersed water may be removed when solvents are filtered through membranes to remove waste in the solvent (Howell and Tarrer, 1994).

In dry-cleaning operations, on-site distillation of perchloroethylene was commonly performed to remove impurities and water. Azeotropic reactions[*] typically allow for certain amounts of boil-over, causing moisture and other impurities to recirculate through the water separator in the solvent/water interphase and return to the clean tank. Distillation degrades and ultimately destroys solvent stabilizers, which allows perchloroethylene to become acidic and corrosive (Laidlaw Company, 2004). The cause of stabilizer loss is most likely related to differences in water solubility between the stabilizers and perchloroethylene.

To minimize the effect of water on methyl chloroform, some solvent maintenance approaches capitalized on the miscibility of water in alcohol by using compounds known as polyhydroxy alcohols (i.e., propylene glycol). Methyl chloroform is treated by drawing off water accumulated atop the cooled liquid solvent and then adding a layer of propylene glycol, which also floats atop the solvent. The remaining water and dissolved acids preferentially partition into the alcohol, which can then be easily removed because it remains as a free liquid phase atop the solvent (Cormany, 1977). Adding alcohols and ethers to solvents can change the solubility of substances in the solvent, sometimes with deleterious effects. For example, the addition of alcohols to TCE to stabilize metal reactions was known to increase the solubility of hydrochloric acid in TCE, which led to corrosion problems (Klabunde, 1949).

[*] An azeotrope is a liquid mixture of two or more substances that retains the same composition in the vapor state as in the liquid state when distilled or partially evaporated.

TABLE 1.29
Boiling-Point Contrasts for Stabilizers with Trichloroethylene

Stabilizer	Maximum Stabilizer (wt%)	Boiling Point (°C)	Boiling-Point Difference[a] (°C)	Rank
Concentrated in Still Bottoms				
Triethanolamine	1	335.4	248.4	1
p-t-Amyl phenol	0.01	262.5	175.5	2
Thymol	0.5	233	146	3
2-Isopropyl-5-methyl-phenol	—	191	104	4
Phenol	0.5	181.75	94.75	5
Isoamyl alcohol	—	132.5	45.5	6
Epichlorohydrin	5	117.9	30.9	7
Pyridine	0.5	115.2	28.2	8
2,2,4-Trimethylpentene-1 (Diisobutylene)	—	101.4	14.4	10
Lost to Vapor				
Triethylamine	—	89.30	2.3	11
Ethyl acetate	—	77	−10	12
1,2-Butylene oxide	—	63.3	−23.7	13
Isobutyl alcohol	—	−88.6	−175.6	14

Source: See Appendix 2 for a tabulation of physical and chemical properties of solvent-stabilizer compounds with full citations.

[a] Trichloroethylene's boiling point is 87°C.

TABLE 1.30
Boiling-Point Contrasts for Stabilizers with Perchloroethylene

Stabilizer	Maximum Stabilizer (wt%)	Boiling Point (°C)	Boiling-Point Difference[a] (°C)	Rank
Concentrated in Still Bottoms				
2,6-di-*tert*-Butyl-*p*-cresol	0.00008	265	144	1
Thymol	0.003	233	112	2
Benzotriazole	—	204	83	3
Butoxymethyl oxirane	0.004	164	43	4
Cyclohexene oxide	—	129.5	8.5	5
Epichlorohydrin	—	117.9	−3.1	6
n-Methylmorpholine	—	113	−8	7
Lost to Vapor				
Diallylamine	—	111	−10	8
Cyclohexane	—	80.7	−40.3	9
Ethyl acetate	—	77	−44	10

Source: See Appendix 2 for a tabulation of physical and chemical properties of solvent-stabilizer compounds with full citations.

[a] Perchloroethylene's boiling point is 121.4°C.

1.2.7.3 Vapor Pressure Differences

A stabilizer with a vapor pressure that is significantly lower than the host solvent will be partitioned into the still bottoms. Stabilizers whose vapor pressures are significantly higher than the host solvent will be progressively depleted as vapor is lost during operation of the degreaser or dry cleaner. Vapor pressure differences at ambient temperatures may contribute to slow loss or enrichment of stabilizers in cold cleaning operations where the solvent is used at room temperature. The literature usually has experimental vapor pressure values available for chemicals at standard temperature and pressure, that is, at 25°C and 1 atm pressure. The vapor pressures of substances change with temperature. The relevant indicator of the potential for stabilizer partitioning is the vapor pressure of the stabilizer at the boiling temperature of the solvent. Obtaining stabilizer vapor pressures at solvent boiling temperatures is more difficult as less experimental data are available.

Several methods are available for calculating vapor pressure at different temperatures, however. The most common method uses the Antoine equation:

$$\log_{10} P = A - [B/(T + C)], \tag{1.23}$$

where P is the vapor pressure (in mm Hg), T is the temperature (in °C), and A, B, and C are the Antoine constants for a particular chemical. The Antoine constants are available in *Yaws' Handbook of Antoine Coefficients for Vapor Pressure* (Knovel Corporation, 2006). The Antoine equation does not apply to all compounds and only applies to a specific temperature range for each compound. Figure 1.3 shows the vapor pressure of 1,4-dioxane and the vapor pressure of methyl chloroform; the plot graphically presents the tendency of 1,4-dioxane to concentrate.

1.2.7.4 Carbon Adsorption

The use of activated charcoal filters for on-site purification of solvents to remove oily waste can contribute to the depletion of stabilizers (Dow Chemical Company, 2006b). Acid acceptors in particular are prone to remain adsorbed to carbon filters used to filter spent solvents for reuse (Howell and Tarrer, 1994). The affinity of an organic compound to adsorb to carbon is commonly taken as the ratio of the chemical adsorbed per unit weight of organic carbon, referred to as a

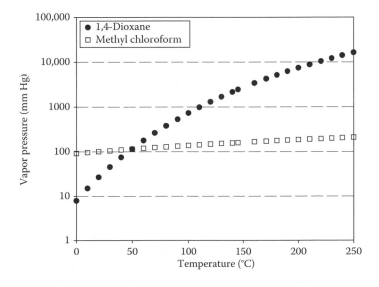

FIGURE 1.3 Vapor pressures of 1,4-dioxane and methyl chloroform. [1,4-Dioxane data from Vinson, C.G., Jr. and Martin, J.J., 1963, *Journal of Chemical Engineering Data* 8(1): 74–75. Methyl chloroform values calculated by using Antoine equation and constants from Knovel Corporation, 2006.]

compound-specific parameter, K_{oc}. Tables 1.31 through 1.34 contrast the K_{oc} values for stabilizers with their host solvents, to predict which stabilizers would be lost to carbon filtration. These tables are also useful for predicting the fate of stabilizers in contact with soil organic matter as they migrate through the soil profile following a release. Fate and transport of stabilizers in the environment are described in detail in Chapter 3.

In addition to carbon filtration, physical filtration was also used to remove soil from solvents in degreasing, cold cleaning, dry cleaning, and solvent reclamation operations. Conventional filters, made of paper and other fibrous materials, were used to remove particles down to 0.5 μm in diameter. Solvent-resistant polymeric membranes were used to remove particles smaller than 1 μm. These filters remove high-molecular-weight matter, such as dissolved oil and grease, but allow molecules smaller than 150–200 Da (daltons) to pass through (Howell and Tarrer, 1994). Membrane separation technology in solvent purification can remove stabilizers with higher molecular weights, such as triethanolamine, and impurities with higher molecular weights, such as perchloroethylene and carbon tetrachloride.

1.2.7.5 Thermal Stability

Stabilizers added to chlorinated solvents must be stable within the operating temperature range of the process for which the solvent is used. Stabilizers should be immune to pyrolysis at temperatures below the boiling point of the solvent and below the temperatures used for distilling solvents in reclamation processes. Solvent boiling temperatures may become slightly elevated when oily waste accumulates in the boiling sump; higher temperatures may be reached where metal fines build up

---

TABLE 1.31

Organic Carbon Partition Coefficient Contrasts for Stabilizers with Methyl Chloroform

Stabilizer	log K_{oc}	K_{oc} Ratio (Stabilizer:Solvent)	Rank
Retained in Filter			
n-Methyl pyrrole	4	54.95	1
Decreasing Retention in Carbon Filter			
Resorcinol	1.8	0.35	2
sec-Butyl alcohol	1.7	0.28	3
tert-Butyl alcohol	1.57	0.20	4
Toluene	1.56	0.20	5
2-Butanone	1.5	0.17	6
Nitroethane	1.48	0.17	7
1,4-Dioxane	1.23	0.09	8
Acetonitrile	1.2	0.09	9
1,3-Dioxolane	1.17	0.08	10
Isobutyl alcohol	0.95	0.05	11
1,2-Butylene oxide	0.90	0.04	12
2-Methyl-3-butyn-2-ol	0.54	0.02	13
t-Amyl alcohol	0.46	0.02	14
Nitromethane	0.28	0.01	15
Trioxane	−1.6	0.00	16

Source: See Appendix 2 for a tabulation of physical and chemical properties of solvent-stabilizer compounds with full citations.

Note: For methyl chloroform, log $K_{oc} = 2.26$.

TABLE 1.32
Organic Carbon Partition Coefficient Contrasts for Stabilizers with Dichloromethane

Stabilizer	Log K_{oc}	K_{oc} Ratio (Stabilizer:Solvent)	Rank
Retained in Filter			
Cyclohexane	2.2	6.61	1
Amine	2.16	6.03	2
2-Methyl-2-butene	1.83	2.82	3
Cresol	1.7	2.09	4
Decreasing Retention in Carbon Filter			
1,4-Dioxane	1.23	0.71	5
Phenol	1.2	0.66	6
Ethanol	1.2	0.66	6
1,3-Dioxolane	1.17	0.62	8
Thymol	0.52	0.14	9
tert-Butylamine	0.46	0.12	10
Nitromethane	0.28	0.08	11
Propylene oxide	0.03	0.04	12
Methanol	−0.77	0.01	13

Source: See Appendix 2 for a tabulation of physical and chemical properties of solvent-stabilizer compounds with full citations.

Note: For dichloromethane, log K_{oc} = 1.38.

around the heating coils in the boiling sump. In general, stabilizer chemicals are relatively immune to pyrolysis within the temperatures encountered in normal degreasing, dry cleaning, and other solvent applications. Higher temperatures may be encountered in transformers, for which the selection of suitable stabilizers requires special attention to their thermal stability.

1.2.7.6 Replenishing Solvent Stabilizers

Stabilizers were replenished in degreasers and in recycled solvent in two ways. For degreaser operators, the solvent distributors and degreasing equipment manufacturers recommended that the solvent be periodically refreshed by adding a quantity of fresh solvent. Operators were advised to purchase new solvent containing higher than usual levels of stabilizer to refortify the stabilizer-depleted solvent. The new solvent was added when the degreaser was not operating and the solvent was cold and provided enough stabilizers to restabilize the solvent (Howell and Tarrer, 1994).

In experiments conducted to simulate operating conditions in a degreaser for the purpose of testing a solvent maintenance system, 4 L of fresh solvent were added weekly to a degreaser whose solvent capacity was 45.4 L.* The weekly addition of almost 10% of the solvent volume was apparently a common practice (Cormany, 1977). Testing the solvent for acid acceptance and performing the aluminum scratch test before and after the addition of fresh solvent ensured the success of this approach.

In the process of recycling solvents, steam stripping was used to separate the oils and other materials solubilized in the solvent during the degreasing process. Many of the stabilizer compounds, particularly those added to methyl chloroform, are hydrophilic and are therefore siphoned off in

---

* A model MLW-120 stainless steel Baron-Blakeslee vapor immersion degreaser unit was used for the experiment.

TABLE 1.33

Organic Carbon Partition Coefficient Contrasts for Stabilizers with Trichloroethylene

Stabilizer	Log K_{oc}	K_{oc} Ratio (Stabilizer:Solvent)	Rank
Retained in Filter			
Diisobutylene	2.4	2.75	1
Pentanol-2-triethanolamine	2.2	1.45	2
Decreasing Retention in Carbon Filter			
Diisopropylamine	2.2	1.41	3
Isobutyl alcohol	1.8	0.63	4
Ethyl acetate	1.8	0.59	5
Styrene oxide	1.7	0.52	6
Pyridine	1.7	0.50	7
Epichlorohydrin	1.6	0.40	8
2-Methyl-3-butyn-2-ol	1.6	0.35	9
2-Butanone	1.5	0.32	10
Propylene oxide	1.4	0.25	11
Isopropyl alcohol	1.4	0.25	11
Phenol	1.2	0.16	13
Isopropyl acetate	1.2	0.15	14
1,2-Butylene oxide	0.90	0.08	15
Thymol	0.52	0.03	16
Triethanolamine	0.48	0.01	17
Triethylamine	0.48	0.03	17
t-Amyl alcohol	0.46	0.03	18

Source: See Appendix 2 for a tabulation of physical and chemical properties of solvent-stabilizer compounds with full
citations.

Note: For trichloroethylene, log $K_{oc} = 2.00$.

steam cleaning or where water is used in the recovery process. Guidance for replenishing stabilizers calls for blending one part of reclaimed solvent with four parts of virgin solvent.

Stabilizers were also added directly back to solvents during ongoing solvent distillation in degreasing operations and in solvent-recycling operations. Operators were cautioned that (1) direct addition of stabilizers in pure form requires more skill and training to determine the correct amounts of acid acceptor, metal stabilizer, and antioxidant and (2) stabilizers are highly flammable, toxic, and possibly carcinogenic. Nevertheless, direct addition of solvent stabilizers was practiced. For example, technicians at Warner Robins Air Logistics Center in Georgia maintained a supply of 1,2-butylene oxide to replenish acid acceptors in reclaimed solvents (Howell and Tarrer, 1994).

Dow Chemical markets MaxiStab, a stabilizer concentrate used to replenish stabilizers in solvents. In low-emission degreasing systems, stabilizers are depleted faster than they are replenished by the addition of fresh solvent. The MaxiStab line of stabilizer concentrates are tailored to each of Dow's chlorinated solvent lines, underscoring the specialization of stabilizer packages for specific applications (Dow Chemical Company, 2006b). A stabilizer concentrate is currently sold for newer lines of degreaser solvents, such as *n*-propyl bromide; the contents may include 1,2-butylene oxide, dimethoxymethane, and 2-methyl-2-propanol, as well as other compounds (Petroferm, Inc., 2005).

Another means of replenishing solvent stabilizers is accomplished through solvent conditioning. For dry cleaning, a deodorizing and antistatic powdered additive is used that also coagulates impurities

TABLE 1.34
Organic Carbon Partition Coefficient Contrasts for Stabilizers with Trichloroethylene

Stabilizer	Log K_{oc}	K_{oc} Ratio (Stabilizer:Solvent)	Rank
Retained in Filter			
2,6-di-*tert*-Butyl-*p*-cresol	3.14	6.607	1
Decreasing Retention in Carbon Filter			
Cyclohexane	2.2	0.759	2
Benzotriazole	2.16	0.692	3
Triethylamine	2.16	0.692	3
Diisopropylamine	2.15	0.676	5
Diallylamine	1.98	0.457	6
Ethyl acetate	1.77	0.282	7
Epichlorohydrin	1.6	0.191	8
1,2-Butylene oxide	0.90	0.038	9
Thymol	0.52	0.016	10
n-Methylmorpholine	0.474	0.014	11
Butoxymethyl oxirane	0.24	0.008	12

Source: See Appendix 2 for a tabulation of physical and chemical properties of solvent-stabilizer compounds with full citations.

Note: For perchloroethylene, log $K_{oc} = 2.32$.

to form a filterable residue, thereby decreasing the contact time with agents that could break down perchloroethylene. The powder also buffers the pH to increase the effectiveness of detergents added to perchloroethylene to remove water-soluble soil from clothing (Laidlaw Company, 2004).

The idea of adding stabilizers back to solvents predates vapor degreasing. A 1937 patent notes that solvent life may be indefinitely extended by repeated additions of stabilizing agents (Dinley, 1937). The concept of developing a concentrated stabilizer package was first described in a 1958 patent application (Kauder, 1960).

ACKNOWLEDGMENTS

The review and comments by Richard Doherty of Engineering and Consulting Resources Inc. in Acton, Massachusetts, are much appreciated.

BIBLIOGRAPHY

AFCEE, 1999, Pro-ACT fact sheet: Parts washing technologies. TI#20835. Air Force Center for Environmental Excellence. http://www.afcee.brooks.af.mil/pro-act/factsheets.asp (accessed July 14, 2006).

AFSCME (American Federation of State, County and Municipal Employees), 2006, Welding hazards. http://www.afscme.org/health/faq-weld.htm (accessed July 28, 2006).

Archer, W.L., 1979, Comparison of chlorinated solvent-aluminum reaction inhibitors. *Industrial & Engineering Chemistry, Product Research and Development* 18: 131–135.

Archer, W.L., 1982, Aluminum—1,1,1-trichloroethane: Reactions and inhibition. *Industrial & Engineering Chemistry, Product Research and Development* 21: 670–672.

Archer, W.L., 1984, A laboratory evaluation of 1,1,1-trichloroethane-metal-inhibitor systems. *Werkstoffe und Korrosion (Materials and Corrosion)* 35: 60–69.

Archer, W.L., Simpson, E.L., and Gerard, R.R., 1977, United States Patent 4,018,837: Stabilized methylchloroform. Assignee: Dow Chemical Company, Midland, MI.

Archer, W.L. and Stevens, V.L., 1977, Comparison of chlorinated aliphatic, aromatic, and oxygenated hydrocarbons as solvents. *Industrial & Engineering Chemistry, Product Research and Development* 16(4): 319–326.

Arena, J.M. and Drew, R.H. (Eds), 1986, *Poisoning—Toxicology, Symptoms, Treatments*, 5th Edition. Springfield, IL: Charles C. Thomas Publishers.

Ashland Chemical Company, 1995, Material Safety Data Sheet for trichloroethane 111 degrs cold/V, issued October 28, 1995, Columbus, OH. MSDS No. 001466-014.001.

Ashland Chemical Company, 1996, Material Safety Data Sheet for trichloroethane 111 w/o dioxan, issued January 15, 1996, FSC: 6810, NIIN: 00072116.

ASTM, 1962, *Handbook of Vapor Degreasing*. American Society for Testing and Materials Special Technical Publication No. 310.

ATSDR, 2000, Toxicological profile for methylene chloride. U.S. Department of Health and Human Services Public Health Service, Agency for Toxic Substances and Disease Registry. http://www.atsdr.cdc.gov/ 2p-publications.html (accessed June 2006).

ATSDR, 2005, Toxicological profile for carbon tetrachloride. U.S. Department of Health and Human Services Public Health Service, Agency for Toxic Substances and Disease Registry. http://www.atsdr.cdc.gov/ 2p-publications.html (accessed June 2006).

Aviado, D.M., Zakhari, S., Simaan, J.A., and Ulsamer, A.G., 1976, *Methyl Chloroform and Trichloroethylene in the Environment*. Cleveland: CRC Press.

Bachtel, H.J., 1957, United States Patent 2,811,252: Methyl chloroform inhibited with dioxane. Assignee: Dow Chemical Company, Midland, MI.

Beckers, N.L., 1959, United States Patent 2,903,488: Chemical method and composition. Assignee: Diamond Alkali Company, Cleveland, OH.

Beckers, N.L., 1973, United States Patent 3,923,912: Methylene chloride stabilized with ketones. Assignee: Diamond Shamrock Corporation, Cleveland, OH.

Beckers, N.L., 1974, United States Patent 3,796,755: Stabilization of chlorinated hydrocarbons. Assignee: Diamond Shamrock Corporation, Cleveland, OH.

Blake, D., Jacoby, W., Nimlos, M., and Noble, R., 1993, Identification and quantification of by-products and intermediates in the photocatalytic oxidation of gas phase trichloroethylene. In: M. Macias and M. Sanchez (Eds), *Proceedings, Volume 2, 6th International Symposium on Solar Thermal Concentrating Technologies*, pp. 1223–1231. Madrid, Spain.

Blum, K., 1984, United States Patent 4,469,906: Process for stabilizing epoxide-containing perchloroethylene. Assignee: Wacker-Chemie GmbH, Munich, Germany.

Brown, W.E., 1962, United States Patent 3,049,571: Stabilized degreasing solvent. Assignee: Dow Chemical Company.

Budavari, S. (Ed.), 1996, *The Merck index—An Encyclopedia of Chemicals, Drugs, and Biologicals*. Whitehouse Station, NJ: Merck and Co., Inc.

California Air Resources Board (CARB), 1996, Curriculum for the environmental training program for perchloroethylene dry cleaning operations. California Environmental Protection Agency.

Campbell, D.H., 1962, United States Patent 3,040,108: Stabilization of perchloroethylene. Assignee: Hooker Chemical Corporation, Niagara Falls, New York.

Cantin, J., 1992, Overview of exposure pathways. In *Proceedings: International Roundtable on Pollution Prevention and Control in the Drycleaning Industry*. U.S. Environmental Protection Agency, Office of Pollution Prevention and Toxics. EPA/774/R-92/002.

Carey, F.A., 1987, *Organic Chemistry*. New York: McGraw Hill.

Carlisle, P.J. and Levine, A.A., 1932, Stability of chlorohydrocarbons. *Industrial & Engineering Chemistry* 24: 1164–1168.

Cheap Solvents, 2002, Material Safety Data Sheet for stabilizer solvent. http://www.cheapsolvents.com (accessed March 1, 2003).

Cohen, W.L. and Izzo, V.J., 1992, Degradation of ground water by tetrachloroethylene. In *Proceedings of the Irrigation and Drainage Sessions at Water Forum '92, ed.*, pp. 63–69. New York: American Society of Civil Engineers.

Considine, D.M. (Ed.), 1974, *Chemical and Process Technology Encyclopedia*, New York: McGraw-Hill Book Company.

Copelin, H.B., 1957, United States Patent 2,297,250: Stabilization of chlorinated hydrocarbons. Assignee: E. I. du Pont de Nemours and Company, Wilmington, DE.

Copelin, H.B., 1959, United States Patent 2,904,600: Stabilization of chlorinated hydrocarbons. Assignee: E. I. du Pont de Nemours and Company, Wilmington, DE.

Cormany, C., 1977, United States Patent 4,065,323: Degreasing process using stabilized methylchloroform solvent. Assignee: PPG Industries, Inc., Pittsburgh, PA.

Cornell Dubilier, 2003, V-chip cleaning and coating guide. http://www.cornell-dubilier.com (accessed November 9, 2003).

Daras, N., 1960, United States Patent 2,935,537: Process for the stabilisation of chlorinated compounds. Assignee: Solvay et Cie, Brussels, Belgium.

Daubert, T.E. and Danner, R.P., 1989, *Physical and Thermodynamic Properties of Pure Chemicals Data Compilation*, Washington, DC: Taylor & Francis.

Dawson, B.R., 1989, Commercial off-site solvent reclamation, Chapter 11: U.S. Environmental Protection Agency (USEPA), September 1989. EPA/625/4-89/021.

DeGroot, R., 1998, Stabilizing halogenated solvents thwarts undesirable reactions. *Precision Cleaning 6(8)*: 25–27. http://www.p2pays.org/ref/02/01815.htm (accessed August 6, 2006).

Dempf, D., Fruhwirth, O., and Schmidhammer, L., 1977, United States Patent 4,034,051: Stabilization of per-chloroethylene. Assignee: Wacker-Chemie GmbH, Munich, Germany.

Dempf, D., Knabl, R., Schmidhammer, L., and Mack, W., 1980, United States Patent 4,220,607: Stabilized perchloroethylene. Assignee: Wacker-Chemie GmbH, Munich, Germany.

de Nevers, N., 1986, A fatal fire with "nonflammable" methyl chloroform. *Archives of Environmental Health* 41(5): 279–281.

Derrick, E.H. and Johnson, D.W., 1943, Three cases of poisoning by irrespirable gases—Phosgene from trichlo-roethylene, nitrogen dioxide, carbon dioxide with reduction of oxygen. *Medical Journal of Australia* 2: 355–358, as cited in NIOSH; Criteria Document: Phosgene, p. 33 (1976) DHEW Pub. NIOSH 76-137.

Dial, W.R., 1957, United States Patent 2,781,406: Stabilization of halogenated hydrocarbons. Assignee: Columbia Southern Chem Corp., Pittsburgh, PA.

Dinley, C.F., 1937, United States Patent 2,093,736: Stabilized chlorinated solvents and method of stabilizing such solvents. Assignee: James H. Bell, Philadelphia, PA.

Doherty, R.E., 2000a, A history of the production and use of carbon tetrachloride, tetrachloroethylene, trichlo-roethylene and 1,1,1-trichloroethane in the United States: Part 1—historical background; carbon tetra-chloride and tetrachloroethylene. *Journal of Environmental Forensics* 1: 69–81.

Doherty, R.E., 2000b, A history of the production and use of carbon tetrachloride, tetrachloroethylene, trichloroethylene and 1,1,1-trichloroethane in the United States: Part 2—trichloroethylene and 1,1,1-trichloroethane. *Journal of Environmental Forensics* 1: 83–93.

Douthitt, C.A., 1990, Resource recovery through solvent reclamation. Presentation to the U.S. Department of Energy Industrial Solvent Recycling Conference October 15, 1990, Charlotte, NC. NASR (National Association of Solvent Recyclers), www.p2pays.org/ref/26/25692.pdf (accessed February, 2006).

Dow Chemical Company, 1980, Evaluation of Chloroethane VG and its components in the Ames Salmonella/ Mammalian Microsome Mutagenicity Assay with attachments. EPA Doc. No. 86-890001189, Fiche No. OTS0520701.

Dow Chemical Company, 1981, Material Safety Data Sheet for Neu-Tri solvent: Trichloroethylene; #56530, December 14, 1981. FSC: 6810, NIIN: 00-924-7107.

Dow Chemical Company, 1985a, Material Safety Data Sheet for trichloroethylene-E; January 1, 1985. FSC: 6810, NIIN: 00-754-2813.

Dow Chemical Company, 1985b, Material Safety Data Sheet for 1/1/85. FSC: 6810, NIIN: LIIN: 00N003051 DOWPER CS SOLVENT.

Dow Chemical Company, 1986, Material Safety Data Sheet for perchloroethylene SVG, April 2, 1986. FSC: 6850, NIIN: LIIN: 00F049684.

Dow Chemical Company, 1987, Material Safety Data Sheet for perchloroethylene SVG, January 1, 1987. FSC: 6810, NIIN: 01-097-2020.

Dow Chemical Company, 1990, Material Safety Data Sheet for 1,1,1-trichloroethane, inhibited. FSC: 6810, NIIN: 00-476-5612.

Dow Chemical Company, 1992, Material Safety Data Sheet for 1,1,1-trichloroethane, aerosol grade, wide-spec.

Dow Chemical Company, 1995, Material Safety Data Sheet for dichloromethane, technical, December 7, 1995. FSC: 6810, NIIN: 00-616-9188.

Dow Chemical Company, 1999a, Material Safety Data Sheet for methylene chloride, technical grade, cyclo-hexane-inhibited. FSC: 6810, NIIN: 00-616-9188.

Dow Chemical Company, 1999b, Degreasing: Economical and efficient degreasing with chlorinated solvents from DOW.

Dow Chemical Company, 1999c, Maintaining stabilization of chlorinated solvents. *Choices & Solutions Newsletter* 3: 2. http://www.dow.com/gco/na/lit/n_letter.htm (accessed December 14, 2002).

Dow Chemical Company, 2000, Features and benefits for surface cleaning—Product Information Sheet. Form No. 100-06959. http://www.chlorinatedsolvents.com (accessed May 29, 2006).

Dow Chemical Company, 2001, Chlorinated solvents: Aluminum cleaning. Form No. 100-07024. http://www.chlorinatedsolvents.com (accessed May 29, 2006).

Dow Chemical Company, 2002, Chlorinated solvents solubility parameters. Form No. 100-07003 (revised). http://www.chlorinatedsolvents.com (accessed May 29, 2006).

Dow Chemical Company, 2006a, Chlorinated solvents product information: VOC content. Form No. 100-07056-0702. http://www.chlorinatedsolvents.com (accessed May 29, 2006).

Dow Chemical Company, 2006b, Maxistab Stabilizers Product Information Sheet. Form No. 100-06825-198QRP. http://www.chlorinatedsolvents.com (accessed May 29, 2006).

Dow Chemical Company, 2006c, Chlorinated solvents—physical properties. Form No. 100-06358. http://www.chlorinatedsolvents.com (accessed May 29, 2006).

Dow Chemical Company, 2006d, Perchloroethylene SVG double stabilized solvent Product Information Sheet. Form No. 100-07079-0902. http://www.chlorinatedsolvents.com (accessed May 29, 2006).

Dreisbach, R.H., 1987, *Handbook of Poisoning*, 12th Edition. Norwalk, CT: Appleton and Lange.

Ellis, C., 1925, United States Patent 1,557,520: Cleaner. Assignee: Carlton Ellis, Montclair, NJ.

Erdmann, E., 1911, Autoxydation von Trichloräthylen. *Journal für Praktische Chemie* 85(1): 78–89.

EuroChlor, 2003, Aluminium degreasing: An update on best practices. http://www.eurochlor.org/chlorsolvents/publications/digest20.htm (accessed November 17, 2003).

European Chemicals Bureau, 2000a, Evaluation and control of the risks of existing substances, IUCLID data set, substance ID 71-55-6, 1,1,1-trichloroethane. http://ecb.jrc.it/DOCUMENTS/Existing-Chemicals/IUCLID/DATA_SHEETS/71556.pdf (accessed July 29, 2006).

European Chemicals Bureau, 2000b, Evaluation and control of the risks of existing substances, IUCLID data set, substance ID 56-23-5, carbon tetrachloride. http://ecb.jrc.it/DOCUMENTS/Existing-Chemicals/IUCLID/DATA_SHEETS/56235.pdf (accessed July 29, 2006).

European Chemicals Bureau, 2000c, Evaluation and control of the risks of existing substances, IUCLID data set, substance ID 79-01-6, trichloroethylene. http://ecb.jrc.it/DOCUMENTS/Existing-Chemicals/IUCLID/DATA_SHEETS/79016.pdf (accessed July 29, 2006).

European Chemicals Bureau, 2000d, Evaluation and control of the risks of existing substances, IUCLID data set, substance ID 75-09-2, dichloromethane. http://ecb.jrc.it/DOCUMENTS/Existing-Chemicals/IUCLID/DATA_SHEETS/75092.pdf (accessed July 29, 2006).

European Chemicals Bureau, 2004, European Union risk assessment report: Trichloroethylene. 348. Environment Agency Chemicals Assessment Section Ecotoxicology and Hazardous Substances National Centre Isis House, Howbery Park, Wallingford, Oxfordshire.

European Chemicals Bureau, 2005, European Union risk assessment report: Tetrachloroethylene. Environment Agency Chemicals Assessment Section Ecotoxicology and Hazardous Substances National Centre Isis House, Howbery Park, Wallingford, Oxfordshire.

European Patent Office (EPO), 1989, European Patent Office Board of Appeals: Case number T 0131/87–3.3.2; Wacker-Chemie vs. Occidental Electrochemicals Company, European Patent No. 0 041 220, 1985.

Ferri, A. and Patron, G., 1959, United States Patent 2,910,512: Method for stabilizing trichloroethylene. Assignee: Sicedison, S.P.A., Milan, Italy.

Fullhart, L. and Swalheim, D.A., 1962, United States Patent 3,051,595: Non-aqueous phosphatizing solution. Assignee: E. I. du Pont de Nemours and Company, Wilmington, DE.

Goodner, W.R., Smith, J.N., and Horvath, J., 1977, United States Patent 4,046,820: Stabilization of 1,1,1-trichloroethane. Assignee: Motorola, Inc., Schaumberg, IL.

Graham, G.W., 1967, United States Patent 3,314,892: Stabilization of halohydrocarbons. Assignee: Canadian Industries Ltd., Montreal, Quebec, Canada.

Great Western Chemical Company, 1990, Material Safety Data Sheet for 1,1,1-trichloroethane ("dioxane free"). FSC: 6810, NIIN: 00-476-5612.

Gregersen, P. and Hansen, T., 1986, Organic solvents—documentation of the neurotoxic effects in humans exposed to solvents. Miljøproject Nr. 72. National Agency of Environmental Protection, Denmark. http://www.mst.dk/homepage/ (accessed July 19, 2006).

Halogenated Solvents Industry Alliance (HSIA), 1994, White paper on methyl chloroform (1,1,1-trichloroethane). http://www.hsia.org/white_papers/paper.html (accessed April 2002).

Halogenated Solvents Industry Alliance (HSIA), 1999, Perchlorethylene white paper. http://www.hsia.org/white_papers/paper.html (accessed April 2002).

Halogenated Solvents Industry Alliance (HSIA), 2001, Trichlorethylene white paper. http://www.hsia.org/white_papers/paper.html (accessed April 2002).

Halogenated Solvents Industry Alliance (HSIA), 2003, Methylene chloride white paper. http://www.hsia.org/white_papers/paper.html (accessed April 2002).

Hardies, D.E., 1966, United States Patent 3,281,480: Stabilization of methyl chloroform. Assignee: Pittsburgh Plate and Glass Company, Pittsburgh, PAlvania.

Harris, C.R., 1933, United States Patent 1,904,450: Stabilization of halogenated hydrocarbons. Assignee: Roessler & Hasslacher Chemical Co., New York.

Henschler, D., Elasser, H., Romer, W., and Eder, E. 1984, Carcinogenicity study of trichloroethylene, with and without epoxide stabilizers in mice. *Journal of Cancer Research and Clinical Oncology* 107: 149–156.

Hoenig, D.R., 2002, Dry cleaning tenants: Understanding and minimizing the risks. http://www.claytongrp.com/dryclean art.htm (accessed 2002).

Howell, S.G. and Tarrer, A.R., 1994, Minimizing pollution in cleaning and degreasing operations. EPA/600/SR-93/191.

Huntingtons Chemicals, 2006, MSDS for methylene chloride. http://www.huntingtons.fsworld.co.uk/tdsmecl2.htm (accessed August 2006).

Hygnstrom, J., 1995, Pollution prevention: A tool kit for dry cleaners. University of Nebraska Cooperative Extension and Biological Systems Engineering, EC 95-741-S, No. 1.

INEL, 2006, Mass estimate of organic compounds in 743-series waste buried in the subsurface disposal area for operable units 7-08 and 7-13/14. U.S. Department of Energy, Idaho National Engineering Laboratory, Idaho Cleanup Project Report 23256.

Inoue, T., Takeuchi, Y., Hisanaga, N., et al., 1983, A nationwide survey on organic solvent components in various solvent products: 1. Homogeneous products such as thinners, degreasers and reagents. *Industrial Health* 21(3): 175–184.

International Agency for Research on Cancer (IARC), 1979, *Monographs on the Evaluation of the Carcinogenic Risk of Chemicals to Man*, Vol. 20, p. 516. Geneva: World Health Organization.

Irani, M.R., 1977, United States Patent 4,032,584: Stabilized methylene chloride. Assignee: Stauffer Chemical Company, Westport, CT.

Irwin, R.J., Van Mouwerik, M., Stevens, L., Seese, M.D., and Basham, W., 1997, *Environmental Contaminants Encyclopedia*. National Park Service, Water Resources Division, Fort Collins, CO.

Ishibe, N. and Metcalf, T.G., 1982, United States Patent 4,309,301: Methylchloroform stabilizer composition employing an alkynyl sulfide. Assignee: Dow Chemical Company, Midland, MI.

Izzo, V.J., 1992, Dry cleaners—a major source of PCE in ground water. Sacramento, California Regional Water Quality Board, Central Valley Region.

Jeffers, P.M., Ward, L.M., Woytowitch, L.M., and Wolfe, N.L., 1989, Homogeneous hydrolysis rate constants for selected chlorinated methanes, ethanes, ethenes, and propanes. *Environmental Science and Technology* 23: 965–969.

Johnson, J.C. and Wedmore, L.K., 1983, Metal cleaning by vapor degreasing. *Metal Finishing* 81(9): 59–63.

Johnson, K., 1973, *Dry Cleaning and Degreasing Chemicals and Processes*. Park Ridge, NJ: Noyes Data Corporation.

Jones, G.W., Seaman, H., and Kennedy, R.E., 1933, Explosive properties of dioxan-air mixtures. *Industrial & Engineering Chemistry* 25(11): 1283–1286.

Joshi, S.B., Donahue, B.A., Tarrer, A.R., Guin, J.A., Rahman, M.A., and Brady, B.L., 1989, Methods for monitoring solvent condition and maximizing its utilization. In: R.A. Conway, J.A. Frick, D.J. Warner, C.C. Wiles, and E.J. Duckett (Eds), *Hazardous and Industrial Solid Waste Minimization Practices: Proceedings of the 8th Symposium on Hazardous and Industrial Solid Waste Testing and Disposal held 12–13 November 1987 at Clearwater, FL*. American Society for Testing and Materials STP 1043, ASTM, Philadelphia, PA.

Kauder, O.S., 1960, United States Patent 2,944,088: Stabilization of chlorinated solvents. Assignee: Argus Chemical Corporation, New York.

Kavaler, A.R. (Ed.), 1995, Chemical profile: 1,1,1-Trichloroethane. *Chemical Marketing Reporter* 247 (February 27): 9, 37.

Klabunde, W., 1949, United States Patent 2,492,048: Stabilization of trichloroethylene and tetrachloroethylene. Assignee: E. I. du Pont de Nemours and Company, Wilmington, DE.

Knovel Corporation, 2006, *Yaws' Handbook of Antoine Coefficients for Vapor Pressure—Electronic Edition*. Available by subscription at http://www.knovel.com (accessed June 2006).

Laidlaw Company, 2004, Technical product extra on LAIDLAW BUFF™. http://www.laidlawcorp.com/industrial/chemicals/buff.htm (accessed March, 2004).

Lane, V. and Smith, J., 2006, Fact or fiction: The source of perchloroethylene contamination in groundwater is a manufacturing impurity in chlorinated solvents. Presented to the National Groundwater Association Ground Water and Environmental Law Conference, July 7, 2006. http://www.ngwa.org (accessed August 2, 2006).

Larchar, A.W., 1950, United States Patent 2,517,894: Metal degreasing composition. Assignee: E. I. du Pont de Nemours and Company, Wilmington, DE.

Levine, A.A. and Cass, O.W., 1939, United States Patent 2,155,723: Stabilization of Trichloroethylene. Assignee: E. I. du Pont de Nemours and Company, Wilmington, DE.

Lim, S-C., Yang, J-Y., Jang, A-S., Park, Y-U., Kim, Y-C., Choi, I-S., and Park, K-O., 1996, Acute lung injury after phosgene inhalation. *Korean Journal of Internal Medicine* (English Edition) 11(1): 87–92.

Linn, W., 2002, Chemicals used in drycleaning operations. http://www.drycleancoalition.org/chemicals/ChemicalsUsedInDrycleaningOperations.htm (accessed 2002).

Linn, W., Appel, L., Davis, G., et al., 2004, Conducting contamination assessment work at drycleaning sites. State coalition for remediation of dry cleaners. http://www.drycleancoalition.org (accessed January 2004).

Lourenço, M.A.-V., 1863, Recherches sur les composés polyatomiques [Research on the polyatomic compounds]. *Annales de Chimie et de Physique* (Paris) 67: 257–339.

Mabey, W. and Mill, T., 1978, Critical review of hydrolysis of organic compounds in water under environmental conditions. *Physical Chemistry Reference Data* 7: 383–415.

Manner, J.A., 1977, United States Patent 4,026,956: Storage stabilized methylchloroform formulations. Assignee: PPG Industries, Inc., Pittsburgh, PA.

McKinney, L.L., Uhing, E.H., White, J.L., and Picken, J.C., Jr., 1955, Vegetable oil extraction: Auto-oxidation products of trichloroethylene. *Agricultural and Food Chemistry* 3(5): 413–419.

MDEQ, 2004, Michigan dry cleaning environmental compliance workbook. Michigan Department of Environmental Quality. http://www.deq.state.mi.us/documents/deq-ess-caap-drycleaners-workbook.pdf (accessed July 29, 2006).

Meike, A., 1993, Chemical and mineralogical concerns for the use of man-made materials in the post-emplacement environment. Lawrence Livermore National Laboratory, UCRL-ID-113383. http://www.osti.gov/energycitations/servlets/purl/139481-VIxX6s/139481.PDF (accessed August 2006).

Mertens, J.A., 2000a, Vapor degreasing with chlorinated solvents. *Metal Finishing* 98(6): 43–51.

Mertens, J.A., 2000b, Keeping metal machines and the environment clean. *Machine Design*, 72(5): 198.

Michigan Office of Regulatory Reform, 2000, Air Pollution Control, Natural Resources and Environmental Protection Act. Michigan Office of Regulatory Reform, Department of Environmental Quality, Air Quality Division, Air Pollution Control. *Michigan Register*, R 336.1122.

Minford, J.D., Brown, M.H., and Brown, R.H., 1959, Reaction of aluminum and carbon tetrachloride I. *Journal of the Electrochemical Society* 106(3): 185–191.

Missbach, E.C., 1936, United States Patent 2,043,258: Stabilized carbon tetrachloride. Assignee: Stauffer Chemical Company, Richmond, CA.

Mohr, T.K., 2001, *Solvent Stabilizers White Paper*. Santa Clara Valley Water District, San José, CA.

Mohr, T.K., Tulloch, C., Giri, S., Chan, S., Cook, G., and Crowley, J. 2007, *Study of Potential for Groundwater Contamination from Past Dry Cleaner Operations in Santa Clara County*. Santa Clara Valley Water District, San José, CA.

Monroe, R.F. and Rapp, D.E., 1959, United States Patent 2,906,783: Stabilization of chlorinated hydrocarbon solvents with azines. Assignee: Dow Chemical Company, Midland, MI.

Morrison, R., 2003, PCE contamination and the dry cleaning industry. *Environmental Claims Journal* 15(1): 93–106.

Morrison, R.D., 2002a, Environmental toolbox: What is solvent mileage? http://www.rmorrison.com/downloads/spr 02.pdf (accessed 2002).

Morrison, R.D., 2002b, Environmental toolbox: PCE contamination from dry cleaning operations http://www.rmorrison.com/downloads (accessed 2002).

Morrison, R.D., Murphy, B.L., and Doherty, R.E., 2006, Chlorinated solvents. In: R.D. Morrison and B.L. Murphy (Eds), *Environmental Forensics, Contaminant Specific Guide*. New York: Academic Press.

NACE International, The Corrosion Society, 2002, Corrosion survey database (COR·SUR). http://www.knovel.com/knovel2/Toc.jsp?BookID=532&VerticalID=0 (accessed 2006).

National Institutes of Health (NIH), 1986, *Toxicology and Carcinogenesis Studies of Tetrachloroethylene in F344/N Rats and B6C3F1 Mice (Inhalation Studies)*. Research Triangle Park, NC: U.S. Department of Health and Human Services, National Institutes of Health, National Toxicology Program, NTP TR 311; NIH Publication No. 86-256.

National Library of Medicine, 2006, Hazardous Substances Data Bank (HSDB). National Institutes of Health. http://toxnet.nlm.nih.gov/ (accessed March 15, 2006).

New Zealand Department of Labour (NZDL), 1981, Code of practice—vapour degreasing operations. http://www.osh.govt.nz/order/catalogue/archive/vapourdegrease.pdf (accessed August 2006).

North Carolina Division of Health Services, Environmental Health Section (NCDHS), 1983, Alternatives for handling waste solvents in North Carolina. http://www.p2pays.org/ref/23/22202.pdf (accessed February, 2006).

Occidental Chemical Corporation, 1989, Material Safety Data Sheet for 1,1,1-trichloroethane vapor degreasing grade. FSC: 6810, NIIN: 00-476-5613.

Occidental Chemical Corporation, 1991, Material Safety Data Sheet for perclene drycleaning grade. FSC: 6850, NIIN: LIIN: 00D002823.

Office of Environmental Health Hazard Assessment California Environmental Protection Agency (OEHHA), 1999, Public health goal for trichloroethylene. Pesticide and Environmental Toxicology Section. http://www.oehha.ca.gov (see link for Public Health Goals) (accessed November 30, 2003).

Pankow, J.F. and Cherry, J.A., 1996, *Dense Chlorinated Solvents and Other DNAPLs in Groundwater*. Waterloo, Ontario, Canada: Waterloo Press.

Petering, W.H. and Aitchison, A.G., 1945, United States Patent 2,371,645: Degreasing process. Assignee: Westvaco Chlorine Products Corporation, South Charleston, WV.

Petroferm, Inc., 1997, Material Safety Data Sheet for Lenium CP, October 8, 1997. FSC: 6850, LIIN: 00N088108.

Petroferm, Inc., 2005, Material Safety Data Sheet for *n*PB Super Booster. http://www.petroferm.com/static/files/cleaning/msds (accessed July 30, 2006).

Pitman, A.L., 1933, United States Patent 1,910,962: Stabilized trichloroethylene. Westvaco Chlorine Products Corporation, South Charleston, WV.

Pitman, A.L., 1943, United States Patent 2,319,261: Stabilizing chlorinated hydrocarbons. Assignee: Westvaco Chlorine Products Corporation, South Charleston, WV.

Pray, B.O. and Chisholm, R.S., 1960, United States Patent 2,959,623: Stabilization. Assignee: Columbia-Southern Chemical Corporation.

PRO-ACT, 1999, Parts washing technologies (TI#20835). http://www.p2pays.org/ref/07/06053.htm (accessed July 14, 2006).

Quast, J.F., Calhoun, L.L., and Frauson, L.E., 1988, 1,1,1-Trichloroethane formulation: A chronic inhalation toxicity and oncogenicity study in Fischer 344 rats and B6C3F1 mice. *Fundamentals of Applied Toxicology* 11: 611–625.

Ramos, H., 1974, Health hazard evaluation toxicity determination report, No. HHE-73-5-110. Prepared by Cummins Northeastern Inc., Dedham, MA. National Institute for Occupational Safety and Health (NIOSH), Hazard Evaluations and Technical Assistance Branch, Cincinnati, OH.

Reid, E.W. and Hofman, H.E., 1929, 1,4-Dioxan. *Industrial and Engineering Chemistry* 21(7): 695–697.

Reuter, G., 2002, TCE and the right technology. *Clean Tech* September.

Reynolds, J.E.F. and Prasad, A.B. (Eds), 1982, *Martindale—The Extra Pharmacopoeia*, 28th Edition, p. 106. London: The Pharmaceutical Press.

Rice, C.V. and Raftery, D., 1999, Photocatalytic oxidation of trichloroethylene using TiO_2 coated optical microfibers. *Chemical Communications* 1999: 895–896.

Richtzenhain, H. and Stephan, R., 1975, United States Patent 3,878,256: Stabilization of 1,1,1-trichloroethane with a four component system. Assignee: Dynamit Nobel AG, Troisdorf, Germany.

Roehl, E.O., 1981, United States Patent 4,289,542: Method of vapor degreasing. Assignee: Rho-Chem Corporation, Inglewood, CA.

Rowe, E.A., Jr. and Cawley, W.H., 1977, United States Patent 4,008,101: Methylene chloride phosphatizing. Assignee: Diamond Shamrock Corporation, Cleveland, OH.

Schmidt, R., DeZeeuw, R., Henning, L., and Trippler, D., 2001, State programs to clean up dry cleaners. State Coalition for Remediation of Dry Cleaners. http://www.drycleancoalition.org/survey/ (accessed 2002).

Sclar, G., 1999, Case report: Encephalomyeloradiculoneuropathy following exposure to an industrial solvent. *Clinical Neurology and Neurosurgery* 101: 199–202.

Shepherd, C.B., 1962, Perchloroethylene and trichloroethylene. In: S.J. Sconce (Ed.), *Chlorine: Its Manufacture, Properties and Uses*, American Chemical Society Monograph Series, No. 154, Chapter 13. American Chemical Society, Washington, DC.

Skeeters, M.J., 1959, United States Patent 2,868,851: Stabilization of chlorinated hydrocarbons with acetylenic ethers. Assignee: Diamond Alkali Company, Cleveland, OH.

Skeeters, M.J., 1960a, United States Patent 2,958,711: Stabilization of chlorinated hydrocarbons with carboxylic acid esters of acetylenic alcohols. Assignee: Diamond Alkali Company, Cleveland, OH.

Skeeters, M.J., 1960b, United States Patent 2,947,792: Stabilization of tetrachloroethylene with a mixture of hydroxyl alkyne and isoeugenol. Assignee: Diamond Alkali Company, Cleveland, OH.

Smallwood, I., 1993, *Solvent Recovery Handbook*. New York: McGraw Hill.

Snelling, W.O., 1914, United States Patent 1,097,145: Fire extinguishing compound. Assignee: Walter O. Snelling, Pittsburgh, PA.

Solvay SA, 2002a, Chlorinated solvents stabilisation. GBR-2900-0002-W-EN Issue 1—19.06.2002, Rue du Prince Albert, 33 1050 Brussels, Belgium. http://www.solvaychemicals.com (accessed November 9, 2003).

Solvay SA, 2002b, Chlorinated solvents—physical and global properties. PROP-2900-0001-W-EN Issue 1—19.06.2002, Rue du Prince Albert, 33 1050 Brussels, Belgium. http://www.solvaychemicals.com (accessed November 9, 2003).

Spencer, D.R. and Archer, W.L., 1981, British Patent 1,582,803: Stabilization of 1,1,1-trichloroethane. Assignee: Dow Chemical Company, Midland, MI.

Spolyar, L.W., Harger, R.N., Keppler, J.F., and Bumsted, H.E., 1951, Generation of phosgene during operation of a trichloroethylene degreaser. *Archives of Industrial Hygiene and Occupational Medicine* 4: 156–160.

Starks, F.W., 1957, United States Patent 2,795,623: Stabilization of chlorinated hydrocarbons. Assignee: E. I. du Pont de Nemours and Company, Wilmington, DE.

Starks, F.W. and Kenmore, N.Y., 1960a, United States Patent 2,945,070: Stabilization of chlorinated hydrocarbons. Assignee: E. I. du Pont de Nemours and Company, Wilmington, DE.

Starks, F.W. and Kenmore, N.Y., 1960b, United States Patent 2,958,712: Stabilization of chlorinated hydrocarbons. Assignee: E. I. du Pont de Nemours and Company, Wilmington, DE.

Stauffer, W.O., 1956, United States Patent 2,751,421: Stabilization of trichloroethylene. Assignee: E. I. du Pont de Nemours and Company, Wilmington, DE.

Stern, M. and Uhlig, H.H., 1952, Effect of oxide films on the reaction of aluminum with carbon tetrachloride. *Journal of the Electrochemical Society* 99(10): 389–392.

Stern, M. and Uhlig, H.H., 1953, Mechanism of reaction of aluminum and aluminum alloys with carbon tetrachloride. *Journal of the Electrochemical Society* 100(12): 543–552.

Stevens, H.C., 1955, United States Patent 2,721,883: Stabilization of halogenated hydrocarbons. Assignee: Columbia-Southern Chemical Corp.

Stewart, L.C. and DePree, L., 1933, United States Patent 1,917,073: Method of stabilizing chlorinated aliphatic hydrocarbons. Assignee: Dow Chemical Company, Midland, MI.

Stoddard, D.L. and Wells, W.R., 1962, In-service solvent cleaning of electric motors. *Industrial Hygiene Association Journal* 23: 62–66.

Strain, F. and De Witt, B.J., 1956, United States Patent 2,737,532: Stabilization of perchloroethylene with alkoxyalkylnitriles. Assignee: Columbia-Southern Chemical Corporation.

Sunshine, I. (Ed.), 1969, *CRC Handbook of Analytical Toxicology*. Cleveland: The Chemical Rubber Company.

Swanson, M.B., Geibig, J.R., and Kelly, K.E., 2002, Alternative adhesive technologies in the foam furniture and bedding industries: A cleaner technologies substitutes assessment. Volume 2: Risk screening and comparison. U.S. Environmental Protection Agency Office of Pollution Prevention Technology, Design for the Environment Program.

Tarrer, A.R., Donahue, B.A., Dhamavaram, S., and Joshi, S.B., 1989, *Reclamation and Reprocessing of Spent Solvents*. Park Ridge, NJ: Noyes Data Corporation.

Thomas, T., 1995, Clean Air Act compliance for solvent degreasers—regulatory strategies for manufacturers affected by Clean Air Act amendments NESHAP for halogenated solvent cleaners. University of Tennessee Center for Industrial Services, Tennessee Department of Environment and Conservation.

Totonidis, S., 2005, A role for trichloroethylene in developing nation anaesthesia. *Kathmandu University Medical Journal* 3(2,10): 181–190.

Tsuruta, H.I. and Fukuda, K., 1983, Analysis of trace impurities in reagent and technical grade trichloroethylene. *Industrial Health* 21(4): 293–295.

United Nations Environment Programme (UNEP), 1984, International Programme on Chemical Safety, Environmental health criteria 32: Methylene chloride. http://www.inchem.org/documents/ehc/ehc/ehc32.htm (accessed July, 2006).

United Nations Environment Programme (UNEP), Ozone Secretariat, 1999, 1998 Report of the Solvents, Coatings, and Adhesives Technical Options Committee. Nairobi, Kenya.

Unocal (Union Oil Company of California), 1989, Material Safety Data Sheet for 1,1,1-trichloroethane-15620. FSC: 6850, NIIN: -00011560.

U.S. Environmental Protection Agency (USEPA), 1980, RACT compliance guidance for carbon adsorbers on perchloroethylene dry cleaners. U.S. Environmental Protection Agency, Office of General Enforcement. EPA/340/1-80/007.

U.S. Environmental Protection Agency (USEPA), 1989a, Locating and estimating air emissions from trichloroethylene and perchloroethylene. U.S. Environmental Protection Agency, Office of Air Quality Planning and Standards, Research Triangle Park, NC. EPA/450/2-89/013.

U.S. Environmental Protection Agency (USEPA), 1989b, Waste minimization in metal parts cleaning. U.S. Environmental Protection Agency, Office of Solid Waste. EPA/530/SW-89/049.

U.S. Environmental Protection Agency (USEPA), 1991, Conservation and recycling practices for CFC-113 and methyl chloroform. U.S. Environmental Protection Agency, Office of Air and Radiation. EPA/400/1-91/017.

U.S. Environmental Protection Agency (USEPA), 1994a, Locating and estimating air emissions from methyl chloroform. U.S. Environmental Protection Agency, Office of Air Quality Planning and Standards. EPA/454/R-93/045.

U.S. Environmental Protection Agency (USEPA), 1994b, Chemical summary for perchloroethylene. U.S. Environmental Protection Agency, Office of Pollution Prevention and Toxics. EPA/749/F-94/020a.

U.S. Environmental Protection Agency (USEPA), 1994c, Chemical summary for methylene chloride (dichloromethane). U.S. Environmental Protection Agency, Office of Pollution Prevention and Toxics. EPA/749/F-94/018a.

U.S. Environmental Protection Agency (USEPA), 1995, EPA Office of Compliance Sector Notebook Project: Profile of the dry cleaning industry. U.S. Environmental Protection Agency, Office of Enforcement and Compliance Assurance. EPA/310/R-95/001.

U.S. Environmental Protection Agency (USEPA), 2001, Sources, emission and exposure for trichloroethylene (TCE) and related chemicals. U.S. Environmental Protection Agency, National Center for Environmental Assessment, Office of Research and Development. EPA/600/R-00/099.

van Gemert, B., 1982, Role of stabilizer-aluminum reactions in methylchloroform stabilization. *Industrial Engineering and Chemical Production Research and Development* 21: 296–299.

Van Waters and Rogers, Inc., 1990, Material Safety Data Sheet for tetrachloroethylene, technical, grade a, specification ASTM D 4081, 55 gallon drum. FSC: 6810, NIIN: 00-270-9982.

Vinson, C.G., Jr. and Martin, J.J., 1963, Heat of vaporization and vapor pressure of 1,4-dioxane. *Journal of Chemical Engineering Data* 8(1): 74–75.

Vogel, T.M. and McCarty, P.L., 1987, Abiotic and biotic transformations of 1,1,1-trichlorethane under methanogenic conditions. *Environmental Science and Technology* 21: 1208–1213.

Vulcan Chemicals, 1987, Material Safety Data Sheet for methylene chloride, degreasing grade. FSC: 6810, NIIN: 00-616-9188.

Wadden, R.A., Scheff, P.A., and Franke, J.E., 1989, Emission factors for trichloroethylene vapor degreasers. *American Industrial Hygiene Association Journal* 50: 496–500.

Watson, J.E. and Rapp, D.E., 1959, United States Patent 2,878,297: Stabilization of chlorinated solvents with aziridines. Assignee: Dow Chemical Company.

Willis, G.C. and Christian, C.A., 1957, United States Patent 2,803,676: Trichloroethylene stabilized with propargyl alcohol and pyrrole. Assignee: Dow Chemical Company, Midland, MI.

World Health Organization, 1984a, Perchloroethylene. International Program on Chemical Safety, Environmental Health Criteria 31. Geneva: Published under the joint sponsorship of the United Nations Environment Program, the International Labor Organization, and the World Health Organization. http://www.inchem.org/documents/ehc/ehc/ehc31.htm (accessed February 17, 2003).

World Health Organization, 1984b, Methylene chloride. International Program on Chemical Safety, Environmental Health Criteria 32. Geneva: Published under the joint sponsorship of the United Nations Environment Program, the International Labor Organization, and the World Health Organization. http://www.inchem.org/documents/ehc/ehc/ehc32.htm (accessed February 17, 2003).

World Health Organization, 1985, Trichloroethylene. International Program on Chemical Safety, Environmental Health Criteria 50. Geneva: Published under the joint sponsorship of the United Nations Environment Program, the International Labor Organization, and the World Health Organization. http://www.inchem.org/documents/ehc/ehc/ehc50.htm (accessed July 3, 2003).

World Health Organization, 1990, 1,1,1-Trichloroethane. International Program on Chemical Safety, Environmental Health Criteria 136. Geneva: Published under the joint sponsorship of the United Nations Environment Program, the International Labor Organization, and the World Health Organization. http://www.inchem.org/documents/ehc/ehc/ehc136.htm (accessed July 3, 2003).

Wurtz, A., 1863, Mémoire sur l'Oxyde d'Ethyléne et les Alcools Polyéthyleniques [Memorandum on the ethylene oxide and polyethylene alcohols]. *Annales de Chimie et de Physique* 69: 317–355.

2 1,4-Dioxane
Chemistry, Uses, and Occurrence

Thomas K.G. Mohr

1,4-Dioxane has many uses beyond its key role as a stabilizer for methyl chloroform. It is used directly in several industrial and commercial processes and is found in a wide range of consumer products. 1,4-Dioxane also occurs as a by-product in the production of certain surfactants, synthetic textiles, plastics, and resins. As a result of its widespread use, 1,4-dioxane is found to a limited extent in ambient air, surface water, and groundwater, but it occurs more commonly in wastewater, landfill leachate, and landfill gas. Documenting the uses, occurrence, and history of 1,4-dioxane production provides practical information for those investigating 1,4-dioxane releases in soil and groundwater. Knowledge of the many uses of 1,4-dioxane may also be helpful in reconstructing release histories by forensic investigative techniques, as discussed in Chapter 9.

1,4-Dioxane possesses some unique and surprising chemical properties, upon which its industrial utility and environmental fate depend. A comprehension of both the basic chemical structure of 1,4-dioxane and the many implications of that structure provides the foundation for (1) developing analytical methods capable of reliably separating low levels of this hydrophilic contaminant from water samples and (2) designing remedial technologies capable of removing the contaminant from affected water supplies. The chemical properties governing environmental fate and transport are described further in Chapter 3. Analytical methods for quantifying 1,4-dioxane in soil, water, air, and other media are described in detail in Chapter 4. Remedial engineering considerations for designing treatment technologies are reviewed in Chapter 7.

2.1 CHEMISTRY OF 1,4-DIOXANE

1,4-Dioxane was first described by A. V. Lourenço in 1863 as the product of reacting ethylene glycol and 1,2-dibromoethane (Stumpf, 1956; Flick, 1998). Lourenço published his discovery of 1,4-dioxane in *Annales de Chimie et de Physique* (Lourenço, 1863). In the same year, the derivation of 1,4-dioxane from ethylene oxide was described by A. Wurtz in a later issue of the same journal (Wurtz, 1863).

1,4-Dioxane is also known by the synonyms *p*-dioxane, diethylene oxide, 1,4-diethylene dioxide, and glycol ethylene ether. A complete list of synonyms is given in Table 2.1, and Table 2.2 lists identifiers for 1,4-dioxane as assigned by various organizations and agencies. 1,4-Dioxane is a cyclic ether that has four carbon atoms and two oxygen atoms, resulting in two ether functional groups in the same molecule. Cyclic ethers have a ring structure in which the oxygen has become part of the ring. In 1,4-dioxane, the oxygen atoms occur directly opposite each other to form symmetrical ether linkages. This structure makes 1,4-dioxane highly stable and relatively immune to reaction with acids, oxides, and oxidizing agents. The 1,4-dioxane ring does not rupture except in the presence of concentrated acids and strong oxidizing agents, under the influence of high temperature and pressure (Reid and Hoffman, 1942). The symmetry of the 1,4-dioxane ring should impart only a very small dipole moment to the molecule; nevertheless, it has good solvency (Stoye, 2005). The solvency and solubility of 1,4-dioxane are attributed to the polarity it acquires when a pair of 1,4-dioxane rings forms a dimer with two intermolecular hydrogen bonds (a dimer refers to a molecule

TABLE 2.1
Synonyms, Translations, and Identification of 1,4-Dioxane

Dioxane	Glycol Ethylene Ether	Diethylene Dioxide	Diethylene Ether
p-Dioxane	Diethylene-1,4-dioxide	Dioxyethylene ether	1,4-Dioxacyclohexane
1,4-Diethylenedioxide	1,4-Dioxacyclohexane	Tetrahydro-1,4-dioxane	Tetrahydro-p-dioxane
Di(Ethylene oxide)	Tetramethylene 1,4-oxide	NE 220	Dioksan (Polish)
Diossano-1,4 (Italian)	Dioxaan-1,4 (Dutch)	Dioxan; dioxan-1,4 (German)	1,4-Dioksaani (Finnish)
1,4-Dioxano (Spanish)	Dioxane; dioxyethylene ether; p-dioxan (Czech)	1,4-Dioxanne; dioxane-1,4; dioxanne (French)	をウェブから探す (Japanese)
1,4-Dioxan (Danish)	1,4-Dioxano (Portuguese)	1,4-διοξάν (Greek)	1,4-Dioxan (Swedish)

TABLE 2.2
Identification of 1,4-Dioxane

Identification Type	Identifier
Chemical Abstracts Service Registry Number (CASRN)	123-91-1
Registry of Toxic Effects of Chemical Substances (RTECS)	JG8225000
United Nations Number (UN)	1165
National Institute of Occupational Safety and Health (NIOSH)	JG8225000
EPA Hazardous Waste Identification Number	U108
Department of Transportation Number	1165; Guide 127
Standard Transportation Number	49 091 55; Dioxane
European Inventory of Existing Chemical Substances (EINICS Number)	204-661-8
Beilstein Registry Number	102551

composed of two similar subunits or monomers linked together). The two remaining oxygen atoms are available for interaction with the water molecules (Mazurkiewicz and Tomasik, 2006).

1,4-Dioxane has a slight odor similar to butanol. It is fully miscible with water and most organic solvents, and water is fully soluble in dioxane. 1,4-Dioxane is favored as a solvent for cellulose derivatives, polymers, chlorinated rubber, and resins, and it has many other uses, as described further in this chapter. 1,4-Dioxane remains inert as a reagent in some chemical reactions, such as hydrogenation and sulfonation (Stoye, 2005).

There are two other dioxane isomers, 1,2-dioxane (CASRN: 5703-46-8) and 1,3-dioxane (CASRN: 505-22-6); in each designation, the two numbers separated by a comma indicate the oxygen positions in the ring structure, as shown in Figure 2.1. 1,3-Dioxane should not be confused with 1,3-dioxolane, which has a pentagonal ring structure similar to tetrahydrofuran and is used as a stabilizer of methyl chloroform.

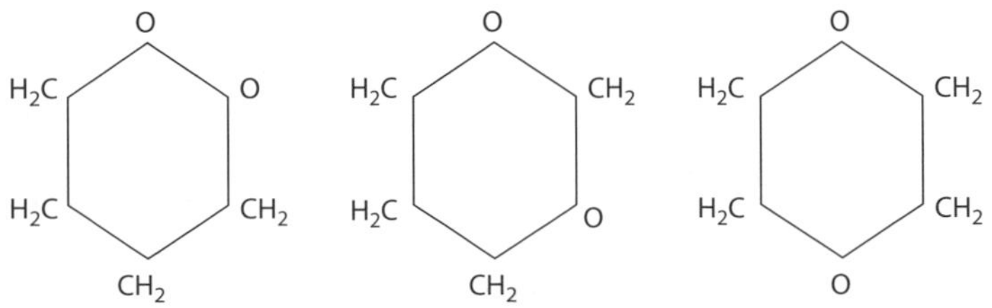

FIGURE 2.1   Structures of 1,2-dioxane, 1,3-dioxane, and 1,4-dioxane.

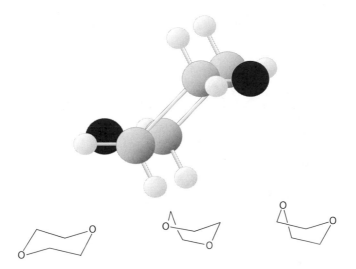

FIGURE 2.2 1,4-Dioxane in the chair conformation (black = oxygen, gray = carbon, white = hydrogen). The three conformations of 1,4-dioxane are shown below the ball and stick molecule. (After Mazurkiewicz, J. and Tomasik, P., 2006, *Journal of Molecular Liquids* 126: 111–116.)

1,4-Dioxane is a Lewis base because its oxygen molecules have electrons available for sharing (a base is a proton acceptor; a Lewis base is an electron-pair donor). The molecular structure of 1,4-dioxane is shown in Figure 2.1. Its two oxygen atoms (in black) make it hydrophilic and infinitely soluble in water. Conventional rankings of solvency for different solvents predict that pure 1,4-dioxane should have a medium degree of solvency; however, its anomalously high dipole moment in aqueous solutions allows it to act as an efficient water-structure breaker, which gives 1,4-dioxane a higher solvency than expected. 1,4-Dioxane is most stable in the "chair conformation," shown in Figure 2.2 (Mazurkiewicz and Tomasik, 2006), but can shift between two boat and two chair conformations, occupying a different energy state in each conformation (Caulkins et al., 2006).

Table 2.3 summarizes structural properties of the 1,4-dioxane molecule, and Table 2.4 provides a list of chemical and physical properties compiled from a variety of sources. The key property controlling the fate and transport of 1,4-dioxane is its solubility. A tabulation of environmental fate and transport properties is presented in Chapter 3.

TABLE 2.3
Structural Properties of the 1,4-Dioxane Molecule

Bond	Bond Length (Å)	Bond Angle Components	Bond Angle (°)	Structure
C–H	1.112			
		C–C–O	109.2	
C–O	1.423			Chair
		C–O–C	112.45	
C–C	1.523			

Source: Lide, D.R., 2007, In: D.R. Lide, (ed.), *CRC Handbook of Chemistry and Physics, Internet Version* 2007, 8th Edition, Boco Raton, FL: Taylor & Francis.

Note: Determined on vapor phase 1,4-dioxane.

TABLE 2.4
Properties of 1,4-Dioxane

Property	Value	Notes
Molecular weight	88.106 Da	Daltons
Density	1.028 g/cm³	~8.6 lb/gallons at 20°C
Composition	C 54.53%; H 9.15%; O 36.32%	
Boiling point	101.2°C	Range in citations: 101.1–101.3°C
Heat of vaporization	98.6 cal/g	
Freezing point	11.85°C	
Heat of fusion	33.8 cal/g	
Specific heat	36.01 cal/(mol K)	0.420 cal/g at 20°C
Vapor pressure	5.08 kPa at 25°C	38.09 mm Hg at 25°C
	15.8 kPa at 50°C	29 mm Hg at 20°C
	42.2 kPa at 75°C	
	97.2 kPa at 100°C	
	200 kPa at 125°C	
Antoine vapor coefficients	A: 7.94016	$P = 0.13332210^{[A-(B/(C+T))]}$
		P is vapor pressure (kPa) and T is temperature (°C); used for determining estimates of vapor pressure at different temperatures
	B: 1906.23	
	C: 275.577	
Vapor density	3.06	Air = 1; other citations: 3 and 3.3
Evaporation rate	2.42	Relative to butyl acetate = 1
Viscosity at 20°C	0.012 poise	
Surface tension (pure liquid)	32.8 mN/m	36.9 dyne/cm at 25°C; 40 dyne/cm at 20°C
Refractive index (at 20°C)	1.4224	
Auto-ignition temperature	180°C	
Flash point	12°C	Closed cup method
Lower explosive limit	2%	
Upper explosive limit	22%	
Dielectric constant	2.209	
Molar volume	85.8 cm³/mol at 25°C	
Kinetic diameter	5.3 Å	
Dipole	0.4 D	D = debyes; 1 D = 1 × 10⁻¹⁸ esu cm; esu = electrostatic units of molecular charge; cm = intramolecular distance between positive and negative charge in the molecule in centimeters
Polarity	16.4	Water = 100
Hildebrand solubility parameter (δ)	10.13 cal¹ᐟ² cm⁻³ᐟ² or 10 MPa	$\delta/MPa^{1/2} = 2.0455 \times \delta/cal^{1/2}\ cm^{-3/2}$
Solubility	Miscible	
Ultraviolet light absorption maximum	180 nm	
Acid dissociation constant	$pK_a = -2.92$	

Sources: Knovel, 2006, *Yaws' Handbook of Antoine Coefficients for Vapor Pressure—Electronic Edition.* http://www.knovel. com (accessed June 2006); Lide, D.R., 2007, In: D.R. Lide (ed.), CRC *Handbook of Chemistry and Physics, Internet Version* 2007, 87th Edition. Boca Raton, FL: Taylor & Francis; Mackay, D., Shiu, W.-Y., and Ma, K.-C., 1993, *Illustrated Handbook of Physical-Chemical Properties and Environmental Fate for Organic Chemicals: Volume III. Volatile Organic Chemicals.* Boca Raton, FL: Lewis Publishers (Taylor & Francis); Merck and Company, Inc., 1983, *The Merck Index.* Rahway, NJ: Merck and Company, Inc.; Monneyron, P.E., Manero, M.-H., and Foussard, J.-N., 2003, *Environmental Science and Technology* 37(11): 2410–2414; Smallwood, I.M., 1996, Handbook of Organic Solvent Properties. London: Arnold; Surprenant, K.S., 2005, In: W. Gerhartz, Y.S. Yamamoto, F.T. Campbell, et al., (Eds). Ullmann's Encyclopedia of Industrial Chemistry. Weinheim, Germany: Wiley Interscience; and Vinson, C.G. and Martin, J.J., 1963, Journal of Chemical and Engineering Data (1): 74–75.

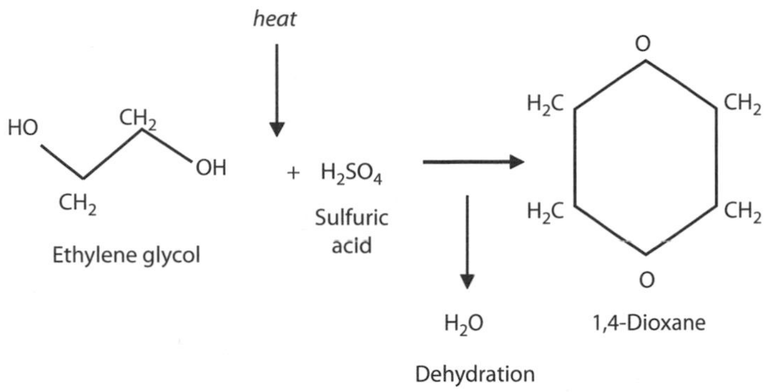

FIGURE 2.3 Conceptual representation of a common 1,4-dioxane production method involving the dehydration and ring closure of ethylene glycol with a strong acid catalyst (sulfuric acid).

2.2 HISTORY OF 1,4-DIOXANE PRODUCTION

1,4-Dioxane is produced from ethylene glycol. The most commonly used process, shown in Figure 2.3, involves heating ethylene glycol to 160°C and reacting it with concentrated sulfuric acid under a vacuum. 1,4-Dioxane forms when a catalyst—usually a strong acid—drives off water from ethylene glycol and reconfigures it to a ring structure. The reaction may have a 90% yield and is usually carried out on a continuous basis. The 1,4-dioxane/water-vapor azeotrope is collected and distilled with an acid trap to remove water and sulfuric acid (ATSDR, 2005; Surprenant, 2005). 1,4-Dioxane and its derivatives can also be produced with ethylene oxide as feedstock. A catalysis reaction on an acid ion exchange resin or on zeolites will dimerize ethylene oxide to diethylene dioxide, that is, 1,4-dioxane. Another method produces 1,4-dioxane and substitutes dioxane derivatives from bis(2-chloroethyl)ether (also called 2-chloro-2′-hydroxyethyl ether) by heating and treating with a 20% sodium hydroxide reactant (ATSDR, 2005; Surprenant, 2005).

2.2.1 1,4-DIOXANE PRODUCTION FOR COMMERCIAL APPLICATIONS

1,4-Dioxane was first produced for commercial sale and use in 1929 (ATSDR, 2004). 1,4-Dioxane could be purchased in New York for $0.50/lb in 1929 (American Chemical Society, 1929); the price lowered to $0.25/lb by 1934 (American Chemical Society, 1935). The United States began larger-scale production of 1,4-dioxane in 1951 (National Cancer Institute, 1985). A company in Japan began production in 1958 (Osaka Organic Chemical Industry, Ltd., 2006). The largest demand for 1,4-dioxane arose in the late 1950s and early 1960s with its use in stabilizing methyl chloroform. Thereafter, 1,4-dioxane production closely tracked the production of methyl chloroform (Figure 2.4).

In 1985, the worldwide production of 1,4-dioxane was estimated at 30.8 million pounds (15,400 tons or 14,000 metric tons) (Surprenant, 2005). Production in the United States in 1985 was estimated at 25 million pounds (12,500 tons or 11,300 metric tons). In the same year, the total methyl chloroform consumption in the United States—accounting for the total U.S. production, imports, and exports—was 450,000 tons. The approximate ratio of 1,4-dioxane production for domestic consumption to the comparable figure for methyl chloroform is about 3%. In the mid-1980s, about 90% of the 1,4-dioxane produced annually was used to stabilize methyl chloroform (National Cancer Institute, 1985). By 1995, the year in which ozone-depleting substance regulations severely curtailed the use of methyl chloroform, the production of 1,4-dioxane decreased to 22 million pounds (11,000 tons or 10,000 metric tons) (European Chemicals Bureau, 2002). The U.S. Environmental Protection Agency (USEPA) reports that the range of 1,4-dioxane production was from 10 to 50 million pounds between 1986 and the early 1990s; after 1994, production ranged from 1 to 10 million pounds (USEPA, 2008), as shown in Figure 2.4.

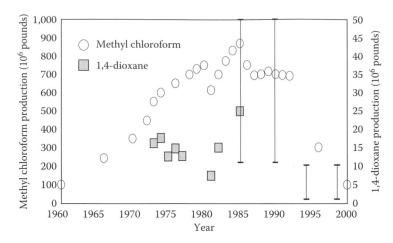

FIGURE 2.4   Annual U.S. production of methyl chloroform and 1,4-dioxane. Brackets indicate range of 1,4-dioxane production. (After Doherty, R.E., 2000, *Journal of Environmental Forensics* 1: 11. With permission; Data from Stanford Research Institute, 1989, 1989 Directory of chemical producers—United States of America. Menlo Park, CA: SRI International; Hazardous Substances Data Bank (HSDB), 2006, Online database. National Library of Medicine, National Institutes of Health. http://toxnet.nlm.nih.gov/ (accessed July 2, 2005); and U.S. Environmental Protection Agency (USEPA), 2008, *Non-confidential Inventory Update Reporting Production Volume Information.* http://www.epa.gov/oppt/iur/tools/data/2002-vol.htm (accessed January 2, 2008).)

1,4-Dioxane was produced at 10 locations in the 1980s, as summarized in Table 2.5. By 2000, fewer companies produced this chemical. The primary ones were Ferro Chemical Corporation (Baton Rouge, Louisiana), Dow Chemical (Freeport, Texas), BASF (Ludwigshafen, Germany), Osaka Yuki (Osaka, Japan), and Toho Chemical (Tokyo, Japan) (European Chemicals Bureau, 2002).

2.2.2 Impurities and Stabilizers of 1,4-Dioxane

1,4-Dioxane is available in several different grades that vary in purity, as well as the nature of impurities present due to different production methods. Table 2.6 lists the common impurities of 1,4-dioxane and

TABLE 2.5
Producers of 1,4-Dioxane (ca. 1985)

Company	Location
Ferro Corporation/Grant Chemical Division	Baton Rouge, Louisiana
Union Carbide	South Charleston, West Virginia
CPS Chemical Company	Old Bridge, New Jersey
Dow Chemical Company	Freeport, Texas
Ugine-Kuhlmann	Frankfurt, Germany
BASF Aktiengesellschaft	Ludwigshafen, Germany
Dow Chemical Company	Terneuzen, Netherlands
Osaka Organic Chemical Industry, Ltd.	Osaka, Japan
Sanraku-Ocean Company	Tokyo, Japan
Toho Chemical Industry Company	Tokyo, Japan

Source: Surprenant, K.S., 2005. In: W. Gerhartz, Y.S. Yamamoto, F.T. Campbell, et al., (Eds). *Ullmann's Encyclopedia of Industrial Chemistry.* Weinheim, Germany: Wiley Interscience.

TABLE 2.6
Common Impurities of 1,4-Dioxane

Impurity	Concentration
bis(2-Chloroethyl) ether	<0.1%
Water	<0.1%
2-Methyl-1,3-dioxane	<0.1%
2-Ethyl-1,3-dioxane	≤0.03%
Hydrogen peroxide	≤0.001%
Lead	≤0.25 ppm
Iron	<0.25 ppm

Sources: European Chemicals Bureau, 2002, European Union risk assessment report: 1,4-Dioxane. Luxembourg, European Union, Institute for Health and Consumer Protection, Report 21; DeRosa, C.T., Wilbur, S., Holler, J., et al., 1996, *Toxicology and Industrial Health* 12: 43; U.S. Environmental Protection Agency (USEPA), 1981, Phase I risk assessment of 1,4-dioxane. U.S. Environmental Protection Agency, EPA Contract No. 68-01-6030; and VROM, 1999, Risk assessment: 1,4-Dioxane. Bilthoven, Netherlands: S.P. Netherlands Ministry of Housing and the Environment (VROM), Chemical Substances Bureau 153.

their typical concentrations. Other impurities may also be present in 1,4-dioxane in unspecified amounts, such as acetaldehyde, acetic acid, glycol acetal paraldehyde, and crotonaldehyde (Wypych, 2001).

Low-molecular-weight ethers, including 1,4-dioxane, react with oxygen to form explosive peroxides. When 1,4-dioxane comes into contact with oxygen or ultraviolet light, it can form the peroxide compound 1,4-dioxanyl-2-hydroperoxide (Gierer and Pettersson, 1968). Prolonged exposure to air and sunlight may form unstable peroxides, particularly when water is absent. Peroxide compounds can form and accumulate in 1,4-dioxane, which may then explode when subjected to heat or shock. Handling 1,4-dioxane is most hazardous when peroxide levels are concentrated by distillation or evaporation. To prevent peroxide formation, 1,4-dioxane is stabilized by adding small amounts of reducing agents such as stannous chloride or ferrous sulfate. In addition, it can be stabilized with 25 mg/kg 2,6-*tert*-butyl-*p*-cresol, a phenolic antioxidant also used to stabilize chlorinated solvents, or with 25 mg/kg butyl hydroxy toluene (BHT), an antioxidant commonly found in baked goods (BASF, 1986, 2005; European Chemicals Bureau, 2002). Another method to prevent peroxide formation is filling container headspace with nitrogen. When stabilized and stored under nitrogen in original containers, 1,4-dioxane has a shelf life of 24 months (BASF, 2005).

2.3 DIRECT USES OF 1,4-DIOXANE

The leading U.S. producer of 1,4-dioxane, Ferro Corporation, lists the following uses of 1,4-dioxane on its web site (Ferro Corporation, 2006):

- Wetting and dispersing agent in textile processing
- Dye baths and stain and printing compositions
- Some cleaning and detergent preparations, adhesives, cosmetics, deodorants, fumigants, emulsions, and polishing agents
- Some lacquers, paints, varnishes, and paint and varnish removers
- Solvent for fats, oils, waxes, and natural and synthetic resins
- Purification of drugs
- Reaction media in various organic synthesis reactions
- Stabilizer for chlorinated solvents

TABLE 2.7
Emissions Sources for 1,4-Dioxane in Industrial Production Operations from Direct Use or Formation of 1,4-Dioxane as a By-Product

Emissions Sources for 1,4-Dioxane

Vapor degreasing
Surface coating
Production of adhesives and sealants
Production of dyes
Production of miscellaneous printing inks
Production of polishes and waxes
Production of surface active agents
Reagent in university research facilities
Textile finishing
Paint and varnish removal
Wood finishing
Drug and pharmaceutical purification
Production of inorganic pigments
Paint production
Detergent manufacturing
Manufacture of membrane filters
Textile mills production
Weaving mills—synthetics, cotton, and wool

Source: California Air Resources Board (CARB), 2006, "Hot spots" inventory guidelines. California Air Resources Board, Health and Safety Code Section 44300 et seq.

1,4-Dioxane is also used for making photosensitive resins and in magnetic tape production as a base for applying magnetic pigments to tape audio and data recording (Royal Society of Chemistry, 1992). The solvents used in the manufacturing of certain varieties of magnetic tapes contain as much as 40% 1,4-dioxane. During formulation of some products, the pure substance is used (European Chemicals Bureau, 2002). The California Air Resources Board (CARB) lists manufacturing processes that use 1,4-dioxane directly or produce 1,4-dioxane as a by-product in regulatory guidance documentation for the Air Toxics "Hot Spots" Information and Assessment Act of 1987 (California Assembly Bill 2588), as summarized in Table 2.7 (CARB, 2006).

A study of solvent uses by USEPA included applications of 1,4-dioxane (USEPA, 1996c). In addition to the uses just listed, possible uses of 1,4-dioxane mentioned in the study but not expressly confirmed included petroleum refining, petrochemicals, pulp and paper, explosives, electroplating, polishing, semiconductors, electronic components, photographic equipment, polymers, plastics, rubber manufacture, and organic and inorganic chemical manufacture. Use of 1,4-dioxane was confirmed in some industries in the USEPA study by questionnaire responses. Sixteen pharmaceutical facilities used 1,4-dioxane, primarily in small amounts for laboratory operations such as quality control of finished goods and laboratory experiments. Two pharmaceutical firms used 1,4-dioxane as a medium for crystallization and for the distillation and dissolution of an intermediate product. One laminated-paper firm identified 1,4-dioxane as a chemical to dissolve polymers to produce a coated paper. One business in the office-machines industry used 1,4-dioxane for the dissolution of pigment for an interface spray. Two photographic industry facilities surveyed in the USEPA study used 1,4-dioxane. One facility used it to dissolve resins and polymers for film coating. The other facility used 1,4-dioxane as a photochemical reaction medium and for miscellaneous research and development projects (USEPA, 1996c). Other surveys of 1,4-dioxane uses list pulping of wood, deodorants, production of fumigants, reagents to provide the mobile phase in some laboratory chromatography applications, reagents for measuring optical activity, reagent for cryoscopic analysis, and as a polymerization catalyst (ATSDR, 2005).

This chapter explores the various uses of 1,4-dioxane and its occurrence as a by-product in different industries by providing examples of how 1,4-dioxane was used or generated, with discussion of the relative quantities involved. Appreciation for the many ways 1,4-dioxane has been used or generated and released to the environment can assist with site investigations in locations where multiple industries have operated.

2.3.1 Cellulose Acetate Membrane Production for Reverse Osmosis and Kidney Dialysis Filters

Cellulose acetate membranes and cellulose triacetate membranes are used to filter small particles from water in reverse osmosis water treatment applications, from blood in kidney dialysis applications, and from water and other liquids in a variety of additional technological applications. Austrian chemist Karl Weissenberg recognized that cellulose acetate will flow when dissolved in 1,4-dioxane in rheological studies published in 1929 (Philippoff, 2006). 1,4-Dioxane was used to produce artificial silk in England in 1932 by treating cellulose acetate yarn with 1,4-dioxane (Barber, 1934, in USEPA, 2002).

The concept of reverse osmosis was first demonstrated in the late 1950s with cellulose acetate membranes. Reverse osmosis membranes are produced via solution deposition and hollow fiber extrusion. Solution deposition involves spreading a mixture of cellulose, a solvent, and a swelling agent over a smooth surface. As the solvent evaporates, the air-dried surface hardens and pores form within the cellulose layer. Reverse osmosis membranes are made with a very thin layer of polymer film applied on top of the porous cellulose acetate substrate. After the solvent evaporates from the cellulose membrane surface, the membrane is quenched by immersing in water, which precipitates the polymer (Klein and Smith, 1972).

The solvents used in this process are selected for their solubility in water. The solvent was commonly a mixture of 1,4-dioxane and acetone in ratios ranging from 1.5:1 to 2.5:1. Swelling agents include methanol and ethanol, used at between 2 wt% and 10 wt% of the final solution (Yashushi, 1977). Wastewater effluent from cellulose acetate and triacetate membrane production facilities has often included 1,4-dioxane. Several 1,4-dioxane groundwater contamination incidents have been caused by effluent from membrane production facilities, as described further in the case studies presented in Chapter 8.

Cellulose acetate and triacetate membranes produced by using 1,4-dioxane as a deposition solvent may contain residual 1,4-dioxane and thus require a California Proposition 65 warning label. Manufacturers acknowledge that trace amounts of 1,4-dioxane may be present and recommend flushing with water to purge 1,4-dioxane from the system. Because 1,4-dioxane is infinitely soluble, it is quickly removed from filters by flushing with water. Nearly all residual 1,4-dioxane is now flushed out in the manufacturing process, though this may not have always been the case. Most manufacturers no longer use 1,4-dioxane for producing reverse osmosis or kidney dialysis filters.

Cellulose acetate fibers are also used to produce some varieties of cigarette filters. 1,4-Dioxane is among the solvents listed in the patent literature for the process used to create the fiber bundles that remove tar from cigarette smoke (Matsumura et al., 1997).

2.3.2 Liquid Scintillation Cocktails

Numerous experiments and routine measurements in the medical and life sciences are made by injecting or orally administering a radio-labeled solution containing β-emitting nuclides and then counting the radioactivity in blood, tissue, excreted urine, or other media. The process of counting levels of radiation often involves the use of liquid scintillation "cocktails." Samples selected for radiation measurement are dissolved in a solvent mixture, called a liquid scintillation cocktail, which includes fluorescent compounds or fluors. Energy from emitted β particles is transferred to solvent molecules, which then achieve an excited state and transfer excess energy to adjacent solvent molecules. The solvents selected for liquid scintillation cocktails remain in an excited state and return to the zero state without emitting light. Energy from excited solvent molecules is captured by the fluorescent solutes, and excited fluors dissipate the energy by emitting light photons, which permits radiation counting in the visible light range. Ideally, each β emission results in a pulse of light, though β-counting efficiencies are generally below 100% and vary by the strength of the β emitter. The liquid scintillation counter instrument consists of two photomultiplier tubes in a circuit. Only those light pulses reaching both tubes are counted. Each fluorescence event is proportional to a radioactive decay event, and the frequency of these events is directly proportional to the number of ^{14}C, ^{3}H, ^{32}P, ^{35}S, or other radionuclides present in the sample.

1,4-Dioxane was favored for liquid scintillation cocktails because it is fully miscible and holds a large amount of water in a homogeneous solution. Therefore, 1,4-dioxane is an ideal solvent for samples containing various amounts of water. Impurities in 1,4-dioxane, such as peroxides formed during storage, will quench the energy transmitted from β particles and must be avoided. The first widely used liquid scintillation formulation, today known as Bray's solution, was a mixture of naphthalene and 1,4-dioxane, with ethylene glycol as an antifreeze to allow long-term storage of counting vials in freezers (Bray, 1960). Bray's solution consisted of the following mixture dissolved in 1,4-dioxane; the 1,4-dioxane comprised more than 80% of the solution (Perkin Elmer, 2006):

Naphthalene	60 g/L
Methanol	100 mL/L
Ethylene glycol	20 mL/L
PPO = diphenyloxazole	4 g/L
POPOP = 1,4-bis-(5-phenyloxazolyl)benzene	0.2 g/L
(note that 5-phenyloxazolyl = phenyloxazolyl-phenyl-oxazolyl-phenyl)	

MSDSs indicate that scintillation-grade 1,4-dioxane used to make the cocktail has a composition of 95–99% 1,4-dioxane (National Diagnostics, Inc., 1987a, 1987b; Fisher Scientific, 2006). There are many other formulations for liquid scintillation cocktails; toluene and xylene are the common components. Larger laboratories have been known to recycle their scintillation wastes by distillation in a rotary evaporator and desiccation with excess potassium hydroxide. These steps were also followed to reduce waste volumes prior to disposal (Miyatake and Saito, 1984). Because the half-lives of the radioactive components of scintillation wastes are relatively short, wastes are often stored long enough for sufficient decay to occur until radioactivity is no longer measurable. Scintillation wastes have also been managed by incineration and by incorporating flammable liquid scintillation fluids into fuels; however, this practice releases the remaining radioactive components to the environment. The University of Illinois once burned liquid scintillation cocktails as a fuel supplement at its campus power plant with approval by the Nuclear Regulatory Commission, which conducted inspections to confirm that the levels of radioisotopes at the point of release were within regulatory limits. Because only small amounts were burned, the practice was exempt from RCRA (Resource Conservation and Recovery Act) Part B permit requirements. A firm in Gainesville, Florida, has processed liquid scintillation wastes by removing enough radiative components to allow burning of the residual liquid in cement kilns (USEPA, 1996b).

A 1990 survey of more than 1000 facilities producing liquid scintillation fluid wastes showed that the leading producers were industrial facilities, followed by academic facilities, government facilities, and medical facilities. The total quantity of all liquid scintillation fluids produced at the facilities surveyed was approximately 100,000 ft^3 (approximately 748,000 gallons) (USEPA, 1996b).

In the 1960s, landfills serving universities that performed biological research involving liquid scintillation counting were highly likely to have been contaminated with 1,4-dioxane. At Duke University in North Carolina, the campus landfill predates hazardous waste regulations. Laboratory wastes were buried in the landfill in the 1960s, and releases of 1,4-dioxane to groundwater at problematic levels have been reported (University of North Carolina, 2002). In 1964, the University of Kansas began operating a landfill for disposal of low-level radioactive wastes, including radionuclides contained in toluene and 1,4-dioxane; the landfill produced 1,4-dioxane contamination in off-site groundwater (Shamberg, 2002). The University of Michigan Landfill, the University of Nebraska Landfill, and the Gloucester Landfill located south of Ottawa in Ontario, Canada, are additional examples where liquid scintillation wastes have caused 1,4-dioxane contamination in groundwater. Municipal landfills receiving hospital wastes and commercial medical and veterinary laboratory wastes may also be sources of 1,4-dioxane contamination. As discussed in Section 2.7, landfills of all types have been found to be sources of 1,4-dioxane, though most landfill groundwater-monitoring programs do not analyze groundwater, landfill gas, or landfill gas condensate for 1,4-dioxane.

2.3.3 TISSUE PRESERVATIVE IN HISTOLOGY AND OTHER LABORATORY MICROSCOPY METHODS

1,4-Dioxane has been used as a laboratory solvent for a variety of purposes, including study of tissue samples, as a carrier for biocides to eliminate biodegradation of analytes in water samples, for analysis of plant and wood samples, for purification/isolation of DNA elements, and for drying soil samples prior to analysis. Human and animal tissue samples are prepared as thin sections on microscope slides for examination by medical technologists and research scientists. A wide variety of methods are used, depending on the purpose for which the slide is prepared. Dehydration and staining are common steps in preparing tissue slides. 1,4-Dioxane has been used as a dehydrating agent for tissues in the preparation of histological slides since the 1930s (Mossman, 1937). It is also used by biologists to impregnate tissue sections with paraffin, and it has advantages in the preparation of histological slides for electron microscopy (Shearer and Hunsicker, 1980). 1,4-Dioxane is preferred over ethanol and xylene for dehydrating tissues because it removes water without causing substantial shrinkage or hardening in the tissue (Ralph, 1938).

1,4-Dioxane is also useful for dissolving and removing mercuric chloride, a biocide formerly used to preserve tissue samples. Tissue slides are also treated with stains or fixatives including dichromate (orange) and chromium trioxide (sometimes called "chromic acid"—red). When treated with a 70% 1,4-dioxane solution, both chrome-based fixatives reduce to chromic oxide, which is green-brown (Winsor, 2006). Because 1,4-dioxane is an expensive reagent (about $75/L for histological grade in 2006), some laboratories reclaim it by drying the 1,4-dioxane solution over a layer of calcium oxide or anhydrous cupric sulfate or by freezing the 1,4-dioxane with water in a spark-proof refrigerator (Winsor, 2006).

1,4-Dioxane is used for extraction and analysis of plant material, for example, lignin in wood samples. Analysis of wood samples involves hydrolyzing lignins and cellulose with a 1,4-dioxane solution. Some procedures use 100% 1,4-dioxane for lignin extraction (Agarwal and Ralph, 1997).

The study of soil samples by x-ray diffraction and other applications requires drying. Drying can be accomplished by introducing a hydrophilic solvent such as acetone or 1,4-dioxane. In some procedures, 1,4-dioxane is used to displace water; the 1,4-dioxane is subsequently displaced with a resin to allow study of friable soil structures. 1,4-Dioxane is preferred for soil drying because it preserves the soil structure better than acetone (Venediktova and Rudnyi, 1965; Moran et al., 1986, 1989; Chartres et al., 1989).

Preparation of water samples for the analysis of biodegradable analytes by ion-selective electrodes may involve treating with a biocide carried by 1,4-dioxane. One recommended protocol uses phenyl mercuric acetate as a biological inhibitor to treat water samples; 0.1 g is added to 20 mL of 1,4-dioxane, which is then diluted with deionized water to 100 mL of solution. A few drops of this solution are added to the sample prior to storage and analysis (NICO2000, Ltd., 2006). 1,4-Dioxane has been used as a reagent in various laboratory analytical methods, including detection of iodide in iodized salt (Saifer and Hughes, 1937). Its use for other iodide analysis methods dates back to 1926 (Anschutz and Broeker, 1926).

2.3.4 PRINTING INKS AND PAINTS AND RELATED USES

2.3.4.1 Inks and Printing Operations

1,4-Dioxane has been used as a solvent in printing inks since the early 1950s (West, 1959). Although many general references cite 1,4-dioxane as a constituent of inks and paints, few citations were found to document the direct use of 1,4-dioxane in printing inks. The presence of 1,4-dioxane in printing inks may be due to the minor use of methyl chloroform as a solvent in solvent-based printing inks and in printing operations. Inks and coatings were listed as the destination for up to 5% of the 715 million pounds of methyl chloroform produced in 1989. The most common solvent component of ink formulations was toluene. Methyl chloroform was also used for contact cleaning of printing equipment. The USEPA's 1993 Toxic Releases Inventory (TRI) reported that 37 printing operations released a total of about 1.4 million pounds of methyl chloroform that year. Toxic Release Inventory data are inherently problematic; for example, the printing facilities reporting represented less than 0.15% of the 70,000 printing operations in the United States at the time (USEPA, 1995a). A 1997 update to USEPA's Sector Notebook Series (which gives surveys of chemical releases in different industries or sectors) provides 1995 figures for 1,4-dioxane releases from two printing facilities. The average release per facility was 8500 pounds per year (USEPA, 1997a).

Another possible source of 1,4-dioxane in inks and printing is the use of propylene glycol in both paint and ink formulations. Formulations for both paints and inks commonly contain up to 20% propylene glycol (Flick, 2005). 1,4-Dioxane is a known production by-product of propylene glycol (Johnson, 2001; Pundlik et al., 2001).

1,4-Dioxane is used in the office machines industry for the dissolution of pigment via an interface spray in photocopying equipment. 1,4-Dioxane is used to help inks adhere to or distribute in plastics. Some inks used to color thermoplastic fluoropolymers, such as electrical wiring insulation plastics, are dissolved in 1,4-dioxane (West, 1959). Inks containing resins to ensure durability—for example, to apply permanent ink markings to golf balls—use a variety of solvents that may include 1,4-dioxane (Kikuchi and Tanaka, 2003). 1,4-Dioxane has been used as an ingredient of deinking solvents for cleaning fingerprint rollers (Sirchie Finger Print Labs, 1989).

2.3.4.2 Historical Restoration

1,4-Dioxane is used to remove fungal stains from paper to preserve historical papers and art works. The solvent is also used in the conservation of oil-on-canvas and other paintings.

2.3.4.3 Painting, Coating, and Paint Stripping

Solvent-borne paints and coatings are made with solvents, binders, pigments, and additives. The combination of binder and solvent is referred to as the paint "vehicle"; pigment and additives are dispersed within the vehicle. Solvents comprise about 60% of the total formulation; binders, 30%; pigments, 7–8%; and additives, 2–3%. The primary paint solvents include toluene, xylene, methyl ethyl ketone, and methyl isobutyl ketone (Northeast Waste Management Officials Association, 1998). Chlorinated solvents such as trichloroethylene and perchloroethylene were used in phosphatizing applications to prepare car bodies and other metal surfaces for coating and painting, whereas methyl chloroform was often used as a solvent for paints formulated with alkyd and polyamide resins.

1,4-Dioxane can occur as a by-product of resin production and may be carried with resins into paint as an impurity. Resins commonly used in paints and coatings have included vegetable oils, alkyds, polyesters, polyamides, phenolics, polyurethanes, epoxies, silicones, acrylics, vinyls, cellulosics, and fluorocarbons. Coatings for metal surfaces are dominated by alkyd resins; however, water-based acrylics, epoxies, polyurethanes, and polyesters also are used for certain applications.

Methyl chloroform was the favored solvent for alkyd resin coatings and for cleaning paint and coating spray guns and fluid hoses (USEPA, 1997b).

Methyl chloroform and dichloromethane could contribute 1,4-dioxane to paint compositions in minor amounts when 1,4-dioxane is present as a stabilizer. 1,4-Dioxane has been documented as a trace ingredient in paints that used methyl chloroform as a drying solvent. For example, polyurethane coatings contained less than 1% 1,4-dioxane (Glidden Paint, 1992). Resin formulations used large amounts of solvents to facilitate application and drying. A polyamide resin formula included 70% methyl chloroform with less than 5% 1,4-dioxane (DeSoto, Inc., 1991). Some lacquer formulations contained only 8% solids; the balance was composed of solvents and thickeners, including as much as 60% 1,4-dioxane (American Cyanamid, 1987). Pervo Paint Company (1993) stated that white traffic paints for striping lanes contained as much as 24% methyl chloroform and 0.5% 1,4-dioxane; yellow and red traffic paints had less than 0.1% 1,4-dioxane. However, according to the USEPA Toxic Release Inventory database, Pervo Paint Company removed 1,4-dioxane from its products in 1994.

"Dioxane purple" is a color name for an acrylic paint used by hobbyists and artists. 1,4-Dioxane has been found in the low part per million range in European felt tip pen inks and fabric dyes (Hansen, 2005). Mixtures of cyclohexanone with 1,4-dioxane or tetrahydrofuran and 1,4-dioxane are preferred for painting the primer layer on magnetic tapes; the outer magnetic layer is applied by using xylene and toluene. 1,4-Dioxane comprises about 10% of the magnetic tape paint mix (Roller et al., 1977). Magnetic tape producers have historically discharged 1,4-dioxane in wastewater.

The dominant paint-removal solvent is dichloromethane, which in some instances includes 1,4-dioxane as a stabilizer. The most common paint-stripper formulation contains 60–65% dichloromethane. This solvent is particularly effective at penetrating the coating and causing it to swell and separate from the substrate (USEPA, 1996a). In 1989, more than 100 million pounds of dichloromethane was used for paint stripping, which was about a quarter of all dichloromethane produced in that year. About 40% of dichloromethane applied for paint stripping was used in the maintenance sector for stripping aircraft and other large vehicles and parts (SRRP, 1991). Because paint stripping often includes a pressure-washing step following treatment with the paint-stripping solvent, water may become contaminated with dichloromethane. USEPA estimates that the discharge of dichloromethane to water from paint-stripping operations in 1989 exceeded 8 million pounds (SRRP, 1991). Several citations in publications focusing on toxicology, risk, or summary fact sheets for 1,4-dioxane have listed 1,4-dioxane as a solvent used directly for paint stripping; however, none of the sources checked corroborated this assertion.

2.3.5 FLAME RETARDANT PRODUCTION

1,4-Dioxane is used as a solvent in the production of brominated fire retardants. These are produced by reacting cyclododecatriene with bromide in the presence of a solvent, generally an alcohol such as isobutanol or 1,4-dioxane. 1,4-Dioxane is favored over alcohol because alcohol causes reaction intermediates to precipitate out of solution before they can be completely reacted and because alcohols directly consume the brominating agent. Using 1,4-dioxane eliminates the problem of intermediates remaining at the end of the reaction, and 1,4-dioxane does not react with the brominating agent (Business Communications Company, 2002). Waste 1,4-dioxane is recovered for off-site use as a fuel (USEPA, 2006).

2.3.6 RUBBER AND PLASTICS INDUSTRY

The USEPA's surveys of chemical releases in different industries, known as Sector Notebooks, group the rubber and plastics industries together. The Rubber and Plastics Industry Sector Notebook lists 1,4-dioxane as a waste emitted from four facilities out of more than 400 surveyed; however, the Toxics Release Inventory data cited are probably inconsistent in reporting 1,4-dioxane if it was not directly used. The Toxics Release Inventory data cited for 1993 include about 200 facilities releasing about 11 million pounds of methyl chloroform to the atmosphere and transferring about 1 million pounds to waste-handling facilities (USEPA, 1995b). The quantities of 1,4-dioxane reported in the 1993 Toxics Release Inventory, about 11,000 lb, probably understate the releases associated with its likely presence as a stabilizer of methyl chloroform. The 1995 Toxics Release Inventory also reports 1,4-dioxane for only four facilities, but lists more than 100,000 lb of 1,4-dioxane transferred and about 5000 lb released.

2.3.7 AIRCRAFT DEICING FLUID AND ANTIFREEZE

Before 2000, 1,4-dioxane was an ingredient or impurity in aircraft deicing fluids and antifreeze solutions. The Air Transport Association reportedly listed the following components of deicing fluid in 1994: ethylene glycol or propylene glycol, water, surfactants (wetting agents), corrosion inhibitors (including flame retardants), pH buffers, dyes, 1,4-dioxane, and complex polymers as thickening agents (USEPA, 2000). Glycol deicing formulations were previously used to treat airport runways. Because 1,4-dioxane was an impurity as residual solvent carried over from production of glycol, past uses of glycol deicing agents could leave trace levels of 1,4-dioxane in soil and ground-water near runways. MSDS for aircraft deicing fluids list a trace presence of 1,4-dioxane (0.0022% or 22 mg 1,4-dioxane per liter of deicing fluid) as an impurity of glycol compounds (Union Carbide, 1989). 1,4-Dioxane was also used as a wetting and dispersing agent but composed less than 0.5 mg/L of deicing fluids (USEPA, 2000). Others report that past levels of 1,4-dioxane in deicing fluids were on the order of 3 mg/L (Gelman Sciences, 1989a). 1,4-Dioxane can be detected in glycols in the low part per million range (Pundlik et al., 2001).

To reduce groundwater contamination, deicing fluid manufacturers have removed 1,4-dioxane from their formulations (USEPA, 2000). The main deicing agents used today are urea and calcium-magnesium acetate, more commonly known as "road salt."

Many brands of automotive antifreeze list 1,4-dioxane as an ingredient or a component at less than 0.0086% (86 mg 1,4-dioxane per liter of antifreeze fluid) (Old World Automotive Products, Inc., 2001). 1,4-Dioxane is also listed as a trace component of recycled antifreeze at less than 0.004% (40 mg/L) (FPPF Chemical Company, Inc., 2006). Analysis of consumer antifreeze products found 1,4-dioxane present at concentrations ranging from 0.1 to 3.4 mg/L (Gelman Sciences, 1989c). Analysis of radiator fluid boil-over verified the presence of 1,4-dioxane at up to 22 mg/L (ATSDR, 2005). The higher concentration of 1,4-dioxane in radiator boil-over compared to the concentration in antifreeze products suggests that heating ethylene glycol in a radiator, coupled with possible acidification from metal oxides, could lead to formation of 1,4-dioxane, as described in Figure 2.3.

In an effort to draw attention to the potential widespread contamination by 1,4-dioxane from antifreeze and deicing fluids, Charles Gelman, the CEO of Gelman Sciences, testified at a State of Michigan hearing that sample results from puddles of car radiator boil-over contained high levels of 1,4-dioxane (Judge, 1988). Gelman's letters cautioned that 1,4-dioxane could pollute groundwater at rest stops and wherever winterization practices for plumbing systems in summer cottages included filling pipes with ethylene glycol to prevent freezing during the winter. Flushing out the lines the following summer would cause 1,4-dioxane to flow through septic systems. The same issue was raised for winterizing recreational vehicles and large boats (Gelman Sciences, 1989b).

2.3.8 ADHESIVES

A number of glues and adhesives were prepared with a chlorinated solvent base to facilitate spreading and drying. 1,4-Dioxane is present in some glues as a solvent and in others as a stabilizer of methyl chloroform. The National Library of Medicine Household Products Database lists a brand of wood parquet floor paste as containing 0.5% 1,4-dioxane; an adhesive for affixing trim and detailing to automotive surfaces is listed as containing 1–3% 1,4-dioxane (DeLima Associates, 2004; HSDB, 2006). Adhesives containing methyl chloroform can also be expected to contain trace amounts of

1,4-dioxane. One formulation of wood glue contained from 1% to 5% 1,4-dioxane (DAP, Inc., 1994). Some formulations of contact cements had ratios of 1,4-dioxane to methyl chloroform at considerably higher proportions than those found in degreasing compositions of stabilized methyl chloroform (Wilke et al., 2004). DAP's Smooth Spread Contact Cement had 2.5% 1,4-dioxane and only 15–20% methyl chloroform (DAP, Inc., 1990).

Adhesives used in architectural coatings are thought to be the primary cause of 1,4-dioxane detections in indoor air samples (Hodgson and Levin, 2006). In offices equipped with cubicles, adhesives are used to join the many components of cubicle furniture; coatings and varnishes may also contain compounds bearing 1,4-dioxane. Testing of indoor air in offices using cubicles detected 1,4-dioxane as well as naphthalene, benzene, trichloroethylene, and perchloroethylene. Newer office furniture is designed to avoid emissions of these compounds (Betts, 2005). Draperies and drapery linings also emit measurable concentrations of 1,4-dioxane to indoor air (Smith and Bristow, 1994).

Film cements for joining or mounting sections of cellulose triacetate movie film use acetone, methylene chloride, and 1,4-dioxane as solvents. 1,4-Dioxane was a significant component in commercial film cement, exceeding 10% in some formulations (Thompson-Hayward Chemical Company, 1993). Another variety of film cement is composed of 50% 1,4-dioxane (Eastman Kodak Company, 1998). 1,4-Dioxane is used to dissolve resins and polymers into solution for film coating. It is also used as a photochemical reaction/synthesis medium as well as for miscellaneous photographic research and development projects (USEPA, 1996c).

2.3.9 Polyurethane Materials for Medical Devices

Polyurethanes are durable elastomers used for a wide variety of applications, including flexible and rigid foams, adhesives, seals, gaskets, carpet padding, and hard plastic parts. Polyurethanes are biocompatible, which makes them suitable for medical devices. Medical devices made with polyurethanes include pacemaker leads, catheters, feeding tubes, angioplasty balloons, condoms, surgical gloves, wound dressings, and many other products. Medical devices are made by injection molding, extrusion molding, and solution processing. Mold-release solvents include 1,4-dioxane, tetrahydrofuran, dichloromethane, methyl ethyl ketone, N,N-dimethylformamide, N-methylpyrrolidone, cyclohexanone, and chloroform. These solvents contribute to highly transparent polyurethanes, a desired feature for medical devices. 1,4-Dioxane and other polyurethane solvents dissolve polymers and form clear solutions. Criteria for solvent selection include ease of solvent removal from the product; all traces of solvents must be eliminated because even trace amounts may interfere with the treatment and the patient's health (Wypych, 2001).

2.3.10 Solvent-Based Cleaning Agents

Where methyl chloroform was a constituent of various household and automotive consumer and commercial contact cleaning agents, 1,4-dioxane was often present in the low percentage range as a stabilizer. Some cleaning agents are better known for their main ingredient, which may mask the presence of methyl chloroform and 1,4-dioxane. For example, brake-cleaning formulations used in automotive shops use tetrachloroethylene as the primary solvent (60–65%), but also contain methyl chloroform (30–35%) and 1,4-dioxane (2.5%) (Valvoline Oil Company, 1991). Some aerosol-spray brake-cleaner compositions were primarily methyl chloroform (89%), with higher percentages of 1,4-dioxane than normally required for vapor degreasing (up to 6.8%) (CRC Chemicals, 1987).

Other solvent-based consumer products contained only small amounts of methyl chloroform but larger amounts of 1,4-dioxane. Some consumer rust-removing agents contained up to 10% 1,4-dioxane, but only 4% methyl chloroform, with butyl alcohol and butyl cellosolve as the primary ingredients (Watsco Components, Inc., 1991). Rust inhibitors have been formulated with as much as 50% methyl chloroform and 0.8% 1,4-dioxane (Crown Industrial Products Company, 1989). Commonly used solvents for loosening frozen hardware such as nuts and bolts contained 1–3% 1,4-dioxane (Loctite Corporation, 1995).

2.3.11 ETHER SUPPLEMENTS IN FUELS

Cyclic ethers including 1,4-dioxane and 1,3-dioxolane have been considered for addition to fuels to enhance the octane value and to reduce the formation of carbon monoxide in automobile exhaust (Maurer et al., 1999). Ether compounds are sometimes added to racing fuel by karters (hot rodders) to increase fuel octane. Karters report that it takes 10–15% 1,4-dioxane added to racing fuel to produce increased octane and yield (Pro-Systems, 2002). Racing rules now prohibit using 1,4-dioxane, dinitrotoluene, propylene oxide, or nitropropane in racing fuels. 1,4-Dioxane is also banned as a racing motor oil additive in "XTREME Stock Car" and "Women on Wheels" racing (Susquehanna Speedway Park, 2006). However, ethers elude detection by the digital fuel meters commonly used at race tracks to verify fuel composition compliance. 1,4-Dioxane, propylene oxide, nitromethane, and nitropropane have all been used as "illegal" racing fuel additives. The quantities added vary according to the whims of the racer.

Amateur racers in the karting community have cautioned against the use of 1,4-dioxane because it is listed as a carcinogen and is easily absorbed through the skin (Copeland, 2004). Racers developed test methods to check racing fuels for the presence of banned substances including 1,4-dioxane, which is not detected by the widely used Godman DT-15 Digatron meter (a dielectric testing device used to check racing fuels for polar substances). The method involves adding an acid reagent and checking for the presence of a white precipitate (the "acid drop test").

1,4-Dioxane is a flammable liquid and tends to form explosive peroxides. Its development in glycols (low parts per million levels), which are used as dehumidifying agents in refineries, may take place by condensation. 1,4-Dioxane thus formed gets distilled over with benzene in the refinery process. Therefore, it is necessary to identify and determine the levels of 1,4-dioxane in glycols as well as benzene. Gas chromatography (GC) is probably the best technique for this purpose. GC analysis may be carried out with a flame ionization detector. Results show that 1,4-dioxane can be reliably determined down to 2 ppm in glycols and benzene and 1 ppm in toluene (Cortellucci and Dietz, 1999).

2.4   1,4-DIOXANE AS A BY-PRODUCT OF MANUFACTURING

1,4-Dioxane occurs as a reaction by-product in several chemical processes used to produce polyester, soaps, and plastics. Consequently, waste streams from facilities that produced these products have included 1,4-dioxane, and some have caused contamination of soil and groundwater in past decades.

2.4.1   ETHOXYLATED SURFACTANTS

1,4-Dioxane is produced as a by-product during the sulfonation reaction with alcohol ethoxylates, a process used to produce surfactants in a wide variety of soaps and detergents. Ethoxylated alcohols arc uscd as surfactants, dctcrgcnts, foaming agcnts, cmulsificrs, and wctting agcnts. During alcohol ethoxylation, ethylene oxide is combined and rearranged to form the polymer of ethylene oxide. This process allows ethylene oxide to dimerize to form diethylene dioxide, that is, 1,4-dioxane. If no effort is made to control formation of 1,4-dioxane, levels may approach 500 ppm or higher. By controlling mixing ratios, temperature, and other reaction parameters, formation of 1,4-dioxane has been limited to 30–200 ppm (Stepan Company, 2006). Since 1990, MSDS have reported 1,4-dioxane at an "upper bound concentration" or "typical maximum" of less than 15 ppm (Talmage, 1994). Analyses of surfactants and products containing surfactants have shown rather high levels of 1,4-dioxane associated with surfactants and associated products in the 1980s and 1990s.

Today, the 1,4-dioxane impurity in ethoxylated surfactants is removed through a stripping process employed during production (ATSDR, 2004). Between 1992 and 1997, the average concentration of 1,4-dioxane in cosmetic finished products was reported to fluctuate from 14 to 79 ppm (mg/kg) (ATSDR, 2004). As described in the next section, producers of nonionic surfactants also used

1,4-dioxane for routine quality-control testing, which could produce a laboratory waste stream with a substantial volume of 1,4-dioxane over time.

2.4.1.1 Personal Care Products: Detergents, Shampoos, and Sundries

1,4-Dioxane was historically present in the part per million range in alcohol ethoxy sulfate (AES) compounds. Sodium laureth sulfate [the compound sodium 2-(2-dodecyloxyethoxy) ethyl sulfate, CASRN 3088-31-1; also called sodium lauryl ether sulfate] in particular had elevated levels of 1,4-dioxane, as high as 500 ppm. Sodium laureth sulfate is a component of anionic surfactants in detergents and shampoos. Powdered detergents typically used as much as 12% AES, whereas liquid detergents used as much as 18% AES, and dish-washing detergents may contain as much as 27% AES (Greek and Layman, 1989; Stepan Company, 2006). In addition to sodium laureth sulfate, the common shampoo, detergent, and dish-washing soap ingredients containing ammonium laureth sulfate and triethanolamine laureth sulfate may also contain 1,4-dioxane as a trace contaminant. In the 1980s, these ingredients contained elevated levels of 1,4-dioxane. Ammonium laureth sulfate contained 1,4-dioxane from 288 to 1282 ppm; sodium laureth sulfate had from 69 to 340 ppm. In 1988, 400 million pounds of ethoxylated alcohol surfactants were used in household detergents in the United States (Talmage, 1994). Some earlier formulations of laundry presoak spray had particularly high levels of 1,4-dioxane. In one of six samples of laundry presoak spray analyzed, 1,4-dioxane was detected at a concentration of 15.0 wt% (USEPA, 1992).

Since the 1950s, quality-control testing for surfactant production included determining the hydrophilic versus lipophilic balance or relative solubility number of nonionic surfactants. The relative solubility number is the volume in milliliters of distilled water necessary to produce persistent turbidity in a benzene and 1,4-dioxane solvent system consisting of 1 g of surfactant sample and 30 mL of solvent. To avoid using toxic solvents, the test now uses toluene and ethylene glycol dimethyl ether (Wu et al., 2004).

In 1979, the U.S. Food and Drug Administration (FDA) announced that hundreds of cosmetics products contained 1,4-dioxane. The announcement cautioned that the potential associated risk was unknown (Washington Post, 1979). In the early 1980s, the FDA tested 100 samples of ethoxylated surfactant ingredients used in personal care products and found that 81% of ingredients tested contained detectable levels of 1,4-dioxane. Results of the FDA study are summarized in Table 2.8 (FDA, 1981). In finished products in the sundries category, the same study found that 2 of 11 samples contained greater than 100 ppm 1,4-dioxane; 5 of 11 contained 1,4-dioxane between 10 and 100 ppm; one sample had less than 10 ppm; and three samples did not detect 1,4-dioxane at a reporting limit of 0.5 ppm (FDA, 1980, 1981). Table 2.9 summarizes levels of 1,4-dioxane in shampoos, bath, and sundry products.

Following the realization that 1,4-dioxane is present in cosmetics, shampoos, and detergents as an impurity of surfactants and emulsifiers, surfactant producers sought to reduce the amount of 1,4-dioxane formed. The degree of reduction in the levels of 1,4-dioxane present varied considerably in

---

TABLE 2.8

Summary of 1,4-Dioxane Analysis of Ethoxylated Raw Materials as of September 30, 1981

Range of Dioxane Detections	Number of Samples in This Range (Out of 100)
Levels above 100 ppm	22
Levels between 10 and 100 ppm	36
Levels between 0.5 and 10 ppm	23
Not detected above 0.5 ppm	19

Source: Food and Drug Administration (FDA), 1981, Progress report on the analysis of cosmetics raw materials and finished cosmetics products for 1,4-dioxane. Division of Cosmetics Technology, Food and Drug Administration 15.

TABLE 2.9
Partial Results for 1981 1,4-Dioxane Analyses in Single Samples of Personal Care Products

Cosmetic or Sundry Product	Manufacturer	Brand Name	1,4-Dioxane Concentration (ppm)
Shampoo	Estée Lauder	Azuree Natural Shampoo	66
Shampoo	Aramis	Aramis Malt Enriched Shampoo Concentrate	279
Bath gel	Ritz Group	Jean Naté Bath Gel	36
Bath gel	Peoples Drug Store	Peoples Peach Concentrated Foam Bath	64
Foundation	Cosmetic Sciences, Inc.	Liquid Jojoba Foundation	ND (<0.5)
Foundation or cream	Hasbro Industries, Inc.	Fresh'n Fancy Red/White Mixing Cream	ND (<0.5)
Skin cleanser	Winthrop Labs	Phisoderm	140
Lotion	Westwood Pharmaceuticals	Keri Lotion for Hands, Face, Body	ND (<0.5)
Lotion	Miller-Morton	Skin Quencher Hand and Body Lotion	30

Sources: FDA (1980, 1981; see also USEPA, 1989b, and Gelman Sciences, 1989c; the information presented here is from the summary tables in the FDA interim update memo included on EPA's OPPT TSCA (Office of Pollution Prevention and Toxics, and Toxic Substances Control Act) Web site (USEPA, 1989b).

Notes: ND = not detected. These data, produced 25 years before this publication, have little bearing on the safety of similar products produced today. All of the companies listed above and other companies producing similar products are regulated by the FDA and must comply with an extensive set of health-based protective standards. Technology to avoid formation of 1,4-dioxane and to strip any 1,4-dioxane and ethylene oxide formed during production before packaging has been employed for more than a decade. The information presented here is of interest primarily for the study of potential past modes of 1,4-dioxane introduction to the environment through discharge from septic tanks, leaking sewer lines, and treated wastewater. For this publication, no laboratory test records or validated quality control records for the analyses presented were inspected or verified.

the 1980s and 1990s. A 1979 review of Colgate Palmolive products found levels of 1,4-dioxane up to 0.423 wt%; the permissible upper limit for by-products recommended by the National Institute of Occupational Safety and Health at that time was 1%, or 10,000 ppm (Belanger, 1980). In 1986, an analytical survey of German antidandruff shampoos for 1,4-dioxane found concentrations ranging from 10 to 390 mg/kg; a 1990 follow-on survey found the same German antidandruff shampoos free of 1,4-dioxane, a result that reflected improvements to surfactant production methods (Anonymous, 1987). However, a 1987 study of German personal care products including shampoos and liquid soaps reported 1,4-dioxane present in 22 products at concentrations ranging from less than 50 to 300 mg/kg. The same study found that 1,4-dioxane concentrations in German cosmetics were generally below 100 mg/kg and more commonly below 10 mg/kg. Shower gels, bubble baths, and hair care products containing alkyl ethers commonly contained up to 500 ppm 1,4-dioxane (Rümenapp and Hild, 1987). In 1989, West Germany produced 227,000 metric tons of surfactants in detergents and cleaners, of which 3 metric tons were estimated to be 1,4-dioxane, or a bulk concentration for all surfactant products of about 13 ppm.

A 1990 Danish study determined that 82% of the personal care products analyzed contained from 0.3 to 96 ppm 1,4-dioxane. The same study showed that 85% of dish-washing detergents contained from 1.8 to 65 ppm 1,4-dioxane. Of 76 products tested, four were in the range of 50–100 ppm; 27, from 20 to 50 ppm; 11, from 10 to 20 ppm; 22, from 0.3 to 10 ppm; and 12 did not detect 1,4-dioxane at a reporting limit of 0.3 ppm (Rastogi, 1990). A 1991 review found that 37 of 70 Japanese shampoos analyzed contained 1,4-dioxane greater than the study's 2 ppm detection threshold. The average concentration was 9 ppm; the maximum, 67 ppm (Fox, 1993). A review of shampoos in the U.S. market found 1,4-dioxane present in all the 12 shampoos containing ethoxylated surfactants. The range was from 6 to 144 ppm (Italia and Nunes, 1991). A Polish study in 1998 found that

TABLE 2.10

Detections of 1,4-Dioxane in Taiwanese Surfactants and Soaps

Sample Type	1,4-Dioxane (ppm)
Surfactant: Polyethylene oxide	72
Surfactant: Poly(ethylene/propylene) oxide	12
Surfactant: Polyhydric alcohol	65
Shampoo[a]	12–41 (in 3 of 9 samples; 6 ND)
Liquid soap[b]	8 (in 1 of 9 samples; 8 ND)
Dish-washing detergent[c]	6 (in 1 of 9 samples; 8 ND)

Source: Fuh, C.B., Lai, M., Tsai, H.Y., and Chang, C.M., 2005, *Journal of Chromatography A* 1071(1): 141–145.

ND = not detected.

low-quality ethoxylates were still being used to produce sundries, with 2–7 ppm 1,4-dioxane detected in shampoos, including baby shampoos, 2–8 ppm in body lotions, and 2–10 ppm in dish-washing soaps (Wala-Jerzykiewicz and Szymanowski, 1998).

An analytical survey of ethoxylated raw materials in the U.S. market found 1,4-dioxane at levels as high as 1410 ppm and at levels as high as 279 ppm in personal care products. Children's shampoos contained levels of 1,4-dioxane in excess of 85 ppm (Black et al., 2001). In Taiwan, 1,4-dioxane residue was still present in 2005 in some nonionic surfactants, as well as in shampoos, liquid soaps, and dish-washing detergents (Fuh et al., 2005). Table 2.10 summarizes findings from analysis of surfactants and soaps in Taiwan.

Liquid laundry soaps often include nonylphenol ethoxylate (CASRN 127087-87-0), which may include trace residuals of 1,4-dioxane at levels sufficiently high to require reporting under California Proposition 65 and 40 CFR Section 302.4 rules. 1,4-Dioxane remaining in detergents and personal care products is released through use to publicly owned treatment works (POTWs) to produce a diffuse presence of 1,4-dioxane in wastewater and POTW effluent.

In 2007, the Campaign for Safe Cosmetics released results from a survey of 1,4-dioxane in children's shampoos and bubble baths. More than two dozen baby and children's consumer products were analyzed by a commercial laboratory by isotope-dilution, headspace/GCMS and a reporting limit of 0.2 ppm. Four of the products tested exceeded 10 ppm, the FDA recommendation for the maximum 1,4-dioxane content in cosmetic products. The median-detected 1,4-dioxane value (excluding eight samples with no detection) was 5 ppm; the average was 7 ppm, and the standard deviation was 7 ppm. See Table 6.17 for a list of product testing results.

2.4.1.2 Personal Care Products: Cosmetics

The *Cosmetic Handbook* (FDA, 1992) advises that cosmetics containing ethoxylated surfactant agents that may be contaminated with 1,4-dioxane—such as detergents, foaming agents, and emulsifiers—are identifiable by the words, acronyms, or suffixes "PEG," "polyethylene," "polyethylene glycol (PEG)," "polyoxyethylene," "–eth–," or "–oxynol–." A survey of cosmetic products in 1979 found that 31 of 65 products analyzed contained greater than 10 ppm 1,4-dioxane (Fishbein, 1981). An analytical survey of Belgian cosmetic products found 1,4-dioxane concentrations from less than 2 to 613 mg/kg (Beernaert et al., 1987). A survey of Italian cosmetic products found that 48% contained from 7.3 to 85.9 ppm 1,4-dioxane (Scalia and Menegatti, 1991; Scalia et al., 1992). In 1985, the FDA instituted a formal policy that cosmetic products should not contain 1,4-dioxane at concentrations greater than 10 ppm (mg/kg) (ATSDR, 2004); the same standard is currently employed by the German Cosmetic, Toiletry, Perfumery and Detergent Association (IKW) (Fruijtier-Pölloth, 2005).

TABLE 2.11
FDA Analytical Survey Results for 1,4-Dioxane in U.S. Ethoxylated Raw Materials and Finished Cosmetics

	1,4-Dioxane Concentration (ppm)			
	Raw Ethoxylated Surfactants/Alkyl Sulfate Surfactants		Finished Cosmetic Products	
Year	Average/Number of Analyses	Range	Average	Range
1979	49/229	71–580	—	—
1980	280/226	6–1410	—	—
1981	—	—	50	2–279
1982	—	—	19	2–36
1983	—	—	2	1–8
1991	—	—	—	3–108
1992	—	—	41	5–141
1993	71/80	16–243	79	50–112
1994	—	—	45	20–107
1995	—	—	74	42–90
1996	180/188	20–653	14	6–34
1997	348/NA	45–1102	19	6–34

Sources: Food and Drug Administration (FDA), 1981, Progress report on the analysis of cosmetics raw materials and finished cosmetics products for 1,4-dioxane. Division of Cosmetics Technology, Food and Drug Administration 15; Italia, M.P. and Nunes, M.A., 1991, *Journal of the Society of Cosmetics Chemistry* 42: 97–103; Black, R.E., Hurley, F.J., and Havery, D.C., 2001, *Journal of AOAC [Association of Official Analytical Chemists] International* 84(3): 666–670; and ATSDR, 2004, Draft toxicological profile for 1,4-dioxane. Division of Health Assessment and Consultation, U.S. Department of Health and Human Services.

NA = not available.

The FDA conducted periodic surveys of 1,4-dioxane levels in ethoxylated raw materials; results are summarized in Table 2.11.

Upon learning that 1,4-dioxane was a contaminant of nonionic surfactants, the U.S. Cosmetic, Toiletry and Fragrance Association commissioned a pharmacology and toxicology committee review of 1,4-dioxane contamination in cosmetics and concluded that the levels found are below a "virtually safe dose." The association nevertheless conducted a survey of suppliers of ethoxylated products to obtain data and information on specifications and processing. The goal was to work toward achieving acceptable levels and controls for 1,4-dioxane and to establish a standard for 1,4-dioxane-containing ingredients (Jass, 1989).

AES compounds and related surfactants have many varieties and many applications beyond detergents and personal care products. Secondary alcohol ethoxylate is used in crop protection agents, metal working and processing, textile processing, paints and coating compositions, and as a conditioning agent for pulp and paper processing (Kosswig, 2002). 1,4-Dioxane residuals may be present in ethoxylated C_{12}–C_{14} secondary alcohols, according to available MSDS. 1,4-Dioxane is also reportedly used in the dissolution of polymers to produce a coating for laminated paper products (USEPA, 1996c). Soil surfactants use ethoxylated surfactant compounds to improve irrigation of golf courses, athletic fields, and other turf and agricultural applications. MSDS listings for some varieties of soil surfactants note the trace presence of 1,4-dioxane.

Manufacturers of ethoxylated surfactants were challenged to both optimize surfactant production and minimize formation of 1,4-dioxane. The amount of 1,4-dioxane formed in AES compounds is a function of the mole ratio of sulfur trioxide fed to the reactor to the ethoxylated alcohol stock (e.g., alcohol) in the reactor. 1,4-Dioxane forms when two molecules of ethylene oxide are cleaved from the parent ethoxylated alcohol. 1,4-Dioxane formation is favored by an excess of sulfur trioxide, high temperatures, and moisture and longer ethylene oxide chains in the ethoxylated alcohol feedstock (Kosswig, 2002; Dado et al., 2006). The level of 1,4-dioxane in the product remains relatively low at 20–30 ppm until a critical point of oversulfation occurs. Once oversulfation occurs and the mole ratio exceeds 1.04, 1,4-dioxane production increases rapidly to values measured in hundreds of parts per million (Foster, 1997).

Vacuum stripping, steam stripping, drying, and other solutions were eventually successful in achieving substantial reductions in the 1,4-dioxane content of finished surfactant products (Sachdeva and Gabriel, 1997). The recovered 1,4-dioxane is often condensed and used as a fuel for on-site boilers or in other energy-recovery operations. Equipment specifically designed for removal of 1,4-dioxane in the production of lauryl ether sulfate achieves an eightfold decrease. Additional equipment is designed to subsequently destroy 1,4-dioxane in the condensate. Destruction methods include a catalyzed reaction with hydrogen peroxide to oxidize 1,4-dioxane, producing carbon dioxide and water (Chemithon Corporation, 2006).

2.4.1.3   1,4-Dioxane in Contraceptive Sponges and Spermicidal Lubricants

The active spermicide nonoxynol-9 has been commonly used in contraceptive products since the 1970s. Nonoxynol-9 is among the surfactants that may contain 1,4-dioxane as an impurity of production. The levels of 1,4-dioxane present were deemed to be less than the 10 ppm limit recommended for polysorbates in food and applied to contraceptives by the FDA (FDA, 1997a). Protracted debate over the safety of a contraceptive sponge in the 1980s and 1990s focused on the presence of 1,4-dioxane in nonoxynol-9 in the "Today" sponge. The FDA determined that there is no appreciable risk because of lack of vaginal absorption of nonoxynol-9 (Woodcock, 1997). Approximately one-sixth of the nonoxynol-9 contained in the sponge is released during use. The process used to create the contraceptive sponge is thought to drive off 1,4-dioxane, reducing its concentration sevenfold (FDA, 1997b). The level of 1,4-dioxane present in nonoxynol-9, 7 ppm, would probably not appear in the sponge (Medical Economics Publishing, 1983). Nonoxynol-9 is also used as a spermicidal lubricant in condoms and related products.

2.4.1.4   1,4-Dioxane in Polyethylene Glycol

PEG compounds are used in a wide range of products (Union Carbide, 2001):

- Pharmaceuticals
- Cosmetics such as deodorant sticks, lipsticks, shaving creams, toothpastes, and lotions
- Detergents in laundry soaps and dish-washing soaps
- Textiles and leather processing
- Plastics and resins
- Paper
- Printing inks
- Lubricants
- Mold-release agents in the rubber industry
- Metal corrosion inhibitors in the petroleum industry
- Anticracking and preservation in woodworking
- Brake fluid lubricants

Pharmaceutical grades of PEG typically contain less than 10 ppm 1,4-dioxane.

Dow Chemical tested the 1,4-dioxane content of various grades of PEG. Some grades showed no detection; all grades tested less than 1 part 1,4-dioxane per million parts PEG (Dow Chemical Corporation, 1989). In 1988, Union Carbide notified its customers that its Carbowax line of PEG products contained no more than 5 ppm 1,4-dioxane and was therefore in compliance with

California's Safe Drinking Water and Toxics Enforcement Act of 1986 (Proposition 65) and Section 313 of Title III of the Superfund Amendments and Reauthorization Act of 1986 (SARA Title III) (Union Carbide, 1988).

The PEG compounds of interest in the context of personal care products include ethers of propylene glycol, propylene glycol stearate, propylene glycol oleate, and propylene glycol cocoate. These compounds serve as cleansing, solubilizing, and emulsifying agents, skin-conditioning agents such as humectants and emollients, and solvents in cosmetic formulations. Only a minor percentage of cosmetic formulations use these compounds; some are not currently used. PEG, propylene glycol cocoate, and propylene glycol oleate are produced by the esterification of polyoxyalkyl alcohols with lauric acid and oleic acid, respectively. Impurities include ethylene oxide (maximum 1 ppm), 1,4-dioxane (maximum 5 ppm), polycyclic aromatic compounds (maximum 1 ppm), and the heavy metals lead, iron, cobalt, nickel, cadmium, and arsenic (maximum 10 ppm combined) (Johnson, 2001).

2.4.1.5 1,4-Dioxane in Glyphosphate Herbicides (Accord®, Roundup®, Rodeo®, Vision®)

1,4-Dioxane may also be present as an impurity of polyoxyethyleneamine (POEA), the salt of glyphosphate [N-(phosphonomethyl)glycine], which is the major component in widely used herbicide formulations such as Roundup®. The United Nations Food and Agriculture Organization has established a limit of 1 ppm for 1,4-dioxane in POEA. The USEPA considers 1,4-dioxane as an "inert of toxicological concern" in pesticide and herbicide formulations. The most popular herbicide, Roundup®, contained less than 0.03% 1,4-dioxane in 1990 (<300 mg 1,4-dioxane per kilogram of Roundup®) (Diamond and Durkin, 1997). Monsanto, the producer of Roundup®, reported in 1997 that 1,4-dioxane contamination had been further reduced to 23 ppm (U.S. Forest Service, 1997). Although 1,4-dioxane content was probably too low to cause groundwater contamination from application to fields, locations where glyphosphate-based herbicide was loaded into or washed from spray equipment could be sources of low-level 1,4-dioxane contamination (Pesticide Action Network, 2006). The practice of rinsing herbicide containers prior to recycling may also cause detectable 1,4-dioxane contamination in groundwater.

Some urea herbicides could potentially contain 1,4-dioxane as a residue of production. 1,4-Dioxane is reacted with 4-chloroaniline, anhydrous hydrogen chloride, and phosgene at 70–75°C to produce p-chlorophenyl isocyanate, an intermediate used for the production of urea herbicides (Roig et al., 2003). Another herbicide used in plantings of tomatoes and other crops, rimsulfuron, could contain 1,4-dioxane as an impurity of production. Technical rimsulfuron, also called Shadeout, is manufactured in a two-step batch process in which a carbamic acid and an aniline compound are reacted in a 1,4-dioxane solution (California Environmental Protection Agency, 1997). Rimsulfuron is 1-(4,6-dimethoxypyrimidin-2-yl)-3-(3-ethylsulfonyl-2-pyridylsulfonyl)urea (CASRN 122931-48-0; formula: $C_{14}H_{17}N_5O_7S_2$). Agricultural products that contain ethoxylated surfactants may also contain traces of 1,4-dioxane. 1,4-Dioxane is listed as present but below reportable quantities in the insecticide Dursban, which includes surfactants among its components (Dow Chemical Corporation, 1985).

2.4.1.6 1,4-Dioxane in Pesticides

Some insecticidal formulations claimed in patents have used a solvent mixture comprising 30–50% of the formulation. An example solvent mixture claimed in the patent literature is 20% 1,4-dioxane along with methyl chloroform, methanol, and ethanol (Chang, 1999). It is unlikely that 1,4-dioxane was widely used in commercial production as a carrier in insecticides in such high percentages because of its high cost. 1,4-Dioxane was present in some pesticide formulations as a stabilizer of methyl chloroform, which has been documented as an "inert" ingredient in pesticides. In 1989, USEPA's Office of Pesticide Programs issued a notice concerning inert ingredients of pesticide products, in which 1,4-dioxane was included in the list of "inerts of toxicological concern." Methyl chloroform is also a carrier solvent for pesticides; it was included on a list of "potentially toxic inerts"

(USEPA, 1989a). Trace amounts of 1,4-dioxane are often found in pesticides because of the inclusion of ethoxylated surfactants, which include 1,4-dioxane as an impurity (see Section 2.4.1). For example, a solid mosquito larvicide designed to float atop ponds and release its active biocidal ingredient, *Bacillus thuringiensis*, lists 1,4-dioxane as present at less than 20 ppm. Other formulations list 1,4-dioxane present at less than 10 ppm as a by-product with ethoxylated alcohols (Cognis Corporation, 2001). Some agency risk assessments and other reviews that list 1,4-dioxane use as a pesticide may confuse 1,4-dioxane derivatives for 1,4-dioxane. For example, dioxathion [CASRN 78-34-2, is *O,O,O',O'*-tetraethyl *S,S'*-(1,4-dioxane-2,3-diyl) diphosphorodithioate], a pesticide for deciduous fruit, is a derivative of 1,4-dioxane that could be mistaken for 1,4-dioxane used as a pesticide.

1,4-Dioxane was used together with benzene to extract and distill the pesticide hexachlorocyclohexane (also called benzene hexachloride, BHC, Lindane), which was widely used as a contact pesticide for head lice, as a wood preservative, and for other purposes. The solvent mixture was recovered and reused; however, losses to the waste by-products and product were routinely replaced with new solvent. Waste solvents were thermally destroyed in well-controlled production operations; however, there have been examples of extraordinary volumes of waste BHC isomers littering production sites in Europe (Vijgen, 2006). Disposal of BHC wastes could also be a source of 1,4-dioxane releases to soil and groundwater.

2.4.2 Terephthalate Esters (Polyester) and Resins

Polyethylene terephthalate (PET) polyester was first introduced in 1952 (USEPA, 1997b). Polyester fibers are produced from the polycondensation reaction of a dicarboxylic acid, such as terephthalic acid, and a dihydroxy alcohol, such as ethylene glycol. 1,4-Dioxane forms as a by-product during esterification (Ellis and Thomas, 1998). To manufacture polyester fiber, terephthalic acid and ethylene glycol are first passed through primary and secondary esterifiers to form the monomer. The melt is then passed to a polymerizer equipped with a high vacuum to allow excess ethylene glycol to escape. Wastes generated during polymerization may include emissions of volatile organic compounds (VOCs) from leaks, spills, and vents, solid wastes from off-specification polymers, and spent solvent from incomplete polymerization (USEPA, 1997b).

Polymers of PET always contain a certain amount of incorporated diethylene glycol (DEG). DEG is formed in a side reaction during the ester interchange of dimethyl terephthalate (DMT) with ethylene glycol. Alternatively, during direct esterification of terephthalic acid with ethylene glycol, DEG is formed via an unusual type of reaction: ester + alcohol ether + acid = DEG + 1,4-dioxane + methyl cellosolve + methyl carbitol (Hovenkamp and Munting, 1970).

Factories producing PET are generally equipped with wastewater-treatment plants to remove waste chemicals; however, 1,4-dioxane has proved difficult to remove. Treated effluent from an industrial wastewater-treatment plant at a PET manufacturing facility contained 100 mg/L of 1,4-dioxane in 1995 (European Chemicals Bureau, 2002). Nearly 50% of effluent samples from a PET resin plant in Spain had 1,4-dioxane concentrations exceeding 1000 μg/L; the remainder showed 1,4-dioxane present from 100 to 1000 μg/L. The mean concentration of 35 samples was 6400 μg/L; the range was 100 to 31,400 μg/L (Romero et al., 1998). Waste streams from polyester production plants in the eastern United States routinely contained as much as 2000 mg of 1,4-dioxane per liter (Grady et al., 1997). Air emissions created during polyester production are estimated to include 1 mg of 1,4-dioxane per kilogram of polyester fiber produced (Laursen et al., 1997).

The equipment used to produce polyester must be cleaned periodically. The cleaning process consists of dipping parts in triethylene glycol (TEG) to remove accumulated polyester and by-products. Spent TEG is recovered and transported off-site for recycling. Parts are then rinsed with water to remove residual TEG and polyester ingredients. This rinse water represents the primary wastewater stream at a typical polyester production facility and may total as much as 2 million gallons annually. Rinse-water constituents typically include TEG, 1,4-dioxane as a by-product of heating TEG, and polyester by-products such as methanol. Rinse waters may be processed in

TABLE 2.12
USEPA 1995 Toxic Release Inventory Data for 1,4-Dioxane from EPA Sector Notebook

Fugitive Air	Point Air	Water Discharges	Total Releases	Average Releases per Facility
3810	1763	17,246	22,841	4568
Disposal transfers	Treatment transfers	Energy recovery transfers	Total transfers	Average transfer per facility
271	12,655	11,990	24,916	4983

Source: U.S. Environmental Protection Agency (USEPA), 1995b, Profile of the rubber and plastics industry. Office of Enforcement and Compliance Assurance, USEPA.

Note: Values in pounds per year for five plastic-resin manufacturing facilities.

wastewater treatment plants or shipped by rail for off-site treatment (North Carolina Department of Environment and Natural Resources, 1998).

In 1992, chemical manufacturing operations produced 2.56 billion kg of terephthalic acid (2.82 million tons). A summary of 1,4-dioxane emissions, releases, and off-site transfers for five plastic-resin manufacturing facilities in 1995 is given in Table 2.12.

Many photographic films are produced from DMT to create a polyester film base. Polyester waste material such as old x-rays and used movie film is commonly recycled to make DMT available for reuse. The processing of polyester waste produces 1,4-dioxane as a by-product, which has been discharged from operations as wastewater. For example, in 1994, 140,000 pounds of 1,4-dioxane were discharged to the industrial sewer in Rochester, New York, from Kodak, the leading U.S. film products company. Because the industrial wastewater-treatment plant was inefficient for 1,4-dioxane removal, plant engineers have since devised and deployed several innovative means to separate and collect 1,4-dioxane for on-site energy recovery so that the chemical can be used as a fuel for plant boilers. The plant successfully eliminated the discharge of 1,4-dioxane (Eastman Kodak Corporation, 2001).

2.4.3   1,4-DIOXANE AS A BY-PRODUCT OF PETROLEUM REFINING AND GASOLINE ANTIKNOCK AGENT

The literature contains reference to the use and formation of 1,4-dioxane during petroleum refining; however, the information that exists is insufficient to suggest that 1,4-dioxane was present in gasoline at levels that can be detected or cause environmental impacts. The main impurities in refined benzene are nonaromatics with nine or fewer carbon atoms, toluene, 1,4-dioxane, and aromatics containing eight carbon atoms (Wypych, 2001). 1,4-Dioxane may also occur as an impurity of toluene. Some petroleum refining methods for isolating and purifying toluene have reportedly used oxygenated reagents. Analytical methods have been specifically developed to address the possible residual presence of 1,4-dioxane in toluene (Cortellucci and Dietz, 1999).

Universal Oil Products Company of Chicago obtained a patent in 1938 for a petroleum refining process to produce 1,4-dioxane from the ethylene by-product of petroleum refining and then to use the 1,4-dioxane to serve as an antiknock agent in gasoline. Ethylene is first converted to ethylene glycol by adding hydrogen peroxide; the ethylene glycol is then converted to 1,4-dioxane by using concentrated sulfuric acid. 1,4-Dioxane acts as a polymerizing agent in the refining process, and excess 1,4-dioxane remaining in the gasoline acts as an antioxidant to preserve the antiknock rating in gasoline stored for 6 months or more (McCaffrey, 1938). 1,4-Dioxane was thus apparently considered a viable oxygenate for gasoline at levels up to 0.05 wt%, although the extent to which this invention was put into practice has not been determined.

2.5   1,4-DIOXANE IN FOOD

The first systematic study designed to determine the levels of 1,4-dioxane in food was conducted in 2003 by a team of Japanese scientists led by Tetsuji Nishimura of Japan Food Research Laboratories in Tokyo (Nishimura et al., 2004). A number of risk assessments and drinking water fact sheets have cited earlier literature to assert that 1,4-dioxane is commonly found in fried chicken, tomatoes, shrimp, and other foods. However, none of the studies cited actually quantified the amount of 1,4-dioxane present, and the studies were all performed in the 1970s and 1980s, before analytical methodologies for 1,4-dioxane detection were improved (ATSDR, 2004). Table 2.13 summarizes early citations of 1,4-dioxane in various food products at unquantified levels.

In Nishimura's landmark study, a market basket approach was taken to analyze groups of food by blending them together to assay 1,4-dioxane content. Food was prepared in the same manner in which consumers cook food for routine consumption, using purified water and clean utensils. Raw and cooked foods in 12 dietary groups were blended together with other foods from the same group and then analyzed by using active carbon solid-phase extraction cartridges (described further in Chapter 4), coupled with GC and mass spectrometry. 1,4-Dioxane was found in all foods except rice and rice noodles; detections were in the low part per billion range. Corresponding dietary intake was determined to be in the nanogram-per-day range, far below the applicable risk thresholds. The study does not specifically explain the cause of the nearly ubiquitous presence of 1,4-dioxane in food; however, the widespread presence of 1,4-dioxane from industrial discharges to river water and groundwater used for irrigation and food processing are suggested as possible reasons. Table 2.14 summarizes the findings from the analysis of 1,4-dioxane in different food groups.

Food additives have been regulated in the United States and Europe for their 1,4-dioxane content. The U.S. Food Chemicals Codex restricts residual 1,4-dioxane content in polysorbates to 10 mg/kg. Polysorbate 60 [polyoxyethylene (20) sorbitan monostearate] is an emulsifier used in ice cream, frozen custard, ice milk, fruit sherbet, and other frozen desserts. Polysorbate 60 is used alone or in combination with polysorbate 65 and/or polysorbate 80. Polysorbates have been found to contain trace levels of 1,4-dioxane and ethylene oxide as impurities, which result from the production processes used for polysorbates (FDA, 1999). Polysorbate 60 and polysorbate 80, which are produced from the polymerization of polyoxyethylene, have historically been found to contain 1,4-dioxane (Birkel et al., 1979). Levels of 1,4-dioxane in these compounds have been reported to range from 5 to 6 ppm (mg/L) (ATSDR, 2004).

1,4-Dioxane was detected in Tween 20 and Tween 80 (polysorbate 20 and polysorbate 60) at approximately 100 and 200 ppm, respectively (Guo and Brodowsky, 2000). Table 2.15 summarizes the varieties of polyoxyethylene-based emulsifiers in which 1,4-dioxane may be present in trace amounts.

Eating ice cream and other frozen desserts that use polysorbates as an emulsifier could cause a maximum exposure to 1,4-dioxane of 19 ng per person per day (FDA, 1999). For a 70 kg person,

---

TABLE 2.13
Studies Reporting the Identification of 1,4-Dioxane in Food Products at Unquantified Levels Prior to Japanese Study by Nishimura et al. (2004)

Food Product	Reference
Chicken flavor and meat volatiles	Shahidi et al. (1986)
Fried chicken volatile flavor compounds	Tang et al. (1983)
Tomato fruit juice volatiles	Chung et al. (1983)
Fat oil compound (tillnoloin) formed in deep frying	Chang et al. (1978)
Cooked shrimp volatile flavor compounds	Choi et al. (1983)
Patis (Philippine fermented fish sauce)	Sanceda et al. (1984)
Brazilian coffee	Spadone et al. (1990)

TABLE 2.14
1,4-Dioxane in Japanese Dietary Food Groups

Dietary Group	Foods	1,4-Dioxane Content (μg/kg)	Total Dietary Intake (μg)
I	Rice, rice noodles	ND	0
II	Oatmeal, wheat flour, bread, buckwheat, cornflakes, sweet potato, almonds, instant noodles, potato, taro root	6	0.057
III	Sugar, jam, caramel, rice cracker, pound cake, chocolate, others	6	0.010
IV	Butter, margarine, soybean oil, lard, dressing	8	0.008
V	Soybean paste, tofu, frozen tofu, deep fried tofu	3	0.006
VI	Orange, apple, banana, strawberry, tomato juice, watermelon, apricot, loquat	4	0.0026
VII	Carrot, spinach, green pepper, tomato, broccoli, celery, okra	3	0.015
VIII	Daikon radish, cucumber, onion, cabbage, bean sprouts, pickles, pickled relish, mushroom, green seaweed	8	0.076
IX	Soy sauce, ketchup, salt, sauce, sake, beer, wine, tea, soft drinks	7	0.070
X	Salmon, tuna, mackerel, flatfish, scallop, salted cod, bonito, sardine	5	0.022
XI	Beef, pork, chicken, lamb, ham, chicken, eggs	6	0.055
XII	Milk, cheese, yoghurt, sake sediment	13	0.095

Source: Nishimura, T., Iizuka, S., Kibune, N. et al., 2004, *Journal of Health Science* 50(1): 101–107.
ND = not detected

this daily value equates to 2.7×10^{-7} mg/kg, which is about 40,000 times *less* than the suggested Integrated Risk Information System daily oral slope factor for estimated carcinogenic risk, that is, 1.1×10^{-2} mg/kg. The FDA concluded that there is reasonable certainty that no harm from exposure to 1,4-dioxane would result from the daily consumption of ice cream (FDA, 1999).

Glycerides and polyglycides of hydrogenated vegetable oils, also known as PEG esters, may also contain 1,4-dioxane as an impurity. Glycerides include mixtures of mono-, di-, and triglycerides, whereas polyglycides include PEG mono- and diesters of fatty acids. The Code of Federal Regulations Title 21, Food and Drugs, sets limits for allowable 1,4-dioxane in food and excipients of dietary supplements at no greater than 10 mg/kg (FDA, 2006). Consumption of glycerides and polyglycides in dietary supplement tablets, capsules, and liquid formulations produces an exposure

TABLE 2.15
Emulsifiers in Which 1,4-Dioxane May Form at Trace Levels

Emulsifier Name	Chemical Description	Uses
Polysorbate 20	Polyoxyethylene sorbitan monolaurate	Emulsifier
Polysorbate 80	Polyoxyethylene sorbitan mono-oleate	Emulsifier, flavoring, surfactant, defoaming agent
Polysorbate 40 (BAN)	Polyoxyethylene sorbitan monopalmitate	Emulsifier (from animal fatty acids)
Polysorbate 60	Polyoxyethylene sorbitan monostearate	Emulsifier
Polysorbate 65	Polyoxyethylene sorbitan tristearate	Emulsifier

Source: Food and Drug Administration (FDA), 1999, *Federal Register* 64: 57974–57976.

to 1,4-dioxane that equates to not more than 800 ng per person per day (FDA, 2006). This exposure level is about 1000 times lower than the suggested oral slope factor for cancer risk for 1,4-dioxane.

2.6　1,4-DIOXANE USE IN THE PHARMACEUTICAL INDUSTRY

The pharmaceutical industry fabricates, processes, and formulates medicinal chemicals and pharmaceutical products and grinds, grades, and mills botanical products. Pharmaceutical products may include agents of natural origin, hormonal products, basic vitamins, and isolated alkaloids from botanical drugs and herbs (USEPA, 1997c). The majority of pharmaceutical production facilities in the United States are located in California, New Jersey, New York, and Puerto Rico. Formulated pharmaceutical products are manufactured by chemical synthesis, fermentation, or isolation and recovery from natural sources, or a combination of these processes. Fermentation is economically the most important process; however, by tons produced, chemical synthesis dominates. Fermentation processes are used to produce antibiotics such as penicillin and tetracycline. Chemical synthesis processes are used to manufacture psychotropic and antihistamine drugs. Extracts from animal products provide hormone-based drugs, biological processes are used to produce vaccines and serums, and vegetable extraction processes yield steroids and alkaloids (Chenier, 2002). Production of some pharmaceutical products incorporates the use of various solvents, including 1,4-dioxane.

Solvents are used primarily as a vehicle or reagent and to clean residues from equipment between production runs of different drugs. Solvent wastes, primarily as vapors, may be generated at most steps in the production of pharmaceuticals, including

- Cleaning reaction vessels and condensers
- Separation by extracting, decanting, centrifugation, filtration, or crystallization
- Purification by recrystallization, centrifugation, or filtration
- Product drying

Solvent-laden wastewater may be generated in the reaction, separation, extraction, and purification steps. Devices to control air emissions and equipment for wastewater treatment are designed to remove residual solvent and comply with air quality and discharge elimination regulations (USEPA, 1997c).

Many bulk pharmaceutical reactions require organic solvents to dissolve chemical intermediates and reagents. To prevent reaction with reagents and intermediates, nonreactive solvents are preferred. Dichloromethane is often the optimum choice for pharmaceutical reactions. The most commonly used solvents in the pharmaceutical industry include methanol, ethanol, acetone, isopropanol, and other oxygenated organic solvents (USEPA, 1997c). 1,4-Dioxane is used in some processes but has not been a major solvent used in production pharmaceuticals. In addition to 1,4-dioxane, other compounds used in the pharmaceutical industry that also serve as solvent stabilizers include acetonitrile, *n*-butyl alcohols, cyclohexane, pyridine, and tetrahydrofuran. Specific uses for 1,4-dioxane include chemical synthesis and biological and natural product extraction (USEPA, 1997c). 1,4-Dioxane, toluene, xylene, and pyridine are routinely monitored by the pharmaceutical industry (Kauffman, 2003).

In 1995, TRI participation involved 200 out of the 916 pharmaceutical production facilities listed in the 1992 census of manufacturers. Of those 200 facilities, two reported releases of 1,4-dioxane (USEPA, 1997c). This low percentage may be because very few facilities used 1,4-dioxane, but it is more likely the result of a tendency for TRI responses to exclude chemicals used in only minor quantities. The two facilities that did report 1,4-dioxane releases in 1995 summarized the quantities released, in pounds per year, as follows:

Fugitive Air Emissions	Point Air Emissions	Water Discharges	Total Releases	Average Releases per Facility
270	260	0	530	265

The same two facilities reported larger quantities transferred for treatment at wastewater facilities (4170 lb) and energy-recovery facilities such as cement kilns (47,916 lb) (USEPA, 1997c). Following the Pollution Prevention Act of 1990, the pharmaceutical industry engaged in a large effort to minimize solvent use and emissions; the industry developed alternative, solvent-free methods for producing pharmaceuticals. In earlier decades (i.e., the 1980s and earlier), pharmaceutical facilities *potentially* released 1,4-dioxane from on-site handling of wastewater, burial of solvent-laden sludge from the solid residues of botanical extraction processes (or landfills that received such sludge), and energy-recovery operations that handled solvents from pharmaceutical facilities.

An indirect means of establishing the use of 1,4-dioxane in the pharmaceutical industry is found in the literature for analytical methods developed to detect organic volatile impurities (OVIs) in finished drug products. U.S. Pharmacopeia (USP) analytical method 467 for analysis of OVIs also specifies the maximum permissible levels of five solvents of interest, including methylene chloride, benzene, trichloroethylene, 1,4-dioxane, and chloroform. Literature from the commercial laboratory sector, the analytical instrument sector, peer-reviewed journals on laboratory analysis, and the pharmaceutical industry literature all cite methods for improving the detection of 1,4-dioxane as an OVI in finished pharmaceutical products to ensure compliance with U.S. Food and Drug restrictions on residual solvents. For example, one gas chromatograph manufacturer's technical literature notes that analysis of ascorbic acid (vitamin C) and acetaminophen (a common pain killer) showed all solvents well below the regulated levels. Vitamin C had benzene at 35 ppb, and acetaminophen had dichloromethane at 407 ppb; neither product had 1,4-dioxane above a method detection limit of just less than 1 ppm (Rankin, 1996).

In the United States and Europe, the permitted daily exposure to 1,4-dioxane through consumption of pharmaceuticals is 3.8 mg per person. The permissible concentration limit in the finished drug product is 380 ppm (FDA, 1997a; European Medicines Agency, 1998). The same levels are considered permissible in U.S. veterinary medicinal products (FDA, 2001).

2.7   DETECTIONS OF 1,4-DIOXANE IN AMBIENT SURFACE WATER, GROUNDWATER, AND AIR

There has not been a great deal of analysis for 1,4-dioxane in ambient environmental monitoring samples. The most comprehensive monitoring program for ambient water quality in the United States, the U.S. Geological Survey's National Ambient Water-Quality Assessment Program (NAWQA), excluded 1,4-dioxane from its analyte list because of analytical challenges for quantifying 1,4-dioxane by using the purge-and-trap GC methods preferred for the majority of the organic compounds sought by the program. California's Groundwater Ambient Monitoring and Assessment Program (GAMA) targeted 1,4-dioxane in a study of groundwater in the San Diego, California, area. Groundwater samples were analyzed from 24 wells; at a reporting limit of 2 ppb, no 1,4-dioxane was detected in any of the samples (Wright et al., 2004). Lawrence Livermore National Laboratory conducted a similar GAMA study, using a reporting limit of 0.2 ppb, for groundwater in the Sacramento, California, area; samples from 108 wells were analyzed for 1,4-dioxane. Three samples had detections below 1 ppb (Moran et al., 2003). Another aquifer-vulnerability assessment tested 60 municipal wells in Santa Clara County, California ("Silicon Valley"), where widespread solvent contamination of shallow aquifers had occurred. 1,4-Dioxane was included in the list of analyses performed on the municipal well samples; however, none of the samples had detections using a reporting limit of 0.15 ppb (J.E. Moran, personal communication, 2003).

Levels of 1,4-dioxane between 0.2 and 1.5 μg/L were detected in tap water samples collected during 1995 and 1996 in Kanagawa Prefecture, Japan, an 870-square-mile province adjacent to and southwest of Tokyo (Abe, 1997). An additional comprehensive survey of the Kanagawa Prefecture included multiple sampling events in three rivers and 20 wells and found 1,4-dioxane to be ubiquitous at low levels; a few locations had relatively elevated levels. 1,4-Dioxane was detected in 90% of the wells, with two-thirds of the detections falling below 1 μg/L, 20% between 1 and 10 μg/L, and 10%

above 10 μg/L; the highest detections ranged from 50 to 95 ppb (Abe, 1999). All of the seven sampling stations in the three rivers had 1,4-dioxane detections in each of five sampling events. Three river sampling stations had all five analyses falling in the range 0.1–0.99 ppb; three fell in the range 0.2–3.3 ppb, and one river sampling station fell in the range 2–16 ppb (Abe, 1999). Sources of 1,4-dioxane were attributed to legacy contamination by methyl chloroform and industrial and domestic wastewater discharges.

1,4-Dioxane has been detected in ambient air monitoring samples in a few isolated studies. USEPA compiled 1,4-dioxane measurements of ambient air performed at 45 locations in 12 cities taken between 1979 and 1984. The results indicated concentrations ranging from below detection limits to 30 μg/m^3. The mean ambient concentration of 1,4-dioxane was 0.44 μg/m^3 (0.12 ppbv) (USEPA, 1993). Air samples in industrial areas in Newark, Elisabeth, and Camden, New Jersey, were collected from July 6 to August 16, 1981. The geometric mean 1,4-dioxane concentrations ranged from 0.01 to 0.02 ppbv (0.04 to 0.07 μg/m^3). 1,4-Dioxane was detected in 51% of the samples (Harkov et al., 1981).

Washington State's Northwest Air Pollution Authority (NWAPA) performed a study of ambient concentrations of hazardous air pollutants at five different sampling locations in Bellingham, Washington (north of Seattle), on 107 sampling days between July 1995 and August 1999. Data quality in this study was compromised: detection of 1,4-dioxane in control blanks left some of the results unusable. 1,4-Dioxane was detected above reporting limits at all five sites. The geometric mean of all samples at each site, including suspect values, was about 5 μg/m^3 at two sites and about 10 μg/m^3 at a third site. No known major emission sources are present in the area (Keel and Franzmann, 2000).

In the 1980s, USEPA compiled a National Ambient Air Database of indoor and outdoor measurements of 320 VOCs. The database includes 585 analyses for 1,4-dioxane, for which the average concentration was 1.03 ppbv; 75% of measured values were less than 0.09 ppbv, and more than half showed no detection (Shah and Singh, 1988).

EXTRATERRESTRIAL FORMATION OF 1,4-DIOXANE?

1,4-Dioxane may be among the organic compounds found in trapped volatiles in carbonaceous chondrites from meteorites formed in the solar nebula (Studier et al., 1965). A NASA study investigated the production of aliphatic and aromatic hydrocarbons from reactions between carbon monoxide and hydrogen, the principal components of cosmic gas, catalyzed by iron and stony meteorites at temperatures between 25°C and 580°C. Experiments involved exposing Canyon Diablo powder (a ground iron meteorite) to a 1:1 mixture of carbon monoxide and hydrogen. After 4 h and 20 min at 115°C, mass spectral peaks at 88 and 58 m/z were observed, which indicated formation of 1,4-dioxane (Studier et al., 1965).

2.8 ESTIMATED RELEASES OF 1,4-DIOXANE TO THE ENVIRONMENT

The amount of 1,4-dioxane released to the environment since 1988 can be estimated by tracking the quantity of 1,4-dioxane reported in USEPA's TRI database. 1,4-Dioxane is reportable in waste streams at concentrations above 0.1% because it is listed by the Occupational Safety and Health Administration as a probable carcinogen, based on reports from the National Toxicology Program and the International Agency for Research on Cancer. For each year, the TRI database presents data compiled from hundreds of thousands of reports from hazardous waste generators; the data are categorized into a variety of sectors of treatment and disposal. It is not surprising that the data available from USEPA's TRI database, particularly in the first few years of TRI data collection and reporting, are sometimes imprecise, given the monumental scope of the task. The TRI data sets are nonetheless useful for providing the approximate quantities, locations, and final disposition of 1,4-dioxane releases.

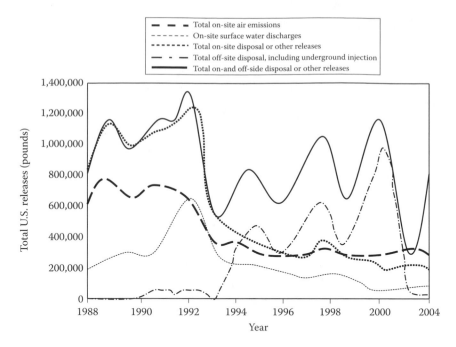

FIGURE 2.5 Estimated releases of 1,4-dioxane from 1988 through 2004. [U.S. Environmental Protection Agency (USEPA), 2000, Preliminary data summary: Airport deicing operations. Office of Water, USEPA.]

Because 90% of all 1,4-dioxane produced in the United States was reportedly used as a stabilizer for methyl chloroform, tracking chemical market reports for methyl chloroform provides a means of corroborating TRI data for 1,4-dioxane. Prior to 1988, 1,4-dioxane production, consumption, and releases to the environment can only be inferred from chemical market reports. The release of 1,4-dioxane as a by-product of manufacture can be estimated from total production in the plastics, synthetic textiles, and ethoxylated surfactants industries. The trends apparent from TRI data for 1,4-dioxane releases to air, surface water, landfills, and injection wells are presented in Figure 2.5. Trends in the TRI database for transfers of 1,4-dioxane wastes for treatment and energy recovery are shown in Figure 2.6.

During the 16-years of waste handling records depicted in Figures 2.5 and 2.6, 28 million pounds were transferred to treatment facilities, and 15 million pounds were released. Releases of 1,4-dioxane to air and surface water have decreased substantially, while recovery and transfer to treatment has increased. For example, U.S. facility emissions of 1,4-dioxane to air reported in the TRI database have decreased from a high of about 840,000 lb in 1989 to approximately 115,000 lb in 2004. Discharges to surface water reported in the TRI database increased from about 200,000 lb in 1988 to 650,000 lb in 1993, and then decreased to about 90,000 lb in 2004. Overall discharges have decreased approximately sixfold from 1993 to 2004. Some of the apparent trends in the TRI database may be reporting artifacts, but the overall trend shows decreases in discharges and increased recovery and treatment of 1,4-dioxane from facilities that use and generate it. A cursory review of records for facilities generating 1,4-dioxane shows that since methyl chloroform was banned by the Montreal Protocol Clean Air Act Amendments, 1,4-dioxane emissions and releases to surface water have been eliminated or significantly decreased. 1,4-Dioxane occurrence at hazardous waste sites is the focus of Chapter 8.

2.8.1   1,4-DIOXANE IN WASTEWATER

1,4-Dioxane has been detected in raw wastewater in Michigan and Japan. Treated effluent from a wastewater-treatment plant serving several apartment complexes in Japan's Kanagawa Prefecture

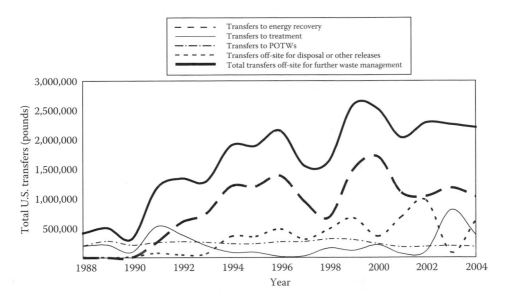

FIGURE 2.6 Estimated transfers and treatment of 1,4-dioxane from 1988 through 2004. [U.S. Environmental Protection Agency (USEPA), 2006, Envirofacts data warehouse. http://www.epa.gov/enviro/ (accessed July 22, 2007).]

was found to contain an average of 0.25 mg 1,4-dioxane per person per day. Sources of 1,4-dioxane in households discharging to the treatment plant were presumed to be shampoos and liquid dishwashing and laundry soaps, which in Japan contain from 0.2 to 0.56 mg/L 1,4-dioxane (Abe, 1999).

The City of Ann Arbor, Michigan, tested raw wastewater and treated wastewater effluent for 1,4-dioxane. In each of three sampling events, 1,4-dioxane was present in influent at an average concentration of 3 ppb; 1,4-dioxane remained present in all the three corresponding treated wastewater effluent samples at an average concentration of 2 ppb. The source of 1,4-dioxane in raw wastewater was not identified in the study (Skadsen et al., 2004). A similar comparison of wastewater influent and treated wastewater effluent from Japan showed 16 ppb in influent and 11 ppb in effluent for one sampling event; a subsequent sampling event showed 3 ppb in influent and 3 ppb in effluent. The treatment plant used an activated sludge process. Sources of 1,4-dioxane in the Japanese study were indicated as industrial pollution and surfactants (Abe, 1999). A 1988 survey of organic compounds in U.S. sewage sludge samples detected 1,4-dioxane in three samples out of more than 100 samples analyzed (USEPA, 1990). The ability of conventional aerobic treatment technology to remove 1,4-dioxane is discussed in Chapter 7. A case study of 1,4-dioxane in recycled water is presented in Chapter 8.

2.8.2 1,4-Dioxane in Landfills

1,4-Dioxane is frequently detected in landfill leachate, groundwater beneath municipal and industrial landfills, and in landfill gas and landfill gas condensate. Not all landfills release 1,4-dioxane. Landfills receiving vapor degreasing still bottoms, solvent wastes, paint filters, scintillation and other laboratory wastes, ink sludge, pesticide containers, household products with methyl chloroform as an ingredient, and industrial sludges from textile production, resin production, and cellulose acetate membrane production have a higher likelihood of 1,4-dioxane presence in leachate, gas, and affected groundwater. 1,4-Dioxane has been detected in landfill gas at a landfill in Westphalia, Germany (Bruckman and Mulder, 1982).

An USEPA survey of U.S. landfills reviewed the chemistry of leachates. The mean concentration of 1,4-dioxane in leachate from all municipal landfills, including landfills not regulated under Subtitle D, was 118 ppb. The mean concentration of 1,4-dioxane at a construction and demolition landfill was

49 ppb; four other construction and demolition landfills had no detections. The median concentration of 1,4-dioxane in municipal landfill leachate at Subtitle D landfills was 11 ppb, the mean was 118 ppb, and the maximum detected concentration was 323 ppb (USEPA, 2000). At hazardous waste landfills, the median concentration was 466 ppb and the maximum was 7611 ppb (USEPA, 2000).

In Japanese landfills, 1,4-dioxane was detected in 87.5% of leachate samples analyzed. Concentrations ranged from 1.1 to 109 ppb with a reporting limit of 0.9 ppb (Yasuhara et al., 1997). 1,4-Dioxane was detected in leachates from hazardous wastes landfills in Japan in the range 20.7–1370 μg/L (Yasuhara et al., 1999).

1,4-Dioxane was detected in landfill leachate at concentrations as high as 19,000 ppb at the Operating Industries, Inc. (OII) site, a 190-acre hazardous waste landfill located in Monterey Park, California, near Los Angeles (USEPA, 1987). The Department of Energy's Pacific Northwest Laboratory analyzed 1,4-dioxane in industrial landfill leachate at the Hanford site in Washington and found that concentrations ranged from 40 to 180 ppb; most results fell in the range from 50 to 100 ppb (Department of Energy, 2000).

A study of the origins of 1,4-dioxane in leachate at a landfill accepting only fly ash and bottom ash from two incineration facilities sought to determine what chemical processes during incineration might cause 1,4-dioxane concentrations to be as high as 134 μg/L. Investigators concluded that the thermal treatment of foam plastics, such as incineration of polyurethane foam upholstery material in automobile-shredder wastes, may produce 1,4-dioxane as a by-product (Watanabe et al., 2006). Bench-scale laboratory tests partially demonstrated that this synthesis can occur. A leaching test on the incineration residue of isolated foam plastic waste produced 0.2 ppm 1,4-dioxane, whereas leachate from unincinerated foam waste released no 1,4-dioxane (Fujiwara et al., 2006). These findings were corroborated by a second study that analyzed 38 leachate samples from Japanese municipal solid-waste landfills. The maximum detected concentration of 1,4-dioxane in leachate was 340 ppb. Of the leachate samples from landfills that received incinerator wastes and shredded solid waste mixed with incinerator waste (as opposed to noncombustible waste), 70% had 1,4-dioxane detections. However, just 38% of the leachate samples from landfills receiving only noncombustible waste had 1,4-dioxane detections (Kurata et al., 2006).

BIBLIOGRAPHY

Abe, A., 1997, Determination method for 1,4-dioxane in water samples by solid phase extraction GC/MS. *Journal of Environmental Chemistry* 7: 95–100.

Abe, A., 1999, Distribution of 1,4-dioxane in relation to possible sources in the water environment. *The Science of the Total Environment* 227: 41–47.

Agarwal, U.P. and Ralph, S.A., 1997, FT-Raman spectroscopy of wood: Identifying contributions of lignin and carbohydrate polymers in the spectrum of Black Spruce (*Picea mariana*). *Applied Spectroscopy* 51(11): 1641–1648.

American Chemical Society, 1929, Market report—October, 1929. *Industrial and Engineering Chemistry* 21(11): 1153–1154, DOI: 10.1021/ie50239a051.

American Chemical Society, 1935, Market report. *Industrial and Engineering Chemistry* 27(6): 733–734, DOI: 10.1021/ie50306a028.

American Cyanamid, 1987, Material Safety Data Sheet for FM 47 identification lacquer, 8% solids, issued January 1, 1987, Wayne, NY. FSC: 8010, LIIN: 00F003660.

Anonymous, 1987, Testergebnisse: Von Appretur bis Zahnbürste [Test results: From glaze to toothbrushes]. *Der Okotest* 2: 129–135.

Anschutz, L. and Broeker, W., 1926, Dioxan als Lösungsmittel, insbesondere bei Molekulargewichts-Bestimmungen [Dioxane as a solvent, particularly for determining molecular weights]. *Berichte der Deutschen Chemische Gesellschaft* 59: 2844–2847.

ATSDR, 2004, Draft toxicological profile for 1,4-dioxane. Division of Health Assessment and Consultation, U.S. Department of Health and Human Services.

ATSDR, 2005, Public health implications of 1,4-dioxane-contaminated groundwater at the Durham Meadows Site, Durham, CT. Division of Health Assessment and Consultation, Agency for Toxic Substances and Disease Registry, U.S. Department of Health and Human Services.

Barber, H., 1934, Haemorrhagic nephritis and necrosis of the liver from dioxane poisoning. *Guy's Hospital Reports* 84: 267–280.

BASF, 1986, Datenblatt Dioxan [Dioxane datasheet]. BASF (Badische Anilin- & Soda-Fabrik).

BASF, 2005, Dioxane—Stab Technical Data Sheet. Ludwigshafen, Germany: BASF.

Beernaert, H., Herpol-Borremans, M., and De Cock, F., 1987, Determination of 1,4-dioxane in cosmetic products by headspace gas chromatography. *Belgian Journal of Food Chemistry and Biotechnology* 42(5): 131–135.

Belanger, P.L., 1980, Health hazard evaluation determination, Report No. HE 80-21-721. National Institute for Occupational Safety and Health, U.S. Department of Health and Human Services, Center for Disease Control.

Betts, K.S., 2005, The changing chemistry of office cubicles. *Environmental Science and Technology Online—Technology News*, June 22, 2005. http://pubs.acs.org/subscribe/journals/esthag-w/2005/jun/tech/kb_cubicles.html (accessed July 5, 2005)

Birkel, T.J., Warner, C.R., and Fazio, T., 1979, Gas chromatographic determination of 1,4-dioxane in polysorbate 60 and polysorbate 80. *Journal of the Association of Official Analytical Chemists* 62(4): 931–935.

Black, R.E., Hurley, F.J., and Havery, D.C., 2001, Occurrence of 1,4-dioxane in cosmetic raw materials and finished cosmetic products. *Journal of AOAC [Association of Official Analytical Chemists] International* 84(3): 666–670.

Bray, G.A., 1960, A simple efficient liquid scintillator for counting aqueous solutions in a liquid scintillation counter. *Analytical Biochemistry* 1: 279–285.

Bruckman, P. and Mulder, W., 1982, Der Gehalt an organischen Spurenstoffen in Deponiegasen [The trace organic composition of landfill gases]. *Mull und Abfall* 14: 339–345.

Business Communications Company, 2002, Process makes purer plastics retardant. *Flame Retardancy News* 12(8): 1.

California Air Resources Board (CARB), 2006, "Hot spots" inventory guidelines. California Air Resources Board, California Air Resources Board, Health and Safety Code Section 44300 et seq.

California Environmental Protection Agency, 1997, Assessment of the tolerances for Section 3 registration of Shadeout (rimsulfuron) on tomatoes. Department of Pesticide Regulation.

Caulkins, J., Labbe, N., Luth, C., Valleries, P., Wilcox, J., and Thompson, R., 2006, Using computational chemistry to understand effective adsorption strategies for separating contaminants from water. American Society of Environmental Engineers Student Poster Competition, Worcester Polytechnic Institute, March 17, 2006.

Chang, B., 1999, United States Patent 5,912,003: Spray-type insecticidal paint and manufacturing process thereof. Assignee: Kukbo Pharma Co., Ltd., Seoul, Korea.

Chang, S.S., Peterson, R.J., and Ho, C.-T., 1978, Chemical reactions involved in the deep-fat frying of foods. *Journal of the American Oil Chemist's Society* 1: 718–727.

Chartres, C.J., Ringrose-Voase, A.J., and Raupach, M., 1989, A comparison between acetone and dioxane and an explanation of their role in water replacement in undisturbed soil samples. *Journal of Soil Science* 40(4): 849–850.

Chemithon Corporation, 2006, DRS™ dioxane reduction system. http://www.chemithon.com/drs_diox_reduction.html (accessed March 22, 2006).

Chenier, P.J., 2002, *Survey of Industrial Chemistry*. New York: Kluwer Academic Publishing.

Choi, S.H., Kobayashi, A., and Yamanishi, T., 1983, Odor of cooked small shrimp, *Acetes japonicus* Kishinouye: Difference between raw material and fermented product. *Agricultural and Biological Chemistry* 47(2): 337–342.

Chung, T.-Y., Hayase, F., and Kato, H., 1983, Volatile components of ripe tomatoes and their juices, purees and pastes. *Agricultural and Biological Chemistry* 47: 343–351.

Cognis Corporation, 2001, Material Safety Data Sheet for Agnique MMF mosquito larvicide & pupicide, issued September 4, 2002, Cincinnati, OH.

Copeland, J., 2004, The question of fuel—Part 4. Oxygenators are the "hottest" topic in fuel chemistry. http://www.foxvalleykart.com/fuel4.html (accessed March, 2004).

Cortellucci, R. and Dietz, E.A., Jr., 1999, Determination of 1,4-dioxane in toluene by gas chromatography. *Journal of High Resolution Chromatography* 22(4): 201–204.

CRC Chemicals, 1987, Material Safety Data Sheet for 5089 Brakleen Aerosol, issued April 10, 1987, Warminster, PA. FSC: 6850, LIIN: 00F024407.

Crown Industrial Products Company, 1989, Material Safety Data Sheet for rust inhibitor 6007, issued February 28, 1989, Hebron, IL. FSC: 6850, LIIN: 00N003634.

Dado, G.P., Knaggs, E.A., and Nepras, M.J., 2006, Sulfonation and sulfation. *Kirk-Othmer Encyclopedia of Chemical Technology Online Edition*. http://mrw.interscience.wiley.com/emrw/9780471238966/home/ WileyInterscience (accessed October 3, 2007).

DAP, Inc., 1990, Material Safety Data Sheet for smooth spread contact cement, issued February 22, 1990, Dayton, OH. FSC: 8040, LIIN: 00F017978.

DAP, Inc., 1994, Material Safety Data Sheet for weldwood contact cement, issued January 4, 1994, Dayton, OH. FSC: 8040, LIIN: 00N056641.

DeLima Associates, 2004, Consumer Product Information Database. http://hpd.nlm.nih.gov/cgi-bin/household (accessed July 2, 2005).

Department of Energy, 2000, *Hanford Environmental Report 2000*. Hanford, Washington, DC: Pacific Northwest National Laboratory.

DeRosa, C.T., Wilbur, S., Holler, J., Richter, P., and Stevens, Y., 1996, Health evaluation of 1,4-dioxane. *Toxicology and Industrial Health* 12: 43.

DeSoto, Inc., 1991, Material Safety Data Sheet for polyamide resin component 910X746, issued February 21, 1991, Berkeley, CA. FSC: 8010, LIIN: 00N019667.

Diamond, G.L. and Durkin, P.R., 1997, Effects of surfactants on the toxicity of glyphosate, with specific reference to RODEO. U.S. Department of Agriculture Animal and Plant Health Inspection Service Report No. 53-3187-5-12, prepared by Syracuse Research Corporation, Syracuse, NY.

Doherty, R.E., 2000, A history of the production and use of carbon tetrachloride, tetrachloroethylene, trichloroethylene and 1,1,1-trichloroethane in the United States: Part 1—Historical background; carbon tetrachloride and tetrachloroethylene. *Journal of Environmental Forensics* 1: 11.

Dow Chemical Corporation, 1985, Material Safety Data Sheet for Dursban, issued October 16, 1985, Midland, MI. FSC: 6840, LIIN: 00D001801.

Dow Chemical Corporation, 1989, TSCA 8(e) letter: Analyses of polyethylene glycol for 1,4-dioxane, January 20, 1989. Office of Toxic Substances Report No. 89-890000079, USEPA. http://yosemite.epa.gov/oppts/ epatscat8.nsf (accessed April 9, 2004.)

Eastman Kodak Company, 1998, Material Safety Data Sheet for Kodak film cement, issued June 24, 1998, Rochester, NY. MSDS No. CKRRX.

Eastman Kodak Corporation, 2001, 2000 Kodak Park Environmental Update. Rochester, NY: Eastman Kodak Corporation 8.

Ellis, R.A. and Thomas, J.S., 1998, United States Patent 5,817,910: Destroying 1,4-dioxane in by-product streams formed during polyester synthesis. Assignee: Wellman, Inc., Shewsbury, NJ.

Elmer, P., 2006, The evolution of liquid scintillation counter cocktails and safety. http://www.npl.co.uk/ lsuf/2003/the_evolution_of_lsc_cocktails_and_safety.pdf (accessed August 3, 2006).

European Chemicals Bureau, 2002, European Union risk assessment report: 1,4-Dioxane. Luxembourg, European Union, Institute for Health and Consumer Protection, Report 21.

European Medicines Agency, 1998, ICH harmonised tripartite guideline: ICH topic Q 3 C (R3) impurities: Residual solvents. In *Note for Guidance on Impurities: Residual Solvents*. London, UK: European Medicines Agency. http://www.emea.eu.int (accessed May 10, 2006).

Ferro Corporation, 2006, Ferro fine chemicals: 1,4-Dioxane. http://www.ferro.com/Our+Products/Fine+Chemicals/ Products+and+Markets/1+4-Dioxane/ (accessed November 5, 2006).

Fishbein, L., 1981, Carcinogenicity and mutagenicity of solvents: I. Glycidyl ethers, dioxane, nitroalkanes, dimethylformamide and allyl derivatives. *Science of the Total Environment* 17(2): 97–110.

Fisher Scientific, 2006, Material Safety Data Sheet for OptiDry ACC# 26970, issued June 6, 1999, Fair Lawn, NJ. https://www1.fishersci.com/Browse?bp=1336H%7CL2884706H (accessed October 16, 2006).

Flick, E.W., 1998, *Industrial Solvents Handbook*. Westwood, NJ: Noyes Data Corporation.

Flick, E.W., 2005, *Paint and Ink Formulations Database*. Norwich, NY: William Andrew Publishing. http:// www.knovel.com/knovel2/Toc.jsp?BookID=1197&VerticalID=0 (accessed November 5, 2006).

Food and Drug Administration (FDA), 1980, Progress report on the analysis of ethoxylated ingredients for 1,4-dioxane. Division of Cosmetics Technology, Food and Drug Administration. http://yosemite.epa.gov/ oppts/epatscat8.nsf/ALLIDS/8FC76CFF09DF768485256F0800556E8A/$FILE/89890000081. pdf?OpenElement (accessed April 2, 2008).

Food and Drug Administration (FDA), 1981, Progress report on the analysis of cosmetics raw materials and finished cosmetics products for 1,4-dioxane. Division of Cosmetics Technology, Food and Drug Administration 15.

Food and Drug Administration (FDA), 1992, *Cosmetic Handbook*. Center for Food Safety and Applied Nutrition. http://www.cfsan.fda.gov/~dms/cos-hdb3.html (accessed November 29, 2007).

Food and Drug Administration (FDA), 1997a, Guidance for industry: Q3C impurities: Residual solvents. Rockville, MD: U.S. Department of Health and Human Services, Food and Drug Administration, Center for Drug Evaluation and Research ICH 16.

Food and Drug Administration (FDA), 1997b, Letter from FDA regarding dioxane in contraceptive sponges. Center for Drug Evaluation and Research 10.

Food and Drug Administration (FDA), 1999, Food additives permitted for direct addition to food for human consumption: Polysorbate 60. *Federal Register* 64: 57974–57976.

Food and Drug Administration (FDA), 2001, Impurities: Residual solvents in new veterinary medicinal products, active substances and excipients. Rockville, MD: Center for Veterinary Medicine, U.S. Food and Drug Administration, http://www.fda.gov/cvm/Guidance/guide100.doc (accessed June 8, 2006).

Food and Drug Administration (FDA), 2006, Food additives permitted for direct addition to food for human consumption: Glycerides and polyglycides. *Federal Regulations* 21: 75–56.

Foster, N.C., 1997, *Sulfonation and Sulfation Processes*. Seattle, Washington, DC: Chemithon. www.chemithon.com (accessed November 5, 2006).

Fox, C., 1993, Hair care literature and patent review. *Cosmetics and Toiletries* 108(3): 29–50.

FPPF Chemical Company, Inc., 2006, Material Safety Data Sheet for glyclean recycled antifreeze 6-6. FPPF Chemical Company, Inc., issued February 22, 1999, Buffalo, NY. FSC: 6850, NIIN: LIIN: 00N092811.

Fruijtier-Pölloth, C., 2005, Safety assessment on polyethylene glycols (PEGs) and their derivatives as used in cosmetic products. *Toxicology* 214: 1–37.

Fuh, C.B., Lai, M., Tsai, H.Y., and Chang, C.M., 2005, Impurity analysis of 1,4-dioxane in nonionic surfactants and cosmetics using headspace solid-phase microextraction. *Journal of Chromatography A* 1071(1): 141–145.

Fujiwara, T., Fukuda, N., Tamada, T., Ono, Y., Kurata, Y., Ono, Y., Nishimura, F., Ohtoshi, K., 2006, Investigation of the origins of 1,4-dioxane in leachate from municipal solid waste landfills. In: *Proceedings of the 4th Intercontinental Landfill Research Symposium*, June 14–16, 2006, Gällivare, Sweden.

Gelman Sciences, 1989a, Information on the dioxane content of polyethylene glycol, raw materials and consumer products with cover letter dated 011789 [submitted to USEPA under TSCA Section 8E, OTS0516624-1].

Gelman Sciences, 1989b, Reports on 1,4-dioxane and polyethylene glycol in products with cover letter dated 011489 [submitted to USEPA under TSCA Section 8E, OTS0516624-1].

Gelman Sciences, 1989c, Letter from Gelman Sciences to USEPA regarding contamination of 1,4-dioxane at rest stops with attachments [submitted to USEPA under TSCA Section 8E, OTS0516624].

Gierer, J. and Pettersson, I., 1968, On the preparation and characterization of *p*-dioxanyl hydroperoxide. *Acta Chemica Scandinavica* 22: 3183–3190.

Glidden Paint, 1992, Material Safety Data Sheet for Glid-Thane One polyurethane coating 6124, issued June 1, 1992, Cleveland, OH.

Grady, C.P.L., Jr., Sock, S.M., and Cowan, R.M., 1997, Biotreatability kinetics. In: G.S. Sayler, J. Sanseverino, and K.L. Davis (Eds), *Biotechnology in the Sustainable Environment*, Vol. 54, p. 14. New York: Plenum Press.

Greek, B.F. and Layman, P.L., 1989, Higher costs spur new detergent formulations. *Chemical and Engineering News* 67(4): 29–38.

Guo, W. and Brodowsky, H., 2000, Determination of the trace 1,4-dioxane. *Microchemical Journal* 64: 174–180.

Hansen, O.C., 2005, Screening for health effects from chemical substances in textile colorants. In: F. Jensen and O.C. Hansen (Eds), *Survey of Chemical Substances in Consumer Products*. Danish Ministry of the Environment, Danish Technological Institute, Report No. 57. http://miljoestyrelsen.dk/homepage/ (accessed October 9, 2006).

Harkov, R., Katz, R., Bozzelli, J., and Kebbekus, B., 1981, Toxic and carcinogenic air pollutants in New Jersey—volatile organic substances. In: J.J. McGovern (Ed.), *Proceedings of the International Technical Conference of Toxic Air Contaminants*. Pittsburgh, PA: Air Pollution Control Association.

Hazardous Substances Data Bank (HSDB), 2006, Online database. National Library of Medicine, National Institutes of Health. http://toxnet.nlm.nih.gov/ (accessed July 2, 2005).

Hodgson, A.T. and Levin, H., 2006, Classification of measured indoor volatile organic compounds based on noncancer health and comfort considerations. http://eetd.lbl.gov/ie/pdf/LBNL-53308.pdf (accessed November 5, 2006).

Hovenkamp, S.G. and Munting, J.P., 1970, Formation of diethylene glycol as a side reaction during production of polyethylene terephthalate. *Journal of Polymer Science: Part A-1. Polymer Chemistry* 8(3): 4.

Italia, M.P. and Nunes, M.A., 1991, Gas chromatographic determination of 1,4-dioxane at the parts-per-million level in consumer shampoo products. *Journal of the Society of Cosmetics Chemistry* 42: 97–103.

Jass, H.E., 1989, CTFA today—despite numerous personnel changes, the workload continues heavy. *Cosmetics and Toiletries* 1989(9): 10–12.

Johnson, W., Jr., 2001, Final report on the safety assessment of PEG-25 propylene glycol stearate, PEG-75 propylene glycol stearate, PEG-120 propylene glycol stearate, PEG-10 propylene glycol, PEG-8 propylene glycol cocoate, and PEG-55 propylene glycol oleate. *International Journal of Toxicology* 20(4): 13–26.

Judge, P., 1988, Gelman cites other dioxane sources. *Ann Arbor News* 1.

Kauffman, J.S., 2003, Trends in impurity analysis: Determination of extractables, leachables, residual solvents, and unknowns by mass spectrometry. *Pharmaceutical Technology North America* 2003: 24–27.

Keel, L. and Franzmann, A., 2000, Downtown Bellingham air toxics screening project: 1995–1999. Northwest Air Pollution Authority (NWAPA).

Kikuchi, M. and Tanaka, H., 2003, United States Patent 6,646,021: Golf ball printing ink, golf ball bearing a mark printed with the ink, and method of manufacturing the golf ball. Assignee: Sumitomo Rubber Industries, Ltd., Kobe, Japan.

Klein, E. and Smith, J.K., 1972, Asymmetric membrane formation: Solubility parameters for solvent selection. *Industrial Engineering and Chemical Production Research and Development* 11(2): 207–212.

Knovel, 2006, *Yaws' Handbook of Antoine Coefficients for Vapor Pressure—Electronic Edition.* http://www. knovel.com (accessed June 2006).

Kosswig, K., 2002, Surfactants. In: W. Gerhartz, Y.S. Yamamoto, and F.T. Campbell (Eds), *Ullmann's Encyclopedia of Industrial Chemistry.* Weinheim, Germany: Wiley-VCH Verlag GmbH & Co. KGaA, DOI: 10.1002/14356007.a25_747.

Kurata, Y., Watanabe, Y., Kawasaki, M., Ono, Y., and Ono, Y., 2006, Occurrence of trace organic compounds in leachates from municipal solid waste landfill sites in Japan. *4th Intercontinental Landfill Research Symposium.* Gällivare, Sweden, Swedish Waste Management Association.

Laursen, S.E., Hansen, J., Bagh, J., Jensen, O., and Werther, I., 1997, Environmental assessment of textiles— life cycle screening of textiles containing cotton, wool, viscose, polyester or acrylic fibres. Danish Environmental Protection Agency, Copenhagen, Denmark. http://www.defra.gov.uk/science/project_ data/DocumentLibrary/SD14010/SD14010_3064_FRP.doc (accessed November 5, 2006).

Lide, D.R., 2007, Bond lengths and angles in gas-phase molecules. In: D.R. Lide (ed.), *CRC Handbook of Chemistry and Physics, Internet Version 2007*, 87th Edition. Boca Raton, FL: Taylor & Francis.

Loctite Corporation, 1995, Material Safety Data Sheet for RP 701 LV Anti-Seize 70175, issued January 3, 1995, Rocky Hill, CT. FSC: 8030, NIIN: LIIN: 00N070591.

Lourenço, M.A.-V., 1863, Recherches sur les composés polyatomiques [Research on the polyatomic compounds]. *Annales de Chimie et de Physique* (Paris) 67: 257–339.

Mackay, D., Shiu, W.-Y., and Ma, K.-C., 1993, *Illustrated Handbook of Physical-Chemical Properties and Environmental Fate for Organic Chemicals: Volume III. Volatile Organic Chemicals.* Boca Raton, FL: Lewis Publishers (Taylor & Francis).

Matsumura, S., Shimamoto, S., and Shibata, T., 1997, United States Patent 5,692,527: Tobacco smoke filter materials, fibrous cellulose esters, and production processes. Assignee: Daicel Chemical Industries, Ltd., Himeji, Japan.

Maurer, T., Hass, H., Barnes, I., and Becker, K.H., 1999, Kinetic and product study of the atmospheric photo-oxidation of 1,4-dioxane and its main reaction product ethylene glycol diformate. *Journal of Physical Chemistry A* 103: 8.

Mazurkiewicz, J. and Tomasik, P., 2006, Why 1,4-dioxane is a water-structure breaker. *Journal of Molecular Liquids* 126: 111–116.

McCaffrey, J.W., 1938, United States Patent 2,119,110: Treatment of hydrocarbon oils. Assignee: Universal Oil Products Company, Chicago, IL.

Medical Economics Publishing, 1983, Contraceptive sponge holds up under criticism. *Mark Drug Topics* 127(August 15) 23.

Merck and Company, Inc., 1983, *The Merck Index.* Rahway, NJ: Merck and Company, Inc.

Miyatake, H. and Saito, K., 1984, Distillation of used liquid scintillation cocktail and the possibility of re-utilization of the recovered solvent. *Radioisotopes* 33(3): 146–149.

Monneyron, P.E., Manero, M.-H., and Foussard, J.-N., 2003, Measurement and modeling of single- and multi-component adsorption equilibria of VOC on high-silica zeolites. *Environmental Science and Technology* 37(11): 2410–2414.

Montgomery, J.H., 1996, *Groundwater Chemicals Desk Reference*, 2nd Edition. Chelsea, MI: Lewis Publishers.

Moran, C.J., McBratney, A.B., and Ringrose-Voase, A.J., 1986, A method for the dehydration and impregnation of clay soil. *European Journal of Soil Science* 40(3): 569–575.

Moran, C.J., McBratney, A.B., Ringrose-Voase, A.J., and Chartres, C.J., 1989, A method for the dehydration and impregnation of clay soil. *Journal or Soil Science* 40: 569–585.

Moran, J.E., Hudson, G.B., Eaton, G.F., and Leif, R., 2003, Ambient groundwater monitoring and assessment: A contamination vulnerability assessment for the Sacramento area groundwater basin. Lawrence Livermore National Laboratory internal report UCRL-TR-203258.

Mossman, H.W., 1937, The dioxan technic. *Stain Technology* 12(4): 147–156.

National Cancer Institute, 1985, *Monograph on Human Exposure to Chemicals in the Workplace: 1,4-Dioxane*. Bethesda, MD: National Cancer Institute.

National Diagnostics, Inc., 1987a, Material Safety Data Sheet for Bray's solution, issued October 7, 1987, Manville, NJ. FSC: 6810, NIIN: LIIN: 00N067365.

National Diagnostics, Inc., 1987b, Material Safety Data Sheet for Dioxscint LS-171, issued October 9, 1987, Manville, NJ. FSC: 6810, NIIN: LIIN: 00N067374.

NICO2000, Ltd., 2006, Ion selective electrode analysis. http://www.nico2000.net/index.htm (accessed October 28, 2006).

Nishimura, T., Iizuka, S., Kibune, N., and Ando, M., 2004, Study of 1,4-dioxane intake in the total diet using the market-basket method. *Journal of Health Science* 50(1): 101–107.

North Carolina Department of Environment and Natural Resources, 1998, Focus: Waste minimization. *The North Carolina Division of Pollution Prevention and Environmental Assistance Newsletter* 7(3). http://www.p2pays.org (accessed November 5, 2006).

Northeast Waste Management Officials Association, 1998, Metal painting and coating operations. http://www.p2pays.org/ref/01/00777/toc.htm (accessed July 18, 2007).

Old World Automotive Products, Inc., 2001, Material Safety Data Sheet for full force antifreeze and coolant. Northbrook, IL.

Osaka Organic Chemical Industry, Ltd., 2006, History of Osaka Organic Chemical Industry, Ltd. http://www.ooc.co.jp/00english/03company/company_01.html (accessed July 18, 2007).

Pervo Paint Company, 1993, Material Safety Data Sheet for White R D traffic paint, 5950, issued January 22, 1993, Los Angeles, CA. FSC: 8010, LIIN: 00N065622.

Pesticide Action Network, 2006, Glyphosphate fact sheet. *Pesticide News* http://www.pan-uk.org/pestnews/actives/glyphosa.htm (accessed October 22, 2006).

Philippoff, W., 2006, Weissenberg's contributions to rheology. *Karl Weissenberg—The 80th birthday celebration essays.* http://innfm.swan.ac.uk/bsr/frontend/home.asp (accessed October 29, 2006).

Pro-Systems, 2002, *Under the Scoop: Carburetors and Concepts.* Spring Lake, MI. http://www.pro-system.com/pjames072702.htm (accessed November 25, 2006).

Pundlik, M.D., Sitharaman, B., and Kaur, I., 2001, Gas chromatographic determination of 1,4-dioxane at low parts-per-million levels in glycols. *Journal of Chromatographic Science* 39(2): 4.

Ralph, P.A., 1938, Comparative study of some dehydration and clearing agents. *Stain Technology* 13: 9–15.

Rankin, T., 1996, Residual solvents in drugs by static headspace analysis. In: *Pharmaceutical Analysis: GC Technical Report No. TR9115/99.04.* Austin, Texas: ThermoQuest. http://www.thermo.com (accessed November 25, 2006).

Rastogi, S.C., 1990, Headspace analysis of 1,4-dioxane in products containing polyethoxylated surfactants by GC-MS. *Chromatographia* 29(9–10): 441–444.

Reid, E.W. and Hoffman, H.E., 1942, 1,4-Dioxan. *Industrial and Engineering Chemistry* 21(7): 695–697.

Roig, B., Allan, I.J., and Greenwood, R., 2003, A toolbox of existing and emerging methods for water monitoring under the Water Framework Directive. In: B. Roig, I.J. Allan, and R. Greenwood (Eds), *Water Framework Directive.* St.-Etienne, France: Ecologic Institute for International and European Environmental Policy. www.ecologic-events.de/swift-germany/documents/Emerging_tools_manual.pdf (accessed November 5, 2006).

Roller, H., Senkpiel, W., Wunsch, et al., 1977, United States Patent 4,018,967: Magnetic recording tape. Assignee: BASF Aktiengesellschaft, Ludwigshafen, Germany.

Romero, J., Ventura, F., Caixach, J., Rivera, J., Godé, L.X., and Niñerola, J.M., 1998, Identification of 1,3-dioxanes and 1,3-dioxolanes as malodorous compounds at trace levels in river water, groundwater, and tap water. *Environmental Science and Technology* 32(2): 206–216.

Royal Society of Chemistry, 1992, Union Carbide sells dioxane business to division of Ferro. *American Paint and Coatings Journal* 21: 1.

Rümenapp, J. and Hild, J., 1987, Dioxan in kosmetischen Mitteln [Dioxane in cosmetic preparations]. *Lebensmittelchemie und Gerichtliche Chemie* 41(3): 59–62.

Sachdeva, Y.P. and Gabriel, R.P., 1997, Apparatus for decontaminating a liquid surfactant of dioxane. United States Patent and Trademark Office, Pharm-Eco Laboratories, Lexington, Massachusetts.

Saifer, A. and Hughes, J., 1937, Dioxane as a reagent for qualitative and quantitative determination of small amounts of iodide: Its application to the detection of iodide in iodized salt. *The Journal of Biological Chemistry* 118: 241–245.

Sanceda, N.G., Kurata, T., and Arakawa, N., 1984, Fractionation and identification of volatile compounds in Patis, a Philippine fish sauce. *Agricultural and Biological Chemistry* 48: 3047–3052.

Scalia, S. and Menegatti, E., 1991, Assay of 1,4 dioxane in commercial cosmetic products by HPLC. *Farmaco* 46(11): 5.

Scalia, S., Testoni, F., Frisina, G., and Guarnerij, M., 1992, Assay of 1,4-dioxane in cosmetic products by solid-phase extraction and GC-MS. *Journal of the Society of Cosmetic Chemists* 43: 6.

Shah, J.J. and Singh, H.B., 1988, Distribution of volatile organic chemicals in outdoor and indoor air—a national VOCs data base. *Environmental Science and Technology* 22(12): 1381–1388.

Shahidi, F., Rubin, L.J., and D'Souza, L.A., 1986, Meat flavor volatiles: A review of the composition, techniques of analysis, and sensory evaluation. *CRC Critical Reviews in Food Science and Nutrition* 24: 141–252.

Shamberg, J.B., 2002, State of Kansas settles litigation over pollution from KU landfill. http://www.sjblaw.com/headlines/landfill.html (accessed March 2002).

Shearer, T.P. and Hunsicker, L.G., 1980, Rapid method for embedding tissues for electron microscopy using 1,4-dioxane and polybed 812. *Journal of Histochemistry and Cytochemistry* 28(5): 465–467.

Sirchie Finger Print Labs, 1989, Material Safety Data Sheet for lab/roller cleaner, 234T, issued June 8, 1989, Raleigh, NC. FSC: 7930, NIIN: 00N047242.

Skadsen, J.M., Rice, B.L., and Meyering, D.J., 2004, The occurrence and fate of pharmaceuticals, personal care products, and endocrine disrupting compounds in a municipal water use cycle: A case study in the City of Ann Arbor, Michigan. Water Utilities and Fleis & VandenBrink Engineering, Inc., Ann Arbor, MI. http://www.ci.annarbor.mi.us/PublicServices/water/WTP/EndocrineDisruptors.pdf (accessed October 9, 2006).

Smallwood, I.M., 1996, *Handbook of Organic Solvent Properties*. London: Arnold.

Smith, B., and Bristow, V., 1994, Indoor air quality and textiles: An emerging issue. *American Dyestuff Reporter*, 83(1): 37–46.

Spadone, J.C., Takeoka, G., and Liardon, R., 1990, Analytical investigation of Rio off-flavor in green coffee. *Journal of Agricultural and Food Chemistry* 38(1): 226–233.

SRRP, 1991, Paint removal. Source Reduction Research Partnership: Metropolitan Water District of Southern California and the Environmental Defense Fund.

Stanford Research Institute, 1989, 1989 Directory of chemical producers—United States of America. Menlo Park, CA: SRI International.

Stepan Company, 2006, Test plan for sodium 2-(2-dodecyloxyethoxy) ethyl sulfate. High Production Volume Program, USEPA. http://www.epa.gov/chemrtk/pubs/summaries/sodium22/c16316tc.htm (accessed November 8, 2006).

Stoye, D., 2005, Solvents. *Ullman's Encyclopedia of Industrial Chemistry*. Weinheim, Germany: Wiley-VCH Verlag GmbH & Co. KGaA.

Studier, M.H., Hayatsu, R., and Anders, E., 1965, Origin of organic matter in early solar system: I. Hydrocarbons. Chicago, IL: Argonne National Laboratory and Enrico Fermi Institute, University of Chicago, NASA Grant NsG-366 Research EFINS 65-115:22.

Stumpf, W., 1956, Chemie und Anwendungen des 1,4-dioxans [Chemistry and applications of 1,4-dioxane]. *Monographien Zu Angewandte Chemie und Chemie-Ingenieur-Technik*.

Surprenant, K.S., 2005, Dioxane. In: W. Gerhartz, Y.S. Yamamoto, F.T. Campbell, R. Pfefferkorn, and J.F. Rounsaville (Eds). *Ullmann's Encyclopedia of Industrial Chemistry*. Weinheim, Germany: Wiley Interscience.

Susquehanna Speedway Park, 2006, 2006 SSP XTREME Stock Car and Women on Wheels (W.O.W.) specification rules. http://www.sspracing.net/docs/2006xtreme.pdf (accessed November 28, 2007).

Talmage, S.S., 1994, *Environmental and Human Safety of Major Surfactants: Alcohol Ethoxylates and Alkylphenol Ethoxylates*. Boca Raton, FL: CRC Press.

Tang, J., Jin, Q.Z., and Shen, G.-H., 1983, Isolation and identification of volatile compounds from fried chicken. *Journal of Agricultural and Food Chemistry* 31: 1287–1292.

Thompson-Hayward Chemical Company, 1993, Material Safety Data Sheet for cement-o-film D-139, issued September 7, 1993, Chicago, IL. FSC: 6850, LIIN: 00N052426.

Union Carbide, 1988, Right-to-know information for polyethylene glycol products. USEPA.

Union Carbide, 1989, Material Safety Data Sheet for UCAR aircraft deicing fluid 2-D, PM-6435, issued November 14, 1989. Bound Brook, NJ. FSC: 6850, NIIN: LIIN: 00F015044.

Union Carbide, 2001, Polyethylene glycols fact sheet. Belgium: INEOS Oxides.

University of North Carolina, 2002, UNC hazardous waste dump ordered removed immediately. *The Hill Online*. http://alumni.unc.edu/car/weekly/story.asp?sid=206 (accessed July 2007).

U.S. Environmental Protection Agency (USEPA), 1981, Phase I risk assessment of 1,4-dioxane. U.S. Environmental Protection Agency, EPA Contract No. 68-01-6030.

U.S. Environmental Protection Agency (USEPA), 1987, Superfund record of decision: Operating Industries, Inc. Superfund Program, USEPA.

U.S. Environmental Protection Agency (USEPA), 1989a, Inert ingredients in pesticide products: Policy statement; revision and modification of lists. Office of Pesticide Programs, USEPA.

U.S. Environmental Protection Agency (USEPA), 1989b, Memo from Gelman Sciences to USEPA. Office of Pollution Prevention and Toxics, and Toxic Substances Control Act, USEPA. http://yosemite.epa.gov/oppts/epatscat8.nsf/ALLIDS/8FC76CFF09DF768485256F0800556E8A/$FILE/89890000081. pdf?OpenElement (accessed April 2, 2008).

U.S. Environmental Protection Agency (USEPA), 1990, National sewage sludge survey: Availability of information and data and anticipated impacts on proposed regulations. USEPA, *Federal Register* 55: 47210–47283.

U.S. Environmental Protection Agency (USEPA), 1992, Project summary: Indoor air pollutants from household product sources. Las Vegas, NV: USEPA, EPA Contract No. 600/S-49-1025: 4.

U.S. Environmental Protection Agency (USEPA), 1993, Final report on ambient concentration summaries for Clean Air Act Title III hazardous air pollutants. Research Triangle Park, North Carolina: USEPA, EPA Contract No. 600/R94/090.

U.S. Environmental Protection Agency (USEPA), 1995a, Profile of the printing and publishing industry. Office of Enforcement and Compliance Assurance, USEPA.

U.S. Environmental Protection Agency (USEPA), 1995b, Profile of the rubber and plastics industry. Office of Enforcement and Compliance Assurance, USEPA.

U.S. Environmental Protection Agency (USEPA), 1996a, Pollution prevention in the paints and coatings industry. National Risk Management Research Laboratory Office of Research and Development, Center for Environmental Research Information, USEPA.

U.S. Environmental Protection Agency (USEPA), 1996b, Profile and management options for EPA laboratory generated mixed waste. Center for Remediation Technology and Tools; Radiation Protection Division; Office of Radiation and Indoor Air, USEPA. EPA Contract No. 402-R-96-015:120.

U.S. Environmental Protection Agency (USEPA), 1996c, Solvents study. Office of Solid Waste, USEPA.

U.S. Environmental Protection Agency (USEPA), 1997a, Sector notebook data refresh—1997. Office of Enforcement and Compliance Assurance. USEPA.

U.S. Environmental Protection Agency (USEPA), 1997b, Profile of the plastic resin and manmade fiber industries. Office of Enforcement and Compliance Assurance, USEPA.

U.S. Environmental Protection Agency (USEPA), 1997c, Profile of the pharmaceutical manufacturing industry. Office of Enforcement and Compliance Assurance, USEPA.

U.S. Environmental Protection Agency (USEPA), 2000, Preliminary data summary: Airport deicing operations. Office of Water, USEPA.

U.S. Environmental Protection Agency (USEPA), 2002, 1,4-Dioxane: Proposed acute exposure guideline levels (AEGLs). Office of Pollution Prevention and Toxics, USEPA 76.

U.S. Environmental Protection Agency (USEPA), 2006, Envirofacts data warehouse. http://www.epa.gov/enviro/ (accessed July 22, 2007).

U.S. Environmental Protection Agency (USEPA), 2008, Inventory update reporting. *Non-confidential Inventory Update Reporting Production Volume Information*. http://www.epa.gov/oppt/iur/tools/data/2002-vol.htm (accessed January 2, 2008).

U.S. Forest Service, 1997, Glyphosphate herbicide information profile. U.S. Forest Service.

Valvoline Oil Company, 1991, Material Safety Data Sheet for Mac's Brake and Electric Motor Cleaner 4700, Valvoline Oil Company. LIIN: 00F029811.

Venediktova, R.I. and Rudnyi, N.M., 1965, Study of mass transfer in the extraction of moisture from capillary-porous materials. *Journal of Engineering Physics and Thermophysics* 9(3): 273–274.

Vijgen, J., 2006, The legacy of lindane HCH isomer production—main report: A global overview of residue management, formulation and disposal. Copenhagen, Denmark: International HCH & Pesticides Association. www.ihpa.info/library_access.php (accessed November 28, 2007).

Vinson, C.G. and Martin, J.J., 1963, Heat of vaporization and vapor pressure of 1,4-dioxane. *Journal of Chemical and Engineering Data* 8(1): 74–75.

VROM, 1999, Risk assessment: 1,4-Dioxane. Bilthoven, Netherlands: S.P. Netherlands Ministry of Housing and the Environment (VROM), Chemical Substances Bureau 153.

Wala-Jerzykiewicz, A. and Szymanowski, J., 1998, Headspace gas chromatography analysis of toxic contaminants in ethoxylated alcohols and alkylamines. *Chromatographia*, 48(3/4): 299–304.

Washington Post, 1979, *A Suspect Chemical is Found in Cosmetics.* July 20, 1979, p. A16.

Watanabe, Y., Kurata, Y., Kawasaki, M., and Ono, Y., 2006, Leaching of the chemical substance from MSW incineration ashes and the shredded residues. *4th Intercontinental Landfill Research Symposium*, Gällivare, Sweden: Swedish Waste Management Association.

Watsco Components, Inc., 1991, Material Safety Data Sheet for 6404 rust eliminator DR-1, issued August 1, 1991, Hialeah, FL. LIIN: 00F025818.

West, F.W., 1959, United States Patent 2,915,416: Ink composition. Assignee: Minnesota Mining and Manufacturing, St. Paul, MN.

Wilke, O., Jann, O., and Brodner, D., 2004, VOC- and SVOC-emissions from adhesives, floor coverings and complete floor structures. *Indoor Air* 14(s8): 98–107.

Winsor, L., 2006, Tissue processing. http://home.primus.com.au/royellis/tp/tp.htm (accessed October 28, 2006).

Woodcock, J., 1997, Letter to associated pharmacologists and toxicologists, Washington, D.C., and to Empire State Consumer Association, Rochester, NY. Rockville, MD: Center for Drug Evaluation and Research, U.S. Food and Drug Administration.

Wright, M.T., Belitz, K., and Burton, C.A., 2004, California GAMA program: Ground-water quality data in the San Diego drainages hydrogeologic province, California, 2004. U.S. Geological Survey Data Series 129.

Wu, J., Xu, Y., Dabros, T., and Hamza, H., 2004, Development of a method for measurement of relative solubility of nonionic surfactants. *Colloids and Surfaces A: Physicochemical and Engineering Aspects* 232: 229–237.

Wurtz, A., 1863, Mémoire sur l'oxyde d'ethyléne et les alcools polyéthyleniques [Memorandum on the ethylene oxide and polyethylene alcohols]. *Annales de Chimie et de Physique* 69: 317–355.

Wypych, G. (Ed.), 2001, *Handbook of Solvents.* Toronto, Ontario: Chemtec Publishing.

Yashushi, J., 1977, United States Patent 4,323,627: Hollow fiber and method of manufacturing the same. Assignee: Nippon Zeon Co., Ltd., Tokyo, Japan.

Yasuhara, A., Shiraishi, H., Nishikawa, M., Yamamoto, T., Nakasugi, O., Okumura, T., Kenmotsu, K., Fukui, H., Nagase, M., and Kawagoshi, Y., 1999, Organic components in leachates from hazardous waste disposal sites. *Waste Management and Research* 17(3): 186–197.

Yasuhara, A., Shiraishi, H., Nishikawa, M., Yamamoto, T., Uehiro, T., Nakasugi, O., Okumura, T., Kenmotsu, K., Fukui, H, Nagase, M., Ono, Y., Kawagoshi, Y., Baba, K., and Noma, Y., 1997, Determination of organic components in leachates from hazardous waste disposal sites in Japan by gas chromatography–mass spectrometry. *Journal of Chromatography A* 774(1–2): 11.

3 Environmental Fate and Transport of Solvent-Stabilizer Compounds

Thomas K.G. Mohr

Industrial waste discharges to the environment are universally associated with degradation of water quality and risk to human and ecological health. Many chemicals in industrial effluents do not persist because of physical removal, chemical reaction, or biodegradation to less harmful by-products. The factors that differentiate compounds that are recalcitrant and compounds that are assimilated into environmental systems without appreciable consequences are called *contaminant fate and transport properties*. Fate and transport properties describe the propensity for a chemical to volatilize, photo-oxidize, adsorb, hydrolyze, or biodegrade and, if the chemical survives these fates, how it may advect, diffuse, or otherwise migrate through soil, groundwater, surface water, and air. These processes not only eliminate some compounds and retard the movement of others, but also affect the phase in which the chemical is present (gas, liquid, or solid). Some compounds are relatively immune to degradation and will persist in the environment for long periods of time. If released in sufficient mass, those contaminants that are persistent, mobile, and also toxic constitute the greatest potential threat to drinking water resources.

Chapter 1 profiled the industrial fate of solvent-stabilizer compounds. The environmental fate of the stabilizers discussed in Chapter 1 is summarized here in brief descriptions and tables.

In this chapter, the environmental fate and transport of stabilizer compounds is examined by profiling their potential to persist in air, surface water, soil, and groundwater. The focus is on those stabilizer compounds that become concentrated by the industrial processes that precede their discharge. A more detailed review is provided of the fate and transport of 1,4-dioxane, a solvent stabilizer that may pose an elevated threat to drinking water resources because of its persistence and mobility.

3.1 FATE AND TRANSPORT PROCESSES

The processes governing the fate and transport of contaminants following their release to the environment are the subject of a number of books and thousands of peer-reviewed papers. Authoritative works on this subject include Freeze and Cherry (1979), Lyman (1990), Dragun (1988), Howard et al. (1991), Chappelle (1993), Mackay et al. (1993), Schwarzenbach et al. (1993), Thibodeaux (1996), Hemond and Fechner (1994), Verschueren (1996), Domenico and Schwarz (1997), Suthersan (2002), and many others. This section briefly reviews the key processes that determine the environmental fate and transport properties for 1,4-dioxane and some of the other solvent-stabilizer compounds.

Fate and transport properties of stabilizer compounds are estimated through a variety of approaches where literature citations of measured or estimated values are not available. The leading approach to property estimation is quantitative structure–activity relationships (QSARs); related approaches include quantitative property–property relationships (QPPRs), quantitative structure–property relationships (QSPRs), group contribution models (GCMs), and similarity-based models (Reinhard and Drefahl, 1999). These methods relate chemical structure to chemical behavior. For example, the measured relationship between molecular weight and boiling point for many compounds permits estimation of boiling points for compounds whose boiling points have not been measured. The EPIWIN Suite[*] from USEPA and Syracuse Research Corporation was used in this chapter to calculate some physicochemical properties for solvent-stabilizer compounds.

3.1.1 ATMOSPHERIC FATE AND TRANSPORT PROCESSES

Solvent-stabilizer compounds can enter the atmosphere through the following industrial processes:

- Direct vapor emission from vapor degreasers, dry cleaners, solvent-charged steam cleaners, and other applications in which solvents are vaporized
- Volatilization from open-top vapor degreasers, vented solvent storage vessels such as tanks and drums, and waste lagoons or from solvent wastes poured on the ground for disposal
- Volatilization from aqueous solutions discharged as condensation water in degreaser or dry cleaner water traps, solvent-laden steam-cleaning wastewater, surface-water bodies to which solvent wastes were discharged, or air-stripping towers used to treat solvent-contaminated groundwater
- Emissions of uncombusted stabilizer compounds from solvent incineration operations at cement kilns and waste incinerators

As discussed in the next section, contaminated soil and groundwater can serve as secondary release sources to the atmosphere.

3.1.2 VOLATILIZATION FROM DRY SOIL

Estimating the rate of chemical transfer from soil to air is complicated by the wide range of possible fates for the stabilizer compound when discharged to soil. In addition to volatilizing directly into air, the compound may be subjected to some or all of the following abbreviated list of possible fates:

- Sorption to soil mineral and organic matter surfaces
- Migration under matric suction into the soil
- Penetration into the soil as a liquid and then volatilization into a vapor that sinks or raises depending on vapor density and soil-air pressures
- Dissolution into soil moisture and downward migration to groundwater
- Absorption into plant roots
- Microbial metabolic processes

An effective strategy for estimating the transfer of a particular chemical from dry soil to air alone is to compartmentalize the system and assess the degree to which partitioning between the two media is expected to occur. This simplification assumes that chemicals in the environment are present in a pure state; however, contaminant releases usually occur as complex mixtures (Kinerson, 1987). This approach forms the basis for broad estimates of volatilization and other contaminant fate and transport processes.

The rate of evaporation of a chemical from dry soil is a function of soil, chemical, and air temperatures, wind speed, air turbulence and surface roughness, air humidity, solar radiation, and the relevant properties of the chemical, including total mass released, molecular weight, vapor pressure, vapor density, and diffusivity in air (Mackay et al., 1993; Thibodeaux, 1996). In particular, vapor pressure describes the concentration of a chemical as its partial pressure in the gas phase in air above a liquid sample, usually measured at 20°C or 25°C and reported in pressure units (e.g., mm Hg). Vapor density is the mass per unit volume of a chemical in the vapor phase at a fixed temperature, usually expressed as a ratio to air density. Vapors heavier than air are reported as a multiple of air density and measured at 25°C (e.g., perchloroethylene, vapor density = 5.7). Diffusion is the

---

[*] Estimations Programs Interface for Windows (EPI Suite) (USEPA, 2007a).

average rate of migration of a chemical in air in response to temperature, pressure, and concentration gradients exclusive of any chemical movement in response to advection. The air diffusion constant is sometimes called air diffusivity, often denoted as D_a (expressed in units of cm²/s). Temperature affects the air diffusion constant, which affects the volatilization rate.

Values for the air diffusion constant, D_a, are not widely available from experimental data. Methods to estimate D_a include empirical equations based on a compound's molecular weight and specific gravity. The following equation has been used to estimate D_a (USEPA, 2001a):

$$D_a = \frac{0.0029(T + 273.16)^{3/2}\sqrt{0.034 + (1/MW)}(1 - 0.00015MW^2)}{\left[(MW/2.5\rho)^{1/3} + 1.8\right]^2}, \tag{3.1}$$

where D_a is the diffusion coefficient of the chemical in air (in cm²/s), T is the temperature (in °C), MW is the molecular weight (in g/mol), and ρ is the density (in g/cm³).

A useful index for contrasting vapor densities of single compounds is relative vapor density (RVD). A compound's RVD is the ratio of the density of dry air saturated with that compound at 25°C and 1 atm total pressure to the density of dry air. The RVD may be calculated by using Equation 3.2 (Pankow and Cherry, 1996):

$$RVD = \frac{(p°/760)MW + [(760 - p°)29.0]/760}{29}, \tag{3.2}$$

where $p°$ is the saturated vapor pressure, MW (in g/mol) is the molecular weight, and 29 g/mol is the mean molecular weight of dry air. RVD is dependent on temperature, so the temperature at which the RVD is calculated must also be stated (usually at 25°C). The saturated vapor pressure $p°$ can be obtained from the Antoine equation.[*] At a compound's boiling point, RVD = MW/29.

Table 3.1 provides vapor pressure values cited in the literature and estimated values of vapor pressure, vapor density, and air diffusion constants for 1,4-dioxane and other commonly used solvent stabilizers likely to be present in vapor degreasing and other solvent wastes. For comparison, data for the major chlorinated solvents are included.

Chemicals on dry soil partition to the vapor phase according to the parameters in Table 3.1. Once in the vapor phase, advection causes vapor to move from soil to turbulent air, and diffusion causes upward movement of vapor into still air (Dragun, 1988). A number of empirical equations are available to estimate the rate of volatilization of a pure chemical from dry soil. The rates at which chemicals diffuse at a given temperature are inversely proportional to the square roots of their molecular weights:

$$\frac{D_{a_1}}{D_{a_2}} = \left(\frac{MW_1}{MW_2}\right)^{1/2}. \tag{3.3}$$

An equation for estimating the rate of vapor generation of a pure chemical from dry soil under steady-state conditions was proposed by Shen (1981):

$$E = 2P_v W_A \left[\frac{L_A D_a v}{(\pi f)}\right]^{1/2}\left(\frac{W_c}{W}\right), \tag{3.4}$$

---

[*] The Antoine equation describes the relationship between saturated vapor pressure of pure substances and temperature by the relationship $P = 10[A - (B/(C+T))]$, whereby historic convention P is the pressure (in mm Hg), T is the temperature (in °C), and A, B, and C are the Antoine equation coefficients available from the National Institute of Standards and Technology (http://webbook.nist.gov/chemistry/).

TABLE 3.1

Properties Affecting Volatilization Rates from Dry Soil for Chlorinated Solvents and Selected Stabilizer Compounds

Compound	Molecular Weight (g/mol)	Vapor Pressure (mm Hg, 25°C)	Vapor Density Relative to Air (g/cm³)	Diffusivity (cm²/s)	Evaporation Rate Relative to Evaporation Rate of Butyl Acetate
Stabilizers					
1,4-Dioxane	**88.1**	**38.09 [1]**	**3.03 [2]**	**0.229 [3]**	**2.7; 2.42 [4]**
1,3-Dioxolane	74.08	70 at 20°C [5]	2.6 [7]	*0.146*[a]	0.29 [18]
		79 at 20°C [6]			
tert-Butyl alcohol	74.12	40.7 [1]	2.55 [8]	*0.115*[a]	1.05 [19]
sec-Butyl alcohol	74.12	18.3 [1]	2.6 [7]	*0.101*[a]	1.3 [20]
tert-Amyl alcohol	88.15	13.8 [1]	3	*0.102*[a]	0.93 [21]
Epichlorohydrin	92.52	16.4 [1]	3.29 [8]	0.086 [3]	1.35 [6]
Nitroethane	75.08	20.8 [1]	2.58 [8]	*0.165*[a]	1.2 [9]
Nitromethane	61.0	35.8 [1]	2.11 [8]	*0.116*[a]	1.39 [10]
Cyclohexane	84.18	96.9 [11]	2.9 [12]	*0.089*[a]	6.1 [12]
Triethanolamine	149.19	3.59×10^{-6} [1]	5.1 [13]	*0.121*[a]	<1 [13]
		< 0.01 [13]			
1,2-Butylene oxide	72.11	207 [3]	2.2 [7]	0.135 [3]	6.05 [16]
		180 [14]			
Solvents					
Methyl chloroform	133.42	100 [15]	4.63 [2]	0.078 [3]	4.6 [15]
Dichloromethane	84.93	355 [15]	2.93 [16]	0.101 [3]	7.0 [15]
Trichloroethylene	131.39	60 [15]	4.53 [17]	0.079 [3]	3.0 [15]
Perchloroethylene	165.83	14 [15]	5.7 [17]	0.072 [3]	1.5 [15]

Sources: [1] Daubert and Danner (1985); [2] Verschueren (1996); [3] USEPA (1998a); [4] USEPA (1981); [5] Riddick et al. (1985); [6] Sax and Lewis (1987); [7] NFPA (1991); [8] Sax (1984); [9] Mackison et al. (1981); [10] NLM (2006); [11] Chao et al. (1983); [12] Fisher Scientific (1999); [13] Baker (2006); [14] Osborn and Scott (1980); [15] Dow Chemical (2006); [16] Kirk and Othmer (1982); [17] Budavari (1996); [18] Beaujean (2007); [19] NIOSH (1978); [20] Smallwood (1996); and [21] Cheremisinoff and Archer (2003).

[a] Air diffusion constants displayed in italic font were estimated by using Equation 3.1 at 25°C.

where E is the emission rate (in cm³/s), P_v is the equivalent vapor pressure (in percent), where P_v = vapor pressure (in mm Hg) divided by 760 mm Hg, W_A is the width of the area occupied by the chemical (in cm), L_A is the length of the area occupied by the chemical (in cm), D_a is the diffusion coefficient of the chemical in air (in cm²/s), generally at 25°C, v is the wind speed (in cm/s), W_c/W is the weight fraction of the chemical in soil (in g/g), and f is the correction factor, where $f = (0.985 - 0.00775P_v)$ and the range of P_v is 0–80%. The above volumetric emission rate can be converted to a mass emission rate:

$$Q = \frac{E(MW)}{G},\qquad(3.5)$$

where Q is the mass emission rate (in g/s), E is the volumetric emission rate (in cm³/s), MW is the molecular weight (in g/mol), and the empirical conversion factor G is 24,860 cm³/mol (Dragun, 1988).

Many other empirical equations are available for estimating rates of volatilization; however, given the many unknown aspects of ground surface conditions, the best use of such equations is to determine the relative volatility and transfer rates among compounds instead of seeking absolute transfer rates for an individual compound. Several equations for estimating rates of volatilization

TABLE 3.2

Calculated Relative Mass Flux Estimates (Chemical Volatilization Rates) of Chlorinated Solvents and Stabilizer Compounds from Dry Soil for a 10 L Spill in a 1.5 m² Area

Compound[a]	Molecular Weight (g/mol)	Vapor Pressure (mm Hg, 25°C)	Density (g/cm³)	10 L Spill Mass (g)	Mass Fraction in Soil[b] (%)	Diffusion Constant[c] D_a	Mass Flux[d] Q (g/s)
Dichloromethane	84.93	355	1.3255	13,255	0.79	0.101	22.0
1,2-Butylene oxide	72.11	207	0.8297	8297	0.50	0.135	9.1
Methyl chloroform	133.42	100	1.3376	13,376	0.80	0.078	7.6
Trichloroethylene	131.39	60	1.4642	14,642	0.88	0.079	5.0
1,3-Dioxolane	74.08	70	1.06	10,600	0.63	*0.146*	4.4
1,4-Dioxane	**88.1**	**38.09**	**1.0329**	**10,329**	**0.62**	**0.229**	**4.3**
Cyclohexane	84.18	96.9	0.779	7790	0.47	*0.089*	3.1
Nitromethane	61	35.8	1.1322	11,322	0.68	*0.116*	1.6
tert-Butyl alcohol	74.12	40.7	0.78581	7858	0.47	*0.115*	1.5
Perchloroethylene	165.83	14	1.6227	16,227	0.97	0.072	1.5
Nitroethane	75.08	20.8	1.0448	10,448	0.63	*0.165*	1.5
Epichlorohydrin	92.52	16.4	1.175	11,750	0.70	0.086	0.8
sec-Butyl alcohol	74.12	18.3	0.8063	8063	0.48	*0.101*	0.6
tert-Amyl alcohol	88.15	13.8	0.8096	8096	0.48	*0.102*	0.5
Triethanolamine	149.19	3.59×10^{-06}	1.1242	11,242	0.67	*0.121*	4×10^{-07}

[a] Nonitalics compounds are stabilizers. Italics compounds are chlorinated solvents; these are listed for comparison to show whether the stabilizer is likely to evaporate with the solvent or remain behind in the soil where it can be leached down to groundwater.

[b] Mass fraction = grams of compound per gram of soil, if soil bulk density is assumed to be 1.85 g/cm³.

[c] Air diffusion constants shown in italic font were estimated by using Equation 3.1, with temperature set to 25°C.

[d] Mass flux was calculated by using the Shen (1981) equation (Equation 3.4 in the text).

from soil are found in Dragun (1988). Calculated estimates of the relative mass flux of chlorinated solvents and stabilizer compounds from dry soil to air are provided in Table 3.2.

The Shen estimate of relative volatilization rates from dry soil isolates one mode of transfer, but discharge of a single chemical to dry soil is an uncommon occurrence. Generally, wastes are discharged as complex mixtures of chlorinated solvents with concentrated stabilizers, oil, grease, soldering flux, acids, and water. Soil-water plays an important role in the fate and transport of chemicals discharged to soil. Mixtures of wastes will partition from moist soil to air according to their solubilities, vapor pressures, chemical stability, and capacity to adsorb to soil minerals and organic matter. As discussed in the next section, the interaction of the chemical at the water–air interface plays an important role in fate and transport processes.

3.1.3 VOLATILIZATION FROM WATER

Several possible fate and transport mechanisms act on a chemical discharged to wet soil, groundwater, or surface water. The propensity for a chemical to partition between aqueous and vapor phases is a well-studied phenomenon, as Henry's law mathematically describes.

3.1.3.1 Henry's Law

The constant describing the equilibrium concentration of a compound in both the aqueous and vapor phases at the liquid–air interface at a fixed temperature is called the Henry's law constant.

The general principle is that the higher that constant, the greater the potential for the compound to volatilize from the water surface. The Henry's law constant is a function of molecular structure and air and water temperatures (Goss, 2006).

The Henry's law constant, H, is usually defined at a particular temperature and at equilibrium as

$$H = \frac{C_a}{C_w}, \tag{3.6}$$

where C_a is the concentration of a compound in air and C_w is the concentration of that compound in water. H is conventionally defined in terms of gas concentrations (in atm) and liquid concentrations (in mol/m^3); therefore, typical units for H are (atm·m^3)/mol. H is usually reported from measurements made at 25°C or 20°C and 1 atm pressure. It is important to take note of the temperature at which H was measured to ensure that values for different compounds can be compared.

The dimensionless form of the Henry's law constant, noted as H_C, is obtained by converting gas concentrations from partial pressures in atmospheres to moles per cubic meter. The ideal gas law relates pressure, volume, temperature, and number of moles:

$$\frac{n}{V} = \frac{P}{RT}, \tag{3.7}$$

$$H_C = \frac{H}{RT}, \tag{3.8}$$

where R is the universal gas constant and is equal to 8.2×10^{-5} (m^3 atm)/(mol K); T is the temperature (in K). Note that the dimensionless H_C actually has the units of concentration of the gas (in mol/m^3) per concentration of the liquid (in mol/m^3); the units can also be volume (in m^3 of liquid) per volume (in m^3 of gas).

The Henry's law constant is often stated as the simple ratio of a compound's vapor pressure to its aqueous solubility. A limitation to this interpretation of Henry's law is the inherent assumption that water does not dissolve into the compound. The vapor pressure of the pure compound is used in this statement of Henry's law, but the solubility used is for the compound when saturated with water. The Henry's law constant estimates that are calculated from the compound's vapor pressure and solubility fail where the solubility of water in the chemical exceeds a few percent (Corsi, 1998). An exhaustive compilation of measured Henry's law constants is found in Sander (1999).

The Henry's law constant describes the tendency of a compound to escape the liquid phase and move into the gaseous phase. The term for the potential to transfer from one medium to another is *fugacity*, or escaping tendency, and is a function of a compound's activity coefficient (Arbuckle, 1983). Compounds such as dichloromethane with high vapor pressures (i.e., with high fugacity) and high aqueous-phase activity coefficients will tend to partition to the gas phase to equalize the chemical potentials for the compound in water and air; thus such compounds have high values of H. Conversely, compounds such as alcohols with low vapor pressures and high solubilities (i.e., low activity coefficients in water) will have low values of H and will tend to partition into the aqueous phase (Schwarzenbach et al., 1993).

Where the ionic strength of a solution increases, as may occur when polluted water is discharged to marine waters, a compound's solubility in the saline solution will be lower, while its vapor pressure remains constant, leading to higher values of H and increased rates of volatilization. This phenomenon, called electrostriction or "salting out," is leveraged in laboratory analysis, as described in Chapter 4. Table 3.3 provides Henry's law constants in two forms for solvent-stabilizer and chlorinated solvent compounds.

3.1.3.2 Mass Transfer Rates from Water to Air: Flux Density

Henry's law values can be used to estimate the rate of mass transfer, or flux density, from a surface-water body by evaluating the volatility of an aqueous solution in isolation from other factors. The air–water interface can be idealized as a thin layer of static air above the water surface, above which

TABLE 3.3
Henry's Law Constants for Stabilizer and Chlorinated Solvent Compounds

Compound	H [(atm m³)/mol]	H_C (dimensionless)[b]	References
Cyclohexane	1.50×10^{-1}	6.14	Bocek (1976)
Perchloroethylene	1.77×10^{-2}	7.24×10^{-1}	Gossett (1987)
Trichloroethylene	9.85×10^{-3}	4.03×10^{1}	Leighton and Calo (1981)
Methyl chloroform	8.00×10^{-3}	3.27×10^{-1}	Lyman (1990), Dilling et al. (1975)
Dichloromethane	3.25×10^{-3}	1.33×10^{-1}	Leighton and Calo (1981)
tert-Amyl alcohol	7.30×10^{-4}	2.99×10^{-2}	Butler et al. (1935)
1,2-Butylene oxide	1.80×10^{-4}	7.37×10^{-3}	Bogyo et al. (1980)
Nitroethane	4.76×10^{-5}	1.95×10^{-3}	Gaffney et al. (1987)
Epichlorohydrin	3.00×10^{-5}	1.23×10^{-3}	Lyman (1990)
Nitromethane	2.59×10^{-5}	1.06×10^{-3}	Lyman (1990)
1,3-Dioxolane	2.40×10^{-5}	9.82×10^{-4}	Hine and Mookerjee (1975)
sec-Butyl alcohol	9.06×10^{-6}	3.71×10^{-4}	Snider and Dawson (1985)
tert-Butyl alcohol	9.05×10^{-6}	3.70×10^{-4}	Altschuh et al. (1999)
1,4-Dioxane	**4.80×10^{-6}**	**1.96×10^{-4}**	**Park et al. (1987)**
Triethanolamine	1.00×10^{-7}	4.09×10^{-6}	Hine and Mookerjee (1975)

[a] Nonitalics compounds are stabilizers. Italics compounds are chlorinated solvents; these are listed for comparison to show whether the stabilizer is likely to evaporate with the solvent or remain behind in the soil where it can be leached down to groundwater.

[b] Dimensionless values of Henry's law constants calculated from literature values by using Equation 3.8.

turbulent air flows, and a thin layer of stagnant water beneath the water surface, below which water flows. In "thin-film theory," the thickness of the static air layer above the water surface is approximated as 1 mm, whereas the stagnant water layer is taken to be 0.1 mm thick (Schwarzenbach et al., 1993). Molecular diffusion is the dominant transport mechanism in the thin films above and below the water surface, whereas turbulent diffusion is active in the fluids above and below the thin films. Equation 3.9 provides an expression for flux density (Hemond and Fechner, 1994):

$$J = -\left[\frac{1}{\delta_w / D_w + \delta_a / (D_a \cdot H)}\right]\left[C_w - \frac{C_a}{H}\right], \tag{3.9}$$

where J is the flux density($M/L^2 T$), δ_w is the thickness of the hypothetical thin water layer (L), D_w is the molecular diffusion coefficient for the chemical in water (L^2/T), δ_a is the thickness of the hypothetical thin air layer (L), D_a is the molecular diffusion coefficient for the chemical in air (L^2/T), H is the Henry's law constant (dimensionless), C_w is the chemical concentration in water (M/L^3), C_a is the chemical concentration in air (M/L^3), and L, M, and T represent any consistently applied units of length, mass, and time.

Because wind speed affects the thickness of the air film above water and water flow or circulation affects the thickness of the water film below the interface, this flux density equation should only be used for rough approximations or to compare relative mass flux rates. The air film is 10 or more times thicker than the water film, and the molecular diffusion coefficient for a compound in air is about 10,000 times higher than its molecular diffusion coefficient in water. If the air concentration of the compound of interest is zero, then the flux density is directly proportional to the magnitude of the compound's Henry's law constant and its concentration in water (Thomas, 1990). Table 3.3 can therefore be interpreted to provide an approximation of the relative magnitude of mass transfer from aqueous solutions to air for fixed concentrations. Comparison of relative mass transfer rates inferred from ratios of the Henry's law constants for chlorinated solvents and their stabilizers suggests that water-to-air partitioning after a waste is discharged to a surface-water body may favor the ether, epoxide, nitroalkane, and alcohol stabilizers remaining in water, whereas the solvents and cyclohexane volatilize to the atmosphere. The Henry's law constant is also a key parameter for designing groundwater and wastewater treatment systems where air stripping is viable. Chapter 7 discusses the suitability of air stripping for remediation of 1,4-dioxane.

A more tangible means of relating the Henry's law constant to the environmental fate of stabilizers discharged to surface water is to estimate their rates of volatilization from rivers and lakes. Models are available to predict the volatilization half-life of a compound in water. Modeled half-lives for stabilizer and chlorinated solvent compounds are profiled in Table 3.4. The model used was Estimation Programs Interface Suite, EPI Suite™ (a trademark of ImageWare Systems®, Inc.), which predicts physical–chemical properties of compounds entered by their Chemical Abstracts Service Registry Number or by their SMILES notation.* The model within EPI Suite for estimation of volatilization from water, WVOLNT, or Water Volatilization Program is based on the method outlined in *Handbook of Chemical Property Estimation Methods* (Thomas, 1990).

The accuracy of the properties estimated by EPI Suite and the associated estimates of chemical phase partitioning depends on the chemical's class, the quality of the available chemical data used by the model as a training set, and whether the chemical's properties fall within the range of the 353 chemicals in the training data set. In general, EPI Suite is able to predict the measured property value within an order of magnitude of measured values, provided the chemical is appropriate for the types of regression and other estimation techniques used (USEPA, 2007a).

3.1.4 ATMOSPHERIC FATE OF STABILIZER COMPOUNDS

Chemical reactions in the atmosphere may occur in the vapor phase as gas-phase collisions between molecules, on the surfaces of airborne solid particulate matter, and in aqueous solution in water droplets. Reactions in water droplets are predominately of acid–base type. Reactions on particle surfaces are of minor importance because of their short residence time in the atmosphere. Gas-phase reactions are the dominant chemical reaction that transforms chemicals released to the atmosphere (Seinfeld, 1986).

Physical reactions—including condensation and precipitation, dissolution into water droplets followed by rain, and reactions with sunlight—can also remove, transform, or eliminate stabilizer compounds emitted to the atmosphere. The longevity of a chemical in the atmosphere is referred to as its atmospheric half-life, that is, the length of time before half the mass of the emitted compound is removed, transformed, or eliminated by chemical or physical reactions. Most studies of atmospheric half-lives for chemicals involve laboratory bench-top testing of compounds reacting with simulated sunlight or atmospheric oxidants such as hydroxyl radicals, nitrate, ozone, and chlorine. Modeling studies are also used to predict the atmospheric fate of compounds subjected to chemical and physical reactions.

The most important atmospheric reactions affecting stabilizer compounds are photolysis, in which ultraviolet (UV) light reacts directly with the compound to break it down, and photo-oxidation, a reaction with hydroxyl radicals, chlorine, and other oxidants in the atmosphere. These reactions are described in more detail in the following sections.

3.1.4.1 Photolysis

Photolysis occurs when a chemical absorbs light energy and undergoes chemical transformation. Light is said to excite electrons and destabilize bonds in some compounds. Absorption of light does

---

* SMILES is "Simplified Molecular Input Line Entry System," a notation system for describing molecular structure. See Reinhard and Drefahl (1999) for a brief introduction.

TABLE 3.4
Modeled Volatilization Half-Lives and Volatilization Residence Times for Solvent-Stabilizer Compounds in a Model River and a Model Lake

Chemical Name[a]	Estimated Volatilization from Model River (Half-Life)	Time for 99% Removal from Model River	Estimated Volatilization from Model Lake (Half-Life)	Time for 99% Removal from Model Lake
Cyclohexane[b]	0.9 h	6.3 h	3.6 days	25 days
Isopentane	0.9 h	6.3 h	3.4 days	24 days
Isoprene	0.9 h	6.3 h	3.3 days	23 days
Diisobutylene	1.1 h	7.7 h	4.2 days	29 days
Dichloromethane[b]	1.1 h	7.7 h	3.7 days	26 days
Methyl chloroform[b]	1.2 h	8.4 h	4.6 days	32 days
Trichloroethylene	1.2 h	8.4 h	4.6 days	32 days
Carbon tetrachloride	1.3 h	9.1 h	4.9 days	34 days
Perchloroethylene	1.4 h	9.8 h	5.1 days	36 days
1,2-Butylene oxide[b]	3.6 h	1.1 days	4.6 days	32 days
Cyclohexene oxide	4.5 h	1.3 days	5.5 days	39 days
Triethylamine	5 h	1.5 days	5.8 days	41 days
Ethyl acetate[b]	5.1 h	1.5 days	5.6 days	39 days
Isopropyl acetate[b]	6.2 h	1.8 days	5 days	35 days
Diisopropylamine	7.2 h	2.1 days	6.8 days	48 days
Propylene oxide[b]	7.2 h	2.1 days	5.9 days	41 days
Acetonitrile	12 h	3.5 days	7.5 days	53 days
2-Butanone[b]	16 h	4.7 days	7.3 days	51 days
Nitromethane	17 h	5.0 days	10 days	70 days
Nitroethane	19 h	5.5 days	8.3 days	58 days
Epichlorohydrin	20 h	5.8 days	12 days	84 days
1,3-Dioxolane	1.4 days	9.8 days	13 days	91 days
Styrene oxide	1.7 days	12 days	23 days	161 days
Nonylphenol	1.8 days	13 days	18 days	126 days
tert-Amyl alcohol	2.6 days	18 days	22 days	154 days
Methanol	3.1 days	22 days	35 days	245 days
Pyridine	3.1 days	22 days	25 days	175 days
sec-Butyl alcohol[b]	3.6 days	25 days	29 days	203 days
tert-Butyl alcohol[b]	3.6 days	25 days	29 days	203 days
1,4-Dioxane[b]	**4.8 days**	**34 days**	**56 days**	**1.1 years**
Butoxymethyl oxirane	6.4 days	45 days	74 days	1.4 years
2-Methyl-3-butyn-2-ol	8.7 days	61 days	66 days	1.3 years
2,6-di-*tert*-Butyl-*p*-cresol	16 days	112 days	123 days	2.4 years
p-tert-Amyl phenol	23 days	161 days	173 days	3.3 years
Diepoxybutane	28 days	196 days	303 days	5.8 years
Thymol	62 days	1.2 years	1.9 years	13.3 years
Phenol	71 days	1.4 years	2.1 years	15 years

Notes: The times predict only volatility; biodegradation, photolysis, sorption, hydrolysis, etc., are not addressed. Half-lives shown are exclusive of other fate processes such as photodegradation, hydrolysis, and biodegradation. Model assumptions: river = 1 m deep, flow is 1 m/s, wind speed is 5.0 m/s; lake = 1 m deep, wind velocity is 0.5 m/s, and circulation or current is 0.05 m/s. Modeled using EPI Suite™ version 3.20 (USEPA, 2007a).

[a] Nonitalics compounds are stabilizers. Italics compounds are chlorinated solvents; these are listed for comparison to show whether the stabilizer is likely to evaporate with the solvent or remain behind in the soil where it can be leached down to groundwater.

[b] Chemical is included in the EPI Suite™ training set.

not result in degradation in the majority of compounds; instead, absorbed energy is released through fluorescence, and the compounds return to their beginning energy state (Hemond and Fechner, 1994). Photolysis, also called direct photodegradation, is more likely to occur in compounds that have double carbon bonds, such as alkenes (e.g., perchloroethylene and trichloroethylene) and aromatic rings (e.g., benzene and toluene). Shorter-wavelength light has higher frequency and energy and is therefore the primary agent in photolysis. Visible light in the wavelength range from 280 to 730 nm (nanometers) is primarily responsible for photolysis of chemicals in the upper atmosphere (Seinfeld, 1986). Photolysis of some compounds may occur in both the atmosphere and in surface-water bodies; however, because light in the lower atmosphere has longer wavelengths, many compounds are not directly photolyzed.

The intensity of sunlight passing through the atmosphere is decreased through absorption by ozone and other atmospheric gases and by molecular and aerosol scattering. Essentially no light is transmitted to the Earth's surface atmosphere (the troposphere) at wavelengths less than 295 nm, and there is a sharp decrease in intensity in the 280–320 nm wavelength range due mainly to ozone absorption. Sunlight in this wavelength interval, often called UV-B radiation, causes sunburn as well as direct photolysis of many pollutants residing in the upper layers of surface-water bodies (Zepp and Cline, 1977).

3.1.4.1.1   Photolysis of 1,4-Dioxane

1,4-Dioxane is photolyzed at wavelengths shorter than the wavelengths of light penetrating the troposphere, that is, less than 290 nm; therefore, direct photolysis is inconsequential to the atmospheric fate of 1,4-dioxane. Nevertheless, photolysis experiments hold interest for developing treatment technologies involving UV light (see Chapter 7) and for understanding the various means by which the 1,4-dioxane molecule can be disassembled.

Laboratory experiments on photolysis of 1,4-dioxane produced volatile solid by-products, including p-formaldehyde and trioxane. Proposed pathways (Hentz and Parrish, 1971) for direct photolysis of 1,4-dioxane vapor at 1470 Å (147 nm) include

$$C_4H_8O_2 \xrightarrow{\ hv\ } C_2H_4 + 2CH_2O, \tag{3.10}$$

$$C_4H_8O_2 \xrightarrow{\ hv\ } C_2H_4 + CH_2O + H_2 + CO, \tag{3.11}$$

$$C_4H_8O_2 \xrightarrow{\ hv\ } H_2 + [\text{Unidentified product}]. \tag{3.12}$$

1,4-Dioxane absorbs UV light in the wavenumber range* 52200–60510 cm^{-1} (165–191 nm), producing an absorption curve with a single peak, which may be the superposition of absorption associated with the two symmetrical oxygen atoms on the ether ring (Pickett et al., 1951). A more detailed pathway for direct photolysis of 1,4-dioxane exposed to light at 190 nm for 200 h was elicited in later studies, as shown in Figure 3.1.

The hydrogen abstraction from 1,4-dioxane ([B] in Figure 3.1) is disputed; other researchers find the primary mechanism for 1,4-dioxane photolysis to be scission of the CO bond rather than cleavage of the CH bond (Houser and Sibbio, 1977). Photolysis experiments on liquid 1,4-dioxane conducted at 185 nm produced "an exceedingly complex mixture" including six major components and 50 minor components. The products included ethylene glycol, glycolaldehyde, dioxanone, hydroxymethyldioxane, and a variety of ethers, alcohols, aldehydes, and some carbonyl compounds (Houser and Sibbio, 1977). Vapor-phase photolysis of 1,4-dioxane at 147 nm produced formaldehyde and ethylene (Hentz and Parrish, 1971).

---

* Wavenumber (in cm^{-1}) is the inverse of wavelength.

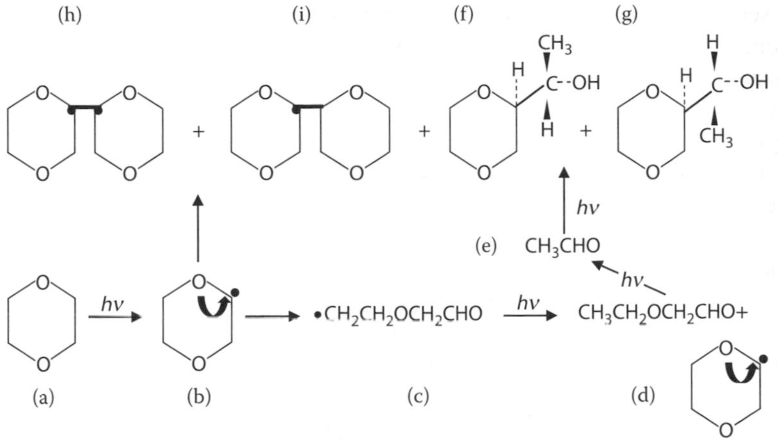

FIGURE 3.1   Laboratory photolysis of 1,4-dioxane [A] produces a dioxyl radical [B] by abstracting a hydrogen from dioxane (indicated by the arrow and dot in [B]). A pair of dioxane dimers ([H] and [I]) is formed by the dimerization of dioxyl radicals. The dioxyl radical also undergoes bond cleavage to open the ring and form radical [C], which then forms ethoxyacetaldehyde [D] and, with further light exposure, acetaldehyde [E]. Acetaldehyde can be photo-reduced to the alcohol isomers [F] and [G]. *Note*: These reactions are not expected to occur in the natural environment. (From Mazzocchi, P.H., and Bowen, M.W., 1975, *Journal of Organic Chemistry* 40(18): 2689–2690. With permission.)

Some experiments have sought to replicate the light range to which 1,4-dioxane would be exposed in the near-surface atmosphere. One experiment simulated UV light with a short-wavelength cutoff of 290 nm and an intensity 2.6 times greater than the noonday sun found in Freeport, Texas (Dilling et al., 1976). Half the mass of 1,4-dioxane disappeared in 3.4 h, giving a corrected half-life of 8.8 h to account for the artificially enhanced light intensity. However, this experiment was conducted in the presence of 5 ppm nitrous oxide, an oxidant involved in photo-oxidation, as discussed further in Section 3.1.4.2.

3.1.4.1.2   Photolysis in Surface Water

The degree to which light penetrates surface water governs the degree to which chemicals in solution are photolyzed. The rates of photolytic reactions in a water body are affected by the solar spectral irradiance at the water surface, radiative transfer of light from air into water, and the transmission of sunlight in the water body.

Indirect or "sensitized" photolysis occurs in surface water when light-absorbing natural organic compounds absorb photons. These compounds, called chromophores, can transfer energy in the form of electrons or hydrogen ions to another compound. Titanium dioxide and other inorganic compounds can also serve as an intermediate in the indirect photolysis of organic contaminants (Hemond and Fechner, 1994). The degree to which contaminants in surface water may be susceptible to photolysis or indirect photodegradation from chromophores is limited by the depth of light penetration in the surface-water body. Light penetration in water is governed by light intensity,* the angle of light entry, turbidity, and dissolved organic matter in water. Light intensity at the water surface decreases with the angle of sunlight entering the water; therefore, intensity decreases from midday to sunset, from summer to winter, and from the tropics to higher latitudes, particularly in the UV-B range (280–320 nm), but also for visible light (UV-A, 320–400 nm) (Zepp and Cline, 1977). Intensity

---

* Light intensity is the number of photons per unit surface area per unit time.

also varies by seasonal changes in atmospheric ozone content and cloud cover. Less than 10% of light from the sun or scattered light from the sky arriving at the water surface is reflected.

Light intensity diminishes with water depth because of scattering and absorption by dissolved humic substances such as humic and fulvic acids and by suspended sediment. The depth range in which photolysis is an active agent for chemical degradation may be on the order of a few centimeters in highly turbid water to several meters in water with greater clarity and lower levels of dissolved organic matter. In the ocean, light absorption is due primarily to water itself. Water is most transparent in the blue region, and scattering is relatively independent of wavelength; hence, solar radiation in clear ocean water acquires a blue hue at great depths. Sunlight penetrates deeper into ocean waters than into inland surface waters, where absorption is due mainly to dissolved organic matter (Zepp and Cline, 1977; Hemond and Fechner, 1994).

A compound in solution in a surface-water body will undergo photolysis if its molecules can absorb UV light in the wavelength range penetrating water and if the bond energy is lower than the energy of the absorbed light. 1,4-Dioxane and other ethers are weak absorbers of UV light in the wavelength range that persists through the troposphere and penetrates surface waters (Houser and Sibbio, 1977; Hill et al., 1997). Therefore, direct photolysis of 1,4-dioxane in surface water is not expected to occur. In addition to ethers, the alcohol, amine, and nitrile stabilizer compounds also do not absorb light in the solar spectral region, that is, between 290 and 700 nm (Harris, 1990a).

3.1.4.2 Photo-Oxidation in the Atmosphere

Photo-oxidation is the reaction of an atmospheric or aqueous compound with hydroxyl radicals and other oxidants. Hydroxyl radicals, $OH^\bullet$, are highly reactive and can degrade a wide range of organic contaminants (Edney and Corse, 1987; Atkinson, 1988).[*] They are produced from the reaction of sunlight with water vapor in the atmosphere or in surface water. In air, hydroxyl radicals may form from four related reactions:

$$H_2O \xrightarrow{\;hv\;} H^+ + OH^\bullet, \tag{3.13}$$

$$H^+ + H_2O \xrightarrow{\;hv\;} H_2 + OH^\bullet, \tag{3.14}$$

$$H^+ + O_2 \xrightarrow{\;hv\;} OH^\bullet + O, \tag{3.15}$$

$$O + H_2O \xrightarrow{\;hv\;} 2OH^\bullet. \tag{3.16}$$

In reaction 3.13, the hydroxyl radical is most commonly produced by breaking the O–H bond of the water molecule to form a hydrogen atom (H^+) and a hydroxyl radical ($OH^\bullet$). In reaction 3.14, the resulting hydrogen atom then reacts with another water molecule to form a hydrogen molecule (H_2) and a second hydroxyl radical. Alternatively, via reaction 3.15, the hydrogen atom reacts with an oxygen molecule (O_2) to form a second hydroxyl radical and an oxygen atom. The new oxygen atom can then react with another water molecule via reaction g to form two new hydroxyl radicals. Hydroxyl radical formation by these reactions supplies a concentration of about 5–10 million hydroxyl radicals per cubic centimeter of air at ground level (Seinfeld, 1986).

Photons impinging on hydrogen peroxide can cause it to dimerize into two hydroxyl radicals, $OH^\bullet$. Hydroxyl radicals constitute the most important catalyst for atmospheric reactions. The potential for photodegradation of stabilizer compounds likely to be released to the atmosphere is summarized in Table 3.5.

Hydroxyl radicals are produced in the presence of sunlight. At night, there is a 10-fold decrease in the concentration of hydroxyl radicals. Atmospheric residence times exceeding 12 h may result in

---

[*] In the notation $OH^\bullet$, the dot represents a free electron; hv represents photon energy, where h is Planck's constant, $h = 6.626 \times 10^{-34}$ J s $= 4.136 \times 10^{-15}$ eV s, and v is the light frequency (in Hz).

TABLE 3.5
Calculated and Experimental Atmospheric Half-Lives of Stabilizer Compounds

Compound	Photo-Oxidation in Presence of 5 ppm NO		Photo-Oxidation	
	$t_{1/2}$	Reference	$t_{1/2}$	Reference
n-Methyl pyrrole	0.4 h	1	—	
2,2,4-Trimethyl pentene-1	1.6 h	1	—	
Cyclohexene	2.3 h	1	—	
Triethanolamine	–		4 h	2
sec-Butyl alcohol	17 h	1	—	
1,4-Dioxane	**8.8 h**	**1**	**22.4 h**	**3**
			1–2 days	4
Nitromethane	24 h	1	100 days	5
1,3-Dioxolane	—		26.5 h	6
Cyclohexane	18 h	1	45 h	7, 8
tert-Amyl alcohol	—		3.3 days	7, 9
1,2-Butylene oxide	—		8.4 days	9
tert-Butyl alcohol	90 h	1	14 days	9
Epichlorohydrin	42 h	1	45 days	9
Nitroethane	—		107 days	9

Sources: [1] Dilling et al. (1976); [2] Dow Chemical (1980); [3] Maurer et al. (1999); [4] Platz et al. (1997); [5] Nielsen et al. (1989); [6] Buxton et al. (1988); [7] Bidleman (1988); [8] Chao et al. (1983); and [9] Meylan and Howard (1993).

Notes: Values cited in the literature from the Dilling et al. (1976) study often mistakenly cite the 1,4-dioxane half-life as 3.4 h. The Dilling study used light intensity much higher than sunlight; measured half-lives are multiplied by 2.6 to estimate the half-life under bright sunlight conditions. Values from the Dilling study were measured in the presence of 5 ppm of nitrous oxide (NO); therefore, photo-oxidation is likely the dominant process for values from Dilling et al.

wet deposition of soluble airborne contaminants at night, leading to a low-concentration distribution to surface water and groundwater. Rain also washes out soluble and particulate airborne contaminants. Numerous studies have been conducted to elicit the photo-oxidation pathway for 1,4-dioxane (Maurino et al., 1997; Platz et al., 1997; Geiger et al., 1999; Li and Pirasteh, 2006). The pathway by which 1,4-dioxane reacts with hydroxyl radicals is summarized in Figure 3.2.

In the photo-oxidation pathway for 1,4-dioxane (mapped in Figure 3.2), one of the eight hydrogen atoms on the 1,4-dioxane molecule [A] reacts with OH$^{\bullet}$ followed by addition of oxygen to form a 1,4-dioxyl radical [B]. The dioxyl radical in turn converts NO to NO_2, producing an alkoxyl radical [C]. The alkoxyl radical is rapidly converted to the $HC(O)O(CH_2)_2OCH_2$ radical [D] via a very fast ring-opening reaction. Addition of oxygen forms another peroxyl radical $HC(O)O(CH_2)_2OCH_2O_2$ [E]. This peroxyl radical will convert NO to NO_2, forming $HC(O)O(CH_2)_2OCH_2O$ [F], which reacts with molecular oxygen to produce ethylene-1,2-diformate (EDF—also called ethylene glycol diformate) [G]. EDF will degrade from successive reactions with hydroxyl radicals, as delineated by Maurer et al. (1999).

Atmospheric oxidation of 1,4-dioxane in the presence of nitrogen oxide (NO) produces EDF with an approximately 100% molar yield. If the annual average tropospheric OH$^{\bullet}$ concentration of 1×10^6 molecules per cubic centimeter is used, the total residence times of 1,4-dioxane and EDF in the atmosphere will be 22.4 h and 24 days, respectively. EDF can therefore be expected to travel far from its source; however, because EDF is highly soluble in water, removal by wet precipitation or "rain-out" will shorten its atmospheric lifetime in wet climatic regions (Maurer et al., 1999).

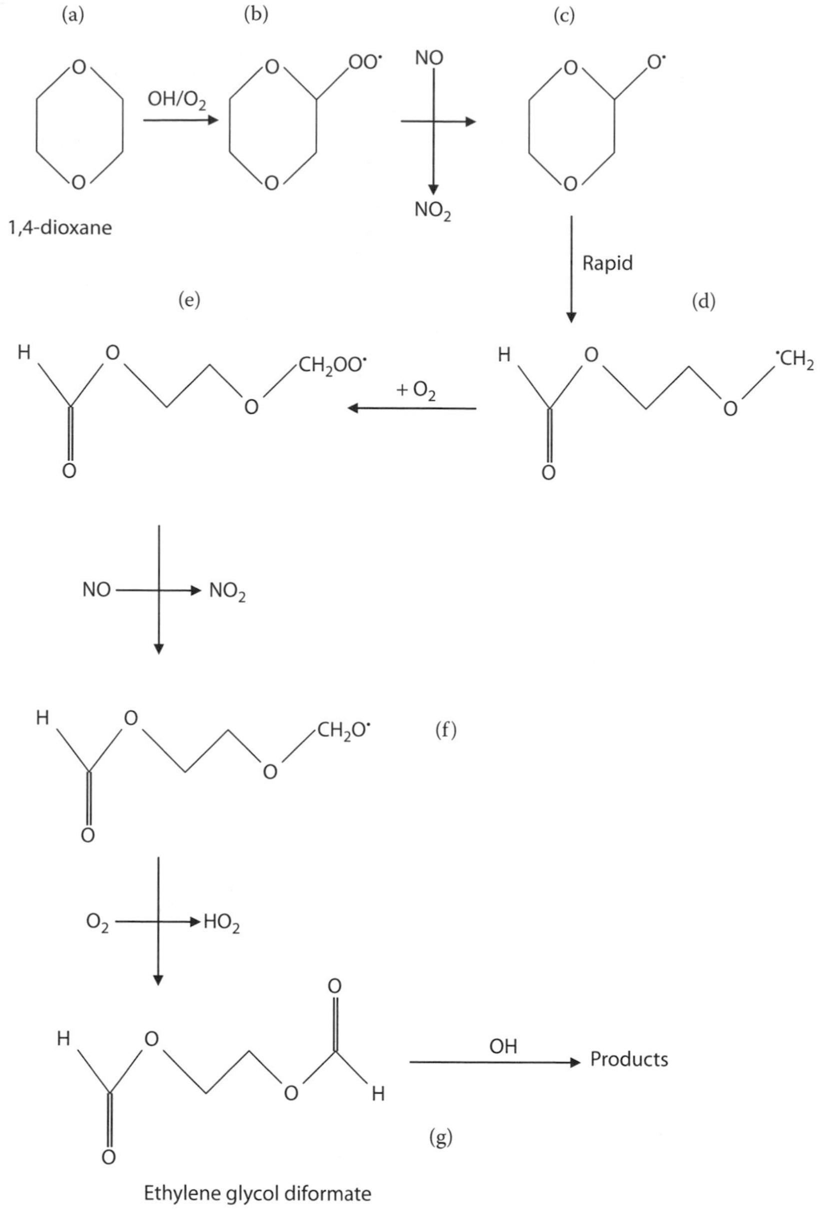

FIGURE 3.2 Photo-oxidation pathway for 1,4-dioxane in the presence of NO$_x$. Note: the product of step (f), ethylene glycol diformate, is also called ethylene-1,2-diformate (EDF). (From Geiger, H., Maurer, T., and Becker, K.H., 1999, *Chemical Physics Letters* 314: 465–471. With permission.)

Although the OH$^\bullet$ concentration in the atmosphere varies with location, time of day, season, and meteorological conditions, a reasonable 24-hour global average atmospheric lifetime for 1,4-dioxane is 1–2 days (Platz et al., 1997). 1,3-Dioxolane follows a similar hydroxyl radical oxidation pathway in which ethylene carbonate and methylene glycol diformate are produced (Freitas-Dinis et al., 2001).

Atomic chlorine is another important atmospheric oxidant capable of oxidizing 1,4-dioxane. Chlorine atoms are introduced into the atmosphere through reactions of HCl with OH and by photolysis of chlorine molecules from industrial emissions and sea salt. The chlorine atoms play a significant role in removal of volatile organic compounds from the atmosphere in coastal regions where marine boundary layers occur (Li and Pirasteh, 2006). The reaction of atomic chlorine and 1,4-dioxane follows the general form

$$Cl + C_4H_8O_2 \rightarrow HCl + C_4H_7O_2 \qquad (3.17)$$

The product of the above reaction, $C_4H_7O_2$, is a dioxane alkyl peroxy radical, which degrades to EDF, a toxic by-product with an atmospheric residency of 24 days (Maurer et al., 1999).

3.1.4.3 Photo-Oxidation in Water

Hydroxyl radicals are present at concentrations on the order of 10^{-17} mol/L in illuminated surface waters (Hemond and Fechner, 1994). Hydroxyl radicals form when an organic chromophore absorbs light and reacts with water to form hydrogen peroxide (H_2O_2). The formation and accumulation of H_2O_2 in natural waters are correlated with the concentration of naturally occurring humic substances (Cooper et al., 1988). Hydrogen peroxide results from a reaction with superoxide, O^-, which forms when oxygen is reduced by free electrons generated by light-absorbing substances through photoionization or energy transfer. Superoxide can also form by electron transfer from reduced metals such as titanium dioxide (Cooper et al., 1988). As a source of hydroxyl radicals, hydrogen peroxide plays a significant role in photo-oxidation of contaminants in natural surface waters, and it may react directly with pollutants. The measured hourly photochemical accumulation rate of H_2O_2 in surface water exposed to midday sunlight ranges from 2.7×10^{-7} to 48×10^{-7} mol/L in waters with dissolved organic carbon ranging from 0.53 to 18 mg/L, respectively. The measurements were made at 24.3° northern latitude, where the midday sunlight intensity is 0.4 W/m^2 in the wavelength range 295–385 nm (Cooper et al., 1988). H_2O_2 will break into two hydroxyl radicals (OH$^•$) if it absorbs a sufficiently energetic photon.

The aqueous photo-oxidation half-life of 1,4-dioxane in water has been estimated to be as low as 67 days and as high as 9.1 years (Dorfman and Adams, 1973; Howard et al., 1991). Estimated photo-oxidation half-lives for additional stabilizer compounds and chlorinated solvents are tabulated in Table 3.6.

3.2 SURFACE-WATER FATE AND TRANSPORT PROCESSES

Solvent wastes discharged to surface water enter a dynamic environment dominated by moving water, biological processes, sedimentation, sunlight, and a variety of chemical reactions. Although discharge of solvent wastes to surface water has been effectively eliminated with the adoption and successful implementation and enforcement of regulations to protect streams, rivers, and lakes, the discharge of 1,4-dioxane to surface water continues today. Untreated 1,4-dioxane may enter surface-water bodies as the effluent of groundwater treatment systems designed to remove chlorinated solvents such as air strippers and granular activated carbon treatment vessels, which are not effective at removing 1,4-dioxane. Other stabilizer compounds that are not effectively removed by conventional pump-and-treat technologies employed at solvent release sites may also be discharged to surface-water bodies. Therefore, the surface-water fate and transport processes for solvent-stabilizer compounds may warrant evaluation for managing the remediation of chlorinated solvent sites.

The transport processes governing contaminants in surface water are advection, diffusion, and sorption to suspended sediment. Diffusion, wind-induced currents, and thermal stratification are important in lakes, whereas advective flow governed by gravity and water velocity is important in streams and rivers. Advection of low to moderate concentrations of dissolved organic compounds generally proceeds independently of chemical properties; however, diffusion is limited by compound-specific properties. Surface-water transport processes are well studied and will not be described further; a concise

TABLE 3.6
Ranges of Measured and Calculated Aqueous Photo-Oxidation Half-Lives for Solvent-Stabilizer Compounds

Compound	Aqueous Photo-Oxidation, $t_{1/2}$		References
	High	Low	
1,4-Dioxane	**9.1 years**	**67 days**	**Anbar and Neta (1967), Dorfman and Adams (1973)**
sec-Butyl alcohol	23 years	129 days	Anbar and Neta (1967), Dorfman and Adams (1973)
1,2-Butylene oxide	55 years	1.4 years	Howard et al. (1991)
tert-Butyl alcohol	64,500 years	2.1 years	Anbar and Neta (1967), Dorfman and Adams (1973)
Ethyl acetate	110 years	2.75 years	Dorfman and Adams (1973), Howard et al. (1991)
Pyridine	24.4 years	14.7 years	Dorfman and Adams (1973), Howard et al. (1991)
Acetonitrile	12,560 years	314 years	Dorfman and Adams (1973), Howard et al. (1991)
Methyl ethyl ketone	81.4 years	48.8 years	Anbar and Neta (1967), Howard et al. (1991)
Cyclohexane	7,800,000 years	160,000 years	Howard et al. (1991)

summary is given in Hemond and Fechner (1994), and a detailed, analytical treatment is given in Thibodeaux (1996). The potential for photolysis and photo-oxidation of stabilizer compounds and their relative potential rates of volatilization from surface-water bodies is discussed in the previous sections of this chapter. This section focuses on the physical and chemical fate of stabilizer compounds in surface-water bodies. Most of these processes are common to the groundwater fate and transport of stabilizer compounds; adsorption and biological processes are discussed in Section 3.3.

3.2.1 HYDROLYSIS

Hydrolysis is a form of nucleophilic substitution reaction in which water is the nucleophile (Hemond and Fechner, 1994; Harris, 1990b). More specifically, hydrolysis is a chemical reaction in which an organic molecule, RX,[*] reacts with hydroxide from the ionization of a water molecule, forming a new carbon—oxygen bond and cleaving a carbon—X bond in the organic molecule, producing H^+X^-. Both the organic compound and the water molecule are split in a hydrolysis reaction, which is essentially a displacement of X by a hydroxyl group (-OH):

$$RX + H_2O \rightarrow ROH + X^- + H^+. \tag{3.18}$$

Hydrolysis acts on organic molecules that possess electrophiles, that is, electron-seeking atoms such as carbon and phosphorous. The nucleophile (water or hydroxide) attacks the electrophile (e.g., H^+) and displaces the functional group that leaves the molecule, often chloride or phenoxide. Some groups of chemicals are resistant to hydrolysis, including many of the solvent-stabilizer compounds,

---

[*] In the RX notation, R usually represents a hydrocarbon, such as an alkyl group, and X represents an organic functional group, such as an ester, amine, or aldehyde.

such as ethers, alcohols, alkanes, alkenes, aromatics, and heterocyclic compounds. Compounds susceptible to hydrolysis include amines, amides, epoxides, nitriles, esters, and several other compounds not found among the stabilizer or chlorinated solvent compounds (Harris, 1990b). There are no hydrolysable functional groups on the 1,4-dioxane molecule; therefore, 1,4-dioxane is not expected to hydrolyze significantly (NICNAS, 1998).

Hydrolysis is a first-order reaction whose rate varies directly with the concentration of the organic species, RX, as shown:

$$\frac{d[RX]}{dt} = -k[RX],$$
(3.19)

where k is the first-order hydrolysis rate constant in units of inverse time (in s^{-1}), giving a disappearance rate of moles per liter per second. The first-order rate constant describes the loss of RX at any concentration of RX and at constant pH, pressure, and temperature. At any time t, the concentration of RX, $[RX]_t$, is given by

$$[RX]_t = [RX]_0\, e^{-kt},$$
(3.20)

where $[RX]_0$ is the starting concentration of RX, t is time in seconds, k is the first-order rate constant, and e is the natural logarithm (USEPA, 1998a; Bennett, 2004a). The time needed to decrease the concentration by one-half, that is, the half-life ($t_{1/2}$), is

$$t_{1/2} = \frac{\ln 2}{k} = \frac{0.693}{k}.$$
(3.21)

The hydrolysis reaction described above is the unimolecular form. A bimolecular form of the hydrolysis reaction involves kinetics that depend on both the concentration of the organic molecule RX and the nucleophile, H_2O, H^+, and OH^-. Hydrolysis reactions are therefore sensitive to the pH of the water, so that a difference of 2 pH units would cause the reaction rate to change 100-fold. Depending on whether hydroxide (OH^-) or hydrogen (H^+) is the dominant species at a given pH, bimolecular hydrolysis reactions are characterized as base-catalyzed hydrolysis or acid-catalyzed hydrolysis (Hemond and Fechner, 1994). Catalysis by microbial or other agents may further complicate the kinetics of hydrolysis; therefore, citations of hydrolysis rate constants from the literature or from predictive modeling must be used with caution, as natural systems are likely to be a great deal more complex than the system measured or modeled.

Equations and software to estimate hydrolysis rate constants for those compounds susceptible to acid-catalyzed or base-catalyzed hydrolysis are freely available (USEPA, 2000a). The HYDROWIN module in the EPIWIN Suite only estimates aqueous hydrolysis rate constants for esters, carbamates, epoxides, halomethanes, and selected alkyl halides (USEPA, 2000a). HYDROWIN does not estimate neutral hydrolysis rate constants, which may be the dominant hydrolysis rate for some epoxide compounds, leading to underestimation of the rate constant. Molecular structure features are used to estimate the acid- and base-catalyzed rate constants, which are then used to calculate hydrolysis half-lives (Mill et al., 1987; USEPA, 2000a). Table 3.7 summarizes estimates for those solvent-stabilizer compounds susceptible to hydrolysis. Literature values for two chlorinated solvents are included to indicate their immunity to hydrolysis. Ethers are generally resistant to hydrolysis; HYDROWIN does not estimate hydrolysis rate constants for ether compounds, including 1,4-dioxane.

3.2.2 Acid Dissociation Potential

Organic compounds that are prone to form ions by dissociating when dissolved in water will behave differently when in pure form. Ionization of an organic compound will affect its solubility and

TABLE 3.7
Estimated Surface-Water Hydrolysis Half-Lives and Residency Times (Time for >99% Removal by Hydrolysis Acting Alone) for Chlorinated Solvents and Solvent Stabilizers Prone to Hydrolysis in Water

Chemical Name	Hydrolysis Half-Lives (Literature Values)	References	Hydrolysis Half-Lives (Estimated)	Residency Time
Solvents				
Dichloromethane	260,000 days	5	Very long	Persistent
Methyl chloroform	350 days at 25°C	4	Very long	Persistent
Stabilizers				
Epichlorohydrin	8.2 days in distilled water	3	350 years	Persistent
Diepoxybutane[a]	—		107 years	Persistent
Glycidol[a]	—		63 years	Persistent
Butoxymethyl oxirane[a]	—		62 years	Persistent
1,2-Butylene oxide[a]	11.6 days; converts to ethylene glycol	1	3 years	21 years
Propylene oxide[a]	11.6 days	1	2.9 years	20 years
Isopropyl acetate	2.4 years at pH 7; 88 days at pH 8	2	2.4 years	17 years
Ethyl acetate	2 years at pH 7 and 25°C	3	1.82 years	13 years
Cyclohexene oxide[a]	—		37.5 days	260 days
Styrene oxide[a]	—		9 days	63 days

Sources: [1] Bogyo et al. (1980) and Hoechst Celanese (2006); [2] Harris (1990b); [3] NLM (2006) (note: the Hazardous Substances Data Bank includes errors in citation and transcription of values from journal articles to online summaries; where available, original sources were checked); [4] Haag and Mill (1988); and [5] Capel and Larson (1995).

Notes: Estimates made with HYDROWIN (USEPA, 2000a), a module in the EPIWIN Suite (USEPA, 2007a). Stabilizer compounds not appearing in this table (e.g., 1,4-dioxane) are not susceptible to hydrolysis. "Very long" describes modeled estimates in excess of 1000 years.

[a] Epoxides estimated for only acid- or base-catalyzed hydrolysis, not for neutral hydrolysis; underestimation may result.

adsorption. This section reviews the potential for stabilizer compounds' participation in acid–base interactions in aqueous solution.

The primary chemical attribute describing a compound's potential to participate in acid–base chemical reactions is its acid dissociation constant, K_a, the equilibrium constant for the reaction

$$HA + H_2O \rightarrow H_3O^+ + A^-, \text{ (j)}$$

where HA is an organic chemical associated with a proton H, H_3O^+ is hydronium, and A^- is the organic compound remaining after splitting off one or more hydrogen atoms. The equilibrium constant K_a is the ratio of the activities of the reactants as follows (Lyman et al., 1990):

$$K_a = \frac{a_{H_2O^+}a_{A^-}}{a_{HA}a_{H_2O}}, \quad a_{H_2O} = 1 \quad \therefore K_a = \frac{a_{H_3O^+}a_{A^-}}{a_{HA}} = \frac{(\gamma_{H_3O^+}M_{H_3O^+})(\gamma_{A^-}M_{A^-})}{\gamma_{HA}M_{HA}}. \tag{3.22}$$

The activity of water is taken as one by convention, and activity is defined as the molar activity coefficient, γ, multiplied by the molar concentration, M. The convention for expressing acid

TABLE 3.8
Acid Dissociation Constants for Stabilizer Compounds

Chemical Name	pK_a	Temperature (°C)	Percent Dissociated at pH 7	References
Pyridine	5.25	25	Complete dissociation	Perrin (1972)
Nitroethane	8.46	25	3.47	Perrin (1972)
Phenol	9.994	20	0.13	Perrin (1972)
Nitromethane	10.21	25	0.062	Perrin (1972)
p-tert-Amyl phenol	10.43	25	0.0167	Schultz (1986)
Triethylamine	10.778	25	0.017	Riddick et al. (1985)
Nonylphenol	10.778	25	0.017	Lipnick et al. (1986)
Diisopropylamine	11.05	25	0.009	Perrin (1972)
2,6-di-tert-Butyl-p-cresol	12.23	25	5.89×10^{-6}	Serjeant and Dempsey (1979)
tert-Butyl alcohol	19.2	25	6.31×10^{-13}	Serjeant and Dempsey (1979)

Note: The majority of stabilizer compounds, including 1,4-dioxane, do not contain functional groups capable of dissociation.

dissociation constants is similar to pH; pK_a is the negative log of the acid dissociation constant. The chlorinated solvents and most of the solvent-stabilizer compounds are stable with respect to acid-base reactions, and few have literature values for the acid dissociation constant.

Weak organic acids in aqueous solutions are strongly affected by the pH of the aquatic system:

$$\log \frac{M_{A^-}}{M_{HA}} = pH - pK_a. \tag{3.23}$$

In the normal pH range of aquatic systems, the relevant range of pK_a is 3–10; pK_a values less than 3 are expected to be 99% dissociated, whereas pK_a values greater than 10 are expected to remain completely undissociated (Lyman et al., 1990). The higher the pH, the more a compound with a pK_a between 3 and 10 will dissociate (Schultz, 1986). Most of the pK_a measurements available are made at room temperature (25°C), which is much warmer than temperatures expected in most surface waters and groundwaters. Therefore, laboratory values of pK_a will typically overestimate the tendency for an organic compound to ionize under ambient subsurface conditions. Table 3.8 provides acid dissociation constants for the few stabilizer compounds whose acid dissociation constants have been measured and reported in the literature. The majority of stabilizer compounds do not contain functional groups capable of dissociation.

3.3   SUBSURFACE FATE AND TRANSPORT PROCESSES

Contaminants in the subsurface migrate slowly. The most rapid rates of migration in the unsaturated zone occur by vapor transport through relatively dry soils, particularly when vapors are denser than air. The most rapid migration rates in the saturated zone are equal to the groundwater flow velocity. Contaminant migration is retarded by chemical, physical, and biological processes contributing to the contaminant's elimination, transformation, or otherwise delayed arrival at the groundwater interface or at an observation well down-gradient from the point of release. The following sections focus on the subsurface fate and transport processes in groundwater that favor migration of some stabilizer compounds while eliminating others.

3.3.1 Aqueous Solubility of Stabilizer Compounds and Stabilizer-Solvent–Waste Mixtures

Solubility, the maximum mass of a compound that will dissolve in pure water at a specific temperature before phase separation occurs, is for many compounds the most important physical property governing subsurface fate. Highly soluble compounds are less prone to partition into the vapor phase and are generally less likely to be adsorbed to organic and mineral surfaces. Consequently, hydrophilic and highly soluble compounds are not significantly retarded in groundwater flow and will migrate rapidly in groundwater. Relative retardation of contaminant migration is discussed further in Section 3.4.

In general, liquids obey the "like dissolves like" rule; polar molecules (e.g., alcohols, ketones, and some ethers) are soluble in polar solvents (e.g., water), and nonpolar molecules (e.g., benzene) are soluble in nonpolar solvents (e.g., oil).

Natural water contains mineral salts that may cause a minor change in the capacity of water to dissolve organic compounds. In most groundwater, ionic strength contributed by dissolved minerals will have a negligible effect on organic compound solubility. For groundwater having higher ionic strength, a correction factor derived from experimental data for aromatic compounds can be applied (Hashimoto et al., 1984):

$$\log_{10} S_{sw} = [(0.0298I + 1) \log_{10} S_{w}] - 0.004I, \tag{3.24}$$

where S_{sw} is the solubility in seawater (in mol/L), S_{w} is the freshwater solubility (in mol/L), and I is the ionic strength, defined by

$$I = \frac{1}{2} \sum \left(c_i z_i^2 \right) \tag{3.25}$$

where C_i is the concentration and Z_i is the charge of the ith ionic species, summed over all ionic species in solution (Lewis and Randall, 1921).

Chemical properties of organic compounds and the geochemical characteristics of groundwater are the primary determinants of pollutant solubility, and molecular structure is the main determinant of most chemical properties. Molecular structures that include branching have increased water solubility for many compounds. Ring formation increases water solubility. Inclusion of a double bond in the molecule, ring, or chain also increases water solubility. The presence of a second or third double bond in a hydrocarbon proportionately increases water solubility, and a triple bond in a chain molecule imparts greater solubility than two double bonds (Verschueren, 1996).

The technical literature lists a wide range of solubilities for the same compounds. Older literature typically lists higher solubility values because older products were not as pure as today's chemicals and because analytical methods have improved (Verschueren, 1996). In addition to laboratory experiments, a number of estimation methods have been developed to calculate the expected compound solubility based on molecular structures (QSARs) and a variety of other approaches.

The WATERNT program uses the atom/fragment contribution (AFC) estimation methodology developed by Syracuse Research Corporation. The AFC estimation methodology is based on a "fragment constant" method wherein coefficients for individual fragments and groups were derived by multiple regression of 1000 reliably measured water solubility values. WATERNT retrieves experimental water solubility values from a database containing more than 6200 organic compounds with measured values. When a compound structure matches a database structure via an exact atom-to-atom connection match, the experimental water solubility value is retrieved (USEPA, 2007a). Table 3.9 lists literature values for measured and calculated solubilities for solvent-stabilizer compounds and chlorinated solvents. The large differences between commonly cited solubility

TABLE 3.9
Estimated and Frequently Cited Solubilities of Solvent-Stabilizer Compounds and Chlorinated Solvents

Stabilizer and Solvent Compounds[a]	Commonly Cited Solubility[b] (mg/L)	Estimated Solubility, WATERNT[c] (mg/L)	Temperature for Commonly Cited Value (°C)	References
Ethanol	Miscible	1,000,000	25	Budavari (1989)
Glycidol	Miscible	1,000,000	25	Sax and Lewis (1987)
Methanol	Miscible	1,000,000	25	Riddick et al. (1985)
N-Methylmorpholine	Miscible	1,000,000	—	Fisher Scientific (2007)
Triethanolamine	Miscible	1,000,000	25	Budavari (1996)
Pyridine	Miscible	729,800	20	Kirk and Othmer (1982)
1,3-Dioxolane	Miscible	276,900	25	Lide (2000)
2-Methyl-3-butyn-2-ol	Miscible	239,500	20	Ahlers (1998)
Diepoxybutane	Miscible	220,500	25	Dean (1985)
tert-Butyl alcohol	Miscible	217,500	25	Riddick et al. (1985)
1,4-Dioxane	**Miscible**	**213,900**	**25**	**Reid and Hoffman (1929)**
Acetonitrile	Miscible	136,700	25	Riddick et al. (1985)
sec-Butyl alcohol	Miscible	130,400	25	Hefter (1984)
Trichloroacetic acid	1,200,000	11,990	25	Willis and McDowell (1982)
Propylene oxide	590,000	129,300	25	Bogyo et al. (1980)
2-Butanone	353,000	76,100	10	Verschueren (1996)
1,3,5-Trioxane	212,000	289,800	25	Budavari (1996)
tert-Amyl alcohol	110,000	70,700	25	Hefter (1984)
1,2-Butylene oxide	95,000	24,650	25	Bogyo et al. (1980)
Phenol	82,800	26,160	25	Southworth and Keller (1986)
Hydroquinone	72,000	129,500	25	Granger and Nelson (1921)
Resorcinol	71,700	85,710	25	Yalkowsky and Dannenfelser (1992)
Epichlorohydrin	65,900	50,630	25	Yalkowsky and Dannenfelser (1992)
Ethyl acetate	64,000	29,930	25	Wasik et al. (1981)
Nitroethane	45,000	38,130	20	Yalkowsky and Dannenfelser (1992)
Isopropyl acetate	29,000	9268	25	Yalkowsky and Dannenfelser (1992)
Dichloromethane	20,000	10,950	25	Pankow and Cherry (1996)
Butoxymethyl oxirane	20,000	26,590	20	Patty (1963)
Triethylamine	15,000	68,260	20	Verschueren (1996)
Styrene oxide	13,000	4228	25	Lapkin (1965)
Nitromethane	11,100	111,200	25	Riddick et al. (1985)
Diisopropylamine	11,000	75,310	25	Kirk and Othmer (1982)
Methyl chloroform	1334	666	25	Banerjee et al. (1980)
Trichloroethylene	1280	779	25	Horvath et al. (1999)
Thymol	1000	437	25	Sax and Lewis (1987)
1-Naphthol	866	1126	25	Hassett et al. (1980)
Carbon tetrachloride	757	280	25	Banerjee et al. (1980)
Isoprene	642	339	25	McAuliffe (1966)
Perchloroethylene	200	80	25	Coca and Diaz (1980)
p-tert-Amyl phenol	168	113	25	Verschueren (1996)

continued

TABLE 3.9   (continued)
Estimated and Frequently Cited Solubilities of Solvent-Stabilizer Compounds and Chlorinated Solvents

Stabilizer and Solvent Compounds[a]	Commonly Cited Solubility[b] (mg/L)	Estimated Solubility, WATERNT[c] (mg/L)	Temperature for Commonly Cited Value (°C)	References
Cyclohexane	55	43	25	McAuliffe (1966)
Isopentane	48	185	25	Riddick et al. (1985)
Nonylphenol	6	2	25	Shiu et al. (1994)
2,6-di-*tert*-Butyl-*p*-cresol	0.4	6	20	Verschueren (1996)
Diisobutylene	Insoluble in water	12	—	Lide (2000)

[a] Nonitalics compounds are stabilizers. Italics compounds are chlorinated solvents; these are listed for comparison to show whether the stabilizer is likely to evaporate with the solvent or remain behind in the soil where it can be leached down to groundwater.

[b] Solubility values commonly cited in ChemFate database (SRC, 2007b) and the Hazardous Substances Databank (NLM, 2006).

[c] WATERNT™ (USEPA, 2007a) in "Estimations Programs Interface for Windows" (EPI Suite) v. 3.20; http://www.epa.gov/oppt/exposure/pubs/episuitedl.htm

values and values estimated by using QSAR techniques (shown in Table 3.9) underscore the importance of reviewing the basis for solubility values listed in data compilations.

Table 3.10 lists estimated and measured solubilities for individual compounds. Stabilizer compounds in a solvent waste mixture dissolve differently than individual compounds when released to groundwater. Each component in the mixture will partition between the aqueous phase and the mixture (Verschueren, 1996). Mixtures of organic compounds will dissolve into water according to their effective solubilities, defined by Raoult's law:

$$S_{eff} = XiS_{water},\qquad(3.26)$$

where S_{eff} is the effective solubility of a compound in a mixture (lower than the compound's solubility when measured alone in water), X_i is the mole fraction of compound i in the mixture, and S_{water} is the compound's single-component aqueous solubility (Banerjee, 1984; Jackson and Dwarakanath, 1999; Jackson and Mariner, 1995). Equation 3.26 applies to mixtures without significant cosolvent effects (Payne et al., 2008). Calculations using an assumed solvent waste composition (based on information in the solvent recovery and patent literature) yield the effective solubility of 1,4-dioxane relative to methyl chloroform, nitromethane, *sec*-butyl alcohol, 1,3-dioxolane, 1,2-butylene oxide, and waste cutting oil (Table 3.10).

Table 3.10 shows that the effective solubility of components in a mixture is considerably reduced from the individual compound's solubility when measured as a single-component system in pure water. Also, the effective solubility of 1,4-dioxane is 10,000 times greater than the effective solubility of methyl chloroform. As a result, 1,4-dioxane will be preferentially removed from the waste mixture; in effect, groundwater will "suck" the dioxane out of the waste mixture because of this large difference in effective solubilities. Because 1,4-dioxane will dissolve in groundwater earlier in time and at a much faster rate, the 1,4-dioxane is likely to lead the other components in the waste mixture in the migration front emanating from the point of release. Consequently, 1,4-dioxane, if released in sufficient quantities, can be expected to migrate farthest and can be found at distances several times the length of the methyl chloroform plume. This pattern has been observed in many plumes, as discussed further in Section 3.4.2 and Chapter 8.

TABLE 3.10
Effective Solubilities of Components in a Waste Mixture of Methyl Chloroform Degreasing Solvent from a 1970s-Era Aerospace Industry Degreaser

Component	Molecular Weight (g/mol)	Specific Gravity (g/cm³)	Initial Mass Fraction[a] (%)	Final Mass Fraction[b] (%)	Mass in 1 mol of Mixture (g/mol)	X_i (Mole Fraction)	Single-Component Aqueous Solubility[c] (mg/L)	S_{eff} (Effective Solubility)[d] (mg/L)
Methyl chloroform	113.40	1.35	93.25	53.62	62.71	0.55	1334	740
1,4-Dioxane	88.12	1.03	3.00	15.00	17.54	0.20	10,000,000	1,991,600
sec-Butyl alcohol	74.10	0.80	1.50	7.06	8.25	0.11	10,000,000	1,114,000
1,3-Dioxolane	74.09	1.06	1.00	1.27	1.48	0.02	10,000,000	200,000
1,2-Butylene oxide	72.12	0.83	0.70	0.31	0.36	0.01	95,000	480
Nitromethane	61.04	1.14	0.55	2.75	3.22	0.05	11,100	590
Cutting oil[e]	400.00	0.95	0.00	20.00	23.39	0.06	50	3
Overall mixture[f]	116.96	1.13	100	100	116.96	1.00		

[a] Initial mass fraction represents the composition of methyl chloroform-containing spent-solvent waste listed in Archer (1984).

[b] Final mass fraction represents the roughly estimated composition of spent-solvent waste after 6 weeks of intensive vapor degreasing during which volatile vapors are lost to the atmosphere, "high-boilers" are concentrated in the still bottoms, and cutting oil accumulates. The assumed oil composition and hypothetical aerospace industry setting are based on Jackson and Dwarakanath (1999).

[c] Miscible compounds assigned a solubility of 10^6 mg/L for calculation; values are from frequently cited solubilities in Table 3.9.

[d] The effective solubility is calculated by using the adaptation of Raoult's law explained in Section 3.3.1 and in Jackson and Mariner (1995) (values are rounded).

[e] Molecular weight, specific gravity, and solubility of cutting oil are assumed values.

[f] Properties of overall mixture are for final composition of the degreasing waste at the point it was removed from the degreaser and replaced with new solvent.

3.3.2 ADSORPTION

Adsorption is the attachment of a compound in the aqueous or vapor phase onto a solid surface. A related process, *ab*sorption, involves the transfer of a compound from the aqueous phase *into* a volume of solid material. Because adsorption and absorption are usually not distinguishable in natural systems, the term *sorption* is used to indicate that the specific mechanism of attachment to or movement into a solid surface or volume is not known (Schwarzenbach et al., 1993).

Adsorption of molecules, while technically a partitioning phenomenon, can be represented as a chemical reaction:

$$A + B \leftrightarrow A \cdot B, \qquad (3.27)$$

where A represents the adsorbate, B the adsorbent, and A·B the adsorbed compounds. The "adsorption reaction" involves a variety of intermolecular forces and is often reversible: molecules continue to accumulate onto the soil particle surface until equilibrium is reached and the rate of forward reaction (adsorption) equals the rate of reverse reaction (desorption) (Snoeyink, 1999).[*]

---

[*] The technical term "soil" is typically reserved for the root-zone material; the geologic material comprising aquifers is referred to as "aquifer solids." For convenience, this chapter uses the less precise practice of calling all subsurface solids by the term "soils."

A key determinant of whether a contaminant will be mobile in the subsurface is the kinetics of the sorptive process compared to the rate of chemical transport by advection and dispersion. Chemicals that are adsorbed more quickly than they are advected will be retained in the aquifer matrix (Schwarzenbach and Westall, 1981).

Adsorption of chemicals to organic and mineral surfaces controls retardation of contaminant migration in surface water and groundwater (Brusseau and Reid, 1991).[*] Compounds may also sorb onto micrometer-sized colloids that are advected with groundwater flow. Low-solubility hydrophobic compounds such as chlorinated solvents can be substantially more mobile when sorbed to colloids than they are in the aqueous phase (Huling, 1989). Adsorption of contaminants to carbon surfaces also has bearing on treatability using liquid or granular activated-carbon filters.

The forces governing sorption include electrostatic interactions, van der Waal's interactions, hydrogen bonding, charge transfer, ligand exchange, direct and induced dipole–dipole interactions, chemisorption, and hydrophobic bonding (Suthersan, 2002). Organic compounds can adsorb to both organic matter and mineral surfaces. The degree to which sorption occurs depends on both soil properties such as soil organic matter content, clay type, and properties of the organic compound, including water solubility, molecular volume, the octanol/water partition coefficient (K_{ow}), and the organic carbon partition coefficient (K_{oc}). These terms and their role in adsorption of stabilizer compounds to aquifer solids and suspended solids are described further in this section.

Contaminant residence time in the subsurface also plays an important role in sorption processes, as contaminants tend to become more strongly bound to soil over time. The age of a release is important as it has bearing on enhanced sorption of organic compounds to inorganic soil particles. Generic fate and transport models rely on published K_{oc} values and assumed total organic carbon content to estimate contaminant partitioning between sorbed and aqueous phases, as well as to predict retardation (see Section 3.3.2.4). This approach can significantly underestimate the sorption potential of a contaminant and overpredict its mobility in the soil or groundwater, as discussed in Section 3.5.

3.3.2.1 Soil Properties Affecting Adsorption

Organic matter in soil or sediment is the primary factor in the sorption of nonionic organic contaminants from water by soils. Organic compounds also adsorb to soil minerals but only weakly, at least initially, owing to the strong competitive adsorption of water on polar mineral surfaces. Sorption to soil organic matter has an approximately inverse proportionality to a compound's aqueous solubility (Chiou and Kile, 1994).

Adsorption is a surface phenomenon: it increases with higher surface area, which is in cubic proportion to particle size and is also related to pore size. Micropores increase the adsorption capacity. The distribution of pore sizes in organic matter is relevant to sorption processes, but it is difficult to quantify and is not usually measured. Large molecules whose hydrated radii exceed the diameter of pore sizes will be excluded from absorption into soil organic matter ("steric exclusion") and will be limited to surface adsorption (Bennett, 2004b).

Most organic matter in soil is bound to clay as a clay–organic complex. Two major types of adsorbing surfaces are available to an organic compound: clay–organic and clay alone. The relative contribution of organic and inorganic surface areas to adsorption will depend upon the extent to which the clay is coated with organic matter. The influence of clay on organic chemical adsorption is significant in soils with organic matter contents below 1% (Dragun, 1988). Active agents governing adsorption of organic compounds onto mineral surfaces include silicates and aluminum and iron oxides and hydroxides. Oxide surfaces may be amphoteric hydroxyl sites[†] capable of both donating and accepting a proton, providing anionic, neutral, and cationic sites, depending on pH and

---

[*] Retardation is the rate of chemical migration in a saturated aquifer relative to groundwater velocity; see Section 3.3.2.3 for further discussion of retardation.

[†] Amphoteric substances can react either as acids or bases.

the hydroxyl reaction rate constant, pK. Amphoteric hydroxyl sites strongly attract water, creating a layer of tightly bound water at the surface (vicinal water). Organic molecules interact first with the charged surface of the vicinal water layer; direct adsorption of the compound onto the mineral surface must be preceded by displacement of the water (Bennett, 2004b).

3.3.2.2 Molecular Properties Affecting Adsorption

The size of a molecule influences its potential to be adsorbed to solid surfaces. Large molecules have more sites where van der Waal's forces[*] contribute to attraction of the molecule to charged soil particles, due to temporary dipoles resulting from time-varying electron distribution or polarity induced by charges on soil particles. The positive regions in the temporary dipole attract electrons in the adjacent soil surface, resulting in a net attraction of the compound to the soil surface (Carey, 1987; Dragun, 1988). For organic chemicals with molecular weights greater than 400–500, van der Waal's forces become the dominant adsorption mechanism in nonsandy soils (Dragun, 1988).

Solubility in water, exemplified by hydrophilic compounds, also controls the degree to which organic compounds are adsorbed to soil organic matter. Molecular fragments including carbon, hydrogen, and the halogens chlorine, bromine, and iodine are hydrophobic, whereas fragments including nitrogen, sulfur, oxygen, and phosphorus are generally hydrophilic. The solubility of compounds with both hydrophilic and hydrophobic groups depends on which fragments are dominant and the geometric arrangement of the molecule (Dragun, 1988). In general, highly soluble and miscible compounds are less prone to adsorption.

The molecular charge of an organic chemical can also play an important role in adsorption onto soil surfaces if the molecule is polar. Many of the stabilizer compounds do not carry a sufficient charge to be affected by electrostatic sorption. Some stabilizer compounds including amines and quaternary ammonium compounds may react in acidic waters to form functional groups that carry a positive charge to make them prone to adsorption. Compounds with hydroxyl or carboxyl groups may acquire a negative charge in alkaline waters and become adsorbed by oxide surfaces on alumina-silicate clay minerals (Dragun, 1988).

Hydrogen bonding can also affect organic chemical adsorption onto soil surfaces. Hydrogen bonding occurs where a hydrogen atom serves as a bridge between two electronegative atoms, with a covalent bond on one end and an electrostatic bond on the other. For example, hydrogen bonding may join a polar organic molecule to an adsorbed cation through a water molecule in the cation's primary hydration shell. This type of hydrogen bonding occurs with pyridine as well as with ketones and amides in soils containing smectite clays (Dragun, 1988).

3.3.2.3 Distribution Coefficients and Sorption Isotherms

Distribution coefficients provide a quantitative description of partitioning between chemicals in solution and soils, bottom sediments, suspended sediments, and other solid media. The distribution coefficient, K_d, describes a linear relationship between the concentration of a chemical in solution and its concentration adsorbed to soil surfaces. At low concentrations, adsorption is a linear process, where the amount adsorbed is directly proportional to the amount in solution (Bennett, 2004b). Sorption potential increases with increasing values of K_d. Sorption isotherms become nonlinear and sorption decreases as the concentration increases, particularly for polar and ionizable compounds and for soils low in organic carbon or high in clay (Suthersan, 2002). The value of K_d is restricted to the system for which that value was obtained; it will vary in different soils.

The relationship between the quantity of a chemical absorbed at constant temperature per quantity of adsorbent, q_e, and the equilibrium concentration of adsorbate in solution, C_e, is called the adsorption isotherm (Snoeyink, 1999). The distribution coefficient K_d is the slope of the sorption

---

[*] Van der Waal's forces, also called London dispersion forces, are the weak electrostatic charges within an otherwise neutral molecule resulting from movement of electrons forming temporary dipoles or charge concentrations.

isotherm at the concentration of interest (a constant for linear isotherms). The use of isotherms for estimating mass of contaminant adsorbed to soil assumes that equilibrium between the adsorbent and the adsorbate is attained instantaneously and that the isotherm is reversible.

The Freundlich equation describes adsorption data for organic compounds adsorbing to organic carbon, particularly granular activated carbon in a water treatment setting:

$$q_e = KC_e^{1/n}, \tag{3.28}$$

or in linear form

$$\log q_e = \log K + \frac{1}{n}\log C_e, \tag{3.29}$$

where K and n are constants.

The units of K and n are determined by the units of q_e, the quantity of chemical adsorbed per quantity of adsorbate, and C_e, the concentration of adsorbate in solution (Snoeyink, 1999). The constant K primarily describes the capacity of the adsorbent for the adsorbate, and $1/n$ is a function of the strength of adsorption. When the adsorbent has become saturated with respect to the adsorbent, q_e is constant. The Freundlich isotherm is more applicable at lower concentrations.

The distribution coefficient is described as the ratio of the mass of a given compound sorbed to soil mineral and organic surfaces to the mass remaining in solution:

$$K_d = \frac{q_e}{C_e}, \tag{3.30}$$

$$q_e = \frac{(C_e - C_i)V}{m}, \tag{3.31}$$

where K_d is the distribution coefficient (in mL/g), q_e is the mass of chemical on the solid phase (in mg/g), and C_e is the mass of chemical in solution (in mg/L). The mass of chemical on the solid phase, q_e, is taken as the difference between the mass of chemical in solution and the initial aqueous-phase concentration C_i (in mg/L) multiplied by the volume of solution and divided by the mass of the solid, m (in g).

In general, compounds with lower values of K_d indicate that those compounds will have higher mobilities. A screening-level approach for estimating K_d involves multiplying K_{oc}, the organic carbon partition coefficient, by f_{oc}, the fraction of organic matter: $K_{oc} \times f_{oc}$. As a rule of thumb, compounds whose values of K_d are higher than 20 are considered to be relatively immobile; these compounds are unlikely to leach from soils or migrate through aquifer materials because of their strong potential for sorption to soil surfaces.

K_d Value of Compounds	Mobility (Fetter, 1993)
>20	Relatively immobile
10–20	"Slight"
1–10	"Low"
0–1	"High"

A related parameter, the retardation factor, R_f or R, expresses the rate of chemical migration in a saturated aquifer relative to groundwater velocity:

$$R_f = 1 + \frac{\rho_b K_d}{\Theta_e} = \frac{V_w}{V_c}, \tag{3.32}$$

where ρ_d is the bulk density of the soil (in g/cm^3), K_d is the distribution coefficient, Θ_e is the effective porosity (unitless because it is vol/vol), V_w is the velocity of water, and V_c is the velocity of the contaminant. In the literature, K_d is sometimes represented as K_p, the mass-specific partition (or distribution) coefficient. The retardation factor assumes that sorption is reversible, instantaneous, and linear (i.e., not proportional to solute concentration or organic matter) (Freeze and Cherry, 1979).

3.3.2.4 Organic Carbon Partition Coefficient, K_{oc}

The sorption of hydrophobic organic compounds is strongly controlled by the presence of soil organic material. K_{oc} is the partitioning coefficient for a solute between the aqueous and organic phases and is defined as the ratio of the amount of chemical adsorbed (in micrograms) per unit mass of organic carbon (oc, in grams) in the soil or sediment to the concentration (in micrograms per milliliter of solution) of the chemical in solution at equilibrium (Lyman et al., 1990). K_{oc} is unitless[*] and is represented as

$$K_{oc} = \frac{K_d}{f_{oc}}, \tag{3.33}$$

where f_{oc} is the fraction of organic carbon in soil (mass ratio).

K_{oc} is not a primary physicochemical compound property because the degree to which a compound is adsorbed is in relation to its K_{oc} and dependent upon the soil organic matter content, organic matter properties, and groundwater chemistry. Measured values of K_{oc} are obtained for particular conditions of the measured soil and solution; however, because K_{oc} is the K_d value normalized for an aquifer's organic carbon content, f_{oc}, use of K_{oc} values to contrast contaminant mobility reduces the variability of partition-coefficient measurements in different soils (Dragun, 1988). K_{oc} values derived from linear isotherms are therefore considered to be largely independent of the properties of soil or sediment. Nevertheless, variability in soil properties including clay content, the particular clay mineral(s), surface area, soil pH, soil temperature, and type of soil organic matter may contribute to different measured organic carbon partition coefficients. Values of K_{oc} for the same compound measured on different soils have uncertainties ranging from 10% to 140% (Lyman, 1990). When estimating the distribution coefficient K_d from the organic carbon partition coefficient K_{oc} and soil organic carbon content f_{oc} (i.e., $K_d = K_{oc} \times f_{oc}$), the following assumptions are made (Sutherson, 2002):

- Sorption is exclusively to the organic component of the soil.
- All soil organic matter has the same sorption capacity per unit mass.
- The sorption–desorption process is in equilibrium.
- The sorption–desorption isotherms are identical, that is, hysteresis does not occur.

Of course, soil organic matter varies in type and sorption capacity among different soils. A significant portion of soil organic matter is composed of humic and fulvic acids. Fulvic acid is composed of organic polymers with molecular weight greater than 2000, whereas humic acid is composed of organic polymers with higher molecular weights, up to 300,000 daltons.[†] Some sources report that compounds have a greater propensity to sorb to one type of organic matter than another (Dragun, 1988), whereas others find that there is only a small variability of the sorption properties of soil organic matter (Nguyen et al., 2005). For humic acid, absorption of organic compounds into the organic matrix plays a greater role than adsorption onto its surfaces (Niederer et al., 2006b). Sorption by amorphous soil organic matter involves *ab*sorption *into* instead of *ad*sorption *onto* the

---

[*] Note that K_{oc} is unitless, which assumes that 1 mL of solution = 1 g of solution, that is, organic compounds are present at low concentrations. Adsorption isotherms become nonlinear at high concentrations.

[†] Dalton (Da) is an atomic mass unit equal to one-twelfth of the mass of an unbound atom of carbon-12, that is, the approximate mass of a hydrogen atom.

organic matter and is referenced as the solubility of the organic compound into the soil organic matter, which is generally proportional to the inverse of aqueous solubility (Chiou and Kile, 1994).

Estimated values of K_{oc} are often used in environmental fate assessment because measurement of K_{oc} is expensive and values are generally not transferable to other settings. Most published values for K_{oc} are estimated by using a variety of QSARs. More than 200 different correlations for estimating K_{oc} have been published. The correlations used to estimate K_{oc} are based on compound solubilities, octanol/water partition coefficients, molecular factors, topological indices, solvation energy relationships, and others (Reinhard and Drefahl, 1999). Table 3.11 lists values for K_{oc} estimated by using the PCKOCWIN algorithm as well as values published in various chemical fate and transport property compilations.

The variation among predicted values of K_{oc} listed in Table 3.11 underscores the uncertainty inherent to estimation of retardation due to sorption onto soil organic matter. Calculated values of K_{oc} can provide an order of magnitude estimate of retardation when combined with measured values of the weight fraction of soil organic matter in aquifer solids. However, measured values of K_{oc} must address organic matter variability due to aquifer heterogeneity. In general, less soil organic matter is present in higher-conductivity strata, and sorption to organic matter does not substantially contribute to retardation in sands and gravels. K_{oc} values in clays and silts are similar, but K_{oc} values for sands are generally 25% lower than for clays (McCarty and Reinhard, 1981).

When reviewing literature values of K_{oc}, it is important to distinguish whether experiments used whole soils or grain-size segregates, given the strong influence of particle size. Fine-grained, low-conductivity sediments such as clays and silts are richer in organic matter and play a greater role in retarding contaminant migration. The type of soil organic matter also makes a difference. Organic matter in consolidated sediments that has been altered by compression and elevated temperature (i.e., diagenesis) may behave differently than unaltered soil organic matter (Patterson et al., 1985).

PCKOCWIN (a soil adsorption coefficient program) estimates K_{oc} for organic compounds. Earlier methods for estimating K_{oc} relied upon relationships with the octanol/water partition coefficient. PCKOCWIN uses a molecular connectivity index (MCI) to predict K_{oc} values for hydrophobic organic compounds, but the MCI approach does not reliably estimate K_{oc} values for hydrophilic polar aprotic compounds such as 1,4-dioxane.[*] PCKOCWIN uses a series of group contribution factors to predict K_{oc} for polar compounds (USEPA, 2007a). The group contribution method outperforms the traditional estimation methods based on octanol/water partition coefficients and water solubility (Meylan et al., 1992). A comparison of PCKOCWIN estimates with experimentally determined distribution coefficients shows that (1) the MCI-based prediction model provides unreliable and inconsistent K_{oc} values, for both polar and nonpolar compounds, and (2) the divergence between experimental and calculated distribution coefficients becomes larger with increasing molecular size (Niederer et al., 2006b). K_{oc} values generated by QSAR models should be used with caution. The similarity of the compounds in the calibration data set for the model and the compound of interest should be confirmed (Niederer and Goss, 2007).

3.3.2.5 Predicting K_{oc} from K_{ow}, the Octanol/Water Partition Coefficient

The octanol/water partition coefficient, K_{ow}, is often used to predict the potential for a compound to adsorb to the aquifer matrix. K_{ow} is the ratio of the concentration of a chemical in octanol and in water at equilibrium and at a specified temperature. Octanol is an organic solvent that is used as a surrogate for natural organic matter. The relationship between the mass-specific distribution coefficient, K_d, and the octanol/water partition coefficient, K_{ow}, has been firmly established for nonpolar organic compounds present in concentrations at less than half their solubility limit and in aquifers whose organic carbon fraction is 0.1% or greater (McCarty and Reinhard, 1981; Schwarzenbach and Westall, 1981). The total organic carbon content of the aquifer matrix has less bearing on the potential for sorption in compounds with low octanol/water partition coefficients (Suthersan, 2002).

---

[*] A polar aprotic solvent is one that does not contain an O—H or N—H bond.

TABLE 3.11
Estimated K_{oc} Values for Stabilizer Compounds

Stabilizer Compound	log K_{oc} Estimated by PCKOCWIN	Commonly Cited log K_{oc}	Estimated[a] K_d ($K_{oc} \times f_{oc}$)	Mobility[b]	References
Nonylphenol	4.79	4.49	6089	Immobile	Swann et al. (1983)
p-tert-Amyl phenol	3.58	2.41	380		Lyman et al. (1990)
1-Naphthol	3.48	4.4	304		Deneer et al. (1988)
Resorcinol	2.64	1.81	43.4		Swann et al. (1983)
Hydroquinone	2.64	1.325	43.4		Swann et al. (1983)
Diisobutylene	2.44	2.44	27.55		Swann et al. (1983)
Phenol	2.43	1.48	26.8		Briggs (1981)
Cyclohexane	2.22	2.2	16.55	Low mobility	Swann et al. (1983)
Diisopropylamine	2.04	2.15	10.9		Swann et al. (1983)
Triethylamine	2.03	2.16	10.72		Swann et al. (1983)
Isopentane	1.83	3.0	6.77		MSDS for isopentane
Isoprene	1.83	2.69	6.77		Lyman et al. (1990)
Styrene oxide	1.73	1.72	5.42		Lyman et al. (1990)
Pyridine	1.52	1.7	3.3		Lyman et al. (1990)
Nitroethane	1.20	1.48	1.58		Swann et al. (1983)
Triethanolamine	1.0	0.477	1.0	High mobility	Hansch and Leo (1985)
Isopropyl acetate	0.98	1.18	0.95		Swann et al. (1983)
Nitromethane	0.91	0.28	0.82		Lyman et al. (1990)
Ethyl acetate	0.79	1.77	0.613		Swann et al. (1983)
Epichlorohydrin	0.65	1.6	0.45		Swann et al. (1983)
Acetonitrile	0.65	1.2	0.45		Swann et al. (1983)
1,2-Butylene oxide	0.65	0.65	0.45		Swann et al. (1983)
2-Butanone	0.58	1.5	0.383		Walton et al. (1992)
2-Methyl-3-Butyn-2-Ol	0.47	1.55	0.292		UNEP (1998)
tert-Amyl alcohol	0.47	0.46	0.292		Swann et al. (1983)
Butoxymethyl oxirane	0.45	0.24	0.283		SPI (2006)
Trichloroacetic acid	0.44	0	0.274		Weber et al. (1981)
Propylene oxide	0.37	1.4	0.232		Swann et al. (1983)
sec-Butyl alcohol	0.31	1.7	0.205		Swann et al. (1983)
tert-Butyl alcohol	0.17	1.57	0.16		Swann et al. (1983)
tert-Butyl phenol	0	2.77	0.10		NICNAS (2003)
1,4-Dioxane	**0**	**1.23**	**0.10**		**Lyman et al. (1990)**
1,3-Dioxolane	0	1.18	0.10		Lyman et al. (1990)
Methanol	0	−0.74	0.10		USEPA (1996)
Methyl chloroform	1.687	2.18	15.2	Low	Pankow and Cherry (1996)
Dichloromethane	1.375	0.944	0.88	High	
Trichloroethylene	1.831	2.1	12.6	Low	
Perchloroethylene	2.029	2.56	36.4	IMM	

Notes: PCKOCWIN is a soil adsorption coefficient estimation program and part of the *Estimation Programs Interface Suite* (EPI Suite™) (USEPA, 2007a).

[a] To calculate K_d, the contaminant distribution coefficient, K_{oc} is multiplied by the fraction of soil organic matter, f_{oc}, where f_{oc} is assumed to be 0.1.

[b] Mobility classes from classification scheme proposed by Fetter (1993); IMM = immobile.

For hydrophobic compounds, K_d and K_{ow} are highly correlated for all sorbents except pure mineral substrates, in which micropores may play a role. For nonpolar organic compounds such as chlorinated solvents, the correlation between the logarithms of K_{oc} and K_{ow} values was determined to follow the relationship (Schwarzenbach and Westall, 1981)

$$\log K_{oc} = 0.72 \ (K_{ow} + 0.49), \quad R^2 = 0.95. \tag{3.34}$$

Another commonly used relationship (Karickhoff, 1981) is

$$K_{oc} = 0.41 K_{ow}. \tag{3.35}$$

These two relationships are not reliable for estimating partition coefficients of polar organic compounds such as alcohols and aliphatic ethers. A regression analysis of about two dozen experimentally determined K_{oc} values produced the following correlation (Nguyen et al., 2005):

$$\log K_{oc} = 0.73 \ (\log K_{ow} + 0.52), \quad R^2 = 0.83. \tag{3.36}$$

Equation 3.36 has a rather low correlation coefficient (R^2). The compounds used to derive Equation 3.36 do not include any ether compounds. A more dynamic and reliable approach to estimating K_{oc} uses linear free-energy relationships (LFERs)[*] to correlate compound structure and sorption to soil organic matter (Niederer et al., 2006b).

Numerous other relationships have been developed for different classes of compounds. In general, compounds with octanol/water partition coefficients less than 1000 will migrate easily through the subsurface without appreciable retardation due to sorption (McCarty and Reinhard, 1981).

K_{ow} values in the literature are prone to considerable uncertainty, sometimes approaching two orders of magnitude or more. Experimental determination of K_{ow} values may incorporate method errors; for example, small quantities of emulsified octanol in the aqueous phase can carry high concentrations of the organic compound being tested, leading to erroneously high determinations of the concentration in water (Mackay et al., 1993). K_{ow} sorption models may be unreliable for polar compounds even if experimental K_{ow} values are used for input data (Niederer et al., 2006b).

Using physicochemical data without carefully assessing its reliability can lead to significant error in estimating contaminant fate. Fate and transport experts address poor data quality of K_{ow} values in the literature by using structure–activity estimation methods to compute K_{ow}, by using carefully selected data, or by applying probabilistic assessments to fate and transport modeling (Renner, 2002).

3.3.2.6   Effect of Surface Oxides on Sorption

Increased amounts of oxygen on granular activated-carbon surfaces can decrease the affinity of organic compounds for sorption to carbon. Oxidation of carbon by acids increases the presence of oxygen surface functional groups and decreases the adsorption capacity of organic compounds (Snoeyink, 1999). Decreased adsorptive capacity was observed in activated carbon for several stabilizer compounds (phenol, nitromethane, methyl ethyl ketone, n-butanol, and 1,4-dioxane) as a result of an increase in acidic surface oxides (Vidic and Suidan, 1991).

3.3.2.7   Competitive Sorption

Solvent-stabilizer compounds are usually released as mixtures composed primarily of chlorinated solvents and oily wastes. The rate of migration of individual components of the mixture will depend

---

[*] LFERs, also called the linear Gibbs energy relation, describe a linear correlation between the logarithm of a rate constant or equilibrium constant for one series of reactions and the logarithm of the rate constant or equilibrium constant for a related series of reactions.

primarily on their solubility, but also on their capacity to adsorb to soil organic matter and mineral surfaces. In mixtures, competitive sorption occurs, in which compounds with greater affinities for soil organic matter are preferentially adsorbed and those compounds with lower affinities stay in solution. Competitive sorption occurs between organic compounds in anthropogenic wastes and natural dissolved organic substances such as humic and fulvic acids. Competitive sorption in soil can cause compounds initially adsorbed to organic matter surfaces to become dislodged and replaced by compounds with stronger affinities for sorption (Snoeyink, 1999). Consequently, compounds that migrate more rapidly because of higher solubility and weak capacity to adsorb to soil surfaces may be displaced as compounds with higher K_{oc} values arrive at the soil surface.

The potential adsorption of a compound determined from a pure-component test does not necessarily predict its degree of removal from a dynamic, multicomponent mixture (Verschueren, 1996). The presence of other organic compounds in solution, whether natural or anthropogenic, usually reduces the adsorption capacity of activated carbon or soil organic matter for a particular organic compound.

In a study of competitive sorption of aqueous-phase solvent-stabilizer compounds onto granular activated carbon, nitromethane was found to have an adsorption capacity similar to methyl ethyl ketone (2-butanone), both of which have substantially larger capacities than n-butanol, which in turn has a larger sorption capacity than 1,4-dioxane (McGuire and Suffett, 1978). The adsorption isotherms for the four-compound mixture were significantly lower than those for each individual compound in solution. As the equilibrium concentration was reduced, the mixture isotherms for n-butanol and 1,4-dioxane became identical to the single-compound isotherms. 1,4-Dioxane was progressively desorbed by the three other compounds in the mixture. Nitromethane, n-butanol, and methyl ethyl ketone continued to adsorb to the granular activated carbon after 1,4-dioxane had completely broken through, indicating that they are more successful competitors for adsorption sites than 1,4-dioxane. Initial adsorption followed by displacement by compounds with stronger affinity to adsorb to carbon is referred to as the "chromatographic effect" (McGuire and Suffett, 1978). Figure 3.3 shows the effect of competitive sorption on the retention on granular activated carbon of 1,4-dioxane and its displacement by other solvents.

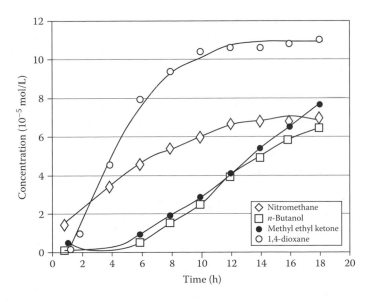

FIGURE 3.3 Breakthrough curves for a mixture of nitromethane, methyl ethyl ketone, n-butanol, and 1,4-dioxane on granular activated carbon (GAC). Influent concentrations were 10^{-4} mol/L each; flow rate = 23 mL/min, pH = 8.0, detention time = 2.1 min. (From McGuire, M.J., and Suffett, I.H., 1978, *Journal of the American Water Works Association* 70(11): 621–635. With permission.)

3.3.2.8   Use of Laboratory Batch Tests to Measure K_d in Soil

Contaminants residing in the vadose zone may leach to underlying groundwater from percolation of rainfall, flood water, or applied irrigation. The degree to which contaminants are prone to dissolution, desorption, leaching, and migration to the saturated zone is the subject of the various estimation methods described in this section. Fate and transport models often rely on published values of K_{oc} and assumed organic matter contents (f_{oc}) in soil to estimate contaminant retardation and partitioning between sorbed and aqueous phases. This approach can significantly underpredict the sorption potential of a contaminant and overpredict its mobility in soil or groundwater, when compared with soil-column leach testing.[*]

Laboratory batch tests can be used to quantify the mobility of contaminants in soil and to estimate the initial concentration of the contaminant in soil leachate and the final concentration as the contaminant enters the saturated zone. Batch tests involve placing a small quantity of soil in buffered, deionized water, agitating the soil slurry for a set period of time, and measuring the soluble fraction of the contaminant. A minimum of three soil samples is generally needed to validate batch test data for each area investigated (HDOH, 2007). The ratio of the mass of a contaminant that remains sorbed to the contaminant mass that goes into solution is referred to as the contaminant's desorption coefficient or K_d value (see additional discussion of K_d—also called the distribution coefficient, but most often called the partition coefficient—in Sections 3.3.2.3, 3.4.1, and 3.4.2).

The key parameter in soil leaching models is the contaminant's K_d value. Lower K_d values imply greater contaminant mobility in soil and a greater leaching potential. Contaminants with K_d values less than 1.0 are considered to be highly mobile, whereas contaminants with K_d values greater than 20 are considered essentially immobile and do not pose a significant leaching concern. Measured values of 1,4-dioxane are typically less than 1.0, as discussed in Section 3.4.2.

The contaminant concentration that partitions into leachable soil moisture can be calculated by using the soil-specific K_d value in an equilibrium partitioning equation:

$$C_{total} = C_{leachate} \times \left[K_d + \left(\frac{\Theta_w + (\Theta_a \times H')}{\rho_b} \right) \right] \times \left(\frac{1\,\text{mg}}{1000\,\mu\text{g}} \right), \tag{3.37}$$

where C_{total} is the total concentration of chemical in the soil sample (in mg/kg), $C_{leachate}$ is the dissolved-phase concentration of chemical (in µg/L), K_d is the estimated or measured distribution coefficient (in L/kg), Θ_w is the water-filled porosity (in L/L or volume of water per volume of soil), Θ_a is the air-filled porosity (in L/L or volume of air per volume of soil), H' is the Henry's law constant at 25°C [in (µg/L)/(µg/L), that is, micrograms of chemical per liter of vapor divided by micrograms of chemical per liter of water], and ρ_b is the soil bulk density (in kg/L).

The final concentration of the contaminant in groundwater as a consequence of its leaching is estimated by assuming a groundwater–leachate dilution factor, DF, usually assumed to be 20 for sites less than 0.5 acre in size (USEPA, 2001b). The DF is the volume of affected groundwater divided by the volume of leachate.

Where site-specific hydrogeological data are available, a site-specific DF can be calculated according to the following equation (USEPA, 2001b):

$$\text{DF} = 1 + \left(\frac{K \times i \times d}{I \times L} \right), \tag{3.38}$$

where K is the aquifer hydraulic conductivity (in m/year), i is the regional hydraulic gradient, d is the assuming mixing zone depth (default is 2 m), I is the surface-water infiltration rate

---

[*] Personal communication with Roger Brewer, PhD, State of Hawaii Department of Health.

(in m/year), and L is the length (in m) of the contaminated soil area that is parallel to groundwater flow (HDOH, 2007).

3.3.2.9　Sorption of Vapor-Phase 1,4-Dioxane

Because 1,4-dioxane is hydrophilic and has a very low Henry's law constant, it is unlikely to migrate significantly in the vapor phase once it has dissolved into water. Vapor transport may be of interest in contamination source areas with thick unsaturated zones where vapor intrusion into occupied buildings is a concern (Little et al., 1992). The extensive literature on sorption includes several studies in which sorption of 1,4-dioxane vapor was measured. A 1994 study by Chiou and Kile measured the vapor adsorption of several solvent-stabilizer compounds dissolving into low-mineral-content samples of peat soil from the Florida Everglades, including nitroethane, 1,4-dioxane, and acetonitrile. The isotherm of acetonitrile was found to be about 4.2 times greater than the isotherm for 1,4-dioxane, and the isotherm for nitroethane was approximately 3.5 times higher greater than the isotherm for 1,4-dioxane (Chiou and Kile, 1994). The comparatively low sorption to peat, which is composed of 86.4% organic carbon,[*] suggests a lower potential for 1,4-dioxane vapors to be sorbed to the much smaller fraction of soil organic matter found in most vadose zone soils.

Scientists at the Swiss Institute of Biogeochemistry and Pollutant Dynamics rigorously measured the humic acid/air partition coefficients for 188 polar and nonpolar organic compounds at different temperatures and relative humidities by using a dynamic flow-through technique. Humic acid/air partition experiment data reflect the concentration sorbed relative to the concentration in air. The measurements were made by using inverse gas chromatography with the humic acid sorbent serving as the chromatographic column stationary phase in a gas chromatograph and the organic compound of interest injected onto the column under isothermal conditions.[†] The measured retention reflects the tendency of the compound to sorb to this particular variety of humic acid at different temperatures and humidities.[‡] The Swiss experiments provide the most comprehensive and reliable data set for sorption into humic material available (Niederer et al., 2006a). The humic acid/air partition coefficients for a few compounds that may be present in solvent wastes as solvent stabilizers are provided in Table 3.12, and Table 3.13 lists humic acid/soil-vapor distribution coefficients at 98% relative humidity. For example, the following equation describes the 1,4-dioxane partitioning between air and humic acid:

$$K_{\text{air/humic acid-1,4-dioxane}} = 4680 \left[\frac{(n_{1,4\text{-dioxane}} / m_{\text{humic acid}})}{(n_{1,4\text{-dioxane}} / V_{\text{air}})} \right] = 4680 \frac{V_{\text{air}}}{m_{\text{humic acid}}}, \qquad (3.39)$$

where n is the number of moles, m is the mass (in kg), and V is the volume (in L).

The 1,4-dioxane aqueous-phase/soil organic carbon partition coefficient is comparatively low, but in the vapor-phase 1,4-dioxane partitions easily into humic acid, the major component of soil organic matter. The variation in humic acid/soil-vapor distribution coefficients with humidity may reflect changes in the polarity of the humic substance with water saturation (Chiou and Kile, 1994).

3.3.2.10　High-Strength 1,4-Dioxane Solutions May Cause Clay Swelling

At high concentrations, some organic compounds, including 1,4-dioxane, can induce swelling and flocculation of smectite (i.e., montmorillonite) clays (Bradley 1945; MacEwan, 1948; Brindley and Tsunashima, 1972). Chemical alteration of clay-mineral arrangement in natural soils and engineered clay liners can affect permeability and diffusive and advective transport (Wu et al., 1994).

---

[*] Measured on a dry-weight basis. The Florida Everglades peat is a reference sample of the International Humic Substances Society.

[†] See Chapter 4 for a brief explanation of chromatographic column stationary phases and gas chromatography.

[‡] The Swiss study used the "leonardite humic acid" with an organic carbon content of 63.8%.

TABLE 3.12
Humic Acid/Air Partition Coefficients at 15°C

Substance	<0.01% K_{HA-air} (L/kg)	45% K_{HA-air} (L/kg)	70% K_{HA-air} (L/kg)	98% K_{HA-air} (L/kg)
Phenol	797,000	1,770,000	967,000	1,920,000
1,4-Dioxane	**4680**	**12,200**	**7000**	**16,900**
Nitroethane	2390	3360	3430	3040
Nitromethane	1560	4270	2650	3110
2-Butanone	2400	1730	1340	2440
n-Propyl Acetate	5030	1220	1480	1930
2-Butanone	2400	1730	1340	2440

Relative Humidity

Source: From Niederer, C., Goss, K.U., and Schwarzenbach, R.P., 2006a, *Environmental Science and Technology* 40(17): 5368–5373. With permission.

Notes: The K_{HA-air} partition coefficients are stated as the liters of air containing the given substance per kilogram of humic acid. Data for 1,4-dioxane are shown in boldface. Calculations are according to Equation 3.39.

Smectite clays have silica and aluminum layers assembled in a 2:1 ratio of silica to aluminum, that is, –Si·Al·Si·Si·Al·Si–. The lattice structure of smectite clays can expand in response to water and other agents, including sorption of organic compounds on both external and internal surfaces. Organic molecules and water can penetrate the layers of the clay-mineral crystal, changing its structure and swelling properties.

Wu et al. (1994) mixed 1,4-dioxane at concentrations of 0.5, 1, and 3 M [4.4%, 8.8%, and 26.4% (i.e., 264,000 mg/L)] in a suspension of deionized water and sodium-saturated smectite. The suspension was deposited on a ceramic filter within a sealed environmental chamber equipped with

TABLE 3.13
Humic Acid/Soil-Vapor Partition Coefficients at 98% Relative Humidity

Substance	5°C K_{HA-air} (L/kg)	15°C K_{HA-air} (L/kg)	25°C K_{HA-air} (L/kg)
Phenol	2,080,000	990,000	416,000
2-Butanone	2430	708	759
Ethyl acetate	1340	247	—
Tetrahydrofuran	1620	504	—
1,4-Dioxane	**15,100**	**6550**	**4380**
Acetonitrile	2630	1460	978
Nitromethane	2900	1880	1070
Nitroethane	2850	1940	—
Pyridine	–	2270	—

Temperature

Source: From Niederer, C., Goss, K.U., and Schwarzenbach, R.P., 2006a, *Environmental Science and Technology* 40(17): 5368–5373. With permission.

Notes: The K_{HA-air} partition coefficients are stated as the liters of soil vapor containing the given substance per kilogram of humic acid. Data for 1,4-dioxane are shown in boldface.

a window to allow x-ray diffraction measurements of spacing between clay layers and a drain to retrieve samples of drained solution. The 1,4-dioxane solution was expressed through the filter by applying pressure with helium; the expressed solution was drained at atmospheric pressure. Successive x-ray diffraction measurements at different helium pressures and for different concentrations of 1,4-dioxane showed that 1,4-dioxane converts fully expanded clay layers with spacing of 30 Å to partially expanded layers with a spacing of 15 Å (Wu et al., 1994).[*]

The researchers in this study postulate that 1,4-dioxane separates from solution and fills the space between the partially expanded layers in the sodium montmorillonite clay, that is, a pure-phase 1,4-dioxane accumulates between the clay layers. 1,4-Dioxane has also been observed to accumulate between layers of calcium montmorillonite (Zhang et al., 1990). 1,4-Dioxane apparently rearranges water molecules between the expanded clay layers to produce an open, coordinated structure that may force layers apart because of the larger volume of the inferred structure (Wu et al., 1994).

Zhang et al. (1990) studied 1,4-dioxane adsorption to sodium and calcium smectite suspensions in the <2 μm size fraction by using ^{14}C-labeled 1,4-dioxane. The 1,4-dioxane sorption isotherm for a calcium smectite followed the Langmuir pattern, wherein the rate of sorption is initially first-order and subsequently zero-order, producing a steep initial slope in the curve of quantity adsorbed versus 1,4-dioxane concentration, followed by an asymptotic tapering off to a zero-slope plateau.[†] This result was markedly different from the sorption isotherm for sodium smectite, for which the millimoles of 1,4-dioxane adsorbed per gram of clay increased monotonically with increasing concentration. The Zhang study found that the 1,4-dioxane adsorption reaction with calcium smectite was exothermic, but bonding was not indicated. Adsorption of most other organic compounds is usually endothermic. Zhang proposed that 1,4-dioxane adsorption onto calcium smectite clay being exothermic is attributable to its unique structure-breaking effect on the tightly held monolayer of water on the clay surface (Zhang et al., 1990). The behavior of 1,4-dioxane as a water-structure breaker has been confirmed; two 1,4-dioxane rings can form a dimer stabilized by two intermolecular hydrogen bonds, leaving two oxygen atoms available for interaction with the water molecules and creating a dipole moment (Mazurkiewicz and Tomasik, 2006).

In solutions with 1,4-dioxane, the interlayer distance in calcium smectite expands from 0.55 to 1.48 nm. In a 1.0 M solution of 1,4-dioxane, the interlayer concentrations in the calcium smectite were found to be 6.0 M (Zhang et al., 1990). Zhang concluded that 1,4-dioxane is not adsorbed because of any specific interaction with the surfaces of the clay layer; instead, 1,4-dioxane is distributed between the bulk phase and the interfacial phase, but prefers the interfacial phase, which facilitates its migration through clays. A different conclusion was reached in an earlier study by Brindley et al. (1969): calcium smectite clays in 1,4-dioxane solutions ranging from 1 to 100 mol% had constant basal spacings, leading to the conclusion that 1,4-dioxane is strongly and preferentially adsorbed by the clay. That study has been superseded by newer research under more controlled conditions, for example, Wu et al. (1994).

3.3.3 Subsurface Vapor-Phase Transport

The migration of contaminants in the vapor phase constitutes an important pathway by which releases of volatile chemicals to soil may manifest as groundwater contamination. Soil vapors are a concern when they migrate to and accumulate in enclosed spaces, where they may lead to either toxic exposure in occupied buildings or create an explosion hazard. Man-made migration pathways for soil vapors—such as permeable sand or gravel fill in trenches for buried electrical lines or rapid

---

[*] Å = angstrom; 1 Å = 10^{-8} m = 0.1 nm.
[†] In first-order processes, the rate is proportional to the concentration of a single reactant or substrate; in a zero-order process, the rate is independent of the substrate concentration.

movement through sewer lines and storm drains—may provide the connection between sources of flammable vapors and sources of ignition.

Vapors accumulating beneath building foundations may intrude into occupied spaces, creating an inhalation exposure hazard. The degree to which vapors may intrude into enclosed spaces depends on the same factors as for volatilization from dry soil described in Section 3.1.2, but also includes factors specific to the building–soil interface. These include the surface area of the building slab or foundation, diffusion coefficients describing the efficiency of vapor transport through cracks in the slab, the thickness of the slab, air exchange rates for the structure, soil–gas flow rates, and other factors. The reader is directed to the authoritative references on the topic of vapor intrusion[*]; this section addresses the intrinsic vapor transport characteristics of 1,4-dioxane and other solvent stabilizers.

The physical and chemical parameters controlling contaminant vapor transport are source-zone concentration, vapor pressure, vapor density, solubility, and sorption characteristics. These parameters determine contaminant equilibrium distribution between the vapor, sorbed, and dissolved phases. Contaminant biodegradability and chemical stability will determine contaminant persistence in each of these subsurface phases, as discussed in Section 3.3.4.

The soil characteristics governing contaminant vapor transport in a homogeneous soil include porosity, soil moisture content, soil-air water vapor content or humidity, and soil-air flow rates governed by pressure and temperature gradients. In the absence of advective flow of soil air, molecular diffusion, characterized by contaminant diffusivity in air as discussed in Section 3.1.2, is the dominant property controlling soil-vapor transport. Soil is rarely homogeneous; soil characteristics such as stratigraphy, structures, organic matter, roots, biota, mineralogy, and density all present challenges to the reliable estimation of soil-vapor transport.

The vapor pressure of a liquid or solid is the pressure of the gas in equilibrium with the liquid or solid at a given temperature. Volatilization, the evaporative loss of a chemical from the liquid to the vapor phase, depends on the vapor pressure of the chemical and on environmental conditions that influence diffusion from the evaporative surface. Chemicals with relatively low vapor pressures, high sorption onto solids, or high solubility in water are less likely to vaporize and become airborne than are chemicals with high vapor pressures or less affinity for solution in water or adsorption to solids and sediments. Because most solvent-stabilizer compounds are highly soluble and have relatively low vapor pressures, they are generally unlikely to partition into the soil-vapor phase and migrate through the soil as vapors.

The probability that a compound occurs in the gas phase depends not only on its vapor pressure but also on its water solubility and adsorption/desorption behavior. Therefore, even substances that have a relatively low vapor pressure (down to 10^{-3} Pa) can be found in soil vapor in measurable quantities (Verschueren, 1996).

Vapor pressures, vapor densities, diffusivities, and relative evaporation rates for selected stabilizer compounds are summarized in Table 3.1. Appendix 3 provides a comprehensive listing of physicochemical parameters for a longer list of solvent-stabilizer compounds. Figure 3.4 contrasts vapor pressures with aqueous solubilities for the solvent and solvent-stabilizer compounds in Appendix 3. In Figure 3.4, compounds plotting in the box in the upper left corner have the highest likelihood of partitioning into the vapor phase and being transported.

3.3.4 BIODEGRADABILITY OF SOLVENT-STABILIZER COMPOUNDS

Information presented in this chapter focuses on laboratory studies geared toward identifying organisms capable of 1,4-dioxane destruction or transformation, identification of degradation pathways,

---

[*] See Johnson and Ettinger (1991), Little et al. (1992), Johnson et al. (1999), American Petroleum Institute (1998), and DTSC (2004).

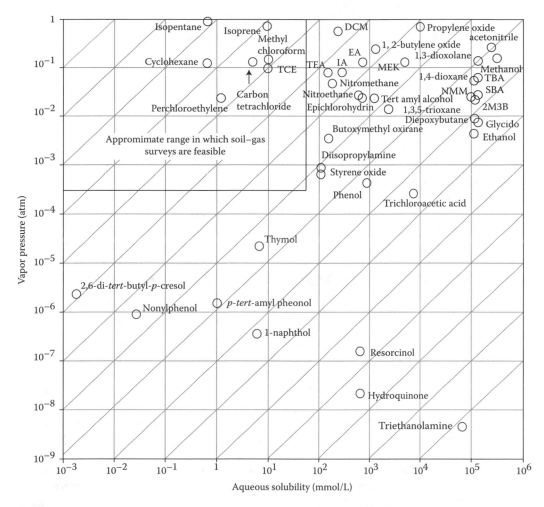

FIGURE 3.4 Plot of vapor pressure versus aqueous solubility for chlorinated solvents and solvent-stabilizer compounds. Diagonal lines represent Henry's law constant equivalents, increasing to the upper left.

and theoretical considerations related to biodegradation. Chapter 7 further assesses various studies related to the possibility of field application of bioremediation as a technology to carry sites toward closure. Although there is some overlap between these two chapters, Chapter 3 primarily addresses the theoretical approaches, and Chapter 7, the applied approaches.

The ability of an organic compound to persist in soil and groundwater is inversely related to the potential for indigenous microflora to dissimilate it (White et al., 1996). Biodegradation refers to the general ability of indigenous soil microbes or fungi to acclimate to and metabolize a contaminant under ambient temperature conditions through microbially induced structural transformation of the parent compound that changes its molecular integrity. Complete or ultimate biodegradation converts a compound to carbon dioxide and water ("mineralization"), whereas acceptable biodegradation is microbial transformation that results in reduced toxicity (Scow, 1982). Microbes can adjust their metabolism to degrade most anthropogenic organic compounds by induction of enzymes, by selecting new metabolic capabilities produced by genetic changes, or by increasing their population (Chappelle, 1993). Although fungi do not increase their population in response to introduction of a contaminant, some can use the contaminant as their sole carbon source.

The term "biodegradable" is qualified by the conditions under which biodegradation occurs. Biodegradable or nonbiodegradable are commonly used terms to indicate whether a compound will be susceptible to microbial destruction under ambient conditions. Some compounds widely considered to be nonbiodegradable under ambient conditions can be degraded under enhanced conditions, which may be created in *ex situ* bioreactors or *in situ* by injection of a substrate, inoculation of acclimated microbes, temperature control, or adding nutrients essential to sustaining microbial populations.

The by-products of microbial metabolism can sometimes produce compounds that are more toxic and regulated at a lower threshold than the original contaminant. The complete mineralization of the contaminant may involve multiple metabolic steps, possibly by different groups of microbes and at different metabolic rates. Biodegradation may be limited by the toxicity of the subject compound to microbes. The metabolic by-product may also be toxic to microorganisms, limiting further biodegradation after an initial step. "Complete biodegradation" occurs in six half-lives, after which only approximately 1.5% of initial contaminant concentration remains; after seven half-lives, less than 1% remains.

Testing for biodegradability often involves using pure cultures of a single strain of microbe to establish the identity of microbes responsible for biodegradation mechanisms and testing for intermediate and end products of biodegradation. Laboratory studies of biodegradation by mixed cultures of microbes are more relevant for determining biodegradability under ambient conditions; however, the conditions under which tests are performed are usually not representative of ambient conditions (Boethling et al., 1994). For example, temperatures in laboratory microcosm studies are elevated, flow rates are increased, growth media and nutrient solutions are enhanced, and high concentrations of the target contaminant are used to accelerate the process and complete the study within a reasonable time frame. When chemicals are present at trace concentrations, microorganisms may not assimilate carbon from the chemicals, and they will not grow or produce the large, acclimated populations needed for enhanced biodegradation (Alexander, 1985). Tests representing ambient conditions may take a long time to complete. Data from different studies may not be comparable where different environmental conditions are used.

Microorganisms in groundwater are more often found attached to grains of aquifer material than in flowing groundwater, and most microbes are larger than the smallest pores in most clays; consequently, hydrophilic contaminants that have diffused into very small pores or intraparticle voids may not be subjected to biodegradation. Microbes generally mediate biotransformation that is energetically favorable. Biotransformation reactions produce a net decrease in the free energy of the chemical system, and the microbes harvest some of the released energy for their own use, stimulating their growth (Hemond and Fechner, 1994).

Microorganisms responsible for biodegradation include heterotrophic bacteria, including actinomycetes, some autotrophic bacteria, fungi including yeasts and basidiomycetes, and some protozoa. Fungi are less able to utilize recalcitrant compounds as growth substrates than are the bacteria, which possess greater metabolic capacity. Fungi function only aerobically, whereas bacteria may evolve to metabolize only in aerobic systems, only in anaerobic systems, or in both environments (i.e., facultative organisms) (Hemond and Fechner, 1994).

Most compounds become usable as an energy or nutrient source only after a period of acclimation during which the population capable of metabolizing the substance grows to a threshold where appreciable degradation becomes noticeable. The acclimation of a microbial community to a chemical may not occur below a threshold concentration; low concentrations of contaminants may be resistant to biodegradation (Alexander, 1994).

Acclimation also involves existing microbes adapting through induction of enzymes that catalyze biodegradation. The number of steps in the biodegradation sequence for a compound will affect the rate of biodegradation (Scow, 1982). Induction is the series of cellular processes by which microbes produce enzymes specific to a particular substrate when that substrate is actually present (Chappelle, 1993). Products generated during the catabolism[*] of one substrate may repress the

---

[*] Catabolism is the metabolic process that breaks down molecules into smaller units.

synthesis of enzymes utilized for the degradation of a second substrate. This effect, called catabolite repression, may suppress the enzymatic activity of a population (Alexander, 1994).

The acclimation of a microbial community to one chemical substrate can result in acclimation to chemical substrates that are structurally similar. Cometabolism is the fortuitous degradation of a compound that microorganisms cannot utilize as growth substrates, that is, the compound does not provide a nutrient or energy source for the degrading organisms (Alexander, 1994; Hyman, 1999). Cometabolic reactions occur when enzymes that metabolize a primary substrate fortuitously transform the cometabolic substrate. For example, enzymes produced during the microbial degradation of the cyclic ether tetrahydrofuran (THF) also degrade 1,4-dioxane. If a synthetic organic compound is transformed by cometabolism, the mass of enzymes involved will be independent of the mass of the compound transformed and will be determined by the mass of the compound that induces the enzymes (Grady et al., 1997).

The rate of biodegradation is controlled by the availability and concentration of the compound to be degraded (e.g., adsorption favors availability) and by the microbe population and temperature of the system, as well as other environmental variables, such as pH, moisture content, and availability of soil organic matter or other organic substrates. Table 3.14 summarizes the conditions that control rates of biodegradation.

The kinetics of microbial processes determine whether biodegradation can play an important role in reducing or eliminating a compound in the subsurface within the time frame of interest, for example, the proposed period of monitored attenuation following source removal in site remediation. The

TABLE 3.14
Conditions Controlling Rates of Biodegradation

Substrate-Related	Influence
State	Liquid> solid> gas
Physicochemical properties	Sorption favors biodegradation; high solubility does not
Concentration	Upper and lower thresholds may restrict biodegradation
Size, shape, ionic charge	Permeation of cell membrane and interaction with enzymes
Toxicity	Specific or general growth-limiting toxicity
Molecular structure	Degree of branching and nature, number, and position of substituents; biodegradability of molecular structural units
Organism-Related	
Species composition of population	High populations favor biodegradation
History of acclimation to substrate	Affects rate; a lag period may be required
Inter- and intraspecies interactions	Competition for resources may favor biodegradation or prevent it if dominant strains do not degrade compound
Enzymatic composition and activity	Enzymes must catalyze contaminant breakdown
Environment-Related	
Temperature	Higher temperatures favor breakdown; upper limits apply
pH	Some pH ranges unfavorable for biodegradation
Moisture	Critical to nearly all biodegradation
Oxygen availability	Necessary to support aerobic biodegradation and fungi
Salinity	May suppress biodegradation above a threshold
Presence of other substances toxic to microbes	Can prevent sufficient population growth for biodegradation

Source: After Scow, K.M., 1982, In: W.J. Lyman, W.F. Reehl, and D.H. Rosenblatt (Eds), *Handbook of Chemical Property Estimation Methods*, pp. 9-1–9-85. Washington, DC: American Chemical Society and Knapp, J.S. and Bromley-Challenor, K.C.A., 2006, Recalcitrant organic compounds. http://www.bmb.leeds.ac.uk/mbiology/ug/ugteach/level3.html (accessed September 9, 2007).

TABLE 3.15
Index of Biodegradability According to BOD$_5$/COD Ratio

BOD$_5$/COD	Rating	Example Compounds
<0.01	Not degradable	1,4-Dioxane, propylene oxide, morpholine, triethanolamine
0.01–0.1	Moderately degradable	n-Butyl acetate, acetonitrile
>0.1	Relatively degradable	n-Propyl alcohol, n-butyl alcohol, pyridine, n-amyl alcohol, phenol

Source: Scow, K.M., 1982 *Handbook of Chemical Property Estimation Methods*, pp. 9-1–9-85. Washington, DC: American Chemical Society.

first-order kinetics follows a linear relationship between concentration reduction by microbial degradation and time, whereas the second-order kinetics follows a hyperbolic relationship (Scow, 1982).

A screening-level index of the biodegradability of organic compounds uses the ratio of biochemical oxygen demand from a 5-day test (BOD$_5$) to chemical oxygen demand (COD) according to the classification scheme shown in Table 3.15.

3.3.4.1 Predicting Biodegradability

Most of the thousands of organic compounds introduced to the marketplace and subsequently released to the environment are subjected to only screening-level tests for biodegradability. Reliable and consistently generated biodegradation data are not widely available. A goal in the assessment of the environmental persistence of organic compounds is the ability to use biodegradation models to predict whether the compounds will be mineralized by ambient soil microbes. Databases of consistently developed biodegradation data—such as the Japanese Ministry of International Trade and Industry (MITI) Test database, which includes more than 800 compounds—are gradually overcoming this limitation to biodegradation modeling, and predictions of biodegradability are improving (Pavan and Worth, 2006). One such biodegradation model is BIOWIN, a commonly used model developed by Syracuse Research Corporation and available as part of the EPIWIN Suite (USEPA, 2007a).

BIOWIN is an algorithm that predicts the potential for aerobic biodegradation from a compound's molecular structure and molecular weight. High-molecular-weight compounds are less prone to aerobic biodegradation. The BIOWIN classification of chemical compounds is based on the presence of structural fragments in a molecule that are also present in compounds whose biodegradability is known. A training set of 200 compounds whose biodegradation fates are well studied is used to predict the biodegradation potential for compounds that include the same structural fragments (Boethling et al., 1994). For example, ester, alcohol, and carboxylic acid groups usually enhance biodegradability, whereas the presence of halogens and nitro groups generally increases resistance to biodegradation (Scow, 1982). Alcohols, aldehydes, acids, esters, and amides are more susceptible to biodegradation than are the corresponding alkanes, olefins, ketones, dicarboxylic acids, nitriles, amines, and chloroalkanes. Ether functions are particularly resistant to biodegradation because of the high dissociation energy of the ether linkage (about 360 kJ/mol) (Kim and Engesser, 2004). Highly branched compounds are also more resistant to biodegradation than are straight chains, and short chains are not as quickly degraded as long chains. Halogenated compounds that resist aerobic degradation may be more rapidly degraded under anaerobic conditions (Aronson et al., 1999). Unsaturated compounds (alkenes or alkynes) are more susceptible to biodegradation than saturated compounds (alkanes). Table 3.16 lists output from the BIOWIN model for predicting biodegradability of solvent stabilizers and chlorinated solvents, and Table 3.17 gives ratings for biodegradability interpreted from literature studies and collated in the Syracuse Research Corporation's BIODEG database (SRC, 2007a).

Because of imperfections in the prediction algorithm, estimates of biodegradability predicted by using BIOWIN must be used conservatively. An evaluation of the reliability of BIOWIN predictions determined that the prediction "*not readily degradable*" is highly accurate (correctly predicted for more

TABLE 3.16
BIOWIN-Predicted Biodegradability of Chlorinated Solvents and Solvent-Stabilizer Compounds

Chemical Name[a]	Readily Biodegradable?	Time Frame	BIOWIN1 Ranking	MITI Probability[b]
Triethanolamine	Yes	Weeks	0.95	0.96
Ethyl acetate	Yes	Weeks	0.88	0.84
Isopropyl acetate	Yes	Weeks	0.87	0.70
2-Butanone	Yes	Weeks	0.72	0.67
Acetonitrile	Yes	Weeks	1.04	0.66
sec-Butyl alcohol	Yes	Weeks	0.87	0.65
1,2-Butylene oxide	Yes	Weeks	0.37	0.58
Cyclohexane	Yes	Weeks	0.71	0.58
Propylene oxide	Yes	Weeks	0.37	0.57
tert-Amyl alcohol	Yes	Weeks	0.52	0.57
tert-Butyl alcohol	Yes	Weeks	0.53	0.56
Butoxymethyl oxirane	Yes	Weeks	0.10	0.56
1,3-Dioxolane	Yes	Weeks	0.02	0.55
Nitroethane	Yes	Weeks	0.71	0.54
Phenol	Yes	Weeks	0.95	0.54
1,4-Dioxane	**Yes**	**Weeks**	**0.01**	**0.53**
Nitromethane	Yes	Weeks	0.72	0.53
2-Methyl-3-butyn-2-ol	Yes	Weeks	0.52	0.53
Diepoxybutane	Yes	Weeks	0.01	0.52
Epichlorohydrin	Yes	Weeks	0.24	0.52
Dichloromethane	*No*	*Months*	*0.48*	*0.51*
1,3,5-Trioxane	Yes	Weeks	−0.34	0.51
Pyridine	No	Weeks	0.56	0.48
Triethylamine	No	Months	0.49	0.48
Diisobutylene	No	Months	0.51	0.45
Nonylphenol	No	Weeks	0.92	0.44
p-tert-Amyl phenol	No	Months	0.60	0.44
Styrene oxide	No	Weeks	0.47	0.41
N-Methylmorpholine	No	Months	0.15	0.41
Methyl chloroform	*No*	*Months*	*0.17*	*0.39*
Diisopropylamine	No	Weeks	0.85	0.34
Trichloroethylene	*No*	*Months*	*0.35*	*0.33*
Carbon tetrachloride	*No*	*Months*	*0.04*	*0.33*
2,6-di-*tert*-Butyl-*p*-cresol	No	Months	0.45	0.32
Perchloroethylene	*No*	*Months*	*0.22*	*0.22*

[a] Nonitalics compounds are stabilizers. Italics compounds are chlorinated solvents; these are listed for comparison to show whether the stabilizer is likely to evaporate with the solvent or remain behind in the soil where it can be leached down to groundwater. Data for 1,4-dioxane are shown in boldface.

[b] "MITI Probability," also called "Inherent Biodegradability," is determined by the Modified MITI Test, an activated sludge assay of biochemical oxygen demand, from the Chemicals Inspection, and Testing Institute, Japan (OECD, 1981). MITI probability values are included in BIOWIN output.

TABLE 3.17
Summary of Biodegradability Test Results from the Literature in Syracuse Research Corporation's BIODEG Database

Compound[a]	Aerobic Summary[b]	Qual[c]	Anaerobic Summary	Qual[c]	Year Filed	Number of References
tert-Amyl alcohol	Slow + Acc	1	—	—	1985	5
1,4-Dioxane	**Slow**	**1**	**Slow**	**3**	**1984**	**3**
tert-Butyl alcohol	BST − 1	1	—	—	1984	15
Nonylphenol	BST − 1	1	—	—	1986	3
Pyridine	BST − 1	1	Fast + Acc	3	1986	6
Acetonitrile	Fast + Acc	1	—	—	1987	6
Triethanolamine	Fast + Acc	1	—	—	1985	9
Dichloromethane	Fast + Acc	1	—	—	1987	5
Diisopropylamine	Fast + Acc	2	—	—	1988	3
Carbon tetrachloride	Fast + Acc	2	Fast	1	1987	9
Phenol	Fast	1	BST − 1	1	1986	66
Isopropyl acetate	Fast	1	—	—	1986	2
2-Butanone	Fast	1	—	—	1986	10
sec-Butyl alcohol	Fast	1	—	—	1987	10
2,6-di-*tert*-Butyl-*p*-cresol	Fast	1	Fast	3	1986	6
Ethyl acetate	Fast	1	Fast	3	1987	9

[a] Nonitalics compounds are stabilizers. Italics compounds are chlorinated solvents; these are listed for comparison to show whether the stabilizer is likely to evaporate with the solvent or remain behind in the soil where it can be leached down to groundwater. Data for 1,4-dioxane are shown in boldface.

[b] BIODEG Evaluation Reliability Codes ("Qual"): Fast = biodegrades fast; Fast + Acc = biodegrades fast with acclimation; Slow = biodegrades slowly; Slow + Acc = biodegrades slowly with acclimation; and BST = biodegrades sometimes.

[c] Reliability ratings: Qual = 1: Chemical tested in three or more tests, consistent results. Qual = 2: chemical tested in two tests, or results in more than two tests are interpretable, but conflicting data. Qual = 3: only one test or uninterpretable, conflicting data.

than 90% of compounds); however, the prediction "*readily degradable*" often *does not agree* with MITI test data. A linear regression of biodegradability predicted by BIOWIN and results from MITI tests had a rather low correlation coefficient, $R^2 = 0.58$ (Arnot et al., 2005). If BIOWIN predicts fast biodegradation, the estimate should be independently verified. If the program predicts slow biodegradation, this result can be reliably used as a confirmation that the compound is not readily biodegradable. The presence of a biodegradation-enhancing fragment generally indicates a *possible* metabolic step, which does not necessarily lead to complete mineralization (Pavan and Worth, 2006).

3.3.4.2   Biodegradability of 1,4-Dioxane: Laboratory Studies

Research into the biodegradability of 1,4-dioxane was conducted first in the context of treating industrial effluents from processes that produce 1,4-dioxane as a by-product, for example, polyethylene terephthalate manufacturing, or processes that use 1,4-dioxane in the manufacture of goods, for example, magnetic tape and cellulose acetate membranes. Recent research into 1,4-dioxane biodegradability is motivated by the desire to develop effective *in situ* remedial technologies or viable *ex situ* treatments. Indigenous soil microbes at contaminated sites are generally considered to be incapable of degrading 1,4-dioxane under ambient conditions (Lesage et al., 1990). Engineered aerobic wastewater treatment systems in Japan decrease but do not completely remove elevated influent concentrations of 1,4-dioxane (1000 µg/L) in activated sludge processes (Abe, 1999). Low influent concentrations of 1,4-dioxane (2–3 µg/L) at the City of Ann Arbor, Michigan, wastewater treatment plant are unchanged in effluent (Skadsen et al., 2004). The general conclusion of the growing body of research into biodegradation of 1,4-dioxane is that it does not biodegrade appreciably under ambient conditions; however, in laboratory-controlled environments, 1,4-dioxane can be degraded completely with enhancements. The "holy grail" in the pursuit of 1,4-dioxane biodegradation is the means to induce significant and sustainable *in situ* biodegradation under ambient temperatures. Table 3.18 summarizes key research on the capacity of microbes to degrade 1,4-dioxane.

Most of the research into 1,4-dioxane biodegradability has focused on aerobic biodegradation. Anaerobic biodegradation of cyclic ethers is not widely reported. Microcosm studies on soils from dioxane-contaminated sites in Maryland and New York showed no degradation under anaerobic conditions (Steffan, 2006). USEPA's guidance document (1999) for implementation of monitored natural attenuation remedies describes protocols for conducting microcosm tests for anaerobic biodegradation. To dissolve hydrophobic or nonaqueous-phase liquids, the guidance recommends using 1,4-dioxane instead of alcohol, because alcohols will biodegrade anaerobically whereas 1,4-dioxane will not (USEPA, 1999).

Aliphatic ethers such as methyl *tert*-butyl ether (MTBE) can be degraded anaerobically (Pruden et al., 2005). MTBE will mineralize in anoxic conditions at slow but measurable rates, beginning with cleavage of the ether linkage (Zenker, 2006). Cyclic ethers like 1,4-dioxane may also be cleavable under strongly anaerobic conditions. Tests of anaerobic biodegradation of the cyclic ether THF with methanotrophic bacteria showed less than 30% of the theoretical gas production (Battersby and Wilson, 1989). Humic acids have many ether linkages in their structure and may be a candidate for stimulating cometabolic breakdown of 1,4-dioxane in both oxic and anoxic conditions (Zenker, 2006). Anaerobic biodegradation of 1,4-dioxane under humic and iron-reducing conditions was recently reported (Pan and Chen, 2006; Shen et al., 2008).

The ether linkage, C—O—C, is commonly encountered in nature (lignins) and in anthropogenic compounds (agrochemicals, detergents) and is generally resistant to biological mineralization (White et al., 1996). The intrinsic refractivity of the ether linkage to biological destruction was recognized early in the history of biodegradation research (Alexander, 1965). Bacterial scission of ether bonds is not thermodynamically favored, as noted in a seminal paper surveying the topic (White et al., 1996): "Microbial cleavage of the ether bond is a remarkable phenomenon, since the C—O bond energy[*] is 360 kJ/mol and necessarily demands an appreciable investment of energy to effect its fission in relation to the sometimes relatively small yield of assimilable carbon." The enzymes responsible for bacterial scission of ether bonds were called etherases in early work and were believed to catalyze a hydrolysis reaction that cleaves the bond; however, catalysis of the ether bond division is now attributed to a heterogeneous group of enzymes that exhibit a variety of mechanisms to break the ether bond. Hydrolysis does not play an important role in cleaving the ether bond (White et al., 1996).

Table 3.18 summarizes the research into metabolism and cometabolism of 1,4-dioxane by pure and mixed cultures of bacteria and fungi. Where biodegradation has been documented, the rate and sustainability must also be determined. Biodegradation that occurs too slowly will not have a significant impact on 1,4-dioxane concentrations in the subsurface environment if migration rates substantially exceed degradation rates. Intermediate products that are toxic to and inhibit the growth of microflora will also limit the degree to which biodegradation lowers concentrations of 1,4-dioxane in groundwater.

Rates of biodegradation of 1,4-dioxane can be described by the commonly used general equation (Chappelle, 1993; Evans et al., 2006):

$$r = \frac{\mu_{max} XS}{Y(K_s + S)},$$ (3.40)

---

[*] kJ = kilojoules; 1 kJ $\approx$ 0.239 kilocalories (kcal); 360 kJ/mol = 86 kcal/mol. Alexander (1994) lists the ether linkage bond energy as 85.5 kcal/mol.

TABLE 3.18
Summary of Studies of Biodegradation of 1,4-Dioxane

Substrate: 1,4-Dioxane Alone

Pure culture

Bernhardt and Diekmann (1991): Six *Rhodococcus* strains tested; none grew on DX alone

Burback and Perry (1993): *Mycobacterium vaccae* Strain JOB5 catabolized 100 ppm DX <50%, that is, growth not supported

Parales et al. (1994): Isolated Strain CB1190 from THF enrichments; DX degraded at 30°C in 18 h from 4 mM to <10 µM without THF

Kim and Engesser (2004): *Rhodococcus* Strain DEE5151 from diethyl ether enrichment cultures did not degrade 1,4-dioxane

Mahendra and Alvarez-Cohen (2005): Isolated *Pseudonocardia dioxanivorans*, the only known pure culture that grows on DX alone. Cells grew at 30°C for up to 28 days

Mahendra and Alvarez-Cohen (2006): DX served as a growth substrate for *Pseudonocardia dioxanivorans* Strain CB1190 and *Pseudonocardia benzenivorans* Strain B5

Mixed culture

Fincher and Payne (1962): No oxygen demand on 1,4-dioxane at 30°C for 72–96 h

Klecka and Gonsior (1986): No oxygen demand in a 20-day BOD test on DX. No biodegradation of DX by sewage cultures at 1 year; aerobic shaker bath at 30°C achieved 44.5% DX removal in 32 days including a 10-day lag

Roy et al. (1994): DX completely degraded at 150 mg/L; sludge microbes utilize DX as sole source of carbon and energy. Incomplete biodegradation at concentrations of >150 mg/L.; by-product toxicity indicated

Taylor et al. (1997): soil microbes did not degrade DX aerobically or anaerobically on representative groundwater samples

Fungi and other

Patt and Abebe (1995), Deshpande (1992): *Aureobasidium pullulans* grew on DX after lag of 36 h, 90% of 50 ppm DX was degraded in the next 36 h

Taylor et al. (1997): fungi from dioxane-contaminated site soils did not degrade DX

Nakamiya et al. (2005): *Cordyceps sinensis* (a fungal insect pathogen) cultivated with DX as sole source of carbon; DX degradation nearly complete (90%) after 3 days. Degradation product was ethylene glycol

Substrate: 1,4-Dioxane with THF as Cometabolite

Pure culture

Bernhardt and Diekmann (1991): six *Rhodococcus* strains tested; 1,4-dioxane completely degraded with 7.5 mM THF

Mahendra and Alvarez-Cohen (2006): See listing below

Vainberg et al. (2006): Biodegradation of DX with THF by *Pseudonocardia* sp. Strain ENV478

Mixed culture

Zenker et al. (1999): Complete biodegradation of DX and THF by sludge microbes at 35°C, biodegradation of DX dependent upon THF presence

Zenker et al. (2000): Biodegradation of DX with THF as cometabolite; DX without THF failed to stimulate biodegradation by soil microbes

Fungi and other

Kelley et al. (2001): phytoremediation with hybrid poplars, CB1190 + THF; 100 mg/L DX removed within 45 days with cosubstrates THF and 1-butanol

Substrate: 1,4-Dioxane with Other Cometabolites

Pure culture

Hyman (1999): *Mycobacterium vaccae* Strain JOB-5 cometabolically degrades cyclic ethers after growth on straight-chain/branched alkanes; catalyzed by alkane monooxygenases

Mahendra and Alvarez-Cohen (2006): Cometabolic transformation of DX observed for monooxygenase-expressing strains induced with methane, propane, tetrahydrofuran, and toluene: *Methylosinus trichosporium* Strain OB3b, *Mycobacterium vaccae* Strain JOB5, *Pseudonocardia* Strain K1, *Pseudomonas mendocina* Strain KR1, *Ralstonia pickettii* Strain PKO1, *Burkholderia cepacia* Strain G4, and *Rhodococcus* Strain RR1

TABLE 3.18   (continued)
Summary of Studies of Biodegradation of 1,4-Dioxane

Vainberg et al. (2006): Biodegradation of DX with sucrose, lactate, yeast extract, 2-propanol, and propane, by
   Pseudonocardia sp. Strains ENV478 and ENV425

Fungi and other

Skinner (2007): *Graphium* sp., a filamentous fungus, degrades DX by cometabolism after growth on THF or gaseous
   alkanes, catalyzed by cytochrome P450 monooxygenase
Kelley et al. (2001): phytoremediation with 1-butanol

Notes: DX = 1,4-dioxane; THF = tetrahydrofuran.

---

where r is the dioxane biodegradation rate, μ_{max} is the maximum growth rate, X is the bacteria concentration, S is the dioxane concentration, Y is the growth yield on dioxane, K_s is the half-saturation constant, and r/X is the specific dioxane biodegradation rate.

The minimum concentration that will maintain growth in a biological treatment system under steady-state conditions ($S_{S,min}$; Alexander, 1994) is

$$S_{S,min} = \frac{K_s b}{\mu_{max} - b}. \tag{3.41}$$

Table 3.19 summarizes published parameters of biodegradation kinetics for bacteria capable of metabolizing 1,4-dioxane.

Information on kinetics is extremely important because it characterizes the concentration of the chemical remaining at any time, permits prediction of the levels likely to be present at some future time, and allows assessment of whether the chemical will be eliminated before it is transported to a site at which exposure will occur (Alexander, 1994).

The relationship between bacterial growth rate and the concentration of the carbon substrate increases, but only at low substrate concentrations. A concentration is reached above which the growth rate falls as the substrate level rises further. This decline is a result of the antimicrobial action (i.e., toxicity to microbes) of the chemical substrate at high concentrations (Alexander, 1994).

The Monod equation describes the growth of a population on a chemical substrate constrained by the toxicity to the population of that substrate in high concentrations:

$$\mu = \frac{\mu_{max} S}{K_s + S + (S^2/K_1)} \approx \frac{\mu_{max} S}{K_s + S}, \tag{3.42}$$

where μ is the specific growth rate of the bacterium, μ_{max} is its maximum specific growth rate, S is the concentration of the chemical substrate, K_s is a constant that represents the concentration of the chemical substrate at which the rate of growth is half the maximum rate, and K_1 is the inhibition constant that represents the suppression of the growth rate by the toxic substrate (Alexander, 1994). This equation may apply to 1,4-dioxane, whose metabolic by-product, 2-hydroxyethoxyacetic acid (2HEAA), is toxic to microorganisms, as described in research by Vainberg et al. (2006) (see Section 7.6.3).

Bacterial growth requires that the substrate be present at sufficiently high concentrations relative to the cell population that acts on the substrate to allow the population to continue doubling. When cell density is higher than the substrate concentration can sustain, growth is limited.

TABLE 3.19
Kinetics of 1,4-Dioxane and 1,3-Dioxolane Biodegradation

Culture	Temperature (°C)	Maximum Growth Rate (μ_{max}/h)	Yield (mg/mg)	Half-Saturation Constant, K_s (mg/L)	Specific Rate, r/X [mg/(g h)]	References
CB1190	30	0.0230	—	—	9.9 at 500 ppm	Adamus et al. (1995)
Mixed	25	0.0100	0.33	13.51	27 at 100 ppm	Grady et al. (1997)
Mixed	35	0.0430	0.64	1.04	66 at 100 ppm	Grady et al. (1997)
Mixed	—	0.141	0.12	9.9	0.12	Grady et al. (1997)
Industrial activated sludge	—	0.062	0.44	1.65	—	Sock (1993) (cited in Zenker et al., 2002)
Nocardioform actinomycete	30	0.033	0.02	—	1.98	Parales et al. (1994)
Industrial activated sludge	—	0.0053	0.22	182	—	Roy et al. (1995)
Pseudonocardia strain ENV478	30				21 at 100 µg/L DX	Vainberg et al. (2006)
Mixed	30	0.083	0.75	13 ± 7.6	0.05 ± 0.003 at 200 ppm	Zenker et al. (2000)
Pseudonocardia dioxanivorans CB1190	30	0.10	0.09	160 ± 44	1.1 ± 0.008 at 100 ppm	Mahendra and Alvarez-Cohen (2006)
Pseudonocardia benzenivorans B5	30	0.07	0.03	330 ± 82	0.1 ± 0.006 at 100 ppm	Mahendra and Alvarez-Cohen (2006)
Amycolata sp. CB1190	~24	—	0.01	—	0.92 ± 0.29 at 100 ppm	Kelley et al. (2001)
Pseudonocardia strain ENV478	30	—	—	—	19	Vainberg et al. (2006)

Source: From Grady, C.P.L., Jr., Sock, S.M., and Cowan, R.M., 1997, In: G.S. Sayler, J. Sanseverino, and K.L. Davis (Eds), *Biotechnology in the Sustainable Environment*. New York: Plenum Press. With permission.

Note: The last entry, for *Pseudonocardia* Strain ENV478, is for biodegradation of 1,3-dioxolane. The other entries are for biodegradation of 1,4-dioxane.

Fungal biomass increases through hyphal lengthening[*] and branching instead of by the splitting and doubling (binary fission) mechanism used by bacteria. Fungal growth occurs primarily at the ends of the filaments, which grow at a constant rate. Because fungi and other filamentous organisms such as actinomycetes may grow radially in three dimensions, growth may follow cubic kinetics; however, growth in the subsurface is constrained by the porous matrix.

The pathway for aerobic degradation of 1,4-dioxane parallels that of THF, which involves oxygenase-mediated hydroxylation of the carbon present in the no. 2 position on the dioxane ring and its subsequent dehydrogenation to form the lactone ring. Hydroxylation is any process that introduces one or more hydroxyl groups (–OH) into a compound (or radical), thereby oxidizing it. In biochemistry, hydroxylation reactions are often facilitated by enzymes called hydroxylases. A lactone is a cyclic ester. After opening the lactone ring, intermediary metabolism pathways may begin (Zenker, 2006).

---

[*] A hypha (plural hyphae) is a long, branching filamentous cell of a fungus and also of unrelated *Actinobacteria*. In fungi, hyphal lengthening is the main mode of growth.

3.3.4.3   Synopsis of 1,4-Dioxane Biodegradability Studies

The key studies of 1,4-dioxane biodegradation listed in Table 3.18 are briefly summarized in this section.

One of the earliest studies of the degradation of ethers was by Fincher and Payne (1962). The study was conducted as part of a waste-stream treatability test for effluent from manufacturing processes that produced ether glycols as a by-product, such as in the manufacturing of explosives, glues, cosmetics, pharmaceuticals, shampoos and sundries, and synthetic detergents. In this study, the authors evaluated utilization of ethers for growth by using a presumed *Pseudomonas–Achromobacter* member (designated TEG-5). The test assayed cells for oxidative activity in cultures maintained at 30°C for 72–96 h in a glycol-basal salt medium. No oxidative activity or other symptoms of growth were observed in the assay for 1,4-dioxane. Fincher and Payne determined that 1,4-dioxane was not utilized by the bacteria for growth and therefore was not directly degraded by the bacteria studied.

Klecka and Gonsior (1986) observed negligible oxygen consumption in a 20-day test for biochemical oxygen demand for 1,4-dioxane. They also did not observe any biodegradation of 1,4-dioxane by mixed cultures of sewage microorganisms exposed for 1 year in a wastewater treatment plant where influent concentrations of 1,4-dioxane ranged from 100,000 to 900,000 μg/L. No toxicity to microorganisms from 1,4-dioxane concentrations up to 300,000 μg/L was observed. Klecka and Gonsior also conducted an aerobic shaker bath test at 30°C and measured 44.5% removal of 1,4-dioxane after 32 days of treatment, including an initial 10-day lag time during which there was little respirometric response. Their study confirmed that aerobic bacteria in sludge could degrade 1,4-dioxane as a primary source of carbon and energy. However, instead of complete destruction, their data suggested conversion of 1,4-dioxane to intermediates and not complete mineralization; they inferred that the metabolic by-products may be toxic to microorganisms and inhibit their growth.

Bernhardt and Diekmann (1991) studied degradation of 1,4-dioxane, THF, and other cyclic ethers by the actinobacterium strain *Rhodococcus* from forest soil or sludge that was obtained from the settling basin of an aerobic wastewater purification plant at a chemical firm. Despite considerable effort, no strains were enriched or isolated when 1,4-dioxane or cyclohexane was the sole carbon substrate. Six strains belonging to the genus *Rhodococcus* that degrade THF were isolated and classified. A strain that degraded dioxane instead of or in combination with THF was further characterized. *Rhodococcus* Strain 219 grows fast and degrades 1,4-dioxane at an optimal temperature of 30°C. The maximum growth rate, μ_{max}, of Strain 219 on 1,4-dioxane in the presence of 7.5 mM THF was 0.019 h^{-1}. Growth occurred in as little as 0.22 mM (16,000 μg/L) THF. In the shaker test, at the end of the logarithmic growth phase, no THF or 1,4-dioxane could be detected by gas chromatography; COD measurement in the supernatant after centrifugation was zero, which was interpreted to mean that no intermediates had accumulated and the substances were totally destroyed. A total of 34 compounds containing nitrogen or oxygen, including the stabilizers 1,3-dioxolane, morpholine, and *n*-methyl morpholine, were tested. Oxygen demands were 0.37 and 2.32 for 1,3-dioxolane and 1,4-dioxane, respectively, and zero for the morpholine compounds. The presumed primary products of THF and 1,4-dioxane degradation were THF-2-ol and 1,4-dioxane-2-ol, but these compounds were not analytically identified.

Burback and Perry (1993) used a pure culture of the propane-oxidizing actinobacterium *Mycobacterium vaccae* [ATCC 29678 (JOB-5)] to demonstrate the biodegradability of 1,4-dioxane. A total of 100 ppm 1,4-dioxane was added to 100 mL of a 1 mg/mL cell suspension and incubated at 30°C in a rotating shaker. Vials were removed at 12, 24, and 48 h, with zero hours as the control, and extracted with ethyl ether for analysis. The bacterium was able to use propane, acetone, and toluene as a sole carbon source, but 1,4-dioxane and the other contaminants evaluated could not sustain growth and were therefore determined to be only cometabolically degraded.

Roy et al. (1994) published a study on the biodegradation of 1,4-dioxane and diglyme in industrial waste. They recorded complete degradation of 1,4-dioxane at 150,000 μg/L based on an oxygen-uptake curve. Roy et al. interpreted the data to mean that microbes utilized 1,4-dioxane as the sole source of carbon and energy below 150,000 µg/L and that there was incomplete biodegradation at concentrations greater than 150,000 µg/L, indicating production of an inhibitory toxic by-product of 1,4-dioxane.

In a follow-on paper (Roy et al., 1995), the growth of microorganisms on 1,4-dioxane was modeled to estimate biodegradation kinetic parameters by using respirometric data. The growth data were obtained from experiments that used microbes from an acclimatized industrial waste to investigate the biodegradability of pure 1,4-dioxane. Growth of the microbes was measured by monitoring oxygen uptake in a respirometer. Experiments were performed on a pure solution of 1,4-dioxane (300,000 and 670,000 µg/L) in electrolytic respirometers as part of a 1989 Master's degree thesis (Anagnostu, 1989). The modeling assumed no stable by-products, no toxicity to microorganisms from dioxane, and Monod kinetics. From a best fit to experimental data of 1,4-dioxane depletion and oxygen uptake, the maximum specific growth rate, μ_{max}, was found to be 0.127 day^{-1}, the saturation constant, K_s, was 182,000 µg/L, and the cell yield was 0.218.

Grady et al. (1997) reported that 1,4-dioxane was very difficult to biodegrade because of the stability of the two ether bonds in the 1,4-dioxane saturated heterocyclic structure. Their study confirmed that 1,4-dioxane could serve as a sole source of carbon and energy through a series of bioreactor experiments using 1,4-dioxane concentrations up to 150,000 µg/L. Growth on dioxane alone was sustained for 5 weeks within a complex bacterial community with several bacterial genera present; however, they were unable to isolate a pure culture capable of converting carbon to energy from 1,4-dioxane alone. Biodegradation of 1,4-dioxane followed Monod kinetics with no indication that either substrate or product inhibition occurred at 1,4-dioxane concentrations up to 2,100,000 µg/L. Grady et al. determined that growth is particularly sensitive to temperature, as profiled in Table 3.20. At low temperatures, the growth rate of microbes on 1,4-dioxane would be too slow to maintain itself without getting flushed out of the bioreactor. This kinetics-based failure to biodegrade 1,4-dioxane occurred especially in winter when bioreactor temperatures decreased to 20°C; it was attributable neither to an inability of bacteria to degrade 1,4-dioxane nor to any inhibitory toxic characteristics.

Hyman (1999) studied the structural features that influence the reactivity of a single, nonspecific monooxygenase enzyme toward a variety of ether compounds. Hyman noted that many ether-bonded compounds are oxidized by microorganisms that cannot further metabolize the products of those reactions and that relatively few microorganisms appear to be able to utilize ether-containing compounds as growth substrates. Hyman's study examined the rates of cometabolism of six ethers and the degradation products obtained by bacteria and fungi. The study also investigated the physiological consequences of ether cometabolism on bacteria. Hyman found that the alkane monooxygenase (AMO) enzyme is largely unreactive to cyclic ethers such as tetrahydropyran (THP) and dioxane isomers, even though this enzyme oxidizes several sulfur analogs of the cyclic ethers. The main focus of Hyman's study was ether oxidation by propane-oxidizing bacteria. AMO is a bacterial enzyme; its capacity for oxidizing specific carbon–hydrogen bonds in some organic compounds leads to cometabolic degradation of recalcitrant contaminants.

Hyman's study used *Mycobacterium vaccae* JOB-5 grown on straight-chain (C3–C8) and branched alkanes to cometabolically degrade the cyclic ethers THF, THP, hexamethylene oxide (HMO), and

TABLE 3.20
Effect of Temperature on 1,4-Dioxane Biodegradation Kinetics in Bioreactor Experiments

Parameter	Temperature			
	40°C	35°C	30°C	25°C
μ_{max}, h^{-1}	0.022	0.043	0.014	0.010
K_S, mg/L	0.30	1.04	9.93	13.51
Y, mg biomass *COD*/mg dioxane *COD*	0.70	0.64	0.23	0.33

Source: From Grady, C.P.L., Jr., Sock, S.M., and Cowan, R.M., 1997, In: G.S. Sayler, J. Sanseverino, and K.L. Davis (Eds), *Biotechnology in the Sustainable Environment*. New York: Plenum Press. With permission.

the dioxane isomers 1,3-dioxane and 1,4-dioxane. The oxidation of all of these compounds was apparently catalyzed by AMO, because all of the transformation reactions were inhibited by acetylene and only occurred in cells grown on alkane substrates. Hyman noted that *Mycobacterium vaccae* cannot utilize cyclic ethers (THF, THP) as growth substrates. However, their lactone (butyrolactone, valerolactone) and hydroxy acid derivatives all supported rapid growth of the bacterium. The cometabolism of ethers produced resultant products that can sustain growth of the organism.

Hyman's research demonstrated some of the potential limitations of cometabolic processes for the degradation of cyclic ethers, including compounds that exert toxic effects on the bacteria that can degrade these compounds. Hyman postulated that the reason ether-bonded compounds are often recalcitrant in the environment may be that latent toxic effects of ether-bonded compounds limit their rate of biodegradation even when appropriate organisms, cosubstrates, and environmental conditions are present.

A very recent study by Kim et al. (2008) isolated and identified a 1,4-dioxane-degrading bacteria (*Mycobacterium sp.* PH-06) from 1,4-dioxane contaminated river sediments in Korea. This bacterial strain was able to degrade 90% of the 1,4-dioxane (at 900 mg/L) in 15 days, using 1,4-dioxane as the sole carbon and energy source, that has been confirmed by only a few researchers (e.g., Roy et al., 1994; Grady et al., 1997; Mahendra and Alvarez-Cohen, 2006). Degradation metabolites (i.e., 1,4-dioxane-2-ol and ethylene glycol) and isotope tagged 1,4-dioxane were used to confirm degradation. The authors identified an initial monooxygenation step that appears to be common to degradation identified by other researchers using different stains of aerobic bacteria.

The ability of the *Rhodococcus* sp. Strain DEE5151 to degrade MTBE and cyclic ethers was studied by Kim and Engesser (2004). *Rhodococcus* Strain DEE5151 was selected because it is able to utilize a broad range of alkyl ethers, including diethyl ether, di-*n*-propyl ether, di-*n*-butyl ether, di-*n*-pentyl ether, di-*n*-hexyl ether, di-*n*-heptyl ether, *tert*-butyl ethyl ether, and other ethers as sole carbon and energy sources. Kim and Engesser isolated 18 ether-degrading *Rhodococcus* strains and two ether-degrading *Sphingomonas* strains from sewage sludge from Stuttgart, Germany. None of the 20 isolates tested was able to utilize dimethyl ether, MTBE, isopropyl ether, THF, THP, or 1,4-dioxane as the sole carbon source.

Researchers with Japan's National Institute for Environmental Studies investigated degradation of 1,4-dioxane and other cyclic ethers by the fungus *Cordyceps sinensis* (Nakamiya et al., 2005). The genus *Cordyceps* is sometimes called the caterpillar fungus and resides in *Hepialus armoricanus* (Lepidoptera: Hepialidae). *C. sinensis* has been identified as mushrooms that grow on the *H. armoricanus* caterpillar in mountainous regions of China. It makes a fruiting body and is consumed as an herbal medicine in Asia.[*] *C. sinensis* exhibits antioxidant activity that could be useful in bioremediation.

The Japanese researchers confirmed that *C. sinensis* (Strain A) can utilize 1,4-dioxane as the sole carbon and energy sources. Strain A has a septum and spores as shown in Figure 3.5 and was identified as the fungus *C. sinensis* from 18S rRNA gene homology (99.75% of a partial sequence of 1617 base pairs).

Nakamiya et al. (2005) found that Strain A can grow in solutions containing up to 0.09 M 1,4-dioxane, but optimal growth was obtained at 0.034 M 1,4-dioxane. The decrease of 1,4-dioxane in the medium was essentially complete after 3 days, and there was a concomitant increase in the concentration of the degradation product, ethylene glycol (0.97×10^{-3} M), after 3 days. After 4 days, the ethylene glycol was almost completely utilized. Rates of 1,4-dioxane degradation and product formation by *C. sinensis* are profiled in Figure 3.6.

---

[*] The herbal medicine is Dong Chong Tsia Tsiao (winter caterpillar summer grass). The caterpillar fungus consists of larvae of *H. armoricanus* (Lepidoptera: Hepialidae) infected with the fungus *C. sinensis* (Clavicipitales, Ascomycotina). The pharmacological properties of the caterpillar fungus are similar to those of ginseng. Dong Chong Tsia Tsiao is mainly used for weak lungs, coughing and shortness of breath, weak kidney, back pain, and other ailments. The wholesale price in China in 2007 was about $700 kg⁻¹ (Steinkraus and Whitefield, 1994; Huang, 2007).

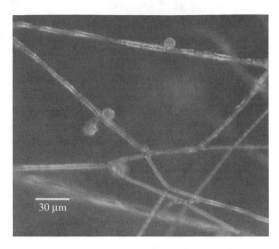

FIGURE 3.5 Light micrograph of *C. sinensis* Strain A. (From Nakamiya, K., et al., 2005, *Applied and Environmental Microbiology* 71(3): 1254–1258. With permission.)

Nakamiya et al. applied an innovative technique, reacting phenyl boronate with the degradation products, to identify the products of *C. sinensis* biodegradation of 1,4-dioxane. Phenyl boronate forms ether bonds with dihydroxy compounds in water–methanol; the derivatives are amenable to gas chromatography/mass spectrometry analysis. By analyzing the breakdown products, Nakamiya et al. were able to propose a degradation pathway for filamentous fungus degradation of 1,4-dioxane, as shown in Figure 3.6. The pathway follows the sequential production of ethylene glycol, glycolaldehyde, glycolic acid, and oxalic acid followed by incorporation of the glycolic acid and/or oxalic acid into the tricarboxylic acid cycle. This study proposed etherases and oxidases as the likely enzymes involved in *C. sinensis* degradation of 1,4-dioxane by Strain A, as shown in Figure 3.7. The

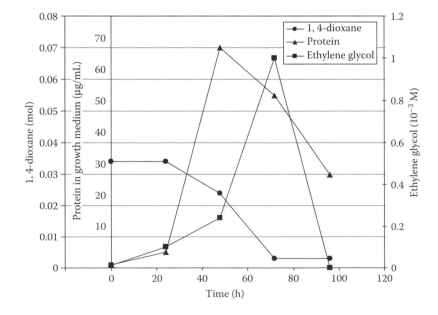

FIGURE 3.6 Rates of 1,4-dioxane degradation and product formation by *C. sinensis*. (From Nakamiya, K., et al., 2005, *Applied and Environmental Microbiology* 71(3): 1254–1258. With permission.)

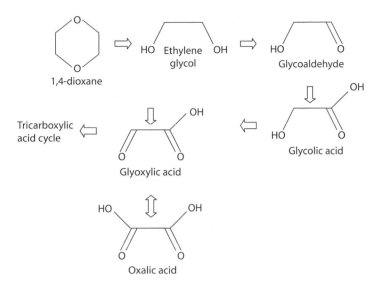

FIGURE 3.7 Proposed pathway for *C. sinensis* degradation of 1,4-dioxane. (From Nakamiya, K., et al., 2005, *Applied and Environmental Microbiology* 71(3): 1254–1258. With permission.)

authors note that ethylene glycol formation may occur via two hemiacetals at one side of the 1,4-dioxane molecule by etherase-type reactions.

Mahendra and Alvarez-Cohen (2005) identified *Pseudonocardia dioxanivorans* as a novel species of actinomycete that grows on 1,4-dioxane in pure culture. Cells of *Pseudonocardia dioxanivorans* Strain CB1190[T] were grown at an optimal temperature of 30°C for up to 28 days. No growth was observed at temperatures higher than 37°C. The cells, 1–3 mm in size, appeared as powdery white colonies floating on the liquid surface and formed biofilms on the inside wall and the bottom of the flask. Figure 3.8 shows an electron micrograph of CB1190[T] cells. The physiological and biochemical

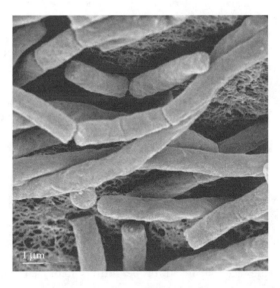

FIGURE 3.8 Scanning electron micrograph of cells of Strain CB1190[T]. (From Mahendra, S., and Alvarez-Cohen, L., 2005, *International Journal of Systematic and Evolutionary Microbiology* 55: 593–598. With permission.)

characteristics of the bacterium are unlike the other species identified in the *Pseudonocardia* genus. To identify the strain, the researchers extracted genomic DNA from CB1190[T] to perform polymerase chain reaction (PCR) amplification of the 16S rRNA gene to obtain gene sequences for comparison with the GenBank database for phylogenetic analyses. Genotypic and phenotypic data support the formal recognition of *Pseudonocardia dioxanivorans* as a unique species.

CB1190[T] was grown aerobically in ammonium BSM at 30°C with 5 mM 1,4-dioxane. Cell growth was determined by measuring total protein and by assaying the enzyme activities of soluble monooxygenases. The observed maximum growth rate over 1 h was 90 µg/L, and the maximum 1,4-dioxane degradation rate, 0.1 mg/mg protein/h, is described by Monod kinetics. Cell yield was 0.09 ± 0.002 g of protein per gram of 1,4-dioxane. Mahendra and Alvarez-Cohen also investigated the ability of CB1190[T] to grow on 1,4-dioxane under anaerobic conditions; however, their data did not show anaerobic growth.

Twenty bacterial isolates, 13 of which were capable of transforming 1,4-dioxane, were evaluated by Mahendra and Alvarez-Cohen (2006). 1,4-Dioxane served as a growth substrate for *Pseudonocardia dioxanivorans* CB1190 and *Pseudonocardia benzenivorans* B5; yields were 0.09 g of protein per gram of 1,4-dioxane and 0.03 g of protein per gram of benzene, respectively. The researchers observed cometabolic transformation of 1,4-dioxane for monooxygenase-expressing bacterial strains induced with methane, propane, THF, and toluene, including *Methylosinus trichosporium* OB3b, *Mycobacterium vaccae* JOB5, *Pseudonocardia* K1, *Pseudomonas mendocina* KR1, *Ralstonia pickettii* PKO1, *Burkholderia cepacia* G4, and *Rhodococcus* RR1. Many of the cometabolic reactions resulted in incomplete degradation of 1,4-dioxane because of product toxicity. Subjecting the bacteria to brief exposure to acetylene, a known monooxygenase inhibitor, prevented oxidation of 1,4-dioxane and toluene in all cases, which supported the hypothesis that monooxygenase enzymes expressed by these strains are the active agent for degrading 1,4-dioxane. The researchers also used a colorimetric assay of naphthol production to rapidly detect monooxygenase activity. The ability of the strains to oxidize naphthalene to naphthol independently confirms the strains' ability to express (i.e., utilize) monooxygenases and that dioxane degradation was positively correlated with monooxygenase activity. Bacterial monooxygenases are similar in structure, function, and reaction mechanisms to mammalian P450 enzymes, which have been documented in the metabolism of 1,4-dioxane during toxicological studies using rats (Woo et al., 1977).

Most of the cultures that acted on 1,4-dioxane cometabolically degraded dioxane incompletely, which suggests that their transformation capacity for dioxane is limited. The primary substrate (e.g., THF) was removed to prevent competition with dioxane during the cometabolic degradation reactions. Incomplete degradation of 1,4-dioxane could be caused by a limited reductant (NADH)[*] supply to fuel the monooxygenase reaction, by toxicity due to dioxane, and/or by toxicity due to dioxane metabolites. 1,4-Dioxane is not likely to exert direct toxicity; CB1190 is able to degrade 1,4-dioxane at concentrations above 1,000,000 µg/L and Strains JOB5, OB3b, and RR1 all tolerate 1,4-dioxane concentrations as high as 500,000 µg/L. Metabolite toxicity is more likely to explain incomplete 1,4-dioxane degradation: the major dioxane metabolite in mammalian cells, 1,4-dioxane-2-one, is more toxic than dioxane. In rats, the lethal dose of 1,4-dioxane-2-one is about one-tenth that of 1,4-dioxane (Woo et al., 1977).

Mahendra and Alvarez-Cohen (2006) found that two of the bacterial strains studied were capable of sustained growth on 1,4-dioxane as the sole carbon and energy source: *Pseudonocardia dioxanivorans* CB1190 and *Pseudonocardia benzenivorans* B5. Rates of oxidation of 1,4-dioxane at 50,000 µg/L were

- 0.01–0.19 mg/h for the metabolic processes
- 0.1–0.38 mg/h for cometabolism by the monooxygenase-induced strains
- 0.17–0.60 mg/h for the recombinant strains per milligram of bacterial protein present

---

[*] NADH is nicotinamide adenine dinucleotide, an important coenzyme found in cells that serve as a carrier of electrons and a participant in metabolic redox reactions.

1,4-Dioxane degradation did not occur in the absence of molecular oxygen. Stoichiometric uptake of dissolved oxygen accompanied 1,4-dioxane degradation by CB1190 within the first minute.

Mahendra and Alvarez-Cohen reported that Monod growth kinetics was observed for CB1190; the maximum dioxane degradation rate (k) and half-saturation concentration (K_s) were computed as 1.1 ± 0.008 mg of dioxane per hour per milligram of protein and $160,000 \pm 44,000$ µg/L, respectively, as calculated by a nonlinear regression fit of the model to the data. Similarly, k and K_s values for Strain B5 were calculated as 0.1 ± 0.006 mg of dioxane per hour per milligram of protein and $330,000 \pm 82,000$ µg/L, respectively. The cell yields for Strains CB1190 and B5 were 0.09 ± 0.002 mg of protein per milligram of dioxane and 0.03 ± 0.002 mg of protein per milligram of dioxane, respectively. The electron equivalents (e^- eq) of these yields were 0.07 e^- eq protein per e^- eq dioxane and 0.008 e^- eq protein per e^- eq dioxane, respectively.

Measured growth rates were relatively low because cell yields of all strains growing on 1,4-dioxane were low. Mahendra and Alvarez-Cohen (2006) noted that the yields were consistently low for all cultures reported in the literature, even on an e^- eq/e^- eq basis. The theoretical dioxane transformation capacity of a microbial consortium was reported to be 1.0 ± 0.36 mg of dioxane per milligram of protein (Zenker et al., 2000). Mahendra and Alvarez-Cohen's study was the first to definitively show the role of monooxygenases in dioxane degradation by using several independent lines of evidence. The complete biodegradation pathway of 1,4-dioxane by monooxygenase-expressing bacteria observed in the study by Mahendra et al. (2007) is presented in Figure 3.9.

Cai et al. (2008) isolated Strain D4 from a polyester production sewage aeration pond in China. They determined that Strain D4 could use 1,4-dioxane as the sole carbon source. Observations of morphology and physiology as well as biochemical tests and 16S rDNA genetic analysis were used to identify Strain D4 as *Bacillus pumilus*. Ideal conditions for 1,4-dioxane degradation were a pH of 7.0, a temperature of 30°C, and an inoculating quantity of 10%. Degradation rates of 83.7% and 85.6% were determined after 24 and 48 h from an initial concentration of 1000 mg/L.

Skinner (2007) completed an extensive study of biodegradation of ether compounds by the eukaryotic alkanotroph *Graphium* sp. (ATCC 58400). *Graphium* sp. does not grow directly on 1,4-dioxane; however, it is able to cometabolize 1,4-dioxane after growth on either THF or alkanes such as propane when mycelia are grown at 27 ± 3°C. Both *n*-alkane- and THF-grown mycelia oxidize 1,4-dioxane. *Graphium* sp. utilizes THF as a sole source of carbon and energy under aerobic conditions via the same THF metabolic pathway used by *Rhodococcus ruber* and two *Pseudonocardia* strains. The cometabolic growth rate of *Graphium* sp. on THF was approximately 10 times greater than on 1,4-dioxane, which itself supports growth about twice the rate of 1,3-dioxane. Growth rates of *Graphium* sp. from Skinner's study are presented in Table 3.21.

Skinner's dissertation explained that physiological and inhibitor studies suggest that THF degradation is initiated by the same cytochrome P450 responsible for oxidizing *n*-alkanes, diethyl ether, and MTBE. There may be an even greater overlap between the enzyme activities involved in *n*-alkane and THF oxidation. Growth of *Graphium* sp. on THF was fully inhibited by acetylene, ethylene, propylene, or propyne (0.5% vol/vol gas phase), whereas growth on either γ-butyrolactone or succinate was unaffected by the presence of these gases at these concentrations.

Although *Graphium* sp. was grown on cyclic ethers containing a single oxygen atom (e.g., THF), no growth was observed on cyclic ethers containing two oxygen atoms (e.g., 1,3-dioxane and 1,4-dioxane), and no growth was observed for any of the nitrogen heterocycles tested as potential growth substrates (morpholine, pyrrolidine, piperazine, and piperadine).

Evidence that the alkane and ether oxidation pathways are superimposed was determined by measuring rates of THF, MTBE, and 1,4-dioxane degradation by *Graphium* mycelia incubated under a spectrum of inductive conditions. Propane-induced mycelia were able to degrade ether compounds (and vice versa) without either a lag phase or the buildup of metabolic intermediates.

The oxidation of alkanes and ethers by *Graphium* sp. is linked through a common catalyst, a cytochrome P450 AMO enzyme that mediates the first step of these pathways. The study

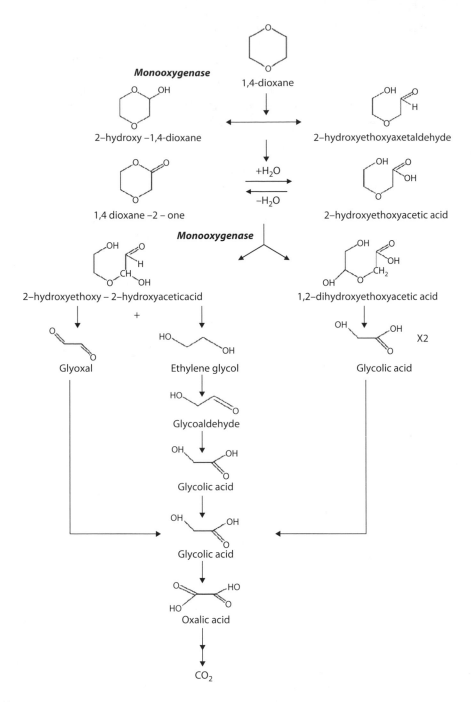

FIGURE 3.9 Complete biodegradation pathway of 1,4-dioxane by monooxygenase-expressing bacteria. (From Mahendra, S., et al., 2007, *Environmental Science and Technology* 41(21): 7330–7336. With permission.)

characterized this enzyme, designated GSPALK1, through molecular, genetic, and biochemical analyses. Skinner's characterization of GSPALK1 through gene encoding is the first description of a cytochrome P450 involved in the terminal oxidation of gaseous *n*-alkanes and cyclic ethers, as well as the first description of a cytochrome P450 involved in MTBE and 1,4-dioxane cometabolism.

TABLE 3.21
Initial Rates of 1,4-Dioxane, THF, and MTBE Oxidation by
Propane-, THF-, and Dextrose-Induced Mycelia of *Graphium* sp.

Induction Compound	1,4-Dioxane	THF
Propane	3.8 ± 1	115 ± 3
THF	9.2 ± 5	123 ± 22
Dextrose	0 ± 6	20.4 ± 25

Source: Skinner, K.M., 2007, *Graphium* species. PhD Dissertation, Oregon State University, Corvallis, OR.
Note: Units: nmol/h per milligram dry weight.

3.4 LABORATORY, FIELD, AND MODELING STUDIES OF 1,4-DIOXANE MOBILITY

The best evidence for 1,4-dioxane's high mobility comes from field studies. A growing number of cleanup case studies have characterized the occurrence of 1,4-dioxane from which interpretations of its mobility relative to solvents can be made. Laboratory column studies of 1,4-dioxane breakthrough in soil columns are also informative. Modeling can also yield an understanding of 1,4-dioxane's high mobility relative to methyl chloroform, its host solvent. This section summarizes laboratory and field studies and presents some screening-level modeling calculations of the relative mobility of 1,4-dioxane at solvent release sites.

3.4.1 LABORATORY STUDIES

A common approach to establishing the retardation rates of contaminants migrating in the subsurface is to perform column tests on saturated soil columns. This approach generally involves sterilizing a column containing a known mass of soil and saturating the soil with water, establishing a pressure gradient and flow rate for the water, then adding a solute to the water and measuring for its presence in the column effluent. Samples of column effluent water are taken at regular intervals and analyzed for the contaminant under study. Breakthrough occurs when the effluent concentration first yields a detection of the subject contaminant, and equilibrium is established when the effluent concentration equals the initial concentration.

Retardation factors (R_f—see Equation 3.32 in Section 3.3.2.3) vary with the flow velocity in column tests. In field tests, the duration of the observation period and the distance over which field measurements are made also affect the retardation factor value. Retardation factors for 1,4-dioxane predicted from field data were twice the values measured in laboratory tests of soil columns (Table 3.22). The difference was attributed to heterogeneities in the aquifer media encountered at the field scale. Highly soluble and miscible compounds are more likely to migrate into low-permeability zones, thereby increasing retardation values derived from field observations (Priddle and Jackson, 1991).

3.4.2 FIELD STUDIES

In a study of 1,4-dioxane migration relative to other contaminants at the Gloucester Landfill near Ottawa, Canada, researchers found that the rate of contaminant migration was inversely proportional to the octanol/water partition coefficient (K_{ow}). The retardation factors and relative mobilities of organic compounds released at the landfill were ascertained by using the ratio of the plume length for each

GLOUCESTER LANDFILL SOIL-COLUMN STUDY
(PRIDDLE AND JACKSON, 1991)

Retardation factors measured from soil core data, presented in Table 3.22, were generated by injecting 1,4-dioxane, THF, diethyl ether, 1.2-dichloroethane, trichloromethane, benzene, and 1,1-dichloroethylene into soil cores at fixed and varying flow rates and using iodide as a conservative tracer to index relative breakthrough. The soil cores were from an aquifer at a well-studied landfill site in Gloucester, near Ottawa, Canada. The contaminants studied were selected because they are found in the aquifer at the site.

Iodide is considered an ideal tracer because it experiences little or no sorption. It has symmetrical breakthrough curves in soil-column tests, indicating equilibrium sorption behavior during transport because any sorption is temporary and reversible and little or no sorption occurs. 1,4-Dioxane breakthrough curves display a similarly symmetrical pattern, whereas the other contaminants display long tailing, that is, the plot of relative concentration to pore volume rises sharply, then slowly decreases as an increasing number of pore volumes passes through the soil core.

For 1,4-dioxane, a retardation factor of 1.1 was obtained from soil core tests at pore-water velocities of 45 and 90 cm/day.

TABLE 3.22
Comparison of Retardation Factors (R_f) Based on Field Data, Equations, and Column Data

| | | | Retardation Factor Value Based on Type of Data | | | | | |
| | | | Field Data | | | Laboratory Data | | |
Compound	Solubility[a] (g/L)	$\log K_{ow}$	Plume Length Estimate[b]	Purge Well Test Result[c]	Correlation Equation[d]	S and W Equation[b,e]	Center of Mass[f]	Maximum Concentration[g]
1,4-Dioxane	M	−0.27	1.6	1.4	1.6	1.0	1.1	1.2
Tetrahydrofuran	M	0.46	2.2	2.2	2.5	1.0	NT	NT
1,2-Dichloroethane	8.7	1.48	7.6	n.p.	5.7	1.2	7.2	4–5
Benzene	1.78	2.04	8.8	n.p.	10.0	1.4	14.3	6–8
1,1-Dichloroethylene	0.40	2.13	n.m.	n.p.	11.0	1.5	10.7	6–7

Source: From Jackson, R.E. and Dwarakanath, V., 1999, *Ground Water Monitoring Review* 19(4): 102–110. With permission.

[a] M = Miscible.

[b] $R_f = L_{Cl}/L_{org}$, the ratio of the length of the chloride plume (L_{Cl}) to that of the organic plume (L_{org}) (Patterson et al., 1985); n.m. = not measured. The chloride originated from the same waste-disposal trench as the organic compounds.

[c] Tracer-labeled contaminants were injected at 140 L/min and extracted 5 m down gradient at the same rate for 6 days; samples were collected at multilevel sampling wells between injection and extraction wells. The retardation factor was determined from the ratio of the times at which $C/C_0 = 0.5$ for the organic compound and the tracer compounds (iodide and fluorinated benzoic acid) (Whiffin and Bahr, 1984); n.p. = contaminant not present.

[d] Correlation equation from field data at the Gloucester landfill site: $\log(R_f - 1) = 0.5 \log(K_{ow}) - 0.065$ (Patterson et al., 1985).

[e] $R_f = 1 + \rho_b K_d/n$, where $K_d = 3.2 f_{oc} \times 0.72(K_{ow})$ (S and W = Schwarzenbach and Westall, 1981); NT = not tested.

[f] In column test data, the center of mass of organic compound versus the center of mass of iodide (Priddle and Jackson, 1991); NT = not tested.

[g] In column test data, the ratio of the time at which $C/C_{max} = 0.5$ for organic plume versus the time at which $C/C_{max} = 0.5$ for iodide plume (Priddle and Jackson, 1991).

DUKE FOREST LANDFILL SOIL SORPTION STUDY (BALL AND BARTLETT, 1992)

A screening-level soil sorption study of 1,4-dioxane transport in native soils was conducted at Duke University. A landfill near the university had received wastes containing 1,4-dioxane, and monitoring wells showed concentrations of several thousand micrograms per liter. Researchers performed soil sorption tests on sieved and homogenized cores of aquifer solids retrieved from below the landfill site to verify that 1,4-dioxane would not bind strongly to soil material. Soil samples sterilized with sodium azide were mixed with water and 1,4-dioxane. For 4 days, the slurry was rotated to maintain active mixing; then it was centrifuged to separate the aqueous-phase solution from the soil for analysis. Because of 1,4-dioxane losses in sample handling (13%) and low analytical recovery on some analyses (e.g., 44%), the distribution coefficient K_d could not be reliably estimated. Nevertheless, the study clearly demonstrated that the potential for 1,4-dioxane to sorb to soil is low. The maximum estimated retardation factor was 4.2, that is, 1,4-dioxane would not move slower than one-quarter the rate of groundwater flow. A related study summarized previous investigations of the Duke Forest landfill site and noted that a consultant report determined an average retardation factor of 1.2 (Liu et al., 2000).

organic compound to the length of the chloride plume, which presumably originated at the same location and time and which is assumed to have a retardation factor of 1.0 (Patterson et al., 1985). A comparison of retardation factors determined with the plume-length approach is given in Table 3.22.

Table 3.22 shows that estimates of retardation factors calculated with the Schwarzenbach and Westall equation can be as little as one-sixth the field-measured values. The Schwarzenbach and Westall equation does not adequately account for adsorption to mineral surfaces; it was derived from data for hydrophobic polar compounds, which do not sufficiently represent the behavior of hydrophilic compounds such as 1,4-dioxane. Table 3.22 also demonstrates that migration of 1,4-dioxane is only minimally retarded. It therefore migrates substantially faster than most other organic contaminants.

3.4.3　Modeling 1,4-Dioxane Transport

Modeling tools can provide useful insights into the relative rates of migration of contaminants under assigned subsurface conditions such as groundwater flow rate, organic matter content of soils and aquifer solids, retardation factors, biodegradation rate constants, abiotic rate constants, source mass strength, release duration, and other factors. There are many analytical approaches and many models available to simulate contaminant transport. The subject of modeling contaminant transport is beyond the scope of this book; however, a brief exercise using a screening-level model is profiled here to provide a sense of the relative calculated rates of migration of 1,4-dioxane, other stabilizer compounds, and methyl chloroform. The screening-level model reviewed here is BIOCHLOR v. 2.2, a spreadsheet model based on modifications to the Domenico analytical solute transport equation, which can simulate one-dimensional advection, three-dimensional dispersion, linear adsorption, and biotransformation via reductive dechlorination for chlorinated solvents (Aziz et al., 2000, 2002). BIOCHLOR is used primarily to evaluate the efficacy of employing monitored natural attenuation as a remedial solution at chlorinated solvent release sites; the model is not intended to accurately predict migration of contaminants. BIOCHLOR includes the assumption that biodegradation occurs only in the aqueous phase; therefore, the model ignores degradation of contaminants sorbed to soil and aquifer solids. When users of BIOCHLOR set longitudinal dispersivity to high values, BIOCHLOR is prone to errors in predicting migration when applied to

TABLE 3.23
Hydraulic Parameters for Modeled Domain

Seepage velocity	111.7 ft/year
Conductivity	0.018 cm/s
Longitudinal dispersivity, α_X	26.9 ft
Transverse dispersivity, α_Y	2.69 ft
Vertical dispersivity, α_Z	0 ft

low-permeability aquifers; however, it is accurate enough for screening-level purposes when applied to advection-dominated transport conditions (Srinivasan et al., 2007; West et al., 2007; USEPA, 2007b).

Relative rates of migration of 1,4-dioxane and methyl chloroform in groundwater were determined by using BIOCHLOR with the first-order biological decay coefficient set to effectively zero biodegradation for 1,4-dioxane. The objective of the screening-level modeling was to predict relative rates of migration at release sites and the relative distances within which regulatory thresholds would be exceeded for 1,4-dioxane and methyl chloroform and its biotransformation products, 1,1-dichloroethane and chloroethane.

The result of this modeling exercise does not necessarily represent the true field behavior of this mixture of compounds. Among other basic limitations, running BIOCHLOR separately for the chlorinated ethanes and 1,4-dioxane ignores any competitive sorption that may occur and thereby possibly underestimates the spatial extent of an actual plume. BIOCHLOR does not account for aquifer heterogeneities such as channels or other preferential pathways. What is most important is that BIOCHLOR does not address matrix diffusion in lenses of fine-grained sediment that occur in heterogeneous aquifers. This application of BIOCHLOR is not intended to simulate migration absolutely; rather, it is used to simulate the relative mobility and persistence of 1,4-dioxane in contrast to methyl chloroform.

The hydraulic and soil properties of an aquifer studied at Cape Canaveral Air Station, Florida—included as a preloaded case study in BIOCHLOR—were used to model transport of methyl chloroform and 1,4-dioxane (Table 3.23). With the exception of redefining the source dimensions, dispersivities, simulation time, and domain length, all other parameters were left as the defaults of the Cape Canaveral case study for methyl chloroform simulations. Table 3.24 summarizes the transport parameters for each compound. The model imposes first-order decay of methyl chloroform and its two degradation products, 1,1-dichloroethane and chloroethane, terminating in sequential fashion with ethane. Sorption is modeled according to K_{oc} values. In cases of multiple contaminants, the median K_{oc} was arbitrarily used.

TABLE 3.24
Regulatory Levels and Transport Properties of Modeled Compounds

	TCA	DCA	Chloroethane	1,4-Dioxane
Regulatory level (μg/L)	200	5	16	3
K_{oc} (L/kg)	426	130	125	17
Retardation	7.13	2.87[a]	2.8	1.1
Degradation, λ (year^{-1})	2.0	1.0	0.7	"0"

Notes: K_{oc} values used are BIOCHLOR model defaults for chlorinated ethanes; $\log K_{oc} = 1.23$ for 1,4-dioxane (17 L/ kg).

[a] Retardation value used in model for all chlorinated ethanes.

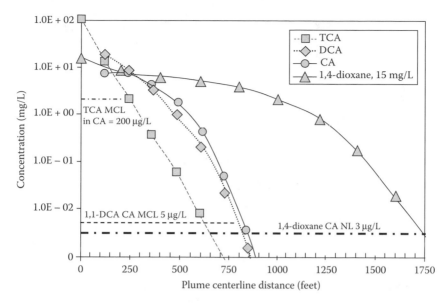

FIGURE 3.10 BIOCHLOR-modeled transport of chlorinated ethanes and 1,4-dioxane, 10-year release scenario. BIOCHLOR parameters include a 10-year continuous source release; source concentrations: 100 mg/L TCA, 15 mg/L 1,4-dioxane. California regulatory and advisory standards are shown for comparison. NL = notification level; MCL = maximum contaminant level (see Chapter 6 for explanation of NL and MCL).

Dispersion is a term inclusive of the physical processes that cause a plume to shear. Fixed values for longitudinal dispersivity, horizontal transverse dispersivity, and vertical transverse dispersivity (Table 3.23) were used in all model runs. The BIOCHLOR model was run several times by using different source durations with an initial aqueous concentration of 100 mg/L methyl chloroform and zero initial concentrations of degradation products. Separate trials were performed for 1,4-dioxane at initial concentrations of 3 and 15 mg/L. The 3 mg/L scenario represents virgin methyl chloroform released to groundwater, whereas the 15 mg/L scenario is intended to represent the release of vapor degreasing still bottoms enriched with 1,4-dioxane because of partitioning in the vapor degreasing process, as described in Chapter 1.

BIOCHLOR only accounts for subsurface movement of dissolved-phase solvents; dense non-aqueous-phase liquids and vapor transport are not considered. The initial concentration of methyl chloroform modeled, 100 mg/L, is less than 10% of the overall solubility of methyl chloroform. 1,4-Dioxane persists over a longer distance from the source than the chlorinated ethanes owing to its infinite solubility, lower sorption, which is incorporated in the lower retardation factor (R_f) reported by Jackson and Dwarakanath (1999), and an assumed lack of biodegradation. Calculated migration using BIOCHLOR shows that 1,4-dioxane concentrations will remain higher than regulatory thresholds over greater distances. Figure 3.10 illustrates the distance from the source at which the contaminant concentration reaches the regulatory level as a function of the source lifetime (see Table 3.23). Figure 3.11 shows the distances that 1,4-dioxane is likely to migrate when released under different duration scenarios and different source concentrations.

3.5 DIFFUSIVE TRANSPORT OF 1,4-DIOXANE AND STORAGE IN FINE-GRAINED SOILS

Transport in fine-grained soils composed predominantly of clay and silt plays an important role in the overall subsurface fate on groundwater contaminants. In clays found in aquitards and in engineered clay liners formerly used as the primary liner for landfills and waste lagoons, contaminant migration by advective transport is very slow. Where the aquitards or liners have low hydraulic

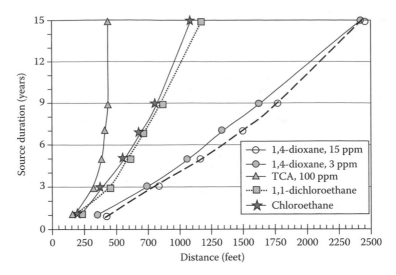

FIGURE 3.11　BIOCHLOR-modeled distance along plume centerline at which contaminant concentration exceeds regulatory levels for increasing release durations (source TCA concentration = 100 mg/L).

conductivity or where hydraulic gradients are insignificant, molecular diffusion through the clay pore fluid is the primary mode of contaminant transport. Predicting contaminant transport through fine-grained strata and engineered clay liners involves modeling diffusive transport (Barone et al., 1992; Hoffman et al., 1998; Young and Ball, 1998; Ball, 2000).

Diffusion-driven contamination is inherently slow. It is controlled by concentration gradients, solubility, and temperature. Limitations on diffusion-driven contamination include the occurrence and heterogeneity of fine-grained media within and adjacent to an aquifer and rates of adsorption, biodegradation, and abiotic degradation. Over long periods of time, contaminants with high concentrations in coarse-grained media will move into adjacent low-permeability zones, including aquitards, fine-grained lenses within an aquifer, and the micropores within individual grains of the aquifer matrix. The process of contaminant diffusion into fine-grained sediments is best articulated by Mackay and Cherry's classic 1989 article "Groundwater Contamination: Pump and Treat Remediation," paraphrased as follows:[*]

> As dissolved contaminants spread through aquifers, they rapidly migrate through more permeable zones while slowly invading less permeable materials by diffusion. Over many years, diffusion can cause dissolved contaminants to occupy large volumes of low-permeability material. To obtain clean water from wells, the lower-permeability components of the aquifer system must be remediated as well as the higher-permeability zones.
>
> Consider a clay lens in the middle of a predominantly sandy hydrostratigraphic unit through which contaminants have migrated for decades. The porosity of the clay lens is often larger than that of the adjacent aquifer, promoting diffusion into the clay. Dissolved contaminants permeate the clay lens by molecular diffusion, which leads to delayed diffusive release. Also, the capacity of a clay lens to sorb contaminants is generally much greater per unit volume than the sorption capacity of the aquifer. When the aquifer is flushed by clean water, the only significant process for release of the contaminant from the clay will be a reversal of the diffusion direction. The relatively slow rate of release of contaminants from the clay by diffusion and the potentially appreciable contaminant mass contained in dissolved and sorbed form in the clay causes a long-term "bleed" or "back-diffusion" of contaminants into the aquifer during remediation. In many aquifers, there are numerous thin beds of silt and clay.
>
> Contaminant concentrations in water extracted for remediation or beneficial uses often remain higher than low regulatory cleanup thresholds for very long periods of time, greatly increasing the

---

[*] See Mackay and Cherry (1989).

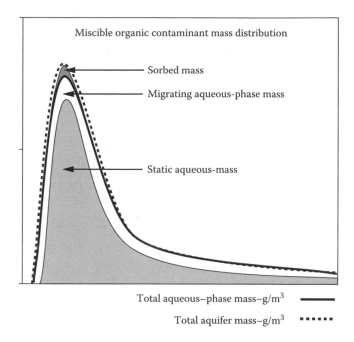

FIGURE 3.12 Distribution of mass of hydrophilic compounds as mass versus distance from site or time at a location down-gradient from the point of release. (From Payne, F.C., et al., 2003, Developing *in situ* reactive zone strategies for 1,4-dioxane. Presented at 1,4-Dioxane and Other Solvent-Stabilizer Compounds in the Environment. December 8, 2003, San Jose, CA, Groundwater Resources Association of California. http://www.grac.org. With permission.)

duration and cost of cleanup. Most investigations fail to delineate site geology in sufficient detail to predict the effects of delayed diffusive release.

Diffusion into fine-grained media is driven by a concentration gradient, away from sustained high concentrations in the higher-mobility aquifer zone supplied by a continuous source of contamination. When that source of contamination is terminated or remediated, the concentration gradient reverses, leading to back-diffusion out of the fine-grained media. The back-diffusion concept is often explained in the context of a sorption–desorption system; however, in the case of high-solubility or miscible hydrophilic compounds with low potential for sorption such as 1,4-dioxane, sorption may not play an active role in the process. Instead, diffusion can lead to long-term storage of 1,4-dioxane in fine-grained media near the source of the release, followed by long-term slow release (Suthersan, 2002; Payne et al., 2003, 2008; Nguyen et al., 2005). Remediation projects commonly cease extraction when the plot of mass removal over time becomes asymptotic; however, a subsequent rebound in concentrations is often observed. Figure 3.12 shows the relative distribution of hydrophilic contaminants in aquifer media.

The diffusion coefficient for a contaminant in water (D_w) can be predicted by using a variety of QSAR equations. A commonly used approach is the Wilke–Chang equation, which incorporates the molar volume of the solute, the viscosity and temperature of water, and several empirical constants (Wilke and Chang, 1955). For many organic compounds, D_w is in the range of 10^{-6}–10^{-4} m^2/day. Through the use of the Wilke–Chang method, the calculated diffusion coefficient for 1,4-dioxane in pure water at 22°C is 10.6×10^{-6} cm^2/s (Barone et al., 1992).

A laboratory study to measure diffusion coefficients and adsorption coefficients for 1,4-dioxane and four other compounds in samples of a clayey soil was conducted by Barone et al. (1992) at the University of Western Ontario.

TABLE 3.25
Characteristics of the Sarnia Clay in 1,4-Dioxane Diffusion Study

Property	Value
Specific gravity	2.73
Saturation	100%
Dry density	1.68 g/cm^3
Calcite/dolomite ratio	0.42
Moisture content	23%
Porosity	39%
Organic carbon content, f_{oc}	0.58
Cation-exchange capacity	10 meq per 100 g dry weight
Minerals in <0.074 mm fraction	
Calcite and dolomite	34%
Quartz and feldspars	15%
Illite	25%
Chlorite	24%
Smectite	2%
Grain-size distribution	
Clay (<0.002 mm)	45%
Silt (0.074–0.002 mm)	43%
Sand (2.0–0.074 mm)	10%
Gravel (>2.0 mm)	2%

Source: Barone, F.S., et al., *Journal of Contaminant Hydrology* 10(3): 225–250.

Distribution coefficients (K_d) were estimated from laboratory diffusion tests in plug samples of a saturated and undisturbed clayey soil using five organic compounds, each tested alone at 300 mg/L (Barone et al., 1992). A K_d value of 0.17 mL/g was estimated for 1,4-dioxane, on the basis of a measured diffusion coefficient of 4×10^{-6} cm^2/s in the clayey soil. The characteristics of the soil tested are listed in Table 3.25. The results of the soil plug diffusion tests are listed in Table 3.26.

The high retardation factor, 2.19, for 1,4-dioxane calculated in Table 3.26 is twice the 1.1 value obtained by Priddle and Jackson (1991). The higher retardation factor calculated by using data from Barone et al. (1992) reflects the low hydraulic conductivity of the Sarnia clay used for diffusion and distribution coefficient testing.

The ratio of the measured effective diffusion coefficient for a water-borne contaminant in soil to the calculated diffusion coefficient of that contaminant through pure water represents the tortuosity of the soil. As a compound diffuses in a porous medium, the rate of diffusion is reduced by the increased length of the molecule's path as it travels around solid particles in the soil. The length of the path traveled by a diffusing molecule relative to the straight-path distance is called the tortuosity factor, τ (Bear, 1972). Tortuosity can be derived from diffusivity as follows:

$$\tau = \frac{L}{L_e} = \frac{D_e}{D_w} \approx \phi^P, \tag{3.43}$$

where L is the length of a straight path through the porous media, L_e is the length of a tortuous path around solid grains, D_e is the effective diffusion coefficient of the water-borne contaminant, and D_w is the calculated diffusion coefficient of the contaminant in pure water (Bear, 1972; Hoffman et al., 1998). The last expression, ϕ^P, provides an empirical way to obtain tortuosity,

TABLE 3.26

Measured Fate and Transport Parameters of Selected Organic Compounds in Clay from Soil Plug Diffusion Tests

Compound	Distribution Coefficient K_d (mL/g)	Calculated Aqueous Diffusion Coefficient in Pure Water[a] D_w (cm²/s)	Measured Effective Aqueous Diffusion Coefficient[b] D_e (cm²/s)	Relative Concentration C/C_0 after 14 days	Calculated Retardation Factor[c] R_f
Acetone	0.19	11.5×10^{-6}	5.6×10^{-6}	0.9	2.33
1,4-Dioxane	0.17	10.6×10^{-6}	4.0×10^{-6}	0.9	2.19
Aniline	1.3	9.8×10^{-6}	7.0×10^{-6}	0.7	10.1
Chloroform	4.5	10.6×10^{-6}	7.0×10^{-6}	0.5	43
Toluene	11	8.9×10^{-6}	5.8×10^{-6}	0.4	183

Source: From Barone, F.S., Rowe, R.K., and Quigley, R.M., 1992, *Journal of Contaminant Hydrology* 10(3): 225–250. With permission.

[a] Diffusion constants calculated by Barone et al. (1992) by using the Wilke–Chang equation for water at laboratory temperature, 22°C.

[b] Barone et al. (1992) noted that the aqueous diffusion coefficients first obtained for chloroform and toluene were unreasonably high owing to the role of nonsettling particles in the soil column.

[c] Retardation factors calculated by the author using soil properties from Barone et al. and the familiar retardation equation (Equation 3.32).

where ϕ is porosity and P is an empirical factor selected for the geological medium under consideration (Millington and Quirk, 1961; Parker et al., 1996). Tortuosity, sometimes represented with the Greek letter χ, accounts for all factors that limit solute diffusion through the porous medium such as the tortuous nature of the diffusion path, dead-end pores, and steric hindrance (Young and Ball, 1998). Higher tortuosity factors indicate shorter flow paths. Well-sorted sands (i.e., those that have essentially uniform grain size) have higher tortuosity factors because smaller grains are not filling pore spaces. Tortuosity factors in well-sorted, fine-grained sands range from 0.6 to 0.8, whereas τ in well-sorted, coarse-grained sands averages 0.4; for poorly sorted sands, τ ranges from 0.2 to 0.8 (Fetter, 1993; Hoffman et al., 1998). Barone et al. (1992) obtained tortuosity factors for the subject Sarnia clay of 0.65 from the calculated and measured diffusivities of chloroform and toluene. A theoretical estimate of tortuosity in unconsolidated materials is 0.67 (Bear, 1972).

Another approach to estimating the tortuosity of a soil involves contrasting measured diffusion coefficients of a compound of interest to the diffusion coefficient of a nonretarded compound such as 1,4-dioxane to obtain a relative approximation of tortuosity. This approach has been used to compare carbon tetrachloride, perchloroethylene, and trichloroethylene to chloroform, whose retardation is considered negligible (Hoffman et al., 1998). Knowledge of the tortuosity of a medium is useful for estimating the role of diffusion in retaining contaminants in fine-grained deposits.

Barone et al. (1992) estimated breakthrough times for vertical transport in clay soil. 1,4-Dioxane was calculated to advance more than 5 m in 100 years, while toluene advanced less than 1 m in the same time. The hypothetical calculation estimates that a leachate containing 1,4-dioxane at 300 mg/L discharged to the subject Sarnia clay would diffuse through a 1-m-thick clay landfill liner in 5 years and contaminate underlying groundwater to concentrations in excess of 30 µg/L (the Canadian drinking water guidance in effect during the 1989 study).

To address the role of sorption in diffusive transport, scientists at Lawrence Livermore National Laboratory (LLNL) developed a hybrid contaminant transport term that accounts for both tortuosity and retardation by sorption (Hoffman et al., 1998). The LLNL scientists noted that effective diffusion is affected by sorption as follows:

$$D_a = \frac{D_e}{R} = \frac{D_w \times \tau}{R}. \qquad (3.44)$$

Hoffman et al. (1998) combined the effects of diffusion and sorption into a single parameter, the diffusion reduction coefficient, Ω:

$$\Omega = \frac{\tau}{R}, \qquad (3.45)$$

where Ω is the diffusion reduction coefficient, τ the tortuosity factor, and R the retardation factor. With these parameters, the rate of migration due to diffusion for a soil with known bulk properties can be estimated.

Diffusion rates in clays are both difficult to measure and difficult to predict. Measured values of diffusion coefficients in clayey materials may only be accurate to within a factor of four, whereas tortuosity factors may be accurate to within a factor of two if controls such as measurement of tritiated water are used. Estimation of diffusion coefficients in low-permeability porous media requires specific measurements for the system of interest (Young and Ball, 1998). Nevertheless, because diffusion plays a critical role in the transport of hydrophilic compounds such as 1,4-dioxane, diffusion coefficients are estimated here by using the approaches of Barone et al. (1992) and Hoffman et al. (1998), described previously in this section, to provide a *relative* sense of the role of molecular diffusion in the transport of solvent stabilizers in aquitards. Table 3.27 provides estimated rates of migration by diffusion for several solvent-stabilizer compounds and chlorinated solvents through a 1-m-thick clay lens by using the bulk soil parameters from the Barone study Sarnia clay described in Table 3.25.

Table 3.27 profiles the potential for compounds to migrate into clays and silts by molecular diffusion and to back-diffuse into the mobile phase when the concentration gradient reverses. The stabilizer compounds in Table 3.27 that have higher apparent diffusion coefficients are candidates for long-term storage in fine-grained material. These stabilizer compounds include the cyclic ethers 1,4-dioxane and 1,3-dioxolane; the alcohols *sec*-butyl alcohol, *tert*-butyl alcohol, and 2-methyl-3-butyn-2-ol; and the epoxy compound butoxymethyl oxirane.

The aqueous and effective diffusivity coefficients presented in Table 3.27 span a relatively small range; therefore, the expected mass transfer by diffusion should be similar among these compounds. Accordingly, concentration ratios of stabilizer compounds in the mobile phase of groundwater should be primarily a reflection of their respective effective solubilities (Feenstra and Guiguer, 1996). However, the apparent diffusion coefficient derived by Hoffman et al., which incorporates sorption and tortuosity, shows a wider range. Hydrophilic and highly soluble ethers, alcohols, and some epoxides can be expected to participate to a greater degree in the aquitard storage and subsequent back-diffusion process than other stabilizers, and 100-fold more than methyl chloroform, trichloroethylene, and perchloroethylene.

The retardation factor for 1,4-dioxane in Table 3.27 is higher than the values measured by Priddle and Jackson (1991; Table 3.22) or Barone et al. (1992; Table 3.25) because it is calculated from the soil parameters for the Sarnia clay used in the Barone et al. study. Where aquifer sands are interbedded with silts and clays or include a clayey matrix, molecular diffusion may play a significant role in the transport, storage, and long-term secondary release back into the aquifer for several solvent-stabilizer compounds. Matrix diffusion from sands into clays and long-term release by back-diffusion can cause persistent presence of 1,4-dioxane at low concentrations, contrary to the expectation that it would quickly flush through the porous reaches of an aquifer owing to its high solubility and low K_{oc} (Payne et al., 2008).

TABLE 3.27
Relative Magnitudes of *Estimated* Effective Aqueous Diffusion Coefficients and Migration of Chlorinated Solvents and Solvent-Stabilizer Compounds in Sarnia Clay

Compound[a]	Distribution Coefficient[b] K_d (Dimensionless)	Aqueous Diffusion Coefficient[c] D_w (cm²/s)	Effective Diffusion Coefficient[d] D_e (cm²/s)	Calculated Retardation Factor[e] R_f (Dimensionless)	Diffusion Reduction Coefficient[f] Ω (Dimensionless)	Apparent Diffusion Coefficient[g] D_a (cm²/s)
1,3-Dioxolane	0.58	2.66×10^{-05}	5.31×10^{-06}	3.50	5.72×10^{-02}	1.52×10^{-06}
1,4-Dioxane	**0.58**	$\mathbf{2.62 \times 10^{-05}}$	$\mathbf{5.24 \times 10^{-06}}$	**3.50**	$\mathbf{5.72 \times 10^{-02}}$	$\mathbf{1.50 \times 10^{-06}}$
tert-Butyl alcohol	0.928	2.14×10^{-05}	4.28×10^{-06}	5.00	4.00×10^{-02}	8.56×10^{-07}
sec-Butyl alcohol	1.189	2.14×10^{-05}	4.28×10^{-06}	6.12	3.27×10^{-02}	6.98×10^{-07}
2-Methyl-3-butyn-2-ol	1.6936	2.19×10^{-05}	4.37×10^{-06}	8.30	2.41×10^{-02}	5.27×10^{-07}
Butoxymethyl oxirane	1.6414	2.10×10^{-05}	4.20×10^{-06}	8.07	2.48×10^{-02}	5.20×10^{-07}
Epichlorohydrin	2.61	2.52×10^{-05}	5.03×10^{-06}	12.24	1.63×10^{-02}	4.11×10^{-07}
Acetonitrile	2.61	2.32×10^{-05}	4.63×10^{-06}	12.24	1.63×10^{-02}	3.78×10^{-07}
1,2-Butylene oxide	2.61	2.20×10^{-05}	4.41×10^{-06}	12.24	1.63×10^{-02}	3.60×10^{-07}
Nitromethane	4.756	2.73×10^{-05}	5.45×10^{-06}	21.49	9.31×10^{-03}	2.54×10^{-07}
Diisopropylamine	63.22	1.94×10^{-05}	3.89×10^{-06}	273.33	7.32×10^{-04}	1.42×10^{-08}
Cyclohexane	95.99	2.13×10^{-05}	4.25×10^{-06}	414.50	4.83×10^{-04}	1.03×10^{-08}
Methyl chloroform	88.16	8.30×10^{-06}	1.66×10^{-06}	381	5.25×10^{-04}	4.36×10^{-09}
Dichloromethane	5.104	1.24×10^{-05}	2.48×10^{-06}	23	8.70×10^{-03}	1.08×10^{-07}
Trichloroethylene	73.08	8.60×10^{-06}	1.72×10^{-06}	316	6.33×10^{-04}	5.45×10^{-09}
Perchloroethylene	211.1	7.70×10^{-06}	1.54×10^{-06}	910	2.20×10^{-04}	1.69×10^{-09}

Source: From Barone, F.S., Rowe, R.K., and Quigley, R.M., 1992, *Journal of Contaminant Hydrology* 10(3): 225–250. With permission.

[a] Nonitalics compounds are stabilizers. Italics compounds are chlorinated solvents; these are listed for comparison to show whether the stabilizer is likely to diffuse out of the solvent or remain with the solvent.

[b] K_d calculated by using $K_{oc} \times f_{oc}$ (see Section 3.3.2.3) and soil parameters from the Sarnia clay.

[c] Aqueous diffusion coefficients in pure water calculated by using the Wilke–Chang method ($T = 22°C$).

[d] Effective diffusion coefficient obtained by dividing aqueous diffusion coefficient by an assumed tortuosity factor, 0.2, for a poorly sorted fine-grained material.

[e] Retardation factor calculated by using soil parameters from Sarnia clay and Equation 3.32.

[f] Diffusion reduction coefficient obtained from Equation 3.45.

[g] D_a obtained as the product of the aqueous diffusion coefficient and the diffusion reduction coefficient.

The geometry of the interface between coarse- and fine-grained aquifer strata plays an important role in determining the degree to which matrix diffusion will occur. Compared to thick sand beds, thin sand beds interbedded with clays and silts provide a larger sand/clay contact surface area across which diffusion may occur. In strongly heterogeneous conditions, contaminant flow in sand lenses less than 10 cm thick will be retarded by aqueous-phase diffusion, causing a significant mass of solute to reside in immobile dissolved-phase storage. Most aquifers are characterized as highly heterogeneous and anisotropic and exhibit sharp contrasts of hydraulic conductivity, often on the scale of a few centimeters (Payne et al., 2008).

The concentration gradient that drives diffusion across interfaces from high to low hydraulic conductivities will rapidly dissipate as 1,4-dioxane or other solutes invade fine-grained pores. The flux density will decrease in response to a decreased concentration gradient, according to Fick's law:

$$J = \Theta_{total} \cdot D_{eff} \frac{dC}{dz}, \tag{3.46}$$

where Θ_{total} is total porosity (dimensionless), D_{eff} is effective diffusivity (in m^2/day), dC/dz is the solute concentration gradient [i.e., the change in mass concentration over distance, in $(kg/m^3) \cdot m^{-1} = kg/m^2$], and J is the flux density [in $\mu g/(cm^2 \cdot s)$]. The diffusive flux density will rapidly decay as contaminant concentration increases in the "immobile porosity," that is, in the micropores in the clays and silts, thereby decreasing the solute concentration gradient, dC/dz. In calculations performed by Payne et al. (2008), the diffusive flux for a trichloroethylene gradient with an initial source-side concentration of 200 $\mu g/L$ will decline 10-fold in 100 days and 1000-fold in 1000 days.

There are two main consequences of appreciable diffusive mass flux density (Payne et al., 2008):

- The diffusive loss of solute from the mobile porosity slows the solute velocity relative to groundwater velocity.
- Immobile porosity in clays and silts serves as a repository for dissolved-phase solute that enters via aqueous-phase diffusion

A further consequence of diffusion is contaminant transport to sites unavailable to microorganisms, precluding biodegradation. For example, in a silt loam, half the measured pore volume consists of pores with radii less than 1 μm, whereas the mean diameter of most soil bacteria is 0.5–0.8 μm, and the mean diameter of soil pores occupied by bacteria is 2 μm (Alexander, 1994). Therefore, diffusion of chemical contaminants into pores smaller than those occupied by soil microbes is an important factor controlling the rate of biodegradation of these compounds.

BIBLIOGRAPHY

Abe, A., 1999, Distribution of 1,4-dioxane in relation to possible sources in the water environment. *The Science of the Total Environment* 227: 41–48.

Adamus, J.E., May, H.D., Paone, D.A., Evans, P.J., and Parales, R.E., 1995, United States Patent 5,474,934: Biodegradation of ethers. Assignee: Celgene Corporation, Warren, NJ.

Ahlers, J., 1998, Screening information data set initial assessment report: 2-Methylbut-3-yn-ol; 8th SIDS2 Initial Assessment Meetings, Paris, France. Nairobi, Kenya: UNEP (United Nations Environment Programme) Organization for Economic Cooperation and Development (OECD).

Alarie, Y., Nielsen, G.D., Andonianhaftvan, J., and Abraham, M.H., 1995, Physicochemical properties of non-reactive volatile organic chemicals to estimate RD$_{50}$: Alternatives to animal studies. *Toxicology and Applied Pharmacology* 134(1): 92–99.

Alexander, M., 1965, Biodegradation: Problems of molecular recalcitrance and microbial fallibility. *Advances in Applied Microbiology* 7: 35–80.

Alexander, M., 1985, Biodegradation of organic chemicals. *Environmental Science and Technology* 19(2): 106–111.

Alexander, M., 1994, *Biodegradation and Bioremediation*. San Diego: Academic Press.

Altschuh, J., Berggemann, R., Santl, H., Eichinger, G., and Piringer, O.G., 1999, Henry's law constants for a diverse set of organic chemicals: Experimental determination and comparison of estimation methods. *Chemosphere* 39(11): 1871–1886.

Anagnostu, G., 1989, Biological detoxification of industrial wastes containing dioxane and diglyme. Master of Science Thesis, Louisiana State University, Baton Rouge, LA..

Anbar, M. and Neta, P., 1967, A compilation of specific bimolecular rate constants for the reactions of hydrated electrons, hydrogen atoms, and hydroxyl radicals with inorganic and organic compounds in aqueous solution. *International Journal of Applied Radiation and Isotopes* 18(7): 493–523.

Arbuckle, W.B., 1983, Estimating activity coefficients for use in calculating environmental parameters. *Environmental Science and Technology* 17(9): 537–542.

Archer, W.L., 1984, A laboratory evaluation of 1,1,1-trichloroethane-metal-inhibitor systems. *Werkstoffe und Korrosion (Materials and Corrosion)* 35: 60–69.

Arnot, J., Gouin, T., and Mackay, D., 2005, Development and application of models of chemical fate in Canada—practical methods for estimating environmental biodegradation rates. Report to Environment Canada, Peterborough, Ontario, Canada: Environment Canada Report No. 200503.

Aronson, D., Citra, M., Shuler, K., Printup, H., and Howard, P.H., 1999, Aerobic biodegradation of organic chemicals in environmental media: A summary of field and laboratory studies. Syracuse Research Corporation, North Syracuse, NY. Syracuse Research Corporation Report No. TR 99-002.

Atkinson, R., 1988, Estimation of gas-phase hydroxyl radical rate constants for organic chemicals. *Environmental Toxicology and Chemistry* 7: 435–442.

Aziz, C.E., Newell, C.J., Gonzales, J.R., and Jewett, D.G., 2000, *BIOCHLOR Natural Attenuation Decision Support System, User's Manual Version 1.0*. U.S. Environmental Protection Agency, Office of Research and Development, National Risk Management Research Laboratory.

Aziz, C.E., Newell, C.J., Gonzales, J.R., and Jewett, D.G., 2002, *BIOCHLOR Natural Attenuation Decision Support System Version 2.2, User's Manual Addendum*. San Antonio, TX: Groundwater Services, Inc., and Air Force Center for Environmental Excellence Technology Transfer Division.

Baker, J.T., 2006, Material Safety Data Sheet for triethanolamine, Phillipsburg, NJ.

Ball, W.P., 2000, Diffusion-limited contamination and decontamination in a layered aquitard: Forensic and predictive analysis of field data. Paper presented at Confinement and Remediation of Environmental Hazards meeting, January 15–19, 2000, University of Minnesota.

Ball, W.P. and Bartlett, L., 1992, *Evaluation of Para-dioxane Sorption by Duke Forest Subsurface Solids*. Durham, NC: Duke University Department of Civil and Environmental Engineering.

Banerjee, S., 1984, Solubility of organic mixtures in water. *Environmental Science and Technology* 18(8): 587–591.

Banerjee, S., Yalkowsky, S.H., and Valvani, S.C., 1980, Water solubility and octanol/water partition coefficients of organics: Limitations of the solubility partition coefficient correlation. *Environmental Science and Technology* 14(10): 1227–1229.

Barone, F.S., Rowe, R.K., and Quigley, R.M., 1992, A laboratory estimation of diffusion and adsorption coefficients for several volatile organics in a natural clayey soil. *Journal of Contaminant Hydrology* 10(3): 225–250.

Barton, A.F.M., 1984, *Alcohols with Water*, Solubility Data Series, Vol. 15. Oxford: Pergamon Press.

BASF, 2006, Safety Data Sheet for 2-methylbut-3-yn-2-ol, Florham Park, NJ.

Battersby, N.S. and Wilson, V., 1989, Survey of the anaerobic biodegradation potential of organic chemicals in digesting sludge. *Applied and Environmental Microbiology* 55(2): 433–439.

Bear, J., 1972, *Dynamics of Flow in Porous Media*. New York: American Elsevier Publishing Company.

Beaujean, M., 2007, Acetals as new solvents in paint and coating formulations, paint strippers and graffiti removers. *Pitture E Vernici—European Coatings* 2007(9): 41–48.

Bennett, P., 2004a, Hydrolysis. http://www.geo.utexas.edu/courses/387E/387E_notes_list.htm (accessed April 2, 2007).

Bennett, P., 2004b, Sorption. http://www.geo.utexas.edu/courses/387E/387E_notes_list.htm (accessed April 2, 2007).

Bernhardt, D. and Diekmann, H., 1991, Degradation of dioxane, tetrahydrofuran, and other cyclic ethers by an environmental *Rhodococcus* strain. *Applied Microbiology and Biotechnology* 36(1): 120–123.

Bhattacharya, S.K., Qu, M., and Madura, R.L., 1996, Effects of nitrobenzene and zinc on acetate utilizing methanogens. *Water Research* 30(12): 3099–3105.

Bidleman, T.F., 1988, Atmospheric processes—wet and dry deposition of organic compounds are controlled by their vapor-particle partitioning. *Environmental Science and Technology* 22(4): 361–367.

Bingham, E., Cohrssen, B., and Powell, C.H., 2001, *Patty's Toxicology*. New York: John Wiley and Sons, Inc.

Bocek, K., 1976, Relationships among activity coefficients, partition coefficients and solubilities. *Experientia Supplementum* 23: 231–239.

Boethling, R.S., Howard, P.H., Meyian, W., Stiteler, W., Beauman, J., and Tiradot, M., 1994, Group contribution method for predicting probability and rate of aerobic biodegradation. *Environmental Science and Technology* 28(3): 459–465.

Bogyo, D.A., Lande, S.S., Meylan, W.M., Howard, P.H., and Santodonato, J., 1980, Investigation of selected potential environmental contaminants: Epoxides—final technical report. Prepared by Syracuse Research Corporation for the U.S. Environmental Protection Agency Office of Toxic Substances.

Boublik, T., Fried, V., and Hala, E., 1984, *The Vapor Pressures of Pure Substances: Selected Values of the Temperature Dependence of the Vapor Pressures of Some Pure Substances in the Normal and Low Pressure Region*. New York: Elsevier Science Publishing Company.

Bradley, W.F., 1945, Molecular associations between montmorillonite and some polyfunctional organic liquids. *Journal of the American Chemical Society* 67(6): 975–981.

Briggs, G.G., 1981, Theoretical and experimental relationships between soil adsorption, octanol–water partition coefficients, water solubilities, bioconcentration factors, and the parachor. *Journal of Agricultural and Food Chemistry* 1981(29): 1050–1059.

Brindley, G.W. and Tsunashima, A., 1972, Montmorillonite complexes with dioxane, morpholine, and piperidine: Mechanisms of formation. *Clays and Clay Minerals* 20(4): 233–240.

Brindley, G.W., Wiewiora, K., and Wiewiora, A., 1969, Intracrystalline swelling of montmorillonite in some water–organic mixtures. *The American Mineralogist* 54(November–December): 1635–1644.

Browning, E., 1965, *Toxicity and Metabolism of Industrial Solvents*. New York: American Elsevier.

Brusseau, M.L. and Reid, M.E., 1991, Nonequilibrium sorption of organic chemicals by low organic-carbon aquifer materials. *Chemosphere* 22(3–4): 341–350.

Budavari, S., Ed., 1989, *The Merck Index—An Encyclopedia of Chemicals, Drugs and Biologicals*. Rahway, NJ: Merck and Co., Inc.

Budavari, S., Ed., 1996, *The Merck Index—An Encyclopedia of Chemicals, Drugs, and Biologicals*. Rahway, NJ: Merck and Co., Inc.

Burback, B.L. and Perry, J.F., 1993, Biodegradation and biotransformation of groundwater pollutant mixtures by *Mycobacterium vaccae*. *Applied Microbiology and Biotechnology* 59(4): 1025–1029.

Butler, J.A.V., Ramchandani, C.N., and Thompson, D.W., 1935, The solubility of non-electrolytes: Part I. The free energy of hydration of some aliphatic alcohols. *Journal of the Chemical Society, Faraday Transactions* 58: 280–285.

Buxton, G.V., Greenstock, C.L., Helman, W.P., and Ross, A.B., 1988, Critical review of rate constants for reactions of hydrated electrons, hydrogen atoms and hydroxyl radicals ($^{\bullet}OH/^{\bullet}O^{-}$) in aqueous solution. *Journal of Physical and Chemical Reference Data* 17: 513–886.

Cai, Z.-Q., Yang, G.-H., Li, E.-Y., and Zhao, X.-Y., 2008, Isolation, identification and degradation character of dioxane degradative strain D4. *China Environmental Science* 28: 49–52.

Capel, P.D. and Larson, S.J., 1995, A chemodynamic approach for estimating losses of target organic chemicals from water during sample holding time. *Chemosphere* 30(6): 1097–1107.

Carey, F.A., 1987, *Organic Chemistry*. New York: McGraw-Hill.

Chao, J., Lin, C.T., and Chung, T.H., 1983, Vapor pressure of coal chemicals. *Journal of Physical and Chemical Reference Data* 12(4): 1033–1063.

Chappelle, F.H., 1993, *Ground-water Microbiology & Geochemistry*. New York: Wiley.

Cheremisinoff, N.P. and Archer, W.L., 2003, *Industrial Solvents Handbook*, 2nd Edition. Boca Raton, FL: CRC Press.

Chiou, C.T. and Kile, D.E., 1994, Effects of polar and nonpolar groups on the solubility of organic compounds in soil organic matter. *Environmental Science and Technology* 28(6): 1139–1144.

Christie, A.O. and Crisp, D.J., 1967, Activity coefficients of the *n*-primary, secondary and tertiary aliphatic amines in aqueous solution. *Journal of Applied Chemistry* 17: 11–14.

Clayton, G.D. and Clayton, F.E., 1994, *Patty's Industrial Hygiene and Toxicology*. New York: Wiley.

Coca, J. and Diaz, R., 1980, Extraction of furfural from aqueous solutions with chlorinated hydrocarbons. *Journal of Chemical Engineering Data* 25(1): 80–83.

Cooper, W.J., Zika, R.G., Petasne, R.G., and Plane, J.M.C., 1988, Photochemical formation of H$_2$O$_2$ in natural waters exposed to sunlight. *Environmental Science and Technology* 22(10): 1156–1160.

Corsi, R., 1998, Environmental risk assessment. http://www.crwr.utexas.edu/gis/gisenv98/class/risk/lecture/Lect4/Fate.html#chemical#chemical (accessed April 2, 2007).

Daubert, T.E. and Danner, R.P., 1985, *Data Compilation Tables of Properties of Pure Compounds*. New York: American Institute of Chemical Engineers.

Dean, J.A., 1985, *Lange's Handbook of Chemistry*. New York: McGraw-Hill.

Deneer, J.W., Sinnige, T.L., Seinen, W., and Hermens, J.L.M., 1988, A quantitative structure–activity relationship for the acute toxicity of some epoxy compounds to the guppy. *Aquatic Toxicology* 13(3): 195–204.

Department of Toxic Substances Control (DTSC), 2004, Guidance for the evaluation and mitigation of subsurface vapor intrusion to indoor air. California Environmental Protection Agency, Department of Toxic Substances Control. http://www.dtsc.ca.gov (accessed November 11, 2006).

Deshpande, M.S., Rale, V.B., and Lynch, J.M., 1992, *Aureobasidium pullulans* in applied microbiology: A status report. *Enzyme and Microbial Technology* 14(7): 514–527.

Dilling, W.L., Bredeweg, C.J., and Tefertiller, N.B., 1976, Simulated atmospheric photodecomposition rates of methylene chloride, 1,1,1-trichloroethane, trichloroethylene, tetrachloroethylene, and other compounds. *Environmental Science and Technology* 10(4): 351–356.

Dilling, W.L., Tefertiller, N.B., and Kallos, G.J., 1975, Evaporation rates and reactivities of methylene chloride, chloroform, 1,1,1-trichloroethane, trichloroethylene, tetrachloroethylene, and other chlorinated compounds in dilute aqueous solutions. *Environmental Science and Technology* 9(9): 833–838.

Domenico, P. and Schwarz, F.W., 1997, *Physical and Chemical Hydrogeology*. New York: Wiley.

Dorfman, L.M. and Adams, G.E., 1973, Reactivity of the hydroxyl radical in aqueous solutions. National Bureau of Standards, National Standard Reference Data System.

Dow Chemical Company, 1980, *The Alkanolamine Handbook*. Midland, MI: Dow Chemical Company.

Dow Chemical Company, 2006, Chlorinated solvents—physical properties. Form No. 100-06358. www.chlorinatedsolvents.com (accessed May 29, 2006).

Dragun, J., 1988, *The Soil Chemistry of Hazardous Materials*. Silver Spring, MD: The Hazardous Materials Control Research Institute.

Edney, E.O. and Corse, E.W., 1987, *Hydroxyl Radical Rate Constant Intercomparison Study*. Research Triangle Park, NC: U.S. Environmental Protection Agency, Atmospheric Science Research Laboratory, EPA-600/S3-86/056.

ERSTG (Epoxy Resin Systems Task Group), 2002, High production volume (HPV) challenge program test plan and robust summaries for *n*-butyl glycidyl ether. U.S. Environmental Protection Agency, AR201-1400A.

Evans, P.J., Parales, R.E., and Parales, J.V., 2006, Bioreactors for treatment of 1,4-dioxane. Presented at Emerging Contaminants in Groundwater: A Continually Moving Target, June 8, 2006, Concord, CA. Groundwater Resources Association of California. http://www.grac.org.

Feenstra, S. and Guiguer, N., 1996, Dissolution of dense non-aqueous phase liquids (DNAPLs) in the subsurface. In: J.F. Pankow and J.A. Cherry (Eds), *Dense Chlorinated Solvents and Other DNAPLs in Groundwater: History, Behavior, and Remediation*, pp. 203–232. Portland, OR: Waterloo Press.

Fetter, C.W., 1993, *Contaminant Hydrogeology*. Englewood Cliffs, NJ: Prentice-Hall.

Fincher, E.L. and Payne, W.J., 1962, Bacterial utilization of ether glycols. *Applied and Environmental Microbiology* 10(6): 542–547.

Fisher Scientific, 1998, Material Safety Data Sheet for *n*-methyl pyrrole for 96-54-8, Fair Lawn, NJ. https://fscimage.fishersci.com/msds/43291.htm (accessed February 4, 2007).

Fisher Scientific, 1999, Material Safety Data Sheet for cyclohexane 99%, University of Wisconsin Nanotechnology Department, ACC# 05870.

Fisher Scientific, 2003, Material Safety Data Sheet for *tert*-amyl alcohol (2-methyl-2-butanol)—75-85-4, Fair Lawn, NJ.

Fisher Scientific, 2007, Material Safety Data Sheet for *n*-methyl morpholine. https://fscimage.fishersci.com/msds/95430.htm (accessed February 4, 2007).

Freeze, R.A. and Cherry, J.A., 1979, *Groundwater*. Englewood Cliffs, NJ: Prentice-Hall.

Freitas-Dinis, C.M., Geiger, H., and Wiesen, P., 2001, Kinetics of the reactions of OH(X$^2\Pi$) radicals with 1,3-dioxolane and selected dialkoxy methanes. *Physical Chemistry and Chemical Physics* 3: 2831–2835.

Gaffney, J.S., Streit, G.E., Spall, W.D., and Hall, J.H., 1987, Beyond acid rain. Do soluble oxidants and organic toxins interact with SO$_2$ and NO$_x$ to increase ecosystem effects? *Environmental Science and Technology* 21(6): 519–524.

Geiger, H., Maurer, T., and Becker, K.H., 1999, OH-initiated degradation mechanism of 1,4-dioxane in the presence of NO$_x$. *Chemical Physics Letters* 314: 465–471.

Goss, K.U., 2006, Prediction of the temperature dependency of Henry's law constant using poly-parameter linear free energy relationships. *Chemosphere* 64(8): 1369–1374.

Gossett, J.M., 1987, Measurement of Henry's law constants for C1 and C2 chlorinated hydrocarbons. *Environmental Science and Technology* 21(2): 202–208.

Grady, C.P.L., Jr., Sock, S.M., and Cowan, R.M., 1997, Biotreatability kinetics. In: G.S. Sayler, J. Sanseverino, and K.L. Davis (Eds), *Biotechnology in the Sustainable Environment*. New York: Plenum Press.

Granger, F.S. and Nelson, J.M., 1921, Oxidation and reduction of hydroquinone and quinone from the standpoint of electromotive-force measurements. *Journal of the American Chemical Society* 43(7): 1401–1415.

Haag, W.R. and Mill, T., 1988, Effect of a subsurface sediment on hydrolysis of haloalkanes and epoxides. *Environmental Science and Technology* 22(6): 658–663.

Hansch, C. and Leo, A.J., 1985, Medchem Project Issue No. 26. Pomona College, Medchem Project, Claremont, CA.

Harris, J.C., 1990a, Rate of aqueous photolysis. In: W.J. Lyman, W.F. Reehl, and D.H. Rosenblatt (Eds), *Handbook of Chemical Property Estimation Methods: Environmental Behavior of Organic Compounds,*, pp. 8-1–8-41. Washington, DC: American Chemical Society.

Harris, J.C., 1990b, Rate of hydrolysis. In: W.J. Lyman, W.F. Reehl, and D.H. Rosenblatt (Eds), *Handbook of Chemical Property Estimation methods: Environmental Behavior of Organic Compounds*, pp. 7-1–7-48. Washington, DC: American Chemical Society.

Hashimoto, Y., Tokura, K., Kishi, H., and Strachan, W.M.J., 1984, Prediction of seawater solubility of aromatic compounds. *Chemosphere* 13(8): 881–888.

Hassett, J.J., Means, J.C., Banwart, W.L., and Wood, S.G., 1980, *Sorption Properties of Sediments and Energy-Related Pollutants*. Athens, GA: U.S. Environmental Protection Agency, Office of Research and Development, Environmental Research Laboratory, EPA-600/3-80-041.

Hawaii Department of Health (HDOH), 2007, Use of laboratory batch tests to evaluate potential leaching of contaminants from soil (April 2007). Hawai'i Department of Health, Office of Hazard Evaluation and Emergency Response. http://www.hawaii.gov/health/environmental/hazard/eal2005.html (accessed February 21, 2007).

Hawley, G.G., 1977, *The Condensed Chemical Dictionary*. New York: Van Nostrand Reinhold Company.

Hawthorne, S.B., Slevers, R.E., and Barkley, R.M., 1985, Organic emissions from shale oil wastewaters and their implications for air quality. *Environmental Science and Technology* 19(10): 992–997.

Hefter, G.T., 1984, Solubility data. 2-Butanol-water system. In: A.F.M. Barton (Ed.), *Alcohols with Water*, Vol. 15. Research Triangle Park, NC: The International Union of Pure and Applied Chemistry (IUPAC).

Hemond, H.F. and Fechner, E.J., 1994, *Chemical Fate and Transport in the Environment*. New York: Academic Press.

Hentz, R.R. and Parrish, C.F., 1971, Photolysis of gaseous 1,4-dioxane at 1470 angstroms. *Journal of Physical Chemistry* 75(25): 3899–3901.

Hill, R.R., Jeffs, G.E., Morris, M.S., and Rebhun, M., 1997, Photocatalytic degradation of 1,4-dioxane in aqueous solution. *Journal of Photochemistry and Photobiology A: Chemistry* 108: 55–58.

Hine, J.H. and Mookerjee, P.K., 1975, Structural effects on rates and equilibriums. XIX. Intrinsic hydrophilic character of organic compounds. Correlations in terms of structural contributions. *Journal of Organic Chemistry* 40(3): 292–298.

Hoechst Celanese, 2006, Ethylene oxide user's guide. Celanese, Dow, Shell, Sunoco, Equistar. http://www.ethyleneoxide.com (accessed March 11, 2007).

Hoffman, F., Chiarappa, M.L., O'Boyle, J., Fox, K., and Bair, K., 1998, Diffusive transport of dissolved volatile organic compounds in sediments. U.S. Department of Energy, Lawrence Livermore National Laboratory Report UCRL-ID-129845.

Horvath, A.L., Getzen, F.W., and Maczynska, Z., 1999, IUPAC-NIST Solubility Data Series 67: Halogenated ethanes and ethenes with water. *Journal of Physical and Chemical Reference Data* 28(2): 395–629.

Houser, J.J. and Sibbio, B.A., 1977, Liquid-phase photolysis of dioxane. *Journal of Organic Chemistry* 42(12): 2145–2151.

Howard, P.H., Boethling, R.S., Jarvis, W.F., Meylan, W.M., and Michalenko, E.M., 1991, *Handbook of Environmental Degradation Rates*. Chelsea, MI: Lewis Publishers.

Huang, Z., 2007, *Insects in Chinese culture*. http://www.cyberbee.net/~huang/pub/insect.html (accessed September 9, 2007).

Huling, S.G., 1989, *Facilitated Transport: Superfund Groundwater Issue*. Ada, OK: U.S. Environmental Protection Agency, Office of Research and Development, EPA/540/4-89/003.

Huntsman, 2005, *Technical Bulletin: N-Methylmorpholine (NMM) CAS 109-02-4*. The Woodlands, TX: Huntsman Corporation.

Hyman, M., 1999, Final report on aerobic cometabolism of ether-bonded compounds. North Carolina State University, U.S. Environmental Protection Agency, Office of Research and Development, National Center for Environmental Research.

Inui, H., Akutsu, S., Itoh, K., Matsuo, M., and Miyamoto, J., 1979, Studies on biodegradation of 2,6-di-*tert*-butyl-4-methylphenol (BHT) in the environment. Part IV: The fate of ^{14}C-phenyl-BHT in aquatic model ecosystems. *Chemosphere* 8(6): 393–404.

Jackson, R.E. and Dwarakanath, V., 1999, Chlorinated degreasing solvents: Physical–chemical properties affecting aquifer contamination and remediation. *Ground Water Monitoring Review* 19(4): 102–110.

Jackson, R.E. and Mariner, P.E., 1995, Estimating DNAPL composition and VOC dilution from extraction well data. *Ground Water* 33(3): 407–414.

Japan Chemical Industry Ecology-Toxicology and Information Center, 1992, Biodegradation and bioaccumulation data of existing chemicals based on CSCL Japan. Compiled under the Supervision of Chemical Products Safety Division, Basic Industries Bureau MITI, ed. Chemicals Inspection and Testing Institute.

Johnson, P.C. and Ettinger, R.A., 1991, Heuristic model for predicting the intrusion of contaminant vapors into buildings. *Environmental Science and Technology* 25(8): 1445–1452.

Johnson, P.C., Kemblowski, M.W., and Johnson, R.L., 1999, Assessing the significance of subsurface contaminant vapor migration to enclosed spaces: Site-specific alternatives to generic estimates. *Journal of Soil Contamination* 8(33): 389–421.

Karickhoff, S.W., 1981, Semi-empirical estimation of sorption of hydrophobic pollutants on natural sediments and soils. *Chemosphere* 10(8): 833–846.

Kelley, S., Aitchison, E.W., Deshpande, M., Schnoor, J.L., and Alvarez, P.J.J., 2001, Biodegradation of 1,4-dioxane in planted and unplanted soil: Effect of bioaugmentation with *Amycolata* sp. CB1190. *Water Research* 35(16): 3791–3800.

Kim, Y.H. and Engesser, K.H., 2004, Degradation of alkyl ethers, aralkyl ethers, and dibenzyl ether by *Rhodococcus* sp. strain DEE5151, isolated from diethyl ether-containing enrichment cultures. *Applied and Environmental Microbiology* 70(7): 4398–4401.

Kinerson, R.S., 1987, Modelling the fate and exposure of complex mixtures of chemicals in the aquatic environment. In: G.C. Butler, V.B. Vouk, A.C. Upton, D.V. Parke, and S.C. Asher (Eds), *Methods for Assessing the Effects of Mixtures of Chemicals*, pp. 409–421. New York: Wiley.

Kirk, R. and Othmer, D., 1982, *Kirk–Othmer Encyclopedia of Chemical Technology*. New York: Wiley.

Klecka, G.M. and Gonsior, S.J., 1986, Removal of 1,4-dioxane from wastewater. *Journal of Hazardous Materials* 13(2): 161–168.

Knapp, J.S. and Bromley-Challenor, K.C.A., 2006, Recalcitrant organic compounds. http://www.bmb.leeds.ac.uk/mbiology/ug/ugteach/level3.html (accessed September 9, 2007).

Lapkin, M., 1965, Epoxides. In: Kirk, R.E. and Othmer, D.F., Eds, Encyclopedia of Chemical Technology, 2nd ed., Vol. 8, pp. 263–293, New York, John Wiley and Sons Inc.

Leighton, D.T. and Calo, J.M., 1981, Distribution coefficients of chlorinated hydrocarbons in dilute air–water systems for groundwater contamination applications. *Journal of Chemical & Engineering Data* 26(4): 382–385.

Lesage, S., Jackson, R.E., Priddle, M.W., and Riemann, P.G., 1990, Occurrence and fate of organic solvent residues in anoxic groundwater at the Gloucester landfill, Canada. *Environmental Science and Technology* 24(4): 559–566.

Lewis, G.N. and Randall, M., 1921, The activity coefficient of strong electrolytes. *Journal of the American Chemical Society* 43(5): 1112–1154.

Li, Z. and Pirasteh, A., 2006, Kinetic study of the reactions of atomic chlorine with several volatile organic compounds at 240–340 K. *International Journal of Chemical Kinetics* 38(6): 386–398.

Lide, D.R., Ed., 2000, *CRC Handbook of Chemistry and Physics*. Boca Raton, FL: CRC Press LLC.

Lide, D.R. and Kehiaian, H.V., 1994, *CRC Handbook of Thermophysical and Thermochemical Data*. Boca Raton, FL: CRC Press.

Little, J.C., Daisey, J.M., and Nazaroff, W.M., 1992, Transport of subsurface contaminants into buildings. *Environmental Science and Technology* 26(11): 2058–2066.

Liu, W.H., Medina, M.A., Jr., Thoman, W., Piver, W.T., and Jacobs, T.L., 2000, Optimization of intermittent pumping schedules for aquifer remediation using a genetic algorithm. *Journal of the American Water Resources Association* 36(6): 1335–1348.

Lyman, W.J., 1990, Adsorption coefficients for soils and sediments. In: W.J. Lyman, W.F. Reehl, and D.H. Rosenblatt (Eds), *Handbook of Chemical Property Estimation Methods: Environmental Behavior of Organic Compounds*, Chapter 4. Washington, DC: American Chemical Society.

Lyman, W.J., Reehl, W.F., and Rosenblatt, D.H., 1990, *Handbook of Chemical Property Estimation Methods: Environmental Behavior of Organic Compounds*. New York: McGraw-Hill.

MacEwan, D.M.C., 1948, Complexes of clays with organic compounds. I. Complex formation between montmorillonite and halloysite and certain organic liquids. *Transactions of the Faraday Society* 44: 349–367.

Mackay, D.M. and Cherry, J.A., 1989, Groundwater contamination: Pump and treat remediation. *Environmental Science and Technology* 23(6): 630–636.

Mackay, D., Shiu, W.Y., and Ma, K.C., 1993, *Illustrated Handbook of Physical–Chemical Properties and Environmental Fate for Organic Chemicals: Volume III. Volatile Organic Chemicals*. Boca Raton, FL: Lewis Publishers.

Mackison, F.W., Stricoff, R.S., and Partridge, L.J., Jr., Eds, 1981, *NIOSH/OSHA—Occupational Health Guidelines for Chemical Hazards*. Washington, DC: National Institute of Occupational Safety and Health (NIOSH).

Mahendra, S. and Alvarez-Cohen, L., 2005, *Pseudonocardia dioxanivorans* sp. nov., a novel actinomycete that grows on 1,4-dioxane. *International Journal of Systematic and Evolutionary Microbiology* 55: 593–598.

Mahendra, S. and Alvarez-Cohen, L., 2006, Kinetics of 1,4-dioxane biodegradation by monooxygenase-expressing bacteria. *Environmental Science and Technology* 40(17): 5435–5442.

Mahendra, S., Petzold, C.J., Baidoo, E.E., Keasling, J.G., and Alvarez-Cohen, L., 2007, Identification of the intermediates of *in vivo* oxidation of 1,4-dioxane by monooxygenase-containing bacteria. *Environmental Science and Technology* 41(21): 7330–7336.

Mallinckrodt Baker, 2008, Material Safety Data Sheet for *tert*-amyl alcohol—75-85-4, Phillipsburg, NJ.

Maurer, T., Hass, H., Barnes, I., and Becker, K.H., 1999, Kinetic and product study of the atmospheric photooxidation of 1,4-dioxane and its main reaction product ethylene glycol diformate. *Journal of Physical Chemistry A* 103: 5032–5039.

Maurino, V., Calza, P., Minero, C., Pelizzeti, E., and Vincente, M., 1997, Light-assisted 1,4-dioxane degradation. *Chemosphere* 35(11): 2675–2688.

Mazurkiewicz, J. and Tomasik, P., 2006, Why 1,4-dioxane is a water-structure breaker. *Journal of Molecular Liquids* 126: 111–116.

Mazzocchi, P.H. and Bowen, M.W., 1975, Photolysis of dioxane. *Journal of Organic Chemistry* 40(18): 2689–2690.

McAuliffe, C., 1966, Solubility in water of paraffin, cycloparaffin, olefin, acetylene, cycloolefin and aromatic hydrocarbon. *Journal of Physical Chemistry* 70(4): 1267–1275.

McCarty, P.L. and Reinhard, M., 1981, Trace organics in groundwater. *Environmental Science and Technology* 15(1): 40–51.

McGuire, M.J. and Suffett, I.H., 1978, Adsorption of organics from domestic water supplies. *Journal of the American Water Works Association* 70(11): 621–635.

Meylan, W.M. and Howard, P.H., 1991, Bond contribution method for estimating Henry's law constants. *Environmental Toxicology and Chemistry* 10(10): 1283–1293.

Meylan, W.M. and Howard, P.H., 1993, Computer estimation of the atmospheric gas-phase reaction rate of organic compounds with hydroxyl radicals and ozone. *Chemosphere* 26(12): 2293–2300.

Meylan, W.M., Howard, P.H., and Boethling, R.S., 1992, Molecular topology/fragment contribution method for predicting soil sorption coefficients. *Environmental Science and Technology* 26(8): 1560–1567.

Mill, T., Haag, W., Penwell, P., Pettit, T., and Johnson, H., 1987, *Environmental Fate and Exposure Studies Development of a PC-SAR for Hydrolysis: Esters, Alkyl Halides and Epoxides*. Menlo Park, CA: Stanford Research Institute, EPA Contract No. 68-02-4254.

Millington, R.J. and Quirk, J.P., 1961, Permeability of porous solids. *Transactions of the Faraday Society* 57: 1200–1207.

Nakamiya, K., Hashimoto, S., Ito, H., Edmonds, J.S., and Morita, M., 2005, Degradation of 1,4-dioxane and cyclic ethers by an isolated fungus. *Applied and Environmental Microbiology* 71(3): 1254–1258.

NLM (National Library of Medicine), 2006, Hazardous substance data bank: MEDLARS Online Information Retrieval System. National Library of Medicine. http://toxnet.nlm.nih.gov/ (accessed April 12, 2006).

NFPA (National Fire Protection Association), 1991, *National Fire Protection Guide: Fire Protection Guide to Hazardous Materials*. Quincy, MA: National Fire Protection Association.

Nguyen, T.H., Goss, K.U., and Ball, W.P., 2005, Polyparameter linear free energy relationships for estimating the equilibrium partition of organic compounds between water and the natural organic matter in soils and sediments. *Environmental Science and Technology* 39(4): 913–924.

NICNAS (National Industrial Chemicals Notification and Assessment Scheme), 1998, 1,4-Dioxane: Priority Existing Chemical No. 7. Canberra, Commonwealth of Australia: National Occupational Health and Safety Commission.

NICNAS (National Industrial Chemicals Notification and Assessment Scheme), 2003, *Full Report: 2,6-di-t-Butyl-4-Methyl Phenol*. Canberra, Commonwealth of Australia: N. I. C. N. A. S. (NICNAS), National Occupational Health and Safety Commission.

Niederer, C. and Goss, K.U., 2007, Quantum chemical modeling of humic acid/air equilibrium partitioning of organic vapors. *Environmental Science and Technology* 41(10): 3646–3651.

Niederer, C., Goss, K.U., and Schwarzenbach, R.P., 2006a, Sorption equilibrium of a wide spectrum of organic vapors in leonardite humic acid: Experimental setup and experimental data. *Environmental Science and Technology* 40(17): 5368–5373.

Niederer, C., Goss, K.U., and Schwarzenbach, R.P., 2006b, Sorption equilibrium of a wide spectrum of organic vapors in leonardite humic acid: Modeling of experimental data. *Environmental Science and Technology* 40(17): 5374–5379.

Nielsen, O.J., Sidebottom, H.W., O'Farrell, D.J., Donlon, M., and Treacy, J., 1989, Rate constants for the gas-phase reactions of OH radicals and Cl atoms with $CH_3CH_2NO_2$, $CH_3CH_2CH_2NO_2$, $CH_3CH_2CH_2CH_2NO_2$, and $CH_3CH_2CH_2CH_2CH_2NO_2$. *Chemical Physics Letters* 156(4): 312–318.

NIOSH, 1978, *Occupational Health Guideline for tert-Butyl Alcohol*. Washington, DC: National Institute for Occupational Safety and Health.

OECD (Organisation for Economic Cooperation and Development), 1981, Inherent biodegradability: Modified MITI test (II). OECD Protocol 302C, adopted May 12, 1981. Paris, France: Organisation for Economic Cooperation and Development.

Osborn, A.G. and Scott, D.W., 1980, Vapor pressures of 17 miscellaneous organic compounds. *The Journal of Chemical Thermodynamics* 12(5): 429–438.

Pan, S. and Chen, H., 2006, Anaerobic degradation of 1,4-dioxane under humic-reducing or Fe(III)-reducing conditions. Presented at the 231st ACS National Meeting, Atlanta, GA. American Chemical Society, Abstract ENVR 144.

Pankow, J.F. and Cherry, J.A., 1996, *Dense Chlorinated Solvents and Other DNAPLs in Groundwater: History, Behavior, and Remediation*. Portland, OR: Waterloo Press.

Parales, R.E., Adamus, J.E., White, N., and May, H.D., 1994, Degradation of 1,4-dioxane by an actinomycete in pure culture. *Applied Environmental Microbiology* 60(12): 4527–4530.

Park, J.H., Hussam, A., Couasnon, P., Fritz, D., and Carr, P.W., 1987, Experimental reexamination of selected partition coefficients from Rohrschneider's data set. *Analytical Chemistry* 59(15): 1970–1976.

Parker, B.L., Cherry, J.A., and Gilham, R.W., 1996, The effects of molecular diffusion on DNAPL behavior in fractured porous media. In: J.F. Pankow and J.A. Cherry (Eds), *Dense Chlorinated Solvents and Other DNAPLs in Groundwater: History, Behavior, and Remediation*, pp. 355–393. Portland, OR: Waterloo Press.

Patt, T.E. and Abebe, H., 1995, United States Patent 5,399,495: Microbial degradation of chemical pollutants. Assignee: The Upjohn Company, Kalamazoo, MI.

Patterson, R.J., Jackson, R.E., Graham, B.W., Chaput, D., and Priddle, M.W., 1985, Retardation of toxic chemicals in a contaminated outwash aquifer. *Water Science and Technology* 17: 57–69.

Patty, F., 1963, *Industrial Hygiene and Toxicology: Volume II. Toxicology*. New York: Wiley-Interscience Publishers.

Pavan, M. and Worth, A.P., 2006, Review of QSAR models for ready biodegradation. European Commission Directorate General Joint Research Centre, European Chemicals Bureau.

Payne, F.C., Quinnan, J.A., and Potter, S.T., 2008, *Remediation Hydraulics*. Boca Raton, FL: Taylor & Francis Group.

Payne, F.C., Suthersan, S.S., Molnaa, B., and Davis, S., 2003, Developing in situ reactive zone strategies for 1,4-dioxane. Presented at 1,4-Dioxane and other solvent stabilizer compounds in the environment, December 8, 2003, San Jose, CA. Groundwater Resources Association of California. http://www.grac.org.

Perrin, D.D., 1972, *Dissociation Constants of Organic Bases in Aqueous Solution: Supplement*. London: Butterworths.

Pickett, L.W., Hoeflich, N.J., and Liu, T.C., 1951, The vacuum ultraviolet absorption spectra of cyclic compounds. II. Tetrahydrofuran, tetrahydropyran, 1,4-dioxane and furan. *Journal of the American Chemical Society* 73(10): 4865–4869.

Platz, J., Sehested, J., Mgelberg, T., Nielsen, O.J., and Wallington, T.J., 1997, Atmospheric chemistry of 1,4-dioxane: Laboratory studies. *Journal of the Chemical Society, Faraday Transactions* 93(16): 2855–2963.

Priddle, M.W. and Jackson, R.E., 1991, Laboratory column measurement of VOC retardation factors and comparison with field values. *Ground Water* 29(2): 260–266.

Pruden, A., Sedran, M.A., Suidan, M.T., and Venosa, M.D., 2005, Anaerobic biodegradation of methyl *tert*-butyl ether under iron-reducing conditions in batch and continuous-flow cultures. *Water Research* 77(3): 297–303.

Reid, E.W. and Hoffman, H.E., 1929, 1,4-Dioxan. *Industrial and Engineering Chemistry* 21(7): 695–697.

Reinhard, M. and Drefahl, A., 1999, *Handbook for Estimating Physicochemical Properties of Organic Compounds*. New York: Wiley.

Renner, R., 2002, The K_{ow} controversy: Doubts about the quality of basic physicochemical data for hydrophobic organic compounds could be undermining many environmental models and assessments. *Environmental Science and Technology* 36(21): 411A–413A.

Riddick, J.A., Bunger, W.B., and Sakano, T.K., 1985, *Techniques of Chemistry, 4th Edition: Volume II. Organic Solvents*. New York: Wiley.

Roy, D., Anagnostu, G., and Chaphalkar, P., 1994, Biodegradation of dioxane and diglyme in industrial waste. *Journal of Environment Science and Health, Part A: Environmental Science* 29(1): 129–147.

Roy, D., Anagnostu, G., and Chaphalkar, P., 1995, Analysis of respirometric data to obtain kinetic coefficients for biodegradation of 1,4-dioxane. *Journal of Environment Science and Health, Part A: Environmental Science* 30(8): 1775–1790.

Sander, R., 1999, Compilation of Henry's law constants for inorganic and organic species of potential importance in environmental chemistry. Max-Planck Institute of Chemistry, Air Chemistry Department, Mainz, Germany. http://www.mpch-mainz.mpg.de/~sander/res/henry.html (accessed May 6, 2007).

Sax, N.I., 1984, *Dangerous Properties of Industrial Materials*. New York: Van Nostrand Reinhold Company.

Sax, N.I. and Lewis, R.J.S., 1987, *Hawley's Condensed Chemical Dictionary*. New York: Van Nostrand Reinhold Company.

Schultz, T.W., 1986, The use of the ionization constant (pK_a) in selecting models of toxicity in phenols. *Ecotoxicology and Environmental Safety* 14(2): 178–183.

Schwarzenbach, R.P., Gschwend, P.M., and Imboden, D.M., 1993, *Environmental Organic Geochemistry*. New York: Wiley.

Schwarzenbach, R.P. and Westall, J., 1981, Transport of nonpolar organic compounds from surface water to groundwater: Laboratory sorption studies. *Environmental Science and Technology* 15(11): 1360–1367.

Scow, K.M., 1982, Rate of biodegradation. In: W.J. Lyman, W.F. Reehl, and D.H. Rosenblatt (Eds), *Handbook of Chemical Property Estimation Methods*, pp. 9-1–9-85. Washington, DC: American Chemical Society.

Seinfeld, J.H., 1986, *Atmospheric Chemistry and Physics of Air Pollution*. New York: Wiley.

Serjeant, E.P. and Dempsey, B., 1979, *Ionisation Constants of Organic Acids in Aqueous Solution*. New York: Pergamon Press.

Shen, T.T., 1981, Estimating hazardous air emissions from disposal sites. *Pollution Engineering* 13(1981): 31–35.

Shen, W., Chen, H., and Pan, S., 2008, Anaerobic biodegradation of 1,4-dioxane by sludge enriched with iron-reducing microorganisms. *Bioresource Technology* 99(7): 2483–2487.

Shiu, W.Y., Ma, K.C., Varhanickova, D., and Mackay, D., 1994, Chlorophenols and alkylphenols: A review and correlation of environmentally relevant properties and fate in an evaluative environment. *Chemosphere* 29(6): 1155–1224.

Skadsen, J.M., Rice, B.L., and Meyering, D.J., 2004, The occurrence and fate of pharmaceuticals, personal care products, and endocrine disrupting compounds in a municipal water use cycle: A case study in the City of Ann Arbor. Water Utilities, City of Ann Arbor, and Fleis & VandenBrink Engineering, Inc.

Skinner, K.M., 2007, Characterization of the molecular foundations and biochemistry of alkane and ether oxidation in a filamentous fungus, a *Graphium* species. PhD Dissertation, Oregon State University, Corvallis, OR.

Smallwood, I.M., 1996, *Handbook of Organic Solvent Properties*. New York: Halsted Press.

Snider, J.R. and Dawson, G.A., 1985, Tropospheric light alcohols, carbonyls, and acetonitrile: Concentrations in the southwestern United States and Henry's law data. *Journal of Geophysical Research* 90(D2): 3797–3805.

Snoeyink, V.L., 1999, Adsorption of organic compounds. In: F.W. Pontius (Ed.), *Water Quality & Treatment Handbook*, pp. 781–875. New York: McGraw-Hill.

Snyder, R., Ed., 1992, *Ethel Browning's Toxicity and Metabolism of Industrial Solvents*, 2nd Edition, Vol.3: Alcohols and esters. New York: Elsevier.

Sock, S.M., 1993, A comprehensive evaluation of biodegradation as a treatment alternative for the removal of 1,4-dioxane. Master of Science Thesis, Clemson University, Department of Environmental Engineering and Science, Clemson, SC.

Southworth, G.R. and Keller, J.L., 1986, Hydrophobic sorption of polar organics by low organic carbon soils. *Journal of Water, Air, & Soil Pollution* 28(3–4): 239–247.

SPI (Society of the Plastics Industry), 2006, Test plans for alkyl (C_{12}–C_{14}) glycidyl ether (CAS No. 68609-97-2) and *n*-butyl glycidyl ether (CAS No. 2426-08-6). Society of the Plastics Industry/Epoxy Resin Systems Task Group, Chemical Right-to-Know—HPV Challenge Program.

SRC (Syracuse Research Corporation), 2007a, BIODEG. Syracuse Research Corporation.

SRC (Syracuse Research Corporation), 2007b, ChemFate. Syracuse Research Corporation.

Srinivasan, V., Clement, T.P., and Lee, K.K., 2007, Domenico solution—is it valid? *Ground Water* 45(2): 136–146.

Steffan, R.J., 2006, *Biodegradation of Ether-Containing Pollutants*. Arlington, VA: Federal Remediation Technologies Roundtable. http://www.frtr.gov/ (accessed August 2, 2007).

Steinkraus, D.C. and Whitefield, J.B., 1994, Chinese caterpillar fungus and world record runners. *American Entomologist* 40(4): 235–239.

Suthersan, S.S., 2002, *Natural and Enhanced Remediation Systems*. Boca Raton, FL: Lewis Publishers.

Swann, R.L., Laskowski, D.A., McCall, P.J., and Vander Kuy, K., 1983, A rapid method for the estimation of the environmental parameters octanol/water partition coefficient, soil sorption constant, water to air ratio and water solubility. *Residue Reviews* 85: 17–28.

Taylor, S.W., Lange, C.R., and Lesold, E.A., 1997, Biofouling of contaminated ground-water recovery wells: Characterization of microorganisms. *Ground Water* 35(6): 973–980.

Thibodeaux, L.J., 1996, *Environmental Chemodynamics*. New York: Wiley.

Thomas, R.G., 1990, Volatilization from water. In: W.J. Lyman, W.F. Reehl, and D.H. Rosenblatt (Eds), *Handbook of Chemical Property Estimation Methods*, pp. 15-1–15-34. Washington, DC: American Chemical Society.

UNEP (United Nations Environment Programme), 1998, *Screening Information Data Set (SIDS) for 2-Methyl-3-yn-2-ol*. Nairobi, Kenya: United Nations Environment Programme Report 115-19-5. http://www.chem.unep.ch/irptc/sids/OECDSIDS/115195.pdf (accessed June 3, 2007).

UNEP (United Nations Environment Programme), 2002, Screening information data set (SIDS) for 2,6-di-*tert*-butyl-*p*-cresol [betahydroxy toluene or BHT]: Initial assessment report for SIDS initial assessment meeting (SIAM), 14th meeting, Paris, France, March 26–28, 2002.

U.S. Coast Guard, 1984, *CHRIS—Hazardous Chemical Data*, Vol. II. Washington, DC: Department of Transportation, U.S. Government Printing Office.

U.S. Environmental Protection Agency (USEPA), 1981, Phase I risk assessment of 1,4-dioxane. U.S. Environmental Protection Agency Contract No. 68-01-6030.

U.S. Environmental Protection Agency (USEPA), 1996, Requirements for management of hazardous contaminated media: Appendix A. Summary of physical/chemical properties. 40 Code of Federal Regulations Parts 260, 261, 262, 264, 268, 269 and 271, Vol. 61, No. 83. Government Printing Office.

U.S. Environmental Protection Agency (USEPA), 1998a, *Industrial Waste Air Model Technical Background Document*. Washington, DC: U.S. Environmental Protection Agency, Office of Solid Waste, EPA 530-R-99-004.

U.S. Environmental Protection Agency (USEPA), 1998b, *ORD/OSW Integrated Research and Development Plan for the Hazardous Waste Identification Rule*. Washington, DC: U.S. Environmental Protection Agency, Office of Research and Development and Office of Solid Waste, EPA-HQ-RCRA-1999-0074.

U.S. Environmental Protection Agency (USEPA), 1999, Use of monitored natural attenuation at Superfund, RCRA corrective action, and underground storage tank sites. U.S. Environmental Protection Agency, Office of Solid Waste and Emergency Response, Directive 9200.4-17P.

U.S. Environmental Protection Agency (USEPA), 2000a, Aqueous Hydrolysis Rate Program (HYDROWIN). A module in the EPIWIN Suite, U.S. Environmental Protection Agency.

U.S. Environmental Protection Agency (USEPA), 2000b, High production volume challenge program submission for 1,3,5-trioxane by Trioxane Manufacturer's Consortium. Prepared by Toxicology and Regulatory Affairs, Flemington, NJ. http://www.epa.gov/hpv/pubs/summaries/triox/c12863.pdf (accessed January 4, 2009).

U.S. Environmental Protection Agency (USEPA), 2001a, *User's Guide for Water9 Software Version 2.0.0*. Research Triangle Park, NC: U.S. Environmental Protection Agency, Office of Air Quality Planning and Standards.

U.S. Environmental Protection Agency (USEPA), 2001b, Supplemental guidance for developing soil screening levels for Superfund sites (Peer Review Draft, May 2001). U.S. Environmental Protection Agency, Office of Solid Waste and Emergency Response, Publication 9355.4–24.

U.S. Environmental Protection Agency (USEPA), 2007a, Estimations programs interface for windows (EPI Suite), v. 3.20. http://www.epa.gov/oppt/exposure/pubs/episuite.htm (accessed November 3, 2007).

U.S. Environmental Protection Agency (USEPA), 2007b, CSMoS comments on the potential limitations of the Domenico-based fate and transport models. Ground Water and Ecosystems Restoration Research. http://www.epa.gov/ada/csmos/domenico.html (accessed September 3, 2007).

U.S. Environmental Protection Agency (USEPA), 2007c, High Production Volume Information System (HPVIS) online database. http://www.epa.gov/chemrtk/hpvis/index.html (accessed November 3, 2007).

Vainberg, S., McClay, K., Masuda, H., Root, D., Condee, C., Zylstra, G.J., and Steffan, R.J., 2006, Biodegradation of ether pollutants by *Pseudonocardia* sp. strain ENV478. *Applied and Environmental Microbiology* 72(8): 5218–5224.

Verschueren, K., 1996, *Handbook of Environmental Data on Organic Chemicals*, 3rd Edition. New York: Van Nostrand Reinhold Company.

Vidic, R.D. and Suidan, M.T., 1991, Role of dissolved oxygen on the adsorptive capacity of activated carbon for synthetic and natural organic matter. *Environmental Science and Technology* 25: 1612–1618.

Walton, B.T., Hendricks, M.S., and Anderson, T.A., 1992, Soil sorption of volatile and semivolatile organic compounds in a mixture. *Journal of Environmental Quality* 21: 552–558.

Wasik, S.P., Tewari, Y.B., Miller, M.M., and Martire, D.E., 1981, *Octanol/Water Partition Coefficients and Aqueous Solubilities of Organic Compounds*. Washington, DC: U.S. Department of Commerce, National Bureau of Standards, NBS TR81-2406.

Weber, R., Parker, P., and Bowser, M., 1981, Vapor pressure distribution of selected organic chemicals. U.S. Environmental Protection Agency, Office of Research and Development, EPA/600/2-81/021.

West, M.R., Kueper, B.H., and Ungs, M.J., 2007, On the use and error of approximation in the Domenico, 1987 solution. *Ground Water* 45(2): 126–135.

Whiffin, R.B. and Bahr, J.H., 1984, Assessment of purge well effectiveness for aquifer decontamination. *Proceedings of the Fourth National Symposium on Aquifer Restoration and Ground Water Monitoring*. National Water Well Association, Las Vegas, NV, May 23–25.

White, G.F., Russell, N.J., and Tidswell, E.C., 1996, Bacterial scission of ether bonds. *Microbiology and Molecular Biology Reviews* 60(1): 216–232.

Wilke, C.R. and Chang, P., 1955, Correlation of diffusion coefficients in dilute solutions. *AIChE Journal* 1(2): 264–272.

Willis, G.H. and McDowell, L.L., 1982, Pesticides in agricultural runoff and their effects on downstream water quality. *Environmental Toxicology and Chemistry* 1: 267–279.

Windolz, M., Ed., 1983, *The Merck Index*. Rahway, NJ: Merck & Co., Inc.

Woo, Y.T., Argus, M.F., and Arcos, J.C., 1977, Metabolism *in vivo* of dioxane: Effect of inducers and inhibitors of hepatic mixed-function oxidases. *Biochemical Pharmacology* 26(16): 1539–1542.

Wu, J., Low, P.F., and Roth, C.B., 1994, Effect of 1,4-dioxane on the expansion of montmorillonite layers in montmorillonite/water systems. *Clays and Clay Minerals* 42(2): 109–113.

Yalkowsky, S.H. and Dannenfelser, R.M., 1992, The AQUASOL database of aqueous solubility. University of Arizona, College of Pharmacy, Tucson.

Yaws, C.L., 1994, *Handbook of Vapor Pressure: Volume 2. C_5 to C_7 Compounds*. Houston, TX: Gulf Publishing Company.

Young, D.F. and Ball, W.P., 1998, Estimating diffusion coefficients in low-permeability porous media using a macropore column. *Environmental Science and Technology* 32(17): 2578–2584.

Zenker, M.J., 2006, 1,4-Dioxane biodegradation—a review. In: A.R. Gavaskar and A.S.C. Chen (Eds), *Remediation of Chlorinated and Recalcitrant Compounds*. Columbus, OH: Battelle Press.

Zenker, M.J., Borden, R.C., and Barlaz, M.A., 1999, Investigation of the intrinsic biodegradation of alkyl and cyclic ethers. *Proceedings of the Fifth International In Situ and On-Site Bioremediation Symposium*, San Diego, CA, April 19–22. Columbus, OH: Battelle Press.

Zenker, M.J., Borden, R.C., and Barlaz, M.A., 2000, Mineralization of 1,4-dioxane in the presence of a structural analog. *Biodegradation* 11(4): 239–246.

Zenker, M.J., Borden, R.C., and Barlaz, M.A., 2002, Modeling cometabolism of cyclic ethers. *Environmental Engineering Science* 19(4): 215–228.

Zenker, M.J., Borden, R.C., and Barlaz, M.A., 2003, Occurrence and treatment of 1,4-dioxane in aqueous environments. *Environmental Engineering Science* 20(5): 423–432.

Zenker, M.J., Borden, R.C., and Barlaz, M.A., 2004, Biodegradation of 1,4-dioxane using trickling filter. *Journal of Environmental Engineering* 130(9): 926–931.

Zepp, R.G. and Cline, D.M., 1977, Rates of direct photolysis in aquatic environment. *Environmental Science and Technology* 11(4): 359–366.

Zhang, Z.Z., Low, P.F., Cushman, J.H., and Roth, C.B., 1990, Adsorption and heat of adsorption of organic compounds on montmorillonite from aqueous solutions. *Soil Science Society of America Journal* 54(1): 59–66.

4 Sampling and Laboratory Analysis for Solvent Stabilizers

Thomas K.G. Mohr

Investigators characterizing solvent-release sites may overlook 1,4-dioxane because (1) they are not required to look for it and (2) they do not find a compelling reason to sample for chemical compounds that are not among the known or regulated threats to drinking water and human health. Awareness of the widespread occurrence of 1,4-dioxane as a co-contaminant of methyl chloroform is still growing, but is not yet widespread; consequently, most site investigators do not direct the laboratory to analyze for 1,4-dioxane. A sensitive and reliable laboratory analytical method was not available until 1997, when the California Department of Health Services developed a method for low-level detection of 1,4-dioxane (Draper et al., 2000). Other solvent-stabilizer compounds may also be overlooked because (1) most investigations of solvent-release sites focus on analyzing for familiar chlorinated aliphatic hydrocarbon compounds and (2) the methods used are unsuitable for trace detection and quantification of ethers, alcohols, and other highly soluble compounds that have been used as solvent stabilizers.

The challenge to analyze for 1,4-dioxane in water is a direct consequence of its miscibility with water, its low water–air partition coefficient, and its low affinity to organic carbon. Understanding the challenges faced by laboratory chemists helps remedial project managers make better use of available laboratory analytical methods. This chapter is written for the "layperson," that is, engineers, geologists, and other people interested in contaminated site remediation who are laboratory clients but who have relatively limited familiarity with the subtleties of laboratory analysis. This chapter also includes detailed information on 1,4-dioxane analysis for the benefit of laboratory chemists.

Sampling methods can make an important difference in the results of analyzing for hydrophilic compounds. The requirements of sampling for 1,4-dioxane by routine or innovative methods are summarized here. By way of review, the basic principles of gas chromatography (GC), mass spectrometry (MS), and sample preparation are briefly discussed here in the context of 1,4-dioxane analysis.

This chapter demonstrates that in laboratory analysis of environmental samples, the chemist is faced with myriad considerations for how the many available combinations of different options in each analytical step best serve to isolate and quantify the contaminants of interest. This universe of possibilities makes analytical chemistry a fascinating science; indeed, its skillful execution deserves the same appreciation as fine art!

4.1 THE FLAWED PARADIGM OF ANALYTE LISTS

To satisfy regulatory requirements and generate comparable results, project managers often require the use of standard methods for analysis of samples from sites contaminated by hazardous wastes. These methods are usually selected from the U.S. Environmental Protection Agency (USEPA) *Test Methods for Evaluating Solid Waste* [SW-846 (see USEPA, 2006a, and earlier editions)]. These methods are designed to reliably report accurate results for a limited list of compounds that behave similarly and that can be analyzed by using the same general approach and instrument parameters.

Analyzing for solvents from standard lists of priority pollutants will overlook important compounds that may be equally or more toxic and mobile than those that the site investigator seeks. Therefore, more information could be obtained by focusing the laboratory phase of an investigation on a wider range of organic compounds than is included on the standard method list. For example, the typical laboratory report for Environmental Protection Agency (EPA) Method 8260 (USEPA, 2006c), "Volatile Organic Compounds by Gas Chromatography/Mass Spectrometry (GC/MS)," includes far fewer than the 94 compounds listed in Method 8260; often only 33 compounds, or an extended list of 62 compounds, will be reported by commercial laboratories. The target compounds in EPA Method 8260 analyses at commercial laboratories most often exclude 1,4-dioxane. The compounds reported can vary from laboratory to laboratory and project to project. Project managers may specify which target analytes to include for each method requested (Simmons, 1997). The choices of sample preparation, chromatographic column, detector, calibration standards, and mass spectrometer settings all affect the ability of a particular analysis to detect and quantify compounds. In order to favor detection of chlorinated solvents at low concentrations, other compounds are typically excluded from reporting.

For laboratory methods using GC–MS, the chemist has the option of looking at nontarget peaks and making tentative identifications by comparison with a library of mass spectra. Tentatively identified compounds (TICs) are usually reported with an estimated concentration, because the instrument is not calibrated for TICs. The USEPA's Contract Laboratory Program (CLP) requires that laboratories report up to 20 TICs per sample. Some laboratories have fixed prices for reporting the 10 or 20 most commonly encountered nontarget compounds (Simmons, 1997).

A variety of ancillary benefits can be obtained from knowing the full complement of anthropogenic compounds in groundwater samples at solvent-release sites. As discussed in Chapter 3, the chlorinated solvents targeted in an investigation are often recalcitrant and retarded relative to groundwater flow rates. Hydrophilic compounds introduced as solvent stabilizers can be helpful for mapping the full extent of a plume, as they are retarded to a much smaller extent and therefore behave like a conservative tracer. Knowing the migration pathways traversed by the more mobile contaminants can be useful for predicting the likely direction of migration of less mobile contaminants released at the same time. Further, parameters of plume dispersion such as dispersivity can be inferred from the distribution of the most mobile contaminant. Knowledge of these migration parameters aids modeling site history and remedial alternatives. Tracking molar ratios of chlorinated solvents (which biodegrade slowly under certain ambient conditions) to 1,4-dioxane (which is not expected to biodegrade under ambient conditions) can also be instructive for analysis of monitored natural attenuation or enhanced biodegradation as a remediation strategy for chlorinated solvents.

The consequences of *not* analyzing for a wider variety of analytes and *not* requesting that TICs be reported at solvent-release sites may include the following: (1) The site characterization may be incomplete (see Chapter 8); (2) the capture zone and treatment system design may be ineffective in protecting drinking water (see Chapter 8); (3) the risks to drinking water consumers may not be fully addressed (see Chapter 6); (4) adjudication of source allocation in legal disputes may be erroneous (see Chapter 9).

Unfortunately, 1,4-dioxane is not a benign tracer. At sufficiently high concentrations, it can cause health effects when consumed, as explained in Chapter 5. Therefore, an expanded analysis of samples for 1,4-dioxane at solvent-release sites should be considered.

In several conference presentations, Christian Daughton of the USEPA summarized the inherent filtering process that results from following the paradigm of restricting analyses by method analyte lists. Daughton has noted that nearly 28 million organic and inorganic substances had been documented as of April 2006 (as indexed by the American Chemical Society's Chemical Abstracts Service in their CAS Registry); 10 million of these are commercially available. Only 240,000 of these compounds are subjected to any form of regulation. Of these 240,000, a much smaller number of chemicals are subjected to routine laboratory analysis of drinking water or environmental samples. A still smaller number of chemicals have been subjected to sufficient study to provide reliable information about health effects (Daughton, 2005).

Daughton (2003, p. 758) has offered several deceptively simple but profoundly important insights into the limitations of environmental analyses:

- Not everything that can be measured is worth measuring, and not everything worth measuring is measurable
- What one finds usually depends on what one aims to search for
- Only those compounds targeted for monitoring have the potential for being identified and quantified
- Those compounds not targeted will elude detection
- The spectrum of pollutants identified in a sample represents but a portion of those present and they are of unknown overall risk significance

These axioms apply very well to solvent stabilizers and help to explain why after more than 20 years of cleanup at solvent-release sites, 1,4-dioxane is only now being discovered. Figure 4.1 summarizes the manner in which the universe of possible chemicals present in the environment is narrowed to the target analyte list by USEPA methods.

The tendency for 1,4-dioxane to be overlooked by routine approaches for analyzing samples for organic compounds was first noted in 1977. In a review article by William T. Donaldson of the USEPA, the basis for establishing analyte lists was questioned. Donaldson noted that a list of 50 compounds *expected* to occur because of their known use in industry did not match well with a survey of 5500 analyses of water samples. Whereas many of the compounds expected in water were not found, 1,4-dioxane was observed in 15 analyses, and nitromethane was observed in 10 samples (Donaldson, 1977).

4.2 SAMPLE COLLECTION, PRESERVATION, AND HANDLING FOR ANALYSIS OF 1,4-DIOXANE

EPA methods that include 1,4-dioxane as an analyte include EPA Method 8260B, EPA Method 8270C, and EPA Method 1624. Sample-preservation requirements for these methods follow the general protocol for all analytes obtained from those methods, as summarized in Table 4.1.

To ensure data validity, it is best to follow the sample container, preservation, and holding time protocols for the EPA method that will be used for 1,4-dioxane analysis or the EPA method upon which the commercial laboratory's adaptation is based. Although acidic preservatives are required for EPA methods for volatile organic compound (VOC) analysis, they are not necessary for 1,4-dioxane. 1,4-Dioxane extraction and recovery are independent of pH (Yoo et al., 2002). Biodegradation of 1,4-dioxane in samples is not likely to be a concern, given the absence of evidence for inherent biodegradation of 1,4-dioxane under ambient conditions (Vainberg et al., 2006).

4.2.1 SAMPLING FOR 1,4-DIOXANE IN WATER

Conventional groundwater sampling approaches involve a one-stop visit to a well site for sample retrieval with a bailer or a submersible sampling pump, after purging three to five well-casing volumes or enough groundwater so that consecutive readings of pH and conductivity are constant. Other systems include dedicated tubing that pulls samples directly from depth-discrete zones in a monitoring well through the use of a peristaltic pump or a nitrogen lift bladder pump. Another approach for sampling soluble organic compounds is the passive-diffusion concept, which provides the basis for a variety of no-purge sampling devices (Parsons, 2005; O'Neill, 2006). These include the polyethylene diffusion bag (PDB; also called passive-diffusion bag), rigid porous polyethylene (RPP), polysulfone membrane, regenerated cellulose, Snap Sampler™, and Hydrasleeve™. All of

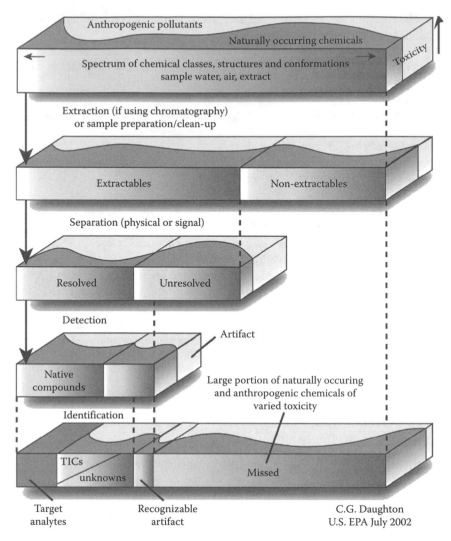

FIGURE 4.1 Limitations and complexity of environmental chemical analysis. TICs = tentatively identified compounds. (From Daughton, C.G., 2006, Pharmaceuticals as environmental contaminants: An overview of the science. Paper presented at the Water Education Foundation Pacific Conference. With permission.)

these cost less than conventional sampling. Performance data for sampling 1,4-dioxane are available for most of these samplers. Polyethylene diffusion bag samplers have been studied extensively and are widely used. Performance data for PDBs can be found online.*

This section briefly examines how some of these passive-diffusion samplers are used and how their results compare for 1,4-dioxane sampling. Diffusion methods determine the concentration from the mass of analyte per volume of sample in the same way as grab samples obtained from a bailer or sampling pump; however, diffusion methods require that the volume within the sampler comes into mass equilibrium with the groundwater outside the sampler. It is therefore necessary to leave the samplers in the monitoring well for as long as two weeks.

---

* See the Interstate Technology & Regulatory Council Web site (http://www.itrcweb.org/) or the EPA Hazardous Waste Cleanup Information site (http://www.clu-in.org).

TABLE 4.1
EPA Recommended Sample Containers, Preservatives, and Holding Times

Method	Container	Preservation	Holding Time	Notes
8260B	Standard 40 mL glass screw-cap VOA vials with Teflon-lined silicone septa	Cool to 4°C and adjust pH to less than 2 with H_2SO_4, HCl, or $NaHSO_4$	14 days	
8270C	1 L amber glass with Teflon-lined silicone septa	Cool to 4°C	Extract within 7 days; analyze within 40 days of extraction	3 mL 10% sodium thiosulfate per gallon if sample is chlorinated
1624	1 L amber glass with Teflon-lined silicone septa	Cool to 4°C; acidify to pH 2 with HCl (1:1) for aromatics	Analyze within 14 days of collection	Sodium thiosulfate (10 mg per 40 mL) recommended if sample of drinking water is chlorinated

Source: U.S. Environmental Protection Agency (USEPA), 2006a, Test Methods SW-846 on line. http://www.epa.gov/epaoswer/hazwaste/test/main.htm (accessed May 4, 2005).

Note: VOA, Volatile organic analysis.

4.2.1.1 Passive-Diffusion Bag Samplers

PDB samplers are narrow plastic sampling bags made of low-density polyethylene (LDPE), typically 18 in. long. The bags are closed at the ends after filling with pure deionized water and then lowered into a monitoring well to a specified depth for a period of 1–2 weeks. During the sampling interval, the deionized water comes into equilibrium with the surrounding well water by diffusion through the semipermeable polyethylene sampling bag, which has a pore size of less than 0.001 μm (10 Å, or 1×10^{-6} mm). Multiple PDBs can be lowered into a single monitoring well with a long screen to characterize vertical stratification of contaminants within a monitoring well screen and to reveal variations in flow rates along different portions of the screen. The primary analytes in PDB sampling programs are VOCs in groundwater monitoring wells. PDBs can also be used to monitor the discharge of contaminated groundwater to surface water (Vroblesky et al., 2000; O'Neill, 2006).

Unfortunately, 1,4-dioxane is not amenable to sampling using PDB samplers. In a field trial comparing different no-flow samplers, results for 1,4-dioxane in a PDB sampler were compared to results for samples obtained by other no-flow samplers and conventional samples. Although 1,4-dioxane was detected in other samples from the same well, it was not detected in the PDB sample. Apparently 1,4-dioxane does not diffuse well through the polyethylene bag (Parsons, 2005).

4.2.1.2 Rigid Porous Polyethylene Samplers

The RPP system is based on a rigid polyethylene tube, about 7 in. long and filled with reagent grade water. The RPP is capped at both ends and lowered into the monitoring well to equilibrate with ambient groundwater. Diffusion occurs through the rigid tube, which has pores ranging in size from 6 to 15 μm (0.006–0.015 mm). RPP samplers are used for VOCs and soluble inorganic compounds (O'Neill, 2006).

1,4-Dioxane recovery from RPP samplers was found to be comparable to recovery from samples obtained by conventional means. In a 14-day test (i.e., the RPP device remained in the monitoring well for 14 days), the RPP sampler yielded 74 parts-per-billion (ppb) 1,4-dioxane, whereas conventional sampling produced 80 ppb. In a 28-day comparison, the RPP device produced a result of 67 ppb; the result from conventional sampling was 64 ppb. These results are within the margin of analytical error (O'Neill, 2006). Another comparison involved 10 wells completed at different depths in a saprolite aquifer formation in the South Carolina Piedmont. RPP samplers are advantageous in this setting because monitoring wells recharge very slowly when pumped or bailed for conventional sampling. A best-fit line to a scatter plot of RPP results versus conventional sampling results falls close to the 1:1 line, with a regression error (R^2) of 0.92 (O'Neill, 2006). A third comparison found that RPP was less reliable for 1,4-dioxane sampling than the Snap Sampler but was comparable to low-flow purge methods (Parsons, 2005).

4.2.1.3 Snap Sampler

The Snap Sampler is a whole-water grab sampler that equilibrates with ambient natural groundwater flowing through the well screen (Britt, 2004). The Snap Sampler is used to retrieve a sample of unfiltered water in 40 mL VOA vials or 125 or 350 mL polypropylene bottles from a discrete depth (S. Britt, personal communication, 2007). The sampler is a double-ended, spring-loaded VOA vial or polypropylene bottle that seals beneath the water surface. Figure 4.2 shows a Snap Sampler. The sample does not come into contact with the atmosphere prior to injection into a gas chromatograph. In field comparison tests, the Snap Sampler has consistently collected higher concentrations of VOCs in samples than are typically found in samples retrieved by using other passive sampling devices, bailers, or sampling pumps (Britt, 2004). Conventional sampling introduces air into the sample by purging monitoring wells before retrieving a sample and by decanting the sample through the air into a laboratory VOA vial.

In a field comparison conducted at a site in Los Angeles, 1,4-dioxane results in samples collected by using the Snap Sampler were about 80% higher than results from purging three well casing volumes (2330 ppb Snap Sampler versus 1310 ppb conventional sampling) (S. Britt, personal

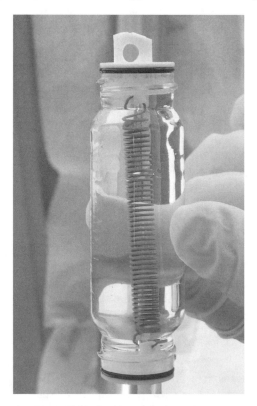

FIGURE 4.2 The Snap Sampler. (From Britt, S., 2004, The Snap Sampler. Paper presented to the Interstate Technology Regulatory Council Diffusion Sampler Team, Albuquerque, NM, October 28, 2004. With permission.)

TABLE 4.2
Summary of Results for 1,4-Dioxane: No-Purge Sampler Demonstration Study, McClellan Air Force Base

Sampler A	Sampler B	Number of Trials	Median RPD	Slope of X–Y Scatter Plot[a]	Regression Error (R^2)[b]	Sampler A		Sampler B
Low-flow	3-Volume	7	−1.5	1.41	0.97	Low flow	=	3-volume
Low-flow	Hydrasleeve®	3	0.0	1.11	1.00	Low flow	=	HSleeve
Low-flow	PSMS	4	7.7	0.94	0.99	Low flow	=	PSMS
Low-flow	RCS	3	−7.4	1.11	1.00	Low flow	<	RCS
Low-flow	RPPS	7	6.1	0.92	0.99	Low flow	=	RPPS
Low-flow	Snap™	3	20.7	1.26	1.00	Low flow	<	Snap
3-Volume	Hydrasleeve®	3	−5.4	1.05	1.00	3-Volume	=	HSleeve
3-Volume	PSMS	4	5.2	0.90	0.99	3-Volume	=	PSMS
3-Volume	RCS	3	30.3	0.74	0.99	3-Volume	=	RCS
3-Volume	RPPS	7	28.6	0.64	0.99	3-Volume	>	RPPS
3-Volume	Snap	3	17.1	0.84	0.99	3-Volume	=	Snap
Hydrasleeve®	PSMS	6	27.6	0.83	1.00	HSleeve	>	PSMS
Hydrasleeve®	RPPS	7	25.0	0.69	0.99	HSleeve	>	RPPS
PSMS	RPPS	9	0.0	0.85	0.97	PSMS	=	RPPS
PSMS	Snap	3	−27.3	1.20	n.a.	PSMS	<	Snap
RCS	RPPS	9	0.0	0.90	0.95	RCS	=	RPPS
RCS	Snap	6	21.8	1.35	0.86	RCS	<	Snap
RPPS	Snap	9	28.6	1.50	0.95	RPPS	<	Snap

Source: Parsons, 2005, Results report for the demonstration of no-purge groundwater sampling devices at former McClellan Air Force Base, CA. Army Corps of Engineers, Omaha, Nebraska. http://www.geoinsightonline.com/resources/mcclellan_report_cover.html (accessed October 12, 2006).

Notes: AFCEE = Air Force Center for Environmental Excellence; RPD = relative percent difference; n.a. = not applicable (too few data points were available for statistical significance). Low-flow sampling removes a small volume of water at a low flow rate from a small portion of the screened interval of a well without mixing water among vertical zones. The 3-volume sampling method rapidly removes 3–5 well-casing volumes from the well to evacuate stagnant water within the well casing and in the surrounding well filter pack. Theoretically, "stagnant" water is replaced with "fresh" groundwater from the surrounding formation with minimal mixing. Hydrasleeve® (abbreviated as HSleeve) is a flat-lying 30 × 2.75-inch polyethylene sleeve, sealed at one end, with a polyethylene reed-valve at the other end and a capacity of 2 L. The 2 × 2-inch polysulfone membrane sampler (PSMS) has 0.2 µm pore size and 108 mL capacity. The 13 × 1-inch regenerated cellulose sampler (RCS) has 0.0018 µm pore size and 400 mL capacity. The 6 × 1.5-inch rigid porous polyethylene sampler (RPPS) has 6–15 µm pore size and 150 mL capacity. The 10 × 1.6-inch Snap Sampler is a spring-loaded volatile organic analysis (VOA) holder that opens a 40 mL VOA or 125/350 mL Polypro bottle at depth.

[a] Scatter-plot slopes between 0.90 and 1.10 indicate that the populations are similar. Slopes less than 0.90 typically indicate that the concentration in sampler A was greater than the concentration in sampler B. Slopes greater than 1.10 typically indicate that the concentration in sampler B was greater than the concentration in sampler A.

[b] R^2 values less than 0.90 indicate a greater degree of scatter and a lower confidence that the slope value is meaningful. R^2 values greater than 0.90 indicate a lower degree of scatter and a higher confidence that the slope value is meaningful.

communication, 2007). In a comprehensive comparison of low- or no-flow sampling devices conducted at McClellan Air Force Base near Sacramento, California, the Snap Sampler provided higher recoveries of 1,4-dioxane than all other types of low- or no-flow samplers. The Snap Sampler™ results were comparable to but slightly lower than those of conventional sampling (Table 4.2). Comparisons of 1,4-dioxane results from different sampling approaches from the McClellan Air Force Base No-Purge Sampler Demonstration Study are summarized in Table 4.2.

4.2.1.4   Gore Sampler

Another approach to passive-diffusion sampling involves the semiquantitative dosimeter introduced by W.L. Gore Company. The Gore Module is a mix of proprietary adsorbents encased in Gore-Tex®, a hydrophobic PTFE* membrane. The sorbents are shielded from microbes, sediment, and water so that contaminants may partition into the enclosed air space according to their Henry's law constants. Contaminants adsorb by diffusion, independent of concentrations or groundwater flow rates. For low concentrations of compounds with low Henry's law constants, such as 1,4-dioxane, the Gore Modules must remain suspended in the monitoring well for a longer period of time than for more volatile compounds. The device is 6 in. long and can be used to profile vertical variation of concentrations within a well screen. Adsorbed contaminants are then thermally desorbed and analyzed by GC–MS (Anderson, 2005).

4.2.2   SAMPLING FOR 1,4-DIOXANE IN AIR AND LANDFILL GAS

A variety of strategies are used to sample for gaseous air contaminants, including whole-air sampling by filling air-sampling vessels and in-field sample concentration by sorption of air contaminants to a variety of specialty sorbent tubes in the field. Air-sampling vessels include specialized bags such as Tedlar® bags or stainless steel canisters such as Summa canisters.

The objectives of air sampling vary widely. For worker health and safety, air sampling is performed to monitor personal breathing zones in the workplace. Air quality studies monitor ambient air quality, and soil vapor studies monitor for vapor intrusion into buildings from underlying soil and groundwater contamination. Air-sampling equipment is also used to monitor for the 1,4-dioxane content of landfill gas as a source of groundwater contamination at landfill sites. Air sampling for 1,4-dioxane may also be necessary for monitoring emissions from industrial facilities. As discussed in Chapter 3, 1,4-dioxane is not likely to persist in the atmosphere owing to its relatively short photodegradation half-life and low Henry's law constant. A detailed discussion of air sampling is beyond the scope of this book. This section provides a brief overview of available methods for sampling 1,4-dioxane in air.

The most common method for sampling 1,4-dioxane in air uses a stainless steel Summa canister for analysis by EPA Method TO-15 (USEPA, 1999). The Summa canister is designed to be evacuated to a near vacuum (0.05 mm Hg) and then opened to the field atmosphere to admit the sample at a fixed rate controlled by a valve. The pressure differential draws the sample into the canister through a flow-restrictive inlet or mass-flow controller. The canister is then sealed and delivered to the laboratory for analysis by GC–MS.

Sampling gas trapped in soil is accomplished by applying a vacuum to a filtered tube and drawing the sample into a Summa canister, a Tedlar bag, a gas syringe, or a gas-tight glass bulb (CalEPA, 2003). The choice of tubing and sample container may have a bearing on 1,4-dioxane results. Some sample tubing has been shown to emit 1,4-dioxane to the sample, whereas other tubing materials retain 1,4-dioxane, on the basis of a few tests with tubing (Hayes et al., 2006). Tubing materials commonly used in soil sampling, and their potential to emit or retain 1,4-dioxane, are listed in Table 4.3.

Similar testing on the contribution of 1,4-dioxane by Tedlar bags to air samples was conducted in the Air Toxics Ltd study (Hayes et al., 2006). Tedlar bags were flushed three times with laboratory grade, humidified "zero-air" samples and then analyzed. 1,4-Dioxane was not found. Testing with laboratory standards showed no difficulties recovering 1,4-dioxane from Tedlar bags. The average recovery was 103%, with a relative standard deviation (RSD) of 5% (Hayes et al., 2006).

---

* PTFE = polytetrafluoroethylene. Gore-Tex is made from a polymer similar to Teflon.

TABLE 4.3
Soil–Gas Sampling Tube Materials and 1,4-Dioxane Emissions and Recovery

Tubing Name	Tubing Material	1,4-Dioxane Emissions	Frequency of Detection (%)	1,4-Dioxane Recovery (%)	RSD[a] (%)
LM Nylaflow	Nylon	0.068 ppbv	11	100	2
PEEK	Polyetheretherketone	<DL[b]	0	79	5
Teflon	Fluoroethylene-propylene (FEP)	<DL[b]	0	77	5
Polyethylene	Low-density polyethylene (LDPE)	<DL[b]	0	113	7

Source: Hayes, H.C., Benton, D.J., and Khan, N., 2006, In: *Vapor Intrusion—The Next Great Environmental Challenge—An Update*. Los Angeles, CA: Air and Waste Management Association. With permission.

[a] Relative standard deviation—all tests were highly reproducible.

[b] Detection limit.

4.2.3 PERSONNEL AIR MONITORING

Personnel air monitoring has typically been performed by using battery-powered portable sampling pumps and sorbent tubes. The National Institute for Occupational Safety and Health (NIOSH) promulgates methods for personnel monitoring, including Method 1602 for 1,4-dioxane. NIOSH Method 1602 uses a solid sorbent tube of coconut shell charcoal. A minimum of 0.5 L to a maximum of 15 L of workplace air is drawn through the tube at a rate of 0.1–0.2 L/min. The sample is then desorbed for 30 min with carbon disulfide as the solvent, and the sample is analyzed by GC with a flame-ionizing detector. The typical limit of detection is 0.010 mg per sample (Kurimo, 1994).

Field instruments for screening VOC vapors operate by using photo-ionization or flame-ionization detectors (FIDs). Photo-ionizing detectors (PIDs) can detect 1,4-dioxane, whose photo-ionization potential is 9.13 electron volts (eV). PID instruments such as the HNU can be calibrated to detect a particular compound when it is the sole compound present within a particular ionization range; however, PIDs are not well suited for distinguishing between detectable compounds in a mixture of gases. PID readings typically provide an integrated response to the gas mixture. Field instruments based on flame ionization can detect 1,4-dioxane. The flame-ionization potential for 1,4-dioxane is 9.41 eV. Another alternative for screening 1,4-dioxane vapors is a portable gas chromatograph (Photovac, 2006).

4.3 SAMPLE PREPARATION

Sample preparation is the first step performed on a sample delivered from the field. There are many established methods for sample preparation, each dependent on the nature of the analyte and the selection of analytical method. This section offers a brief review of the major sample extraction and concentration methods that have been used for 1,4-dioxane in water and soil.

4.3.1 PURGE AND TRAP

EPA Method 5030B describes the most common protocol for the purge-and-trap (PT) technique. The sample is purged by bubbling an inert gas, usually helium or nitrogen, through the aqueous sample at ambient temperature. The volatile components are transferred from the aqueous phase to the vapor phase. The contaminant vapor is introduced at a specified flow rate to a sorbent column where the volatile components are adsorbed. After purging is completed, the sorbent column is heated and back-flushed with inert gas to desorb the components onto a gas chromatographic column

(USEPA, 2006c). Volatile compounds with high air–water partition coefficients, such as the chlorinated solvents, are readily extracted from samples in this manner; however, hydrophilic compounds with low air–water partition coefficients, such as 1,4-dioxane, are not efficiently purged from water.

The high water solubility of 1,4-dioxane results in poor purging efficiency; however, PT systems can be modified to improve 1,4-dioxane recovery, as will be discussed later. The PT device consists of the sample purger, the trap, and the desorber. A typical purging chamber accepts 5 mL samples with a 3-cm-deep water column. The purge gas is introduced close to the sample level in the purging chamber and passes through the water column as finely divided bubbles. Trap dimensions vary by manufacturer; a typical trap may be about 1 foot long, with an inside diameter of about 0.1 in., and may contain adsorbents comprising, for example, approximately equal amounts of 2,6-diphenylene oxide polymer, silica gel, and coconut charcoal. The desorber is designed to be rapidly heated to 180°C for desorption of the target compounds for injection into the chromatographic column. Some configurations include a water trap, and most use a molecular sieve. Extensive maintenance, cleaning, and calibration protocols must be followed to produce reliable and repeatable results from PT equipment (USEPA, 2006a). The carrier-gas flow rate depends on the properties of the trap. A flow rate of 15 mL/min is used for the standard silica gel trap.

4.3.2 HEATED PURGE AND TRAP

1,4-Dioxane is miscible in water and boils at higher temperatures than most GC–MS analytes; therefore, it is more efficiently purged by using a heated PT method. A 1988 USEPA publication explores the benefits of heated PT for polar, highly soluble VOCs, including 1,4-dioxane, nitriles, ketones, and alcohols (Lucas et al., 1988). Although recovery for 1,4-dioxane was poor, use of sodium sulfate to "salt out" dioxane improved recoveries and enabled a preliminary method detection limit (MDL) of 4 ppb and limit of quantitation (LOQ) of 20 ppb. Problems with the salt clogging the columns discouraged further pursuit of this approach (Lucas et al., 1988). PT followed by GC-FID suffers from interferences that prevent attaining low detection limits for 1,4-dioxane (Song and Zhang, 1997).

Many research, water utility, and commercial laboratories have adjusted the purge times, purge and desorption temperatures, and materials used for PT to optimize extraction of the analytes of interest. A number of these innovations were targeted at fuel oxygenates, a suite of compounds whose chemical properties are similar to 1,4-dioxane and other solvent stabilizers in the ether, epoxide, and alcohol groups. Heated PT was optimized for analysis of methyl *tert*-butyl ether (MTBE), *tert*-butyl alcohol, *tert*-amyl alcohol, and other fuel oxygenates by heating the sample to 65°C and bubbling helium through the sample (Rose and Sandstron, 2003). For 1,4-dioxane, heating the sample and increasing the purge flow generally increases the recovery by a factor of two, but poor chromatography results when the purge temperature or purge flow is increased beyond optimal levels. Erratic recoveries may be due to increased amounts of moisture introduced into the PT concentrator at higher temperatures and purge flow rates (Strout et al., 2004b).

Orange County Water District (OCWD) modified the PT extraction and instrumentation used for EPA Method 524.2 to improve the sensitivity and reproducibility for analyzing for 1,4-dioxane in drinking water. Purge efficiency was improved by increasing the purge time from 11 to 20 min. Increasing the purge time was preferred to increasing flow rate, because purge flow rates at 40 mL/min may cause foaming in some samples. Varying the media in the trap, for example, using one that contains more Carbopack™, could improve the response and the shape of the 1,4-dioxane peak (Yoo et al., 2002).

4.3.3 EQUILIBRIUM HEADSPACE EXTRACTION

Equilibrium headspace (i.e., EPA Method 5021) is another approach for transferring the analyte from soil or water to the gaseous phase for analysis. Headspace methods are applicable to organic

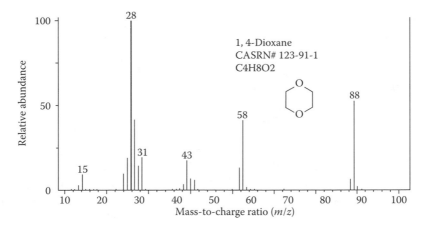

FIGURE 4.3 Mass spectrum for 1,4-dioxane-d_8 and its ion fragments. (From Yoo, L., 2002. Simultaneous determination of 1,2,3-TCP and 1,4-dioxane in drinking water by GC-MS/MS using purge/trap techniques. Orange County Water District. With permission.)

compounds with sufficiently high volatility to allow their transfer from soil samples under equilibrium headspace conditions (USEPA, 2006c). 1,4-Dioxane's hydrophilic nature and low air–water partition coefficient make headspace methods an unsuitable approach for its analysis; however, such methods have been used with modifications. A Danish study of 1,4-dioxane content in cosmetics and sundries adapted the equilibrium headspace method by weighing a 1 g sample in a headspace vial and adding it to 25 µL dichloromethane and 25 µL 0.1% (V/V) 1,4-dioxane-d_8. The headspace vial was then heated for 16–18 h at 80°C, and the gas was routed to a GC–MS operated in selective ion monitoring (SIM) mode for target masses 88 and 96 [for 1,4-dioxane and 1,4-dioxane-d_8 (Figure 4.3); SIM is explained further in Section 4.5.5]. The limit of detection achieved was 300 µg/L (0.3 ppm) (Rastogi, 1990).

A Polish study also focused on the use of headspace methods for 1,4-dioxane analysis in sundries. In this study, 1 g of sample was introduced to headspace vials together with a standard of dioxane dissolved in a solvent; thereafter, the vials were heated to 130°C for 1 h. Careful analysis of partition coefficients of dioxane from the laboratory solvent allowed quantitation to 2 µg/g (2 ppm) for dioxane in shampoos (Wala-Jerzykiewicz and Szymanowski, 1998).

4.3.4 VACUUM DISTILLATION

Vacuum distillation is another means of transferring contaminants from samples to the carrier gas. EPA Method 5032 uses vacuum distillation and a cryogenic trapping procedure followed by GC–MS. The sample is introduced into a sample flask, which is then depressurized to the vapor pressure of water through the use of a vacuum pump. The vapor is passed over a condenser coil chilled to –10°C or less to condense water. The uncondensed distillate is cryogenically trapped on stainless steel tubing chilled with liquid nitrogen (–196°C). The condensate is then thermally desorbed and transferred to the GC by using helium carrier gas (USEPA, 2006b). The 1996 publication of EPA Method 5032 does not identify 1,4-dioxane as suitable for analysis using vacuum distillation; however, a 2004 improvement to the method includes 1,4-dioxane as a suitable analyte (Strout et al., 2004a). EPA Method 8261A uses a vacuum distillation unit to extract volatile and semi-VOCs, including 1,4-dioxane. The lowest MDL obtained for 1,4-dioxane was 2.5 µg/L for low-concentration water samples, but recoveries were variable (Strout et al., 2004a). See Section 4.2.5.4 for more discussion of vacuum distillation in EPA Method 8261A.

4.3.5 Azeotropic Distillation

EPA Method 5031 uses azeotropic distillation to separate nonpurgeable, water-soluble, and VOCs, such as 1,4-dioxane, in aqueous samples or leachates from solid matrices. An azeotrope is a liquid mixture of two or more substances that behaves like a single substance: it boils at a constant temperature, and its vapors have a constant composition. 1,4-Dioxane forms a homogeneous pressure-maximum azeotrope with water in a completely miscible system at 87.8°C where 1,4-dioxane comprises 47.2% of the vapor-phase mass (Lide, 2008). Azeotropic distillation capitalizes on the ability of selected organic compounds to form binary azeotropes with water to facilitate separation of the compounds from a complex matrix. Extraction by azeotropic distillation was used as early as 1992 for the analysis of 1,4-dioxane (Bruce et al., 1992). After 2 min of fractional distillation of the 40 mL samples, the first 100 µL of distillate was collected for analysis by GC–FID. Recoveries were determined by comparison with standards injected directly prior to distillation. An MDL of 7 ppb was achieved for 1,4-dioxane. Azeotropic microdistillation was judged superior to the PT extraction method for analysis of water-soluble VOCs (Bruce et al., 1992). EPA Method 8015 for VOC analysis by GC–MS uses Method 5031, azeotropic distillation, as a preparatory step. Methods reporting limits for 1,4-dioxane analyzed following the standard application of EPA Method 8015B are summarized in Table 4.4 (from USEPA, 2006c).

4.3.6 Sample Extraction

EPA Method 5035 is used in most analyses of solid environmental samples (soil and other solid media) for solvent extraction of organic compounds with methanol. The solvent and surrogates used to calibrate the GC–MS instruments are added to the soil sample at the same time. For soils with low concentrations of the water-soluble target analytes (<200 µg/kg), 5 mL of reagent grade water is added to a weighed soil sample together with surrogates and internal standards, prior to proceeding with PT.

4.3.6.1 Liquid–Liquid Extraction

Liquid environmental samples require a protocol different from that used for solid samples. Method 3510C, separatory funnel liquid–liquid extraction (separatory funnel LLE), and Method 3520C,

---

TABLE 4.4

Results of EPA Method 8015B for Analysis of 1,4-Dioxane at Different Concentrations and in Different Sample Types

Sample Type	Low Concentration (25 µg/L)		Medium Concentration (100 µg/L)		High Concentration (750 µg/L)	
	Recovered (%)	RSD (%)	Recovered (%)	RSD (%)	Recovered (%)	RSD (%)
Groundwater	124	16	96	10	99	8
TCLP leachate	103	20	103	16	102	7
Solid matrices	106[a]	19[a]	105[b]	10[b]	–	–

Source: U.S. Environmental Protection Agency (USEPA), 1996a, Method 8015B: Non-halogenated organics using GC/FID. U.S. Environmental Protection Agency.

Notes: Method detection limits for 1,4-dioxane in 25 µg/L spiked samples of aqueous matrices: reagent water—12 ppb, groundwater—15 ppb; TCLP leachate from aqueous matrix—16 ppb. All values at each concentration of nonpurgeable volatiles are the average from seven replicate analyses by azeotropic microdistillation (EPA Method 5031). TCLP = Toxicity Characteristic Leaching Procedure, an EPA extraction protocol. RSD = relative standard deviation.

[a] Incinerator ash (low concentration; 0.5 mg/kg).

[b] Kaolin (medium concentration; 25 mg/kg).

continuous liquid–liquid extraction (continuous LLE), isolate and concentrate water-insoluble and slightly water-soluble organic compounds from aqueous samples.

In the continuous LLE (Method 3520C), 1 L of sample is placed into a continuous liquid–liquid extractor and extracted with organic solvent for 18–24 h. Then the extract is dried with anhydrous sodium sulfate, concentrated, and transferred to a solvent suitable for the analytical technique.

In the separatory funnel method (Method 3510C), 1 L of sample adjusted to a specified pH is serially extracted with methylene chloride or other solvents by using a separatory funnel. The extract is dried, concentrated by distillation, and exchanged into a solvent compatible with the analytical method to be used. In this extraction, 1 L of sample is reduced to 10 mL for injection into the GC column.

Because 1,4-dioxane and many of the solvent-stabilizer compounds of interest are highly soluble in water, separatory funnel LLE by Method 3510C is not generally used for their analysis. However, chemists at the OCWD were able to use an adaptation of separatory funnel LLE to analyze both 1,4-dioxane and nitrosodimethylamine (NDMA), a compound that is similarly soluble, polar, and volatile (Yoo et al., 2002). Using a separatory funnel saved 80% of the time required for continuous LLE. Three extractions were done with methylene chloride, and deuterated standards for both 1,4-dioxane and NDMA were added during the extraction process. The MS analysis used chemical ionization with methanol to achieve an MDL of 0.2 µg/L for 1,4-dioxane and 0.3 ng/L for NDMA (nanograms per liter, i.e., parts per trillion) (Yoo et al., 2002). Figure 4.4 shows separatory funnels for LLE.

A typical LLE uses a 1 L sample mixed with 100 g of NaCl in a 2 L separatory funnel for three sequential extractions with 60 mL of dichloromethane. The extraction step involves shaking the funnel for a few minutes and then allowing the mix to separate. Dichloromethane is denser than water and ideally settles to the bottom of the funnel in a separate phase. Organic compounds remain dissolved in the solvent, which is concentrated to 1 mL by bubbling nitrogen gas in a tube warmed by a 35°C water bath (Yoo et al., 2002).

In addition to the complete miscibility of 1,4-dioxane and water, LLE of 1,4-dioxane from water is made inherently difficult because of 1,4-dioxane's low octanol–water partition coefficient (K_{ow}). Various approaches have been developed to overcome these characteristics. Addition of salt has been found to improve recovery. Solvents tested for use in LLE have included dichloromethane and MTBE; however, MTBE proved to be an ineffective LLE solvent (Park et al., 2005).

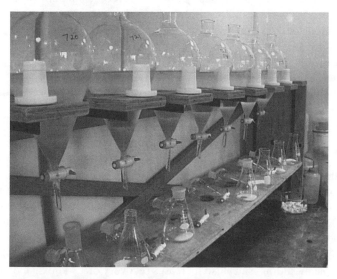

FIGURE 4.4 LLE separatory funnels, OCWD Laboratory. (From Yoo, L., 2002. Simultaneous determination of 1,2,3-TCP and 1,4-dioxane in drinking water by GC-MS/MS using purge/trap techniques. Orange County Water District. With permission.)

4.3.6.2   Solid-Phase Extraction

Solid-phase extraction (SPE), also called liquid–solid extraction in USEPA test methods, is described by EPA Method 3535. SPE involves adsorbing target analytes on a solid sorbent and then desorbing and concentrating them with a solvent before injection into the GC–MS system. A number of researchers have developed viable methods with SPE to achieve low MDLs (Fuh et al., 2005; Isaacson et al., 2006; Shirey and Linton, 2006; USEPA, 2006a). Carbon disks work well for extraction of highly polar, water-soluble compounds, such as 1,4-dioxane and other ethers. For the SPE protocol prescribed by EPA Method 3535, a 1 L sample is treated with 5 mL of methanol, internal standards and surrogates are added, and a pH adjustment is made. An SPE device, usually a coated disk or filter cartridge, is then used to extract the organic analytes. The disks are composed of hydrophobic materials and are often prewetted with a water-miscible solvent such as acetone or acetonitrile. The sample is filtered through the SPE cartridge or disk under vacuum. Vacuum is maintained after the sample has passed through the SPE device to dry the filter. Earlier SPE methods used bulk graphitized carbon or powdered carbon black. Disks reduce the filtering time and cost, allow increased flow rates, improve mass transfer, and eliminate bed channeling (Markell et al., 1991). Dichloromethane or other solvents are used to elute target analytes from disks. The eluted solvent is next dried with anhydrous sodium sulfate. The sample is then concentrated through the use of a nitrogen blow-down and hot water method, and the concentrated extract can be exchanged into a solvent as needed (USEPA, 2006a).

Solid-phase *micro* extraction (SPME) is a variation of SPE in which a much smaller volume of sample is sufficient to complete the analysis. The use of SPME for the analysis of 1,4-dioxane has many advantages over conventional solvent phase extraction and headspace pretreatments, including simplicity, speed, precision, lower detection limit, and minimal solvent consumption (Fuh et al., 2005). In SPME, the solid phase is coated on a fused silica fiber attached to the end of a wire plunger that is pushed through a syringe needle and immersed in the aqueous liquid phase or exposed to the headspace above the liquid. Organic compounds present in the vapor or liquid phase are absorbed into or adsorbed onto the solid phase. After a specified sampling time, the compounds are thermally desorbed from the solid phase in the heated injection port of a gas chromatograph (Black and Fine, 2001). Variants of SPME have evolved, including stir bar sorptive extraction (SBSE) and headspace sorptive extraction (HSSE). These adaptations were developed to increase sorption capacity and to overcome some drawbacks of SPME, such as fiber fragility (Jochmann et al., 2006). Solid-phase *dynamic* extraction (SPDE) is a commercial sample extraction method that is growing in popularity. SPDE uses a 2.5 mL headspace syringe equipped with a needle fabricated with an interior coating similar to a fused-silica GC column (see Section 4.4.1.1) with an immobilized extraction phase (Jochmann et al., 2006). SPDE needle coatings incorporate 4–6 times the amount of sorbent as SPME fibers. The syringe is immersed directly into a sample or the headspace above it, and the plunger is exercised several times to extract the sample, thereby adsorbing the analytes to the internal coating. After several cycles of aspirating and dispensing the sample, analytes are thermally desorbed from the coating inside the needle and purged with nitrogen gas into the GC injector (Jochmann et al., 2006).

As with LLE, the addition of sodium chloride or sodium sulfate salt to the sample improves the recovery of 1,4-dioxane in SPME extraction. The salt helps to decrease the solubility of 1,4-dioxane in water. The amount of 1,4-dioxane extracted from SPME fibers increases linearly up to 15% of sodium chloride (NaCl) concentration. High salt concentrations may shorten the lifetime of SPME fiber coatings. Addition of 10% (w/v) NaCl is considered optimal for 1,4-dioxane determination using SPME while also preserving SPME fiber coatings (Fuh et al., 2005). Increasing the ionic strength of the sample by adding NaCl also increases recovery of 1,4-dioxane in SPDE needles. 1,4-Dioxane recovery increased by approximately 60% when salt content was increased from 10% to 25% (w/w ratios) in samples extracted with SPDE needles (Jochmann et al., 2006). The effect of adding salt to extract more of a water-soluble organic compound is referred to as "salting out" and is quantified by the Setschenow constant, K^S,

$$\log\left(\frac{\gamma_{w,salt}}{\gamma_w}\right) = K^S[salt]_{total}, \qquad (4.1)$$

where γ_w is the activity coefficient of 1,4-dioxane in pure water, $\gamma_{w,salt}$ is the activity coefficient of 1,4-dioxane in saline aqueous solution, and $[salt]_{total}$ is the total molar concentration of NaCl in the solution used to measure 1,4-dioxane activity in an NaCl solution. Setschenow constants for MTBE, tetrahydrofuran (THF), and 1,4-dioxane determined at 70°C were 0.17 L/mol ($R^2 = 0.995$), 0.16 L/mol ($R = 0.988$), and 0.08 L/mol ($R^2 = 0.999$), respectively, and R^2 is the correlation coefficient from four data points (Jochmann et al., 2006). 1,4-Dioxane's Setschenow constant is considered exceptionally low, which means that the addition of NaCl to a sample for extraction by SPDE will have a large increase in sample recovery for 1,4-dioxane.

Selection of the adsorbent coating on the fibers is an important consideration for using SPME. A study comparing the effectiveness of different SPME media determined that carboxen-polydimethylsiloxane (CAR-PDMS) coating is most effective for extraction of 1,4-dioxane (Shirey, 2000). Another study contrasted the recovery of 1,4-dioxane from six different SPE media, each receiving a uniform quality control sample spiked at 10 µg/L. The best recovery was produced by 2OH Diol sorbent. All six media performed well[*] (Song and Zhang, 1997).

The quantity of carbon or other adsorbents used in SPME is also important. When cutting pieces of carbon felt for in-vial elution, at least 1.5 g are needed to produce adequate recovery. Granulated carbon has greater adsorptive power than felt, but may be short-circuited by channeling of the sample through the cartridge (Kawata et al., 2001; Isaacson et al., 2006). Samples extracted onto flexible disks by using SPE may be eluted "in-vial" in a 2 mL autosampler vial and directly analyzed by gas chromatography and tandem mass spectrometry (GC–MS/MS) (Isaacson et al., 2006). Desorption of VOCs from the solid phase may use solvents such as dichloromethane (methylene chloride) or acetone. Desorption is also done in headspace extraction by heating the fiber to an elevated temperature. A comparison of these approaches showed that heated headspace extraction at 55°C is 31% more efficient than direct immersion for the extraction of 1,4-dioxane from CAR-PDMS (Shirey and Linton, 2006). Nonpolar compounds have a higher affinity for the SPME media and may displace polar compounds (Black and Fine, 2001).

Experiments were performed by chemists at Supelco, a manufacturer of chromatographic columns, to determine whether high concentrations of methyl chloroform would compete with 1,4-dioxane for sorption onto CAR-PDMS-coated fibers. Although no displacement or competitive sorption was observed for concentrations as great as 20 mg/L methyl chloroform with 10 µg/L 1,4-dioxane and 25 µg/L 1,4-dioxane-d$_8$, an impurity of methyl chloroform was found to interfere with quantitation of 1,4-dioxane. The impurity was 1,1,2-trichloroethane (Shirey and Linton, 2006; R.E. Shirey, personal communication, 2007). A "solvent effect" on the effectiveness of SPME for 1,4-dioxane extraction was reported in a study of SPME coupled with headspace analysis that analyzed 1,4-dioxane in cosmetics (Fuh et al., 2005). Laboratory solvents were found to decrease the 1,4-dioxane yield from SPME by 30–40% after adding 1%, 5%, and 10% (w/v) of methanol and acetonitrile. The SPME media used in the study were 75 µm CAR-PDMS, 85 µm polyacrylate (PA), and 100 µm PDMS fiber assemblies made by Supelco. To avoid the solvent effect, headspace analysis

---

[*] The six SPE sorbent media, sorbent manufacturers, and their performance with a 10 µg/L standard in seven replicate tests were C$_{18}$ octadecyl (Applied Separations): 10.96 ± 0.49 µg/L; C$_8$ octyl (Analytichem International): 10.9 ± 0.67 µg/L; C$_2$ ethyl (J.T. Baker): 11.0 ± 0.37 µg/L; SI silyca (Analytichem International): 10.7 ± 0.58 µg/L; 2OH Diol (Varian): 9.94 ± 0.58 µg/L; NH$_2$-Aminopropyl (Varian): 9.65 ± 0.17 µg/L. (Kawata et al., 2001.)

was used. In another study, using hexane and dichloromethane as extraction solvents was found to interfere with quantitation of 1,4-dioxane in the mass spectrometer. Use of a lower molecular weight solvent, acetonitrile, was found to avert the interference (Song and Zhang, 1997).

SPE media must be completely air-dried prior to solvent extraction, or the eluted solvent must be dehydrated by using anhydrous sodium sulfate crystals. When water is present in the eluate from the SPE media, the resulting chromatographic peaks for 1,4-dioxane and 1,4-dioxane-d$_8$ become significantly less sharp. With increasing water content, a shoulder develops on the left side of the peak between 13.3 and 13.5 min at 7.6% water content. At 14% water content, the peak broadens dramatically between 13 and 13.5 min (Kawata et al., 2001). To eliminate water from the SPE cartridge, researchers recommend drying with clean laboratory air for 2 min using an aspirator, followed by centrifuging at 1700 g (3000 RPM) for 10 min (Kawata et al., 2001). Figure 4.5 displays the deterioration of the 1,4-dioxane peak with increasing water content.

The temperature at which SPE desorption of target analytes occurs also affects the size of the peak area, that is, the recovery efficiency of the extraction. A German study of SPDE effectiveness determined that increasing the desorption temperature between 30°C and 70°C increased the efficiency of extraction at higher temperatures. A plot of chromatographic peak area versus extraction temperature creates a series of exponential curves. The curve determined for 1.4-dioxane is described by

$$y = (1.73 \times 10^5)e^{0.0376X}, \tag{4.2}$$

where y is the peak area and X is the extraction temperature in Celsius. The correlation coefficient was 0.999. For 1,4-dioxane, extraction recovery at 70°C is 4.5 times greater than at 30°C (Jochmann

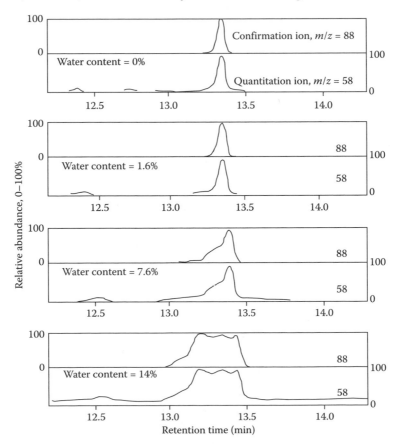

FIGURE 4.5 Deterioration of 1,4-dioxane peak with increasing water content. (After Kawata, K. et al., 2001, *Journal of Chromatography A* 911: 75–83. With permission.)

et al., 2006). The same study found that increases in concentrations of target ether and alcohol analytes in headspace above a liquid sample at higher temperatures are partially offset by a lower sorbent–air partition constant at higher temperatures. Because sorption is an exothermic process, the sorbent–air partition coefficient decreases as the sorbent temperature in the SPDE needle increases. The maximum extraction yields therefore occur at intermediate temperatures, with a decrease in SPDE extraction yield at higher temperatures (Jochmann et al., 2006). For SPDE, it is also important that the selected temperature not be so cool as to allow the condensation of water vapor in the needle, to prevent the 1,4-dioxane peak deterioration with increasing water content already noted (Kawata et al., 2001; Jochmann et al., 2006). The duration of extraction by SPME and the number of cycles of extraction by SPDE also have a significant bearing on the extraction recovery of 1,4-dioxane from water samples. Recovery of 1,4-dioxane approximately doubled when the number of SPDE extraction cycles was increased from 10 to 30; no appreciable change was observed when the number of SPDE extraction cycles was increased from 30 to 50 (Jochmann et al., 2006).

One of the lowest MDLs reported in the peer-reviewed literature for analysis of 1,4-dioxane in water, 130 parts per trillion, was achieved with SPE with solvent desorption (Isaacson et al., 2006). The method was developed at Oregon State University (OSU) and involved the following steps: 25-mm-diameter activated carbon disks with 10 μm particles and 1100 m²/g surface area were fitted to polypropylene disk holders and attached to a vacuum manifold. Sample volumes of 80 mL were spiked with 100 μL of 50 μg/mL dioxane-d$_8$. Samples were pulled through dry disks under a 67 kPa vacuum, followed by 1 h of drying in laboratory air under vacuum. Disks were then transferred into 2 mL GC autosampler vials, to which acetone and instrumental standards were added. Acetone was found to yield greater than 90%, that is, the same recovery as dichloromethane. Acetone is preferred over dichloromethane because it is less hazardous and wastes are more easily managed.

The parameters of GC–MS/MS analysis for the OSU SPE method are summarized in Appendix 3. The pilot study for development of this method also tested the effectiveness of SPE for extraction and identification of two other stabilizer compounds, THF and 1,3-dioxolane. Spike and recovery experiments with THF and 1,3-dioxolane in water yielded greater than 80% recovery for THF and less than 10% for 1,3-dioxolane. Breakthrough of the 1,3-dioxolane was confirmed from the fact that 90% of the added mass was recovered from the water that passed through the disk, as determined by an in-vial LLE method. 1,3-Dioxolane is poorly suited for extraction by SPE using activated carbon. The MDL for 1,4-dioxane was 0.13 μg/L, and the LOQ was 0.31 μg/L. For THF, the MDL was 1.0 μg/L and the LOQ was 3.1 ppb (Isaacson et al., 2006). Chromatograms from the OSU study are shown in Figure 4.6.

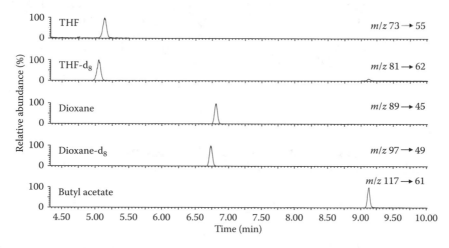

FIGURE 4.6 Chromatograms of THF, THF-d$_8$, dioxane, dioxane-d$_8$, and internal standard butyl acetate. [From Isaacson, C., Mohr, T.K.G., and Field, J.A., 2006, *Environmental Science & Technology* 40(23): 7305–7311; Chromatograms by Oregon State University. With permission.]

CASE STUDY—METHOD DEVELOPMENT FOR 1,4-DIOXANE ANALYSIS OF DRINKING WATER IN A WORKING TESTING LABORATORY

The OCWD, located in Fountain Valley, California, developed a method for time-efficient, low-level analysis of 1,4-dioxane in water. OCWD was challenged with analyzing for 1,4-dioxane because of its low-level presence in treated wastewater used as a seawater intrusion barrier through underground injection in a groundwater basin that is also used by several municipal supply wells. OCWD sought to develop a method that would reliably identify and quantify low levels of 1,4-dioxane following a protocol that would be time efficient and cost effective, would minimize waste generation, and would be robust for a variable sample matrix and presence of other compounds in the sample. OCWD's objective was to achieve a detection limit of 1 ppb with less than 20% RSD for the mean response factor (RF). The objective was achieved by pursuing a modified PT approach with GC–MS/MS. PT allows analysis to be completed with only 25 mL of sample instead of the 1 L sample required by LLE. PT generates minimal solvent wastes, whereas LLE requires 200 mL of dichloromethane per sample.

OCWD confirmed that the extraction efficiency of 1,4-dioxane is independent of pH. Extraction efficiencies ranged from 83% to 87% for spiked 1,4-dioxane samples at pH ranging from 2 to 9. Because 1,4-dioxane is a neutral compound, pH adjustment of the sample is not necessary prior to extraction and analysis.

OCWD increased the purge time from the 11 min recommended by the USEPA for drinking water to 20 min. The baking time for the trap was also increased from 15 min to 20 min. The use of a GC–MS/MS system enhanced the intensity of fragment ions for 1,4-dioxane. The extra MS function coupled with GC–MS helped to verify the sub-ppb levels of 1,4-dioxane in heavy matrix samples and improved the reproducibility and sensitivity of the system. Appendix 3 summarizes the operating parameters for the OCWD method for PT with GC–MS/MS. An MDL of 0.2 ppb was achieved by following this approach (Yoo et al., 2002). The time requirements for this method are comparable to LLE and SPME, but the PT method is less labor intensive.

OCWD has been running between 2500 and 3000 1,4-dioxane analyses annually. OCWD has also experimented with including 1,4-dioxane in a 200 mL methylene chloride (liquid–liquid) extraction method for the analysis of a similarly hydrophilic compound, NDMA, using GC–MS/MS combined with chemical ionization. However, a significant portion of 1,4-dioxane was lost during concentration of the 200 mL methylene chloride extract in the sample concentration system. OCWD is now developing an extraction method to detect nitrosamines and 1,4-dioxane with GC–MS/MS/CI (chemical ionization) and SPE techniques to avoid generating large volumes of laboratory solvent wastes (Yoo, personal communication, 2007). Operating conditions and instrument settings used by OCWD for 1,4-dioxane analysis by PT GC–MS/MS are listed in Table 4.5.

TABLE 4.5
Operating Conditions for OCWD Purge and Trap, GC-MS/MS Method for Low-Level Detection of 1,4-Dioxane

Purge and Trap Conditions[a]	
Purge time	20.0 min
Bake time	15.0 min
Preheat	245°C
Desorb	250°C
Bake	260°C

TABLE 4.5   (continued)
Operating Conditions for OCWD Purge and Trap, GC-MS/MS Method for Low-Level Detection of 1,4-Dioxane

Sample Injection Conditions[b]

Settle	0.3 min
Prepurge	30 s
Sample pressurize	40–60 s
Sample transfer	75 s
Internal standard fill	5 s
Internal standard transfer	50 s
Back flush	Off
Desorb time	4 min
Transfer line rinse	Off

GC and MS Conditions[c]

Initial column temperature	35°C
Hold time	2 min
Final temperature	220°C
Temperature rate	3.1°C/min to 100°C; 100°C; to 220°C at 30°C/min
Hold time	1 min at 30°C/min
Helium linear velocity	1 mL/min
Total run time	27.97 min
Injector temperature	125°C
Detector	Ion trap MS/MS
Transfer line	220°C
Mass range	40–100 m/z
Trap temperature	135°C
Background mass	40 m/z
EI (electron-impact ionization) current	80 μA
Scan time	0.36 s
Filament/multiply delay	14 min
Reagent ion ejection	20 V
Ion storage level	35.0 m/z
Ejection amplitude	20 V
High edge amplitude	15 V
Wave form	Resonance

Source: From Yoo, L.J., Fitzsimmons, S., and Wehner, M., 2002, Improved purge-trap and GC/MS/MS techniques for the trace-level determination of 1,4-dioxane in water. In: *Proceedings, Water Quality Technology Conference*. Seattle, WA: American Water Works Association, CD-ROM. With permission.

[a] Purge and trap by Tekmar ALS 3100.

[b] Sample injector by Tekmar AquaTek 50.

[c] GC—Varian Model 3800; MS—Varian Saturn 2000 (column—DB-VRX fused-silica capillary column, 60 m × 0.32 mm ID with 1.8 μm film thickness).

4.4   LABORATORY METHODS FOR THE ANALYSIS OF 1,4-DIOXANE BY GC–MS

Laboratory methods using GC–MS form the subject of numerous books and university courses. This section presents only a brief synopsis of the key operative elements of these methods, in the context of how they are modified to analyze for 1,4-dioxane.

Identification and quantification of contaminants in soil, water, air, food, and other media involve isolating the contaminants of interest from the sample by

1. Sample extraction and concentration
2. Separating the contaminants on the basis of their rate of passage through a chromatographic column designed with specific properties to retain the contaminants according to their vapor pressure, boiling point, and other properties
3. Detecting and quantifying the contaminants by ionization of the analyte followed by mass separation and detection
4. Identifying contaminants according to their mass spectra by using a computerized library of reference spectra and quantification via calibration standards

The primary approach for analysis of soil and water samples for VOCs has involved PT/GC–MS analytical methods (USEPA, 1986, 1990; Lesage and Jackson, 1992). PT methods are designed to measure as many compounds as possible with a single procedure. Such methods are favored for those compounds that are relatively insoluble in water and that have boiling points below 200°C (USEPA, 1986, 1990) or below 150°C (Lesage and Jackson, 1992). The PT preparation techniques promote low detection limits (ppb range), and the MS detector allows positive compound identification. Compounds that contain polar functional groups such as low molecular weight ketones, alcohols, aldehydes, nitriles, and ethers (i.e., 1,4-dioxane) are generally soluble in water, do not purge well, and produce broad, tailing GC peaks that give poor quantitative estimates and are often difficult to identify by MS (Swallow, 1992). Some polar compounds are included in SW-846 Methods 8240/8260 (USEPA, 1986, 1990), but the recovery of polar compounds is often less than 20% (Swallow, 1992). Azeotropic distillation (SW-846 Method 5031) and closed-system vacuum distillation with cryogenic condensation (SW-846 Method 5032) were introduced as preparation methods for analysis of hydrophilic analytes such as alcohols and ethers in the third update to SW-846 (Lesnik, 1993; USEPA, 1993).

4.4.1 Gas Chromatography

GC involves injecting the mixture to be analyzed into an inert gas stream that sweeps the sample into an open capillary column coated with a thin film or a column coated with a resolving stationary phase. The components in the gas stream absorb and interact with the thin film or stationary coating to varying degrees, which leads to differential separation, causing the components in the gas stream to be swept through the column in a sequence based upon their vapor pressure, boiling points, and other physical properties. As individual components of the mixture elute from the chromatographic column, they are swept by the carrier gas to a detector. The detector generates measurable electrical signals, referred to as peaks, which are proportional to the amount of analyte present. The detector response is plotted as a function of the time required for the analyte to elute from the column after it was injected. The resulting plot is called a chromatogram. The position of the peaks on the time axis may serve to identify the components, and the areas under the peaks provide a quantitative measure of the amount of each component when using a mass spectrometer. To identify the compound, the spectra corresponding to the peaks are compared to the spectra of calibration standards or to a computerized library of spectra. Because several compounds may possess similar retention times, the nature of the resulting peak may require further identification using MS (McMaster and McMaster, 1998).

Mass spectrometers differentiate compounds based on the mass-to-charge ratio (m/z) of their ionized products. Compounds entering a mass spectrometer are ionized in a vacuum. The resulting ions are directed through a mass analyzer, which separates the ions by the m/z ratio. MS alone is useful for resolving the identity of individual compounds, but is less efficient for resolving compound identification in mixtures. The combination of GC and MS, GC–MS, forms a system that is effective at separating mixtures into their individual compounds and determining the amount and structure of the compounds. GC–MS has become the mainstay of environmental chemistry for volatile analytes such as solvents and related organic compounds in soil, air, and water; however, it is not without its limitations (McMaster and McMaster, 1998).

4.4.1.1 Methods in Gas Chromatography

The laboratory analyst must choose operating parameters for the analysis of each suite of analytes by GC. These include selection of carrier gas, injector type, column type, oven temperature, and temperature gradient, that is, the rate at which column temperature is increased and the duration it is held at a given temperature (Rödel and Wölm, 1987). The combinations of sample-preparation methods described in Section 4.3 with the detectors described in Section 4.4.2 differentiate the individual analytical methods for 1,4-dioxane analysis detailed in Section 4.5. Although there is a seemingly infinite number of permutations of instrumental configurations, extraction techniques, operating parameters, chromatographic columns, and detectors, seven methods for 1,4-dioxane analysis are published, and most are commercially available.

4.4.1.2 Selecting Chromatographic Columns for 1,4-Dioxane Analysis

The chromatographic column separates mixed analytes in the sample. Columns are classified according to their diameter, type, and length. Conventional columns range from 6 to 20 feet or more in length and may be of either the packed or open-tube (capillary) design. Packed columns are filled with a porous granular or beaded material, such as a specialized variety of diatomaceous earth, but more often a fused silica material. Capillary columns are longer than conventional columns, 5–100 m long with 30 m being most common, but they have a similar porous or liquid interior coating. Fused silica capillary columns are common for gas chromatographs used for environmental analysis, although a wide variety of columns are available. Capillary columns provide greater resolution of analytes than packed columns when handling small-volume samples. Most capillary columns are designed with a fused silica open tube, to which a liquid stationary phase is chemically bonded.

The liquid stationary phase is selected for its stability at high temperatures and for its low volatility. Polysiloxanes (i.e., "OV-1") and polyethylene glycol (i.e., "CarboWax 20M") are two examples of liquid stationary phases used in GC columns. For a given analyte, the thicker the liquid stationary-phase film in the column, the longer the retention time, and thicker films are used for more volatile analytes.

Oven temperature is the primary factor in determining retention time of an analyte. Temperature is used to separate analytes by programming the GC oven to increase temperature at a specified rate. Increasing the temperature and the rate of heating can cause analytes with higher boiling points to elute earlier from the column to the detector. Temperature programs are recommended for each column type by the EPA method for particular groups of target analytes (USEPA, 2006a). Optimum temperature gradients must be independently determined for each analyte and column combination; however, lowered temperature means an increase in the elution time and the time required to complete the analysis.

The polarity of the stationary phase should be matched to the group of analytes being tested. Components of the mixture with similar polarity and therefore a high degree of affinity for the stationary phase are strongly retained, whereas components with dissimilar polarity will have low affinity for the stationary phase and migrate rapidly through the column. These differing behaviors present a challenge for analyzing samples for both solvents and stabilizers, because many solvents are nonpolar whereas many alcohol, epoxide, and ether stabilizers such as 1,4-dioxane are polar. This challenge can be overcome with a gas chromatograph in which the sample gas flow is split between two columns of different materials to address both polar and nonpolar compounds.

Polysiloxanes, among the most common stationary phases in GC columns, contain various substituent groups to change the polarity of the stationary phase. The nonpolar end of the spectrum is polydimethyl siloxane, which is made more polar by increasing the percentage of phenyl groups on the polymer. For very polar analytes, polyethylene glycol (i.e., Carbowax) is commonly used as the stationary phase. Some gas chromatographers conclude that the column material that best separates 1,4-dioxane from the sample is DB-WAX, a fused silica column bonded with polyethylene glycol (Kawata et al., 2001). After the polymer coats the column wall or packing material, it is often cross-linked to increase the thermal stability of the stationary phase and prevent it from gradually bleeding out of the column.

The stationary phase should be similar in chemical structure to the components of the mixture targeted for analysis. To match chromatographic column selectivity to target analytes, several classification schemes of analyte polarity have been developed, including McReynold's numbers and Rohrschneider constants. McReynold's tables predict the chromatographic separation of various compounds with different stationary phases and provide the best information available for the comparison of the selectivity of the stationary phase. In the McReynold's system, 1,4-dioxane is used as an index analyte for quantifying column selectivity. 1,4-Dioxane is representative of "Class III" compounds of medium polarity. Stationary phases that will more strongly retain Class III compounds include the following:

- All polyesters
- SAIB tricresyl phosphate
- Benzyl cyanide
- Propylene carbonate
- Polyphenylether
- OV-17
- Dibutyl tetrachloro-phthalate
- STAP
- Lexan
- QF-1
- Dimethylsulfolane

The stationary phases recommended by USEPA for the analysis of ethers and other compounds of intermediate polarity ("Class III polarity") include Carbowax 400, Tricresyl phosphate, LAC, Apiezon L, and β–β'-oxydipropionitrile (USEPA, 2006c). Other listings for stationary phases suitable for the separation of ethers include Carbowax 20M, Carbowax 1500, and SE-30 (GC Grade) (Supelco/SigmaAldrich, 1997).

4.4.2 DETECTORS IN GAS CHROMATOGRAPHY

A variety of detectors are used in GC. The thermal conductivity detector (TCD) was the most widely used detector system during the early years of GC. The FID became more popular for common hazardous waste analyses. Specialty detectors were subsequently developed, such as the electron-capture detector (ECD) and the nitrogen–phosphorus detector. All detectors described in this section are nonspecific detectors and have largely been replaced by mass spectrometers. MS is the preferred detection technique because it is specific, sensitive, and quantitative, and good precision can be obtained at low concentrations under the right conditions.

TCDs were widely used in the early years of GC because of their simplicity, universal applicability, and low cost. Analyte detection is based on changes in the conductivity of the column effluent. The TCD is a destructive detector that can be used in series only after nondestructive detectors. The TCD detects gaseous compounds in the ppm range. TCDs are not generally used for analysis of low-concentration samples as the possibility of false identification of analytes is a large problem. Larger sample volumes are required to achieve increased sensitivity, which in turn requires using a large-diameter chromatographic column (USEPA, 2006b). TCDs are not well suited for detection of 1,4-dioxane and are no longer widely used for environmental sample analysis.

The FID is sensitive for most organic compounds containing oxidizable carbon such as aromatic and chlorinated aliphatics, petroleum compounds, semivolatile compounds, and polychlorinated biphenyls (PCBs). The FID is a destructive detector that can be used in series only after nondestructive detectors. The FID is a nonspecific detector, as most carbon-containing compounds are detectable by the FID.

An FID uses a small stainless steel jet positioned at the end of the chromatographic column. As carrier gas exits the column and flows through the jet, it mixes with hydrogen supplied in the jet and burns at the tip of the jet. Hydrocarbons and other molecules in the sample are ionized in the flame and attracted to a metal collector electrode located just to the side of the flame. The resulting electron current is amplified to convert very small currents to voltages that are recorded as the sample signal (USEPA, 2006b). An FID has been used to detect 1,4-dioxane vapors in occupational exposure testing; the detection limits are 5–190 ppmv (20–700 mg/m^3) (Cooper et al., 1971, in NICNAS 1998).

An ECD consists of a sealed stainless steel cylinder that typically contains radioactive nickel-63. The nickel-63 emits beta particles that collide with and ionize carrier-gas molecules; in the process, a stable cloud of free electrons forms in the ECD cell. When a halogenated or other electronegative molecule enters the cell, it is immediately combined with one of the free electrons, causing a temporary but measurable reduction in the number of free electrons in the cell. The ECD is a nondestructive detector that can be used in series before other detectors (USEPA, 2006b). Compared to TCD or FID, the ECD is a more specific detector for halogenated or other electronegative compounds, but it is relatively insensitive for hydrocarbons, alcohols, ketones, and ethers such as 1,4-dioxane.

4.4.2.1 Mass Spectroscopy

A mass spectrometer is a detector used in GC. Mass spectrometers ionize gaseous molecules, separate the ions produced on the basis of the mass-to-charge ratio (m/z), and then record the relative number of different ions produced. The m/z ratio is then plotted as the abscissa with relative intensity as the ordinate. This plot is referred to as a "mass spectrum." The mass spectrum of a compound can be considered its "fingerprint" and can be used to identify a compound through comparison with published reference spectra. MS systems interface with computers that compare experimental spectra to standard spectra and perform the identification automatically. Spectra that do not match calibration standard spectra can be compared with a library of spectra, and a tentative identification can be made.

Analytes can be fragmented by using either electron or chemical ionization. In electron-impact ionization, a 70 eV beam of electrons ionizes the analyte. Many analytes will fragment when exposed to the electron beam. The fragmentation pattern is characteristic of the ionized analyte and produces the spectrum from which the compound is identified. This experimental spectrum is then compared with spectra of compounds of the same molecular weight in a computerized spectra database or a published library of spectra (McMaster and McMaster, 1998).

Chemical ionization uses an ionization gas such as methane, butane, or carbon dioxide mixed with the sample stream. The ionization gas absorbs the initial ionizing electron and transfers the energy to the sample molecule. Chemical ionization creates a lower energy state than electron-impact ionization and thus provides less fragmentation and a larger abundance of molecular ions (McMaster and McMaster, 1998).

The major components of the MS system include a vacuum pump to maintain a vacuum in the mass spectrometer, an ionization chamber, a focusing lens to direct the stream of ions, a mass analyzer, a detector, and a data control system. Following the ionization process already described, the charged-ion stream is focused through a pinhole and a series of electrical focusing lenses to narrow and intensify the ion stream before it enters the mass analyzer. A quadrupole analyzer is a set of four cylindrical metal rods clamped together in a specific configuration spaced so that their electromagnetic fields interact in a precise manner. Direct current and radio-frequency signals are applied across the rods; adjacent rods have opposite charges. The effect of the combined direct current and radio-frequency fields is to force the ion stream entering the quadrupole into a corkscrew-shaped three-dimensional sine wave. The combined electromagnetic fields establish a standing wave for just a single mass-to-charge ratio at a particular frequency, allowing it to be directed to the detector at the end of the quadrupole, while other ion fragments follow unstable paths and collide with the walls of the quadrupole rods. As the combined electromagnetic fields are varied, larger or smaller masses impinge upon the detector (Chapman, 1993; McMaster and McMaster, 1998).

Ionized analytes and their fragments leave the quadrupole and are deflected from a linear path to the ion detector. Fragment ions colliding with the conductive surface of the ion detector induce a cascade of ions in the body of the ion detector, amplifying the signal from the single fragment and producing a signal strong enough to be counted in the data system. The combination of ion detectors and data acquisition systems can record about 25,000 data points per second. The range and scan rate are instrument dependent; for example, mass spectra from 35 to 550 m/z can be scanned up to 10 times per second in some mass spectrometers (McMaster and McMaster, 1998).

4.4.2.2 Mass Spectroscopy and 1,4-Dioxane

To improve analytical sensitivity for detection of 1,4-dioxane and other compounds at low concentrations, MS systems can be run in selective ion mode (SIM). In SIM, the MS can focus on a narrow m/z range, refining the ability to detect ion fragments in the target range. To ensure accurate compound identification using SIM, multiple ions should be monitored for each compound (USEPA, 2006a). For 1,4-dioxane, the mass-to-charge ratio (m/z) of the molecular ion is 88, and the masses of fragment ions of 1,4-dioxane are 58, 43, and 57. For 1,4-dioxane's deuterated internal standard, 1,4-dioxane-d_8, the mass-to-charge ratio (m/z) of the molecular ion is 96, while the m/z ratios for its fragment ions are 64 and 46. Mass spectra for 1,4-dioxane and 1,4-dioxane-d_8 are shown in Figure 4.3.

4.5 ENVIRONMENTAL PROTECTION AGENCY METHODS FOR ANALYSIS OF 1,4-DIOXANE

Commercial laboratories typically offer analysis of 1,4-dioxane in water by four EPA methods: 524.2 for drinking water and 8260 (purgeable), 8260B (SIM), or 8270 (extractable). Because of the high water solubility and poor purging efficiency of 1,4-dioxane, GC–MS methods without modifications produce high detection limits, typically greater than 100 µg/L (ppb) (Strout et al., 2004b).

The development of analytical methods for 1,4-dioxane followed demand for quantifying 1,4-dioxane in the chlorinated solvents, pharmaceutical, cosmetics, and petroleum industries, as well as personnel air monitoring in these industries. Earlier test methods in these applications required quantitation in the percent range, which could be achieved by using various combinations of PT and GC with FIDs. The motivation to analyze water-borne 1,4-dioxane in the part-per-billion range stemmed from risk analyses leading to low-level regulatory guidelines for drinking water.

Among the earlier EPA methods for 1,4-dioxane, a heated PT method was proposed for polar, water-soluble VOCs (Lucas et al., 1988). The method involved purging samples at 90°C. To shift the aqueous phase/vapor phase equilibrium constant toward the vapor phase, salt was added to the sample or the heated PT method was used. Sodium chloride was the salt selected for heated PT because it is readily available and provides the desired high ionic strength. To obtain the preferred "salting out" effect, sodium chloride and sodium or magnesium sulfate are used at concentrations that attain 80% saturation at 65°C. 1,4-Dioxane recovery with heated PT increased 21% without salt and as much as 56% with addition of sodium sulfate salt. The heated PT method provided a preliminary MDL of 4 ppb and limit of quantitation of 20 ppb (Lucas et al., 1988).

USEPA published CLP Method OLM03.1 for 1,4-dioxane in water by SIM GC–MS in 1994 (USEPA, 1994). The method uses a heated PT at 50°C, followed by GC–MS analysis using SIM rather than full-scan analysis. The monitored ions have mass-to-charge ratios (m/z) of 88 ± 0.5 and 58 ± 0.5 for 1,4-dioxane and 96 ± 0.5 and 64 ± 0.5 for 1,4-dioxane-d_8 as an internal standard. Method OLM03.1 could reportedly achieve a practical quantitation limit of 5 µg/L (USEPA, 1994).

In 1997, USEPA's Central Regional Laboratory, operated by EPA Region V in Chicago, published a method for measurement of purgeable 1,4-dioxane in water using wide-bore capillary column GC–MS (Rudinsky et al., 1997). The method uses a heated PT with sodium sulfate and 1,4-dioxane-d_8 added as the internal standard. The method was accurate for samples with 1,4-dioxane concentrations between 20 and 200 µg/L. A summary of instrumentation and operating parameters for this method is included in Appendix 3.

Researchers with the Shaw Group and USEPA in Las Vegas conducted a multilaboratory comparison study of purgeable versus extractable methods and conventional versus isotope-dilution quantitation methods (Strout et al., 2004b). The results show that quantitation by isotope dilution provides higher accuracy and precision than the conventional method of quantitation, regardless of whether the method uses purgeable or extractable sample preparation.

Carryover contamination can occur with the 1,4-dioxane purgeable method, and method blank contamination was observed in the study. High recoveries of low spiked concentrations, that is, 2 µg/L, indicate that carryover was occurring. Carryover contamination must be addressed by running an additional method or instrument blanks, which adds to the method cost and time of analysis. The study concluded that analysis of 1,4-dioxane using a 5 mL heated purge, an accelerated sample purge flow, and SIM MS detection produces the most linear calibration for a concentration range from 0.5 to 200 µg/L, as well as the highest signal-to-noise (S/N) ratio. Running a 25 mL purge volume with the same analytical conditions provided a calibration over a narrower concentration range (0.5–20 µg/L). The optimal sample temperature is approximately 50°C, and the optimal sample purge flow is approximately 50 cm³/min for the analysis of 1,4-dioxane as a purgeable analyte on the SIM MS system (Strout et al., 2004b).

USEPA's CLP developed a draft protocol for the analysis of trace 1,4-dioxane using the SIM method. SIM requires method blanks at the beginning and end of the analytical sequence and continuing calibration verification (CCV) standards. Quantitation is performed by using the average relative response factor (RRF) from the initial calibration. USEPA Method 8260 is often used with SIM for analysis of 1,4-dioxane, as described in Section 4.5.2.

4.5.1 ENVIRONMENTAL PROTECTION AGENCY METHOD 8015B

Method 8015 analyzes nonhalogenated volatile and semivolatile organic compounds by GC and flame-ionization detection. The recommended sample-preparation methods for analysis of 1,4-dioxane by EPA Method 8015B are direct injection and azeotropic microdistillation. Poor purging efficiency is noted for 1,4-dioxane sample preparation by PT, and solvent extraction is not recommended for EPA Method 8015B analysis of 1,4-dioxane (USEPA, 1996a). Four GC columns are recommended for EPA Method 8015B, including a DB-Wax column suitable for 1,4-dioxane retention. The recommended columns are summarized in Table 4.6.

For reagent grade water, groundwater, and leachate, EPA Method 8015B can achieve MDLs of 12, 15, and 16 µg/L, respectively (according to EPA's Toxicity Characteristic Leaching Procedure test; USEPA, 1996c). Detection limits for other media are shown in Table 4.4.

TABLE 4.6
Recommended Columns for EPA Method 8015B

Column Length	Column ID	Column Material	Column Film Thickness (µm)
8 feet	2.54 mm (0.1 inch)	Stainless steel or glass packed, 1% SP-1000 on Carbopack-B 60/80 mesh	—
6 feet	2.54 mm (0.1 inch)	Stainless steel or glass column packed with *n*-octane on Porasil-C 100/120 mesh (Durapak)	—
30 m	0.53 mm	Fused-silica capillary column bonded with DB-Wax (or equivalent)	1
30 m	0.53 mm	Fused-silica capillary column chemically bonded with 5% methyl silicone (DB-5, SPB-5, RTx, or equivalent)	1.5

Source: U.S. Environmental Protection Agency (USEPA), 1996a, Method 8015B: Non-halogenated organics using GC/FID. U.S. Environmental Protection Agency.

4.5.2 Environmental Protection Agency Method 8260B

Method 8260B is a method that uses GC–MS and can therefore achieve lower detection limits for 1,4-dioxane than EPA Method 8015B (FID, i.e., GC–FID). For 1,4-dioxane by 8260B, recommended sample-preparation methods include EPA Methods 5031 and 5032, that is, azeotropic microdistillation and vacuum extraction, respectively. For most of the 107 compounds on the EPA Method 8260B analyte list, the estimated quantitation limit for groundwater samples is 5 µg/L. If azeotropic microdistillation is used for sample preparation and if 1,4-dioxane-d_8 is used as an internal standard, the general reporting limit for 1,4-dioxane by EPA Method 8260B is 12 µg/L (USEPA, 1996b). In practice, 1,4-dioxane reporting limits by EPA Method 8260B are typically in the 70 µg/L range (Lancaster Laboratories, 2004). Some laboratories provide an 8260B analysis for 1,4-dioxane with SIM, which improves the detection limit of the method. For example, commercial laboratories using 8260B-SIM with a heated PT can achieve an MDL of 0.5 µg/L (Lancaster Laboratories, 2004) and a reporting limit of 1 µg/L (Curtis and Tomkins Laboratory, 2006).

The chromatographic columns recommended for EPA Method 8260 are listed in Table 4.7.

TABLE 4.7
Recommended Columns for EPA Method 8015B

Column Length (m)	Column ID (mm)	Column Material	Column Film Thickness (µm)
60	0.75	Capillary column coated with VOCOL (Supelco)	1.5
75	0.53	Capillary column coated with DB-624 (J&W Scientific), Rt-502.2 (RESTEK), or VOCOL (Supelco)	3.0
30	0.25–0.32	Capillary column coated with 95% dimethyl—5% diphenyl polysiloxane (DB-5, Rt-5, SPB-5, or equivalent)	1.0
60	0.32	Capillary column coated with DB-624 (J&W Scientific)	1.8

Source: U.S. Environmental Protection Agency (USEPA), 1996b, Method 8260b: Volatile organic compounds by gas chromatography/mass spectrometry (GC/MS). U.S. Environmental Protection Agency.

TABLE 4.8
**Method 8260C Accuracy and Precision for 1,4-Dioxane by Mean Percent Recovery (*R*)
and Percent Relative Standard Deviation (RSD)**

	25 ppb Spike		100 ppb Spike		500 ppb Spike	
	R	RSD	*R*	RSD	*R*	RSD
1,4-Dioxane-d_8	63	25	55	16	54	13

Source: U.S. Environmental Protection Agency (USEPA), 2006c, Method 8260C: Volatile organic compounds by gas chromatography/mass spectrometry (GC/MS). U.S. Environmental Protection Agency.

Notes: Sample preparation by Method 5031—the microdistillation technique (single laboratory and single operator). Lower RSD values reflect increased precision. Data from analysis of seven aliquots of reagent water spiked at each concentration, using a quadrupole mass spectrometer in the selected-ion monitoring mode.

4.5.3 ENVIRONMENTAL PROTECTION AGENCY METHOD 8260C

EPA Method 8260C is a heated PT method that uses a 1,4-dioxane-d_8 standard. The sample is heated to 90°C, which produces an approximately fivefold increase in 1,4-dioxane recovery over conventional analysis; however, the 1,4-dioxane recovery can nevertheless be as low as 1%. Using heated PT requires careful control on temperature to generate reproducible results. Because the majority of the 1,4-dioxane remains in the sparger, thorough rinsing and bake-out must be performed to prevent carryover (Fitzpatrick and Tate, 2006). The description for EPA Method 8260C requires that the sample be heated to at least 80°C before 1,4-dioxane can be considered a viable analyte by PT. Sample preparation may also include Method 5031, azeotropic distillation, and Method 5032, closed-system vacuum distillation (USEPA, 2006c). The recovery of 1,4-dioxane described in Method 8260C is low, but the recovery is indexed to the isotope standard, allowing quantitation of 1,4-dioxane. Recoveries and RSDs for 1,4-dioxane analyses by Method 8260C at a single laboratory by a single operator are listed in Table 4.8. Method 8260C does not list a lower limit of quantitation for 1,4-dioxane.

4.5.4 ENVIRONMENTAL PROTECTION AGENCY METHOD 8261A

EPA Method 8261A uses vacuum distillation to vaporize compounds and separate VOCs, some low-boiling-point semivolatile organic compounds, and polar nonpurgeable or low-purging compounds from soil, water, and other environmental sample matrices. The volatilized material passes over a chilled condenser coil to condense and remove water. The analytes are condensed by using liquid nitrogen to obtain cryogenic temperatures (down to –196°C). Volatile organic analytes are then introduced by a vacuum distiller into the gas chromatograph, which is interfaced with a mass spectrometer. The combination of cryogenic trapping and vacuum distillation is described by EPA Method 5032.

EPA Method 8261A uses internal standards to measure matrix effects and to compensate analyte responses for matrix effects. This method should be considered for samples in which matrix effects are anticipated to severely affect analytical results (USEPA, 2006d). The inventor of the vacuum distillation technique, Michael Hiatt, is a USEPA scientist; EPA was granted a patent for the method (Hiatt, 1995).

The EPA vacuum distillation method accounts for the physical chemistry of analytes in measuring matrix effects, yielding method performance data by analyte. Results for analyses by EPA Method 8261A are reported with confidence intervals. EPA Method 8261A eliminates the need for matrix spike–matrix spike duplicates as well as calibration of instrumentation by matrix type (Hiatt, 2007).

TABLE 4.9
Laboratory Performance for 1,4-Dioxane Analyses by EPA Method 8261, Vacuum Distillation (VD/GC/MS)

Lab	Sample Size	Matrix	Number of Samples	Spike Concentration (µg/L)	Average Recovery (%)	Relative Standard Deviation[a]	Method Detection Limit (Estimated)
1	5 g	Soil	8	10	186	16.7	9.4 µg/kg
2	5 g	Soil	7	10	87	20.9	5.7 µg/kg
1	5 mL	Water	8	20	79	24.9	11.6 µg/L
2	5 mL	Water	7	10	113	17.9	6.3 µg/L
2	25 mL	Water	7	4	113	17.7	2.5 µg/L

Lab	Number of Samples	Sample Matrix and 1,4-Dioxane Concentration (ppb)	Average Recovery at Different Concentrations				Relative Standard Deviation[a]
			Low (%)	Medium (%)	High (%)	Overall (%)	
1a	9	Water: 1, 10, and 50	128	103	114	116	6.4
2b	12	Water: 1, 3, and 5	327	145	178	217	13.9
1c	9	Soil: 1, 10, and 50	137	103	122	121	21.0
2d	12	Soil: 1, 3, and 5	156	127	166	150	21.0

Matrix of Five Blind Samples (5 mL)	Spike Level (µg/L)	Average Recovery (%)	Average Relative Standard Deviation[a]
Water	375	95	11.4
Water	125	111	7.8
Water	55	102	13.6
Water	7	196	24.2
Water	1	660	24.0

Source: Strout, K., et al., 2004a, Vacuum distillation unit interlaboratory study evaluation. Paper presented at the 20th Annual National Environmental Monitoring Conference, July 19–22, Independent Laboratory Institute, Washington, DC. http://nemc.us/proceedings/2004proceedings.pdf (accessed January 2005). With permission.

[a] Lower relative standard deviation (RSD) values reflect increased precision. Recovery of volatilized 1,4-dioxane may be diminished by condensation in the apparatus (Hiatt, 2007).

In vacuum distillation, the effect of boiling point and relative volatility on the internal standards must be addressed by correcting the GC–MS response. Surrogate recovery is determined to predict the recoveries of analytes within surrogate groups on the basis of relative volatility[*] and boiling point. Once the recoveries of the internal standards have been corrected for boiling-point effects, recoveries then reflect only the matrix effects owing to relative volatility. Relative volatility is also corrected by using a software algorithm for processing the data generated from EPA Method 8261A (USEPA, 2006d). Method 8261A does not list a lower limit of quantitation for 1,4-dioxane. An estimated MDL for 1,4-dioxane on a 25 mL aliquot of water is 2.5 µg/L, whereas the routine injection of 5 mL of sample produces an estimated MDL or 6.3 µg/L (Strout et al., 2004a). Table 4.9 presents data from laboratory performance studies for EPA Method 8261A.

---

[*] "Relative volatility" describes the distillation of one compound relative to that of another compound as a ratio of their respective partition coefficients.

4.5.5 ENVIRONMENTAL PROTECTION AGENCY METHOD 8270C—EXTRACTED SEMIVOLATILE ORGANIC COMPOUNDS BY GAS CHROMATOGRAPHY–MASS SPECTROSCOPY

EPA Method 8270C is a GC–MS method designed to analyze for semivolatile organic compounds on extracted samples. This method targets compounds that are soluble in methylene chloride and capable of being eluted as sharp peaks from a gas chromatographic fused-silica capillary column coated with a slightly polar grade of silicone (USEPA, 1996c). Sample-preparation methods for EPA Method 8270C include LLE. EPA Method 8270C uses a single-column configuration, 30-m-long silicone-coated fused-silica capillary column (J&W Scientific DB-5 or equivalent) with a 0.25 or 0.32 mm internal diameter and a 1 μm film thickness (USEPA, 1996c). 1,4-Dioxane is not a listed analyte in EPA Method 8270C; however, the method can be adapted for compound-specific analysis with isotope dilution to quantify 1,4-dioxane with reporting limits as low as 0.5 μg/L (Weck Laboratories, 2006).

In analyses using extractable sample-preparation methods, 1,4-dioxane elutes immediately after the solvent front,[*] which reduces interference from other compounds. In isotope dilution, samples are spiked with 1,4-dioxane-d_8 and then subjected to routine sample extraction procedures. For improved resolution, the sample extract can be reanalyzed with the mass spectrometer in SIM mode. Isotope dilution in MS analysis involves comparing the mass differences between 1,4-dioxane and its deuterated analog, 1,4-dioxane-d_8. 1,4-Dioxane-d_8 is the ideal internal standard for 1,4-dioxane because it has similar chemical properties as 1,4-dioxane, but its mass spectra are distinctly different (Linton and Alonso, 2006). Samples can be analyzed for 1,4-dioxane by using the full-scan method with other semivolatile analytes, but with higher detection limits (Strout et al., 2004b).

Isotope dilution improves precision and accuracy by reducing the problems with calibration and sample-preparation matrix effects. Some analytes are not easily recovered during chromatography or sample extraction. Calibration problems and matrix effects are usually compensated for in part by using internal standards and surrogate analytes. In 1,4-dioxane-d_8, hydrogen atoms on the 1,4-dioxane molecule are replaced with deuterium atoms.[†] The 1,4-dioxane-d_8 is available as "99 atom% deuterium," meaning that 99% of the sites on the 1,4-dioxane ether ring normally occupied by ^{1}H are occupied by deuterium in the standard. In GC analyses, the error from methods used to inject the sample can be approximately 5%. By adding an internal standard, the error can be reduced by half to approximately 2–3%. With isotope dilution, the error should be half of that again, approximately 1% (Lindsay, 2000).

To achieve low-level reporting limits by EPA Method 8270C, commercial laboratories have adjusted the method by adding a salting-out step that uses large-volume injection of sodium chloride or sodium sulfate and by running the mass spectrometer in single ion mode. For example, a method used by Columbia Analytical Services involves separatory funnel extraction using dichloromethane on a 100 mL sample spiked with 1,4-dioxane-d_8 and prepared with NaCl. The extract is dried and reduced to a final volume of 5 mL. A large-volume injector is used to introduce the sample to the GC–MS, which is run in SIM mode. Columbia Analytical Services reports that a method reporting limit of 0.1 μg/L is routinely achievable through the use of this approach (Grindstaff, 2004).

4.5.6 ENVIRONMENTAL PROTECTION AGENCY METHOD 1624—ISOTOPE-DILUTION GAS CHROMATOGRAPHY–MASS SPECTROSCOPY

EPA Method 1624 is a modification of Method 8270C. Method 1624 also adds 1,4-dioxane-d_8 to the sample as an internal standard prior to extraction. Losses of the target analyte that may occur during the

---

[*] The solvent front is the chromatographic peak corresponding to the carrier solvent, usually dichloromethane, used to extract the analyte from the sample.

[†] Deuterium, written as ^{2}H or D, is the naturally occurring stable isotope of hydrogen. It has one extra neutron and is therefore heavier than hydrogen. Deuterium has the same chemical properties as hydrogen. A "deuterated" standard is prepared so that molecular sites normally occupied by ^{1}H are replaced with deuterium.

sample concentration step in EPA Method 8270C are reduced in EPA Method 1624 by limiting the degree of concentration (2 mL final volume versus 1 mL final volume by the standard procedure) and using high-volume injection into the gas chromatograph (Linton and Alonso, 2006). The detection limit of EPA Method 1624 depends more on the level of interferences than instrumental limitations (USEPA, 2001).

1,4-DIOXANE METHOD COMPARISON MINI-STUDY

A groundwater investigation to delineate 1,4-dioxane occurrence was conducted at a former industrial facility located in Tallevast, Florida. Analyses for 1,4-dioxane initially used a modified Method 8270C. Beginning in 2006, samples were analyzed with both EPA Method 8270C and EPA Method 8260B in SIM mode. Results reported for the Method 8260B (SIM) analysis were consistently and significantly higher than those reported for the Method 8270C analysis, in some cases, as much as two orders of magnitude higher. Environmental scientists with the consulting firm conducting the investigation, Blasland Bouck and Lee Environmental Services (now part of Arcadis), undertook a comparison study to identify differences in analytical methods for 1,4-dioxane (Linton and Alonso, 2006).

Samples were collected in triplicate in two separate sampling events from 13 monitoring wells and two private wells by using low-flow sampling techniques. Analysis was performed by EPA Methods 8270C, 8260B, and 8270 with isotope dilution ("8270-ID") at three different commercial laboratories. The purpose of the comparison study was to obtain a more quantitative statement of a pattern of conflicting results for analyses by different parties using different methods to analyze samples from domestic wells. The conflicting results led to controversy in the affected community in Bradenton County, Florida, and brought the subtleties of GC–MS configurations into the local newspaper. Data used in the method evaluation by Linton and Alonso have not been fully validated, and the study was not intended to be a rigorous or fully validated review of these three analytical methods. Full validation of the data would require an exhaustive review of laboratory quality control procedures, field methods, and so on. Samples were analyzed by Method 8260B in standard scan mode initially and subsequently reanalyzed in SIM mode if results were below the detection limit. To check for analyte loss and potential interference, field samples were spiked at three different concentrations and used as matrix spike–matrix spike duplicate (MS/MSD) samples for all three methods to determine method accuracy (percent recovery) and method precision (relative percent difference, RPD). Recovery was also measured on laboratory control standards, to distinguish instrument accuracy from possible matrix effects attributable to sample characteristics.

Results for samples analyzed by Method 8260B were highest among the three methods; some results were more than an order of magnitude higher than the Method 8270C results. Accuracy was consistent for the three spike concentration levels, and precision increased somewhat with spike concentration. Method 8260B in scan and SIM modes had an average recovery of 73% and an average RPD of 5%. Accuracy increased with increasing spike concentration, ranging from 49% recovery for the low spike level to 98% recovery for the high spike level. Precision was relatively constant at all spike concentration levels.

As shown in Figure 4.7, the results using Method 8260B paralleled results for the isotope-dilution method. Method 8270C had the lowest recoveries. The average recovery was 65%, and the average RPD for recoveries in MS/MSD analyses was 20%. Precision and accuracy of Method 8270C decreased as the concentration of spike addition increased. Overall accuracy for Method 8270ID averaged 100% recovery, and precision averaged 9% RPD in the first sampling event and 12% in the second sampling event. Accuracy was independent of spike concentrations, whereas precision increased somewhat with spike concentration. Figure 4.7 presents a comparison of the concentration trends for the results from the three methods.

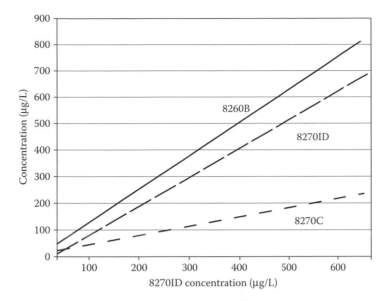

FIGURE 4.7 Paired comparisons of the results of EPA Methods 8260B and 8270C with 8270–Isotope Dilution from Linton and Alonso Method Comparison Study. (From Linton, J. and Alonso, J.C. 2006, 1,4-Dioxane comparative analysis study—EPA Methods 8260B, 8270C and 8270 with isotope dilution. Tampa, FL: Blasland Bouck and Lee Environmental Services [now Arcadis-US]. With permission.)

Results for samples from each well by Methods 8260B and 8270C were compared to results from that well by Method 8270ID. In Figure 4.7, results from a specific well by Methods 8260B and 8270C (*y*-axis) are plotted against the concentration obtained by Method 8270ID (*x*-axis). Regression analysis and a paired *t*-test showed that the results of Method 8260B did not significantly differ from those of the isotope-dilution method, and the methods correlated well. BBL recommended that the EPA consider performing further optimization of Method 8260B to increase the sensitivity of the method. Increasing the sample and purge volumes may decrease the MDL, and using a heated purge would increase the purging efficiency of 1,4-dioxane from the groundwater samples, thereby increasing sensitivity of the method.

CAVEAT EMPTOR—DOUBLE-BLIND STANDARDS CAN HELP TO ENSURE GOOD DATA QUALITY

The majority of contaminated site investigations rely upon laboratories' internal quality control protocols to confirm that results are within acceptable limits of precision and accuracy. The subject of laboratory quality control is beyond the scope of this book; the reader is referred to USEPA's quality pages (http://www.epa.gov/quality/) for a thorough review of the fundamentals of laboratory quality control. Quality control tests include method blanks, initial and continuing calibration standards, matrix spike and recovery analyses, and several others. For emerging contaminants, unconventional analyses, analyses reported at the low end of the method range, and analyses performed infrequently by laboratories, it is sometimes useful to confirm laboratory performance independently by submitting double-blind standards.

Double-blind standards are usually prepared by a third-party laboratory or provider of analytical standards. The subject laboratory's sample container is shipped to the standards

continued

provider, where a whole-volume standard with a value similar to concentrations found at a field site is prepared and filled in the container and then express-shipped back to the laboratory client. The client then labels the sample with a fictitious name similar to names of other samples in the set and ships it in a cooler with other field samples from a sampling event. Laboratory results are then compared with the value of the whole-volume standard given by the standards provider. Laboratory performance can then be gauged to confirm that the laboratory's in-house quality control tests are accurate. Because there is analytical error associated with both the whole-volume standard and the subject laboratory's analysis, the combined error range can be larger than the errors of the individual analyses. Nevertheless, the laboratory client can use double-blind standards to independently verify the reasonableness of reported results.

Standards for 1,4-dioxane and 1,4-dioxane-d_8 can be purchased from laboratory standards providers. If desired, the whole-volume standard can be prepared by using the same matrix as other samples in the set—for example, groundwater from a monitoring well that has consistently tested nondetect for 1,4-dioxane—and then analyzing a second sample from a whole-volume split of that groundwater to confirm absence of 1,4-dioxane. Matrix-matched double-blind standards prepared in this manner are less obvious to laboratory analysts, who might otherwise recognize a double-blind standard by its low electrical conductivity and use extra care when conducting the analysis.

4.5.7 ENVIRONMENTAL PROTECTION AGENCY METHOD 522—SOLID-PHASE EXTRACTION AND GAS CHROMATOGRAPHY–MASS SPECTROSCOPY WITH SELECTIVE ION MONITORING

USEPA's National Exposure Research Laboratory is developing a new method for measuring 1,4-dioxane and other water-soluble solvent-stabilizer compounds in drinking water. The method applies SPE, using coconut charcoal as the solid sorbent and dichloromethane as the eluent, to extract and concentrate hydrophilic volatile organic chemicals from water matrices. Extracts are analyzed by GC–MS with large-volume injection. Preliminary trials achieved recovery of 1,4-dioxane from both reagent and tap water at 89.5% and 95.3%, respectively, with RSDs of less than 6.3% (Munch, 2006). Preliminary data for four other solvent stabilizers—1,2-butylene oxide, 1,3-dioxolane, *t*-butanol, and epichlorohydrin—demonstrate 92.5–114% recovery with RSDs of less than 5.1% (Munch, 2006). USEPA is developing this method as EPA Method 522, which has been drafted and is undergoing evaluation (personal communication with Andrew Eaton, Vice President, MWH Laboratories, Monrovia, California, June 2008).

4.6 LABORATORY SAFETY FOR 1,4-DIOXANE HANDLING AND INSTRUMENT CLEANING

1,4-Dioxane can react with oxygen to form explosive peroxides. Discovery of old bottles of full-strength ethers can lead to expensive removal by demolition experts, or worse, accidental detonation. The peroxide crystals often form around the threads of the bottle top and explode when compressed if an unsuspecting laboratory chemist attempts to open the bottle. This problem is more likely to occur at facilities using pure 1,4-dioxane such as histology laboratories, liquid scintillation counters, and research chemistry laboratories. Several news stories have profiled incidents in which old cans or bottles of dioxane have been discovered, including a hospital in Fort Lauderdale, Florida, a laser facility in Santa Clara, California, and a college laboratory in Salisbury, North Carolina. In each case, these discoveries caused facility shutdown and a bomb squad response to remove the container of 1,4-dioxane to a remote location for detonation. Once peroxides form, controlled detonation is the safest method to remove the hazard. The explosive power of a gallon of peroxidized 1,4-dioxane is probably equivalent to several hand grenades, according to the reported explosive power of other peroxidizable materials such as ethyl ether and isopropyl ether. 1,4-Dioxane is also known to form explosive mixtures with nitromethane, sulfur trioxide, silver perchlorate, and decaborane (NICNAS, 1998). Laboratory standards used for instrument calibration and quality control are generally low concentration and unlikely to form peroxides.

Unstabilized 1,4-dioxane may also form toxic and hazardous formate esters of 1,2-ethanediol over long periods of storage. The mono- and diformates of 1,2-ethanediol are formed by peroxide intermediates and have been found at concentrations as high as 1.8 M in partially consumed bottles of 1,4-dioxane. Purging containers of 1,4-dioxane with nitrogen each time they are opened, keeping containers away from light, and refrigerating 1,4-dioxane can prevent the possible health hazard caused by contamination with peroxide and formate ester (Jewett and Lawless, 1980). As described in Chapter 3, photo-oxidation of 1,4-dioxane in the atmosphere produces ethylene-1,2-diformate.

As discussed in Chapter 2, 1,4-dioxane can be stabilized to prevent peroxide and formate ester formation. Stabilizers include stannous chloride, ferrous sulfate, 2,6-*tert*-butyl-*p*-cresol, or butyl hydroxy toluene (BHT) (BASF, 1986, 2005; European Chemicals Bureau, 2002). Peroxide formation can also be prevented by filling the container headspace with nitrogen. When stabilized and stored under nitrogen in original containers, 1,4-dioxane has a shelf life of 24 months (BASF, 2005).

Absence of 1,4-dioxane to stabilize methyl chloroform has also created laboratory safety problems. A common method for cleaning parts of mass spectrometers is sonication (ultrasound cleaning) in a solvent bath. Before the use of methyl chloroform was banned by the Montreal Protocol, it was commonly used in laboratory cleaning operations. At Purdue University, a laboratory technician performed a routine sonication cleaning operation that involved placing aluminum mass spectrometer parts in a beaker with methyl chloroform. On this occasion, the technician used a new bottle of methyl chloroform. The methyl chloroform in the beaker began "boiling," became dark, and began producing a sharp odor that filled the laboratory, forcing evacuation. The mass spectrometer part was ruined. Most mass spectrometer parts are made from stainless steel. In this instance, the ion source from a Hewlett Packard (Agilent) 5970, 5992, 5995, or 5996 Series GC–MS was made of aluminum and reacted with the methyl chloroform (Purdue University, 2003). It is likely that the particular grade of methyl chloroform used was not intended for cleaning aluminum parts and was therefore not stabilized with 1,4-dioxane.

ACKNOWLEDGMENTS

The critical reviews and helpful comments on this chapter by Carl Isaacson, a recent doctoral student at OSU now with the USEPA in Athens, Georgia, and Bart Simmons, retired Director of the California Department of Toxic Substances Control Hazardous Materials Laboratory, are gratefully acknowledged.

BIBLIOGRAPHY

Anderson, H., 2005, Memo to D. Easley, U.S. EPA Region VII, from Gore Survey Products. EPA Region VII.

BASF, 1986, *Datenblatt Dioxan [Dioxane datasheet]*. Ludwigshafen, Germany: BASF (Badische Anilin- und Soda-Fabrik).

BASF, 2005, *Dioxane—Stab Technical Data Sheet*. Ludwigshafen, Germany: BASF (Badische Anilin- und Soda-Fabrik).

Black, L. and Fine, D., 2001, High levels of monoaromatic compounds limit the use of solid-phase microextraction of methyl *tert*-butyl ether and *tert*-butyl alcohol. *Environmental Science and Technology* 35(15): 3190–3192.

Britt, S., 2004, The Snap Sampler. Paper presented to the Interstate Technology Regulatory Council Diffusion Sampler Team, Albuquerque, NM, October 28, 2004.

Bruce, M.L., Lee, R.P., and Stephens, M.W., 1992, Concentration of water soluble volatile organic compounds from aqueous samples by azeotropic microdistillation. *Environmental Science and Technology* 26(1): 160–163.

CalEPA, 2003, *Advisory—Active Soil Gas Investigations*. Los Angeles: Los Angeles Regional Water Quality Control Board and California Department of Toxic Substances Control.

Chapman, J.R., 1993, *Practical Organic Mass Spectrometry—A Guide for Chemical and Biochemical Analysis*. Chichester: Wiley.

Cooper, C.V., White, L.D., and Kupel, R.E., 1971, Qualitative detection limits for specific compounds utilizing gas chromatographic fractions, activated charcoal and a mass spectrometer. *American Industrial Hygiene Association Journal* 32: 383–386.

Curtis and Tomkins Laboratory, 2006, 1,4-Dioxane by GC/MS with selected ion monitoring—EPA 8270 SIM. http://www.curtisandtompkins.com (accessed February 12, 2006).

Daughton, C.G., 2003, Cradle-to-cradle stewardship of drugs for minimizing their environmental disposition while promoting human health. I. Rationale for and avenues toward a green pharmacy. *Environmental Health Perspectives* 111: 757–774.

Daughton, C.G., 2005, Emerging chemicals as pollutants in the environment: A 21st century perspective. *Renewable Resources Journal* 23(4): 6–23.

Daughton, C.G., 2006, Pharmaceuticals as environmental contaminants: An overview of the science. Paper presented at the Water Education Foundation Pacific Conference.

Donaldson, W.T., 1977, Trace organics in water: Refined methods are needed to document their occurrence and concentrations. *Environmental Science and Technology* 11(4): 348–352.

Draper, W.M., Dhoot, J.S., Remoy, J.W., and Perera, S.K., 2000, Trace-level determination for 1,4-dioxane in water by isotopic dilution GC and GC-MS. *Analyst* 125(8): 1403–1408.

Epstein, P.S., Mauer, T., Wagner, M., Chase, S., and Giles, B., 1987, Determination of parts per billion concentrations of dioxane in water and soil by purge and trap gas chromatography/mass spectrometry or charcoal tube enrichment gas chromatography. *Analytical Chemistry* 59(15): 1987–1990.

European Chemicals Bureau, 2002, European Union risk assessment report: 1,4-Dioxane. Institute for Health and Consumer Protection, Luxembourg, European Union, Report 21.

Fitzpatrick, T.W. and Tate, K., 2006, 1,4-Dioxane environmental and measurement issues. In: *Proceedings, Annual Industry Workshop 2006*. Sarasota, FL: Florida Department of Environmental Protection.

Fuh, C.B., Lai, M., Tsai, H.Y., and Chang, C.M., 2005, Impurity analysis of 1,4-dioxane in nonionic surfactants and cosmetics using headspace solid-phase microextraction. *Journal of Chromatography A* 1071(1): 141–145.

Grindstaff, J., 2004, Low level analysis of 1,4-dioxane by EPA Method 8270C SIM with large volume injection. *CAS Connection* 1: 4. http://www.caslab.com/contact/cc_spring2004_email.pdf

Guo, W. and Brodowsky, H., 2000, Determination of the trace 1,4-dioxane. *Microchemical Journal* 64: 173–179.

Hayes, H.C., Benton, D.J., and Khan, N., 2006, Impact of sampling media on soil gas measurements. In: *Vapor Intrusion—The Next Great Environmental Challenge—An Update*. Los Angeles, CA: Air and Waste Management Association.

Hiatt, M., 1995, United States Patent 5,411,707: Vacuum extractor incorporating a condenser column. Assignee: U.S. Environmental Protection Agency.

Hiatt, M., 2007, An overview of SW-846 Method 8261 chemistry: Volatile organic compounds by vacuum distillation in combination with gas chromatography/mass spectrometry (VD/GC/MS): New method for analyzing organic chemicals in a wide array of samples. http://www.epa.gov/esd/chemistry/vacuum/training/pdf/theory-rev5.pdf (accessed March 23, 2007).

Isaacson, C., Mohr, T.K.G., and Field, J.A., 2006, Quantitative determination of 1,4-dioxane and tetrahydrofuran in groundwater by solid phase extraction GC/MS/MS. *Environmental Science & Technology* 40(23): 7305–7311.

Jewett, D. and Lawless, J.G., 1980, Formate esters of 1,2-ethanediol—major decomposition products of *p*-dioxane during storage. *Bulletin of Environmental Contamination and Toxicology* 25: 118–121.

Jochmann, M.A., Kmiecik, M.P., and Schmidt, T.C., 2006, Solid-phase dynamic extraction for the enrichment of polar volatile organic compounds from water. *Journal of Chromatography A* 1115: 208–216.

Kawata, K., Ibaraki, T., Tanabe, A., et al., 2001, Gas chromatographic–mass spectrometric determination of hydrophilic compounds in environmental water by solid-phase extraction with activated carbon felt. *Journal of Chromatography A* 911: 75–83.

Kurimo, R.W., 1994, NIOSH Method 1602: Dioxane, NIOSH. http://www.cdc.gov/niosh/nmam/pdfs/1602.pdf (accessed May 2004).

Lancaster Laboratories, 2004, 1,4-Dioxane in environmental samples. http://www.lancasterlabs.com/ (accessed May 14, 2004).

Lesage, S. and Jackson, R.E., 1992, *Groundwater Contamination and Analysis at Hazardous Waste Sites*. Boca Raton, FL: CRC Press.

Lesnik, B., 1993, New SW-846 methods for the analysis of conventional and non-conventional volatile organics in solid matrices (abstract). In: R.L. Siegrist and J.J. van Ee (Eds), *Proceedings, National Symposium on Measuring and Interpreting VOCs in Soils: State of the Art and Research Needs*. Las Vegas, NV: U.S. Environmental Protection Agency, January 12–14.

Lide, D.R., 2008, *Handbook of Chemistry and Physics*, 88th Edition. Boca Raton: Taylor & Francis Group.

Lindsay, E., 2000, Isotope dilution. *West Coast Analytical Services Digest* 8(1). http://www.wcaslab.com/tech/tech2.htm (accessed May 2004).

Linton, J. and Alonso, J.C., 2006, 1,4-Dioxane comparative analysis study—EPA Methods 8260B, 8270C and 8270 with isotope dilution. Tampa, FL: Blasland Bouck and Lee Environmental Services (now Arcadis-US). http://www.tallevast.info/pdfs/dioxane_may19.pdf (accessed September 2006).

Lucas, S.V., Burkholder, H.M., and Alford-Stevens, A., 1988, Heated purge and trap method development and testing. U.S. Environmental Protection Agency, Environmental Monitoring and Support Laboratory, EPA/600/4-88/029.

Markell, C., Hagen, D.F., and Bunnelle, V.A., 1991, New technologies in solid-phase extraction. *LC-GC* 9: 332–337.

McMaster, M. and McMaster, C., 1998, *GC/MS: A Practical User's Guide*. New York: Wiley-VCH.

Munch, J.W., 2006, Analysis of 1,4-dioxane and other water soluble volatile organic compounds by solid phase extraction and GC/MS. Paper presented at American Water Works Association Water Quality Technology Conference, Denver, CO.

NICNAS, 1998, Priority existing chemical assessment reports—1,4-dioxane. National Industrial Chemicals Notification and Assessment Scheme (NICNAS), Australian Government.

O'Neill, D., 2006, Passive diffusion samplers: Cost effective sampling techniques for sampling groundwater for organic and inorganic analytes of interest—highlighting rigid porous polyethylene samplers (RPPS). Presented at DOD Chemist Conference, San Antonio, TX, April.

Park, Y.-M., Pyoa, H., Park, S.-J., and Park, S.-K., 2005, Development of the analytical method for 1,4-dioxane in water by liquid–liquid extraction. *Analytica Chimica Acta* 7(548): 109–115.

Parsons, 2005, Results report for the demonstration of no-purge groundwater sampling devices at former McClellan Air Force Base, CA. Army Corps of Engineers, Omaha, Nebraska. http://www.geoinsightonline.com/resources/mcclellan_report_cover.html (accessed October 12, 2006).

Photovac, 2006, VOC detection instruments. http://www.photovac.com/White_Papers/VOC_Compounds_Detected_by_Photovac_Instruments.htm (accessed July 14, 2006).

Purdue University, 2003, The Lilly incident. http://web.ics.purdue.edu/~swihart/safety/LillyIncidentOct2003.htm (accessed August 28, 2005).

Rankin, T., 1996, Residual solvents in drugs by static headspace analysis. In: *Pharmaceutical Analysis: GC Technical Report*. Austin, TX: ThermoQuest CE www.tmqaustin.com (accessed February 2005).

Rastogi, S.C., 1990, Headspace analysis of 1,4-dioxane in products containing polyethoxylated surfactants by GC–MS. *Chromatographia* 29(9/10): 441–445.

Rödel, W. and Wölm, G., 1987, *A Guide to Gas Chromatography*. Berlin, German Democratic Republic: VEB Deutscher Verlag der Wissenschaften.

Rose, D.L. and Sandstron, M.W., 2003, Determination of gasoline oxygenates, selected degradates, and BTEX in water by heated purge and trap/gas chromatography/mass spectrometry—methods of analysis by the U.S. Geological Survey National Water Quality Laboratory. USGS Method O-4024-03:39. http://web1.er.usgs.gov (accessed February 2007).

Rudinsky, R., Tang, C.M., and Elly, C.T., 1997, CRL Method 624 VOC-dioxane: Standard operating procedure for measurement of purgeable 1,4-dioxane in water by wide-bore capillary column gas chromatography/mass spectrometry. U.S. Environmental Protection Agency, U.S. EPA Region V.

Shirey, R.E., 2000, Optimization of extraction conditions for low molecular weight analytes using solid-phase microextraction. *Journal of Chromatographic Science* 38(3): 109–115.

Shirey, R.E. and Linton, C.M., 2006, The extraction and analysis of 1,4-dioxane from water using solid phase microextraction coupled with gas chromatography and gas chromatography–mass spectrometry. *Journal of Chromatographic Science* 44(7): 444–450.

Simmons, B., 1997, Non-target compounds—you can't always get what you want. *Hydrovisions* 6(2): 1. Groundwater Resources Association of California. http://www.grac.org/summer97/non-target.htm (accessed May 2004).

Song, D. and Zhang, S., 1997, Rapid determination of 1,4-dioxane in water by solid-phase extraction and gas chromatography–mass spectrometry. *Journal of Chromatography A* 787: 283–287.

Strout, K., Hedin, C., Zimmerman, M., and Smith, T., 2004a, Vacuum distillation unit interlaboratory study evaluation. Paper presented at the 20th Annual National Environmental Monitoring Conference, July 19–22, Independent Laboratory Institute, Washington, DC. http://nemc.us/proceedings/2004proceedings. pdf (accessed January 2005).

Strout, K., Zimmerman, M., Hedin, C., and Smith, T., 2004b, Comparison of 1,4-dioxane as a volatile and semivolatile analyte in single and multi-laboratory studies. Paper presented at the 20th Annual National Environmental Monitoring Conference, Book of Abstracts, No. 29. July 19–22, Independent Laboratory Institute, Washington, DC. http://nemc.us/proceedings/2004proceedings.pdf (accessed January 2005).

Supelco/SigmaAldrich, 1997, *The Retention Index System in Gas Chromatography: McReynolds Constants.* Bellefonte, PA. http://www.sial.com (accessed January 2005).

Swallow, K.C., 1992, Nonpriority pollutant analysis and interpretation. In: S. Lesage and R.E. Jackson (Eds), *Groundwater Contamination and Analysis at Hazardous Waste Sites.* New York: Marcel Dekker.

U.S. Environmental Protection Agency (USEPA), 1986, *Test Methods For Evaluating Solid Waste, SW-846,* 3rd Edition. Washington, DC: U.S. Environmental Protection Agency, Office of Solid Waste and Emergency Response.

U.S. Environmental Protection Agency (USEPA), 1990, Update II to SW-846. Methods section. U.S. Environmental Protection Agency, Office of Solid Waste.

U.S. Environmental Protection Agency (USEPA), 1993, Behavior and determination of volatile organic compounds in soil: A literature review. U.S. Environmental Protection Agency, Office of Research and Development. EPA/600/R-93/140.

U.S. Environmental Protection Agency (USEPA), 1994, 1,4-Dioxane in water by selective ion monitoring (SIM) gas chromatography/mass spectrometry (GC/MS)—EPA CLP Method OLM03.1. U.S. Environmental Protection Agency.

U.S. Environmental Protection Agency (USEPA), 1996a, Method 8015B: Non-halogenated organics using GC/ FID. U.S. Environmental Protection Agency.

U.S. Environmental Protection Agency (USEPA), 1996b, Method 8260b: Volatile organic compounds by gas chromatography/mass spectrometry (GC/MS). U.S. Environmental Protection Agency.

U.S. Environmental Protection Agency (USEPA), 1996c, Method 8270C semivolatile organic compounds by gas chromatography/mass spectrometry (GC/MS). U.S. Environmental Protection Agency.

U.S. Environmental Protection Agency (USEPA), 1999, Compendium Method TO-15: Determination of volatile organic compounds (VOCs) in air collected in specially prepared canisters and analyzed by gas chromatography/mass spectrometry (GC/MS). U.S. Environmental Protection Agency.

U.S. Environmental Protection Agency (USEPA), 2001, Method 1624 revision B—volatile organic compounds by isotope dilution GC/MS. 40 CFR Chapter I, Part 136, Appendix A, Method 1624, July 1, Edition.

U.S. Environmental Protection Agency (USEPA), 2006a, Test Methods SW-846 on line. http://www.epa.gov/ epaoswer/hazwaste/test/main.htm (accessed May 4, 2005).

U.S. Environmental Protection Agency (USEPA), 2006b, Characterization methods web site. http://www. clu-in.org/char/technologies/gc.cfm (accessed December 2006).

U.S. Environmental Protection Agency (USEPA), 2006c, Method 8260C: Volatile organic compounds by gas chromatography/mass spectrometry (GC/MS). U.S. Environmental Protection Agency.

U.S. Environmental Protection Agency (USEPA), 2006d, Method 8261A: Volatile organic compounds by vacuum distillation in combination with gas chromatography/mass spectrometry (VD/GC/MS). U.S. Environmental Protection Agency.

Vainberg, S., McClay, K., Masuda, H., et al., 2006, Biodegradation of ether pollutants by *Pseudonocardia* sp. strain ENV478. *Applied and Environmental Microbiology* 72(8): 5218–5224.

Vroblesky, D.A., Borchers, J.W., Campbell, T.R., and Kinsey, W., 2000, Investigation of polyethylene passive diffusion samplers for sampling volatile organic compounds in ground water at Davis Global Communication Station, Sacramento, CA, August 1998 to February 1999. U.S. Geological Survey Open-File Report 00-307.

Wala-Jerzykiewicz, A. and Szymanowski, J., 1998, Headspace gas chromatography analysis of toxic contaminants in ethoxylated alcohols and alkylamines. *Chromatographia* 48(3/4): 299–304.

Weck Laboratories, 2006, Method reporting limits. http://www.wecklabs.com/Resources/MethodReportingLimits/ SemivolatileOrganics/tabid/130/Default.aspx (accessed February 12, 2006).

Yasuhara, A., Shiraishi, H., Nishikawa, M., et al., 1999, Organic components in leachates from hazardous waste disposal sites. *Waste Management and Research* 17(3): 186–197.

Yasuhara, A., Tanaka, Y., Tanabe, A., and Katami, T. 2003, Elution of 1,4-dioxane from waste landfill sites. *Bulletin of Environmental Contamination and Toxicology* 71: 641–646.

Yoo, L., 2002. Simultaneous determination of 1,2,3-TCP and 1,4-dioxane in drinking water by GC-MS/MS using purge/trap techniques. Orange County Water District.

Yoo, L.J., Fitzsimmons, S., and Wehner, M., 2002, Improved purge-trap and GC/MS/MS techniques for the trace-level determination of 1,4-dioxane in water. In: *Proceedings, Water Quality Technology Conference*. Seattle, WA: American Water Works Association, CD-ROM.

5 The Toxicology of 1,4-Dioxane

Julie A. Stickney

The risk assessment and regulation of 1,4-dioxane as an environmental contaminant depend on assumptions regarding human exposure and toxic potency. The toxicity of 1,4-dioxane in humans is evaluated by considering studies in laboratory animals and limited data from human exposures. This chapter briefly summarizes the toxicological data for 1,4-dioxane. The uptake and behavior of this chemical within the body are discussed (i.e., toxicokinetics). Toxicity studies in humans are presented; these consist of case reports of occupational poisoning, volunteer studies of acute inhalation exposure, and limited epidemiology studies of occupational workers. The acute effects of high-dose exposures (less than one month in duration) on laboratory animals are also described. These studies are relevant in identifying the potential outcomes of accidental or industrial poisoning from 1,4-dioxane, but do not provide useful information about lower-level environmental exposures. Subchronic and chronic laboratory animal studies are designed to evaluate the potential for health effects to result from exposure to different doses over longer periods of time. These studies are generally used to assess long-term environmental exposures when human data are limited or absent. Brief study summaries are provided for the subchronic and chronic toxicity studies conducted for 1,4-dioxane. Also discussed are mechanistic studies that provide some insight into the potential mechanisms of action leading to toxicological outcomes.

Taken together, the human and laboratory animal data identify possible health effects that may result from exposure to 1,4-dioxane. The primary exposure routes leading to 1,4-dioxane toxicity are inhalation and ingestion. Significant inhalation exposures to 1,4-dioxane are likely to occur only in occupational settings. Oral exposure in humans may result from ingestion of contaminated drinking water. Liver and kidney toxicities are the primary noncancer health effects associated with exposure to 1,4-dioxane in humans and laboratory animals. Liver tumors found in rats and mice in several lifetime drinking-water studies suggest that oral exposure to 1,4-dioxane may result in carcinogenicity in humans. Liver tumors generally occur at dose levels that also produce liver toxicity. Genotoxicity data for 1,4-dioxane are generally negative and do not support a mutagenic mode of action for carcinogenicity. Alternative modes of carcinogenic action may involve cytotoxicity followed by regenerative cell proliferation or growth promotion of liver cells in the absence of significant cytotoxicity (i.e., mitogenesis).

5.1 TOXICOKINETICS

Toxicokinetics describe the absorption, distribution, metabolism, and excretion of a substance from the human body following exposure by oral, dermal, or inhalation routes. The toxicokinetics of 1,4-dioxane are described by using human volunteer studies, laboratory animal studies, *in vitro* test systems, and computer modeling [i.e., physiologically based pharmacokinetic (PBPK) models]. Toxicokinetic data for 1,4-dioxane in humans are limited; however, the absorption, distribution, metabolism, and elimination of this compound are well described in rats exposed via the oral, inhalation, and intravenous (i.v.) routes.

1,4-Dioxane is well absorbed following oral and inhalation exposure. Studies in workers and human volunteers demonstrated rapid absorption following inhalation exposure to 1,4-dioxane by measuring concentrations of 1,4-dioxane and its metabolite, β-hydroxyethoxy acetic acid (HEAA), in blood and urine (Young et al., 1976, 1977). Similar findings were reported for inhalation exposure in rats (Young et al., 1978a,b). Gastrointestinal absorption was nearly complete in male rats administered oral doses of 10–1000 mg/kg,[*] given as a single dose or as 17 consecutive daily doses (Young et al., 1978a,b). Dermal absorption data for 1,4-dioxane are limited. A study in rhesus monkeys showed little dermal penetration from a methanol or skin lotion vehicle (Marzulli et al., 1981); however, significant dermal absorption was suggested by an *in vitro* experiment using excised human abdominal skin and three different vehicle preparations (water, lotion, and isopropyl myristate) (Bronaugh, 1982).

No data are available to describe the distribution of 1,4-dioxane in humans or animals after oral or inhalation exposure to 1,4-dioxane. After an injection of radiolabeled 1,4-dioxane into the abdominal cavity [intraperitoneal (i.p.) injection] of male Sprague Dawley rats, the tissue distribution was generally uniform across liver, kidney, spleen, lung, colon, and skeletal muscle 1, 2, 6, and 12 h later (Woo et al., 1977a). Blood concentrations were higher than tissue concentrations at all times except for 1 h postdosing, when kidney levels were approximately 20% higher than blood. A PBPK modeling study suggested that 1,4-dioxane may be transferred to breast milk in lactating mothers (Fisher et al., 1997).

Covalent binding (i.e., irreversible binding) was measured in liver cells 16 h after dosing by i.p. injection (Woo et al., 1977a). Covalent binding to cellular proteins and lipids can lead to toxicity, and binding to deoxyribonucleic acid (DNA) may lead to mutagenesis. Binding was greatest in the nuclear, microsomal, and mitochondrial fractions of the cell; little covalent binding was seen in the cytoplasmic fluid of liver cells (cytosol of hepatocytes). Pretreatment of rats with mixed-function oxidase[†] inducers [phenobarbital, 3-methylcholanthrene, or polychlorinated biphenyls (PCBs)] did not significantly change the extent of covalent binding in subcellular fractions of hepatocytes (Woo et al., 1977a). This finding suggests that oxidative metabolism of 1,4-dioxane in the liver does not lead to the formation of reactive intermediates that covalently bind to cellular constituents resulting in toxicity.

A proposed metabolic scheme for 1,4-dioxane is presented in Figure 5.1. The major product of 1,4-dioxane metabolism is HEAA. This metabolite could be generated from oxidation of 1,4-dioxane via diethylene glycol (DEG), 1,4-dioxane-2-ol, or 1,4-dioxane-2-one (Woo et al., 1977b). Oxidation to DEG and HEAA appears to be the most likely, because DEG was found as a minor metabolite in Sprague Dawley rat urine following a single 1000 mg/kg gavage[‡] dose of 1,4-dioxane (Braun and Young, 1977). Additionally, i.p. injection of 100–400 mg/kg DEG in Sprague Dawley rats resulted in urinary elimination of HEAA (Woo et al., 1977b). 1,4-Dioxane oxidation was shown to be mediated by the cytochrome P450 (CYP) enzyme system in the liver.[§] CYP induction with phenobarbital or Arochlor 1254 (a PCB mixture) and suppression with 2,4-dichloro-6-phenylphenoxy ethylamine or cobaltous chloride were effective in significantly increasing and decreasing, respectively, the appearance of HEAA in the urine of Sprague Dawley rats (Woo et al., 1977c, 1978). 1,4-Dioxane itself induced the CYP-mediated metabolism of other compounds in laboratory animals (Mungikar and Pawar, 1978). Nannelli et al. (2005) described the induction of several isoforms of CYP in hepatic, renal, and nasal tissues following exposure to 1,4-dioxane in rats.

Metabolism of 1,4-dioxane to HEAA is extensive in humans following inhalation exposure. Workers exposed to a time-weighted average (TWA) concentration of 0.6 ppm 1,4-dioxane for 7.5 h had a ratio of 118:1 for HEAA and 1,4-dioxane concentrations measured in the urine (Young et al., 1976). Similarly, in male volunteers exposed to 50 ppm for 6 h (Young et al., 1977), more than 99% of inhaled 1,4-dioxane appeared in the urine as HEAA. Metabolism in rats is also

---

[*] Milligrams of 1,4-dioxane per kilogram of body weight of the subject animal.

[†] Mixed-function oxidase enzymes catalyze the oxidation of drugs and chemicals.

[‡] Gavage is the introduction of material into the stomach by a tube.

[§] Cytochrome P450 enzymes are a family of intracellular hemoprotein-containing enzymes that function in the oxidative metabolism of drugs and chemicals.

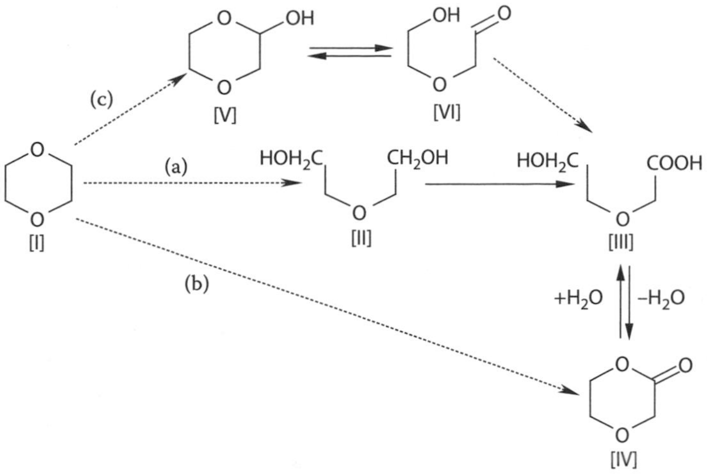

FIGURE 5.1   Suggested metabolic pathways of 1,4-dioxane in the rat. I = 1,4-dioxane; II = DEG; III = β-hydroxyethoxy acetic acid (HEAA); IV = 1,4-dioxane-2-one; V = 1,4-dioxane-2-ol; VI = β-hydroxyethoxy acetaldehyde. (Adapted from Woo, Y., Argus, M.F., and Arcos, J.C., 1977c, *Biochemical Pharmacology* 25: 1539–1542. With permission.)

extensive after inhalation exposure. Urinary levels of HEAA were 3000-fold higher than urinary 1,4-dioxane concentrations after exposure to 50 ppm for 6 h (Young et al., 1978a,b). 1,4-Dioxane metabolism was shown to be a saturable process, as demonstrated by oral and i.v. exposures to various doses of 1,4-dioxane in rats (Young et al., 1978a,b). In rats given radiolabeled 1,4-dioxane via gavage in distilled water as single doses of 10, 100, or 1000 mg/kg or in 17 daily doses of 10 or 1000 mg/kg, the urinary excretion of radiolabeled metabolites decreased significantly with increasing dose, while the radiolabel detected in expired air increased (i.e., exhalation of unmetabolized 1,4-dioxane).

1,4-Dioxane is primarily eliminated as HEAA in the urine; however, the exhalation of 1,4-dioxane in breath increases at higher doses because of metabolic saturation. In workers exposed to a TWA of 0.6 ppm for 7.5 h, 99% of 1,4-dioxane eliminated in urine was in the form of HEAA (Young et al., 1976). The elimination half-life was 59 min in male volunteers exposed to 50 ppm 1,4-dioxane for 6 h (Young et al., 1977). As in humans, the elimination half-life in rats exposed to 50 ppm 1,4-dioxane for 6 h was calculated to be 1.01 h (Young et al., 1978a,b). After oral exposure to rats, urinary elimination ranged from 76% to 99%, depending on the dose (i.e., urinary elimination was decreased at higher doses). Elimination of 1,4-dioxane in expired air increased with increasing dose. Fecal elimination was less than 2% for all doses (Young et al., 1978a,b).

PBPK models are quantitative computer models that describe the movement and fate of chemicals in the body. PBPK models can be used to account for the saturation of 1,4-dioxane metabolism at high doses and adjust for differences in physiology and metabolism between humans and rats. Two PBPK models were previously used for 1,4-dioxane risk assessment (Leung and Paustenbach, 1990; Reitz et al., 1990). These models simulated exposure via i.v., inhalation, and oral pathways. 1,4-Dioxane intake from drinking water was simulated by assuming rapid absorption of the chemical from the gastrointestinal tract directly to the liver. Uptake from air was modeled by assuming rapid equilibration between alveolar air* and pulmonary capillary blood. Blood leaving the lungs was distributed to four compartments including liver, fat, slowly perfused tissues such as muscle and skin, and richly perfused tissues such as kidney, brain, and viscera.[†] The models incorporated

---

* Alveolar air is air held in the alveolar air spaces of the lung.
[†] Perfusion refers to forcing a fluid through an organ or tissue, especially by way of the blood vessels.

kinetic constants for HEAA formation in rats and humans in order to estimate the dose delivered to the liver. New experimental data have recently been generated to reduce the uncertainty associated with the PBPK models for 1,4-dioxane (Sweeney et al., 2008). An updated PBPK model was developed that may be used in future risk assessments for 1,4-dioxane (Sweeney et al., 2008).

5.2   TOXICITY STUDIES IN HUMANS

The human data for 1,4-dioxane consist of case reports of occupational poisoning, volunteer studies of acute inhalation exposure, and epidemiology studies of workers occupationally exposed to 1,4-dioxane. The study details for each of the human studies are presented in Table 5.1. Several fatal cases of acute or chronic inflammation of the kidney (hemorrhagic nephritis) and localized liver tissue death (centrilobular necrosis of the liver) were related to occupational exposure (i.e., inhalation and dermal contact) to 1,4-dioxane (Barber, 1934; Johnstone, 1959). Neurological changes were also reported in one case, including headache, elevation in blood pressure, agitation and restlessness, and coma (Johnstone, 1959). Perivascular widening (enlargement of blood vessels) was observed in the brain of this worker, with localized damage to the nerve sheath (demyelination) in several regions (e.g., cortex and basal nuclei).

Human volunteer studies showed that acute inhalation exposure to high concentrations of 1,4-dioxane (>200 ppm) for several minutes produced irritation of the eyes, nose, and throat (Yant et al., 1930; Fairley et al., 1934; Wirth and Klimmer, 1936; Silverman et al., 1946). Eye irritation was also seen following a 6-h exposure to 50 ppm 1,4-dioxane (Young et al., 1977). No clinical symptoms or change in blink frequency, nasal swelling, pulmonary function, or inflammatory markers in the plasma (i.e., C-reactive protein, interleukin-6) were seen after exposure to 20 ppm 1,4-dioxane for 2 h (Ernstgård et al., 2006). The epidemiology studies in occupational workers exposed to 1,4-dioxane are predominantly negative (Thiess et al., 1976; Buffler et al., 1978); however, only limited conclusions can be drawn from these negative findings because of the small size of the worker populations that were studied and the small number of cases available for identification of low-level excess health risk. In addition, the mean exposure duration of the mortality study (Buffler et al., 1978) was only five years (less than two years for 43% of workers), and the latency period for evaluation was short (less than 10 years for 59% of workers).

5.3   TOXICOLOGY STUDIES IN LABORATORY ANIMALS

5.3.1   Acute and Short-Term Toxicity Studies (<1 Month Exposure)

The acute and short-term toxicity studies of 1,4-dioxane in laboratory animals are summarized in Table 5.2. Several exposure routes were employed in these studies, including dermal application, drinking-water exposure, gavage, vapor inhalation, and i.v. or i.p. injection. Mortality was observed in many acute high-dose studies, and LD_{50} values[*] for 1,4-dioxane were calculated for rats, mice, and guinea pigs (see Table 5.2; Laug et al., 1939; Smyth et al., 1941; Pozzani et al., 1959). Clinical signs of central nervous system (CNS) depression were observed, including staggered gait, narcosis,[†] paralysis, coma, and death (de Navasquez, 1935; Knoefel, 1935; Schrenk and Yant, 1936; Laug et al., 1939; Nelson, 1951). The acute neurotoxicity of 1,4-dioxane was further investigated in several studies (Goldberg et al., 1964; Frantik et al., 1994; Kanada et al., 1994). Frantik et al. (1994) measured the inhibition of the propagation and maintenance of an electrically evoked seizure discharge following acute inhalation exposure to 1,4-dioxane. This endpoint has been correlated with the behavioral effects and narcosis that occur following acute exposure to higher concentrations of organic solvents. Linear regression analysis of the concentration-effect data was used to calculate an isoeffective air

---

[*] LD_{50} is the median lethal dose of a toxin at which 50% of the exposed organisms are killed.
[†] Narcosis is a state of stupor, drowsiness, or unconsciousness produced by drugs or toxic substances.

TABLE 5.1
Human Toxicity Studies for 1,4-Dioxane

Study Type	Exposure Conditions	Concentration	Results	References
Case report	Four workers exposed to 1,4-dioxane in the workplace	Considered to be a high exposure, but concentration not specified	Liver and kidney toxicities—fatal cases of hemorrhagic nephritis and centrilobular necrosis of the liver	Barber (1934)
Case report	One worker exposed via inhalation and dermal contact	Mean air concentrations ranged from 208 to 650 ppm	Toxicity in the liver, kidney, and brain—fatal case of hemorrhagic necrosis of the kidney cortex, centrilobular necrosis of the liver, perivascular widening and small foci of demyelination in several regions of the brain (i.e., cortex and basal ganglia)	Johnstone (1959)
Volunteer study	Six men and six women	20 ppm for 2 h at rest	No clinical symptoms reported; no change was observed in blink frequency, pulmonary function, nasal swelling, or inflammatory markers in plasma (C-reactive protein, interleukin-6)	Ernstgård et al. (2006)
Volunteer study	Four men Six men	1000 ppm for 5 min 2000 ppm for 3 min	No symptoms in most volunteers, constriction in the throat noted for one volunteer	Fairley et al. (1934)
Volunteer study	12 men and 12 women	200 or 300 ppm for 15 min	200 ppm was tolerable; 300 ppm caused irritation to the eyes, nose, and throat	Silverman et al. (1946)

continued

TABLE 5.1 (continued)
Human Toxicity Studies for 1,4-Dioxane

Study Type	Exposure Conditions	Concentration	Results	References
Volunteer study	Five men	>280 ppm for several minutes 1400 ppm for several minutes	Mucus membrane irritation of the nose and throat	Wirth and Klimmer (1936)
Volunteer study	Five men	5000 ppm for 1 min; 1600 ppm for 10 min	Irritation and burning sensation in nose and throat	Yant et al. (1930)
Volunteer study	Four men	50 ppm for 6 h	Eye irritation; odor detection diminished over time	Young et al. (1977)
Pharmacokinetic study in workers	Four workers	1.6 ppm for 7.5 h	No clinical signs reported; urinary excretion of 1,4-dioxane and HEAA was measured (ratio of 1:118)	Young et al. (1976)
Epidemiology (cross-sectional survey)	74 German workers exposed to 1,4-dioxane (24 still actively employed in 1,4-dioxane production at the time of study, 23 employed at other jobs in the plant, 27 retired or deceased)	Measured concentrations 0.06–0.69 ppm; estimated previous concentrations 0.06–7.2 ppm; exposure duration from 3–41 years	Results of medical examinations normal; hematology, clinical chemistry, renal function, and urinalysis normal; no increase in chromosome aberrations observed	Thiess et al. (1976)
Epidemiology (mortality study)	165 workers employed in manufacturing or processing of 1,4-dioxane	Estimated exposure levels for different job categories: low (<25 ppm), intermediate (50–75 ppm), high (>75 ppm); mean exposure duration <5 years	No apparent excess in mortality or deaths due to cancer	Buffler et al. (1978)

TABLE 5.2
Acute and Short-Term Toxicity Studies of 1,4-Dioxane (Less than 1-Month Exposure)

Animal	Exposure Route	Test Conditions	Results	Dose[a]	References
Oral Studies					
Rat (inbred strain and gender unspecified)	Oral via drinking water	Drinking-water exposure for 1–10 days	Ultrastructural changes in the kidney, degenerative nephrosis, hyaline droplet accumulation, crystal formation in mitochondria	11,000 mg/kg-day (5%)	David (1964)
Rat (strain and gender unspecified)	Oral via drinking water	Drinking-water exposure for 5–12 days	Extensive degeneration of the kidney, liver damage, mortality in 8/10 animals by 12 days	11,000 mg/kg-day (5%)	Kesten et al. (1939)
F344/DuCrj rat	Oral via drinking water	Drinking-water exposure for 14 days	Mortality, decreased body weights, histopathological lesions in the nasal cavity, liver, kidney, and brain	2500 mg/kg-day (nuclear enlargement of olfactory epithelial cells), >7500 mg/kg-day for all other effects	JBRC (1998a)
Male Sprague Dawley rat	Gavage	Single-dose gavage	Decreased dopamine and serotonin content of the hypothalamus	1050 mg/kg-day	Kanada et al. (1994)
Female Sprague Dawley rat	Gavage	0, 168, 840, 2550, or 4200 mg/kg by gavage; 21 and 4 h prior to sacrifice	Increased ODC activity, hepatic CYP450 content, and DNA single-strand breaks	840 mg/kg (ODC activity only)	Kitchin and Brown (1990)
Female Carworth Farms–Nelson rat	Gavage	Single-dose gavage	Lethality	LD$_{50}$ = 6400 mg/kg (14,200 ppm)	Pozzani et al. (1959)
Male Wistar rat, guinea pig	Gavage	Single-dose gavage	Lethality	LD$_{50}$ (mg/kg): rat = 7120, guinea pig = 3150	Smyth et al. (1941)
Rat, mouse, and guinea pig	Gavage	Single-dose gavage	Clinical signs of CNS depression, stomach hemorrhage, kidney enlargement, and liver and kidney degeneration	LD$_{50}$ (mg/kg): mouse = 5900, rat = 5400, and guinea pig = 4030	Laug et al. (1939)
Rabbit	Gavage	Single-gavage dose of 0, 207, 1034, or 2068 mg/kg-day	Clinical signs of CNS depression, mortality at 2068 mg/kg, renal toxicity (polyuria followed by anuria), histopathological changes in liver and kidneys	1034 mg/kg-day	de Navasquez (1935)
Rat, rabbit	Gavage	Single dose; mortality after two weeks	Mortality and narcosis	3160 mg/kg	Nelson (1951)

continued

TABLE 5.2 (continued)
Acute and Short-Term Toxicity Studies of 1,4-Dioxane (Less than 1-Month Exposure)

Animal	Exposure Route	Test Conditions	Results	Dose[a]	References
Crj:BDF1 mouse	Oral via drinking water	14-day exposure	Mortality, decreased body weights, histopathological lesions in the nasal cavity, liver, kidney, and brain	10,800 mg/kg-day: hepatocellular swelling	JBRC (1998c)
Dog	Drinking-water ingestion	3–10 days of exposure	Clinical signs of CNS depression and liver and kidney degeneration	11,000 mg/kg-day (5%)	Schrenk and Yant (1936)
Inhalation Studies					
Male CD1 Rat	Vapor inhalation	Serum enzymes measured before and after single 4-h exposure	Increase in ALT, AST, and OCT; no change in G-6-pase	1000 ppm	Drew et al. (1978)
Rabbit, guinea pig, rat, and mouse	Vapor inhalation	3 h for five days; 1.5-h exposure for one day	Degeneration and necrosis in the kidney and liver, vascular congestion in the lungs	10,000 ppm	Fairley et al. (1934)
Male Wistar rats; female H-strain mice	Vapor inhalation	4 h in rats; 2 h in mice	30% decrease in the maximal response to an electrically evoked seizure	1860 ppm in rats; 2400 ppm in mice	Frantik et al. (1994)
Female Carworth Farms–Elias rat	Vapor inhalation	4 h per day, five days per week, for 10 exposure days	Dose-related effect on conditioned avoidance behavior; escape behavior was generally not affected (pole-climb methodology)	3000 ppm	Goldberg et al. (1964)
Rat	Vapor inhalation	5 h of exposure	Mortality and narcosis	6000 ppm	Nelson (1951)
Female Carworth Farms–Nelson rat	Vapor inhalation	Determination of a 4-h inhalation LC50	Lethality	$LC_{50} = 51.3$ mg/L	Pozzani et al. (1959)
Mouse, cat	Vapor inhalation	8 h per day for 17 days	Paralysis and death	8400 ppm	Wirth and Klimmer (1936)

Species	Route	Protocol	Effects	Dose	Reference
Guinea pig	Vapor inhalation	8 h of exposure to 0.1–3 vol%	Eye and nasal irritation, retching movements, altered respiration, narcosis, pulmonary edema and congestion, hyperemia of the brain	0.5 vol%	Yant et al. (1930)
Other Routes					
Male COBS/Wistar rat	Dermal	Nonoccluded technique using shaved areas of the back and flank; single application, 14-day observation	Negative; no effects noted	8300 mg/kg	Clark et al. (1984)
Rabbit, cat	i.v. injection	Single injection of 0, 207, 1034, 1600 mg/kg-day	Clinical signs of CNS depression, narcosis at 1034 mg/kg, mortality at 1600 mg/kg	1034 mg/kg-day	de Navasquez (1935)
Rat (strain not specified); rabbit (strain not specified)	i.p. injection	Single dose	Narcosis, staggering posture	30 mmol/kg (rat); 50 mmol/kg (rabbit)	Knoefel (1935)
Female Sprague Dawley rat	i.p. injection	Single dose; LD_{50} values determined 24 h and 14 days after injection	Increased serum SDH activity at 1/16th of the LD_{50} dose; no change at higher or lower doses	LD_{50} (mg/kg): 24 h = 4848, 14 days = 799	Lundberg et al. (1986)
CBA/J mouse	i.p. injection	Daily injection for seven days; 0%, 0.1%, 1%, 5%, and 10%	Slightly lower lymphocyte response to mitogens	2000 mg/kg-day (10%)	Thurman et al. (1978)

Notes: ND = no data; OCT = ornithine carbamyl transferase; ODC = ornithine decarboxylase; and SDH = sorbitol dehydrogenase.

[a] Lowest effective dose for positive results/highest dose tested for negative results.

concentration that corresponds to the concentration producing a 30% decrease in the maximal response to an electrically evoked seizure. The isoeffective air concentrations for 1,4-dioxane were approximately 1860 ppm in rats and 2400 ppm in mice. Goldberg et al. (1964) evaluated the effect of solvent inhalation on pole-climb performance in rats. Female rats were exposed to 0, 1500, 3000, or 6000 ppm of 1,4-dioxane in air for 4 h per day, five days per week, for 10 exposure days. 1,4-Dioxane exposure produced a dose-related effect on conditioned avoidance behavior in female rats, while escape behavior was generally not affected. Kanada et al. (1994) evaluated the effect of oral exposure to 1,4-dioxane on the regional neurochemistry of the rat brain. 1,4-Dioxane was administered by gavage to male Sprague Dawley rats at a dose of 1050 mg/kg, approximately equal to one-fourth the oral LD_{50}. 1,4-Dioxane exposure was shown to reduce the content of the neurotransmitters dopamine and serotonin[*] in the hypothalamus, while the neurochemical profile of all other brain regions in exposed rats was similar to control rats.

Severe liver and kidney degeneration and necrosis were often seen in acute studies [de Navasquez, 1935; Schrenk and Yant, 1936; Kesten et al., 1939; Laug et al., 1939; David, 1964; Japan Bioassay Research Center (JBRC), 1998a]. JBRC (1998a) additionally reported microscopic changes (histopathological lesions) in the nasal cavity and the brain of rats following two weeks of exposure to 1,4-dioxane in the drinking water.

5.3.2 SUBCHRONIC AND CHRONIC TOXICITY STUDIES

Subchronic and chronic toxicity studies are conducted in laboratory animals to identify potential health effects associated with long-term exposure. Well-conducted animal studies use multiple dose groups with a sufficient number of animals per group. Multiple parameters are evaluated to assess toxicological effects (i.e., survival, body weight, food and water consumption, observation of clinical signs, hematology, clinical chemistry, urinalysis, organ weights, gross pathology, and microscopic histopathology). Dose levels that cause significant mortality or large decreases in animal body weight (>10%) are considered to exceed a maximum tolerated dose (MTD). Because animals at these dose levels are severely ill and may die early (i.e., before the end of the study), useful information pertaining to more subtle organ-system changes that may occur at lower doses given over a longer period of time cannot be obtained. Generally, studies attempt to have the highest-dose group represent an MTD and include several lower-dose groups to assess other adverse health effects. Measured parameters from toxicology studies are taken together to determine the nature and severity of an adverse outcome. For example, changes in hematology, clinical chemistry, and urinalysis measures can support observed histopathological findings of organ-system toxicity. However, small changes in single measures of these parameters (e.g., less than twofold change in serum enzymes) in the absence of other measures of toxicity are not always indicative of an adverse health effect. Histopathological changes are generally considered to be evidence of organ-system toxicity; however, some microscopic changes may simply represent an adaptive response to chemical exposure.

The study summaries provided here give a brief overview of methods and findings. More detailed information on methodology and statistical analysis of the results can be found in the cited study reports or publications. Dose levels—stated on the basis of milligrams of 1,4-dioxane per kilogram of body weight per day—were estimated for the purpose of this summary by using assumptions regarding animal body weight and drinking-water ingestion (USEPA, 1988), if not provided by the study authors. Most of the subchronic and chronic studies conducted for 1,4-dioxane were oral drinking-water studies. Table 5.3 summarizes study details (i.e., species, doses, and duration) for the oral subchronic and chronic studies for 1,4-dioxane. The highest-dose (or highest-exposure) level that produced no observed adverse effect (NOAEL) and the lowest-dose level that produced an observed adverse effect (LOAEL) are indicated, and the toxicological effects that were observed at

---

[*] Dopamine and serotonin are neurotransmitters involved in the regulation of motor function and emotional well-being.

the LOAEL dose are described. Carcinogenic findings, including the target organ, tumor type, and the lowest dose producing a significant increase in tumors, are also indicated for chronic studies. Longer-term inhalation studies consisted of only one subchronic study (Fairley et al., 1934) and one chronic study (Torkelson et al., 1974) (both discussed here).

As reported by Fairley et al. (1934), six rats and six mice (unspecified strains) were given drinking water containing 1.25% 1,4-dioxane for as long as 67 days (approximate daily doses were 1900 and 3300 mg/kg), respectively. Gross pathology and histopathology were evaluated in all animals. Five of the six rats in the study died or were sacrificed at the point of death (in extremis) prior to day 34 of the study. Mortality was lower in mice; five of six mice survived up to 60 days. Kidney enlargement and renal cortical degeneration were observed in rats and mice. Large areas of necrosis were observed in the cortex, whereas cell degeneration in the medulla was slight or absent. Tubular casts[*] were observed and vascular congestion[†] and hemorrhage were present throughout the kidney. Hepatocellular degeneration with vascular congestion was also noted in rats and mice.

In a study by Stott et al. (1981), male Sprague Dawley rats (4–6 per group) were given average daily doses of 0, 10, or 1000 mg/kg 1,4-dioxane in their drinking water, seven days per week for 11 weeks. An increase in the liver-to-body-weight ratio and minimal centrilobular hepatocyte swelling were observed in rats from the high-dose group. Hepatic DNA synthesis, measured by incorporation of the tritium-labeled DNA nucleoside thymidine [(^{3}H)-thymidine], was increased 1.5-fold in high-dose rats. No changes relative to the control were observed for rats exposed to 10 mg/kg per day.

The JBRC administered 1,4-dioxane in the drinking water to groups of six-week-old F344/DuCrj rats (10 per sex per group) and Crj:BDF$_1$ mice (10 per sex per group) for 13 weeks (JBRC, 1998b). The concentrations of 1,4-dioxane in the water for rats and mice were 0, 640, 1600, 4000, 10,000, or 25,000 ppm. On the basis of drinking-water ingestion and body-weight changes, male rats received daily 1,4-dioxane doses of approximately 0, 60, 150, 330, 760, and 1900 mg/kg, and female rats received daily doses of 0, 100, 200, 430, 870, and 2020 mg/kg. Male mice received daily doses of 0, 100, 260, 580, 920, or 1830 mg/kg and female mice received daily doses of 0, 170, 410, 920, 1710, or 2700 mg/kg.

One female rat in the high-dose group (2020 mg/kg per day) died, but cause and time of death were not specified. Food consumption was reduced at the highest dose in rats, and final body weights were reduced at the two highest-dose levels in both male and female rats. A dose-related decrease in water consumption was observed. Hematological effects (i.e., effects on blood) were within 2–15% of control values, and serum biochemistry parameters in treated rats did not differ more than twofold from control values. Absolute and relative kidney weights were increased in females at daily doses of ≥200 mg/kg. No organ weight changes were noted in male rats. Histopathology findings in rats that were related to 1,4-dioxane exposure included alterations in the nasal and tracheal epithelial cells, liver lesions (hepatocyte enlargement and cytoplasmic changes), changes in the kidney (altered appearance of proximal tubule cells), and vacuolar changes in the brain (i.e., changes in vacuole spaces within the cytoplasm of brain cells).

One male mouse in the high-dose group (1830 mg/kg per day) died, but no information was provided regarding the cause or time of the death. Food consumption was not significantly reduced in any exposure group. Final body weights were decreased in high-dose male mice, but were not significantly reduced (i.e., within 10% of controls) in the other male dose groups and in female mice. A dose-related decrease in water consumption was observed in male and female mice. Hematological changes were within 2–15% of control values. With the exception of a threefold increase in the enzymatic activity of alanine aminotransferase (ALT) in male and female mice, serum biochemistry parameters in treated rats did not differ more than twofold from control values. Increases in absolute and relative lung and kidney weights were seen in male and female mice. Histopathology

---

[*] Tubules in the kidneys secrete proteins; when kidney function is impaired, these proteins can precipitate and fill the tubules. The small cylindrical shapes that result and that can be seen under light microscopy are called casts.
[†] Vascular congestion is a condition in which the blood vessels become engorged with blood.

TABLE 5.3
Subchronic and Chronic Oral Toxicity Studies for 1,4-Dioxane

Species	Dose/Duration	NOAEL (mg/ kg·day)	LOAEL (mg/ kg·day)	Noncancer Effects at the LOAEL Dose	Target Organ and Tumor Type (if Applicable)	Lowest Dose with Significant Increase in Tumor Incidence (mg/kg·day)[a]	References
Subchronic Studies							
Rat and mouse (six per species); unknown strain	Rats 0 or 1900 mg/kg-day; mice 0 or 3300 mg/kg-day for 67 days	NA	1900 (rats) 3300 (mice)	Renal cortical degeneration and necrosis, hemorrhage; hepatocellular degeneration	NA	NA	Fairley et al. (1934)
Male Sprague Dawley rat (four to six per group)	0, 10, or 1000 mg/kg-day for 11 weeks	10	1000	Minimal centrilobular hepatocyte swelling; increased DNA synthesis	NA	NA	Stott et al. (1981)
F344/DuCrj rat (10 per sex per group)	Males 0, 60, 150, 330, 760, and 1900 mg/kg-day; females 0, 100, 200, 430, 870, and 2020 mg/ kg-day for 13 weeks	60	150	Nuclear enlargement of nasal respiratory epithelium; hepatocyte swelling	NA	NA	JBRC (1998b)
Crj:BDF₁ mouse (10 per sex per group)	Males 0, 100, 260, 580, 920, or 1830 mg/kg-day; females 0, 170, 410, 920, 1710, or 2700 mg/ kg-day for 13 weeks	170	410	Nuclear enlargement of bronchial epithelium	NA	NA	JBRC (1998b)
Chronic Studies							
Male Wistar rat (26 treated, nine controls)	0 or 640 mg/kg-day for 63 weeks	NA	640	Hepatocytes with enlarged hyperchromic nuclei; glomerulonephritis	Liver (hepatocellular carcinoma)	640	Argus et al. (1965)
Male Sprague Dawley rats (30 per group)	0, 430, 574, 803, and 1032 mg/ kg-day for 13 months	NA	430	Hepatocytomegaly; glomerulonephritis	Nasal cavity (squamous cell carcinoma, liver (hepatocellular carcinoma)	NA[b]	Argus et al. (1973), Hoch-Ligeti et al. (1970)

Species (group)	Dose	NOAEL (mg/kg-day)	LOAEL (mg/kg-day)	Non-neoplastic effects	Neoplastic effects	Tumor dose (mg/kg-day)	Reference
Sherman rat (60 per sex per dose group)	Males 0, 9.6, 94, and 1015 mg/kg-day; females 0, 19, 148, and 1599 mg/kg-day for two years	9.6	94	Degeneration and necrosis of renal tubular cells and hepatocytes	Nasal cavity (squamous cell carcinoma), liver (hepatocellular carcinoma)	1307 (average of male and female dose), 1307 (average of male and female dose)	Kociba et al. (1974)
Osborne-Mendel rat (35 per sex per dose level)	Males 0, 240, and 530 mg/kg-day; females 0, 350, and 640 mg/kg-day for 110 weeks	NA	240	Pneumonia, gastric ulcers, and cortical tubular degeneration in the kidney	Nasal cavity (squamous cell carcinoma); liver (hepatocellular adenoma)	530 (males), 350 (females); 350 (females)	NCI (1978)
B6C3F$_1$ mouse (50 per sex per dose level)	Males 0, 720, and 830 mg/kg-day; females 0, 380, and 860 mg/kg-day for 90 weeks	NA	380	Pneumonia and rhinitis	Liver (hepatocellular adenoma or carcinoma)	720 (males), 380 (females)	NCI (1978)
F344/DuCrj rat (50 per sex per dose level)	Males 0, 16, 81, or 398 mg/kg-day; females 0, 21, 103, or 514 mg/kg-day for two years	81	398	Atrophy of nasal olfactory epithelium; nasal adhesion and inflammation	Nasal cavity (squamous cell carcinoma, sarcoma, rhabdomyo-sarcoma, and esthesioneuro-epithelioma), liver (hepatocellular adenoma or carcinoma)	398 (males), 514 (females); 398 (males), 514 (females)	JBRC (1998c)
Crj:BDF$_1$ mouse (50 per sex per dose level)	Males 0, 66, 251, or 768 mg/kg-day; females 0, 77, 323, or 1066 mg/kg-day for two years	77	323	Nasal inflammation	Liver(hepatocellular adenoma or carcinoma)	768 (males), 77 (females)	JBRC (1998c)
Developmental Studies							
Sprague Dawley rat (18–20 per group)	Pregnant dams 0, 250, 500, or 1000 mg/kg-day on gestation days 6–15	500	1000	Delayed ossification of the sternebrae and reduced fetal body weights	NA	NA	Giavini et al. (1985)

NA = not applicable.

[a] $p \leq 0.05$ by Fisher's exact test pair-wise comparison with controls.

[b] Precise incidences cannot be calculated because the number of rats per group was reported as a range (28–32); incidence in control rats was not reported; no statistical analysis of the results was conducted.

findings in mice that were related to exposure included liver lesions (cell swelling and single-cell necrosis) and alterations in nasal and bronchial epithelial cells.

In a study by Argus et al. (1965), 26 adult male Wistar rats were exposed to 1,4-dioxane in the drinking water at a concentration of 1% for 63 weeks (approximate daily dose of 640 mg/kg). A group of nine untreated rats served as controls. Six of the 26 treated rats developed liver tumors (hepatocellular carcinomas). No liver tumors were observed in control rats. In two rats that died after 21.5 weeks of treatment, preneoplastic changes (histological changes preceding the development of tumors) appeared to involve the entire liver. Animals killed after 60 weeks of treatment showed small neoplastic nodules* or multifocal hepatocellular carcinomas.[†] No cirrhosis was observed in this study.[‡] Many rats had extensive changes in the kidneys often resembling glomerulonephritis.[§] This effect progressed from increased cellularity to thickening of the glomerular capsule followed by obliteration of the glomeruli.

In two later studies (Hoch-Ligeti et al., 1970; Argus et al., 1973), groups of 2–3-month-old male Sprague Dawley rats (28–32 per dose group) were administered 1,4-dioxane in the drinking water for as long as 13 months at concentrations of 0%, 0.75%, 1.0%, 1.4%, or 1.8% (approximate daily doses of 0, 430, 574, 803, and 1032 mg/kg). The progression of liver tumorigenesis was evaluated by an additional group of 10 male rats administered 1% 1,4-dioxane in the drinking water (574 mg/kg per day), five of which were sacrificed after eight months of treatment and five of which were killed after 13 months of treatment.

The first change observed in the liver was an increase in the size of the nucleus of the hepatocytes, mostly in the periportal area.[¶] Precancerous changes were characterized by disorganization of the rough endoplasmic reticulum, an increase in smooth endoplasmic reticulum, and a decrease in glycogen and increase in lipid droplets in hepatocytes. These changes increased in severity in the hepatocellular carcinomas in rats exposed to 1,4-dioxane for 13 months. Cirrhosis of the liver was not observed.

Nasal cavity tumors were observed upon gross examination of six rats with tumors visible either at the tip of the nose, bulging out of the nasal cavity, or on the back of the nose covered by intact or later ulcerated skin. As the tumors obstructed the nasal passages, the rats had difficulty breathing and lost weight rapidly. Neurological signs and compression of the brain were not observed. In all cases, the tumors were squamous cell carcinomas with marked keratinization** and formation of keratin pearls. Bony structure was extensively destroyed in some animals with tumors, but there was no invasion into the brain. Kidney toxicity was observed following exposure to 1,4-dioxane. Effects included glomerulonephritis and pyelonephritis,[††] with characteristic epithelial proliferation of Bowman's capsule,[‡‡] periglomerular fibrosis, and distension of tubules. No kidney tumors were found.

Hoch-Ligeti and Argus (1970) also provided a brief account of the results of exposure of guinea pigs to 1,4-dioxane. 1,4-Dioxane was administered to 22 male guinea pigs in the drinking water for 23–28 months. A group of 10 untreated guinea pigs served as controls. The authors stated that the concentration of 1,4-dioxane was regulated so that normal growth of the guinea pigs was maintained and that the amount of 1,4-dioxane received by the guinea pigs over a 23-month period was

---

* A neoplasm is the abnormal proliferation of cells in tissue or organs to form a tumor.
† Multifocal hepatocellular carcinomas are cancerous cell growths at multiple locations within the liver.
‡ Cirrhosis is the replacement of liver tissue by scar tissue and regenerative nodules, leading to progressive loss of liver function.
§ Glomerulonephritis is a kidney disease characterized by inflammation of the glomeruli, or small blood vessels in the kidneys.
¶ The periportal area surrounds the portal vein of the liver.
** Keratinization is the filling of cells with the fibrous proteins found in hair, nails, and claws.
†† Pyelonephritis is a bacterial infection of the kidney.
‡‡ Bowman's capsule is a capsule-shaped membrane structure surrounding the glomerulus of each mammalian kidney tubule (nephron).

588–635 g (estimated daily dose range of 944–1019 mg/kg). All animals were sacrificed within 28 months, but the scope of the postmortem examination was not reported.

Nine treated guinea pigs showed peri- or intrabronchial epithelial hyperplasia[*] and nodular mononuclear infiltration in the lungs. Four control guinea pigs had peripheral mononuclear cell accumulation in the lungs, and only one had hyperplasia of the bronchial epithelium. Also, two guinea pigs had carcinoma of the gall bladder, three had early hepatomas,[†] and one had an adenoma of the kidney. No further information was presented in the brief narrative of this study.

Kociba et al. (1974) administered 1,4-dioxane to Sherman rats (60 per sex per dose level) via the drinking water at concentrations of 0% (controls), 0.01%, 0.1%, or 1.0% for as long as 716 days (approximate daily doses of 0, 9.6, 94, and 1015 mg/kg for male rats and 0, 19, 148, and 1599 mg/kg for female rats). Treatment with 1,4-dioxane significantly increased mortality among high-dose males and females beginning at about two to four months of treatment. These rats showed degenerative changes in both the liver and kidneys. At termination, the only alteration in organ weights noted by the authors was a significant increase in absolute and relative liver weights in male and female high-dose rats. Histopathological lesions were restricted to the liver and kidney from the mid- and high-dose groups and consisted of variable degrees of renal tubular epithelial and hepatocellular degeneration and necrosis. Rats from these groups also showed evidence of hepatic regeneration, as indicated by hepatocellular hyperplastic nodule formation and evidence of renal tubular epithelial regenerative activity (observed after two years of exposure). An increase in the incidence of liver tumors (hepatocellular carcinomas) and nasal carcinomas (squamous cell carcinoma of the nasal turbinates) occurred in high-dose male and female rats.[†]

The National Cancer Institute (NCI, 1978) made a study of groups of Osborne–Mendel rats (35 per sex per dose) and B6C3F$_1$ mice (50 per sex per dose) that were administered 1,4-dioxane in the drinking water for 110 or 90 weeks, respectively, at levels of 0%, 0.5%, or 1% (estimated daily doses: 0, 240, and 530 mg/kg in male rats; 0, 350, and 640 mg/kg in female rats; 0, 720, and 830 mg/kg in male mice; and 0, 380, and 860 mg/kg in female mice). Mortality was significantly increased in treated rats, beginning at approximately one year into the study. Histopathological lesions associated with 1,4-dioxane treatment were seen in the kidneys, liver, and stomach. Kidney lesions consisted of vacuolar degeneration and/or focal tubular epithelial regeneration in the proximal cortical tubules and occasional hyaline casts. Elevated incidence of liver cell enlargement also occurred in treated female rats, and gastric ulcers occurred in treated males. The incidence of pneumonia was increased above controls in high-dose female rats. 1,4-Dioxane treatment was associated with nasal cavity tumors (squamous cell carcinomas, adenocarcinomas, and one rhabdomyoma)[§] in both sexes and liver tumors (hepatocellular adenomas) in female rats only.

Mortality was significantly increased in female mice beginning at approximately 80 weeks in the study. 1,4-Dioxane produced an increase in the incidence of mice with pneumonia (males and females) and rhinitis[¶] (females only). A dose-related increase was also seen in the incidence of hepatocellular carcinomas or adenomas in male and female mice. Tumors were characterized by parenchymal cells (the cells composing the bulk of the organ) of irregular size and arrangement and were often enlarged with hyperchromatic (densely staining) nuclei. Neoplasms were locally invasive within the liver, but metastasis to the lungs was rarely observed.[**]

In a study reported by JBRC (1998c) and Yamazaki et al. (1994), groups of F344/DuCrj rats (50 per sex per dose level) were exposed to 1,4-dioxane in the drinking water at levels of 0, 200, 1000,

---

[*] Hyperplasia is an increase in the reproduction rate of cells, sometimes seen as an initial stage in the development of cancer.

[†] Hepatomas are liver tumors.

[‡] Nasal turbinates are the spongy bones of the nasal cavity; squamous cell carcinoma is a malignant type of tumor.

[§] Rhabdomyoma is a rare benign tumor of muscle tissue.

[¶] Rhinitis is inflammation of the mucous membrane of the nose (a runny nose).

[**] Metastasis is the development of secondary tumors at a distance from a primary site of cancer.

or 5000 ppm for two years (estimated daily doses were 0, 16, 81, or 398 mg/kg in male rats and 0, 21, 103, or 514 mg/kg in female rats). Groups of Crj:BDF$_1$ mice (50 per sex per dose level) were similarly exposed to 0, 500, 2000, or 8000 ppm of 1,4-dioxane in the drinking water (estimated daily doses were 0, 66, 251, or 768 mg/kg in male mice and 0, 77, 323, or 1066 mg/kg in female mice). Survival was significantly decreased in the high-dose rat groups during the second year of the study. Neither food nor water consumption was significantly affected by treatment; however, terminal body weights were reduced in high-dose rats. Hematological changes observed in rats were generally within 15% of control values, with the exception of a 23% decrease in hemoglobin in high-dose male rats and a 27% increase in platelets in high-dose female rats. Significant changes in serum-chemistry parameters occurred only in high-dose rats [males: increased phospholipids, activity of the enzymes ALT, aspartate aminotransferase (AST), lactate dehydrogenase (LDH), alkaline phosphatase (ALP), γ-glutamyl transpeptidase (GGT), creatinine phosphokinase (CPK), potassium, and inorganic phosphorus and decreased total protein, albumin, and glucose; females: increased total bilirubin,[*] cholesterol, phospholipids, AST, ALT, LDH, GGT, ALP, CPK, and potassium, and decreased blood glucose]. Increases in serum enzyme activities ranged from <2–17-fold above control values; the largest increases were seen for ALT, AST, and GGT. Blood samples were collected only at the end of the two-year study, so altered serum chemistry in high-dose animals may be associated with the tumorigenic changes observed in the liver (see below).

In males, relative liver weights were increased in the high-dose group; in females, relative lung, liver, and kidney weights were increased at the highest dose tested. Microscopic examination of the tissues showed alterations in the nasal cavity, liver, and kidneys mainly in high-dose rats and in a few mid-dose rats. Epithelial cells lining the nasal cavity were changed in size and appearance, and inflammation and proliferation of the nasal gland were observed. Liver lesions included spongiosis hepatis,[†] hyperplasia, cyst formation, and clear and mixed-cell foci.[‡] The only observable change in the kidney was nuclear enlargement of renal proximal tubule cells. Significantly increased incidences of liver tumors (adenomas and carcinomas) and tumors of the nasal cavity occurred in high-dose male and female rats treated with 1,4-dioxane for two years. In addition, significant increases in the incidences of mesothelioma[§] of the peritoneum, mammary gland fibroadenoma, and fibroma of the subcutis[¶] were seen in high-dose male rats. High-dose females showed a significant increase in the incidence of mammary gland adenomas.

In the study in mice, survival was low in all male groups and particularly low in high-dose females. Deaths occurred primarily during the second year of the study. Food consumption was not significantly affected, but water consumption was reduced in high-dose male and female mice. Final body weights were reduced 43% in high-dose males and 15% and 45% in mid- and high-dose females, respectively. Hematological changes were within 15% of control values with the exception of a 60% decrease in platelets in high-dose female mice. Serum activities of the enzymes AST, ALT, LDH, and ALP were significantly increased in mid- and high-dose males, whereas leucine aminopeptidase (LAP) and CPK activities were increased only in high-dose males. AST, ALT, LDH, and ALP activities were increased in mid- and high-dose females, but CPK activity was increased only in high-dose females. Increases in serum enzyme activities ranged from less than two- to sevenfold above control values. Blood samples were collected only at the end of the two-year study, so altered serum chemistry in high-dose animals may be associated with the tumorigenic changes observed in the liver. Glucose and triglycerides were decreased in high-dose males and in mid- and high-dose females. High-dose females also showed decreases in serum phospholipid and albumin concentrations (not

---

[*] Bilirubin is the orange-yellow pigment formed in the liver by the breakdown of hemoglobin and excreted in bile.
[†] Spongiosis hepatis is a cyst-like lesion that arises from the perisinusoidal Ito cells of the liver.
[‡] Such foci represent preneoplastic changes in the liver.
[§] Mesothelioma is a tumor of the pleura, peritoneum, or pericardium.
[¶] Fibroadenoma is a tumor formed of mixed fibrous and glandular tissue; fibroma of the subcutis is a benign fibrous tumor in the loose connective fatty tissue found under the dermis (skin).

reported in males). Blood calcium was lower in high-dose females and was not reported in males. Relative and absolute lung weights were increased in high-dose males and in mid- and high-dose females.

Microscopic examination showed significant alterations in the epithelium of the respiratory tract (i.e., nuclear enlargement, atrophy, and inflammation), mainly in high-dose animals, although some changes occurred in mid-dose mice. Other notable changes included nuclear enlargement of the proximal tubule cells of the kidney and angiectasis in the liver in high-dose males.[*] Treatment with 1,4-dioxane resulted in an increase in the formation of liver tumors (adenomas and carcinomas) in male and female mice. The incidence of male mice with hepatocellular carcinoma or either tumor type (adenoma or carcinoma) was increased in the high-dose group only. In female mice, increased incidence was observed for each liver tumor type in all treatment groups.

Giavini et al. (1985) examined the effects of 1,4-dioxane on rat reproduction. Pregnant female Sprague Dawley rats (18–20 per dose group) were given 1,4-dioxane by gavage in water at concentrations of 0, 0.25, 0.5, or 1 mL/kg per day, corresponding to dose estimates of 0, 250, 500, or 1000 mg/kg per day. The chemical was administered at a constant volume of 3 mL/kg on days 6–15 of gestation. The dams[†] were sacrificed with chloroform on gestation day 21, and the numbers of corpora lutea,[‡] implantations, resorptions, and live fetuses were recorded. Fetuses were weighed and examined for external malformations prior to the evaluation of visceral and skeletal malformations and a determination of the degree of ossification (bone formation). Maternal weight gain was reduced by 10% in the high-dose group (1000 mg/kg per day). Food consumption for this group was 5% lower during the dosing period, but exceeded control levels for the remainder of the study. No change from control was observed in the number of implantations, live fetuses, or resorptions; however, fetal birth weight was 5% lower in the highest-dose group. 1,4-Dioxane exposure did not increase the frequency of major malformations or minor anomalies and variants. Ossification of the sternebrae[§] was reduced in the 1000 mg/kg per day dose group. The study authors suggested that the observed delay in sternebrae ossification combined with the decrease in fetal birth weight indicated a developmental delay related to 1,4-dioxane treatment.

Fairley et al. (1934) also studied rabbits, guinea pigs, rats, and mice (3–6 per species per group) that were exposed to 1000, 2000, 5000, or 10,000 ppm of 1,4-dioxane vapor for 3 h per day, five days per week, and 1.5 h on the sixth day (16.5 h per week). Animals were exposed until death occurred or were sacrificed at varying time periods. At the 10,000 ppm concentration, only one animal (rat) survived a seven-day exposure. The rest of the animals (six guinea pigs, three mice, and two rats) died within the first five exposures. Severe liver and kidney damage and acute vascular congestion of the lungs were observed in these animals. Kidney damage was described as patchy degeneration of the cortical tubules with vascular congestion and hemorrhage. Liver lesions varied from cloudy hepatocyte swelling to large areas of necrosis. At 5000 ppm, mortality was observed in two mice and one guinea pig following 15–34 exposures. The remaining animals were sacrificed following 3–5 weeks of exposure (three rabbits and three guinea pigs). Liver and kidney damage in both dead and surviving animals was similar to that described for the 10,000 ppm concentration. Liver and kidney toxicities were also apparent in animals (four rabbits, four guinea pigs, six rats, and five mice) exposed to 2000 ppm for approximately 2–6 weeks. Cortical kidney degeneration and hepatocyte degeneration and liver necrosis were observed in animals exposed to 1000 ppm for approximately 4–12 weeks (two rabbits, three guinea pigs, three rats, and four mice).

---

[*] Angiectasis is the abnormal dilation of blood vessels.

[†] A *dam* is a pregnant female rat.

[‡] Corpora lutea are temporary endocrine structures in mammals developed from an ovarian follicle and involved in the production of progestogens, the hormones needed to maintain the thick lining (endometrium) of the uterus; they provide an area rich in blood vessels where zygote(s) can be implanted and develop (i.e., so that pregnancy occurs).

[§] The sternebrae is a part of the sternum, the bony plate covering the center of the chest.

Finally, in a study by Torkelson et al. (1974) male and female Wistar rats (288 per sex) were exposed whole-body to 1,4-dioxane vapors at a concentration of 0.4 mg/L (111 ppm), 7 h per day, five days per week for two years. Control rats (192 per sex) were exposed to filtered air. Exposure to 1,4-dioxane vapors had no significant effect on mortality or body-weight gain and induced no signs of eye or nasal irritation or respiratory distress. Slight changes in hematological and clinical chemistry parameters were within the normal physiological limits and were not considered to be of toxicological importance by the investigators. Organ weights were not significantly affected, and microscopic examination of tissues did not reveal any treatment-related effects.

5.3.3 Mechanistic Toxicology Studies

5.3.3.1 Genotoxicity

Weight-of-evidence evaluations of the genetic toxicity data have concluded that 1,4-dioxane is either a very weak genotoxin (Kitchin and Brown, 1990; Ashby, 1994) or not genotoxic (NICNAS, 1998; Netherlands Organization for Applied Scientific Research, 1999). The genotoxicity data for 1,4-dioxane are presented in Table 5.4. Negative findings were reported for mutagenicity in assays of the bacteria *Salmonella typhimurium*, *Escherichia coli*, and *Photobacterium phosphoreum* (Mutatox assay) (Stott et al., 1981; Haworth et al., 1983; Nestmann et al., 1984; Khudoley et al., 1987; Kwan et al., 1990; Hellmer and Bolcsfoldi, 1992; Morita and Hayashi, 1998). Negative results were also indicated for the induction of aneuploidy[*] in yeast (*Saccharomyces cerevisiae*) (Zimmermann et al., 1985) and the sex-linked recessive lethal test in *Drosophila melanogaster* (Yoon et al., 1985); however, positive results were reported for meiotic nondisjunction[†] in *Drosophila* (Munoz and Barnett, 2002).

1,4-Dioxane was also reported as negative in the mouse lymphoma cell mutagenicity assay (McGregor et al., 1991; Morita and Hayashi, 1998). In rat hepatocytes, 1,4-dioxane exposure caused single-strand breaks in DNA at toxic concentrations *in vitro* (Sina et al., 1983). DNA single-strand breaks were also demonstrated in hepatocytes following gavage exposure to female rats (Kitchin and Brown, 1990). 1,4-Dioxane did not affect *in vivo* DNA repair in hepatocytes or the nasal cavity (Stott et al., 1981; Goldsworthy et al., 1991), but did increase hepatocyte DNA synthesis indicative of cell proliferation in several studies (Stott et al., 1981; Goldsworthy et al., 1991; Uno et al., 1994; Miyagawa et al., 1999). 1,4-Dioxane caused a transient inhibition of ribonucleic acid (RNA) polymerase A and B in the rat liver (Kurl et al., 1981). Intravenous administration of 1,4-dioxane at doses of 10 or 100 mg/rat produced inhibition of both polymerase enzymes; a quicker and more complete recovery of activity was noted for RNA polymerase A. 1,4-Dioxane did not covalently bind to DNA under *in vitro* study conditions (Woo et al., 1977a). DNA alkylation[‡] was also not detected in the liver 4 h following a single gavage exposure (1000 mg/kg) in male Sprague Dawley rats (Stott et al., 1981).

1,4-Dioxane did not produce chromosomal aberrations or micronucleus formation in Chinese hamster ovary (CHO) cells (Galloway et al., 1987; Morita and Hayashi, 1998). Results were negative in one assay for sister chromatid exchange in CHO cells (Morita and Hayashi, 1998) and were weakly positive in the absence of metabolic activation in another (Galloway et al., 1987). Studies of micronucleus formation following *in vivo* exposure to 1,4-dioxane produced equivocal results. Negative findings were reported for bone marrow micronucleus formation in B6C3F$_1$, BALB/c, CBA, and C57BL6 mice (McFee et al., 1994; Mirkova, 1994; Tinwell and Ashby, 1994). Negative findings were also indicated for micronucleus formation in peripheral blood of CD-1 mice (Morita, 1994; Morita and Hayashi, 1998). Mirkova (1994) reported an increase in bone marrow micronucleus

---

[*] Aneuploidy is the condition in which the chromosome number of a cell is not an integer multiple of the normal basic (haploid) number.

[†] Nondisjunction is the failure of the chromosomes to properly segregate during meiosis or mitosis, resulting in daughter cells with abnormal numbers of chromosomes.

[‡] DNA alkylation involves the substitution of an alkyl group for hydrogen in DNA that can prevent cell replication by preventing the separation of two DNA chains during cell division.

formation in C57BL6 mice, as compared to negative results for BALB/c mice in the same study. Morita and Hayashi (1998) demonstrated an increase in micronucleus formation in mouse hepatocytes following oral administration of 1,4-dioxane and partial hepatectomy to induce cellular mitosis. Roy et al. (2005) showed an increase in micronuclei in bone marrow erythrocytes and in proliferating hepatocytes [labeled with 5-bromo-2′-deoxyuridine (BrdU)], but not in nondividing liver cells. Gene expression profiling in human hepatoma HepG2 cells was performed by using DNA microarrays to discriminate between genotoxic and nongenotoxic carcinogens (van Delft et al., 2004). The gene expression profile for 1,4-dioxane resulted in classification of this chemical as a nongenotoxic carcinogen.

5.3.3.2 Initiation/Promotion Studies

The development of cancer is considered to be a multistage process involving initiation (a change in the DNA of a cell leading to transformation of a normal cell to a cancerous cell), promotion (the enhanced growth of initiated or transformed cells), and progression (formation of benign tumors that may become malignant and metastasize). Initiation/promotion studies are designed to determine whether a chemical is capable of initiation, promotion, or complete carcinogenesis. Chemical initiators like dimethylbenzanthracene (DMBA) are used to determine whether a chemical is a tumor promoter (i.e., can promote the growth of DMBA-initiated cells). Tumor promoters like tetradecanoylphorbol-13-acetate (TPA) are used to evaluate whether a chemical has initiating properties (i.e., can transform cells that can be promoted to result in tumors).

Bull et al. (1986) tested 1,4-dioxane as a cancer initiator in mice using oral, subcutaneous, and topical routes of exposure. A group of 40 female SENCAR mice was administered a single dose of 1000 mg/kg dioxane by gavage, subcutaneous injection, or topical administration. Two weeks after administration of 1,4-dioxane, TPA (1.0 μg in 0.2 mL of acetone) was applied to the shaved back of mice three times per week for a period of 20 weeks. The yield of papillomas (benign skin tumors) was measured at 24 weeks. Acetone was used instead of TPA in an additional group of 20 mice in order to determine whether a single dose of 1,4-dioxane could induce tumors in the absence of TPA promotion. 1,4-Dioxane did not increase the formation of papillomas compared to mice initiated with vehicle and promoted with TPA, indicating lack of initiating activity under the conditions of the study. Negative results were obtained for all three exposure routes. A single dose of 1,4-dioxane did not induce tumors in the absence of TPA promotion.

1,4-Dioxane was evaluated for complete carcinogenicity and tumor promotion activity in mouse skin (King et al., 1973). In the complete carcinogenicity study, 0.2 mL of a solution of 1,4-dioxane in acetone was applied to the shaved skin of the back of Swiss Webster mice (30 per sex) three times a week for 78 weeks. Acetone was applied to the backs of control mice (30 per sex) for the same time period. In the promotion study, each animal was treated with 50 μg of DMBA one week prior to the 78-week-long topical application of the 1,4-dioxane solution. Acetone vehicle was used in negative control mice (30 per sex). Croton oil was used as a positive control in the promotion study (30 per sex). Weekly counts of papillomas and suspect carcinomas were made by gross examination.

1,4-Dioxane was negative in the complete skin carcinogenicity test using dermal exposure (King et al., 1973). One treated female mouse had malignant lymphoma; however, no papillomas were observed in male or female mice by 60 weeks. Neoplastic lesions of the skin, lungs, and kidney were observed in mice given the promotional treatment with 1,4-dioxane. In addition, the percentage of mice with skin tumors increased sharply after approximately 10 weeks of promotion treatment. Significant mortality was observed when 1,4-dioxane was administered as a promoter, but not as a complete carcinogen. Mice treated with croton oil as a positive control experienced significant mortality, whereas the survival of acetone-treated control mice in the promotion study was not affected. The incidence of mice with papillomas was similar for croton oil and 1,4-dioxane; however, the tumor multiplicity (i.e., number of tumors per mouse) was higher for the croton oil treatment.

Lundberg et al. (1987) evaluated the tumor-promoting activity of 1,4-dioxane in rat liver. Male Sprague Dawley rats (eight per dose group, 19 for control group) underwent a partial hepatectomy

TABLE 5.4
Genotoxicity Studies with 1,4-Dioxane

Test System	Endpoint	Test Conditions	Results		References
			Without Activation	With Activation	
		Bacteria			
Salmonella typhimurium strains TA98, TA100, TA1535, and TA1537	Reverse mutation	Plate incorporation assay	–	–	Haworth et al. (1983)
Salmonella typhimurium strains TA98, TA100, TA 1530, TA1535, and TA1537	Reverse mutation	Plate incorporation assay	–	–	Khudoley et al. (1987)
Salmonella typhimurium strains TA98, TA100, TA1535, and TA1537	Reverse mutation	Plate incorporation and preincubation assays	–	–	Morita and Hayashi (1998)
Salmonella typhimurium strains TA100, TA1535	Reverse mutation	Preincubation assay	–	–	Nestmann et al. (1984)
Salmonella typhimurium strains TA98, TA100, TA1535, TA1537, and TA1538	Reverse mutation	Plate incorporation assay	–	–	Stott et al. (1981)
E. coli K-12 uvrB/recA	DNA repair	Host-mediated assay	–	–	Hellmer and Bolcsfóldi (1992)
E. coli WP2/WP2uvrA	Reverse mutation	Plate incorporation and preincubation assays	–	–	Morita and Hayashi (1998)
P. phosphoreum M169	Mutagenicity, DNA damage	Mutatox assay	–	ND	Kwan et al. (1990)

Yeast/Drosophila

S. cerevisiae D61.M	Aneuploidy	Standard 16-h incubation or cold-interruption regimen	−T	ND	Zimmermann et al. (1985)
D. melanogaster	Meiotic nondisjunction	Oocytes were obtained for evaluation 24 and 48 h after mating	+T	ND	Munoz and Barnett (2002)
D. melanogaster	Sex-linked recessive lethal test	Exposure by feeding and ingestion	−	ND	Yoon et al. (1985)

Mammalian Cells

L5178Y mouse lymphoma cells	Forward mutation assay	Thymidine kinase mutagenicity assay (trifluorothymidine resistance)	−	−	McGregor et al. (1991)
L5178Y mouse lymphoma cells	Forward mutation assay	Thymidine kinase mutagenicity assay (trifluorothymidine resistance)	−	−T	Morita and Hayashi (1998)
Rat hepatocytes	DNA damage; single-strand breaks measured by alkaline elution	3-h exposure to isolated primary hepatocytes	+T	ND	Sina et al. (1983)
Primary hepatocyte culture from male F344 rats	DNA repair	Autoradiography	−	ND	Goldsworthy et al. (1991)
Calf thymus DNA	Covalent binding to DNA	Incubation with microsomes from 3-methylcholanthrene-treated rats	−	−	Woo et al. (1977a)
BALB/3T3 cells	Cell transformation	48-h exposure followed by four-week incubation; 13-day exposure followed by 2.5-week incubation	+T	ND	Sheu et al. (1988)
CHO cells	Sister chromatid exchange (SCE)	5-bromo-2′-deoxyuridine (BrdU) was added 2 h after 1,4-dioxane addition; chemical treatment was 2 h with S9 and 25 h without S9	±	−	Galloway et al. (1987)
CHO cells	Chromosomal aberration	Cells were harvested 8–12 h or 18–26 h after treatment (time of first mitosis)	−	−	Galloway et al. (1987)
CHO cells	SCE	3-h pulse treatment; followed by continuous treatment of BrdU for 23 or 26 h	−	−	Morita and Hayashi (1998)

continued

TABLE 5.4 (continued)
Genotoxicity Studies with 1,4-Dioxane

Test System	Endpoint	Test Conditions	Results Without Activation	With Activation	References
CHO cells	Chromosomal aberration	5-h pulse treatment, 20-h pulse and continuous treatments, or 44-hour continuous treatment; cells were harvested 20 or 44 h following exposure	–	–	Morita and Hayashi (1998)
CHO cells	Micronucleus formation	5-h pulse treatment or 44-h continuous treatment; cells were harvested 42 h following exposure	–	–	Morita and Hayashi (1998)
Laboratory Animal Studies					
Male F344 rat	DNA repair in hepatocytes (autoradiography)	Gavage and drinking-water exposure; thymidine incorporation	–	ND	Goldsworthy et al. (1991)
Male F344 rat	Replicative DNA synthesis (i.e., cell proliferation) in hepatocytes	Gavage and drinking-water exposure; thymidine incorporation	+ (1–2-week exposure)	ND	Goldsworthy et al. (1991)
Male F344 rat	DNA repair in nasal epithelial cells from the nasoturbinate or maxilloturbinate	Gavage and drinking-water exposure; thymidine incorporation	–	ND	Goldsworthy et al. (1991)
Male F344 rat	Replicative DNA synthesis (i.e., cell proliferation) in nasal epithelial cells	Drinking-water exposure; thymidine incorporation	–	ND	Goldsworthy et al. (1991)
Female Sprague Dawley rat	DNA damage; single-strand breaks measured by alkaline elution	Two gavage doses given 21 and 4 h prior to sacrifice	+	ND	Kitchin and Brown (1990)
Male Sprague Dawley rat	RNA synthesis; inhibition of RNA polymerase A and B	i.v. injection; activity measured in isolated hepatocytes	+	ND	Kurl et al. (1981)
Male F344 rat	DNA synthesis in hepatocytes	Gavage; thymidine and BrdU incorporation	+	ND	Miyagawa et al. (1999)
Male F344 rat	DNA synthesis in hepatocytes	Thymidine incorporation	±	ND	Uno et al. (1994)

Male Sprague Dawley rat	DNA synthesis in hepatocytes	Drinking water; thymidine incorporation	+	ND	Stott et al. (1981)
Male Sprague Dawley rat	DNA repair in hepatocytes	Drinking water; thymidine incorporation with hydroxyurea to repress normal DNA synthesis	–	ND	Stott et al. (1981)
Male Sprague Dawley rat	DNA alkylation in hepatocytes	Gavage; DNA isolation and HPLC analysis 4 h after dosing	–	ND	Stott et al. (1981)
Male B6C3F$_1$ mouse	Micronucleus formation in bone marrow	i.p. injection; analysis of polychromatic erythrocytes 24 or 48 h after dosing	–	ND	McFee et al. (1994)
Male and female C57BL6 mice; male BALB/c mouse	Micronucleus formation in bone marrow	Gavage; analysis of polychromatic erythrocytes 24 or 48 h after dosing	+ (C57BL6)– (BALB/c)	ND	Mirkova (1994)
Male CD-1 mouse	Micronucleus formation in peripheral blood	Two i.p. injections (one per day); micronucleated reticulocytes measured 24, 48, and 72 h after the second dose	–	ND	Morita (1994)
Male CD-1 mouse	Micronucleus formation in hepatocytes	Gavage; partial hepatectomy 24 h after dosing; hepatocytes analyzed five days after hepatectomy	+	ND	Morita and Hayashi (1998)
Male CD-1 mouse	Micronucleus formation in peripheral blood	Gavage; partial hepatectomy 24 h after dosing; peripheral blood obtained from tail vein 24 h after hepatectomy	–	ND	Morita and Hayashi (1998)
Male CD-1 mouse	Micronucleus formation in hepatocytes and bone marrow	Gavage; analysis of bone marrow erythrocytes and proliferating BrdU-labeled hepatocytes	+	ND	Roy et al. (2005)
Male CBA and C57BL6 mice	Micronucleus formation in bone marrow	Gavage; analysis of polychromatic erythrocytes from specimens prepared 24 h after dosing	–	ND	Tinwell and Ashby (1994)

Note: + = positive, ± = weak positive, – = negative, T = toxicity, ND = no data.

followed 24 h later by an i.p. injection of 30 mg/kg diethylnitrosamine[*] (initiation treatment). 1,4-Dioxane was then administered daily by saline gavage at doses of 0, 100, or 1000 mg/kg per day, five days per week for seven weeks. Control rats were given saline following diethylnitrosamine initiation. 1,4-Dioxane was also administered to groups of rats that were not given the diethyl-nitrosamine initiation treatment (saline used instead of diethylnitrosamine). Ten days after the last dose, animals were sacrificed, and liver sections were stained for GGT (i.e., preneoplastic enzyme-altered foci). The number and total volume of GGT-positive foci were determined. 1,4-Dioxane did not increase the number or volume of GGT-positive foci in rats that were not given the diethyl-nitrosamine initiation treatment. The high dose of 1,4-dioxane (1000 mg/kg per day) given as a promoting treatment (i.e., following diethylnitrosamine injection) produced an increase in the number of GGT-positive foci and the total foci volume. Histopathological changes were also noted in the livers of high-dose rats. Enlarged, foamy hepatocytes were observed in the midzonal region of the liver; the foamy appearance was due to the presence of numerous fat-containing cytoplasmic vacuoles. These results suggest that cytotoxic doses of 1,4-dioxane may be associated with tumor promotion of 1,4-dioxane in rat liver.

5.3.3.3 Mechanistic Studies Evaluating Mode of Action for Carcinogenicity

Several studies have been performed to evaluate potential mechanisms for the carcinogenicity of 1,4-dioxane (Stott et al., 1981; Kitchin and Brown, 1990; Goldsworthy et al., 1991). Stott et al. (1981) evaluated 1,4-dioxane in several test systems, including *Salmonella typhimurium* mutagen-icity *in vitro* (Ames test), rat hepatocyte DNA repair activity *in vitro*, DNA synthesis determination in male Sprague Dawley rats following acute gavage dosing or an 11-week drinking-water expo-sure and hepatocyte DNA alkylation and DNA repair following a single gavage dose. This study used daily doses of 0, 10, 100, or 1000 mg/kg; the highest dose was considered to be a tumorigenic dose level. Liver histopathology and liver-to-body-weight ratios were also evaluated in rats from acute gavage or repeated-dose drinking-water experiments (only for groups receiving 10 and 1000 mg/kg per day).

The histopathology evaluation indicated that liver cytotoxicity (i.e., centrilobular hepatocyte swelling) was present in rats from the 1000 mg/kg per day dose group that received 1,4-dioxane in the drinking water for 11 weeks (Stott et al., 1981). An increase in the liver-to-body-weight ratio accompanied by an increase in hepatic DNA synthesis was also seen in high-dose animals. No effect on histopathology, liver weight, or DNA synthesis was observed in acutely exposed rats or rats that were exposed to a lower daily dose of 10 mg/kg for 11 weeks. 1,4-Dioxane produced negative findings in the remaining genotoxicity assays conducted as part of this study (i.e., *Salmonella typh-imurium* mutagenicity, *in vitro* and *in vivo* rat hepatocyte DNA repair, and DNA alkylation in rat liver). The lack of genotoxicity in this study suggests that a mutagenic mechanism for 1,4-dioxane hepatocellular carcinoma in rats is unlikely.

Goldsworthy et al. (1991) evaluated potential mechanisms for the nasal and liver carcinogenicity of 1,4-dioxane in the rat. DNA repair activity was evaluated as a measure of DNA reactivity, and DNA synthesis was measured as an indicator of cell proliferation or promotional activity. *In vitro* DNA repair was evaluated in primary hepatocyte cultures from control and 1,4-dioxane-treated rats (1% or 2% in the drinking water for one week). DNA repair and DNA synthesis were also measured *in vivo* following a single gavage dose of 1000 mg/kg, a drinking-water exposure of 1% (1500 mg/kg per day) for one week, or a drinking-water exposure of 2% (3000 mg/kg per day) for two weeks. Liver-to-body-weight ratios and palmitoyl CoA (coenzyme A) oxidase activity were measured in the rat liver to determine whether peroxisome proliferation[†] played a role in the liver carcinogenesis due to 1,4-dioxane. *In vivo* DNA repair was evaluated in rat nasal epithelial cells derived from either the

---

[*] Diethylnitrosamine (CASRN 55-18-5) is also called n-nitrosodiethylamine or NDEA.
[†] Peroxisome proliferation is a mechanism by which some drugs and chemicals induce liver cancer in rodents. Peroxisomes participate in fatty acid metabolism in cells.

nasoturbinate or the maxilloturbinate of 1,4-dioxane-treated rats. These rats received 1% 1,4-dioxane (1500 mg/kg per day) in the drinking water for eight days, followed by a single gavage dose of 10, 100, or 1000 mg/kg 12 h prior to sacrifice. Archived tissues from the NCI (1978) bioassay were reexamined to determine the primary sites for tumor formation in the nasal cavity following chronic exposure in rats. Histopathology and cell proliferation were determined for specific sites in the nasal cavity that were related to tumor formation. This evaluation was performed in rats that were exposed to drinking water containing 1% 1,4-dioxane (1500 mg/kg per day) for two weeks.

1,4-Dioxane and its metabolite 1,4-dioxane-2-one did not affect *in vitro* DNA repair in primary hepatocyte cultures (Goldsworthy et al., 1991). *In vivo* DNA repair was also unaffected by acute gavage exposure or ingestion of 1,4-dioxane in the drinking water for a one- or two-week period. Hepatocyte cell proliferation was not affected by acute gavage exposure, but was increased approximately twofold following drinking-water exposure for 1–2 weeks. A five-day drinking-water exposure to 1% 1,4-dioxane (1500 mg/kg per day) did not increase the activity of palmitoyl CoA or the liver-to-body-weight ratio, suggesting that peroxisome proliferation does not play a role in the hepatocarcinogenesis of 1,4-dioxane. Nannelli et al. (2005) also reported a lack of hepatic palmitoyl CoA induction following 10 days of exposure to 1.5% 1,4-dioxane in the drinking water (2100 mg/kg per day).

Treatment of rats with 1% (1500 mg/kg per day) 1,4-dioxane for eight days did not alter DNA repair in nasal epithelial cells (Goldsworthy et al., 1991). The addition of a single gavage dose of up to 1000 mg/kg 12 h prior to sacrifice also did not induce DNA repair. Reexamination of tissue sections from the NCI (1978) bioassay suggested that the majority of nasal tumors were located in the dorsal nasal septum or the nasoturbinate of the anterior portion of the dorsal meatus (Goldsworthy et al., 1991). The location of these tumors is consistent with the possibility that inhalation of droplets of drinking water may be responsible for nasal lesions observed following chronic exposure to 1,4-dioxane. No histopathological lesions were observed in nasal sections of rats exposed to drinking water containing 1% 1,4-dioxane (1500 mg/kg per day) for two weeks, and no increase was observed in cell proliferation at the sites of highest tumor formation in the nasal cavity.

Sweeney et al. (2008) performed a fluorescent dye experiment to determine whether drinking water directly contacts nasal tissues under bioassay study conditions. Rats (five per group) were exposed to drinking water containing 0 or 0.5% 1,4-dioxane along with a fluorescent dye tracer. The animals were sacrificed approximately 24 h after the start of exposure, and the nasal cavity was examined by fluorescence microscopy. An additional animal was given two gavage doses of water containing dye (30 min between doses) and was sacrificed 5 h later, in order to evaluate the potential for systemic delivery to nasal tissues. Fluorescent dye was readily detectable in the oral and nasal cavities of rats following drinking-water exposure. The dye was observed in several sections of the anterior third of the nose of each rat (nasal vestibule, maxillary turbinates, and dorsal nasoturbinates) and was occasionally detected in the ethmoid turbinate region and nasopharynx. No fluorescence was detected in the nasal cavity of the rat that received a gavage exposure. These results suggest that drinking water directly contacts nasal tissues under bioassay study conditions and may be responsible for the nasal lesions and tumors observed following chronic exposure to 1,4-dioxane.

Female Sprague Dawley rats (3–9 per group) were given 0, 168, 840, 2550, or 4200 mg/kg 1,4-dioxane by corn oil gavage in two doses at 21 and 4 h prior to sacrifice (Kitchin and Brown, 1990). DNA damage (single-strand breaks measured by alkaline elution), ornithine decarboxylase (ODC) activity, reduced glutathione content, and CYP content were measured in the liver. Serum ALT activity and liver histopathology were also evaluated. No changes were observed in hepatic reduced glutathione content or ALT activity. Light microscopy revealed minimal to mild vacuolar degeneration in the cytoplasm of hepatocytes from three of five rats from the 2550 mg/kg dose group. No histopathological lesions were seen in any other dose group, including rats given the highest dose of 4200 mg/kg. 1,4-Dioxane caused 43% and 50% increases in DNA single-strand breaks at dose levels of 2550 and 4200 mg/kg, respectively. CYP content was also increased at the two highest-dose levels (25% and 66%, respectively). ODC activity was increased approximately two-, five-, and eightfold above control values at doses of 840, 2550, and 4200 mg/kg, respectively. The results of this study demonstrated that hepatic DNA damage can occur in the absence of significant cytotoxicity. The fact that parameters associated with tumor promotion (i.e., ODC activity, CYP450 content) were also elevated suggests that promotion may play a role in the carcinogenesis of 1,4-dioxane.

5.4 SUMMARY OF NONCANCER HEALTH EFFECTS

Liver and kidney toxicities were the primary noncancer health effects associated with exposure to 1,4-dioxane in humans and laboratory animals. Several fatal cases of hemorrhagic nephritis and centrilobular necrosis of the liver were related to occupational exposure to 1,4-dioxane (Barber, 1934; Johnstone, 1959). Neurological changes were also reported in one case, including headache, elevation in blood pressure, agitation and restlessness, and coma (Johnstone, 1959). Perivascular widening was observed in the brain of this worker, with small foci of demyelination in several regions (e.g., cortex and basal nuclei). Severe liver and kidney degeneration and necrosis were observed frequently in acute oral and inhalation studies (Fairley et al., 1934; de Navasquez, 1935; Schrenk and Yant, 1936; Kesten et al., 1939; Laug et al., 1939; David, 1964; Drew et al., 1978; JBRC, 1998a).

Liver and kidney toxicities were the primary noncancer health effects of subchronic and chronic oral exposure to 1,4-dioxane in animals. Liver effects included degeneration and necrosis, hepatocyte swelling, cells with hyperchromic nuclei, spongiosis hepatis, hyperplasia, and clear and mixed-cell foci of the liver (Fairley et al., 1934; Argus et al., 1965, 1973; Kociba et al., 1974; NCI, 1978; JBRC, 1998c). Liver hyperplasia and clear and mixed-cell foci are commonly considered precancerous changes and would not be considered evidence of noncancer toxicity. Spongiosis hepatis may also occur in combination with preneoplastic foci or hepatocellular adenoma or carcinoma (Stroebel et al., 1995; Bannasch, 2003). Hepatocellular degeneration and necrosis were seen at high doses in a subchronic study (1900 mg/kg per day in rats) (Fairley et al., 1934) and at lower doses in a chronic study (94 mg/kg per day, male rats) (Kociba et al., 1974).

Kidney damage at high doses was characterized by degeneration of the cortical tubule cells, necrosis with hemorrhage, and glomerulonephritis (Fairley et al., 1934; Argus et al., 1965, 1973; Kociba et al., 1974; NCI, 1978). Renal cell degeneration generally began with cloudy swelling of cells in the cortex (Fairley et al., 1934). Nuclear enlargement of proximal tubule cells was observed at doses below those producing renal necrosis (JBRC, 1998b,c), but is of uncertain toxicological significance. The lowest daily dose reported to produce kidney damage was 94 mg/kg, which produced renal degeneration and necrosis of tubule epithelial cells in male rats (Kociba et al., 1974). Cortical tubule degeneration was seen at higher daily doses in the NCI (1978) bioassay (240 mg/kg, male rats), and glomerulonephritis was reported for rats given daily doses of at least 430 mg/kg (Argus et al., 1965, 1973).

Exposure to 1,4-dioxane in the drinking water resulted in effects on nasal cavity and respiratory tract tissues in mice and rats (NCI, 1978; JBRC, 1998c). Rhinitis and inflammation of the nasal cavity were reported, as well as atrophy of the nasal epithelium and adhesion in rats and mice. Nasal inflammation may be a response to direct contact of the nasal mucosa with drinking water containing 1,4-dioxane (Goldsworthy et al., 1991; Sweeney et al., 2008). Metaplasia and hyperplasia of the nasal epithelium were also observed in high-dose male and female rats (JBRC, 1998c); however, these effects are likely to be associated with the formation of nasal cavity tumors in these dose groups. Nuclear enlargement of the nasal olfactory epithelium was also observed; however, it is unclear whether this alteration represents an adverse toxicological effect. A significant increase in the incidence of pneumonia was reported in mice from the NCI (1978) study. The significance of this effect is unclear, as it was not observed in other studies that evaluated lung histopathology (Kociba et al., 1974; JBRC, 1998b,c). Nuclear enlargement of the tracheal and bronchial epithelium and an accumulation of foamy cells in the lung were also seen in male and female mice (JBRC, 1998c).

5.5 CHARACTERIZATION OF CANCER POTENTIAL

5.5.1 SUMMARY OF HUMAN AND ANIMAL DATA

Human studies of occupational exposure to 1,4-dioxane were generally negative; however, in each case, the cohort size and number of reported cases were small, and the probability of detecting an excess cancer risk was low (Thiess et al., 1976; Buffler et al., 1978). Several carcinogenicity bioassays have been conducted for 1,4-dioxane in mice, rats, and guinea pigs (Argus et al., 1965, 1973; Hoch-Ligeti and Argus, 1970; Hoch-Ligeti et al., 1970; Kociba et al., 1974; Torkelson et al., 1974; NCI, 1978; JBRC, 1998c). Liver tumors have been observed following drinking-water exposure in male Wistar rats (Argus et al., 1965), male guinea pigs (Hoch-Ligeti and Argus, 1970), male Sprague Dawley rats (Hoch-Ligeti et al., 1970; Argus et al., 1973), male and female Sherman rats (Kociba et al., 1974), female Osborne–Mendel rats (NCI, 1978), male and female F344/DuCrj rats (JBRC, 1998c), male and female B6C3F$_1$ mice (NCI, 1978), and male and female Crj:BDF$_1$ mice (JBRC, 1998c). In the earliest cancer bioassays, the liver tumors were described as hepatomas (Argus et al., 1965, 1973; Hoch-Ligeti and Argus, 1970; Hoch-Ligeti et al., 1970); however, later studies made a distinction between hepatocellular carcinoma and hepatocellular adenoma (Kociba et al., 1974; NCI, 1978; JBRC, 1998c). Both tumor types have been seen in rats and mice exposed to 1,4-dioxane.

Kociba et al. (1974) noted evidence of liver toxicity at or below the dose levels that produced liver tumors. Hepatocellular degeneration and necrosis were observed in the mid- and high-dose groups of male and female Sherman rats exposed to 1,4-dioxane, whereas tumors were observed only at the highest dose. Hepatic regeneration was indicated in the mid- and high-dose groups by the formation of hepatocellular hyperplastic nodules. JBRC (1998c) also demonstrated signs of liver hyperplasia in male F344/DuCrj rats at a dose level below the dose that induced a statistically significant increase in tumor formation. The male mouse data from the JBRC (1998c) study were confounded by a high incidence of liver tumors in the male controls (42% incidence).

Nasal cavity tumors were also observed in Sprague Dawley rats (Hoch-Ligeti et al., 1970; Argus et al., 1973), Osborne–Mendel rats (NCI, 1978), Sherman rats (Kociba et al., 1974), and F344/DuCrj rats (JBRC, 1998c). Most tumors were characterized as squamous cell carcinomas. Nasal tumors were not elevated in B6C3F$_1$ or Crj:BDF$_1$ mice. With the exception of the NCI (1978) study, the incidence of nasal cavity tumors was generally lower than that of liver tumors in the same study population. JBRC (1998c) was the only study that evaluated nonneoplastic changes in nasal cavity tissue following prolonged exposure to 1,4-dioxane in the drinking water. Histopathological lesions in female F344/DuCrj rats were suggestive of toxicity and regeneration in this tissue (i.e., atrophy, adhesion, inflammation, nuclear enlargement, and hyperplasia and metaplasia of respiratory and olfactory epithelium). Some of these effects occurred at a lower daily dose (103 mg/kg) than that shown to produce nasal cavity tumors (513 mg/kg). Goldsworthy et al. (1991) suggested that formation of nasal cavity tumors may be related to inhalation of water droplets during drinking and may not be due to systemic exposure to 1,4-dioxane. Reexamination of tissue sections from the NCI (1978) bioassay suggested that the majority of nasal tumors were located in the dorsal nasal septum or the nasoturbinate of the anterior portion of the dorsal meatus. The location of these tumors is consistent with the possibility that inhalation of droplets of drinking water may be responsible for nasal lesions observed following chronic exposure to 1,4-dioxane. Nasal tumors were not observed in an inhalation study in Wistar rats exposed to 111 ppm for five days per week for two years (Torkelson et al., 1974).

5.5.2 WEIGHT-OF-EVIDENCE EVALUATION

The International Agency for Research on Cancer (IARC) classifies 1,4-dioxane as possibly carcinogenic to humans (Group 2B) based on inadequate evidence in humans and sufficient evidence in experimental animals (IARC, 1999). Similarly, the U.S. Environmental Protection Agency (USEPA) has classified 1,4-dioxane as a probable human carcinogen (Group B2) on the basis of inadequate data from human epidemiological studies and sufficient data from laboratory animal studies, including nasal cavity and liver carcinomas in multiple strains of rats, liver carcinomas in mice, and gall bladder carcinomas in guinea pigs (USEPA, 2007). If reevaluated under the 2005 *Guidelines for Carcinogen Risk Assessment* (USEPA, 2005), 1,4-dioxane would be considered *likely to be carcinogenic to humans*, on the basis of adequate evidence of liver carcinogenicity in more than one species and strain of laboratory animal.

5.5.3 Mode of Action

The 2005 cancer guidelines (USEPA, 2005) place a high degree of emphasis on establishing the mode of action for a chemical to produce cancer. The mode of action that is established for a chemical directly affects the quantitative methodology used to assess cancer potency. Chemicals that act through a mutagenic mode of action are assessed with a linear approach to low-dose extrapolation. Other modes of action may be modeled with either linear or nonlinear approaches. USEPA (2005) described a rigorous mode-of-action framework that employs the Hill criteria for causality (Hill, 1965) in determining whether a hypothetical mode of action is plausible. The mode-of-action framework using the modified Hill criteria includes two parts:

1. A *description* of the key events in the mode of action
2. An *evaluation* of (a) the strength, consistency, and specificity of the associated events, (b) the dose response and temporal concordance, and (c) the biological plausibility and coherence of the proposed cancer mode of action

The EPA cancer guidelines (USEPA, 2005) indicate that the determination of the cancer mode of action is a "data rich" determination and suggest that health-protective, default positions (i.e., low-dose linearity) will be employed if data are considered insufficient. A brief description of plausible modes of action for the induction of liver and nasal tumors in laboratory animals is provided next.

1,4-Dioxane was not shown to be mutagenic, that is, a genotoxic mode of action involving an interaction with DNA, causing mutations in critical genes for tumor initiation (i.e., oncogenes or tumor-suppressor genes), is not likely. This conclusion is supported by evidence that 1,4-dioxane is a tumor promoter but not a tumor initiator in mouse skin and rat liver bioassays (King et al., 1973; Lundberg et al., 1987). Possible key events in the mode of action for 1,4-dioxane-induced liver cancer include

1. Saturation of CYP metabolism/clearance leading to accumulation of the parent 1,4-dioxane
2. (a) Liver damage followed by regenerative cell proliferation or (b) cell proliferation in the absence of cytotoxicity (i.e., mitogenesis)
3. Hyperplasia
4. Tumor formation

Liver toxicity appears to be related to the accumulation of the parent compound following metabolic saturation at high doses (Kociba et al., 1975). Nannelli et al. (2005) demonstrated that an increase in the oxidative metabolism of 1,4-dioxane via CYP450 induction using phenobarbital or fasting does not result in an increase in liver toxicity. This result suggested that highly reactive and toxic intermediates did not play a large role in the liver toxicity of 1,4-dioxane, even under conditions where metabolism was enhanced. Dose response and temporal evidence support the occurrence of cell proliferation and hyperplasia prior to the development of liver tumors (Kociba et al., 1974; JBRC, 1998b). However, conflicting data from rat and mouse bioassays (Kociba et al., 1974; JBRC, 1998b) suggest that cytotoxicity may not be a required precursor event for 1,4-dioxane-induced cell proliferation. Mechanistic studies also provide evidence of cell pro-liferation, but do not indicate whether mitogenesis or cytotoxicity is responsible for increased cell turnover (Stott et al., 1981; Goldsworthy et al., 1991; Uno et al., 1994; Miyagawa et al., 1999).

Inhalation of droplets of drinking water may be responsible for the nasal tumors observed in rats following chronic exposure to 1,4-dioxane in drinking water (Goldsworthy et al., 1991; Sweeney et al., 2008). In this case, the mode of action may involve chronic irritation followed by regenerative hyperplasia leading to the formation of nasal tumors. Histopathological lesions in female rats were suggestive of toxicity and regeneration in this tissue (i.e., atrophy, adhesion, inflammation, nuclear enlargement, and hyperplasia and metaplasia of respiratory and olfactory epithelium) (JBRC, 1998b).

5.6   1,4-DIOXANE TOXICOLOGY AND RISK ASSESSMENT

Environmental risk assessment of 1,4-dioxane involves an understanding of potential human expo-sures and toxic potency. Drinking-water contamination with this chemical could result in exposure via ingestion, dermal contact, and inhalation. The limited human and animal data available for the inhalation route suggest that toxic outcomes occur only at high levels of exposure. Toxicity studies for the dermal exposure route are lacking, but limited data suggest that dermal penetration may not be significant. Oral exposure is the primary exposure route of concern for 1,4-dioxane, and most of the laboratory animal studies have used this exposure route.

In characterizing the toxic potency of 1,4-dioxane, human and laboratory animal studies should both be considered. The most useful studies provide an evaluation of the dose–response relationship for the toxic endpoints of concern. These data can be used to quantify the toxic potency and provide a numerical toxicity value for human health risk assessment. Studies of 1,4-dioxane exposure in humans do not provide adequate information for evaluating possible health effects resulting from environmental exposure to this chemical. In this case, laboratory animal studies must be relied upon to provide quantitative toxicity information for use in risk assessment. Data are available to describe the dose–response relationship for both target organ toxicity (i.e., liver and kidney toxicities) and cancer. Noncancer health effects are generally evaluated by using a reference dose (RfD) value, which is defined as an estimate (with uncertainty spanning perhaps an order of magnitude) of a daily oral exposure to the human population (including sensitive subgroups) that is likely to be with-out an appreciable risk of deleterious effects during a lifetime. An RfD value is not currently avail-able for 1,4-dioxane on USEPA's Integrated Risk Information System (IRIS) database (USEPA, 2007); however, an RfD value may be developed in the future by using chronic bioassay data for liver and kidney toxicities (Fairley et al., 1934; Argus et al., 1965, 1973; Kociba et al., 1974; NCI, 1978; JBRC, 1998c).

Oral cancer slope factors (CSFs) or inhalation unit risk values are developed from laboratory animal studies for use in cancer risk assessment. An oral CSF value of 0.011 (mg/kg-day)$^{-1}$ previ-ously published on USEPA's IRIS database (USEPA, 2007) was based on the dose–response data for nasal cavity carcinomas in male Osborne–Mendel rats (NCI, 1978); the USEPA used the default methodology of linear low-dose extrapolation. The verification date for the IRIS assessment was February 3, 1988. Since that time, additional cancer bioassay data have become available (JBRC, 1998c), and several studies have been published that shed light on the possible mode of action for cancer induction in experimental animals (Goldsworthy et al., 1991; Uno et al., 1994; Miyagawa et al., 1999; Nannelli et al., 2005). PBPK models are also available for 1,4-dioxane that estimate the internal dose at the target organ (liver) and take into account species differences (i.e., rats vs. humans) and nonlinear pharmacokinetics (Reitz et al., 1990; Leung and Paustenbach, 1990; Sweeney et al., 2008).

For chemicals that are considered to be possibly carcinogenic to humans, the cancer risk assess-ment is an important component of environmental risk assessment and regulation. The CSF repre-sents the relationship between the dose of a chemical and the incidence of humans or animals with

TABLE 5.5
Safe Drinking Water Concentrations for 1,4-Dioxane

Dose–Response Assessment Method	Cancer Risk	Concentration (mg/L)	References
IRIS CSF—rat nasal tumors	1×10^{-5}	0.030	NCI (1978)
IRIS CSF—rat nasal tumors	1×10^{-6}	0.003	NCI (1978)
PBPK LMS model	1×10^{-5}	20[a]	Reitz et al. (1990)
PBPK LMS model	1×10^{-6}	2[a]	Reitz et al. (1990)
PBPK safety factor approach	NA	120	Reitz et al. (1990)
PBPK LMS model	1×10^{-5}	28[b]	Leung and Paustenbach (1990)
PBPK LMS model	1×10^{-6}	2.8[b]	Leung and Paustenbach (1990)
PBPK safety factor approach	NA	49[b]	Leung and Paustenbach (1990)
Threshold approach	NA	3.5[c]	Stickney et al. (2003)

Source: Adapted from Stickney et al., 2003, *Regulatory Toxicology and Pharmacology* 38: 183–195. With permission.

Notes: IRIS = USEPA's Integrated Risk Information System; PBPK = physiologically based pharmacokinetic; LMS = Linearized Multistage Model; NA = not applicable.

[a] Calculated by assuming a linear relationship between risk level and the estimated cancer slope factor (CSF).

[b] Calculated from risk-specific doses by assuming 70 kg body weight and 2 L per day ingestion rate.

[c] Calculated by dividing the no observed adverse effect (NOAEL) of 10 mg/kg per day by an uncertainty factor (UF) of 100 and by assuming 70 kg body weight and 2 L per day ingestion rate.

tumors. The CSF value can be used to estimate the cancer risk (i.e., the projected incidence of cancer associated with a specified dose), or it can be used to determine "safe" levels of a chemical in the environment by making some assumptions regarding the acceptable level of cancer risk (e.g., 1 in 1 million or 1×10^{-6} cancer risk). Stickney et al. (2003) provided an evaluation of safe drinking-water concentrations of 1,4-dioxane that may be associated with different approaches to carcinogen risk assessment (see Table 5.5). Factors that significantly affect the cancer risk assessment of 1,4-dioxane include the following:

1. Determination of the human relevance of nasal and liver tumors
2. Elucidation of the mode of action for relevant tumor types
3. Identification of an appropriate internal dose metric (i.e., from PBPK modeling) for use in quantification of human cancer risks

Each of these factors is subject to some degree of scientific interpretation. Scientific consensus regarding cancer risk assessment is generally achieved through a peer review process.

ACKNOWLEDGMENTS

The insightful comments and questions provided by Heather Carlson-Lynch of SRC, Inc. are greatly appreciated.

BIBLIOGRAPHY

Argus, M.F., Arcos, J.C., and Hoch-Ligeti, C., 1965, Studies on the carcinogenic activity of protein-denaturing agents: Hepatocarcinogenicity of dioxane. *Journal of the National Cancer Institute* 35: 949–958.

Argus, M.F., Sohal, R.S., Bryant, G.M., Hoch-Ligeti, C., and Arcos, J.C., 1973, Dose-response and ultrastructural alterations in dioxane carcinogenesis: Influence of methylcholanthrene on acute toxicity. *European Journal of Cancer* 9: 237–243.

Ashby, J., 1994, Series: Current issues in mutagenesis and carcinogenesis, No. 45—The genotoxicity of 1,4-dioxane. *Mutation Research* 322(2): 141–150.

Bannasch, P., 2003, Comments on R. Karbe and R.L. Kerlin (2002). Cystic degeneration/spongiosis hepatis (*Toxicologic Pathology* 30(2): 216–227). *Toxicologic Pathology* 31: 566–570.

Barber, H., 1934, Haemorrhagic nephritis and necrosis of the liver from dioxane poisoning. *Guys Hospital Report* 84: 267–280.

Braun, W.H. and Young, J.D., 1977, Identification of β-hydoxyethoxyacetic acid as the major urinary metabolite of 1,4-dioxane in the rat. *Toxicology and Applied Pharmacology* 39: 33–38.

Bronaugh, R.L., 1982, Percutaneous absorption of cosmetic ingredients. In: P. Frost (Ed.), *Principles of Cosmetics for the Dermatologist*, pp. 277–284. St. Louis, MO: C.V. Mosby.

Buffler, P.A., Wood, S.M., Suarez, L., and Kilian, D.J., 1978, Mortality follow-up of workers exposed to 1,4-dioxane. *Journal of Occupational and Environmental Medicine* 20(4): 255–259.

Bull, R.J., Robinson, M., and Laurie, R.D., 1986, Association of carcinoma yield with early papilloma development in SENCAR mice. *Environmental Health Perspectives* 68: 11–17.

Clark, B., Furlong, J.W., Ladner, A., and Slovak, A.J.M., 1984, Dermal toxicity of dimethyl acetylene dicarboxylate, *n*-methyl pyrrolidone, triethylen glycol dimethy ether, dioxane and tetraline in the rat. *IRCS Journal of Medical Science* 12: 296–297.

David, H., 1964, Elektronenmikroskopische Befunde bei der dioxanbedingten Nephrose der Rattenniere [Electron-microscopic findings in dioxane-dependent nephrosis in rat kidneys]. *Beiträge zur pathologischen Anatomie und zur allgemeinen Pathologie* 130: 187–212.

de Navasquez, S., 1935, Experimental tubular necrosis of the kidneys accompanied by liver changes due to dioxane poisoning. *The Journal of Hygiene* 35: 540–548.

Drew, R.T., Patel, J.M., and Lin, F., 1978, Changes in serum enzymes in rats after inhalation of organic solvents singly and in combination. *Toxicology and Applied Pharmacology* 45: 809–819.

Ernstgård, L., Iregren, A., Sjogren, B., and Johanson, G., 2006, Acute effects of exposure to vapours of dioxane in humans. *Human & Experimental Toxicology* 25: 723–729.

Fairley, A., Linton, E.C., and Ford-Moore, A.H., 1934, The toxicity to animals of 1,4-dioxane. *The Journal of Hygiene* 34: 486–501.

Fisher, J., Mahle, D., Bankston, L., Greene, R., and Gearhart, J., 1997, Lactational transfer of volatile chemicals in breast milk. *American Industrial Hygiene Association Journal* 58: 425–431.

Frantik, R., Hornychova, M., and Horvath, M., 1994, Relative acute neurotoxicity of solvents: Isoeffective air concentrations of 48 compounds evaluated in rats and mice. *Environmental Research* 66: 173–185.

Galloway, S.M., Armstrong, M.J., Reuben, C., et al., 1987, Chromosome aberrations and sister chromatid exchanges in Chinese hamster ovary cells: Evaluations of 108 chemicals. *Environmental and Molecular Mutagenesis* 10(Suppl 10): 1, 10, 21, 67.

Giavini, E., Vismara, C., and Broccia, M.L., 1985, Teratogenesis study of dioxane in rats. *Toxicology Letters* 26: 85–88.

Goldberg, M.E., Johnson, H.E., Pozzani, U.C., and Smyth, H.F., Jr., 1964, Effect of repeated inhalation of vapors of industrial solvents on animal behavior. I. Evaluation of nine solvent vapors on pole-climb performance in rats. *American Industrial Hygiene Association Journal* 25: 369–375.

Goldsworthy, T.L., Monticello, T.M., Morgan, K.T., et al., 1991, Examination of potential mechanisms of carcinogenicity of 1,4-dioxane in rat nasal epithelial cells and hepatocytes. *Archives of Toxicology Links—Archives of Toxicology* 65: 1–9.

Haworth, S., Lawlor, T., Mortelmans, K., Speck, W., and Zeiger, E., 1983, *Salmonella* mutagenicity test results for 250 chemicals. *Environmental Mutagenesis* 1(Supp 1): 1, 10.

Hellmer, L. and Bolcsfoldi, G., 1992, An evaluation of the *E. coli* K-12 uvrB/recA DNA repair host-mediated assay I: *In vitro* sensitivity of the bacteria to 61 compounds. *Mutation Research* 272: 145–160.

Hill, A.B., 1965, The environment and disease: Association or causation? *Proceedings of the Royal Society of Medicine* 58: 295–300.

Hoch-Ligeti, C. and Argus, M.F., 1970, Effect of carcinogens on the lung of guinea pigs. In *Morphology of Experimental Respiratory Carcinogenesis*, P. Nettesheim, M.G. Hanna, and J.W. Deatherage (Eds), pp. 267–279. AEC Symposium Series 21, National Cancer Institute and U.S. Atomic Energy Commission. CONF700501.

Hoch-Ligeti, C., Argus, M.F., and Arcos, J.C., 1970, Induction of carcinomas in the nasal cavity of rats by dioxane. *British Journal of Cancer* 24(1): 164–167.

IARC (International Agency for Research on Cancer), 1999, Some chemicals that cause tumours of the kidney or urinary bladder in rodents, and some other substances—1,4-dioxane (group 2B). IARC Monographs on the Evaluation of Carcinogenic Risks to Humans, vol. 71. Lyon, France: World Health Organization. Available at http://www-cie.iarc.fr/htdocs/monographs/vol71/019-dioxane.html

JBRC (Japan Bioassay Research Center), 1998a, Two-week studies of 1,4-dioxane in F344 rats and BDF1 mice (drinking water studies). Kanagawa, Japan: Japan Bioassay Research Center.

JBRC, 1998b, Thirteen-week studies of 1,4-dioxane in F344 rats and BDF1 mice (drinking water studies). Kanagawa, Japan: Japan Bioassay Research Center.

JBRC, 1998c, Two-year studies of 1,4-dioxane in F344 rats and BDF1 mice (drinking water). Kanagawa, Japan: Japan Bioassay Research Center.

Johnstone, R.T., 1959, Death Due to Dioxane? *A.M.A. Archives of Industrial Health* 20: 445–447.

Kanada, M., Miyagawa, M., Sato, M., Hasegawa, H., and Honma, T., 1994, Neurochemical profile of effects of 28 neurotoxic chemicals on the central nervous system in rats (1). Effects of oral administration on brain contents of biogenic amines and metabolites. *Industrial Health* 32: 145–164.

Kesten, H.D., Mulinos, M.G., and Pomerantz, L., 1939, Pathologic effects of certain glycols and related compounds. *Archives of Pathology* 27: 447–465.

Khudoley, V.V., Mizgireuv, I., and Pliss, G.B., 1987, The study of mutagenic activity of carcinogens and other chemical agents with *Salmonella typhimurium* assays: Testing of 126 compounds. *Archiv für Geschwulstforschung* 57: 453–462.

King, M.E., Shefner, A.M., and Bates, R.R., 1973, *Carcinogenesis* bioassay of chlorinated dibenzodioxins and related chemicals. *Environmental Health Perspectives* 5: 163–170.

Kitchin, K.T. and Brown, J.L., 1990, Is 1,4-dioxane a genotoxic carcinogen? *Cancer Letters* 53(1): 67–71.

Knoefel, P.K., 1935, Narcotic potency of some cyclic acetals. *The Journal of Pharmacology and Experimental Therapeutics* 53: 440–444.

Kociba, R.J., McCollister, S.B., Park, C., Torkelson, T.R., and Gehring, P.J., 1974, 1,4-Dioxane. I. Results of a 2-year ingestion study in rats. *Toxicology and Applied Pharmacology* 30: 275–286.

Kociba, R.J., Torkelson, T.R., Young, J.D., and Gehring, P.J., 1975, 1,4-Dioxane: Correlation of the results of chronic ingestion and inhalation studies with its dose-dependent fate in the rat. In *Proceedings of the 6th Annual Conference on Environmental Toxicology*, pp. 345–354. Aerospace Medical Research Laboratory, Wright-Patterson Air Force Base, OH.

Kurl, R.N., Poellinger, L., Lund, J., and Gustafsson, J.A., 1981, Effects of dioxane on RNA synthesis in the rat liver. *Archives of Toxicology* 49(1): 29–33.

Kwan, K.K., Dutka, B.J., Rao, S.S., and Liu, D., 1990, Mutatox test: A new test for monitoring environmental genotoxic agents. *Environmental Pollution* 65: 323–332.

Laug, E.P., Calvery, H.O., Morris, H.J., and Woodward, G., 1939, The toxicity of some glycols and derivatives. *The Journal of Industrial Hygiene and Toxicology Journal of Industrial Hygiene* 21: 173–201.

Leung, H. and Paustenbach, D.J., 1990, Cancer risk assessment for dioxane based upon a physiologically based pharmacokinetic approach. *Toxicology Letters* 51(2): 147–162.

Lundberg, I., Ekdahl, M., Kronevi, T., Lidums, V., and Lundberg, S., 1986, Relative hepatotoxicity of some industrial solvents after intraperitoneal injection or inhalation exposure in rats. *Environmental Research* 40: 411–420.

Lundberg, I., Hogberg, J., Kronevi, T., and Holmberg, B., 1987, Three industrial solvents investigated for tumor promoting activity in the rat liver. *Cancer Letters* 36: 29–33.

Marzulli, F., Anjo, D.M., and Maibach, H.I., 1981, *In vivo* skin penetration studies of 2,4-toluenediamine, 2,4-diaminoanisole, 2-nitro-*p*-phenylenediamine, *p*-dioxane and *N*-nitrosodiethanolamine in cosmetics. *Food and Cosmetics Toxicology* 19: 743–747.

McFee, A.F., Abbott, M.G., Gulati, D.K., and Shelby, M.D., 1994, Results of mouse bone marrow micronucleus studies on 1,4-dioxane. *Mutation Research* 322: 141–150.

McGregor, D.B., Brown, A.G., Howgate, S., McBride, D., Riach, C., and Caspary, W.J., 1991, Responses of the L5178Y mouse lymphoma cell forward mutation assay. V. 27 coded chemicals. *Environmental and Molecular Mutagenesis* 17(3): 196–219.

Mirkova, E.T., 1994, Activity of the rodent carcinogen 1,4-dioxane in the mouse bone marrow micronucleus assay. *Mutation Research* 322: 142–144.

Miyagawa, M., Shirotori, T., Tsuchitani, M., and Yoshikawa, K., 1999, Repeat-assessment of 1,4-dioxane in a rat-hepatocyte replicative DNA synthesis (RDS) test: Evidence for stimulus of hepatocyte proliferation. *Experimental and Toxicologic Pathology* 51: 555–558.

Morita, T., 1994, No clastogenicity of 1,4-dioxane as examined in the mouse peripheral blood micronucleus test. *Mammalian Mutagenesis Study Group Communication* 2: 7–8.

Morita, T. and Hayashi, M., 1998, 1,4-Dioxane is not mutagenic in five *in vitro* assays and mouse peripheral blood micronucleus assay, but is in mouse liver micronucleus assay. *Environmental and Molecular Mutagenesis* 32: 269–280.

Mungikar, A.M. and Pawar, S.S., 1978, Induction of the hepatic microsomal mixed function oxidase system in mice by *p*-dioxane. *Bulletin of Environmental Contamination and Toxicology* 20: 797–804.

Munoz, E.R. and Barnett, B.M., 2002, The rodent carcinogens 1,4-dioxane and thiourea induce meiotic non-disjunction in *Drosophila melanogaster* females. *Mutation Research* 517(1–2): 231–238.

Nannelli, A., Rubertis, A., Longo, V., and Gervasi, P.G., 2005, Effects of Dioxane on Cytochrome P450 enzymes in liver, kidney, lung and nasal mucosa of rat. *Archives of Toxicology* 79(2): 74–82.

National Cancer Institute (NCI), 1978, *Bioassay of 1,4-Dioxane for Possible Carcinogenicity*. Bethesda, MD: National Cancer Institute. National Institutes of Health Publication No. 78-1330 NCICGTR-80.

Nelson, N., 1951, Solvent toxicity with particular reference to certain octyl alcohols and dioxanes. *Medical Bulletin* 11: 226–238.

Nestmann, E.R., Otson, R., Kowbel, D.J., Bothwell, P.D., and Harrington, T.R., 1984, Mutagenicity in a modified *Salmonella* assay of fabric-protecting products containing 1,1,1-trichloroethane. *Environmental Mutagenesis* 6: 71–80.

Netherlands Organization for Applied Scientific Research, 1999, Risk assessment: 1,4-Dioxane. Netherlands Organization for Applied Scientific Research (TNO) and the National Institute of Public Health and the Environment (RIVM). Chemical Substances Bureau, Ministry of Housing, Spatial Planning and the Environment (VROM), Netherlands, Final Version, 5 November, EINECS-No. 204-661-8.

NICNAS, 1998, 1,4-Dioxane: Priority existing chemical no. 7, Full Public Report. National Industrial Chemicals Notification and Assessment Scheme. Commonwealth of Australia.

Pozzani, U.C., Weil, C.S., and Carpenter, C.P., 1959, The toxicological basis of threshold limit values: 5: The experimental inhalation of vapor mixtures by rats, with notes upon the relationship between single dose inhalation and single dose oral data. *Industrial Hygiene Journal* 20: 364–369.

Reitz, R.H., McCroskey, P.S., Park, C.N., Andersen, M.E., and Gargas, M.L., 1990, Development of a physiologically based pharmacokinetic model for risk assessment with 1,4-dioxane. *Toxicology and Applied Pharmacology* 105(1): 37–54.

Roy, S.K., Thilagar, A.K., and Eastmond, D.A., 2005, Chromosome breakage is primarily responsible for the micronuclei induced by 1,4-dioxane in the bone marrow and liver of young CD-1 mice. *Mutation Research* 586: 28–37.

Schrenk, H.H. and Yant, W.P., 1936, Toxicity of dioxane. *The Journal of Industrial Hygiene and Toxicology Journal of Industrial Hygiene* 18: 448–460.

Sheu, C.W., Moreland, F.M., Lee, J.K., and Dunkel, V.C., 1988, *In vitro* BALB/3T3 cell transformation assay of nonoxynol-9 and 1,4-dioxane. *Environmental and Molecular Mutagenesis* 11: 41–48.

Silverman, L., Schulte, H.F., and First, M.W., 1946, Further studies of sensory response to certain industrial solvent vapors. *The Journal of Industrial Hygiene and Toxicology Journal of Industrial Hygiene* 28: 262–266.

Sina, J.F., Bean, C.L., Dysart, G.R., Taylor, V.I., and Bradley, M.O., 1983, Evaluation of the alkaline elution/rat hepatocyte assay as a predictor of carcinogenic/mutagenic potential. *Mutation Research* 113: 357–391.

Smyth, H.F.J., Seaton, J., and Fischer, L., 1941, The single dose toxicity of some glycols and derivatives. *The Journal of Industrial Hygiene and Toxicology Journal of Industrial Hygiene* 23(6): 259–268.

Stickney, J.A., Sager, S.L., Clarkson, J.R., et al., 2003, An updated evaluation of the carcinogenic potential of 1,4-dioxane. *Regulatory Toxicology and Pharmacology* 38: 183–195.

Stott, W.T., Quast, J.F., and Watanabe, P.G., 1981, Differentiation of the mechanisms of oncogenicity of 1,4-dioxane and 1,3-hexachlorobutadiene in the rat. *Toxicology and Applied Pharmacology* 60: 287–300.

Stroebel, P., Mayer, F., Zerban, H., and Bannasch, P., 1995, Spongiotic pericytoma: A benign neoplasm deriving from the perisinusoidal (Ito) cells in rat liver. *The American Journal of Pathology* 146(4): 903–913.

Sweeney, L.M., Thrall, K.D., Poet, T.S., et al., 2008, Physiologically based pharmacokinetic modeling of 1,4-dioxane in rats, mice, and humans. *Toxicological Sciences* 101(1): 32–50.

Thiess, A.M., Tress, E., and Fleig, I., 1976, Industriell-medizinische Untersuchungsentdeckungen von der Arbeiter Einatmung des Dioxans [Industrial-medical investigation results in the case of workers exposed to dioxane]. *Arbeitsmedizin, Sozialmedizin, Präventivmedizin* 11: 35–46.

Thurman, G.B., Simms, B.G., Goldstein, A.L., and Kilian, D.J., 1978, The effects of organic compounds used in the manufacture of plastics on the responsivity of murine and human lymphocytes. *Toxicology and Applied Pharmacology* 44: 617–641.

Tinwell, H. and Ashby, J., 1994, Activity of 1,4-dioxane in mouse bone marrow micronucleus assays. *Mutation Research* 322: 148–150.

Torkelson, R., Leong, B.K.J., Kociba, R.J., Richter, W.A., and Gehring, P.J., 1974, 1,4-Dioxane. II. Results of a 2-year inhalation study in rats. *Toxicology and Applied Pharmacology* 30: 287–298.

Uno, Y., Takasawa, H., Miyagawa, M., Inoue, Y., Murata, T., and Yoshikawa, K., 1994, An *in vivo–in vitro* replications DNA synthesis (RDS) test using rat hepatocytes as an early prediction assay for nongenotoxic hepatocarcinogens: Screening of 22 known positives and 25 noncarcinogens. *Mutation Research* 320(3): 189–205.

U.S. Environmental Protection Agency (USEPA), 1988, Recommendations for and documentation of biological values for use in risk assessment. U.S. Environmental Protection Agency, Office of Research and Development, Cincinnati, OH; EPA 600/6-87/008. Available from National Technical Information Service, Springfield, VA; PB88-179874/AS.

U.S. Environmental Protection Agency (USEPA), 2005, Guidelines for carcinogen risk assessment. Risk Assessment Forum, Washington, DC; EPA/630/P-03/001B. Available at http://www.epa.gov/iris/backgr-d.htm

U.S. Environmental Protection Agency (USEPA), 2007, IRIS summary for 1,4-dioxane. U.S. Environmental Protection Agency, Integrated Risk Information System, Verification Date February 3, 1988. Available at http://www.epa.gov/iris/subst/0326.htm

van Delft, J.H.M., Van Agen, E., van Breda, S.G.J., Herwijnen, M.H., Staal, Y.C.M., and Kleinjans, J.C.S., 2004, Discrimination of genotoxic from non-genotoxic carcinogens by gene expression profiling. *Carcinogenesis* 25(7): 1265–1276.

Wirth, W. and Klimmer, O., 1936, Zur Toxikologie der Lösungsmittel: 1,4-dioxan [On the Toxicology of Organic Solvents: 1,4-Dioxane (diethylene dioxide).] *Archiv für Gewerbepathologie und Gewerbehygiene* 17: 192–206.

Woo, Y.-T., Argus, M.F., and Arcos, J.C., 1977a, Tissue and subcellular distribution of ^{3}H-dioxane in the rat and apparent lack of microsome-catalyzed covalent binding in the target tissue. *Life Sciences* 21(10): 1447–1456.

Woo, Y., Arcos, J.C., Argus, M.F., Griffin, G.W., and Nishiyama, K., 1977b, Structural identification of *p*-dioxane-2-one as the major urinary metabolite of *p*-dioxane. *Naunyn-Schmiedeberg's Archives of Pharmacology* 299: 283–287.

Woo, Y., Argus, M.F., and Arcos, J.C., 1977c, Metabolism *in vivo* of dioxane: Effect of inducers and inhibitors of hepatic mixed-function oxidases. *Biochemical Pharmacology* 25: 1539–1542.

Woo, Y.T., Argus, M.F., and Arcos, J.C., 1978, Effect of mixed-function oxidase modifiers on metabolism and toxicity of the oncogen dioxane. *Cancer Research* 38: 1621–1625.

Yamazaki, K., Ohno, H., Asakura, M., et al., 1994, Two-year toxicological and carcinogenesis studies of 1,4-dioxane in F344 rats and BDF1 mice. Drinking studies. In *Proceedings on the Second Asia-Pacific Symposium on Environmental and Occupational Health, Environmental and Occupational Chemical Hazards (2)*, pp. 193–198. Kobe University.

Yant, W.P., Schrenk, H.H., Waite, C.P., and Patty, F.A., 1930, Acute response of guinea pigs to vapors of some new commercial organic compounds—VI dioxane. *Public Health Reports* 45: 2023–2032.

Yoon, J.S., Mason, J.M., Nalencia, R., Woodruff, R.C., and Zimmering, S., 1985, Chemical mutagenesis testing in drosophila. IV. Results of 45 coded compounds tested for the national toxicology program. *Environmental Mutagenesis* 7: 349–367.

Young, J.D., Braun, W.H., Gehring, P.J., Horvath, B.S., and Daniel, R.L., 1976, Short communication. 1,4-Dioxane and beta-hydroxyethoxyacetic acid excretion in urine of humans exposed to dioxane vapors. *Toxicology and Applied Pharmacology* 38: 643–646.

Young, J.D., Braun, W.H., Rampy, L.W., Chenowith, M.B., and Blau, G.E., 1977, Pharmacokinetics of 1,4-dioxane in humans. *Journal of Toxicology and Environmental Health* 3: 507–520.

Young, J.D., Braun, W.H., and Gehring, P.J., 1978a, The dose-dependent fate of 1,4-dioxane in rats. *Journal of Environmental Pathology and Toxicology* 2: 263–282.

Young, J.D., Braun, W.H., and Gehring, P.J., 1978b, Dose-dependent fate of 1,4-dioxane in rats. *Journal of Toxicology and Environmental Health* 4(5–6): 709–726.

Zimmermann, F.K., Mayer, V.W., Scheel, I., and Resnick, M.A., 1985, Acetone, methyl ethyl ketone, ethyl acetate, acetonitrile and other polar aprotic solvents are strong inducers of aneuploidy in *Saccharomyces cerevisiae*. *Mutation Research* 149: 339–351.

6 Regulation and Risk Assessment of 1,4-Dioxane

Thomas K.G. Mohr

A fundamental maxim of toxicology is "The dose makes the poison."[*] Regulations must protect against exposure to toxic chemicals at doses large enough to result in toxic effects. Exposure prevention extends to multiple routes of exposure. For example, effective regulation will prevent the combination of ingesting contaminated drinking water and inhaling chemical vapors while showering with the same contaminated water. Unfortunately, the regulations promulgated by different agencies pursuing a wide range of protective oversight (such as food quality, air quality, water quality, and drinking water treatment) may not be holistically integrated to produce a framework that accounts for the additive effects of multiple routes of exposure to the same contaminant. For example, perchlorate, the oxidizing salt used in the production of solid rocket motors, first emerged as a drinking water concern. We have subsequently learned that the relative source contribution (RSC) from ingestion of perchlorate in food is much larger than was originally understood; however, perchlorate in food is not regulated and remains the subject of ongoing investigation. Similarly, the concentration in water and air was the primary focus of regulations for trichloroethylene (TCE), whereas haloacetic acids—the breakdown products in the liver—were not. When accounting for total exposure to haloacetic acids from inhaling TCE vapors and ingesting haloacetic acids present in drinking water as disinfection by-products, it may be warranted to adjust the regulatory threshold for TCE downward to minimize the RSC to haloacetic acid exposure. These examples underscore the difficult challenge faced by government scientists who work to establish protective regulations. Their challenge is exacerbated by the large uncertainty associated with correlating laboratory animal data to estimation of toxic effects in humans.

Regulating chemical use, handling, and disposal protects us against risk from exposure to harmful doses of toxic chemicals used widely in industry, consumer products, crop protection, and other applications. Regulations establishing exposure thresholds protect human health by providing regulatory agencies with a framework for evaluating and enforcing the safety of chemical use and associated exposure. The many ways by which chemical exposure may occur is mirrored in the complex web of regulations that have evolved to govern

- Worker health and safety
- Drinking water quality
- Chemical residues in food
- Pharmaceuticals and sundries
- Ecological effects of wastewater discharges
- Ambient air quality
- Transportation safety

---

[*] This adage is attributed to the Swiss scientist "Paracelsus," aka Phillip von Hohenheim, 1493–1541, who said, "Alle Ding' sind Gift und nichts ohn' Gift; allein die Dosis macht, dass ein Ding kein Gift ist." ("All things are poison and nothing is without poison, only the dose permits something not to be poisonous.") The axiom is sometimes wrongly attributed to Pericles, a Greek statesman.

- Consumer right-to-know of the presence of carcinogens or reproductive toxins in consumer products
- Community right-to-know of large quantities of dangerous chemicals stored and used by local industries

This chapter reviews the full complement of regulations and regulatory guidance thresholds for 1,4-dioxane, with an emphasis on the derivation of water quality standards.

6.1   DRINKING WATER REGULATIONS FOR 1,4-DIOXANE

Water quality standards are mandated by the Clean Water Act[*] and preserve the beneficial uses of a water body by designating criteria to protect it against degradation from pollutants. A water quality standard accounts for

- The designated beneficial uses of the water body (i.e., recreation, drinking water supply, aquatic life, and agriculture)
- The water quality criteria to protect beneficial uses (numeric pollutant concentrations and narrative requirements)
- The maintenance and protection of existing beneficial uses and high quality waters (antidegradation policy)
- Practical implementation issues (i.e., low flows, variances, mixing zones) (USEPA, 2006a)

The National Pollution Discharge Elimination System (NPDES), a permit program that controls water pollution by regulating point-source discharges into waters of the United States, is the primary framework for water quality standards from the Clean Water Act.

The Safe Drinking Water Act of 1974 (SDWA) authorizes the U.S. Environmental Protection Agency (USEPA) to establish national health-based standards for drinking water to protect against both naturally occurring and anthropogenic contaminants in drinking water. The SDWA was amended in 1986 to include specific health goals and defined approaches for risk management. The SDWA also outlines criteria for adding chemicals to the list of contaminants regulated by Maximum Contaminant Levels (MCLs). Prior to the 1996 amendments to the SDWA, USEPA was required to regulate an additional 25 contaminants every 3 years. Currently, USEPA has the flexibility to determine whether to regulate a contaminant after completing a required review of at least five contaminants every 5 years.

USEPA uses three criteria in its review to determine whether to regulate a contaminant:

1. The contaminant adversely affects human health
2. The contaminant is likely to occur in public water systems at sufficient frequency and concentrations to create a public health concern
3. Regulation of the contaminant presents a meaningful opportunity for health risk reduction (USEPA, 2006a)

USEPA must also establish criteria for monitoring of unregulated contaminants and must issue an Unregulated Contaminant Monitoring Requirement (UCMR) list of not more than 30 contaminants every 5 years. In order to add a new contaminant to the list of regulated compounds, USEPA must conduct cost-benefit analyses and use the "best available, peer-reviewed science and supporting studies" to set standards.

---

[*] The Clean Water Act or CWA is the Federal Water Pollution Control Act Amendments of 1972 and 1977.

As described in Chapter 4, of the more than 10 million commercially available organic and inorganic chemicals, only about a quarter million are subject to any form of regulation, far fewer are routinely analyzed, and fewer still have been profiled and studied for health effects. Reviewing more chemicals for addition to the list of regulated contaminants is essential to ensure public health protection. At the same time, some compounds on the list of regulated contaminants are being revisited to determine whether to lower, raise, or eliminate their MCLs. For example, in California, revised draft Public Health Goals for several chlorinated solvent compounds including methyl chloroform and *cis*-1,2-dichloroethylene suggest that these compounds may not be as toxic as was first thought when they were originally listed. Similarly, USEPA is reviewing whether to lower the toxicity threshold used to establish the Preliminary Remediation Goal (PRG) for TCE. California has established a Public Health Goal for perchloroethylene at 0.06 μg/L, whereas the MCL remains at 5 μg/L. Setting drinking water standards can be described as a fluid process that is by no means smooth or certain. It is in this uncertain and sometimes controversial regulatory environment that the evaluation of regulating 1,4-dioxane is being made.

The multiyear process of evaluating, listing, and approving additions to the list of regulated contaminants includes many tiers of review and can be subjected to legal challenge and political initiatives. For example, although methyl-*tert* butyl ether (MTBE) and perchlorate have been investigated and remediated for many years, the fact that these compounds are not yet regulated by a federal MCL is due in part to a protracted review process. States may independently adopt and implement regulation of drinking water contaminants on the basis of the prevalence and magnitude of drinking water contaminants within their jurisdictions. To date, 1,4-dioxane and other solvent stabilizer compounds are not regulated by USEPA, except for epichlorohydrin (formerly used to stabilize TCE), which has an MCL goal of zero (USEPA, 2007b). In February 2008, 1,4-dioxane was included in the third contaminant candidate list (CCL) (USEPA, 2008). 1,4-Dioxane has not yet been included among the federal UCMR lists. Colorado became the first state to regulate 1,4-dioxane in March 2005, as discussed further in Section 6.1.2.1.

In 2004, USEPA's National Center for Environmental Assessment in the Office of Research and Development began a review of the Integrated Risk Information System (IRIS) health assessment for 1,4-dioxane (USEPA, 2004c). IRIS is a database of scientific positions on potential adverse human health effects that may result from chronic (or lifetime) exposure to chemicals in the environment. IRIS contains chemical-specific summaries of qualitative and quantitative health information in support of dose–response evaluation. The IRIS database is used to support the development of risk assessments, site-specific environmental decisions, and rule making. USEPA's assessment of the carcinogenic potential of 1,4-dioxane was added to the IRIS database in 1988.

The IRIS Program is preparing an assessment that will incorporate available health effects information for 1,4-dioxane, as well as current risk-assessment methods. The assessment will present reference values for the noncancer effects of 1,4-dioxane (the reference dose [RfD], and the reference concentration [RfC]), where supported by available data, and a cancer assessment. The Toxicological Review and IRIS Summary will be subject to internal peer consultation, USEPA review, and external scientific peer review. The final product will reflect the agency opinion on the toxicity of 1,4-dioxane and is slated for completion by February 2009 (USEPA, 2007h).

This section profiles the existing regulations for 1,4-dioxane and compares the basis for 1,4-dioxane drinking water regulations and guidance adopted by three states.

6.1.1   FEDERAL

There is currently* no federal MCL or MCL Goal for 1,4-dioxane in drinking water. Several of the USEPA regional offices use an advisory guidance level of 6.1 parts per billion (ppb) for limiting 1,4-dioxane concentrations in drinking water. USEPA Region 9 covers the Pacific southwest states

---

* As of early 2009.

and territories and uses a PRG of 6.1 µg/L. Region 3 covers the mid-Atlantic states and uses a *Risk-Based Concentration* of 6.1 µg/L. Region 6—serving Arkansas, Louisiana, New Mexico, Oklahoma, Texas, and 65 tribes—uses a Human Health Tap Water Screening Level of 6.1 µg/L.

The absence of a federal MCL for 1,4-dioxane has left the states to determine whether to promulgate regulatory standards or Advisory Action Levels. Only Colorado has adopted a standard; Michigan held hearings on adopting a state MCL of 35 ppb for 1,4-dioxane, but did not adopt a standard. It is unlikely there will be a federal MCL adopted anytime soon, as the criteria required to justify adopting an MCL have not materialized, and it is unclear whether USEPA will find that there has been enough exposure or that the health risks are sufficiently understood. More familiar contaminants, such as the widely used fuel oxygenate MTBE and the ubiquitous solid rocket fuel oxidizer, perchlorate, still lack federal MCLs. Nevertheless, the author has witnessed the discovery of 1,4-dioxane in drinking water in several communities following his dialog with state and federal regulators working on solvent cleanups in several states. It is likely that there is additional ongoing but undiscovered exposure to 1,4-dioxane in drinking water.

MCLs are usually preceded by Maximum Contaminant Level Goals (MCLGs), which are similar to Public Health Goals or Advisory Action Levels. MCLGs are set at the level that causes no known adverse effects to people and incorporate an adequate margin of safety; MCLs are subsequently set as close as feasible to MCLGs. The MCL may be higher than the MCLG if available analytical methods do not quantify the contaminant in the range at which the MCLG is set. The availability of treatment technologies and the cost of treatment may also factor in the USEPA administrator's selection of the MCL.

The MCLG for an *initiator* carcinogen (i.e., a contaminant with no threshold for adverse effects) is set to zero by default. The MCL for an initiator carcinogen is generally set between the 1 in 10,000 and the 1 in 1,000,000 increased lifetime risk. For a *promoter* carcinogen or for a noncarcinogen (a contaminant showing a threshold for adverse effects), the MCLG is based on the RfD, set to a level below known adverse health effects (Macler, 2006).

A number of events would have to transpire before 1,4-dioxane could be regulated with a federal MCL. First, 1,4-dioxane would have to be identified as a possible health concern in drinking water, leading to its inclusion on the USEPA drinking water CCL. The CCL is compiled by a committee of experts that reviews potential contaminants based on toxicity, potency, or severity, on prevalence and magnitude of occurrence, and on persistence and mobility (NDWAC, 2006). USEPA weighed inclusion of 1,4-dioxane on the CCL in 1997 but decided against adding it. In 2006, USEPA solicited candidates for the third drinking water CCL, and in February 2008, 1,4-dioxane was included on the list (USEPA, 2006b, 2008). Once listed on the CCL, health, occurrence, and exposure information will be assembled to conduct a preliminary risk assessment. To assist with establishing the presence of a contaminant, some contaminants under review for the CCL are included in the UCMRs. If regulators conclude that an opportunity exists to reduce public health risks by regulation, the requisite regulatory elements would have to be developed. These include health risk assessments to dctcrmine potential standards and quantify the benefits derived from regulating 1,4-dioxane. Further, viable analytical methods would have to be available for analyzing 1,4-dioxane in drinking water at the levels of interest for reducing risk. USEPA developed Method 522 for determination of 1,4-dioxane in drinking water in 2008.[*] Review of 1,4-dioxane treatment methods required to analyze costs for compliance and enforcement is also needed to adopt an MCL. The draft MCL and other regulatory elements are then reviewed by USEPA and the President's Office of Management and Budget (OMB). Thereafter, public comments are solicited and addressed, the final National Primary Drinking Water Regulations (NPDWR) document is reviewed by USEPA and OMB again, and then the regulation is promulgated. Implementation and enforcement would follow a schedule outlined in the NPDWR (Macler, 2006).

---

[*] Method 522 (2008) uses solid-phase extraction and gas chromatography–mass spectrometry with selected-ion monitoring.

TABLE 6.1
State, Federal (EPA), and International Regulatory Guidelines, Action Levels, and Remediation Targets for 1,4-Dioxane as of 2007

Concentration of 1,4-Dioxane in Given Amount of Media and Use or Type of Exposure

Alaska: Draft cleanup guidance: Human health risk-based concentrations [1]
Arctic soil in direct contact: 730 mg/kg
Soil in direct contact (>40 ft to groundwater): 540 mg/kg
Soil in direct contact (<40 ft to groundwater): 440 mg/kg
Groundwater: 77 µg/L

Arizona: Soil remediation level [2]
Residential soil: 400 mg/kg
Nonresidential soil: 1700 mg/kg

California: Advisory level [2]
Drinking water: 3 µg/L
Residential or industrial soil: 0.0018 mg/kg

Colorado: Water quality standards [2]
Groundwater (March 2005 to March 2010): 6.1 µg/L
Surface water (after March 2010): 3.2 µg/L

Connecticut: Comparison value for risk assessments [2]
Drinking water: 20 µg/L

Delaware: Remediation standard [2]
Groundwater: 6 µg/L
Soil (critical water resource area): 0.6 mg/kg
Soil (noncritical water resource area): 58–520 mg/kg

Florida: Cleanup target levels [2]
Groundwater: 3.2 µg/L
Surface water: 120 µg/L
Soil (with direct exposure—commercial or industrial): 38 mg/kg
Soil (with direct exposure—residential): 23 mg/kg
Soil (leachable by groundwater): 0.01 mg/kg
Soil (leachable by surface water): 0.5 mg/kg

Hawaii: Action levels [22]
Soil (for releases over groundwater used for drinking water): 3.7 µg/kg
Groundwater (drinking water): 6.1 µg/L
Surface water (nondrinking; aquatic habitat protection): 34,000 µg/L
Surface water (marine): 50,000 µg/L

Illinois: Groundwater remediation objective
Groundwater (acceptable detection limit for carcinogens): 1 µg/L

Iowa [2]
Soil: 280 mg/kg

Maine: Maximum exposure guideline [2]
Drinking water: 32 µg/L

Massachusetts [19]
Guidelines:
  Drinking water: 3 µg/L
Contingency-plan standards:
  Groundwater (guidance value—issued in 2007): 3 µg/L
  Groundwater (shallow aquifer with vapor-intrusion potential): 6000 µg/L
  Groundwater (ecological risk): 50,000 µg/L
  Soil (direct contact and leaching): 0.005 mg/kg

continued

TABLE 6.1   (continued)
State, Federal (EPA), and International Regulatory Guidelines, Action Levels, and Remediation Targets for 1,4-Dioxane as of 2007

Michigan: Cleanup criteria or screening levels [2]
Drinking water (industrial): 350 µg/L
Drinking water (residential): 85 µg/L
Soil (industrial drinking water protection): 7 mg/kg

Minnesota: Recommended allowable limits for drinking water contaminants [13,14]
Drinking water: 30 µg/L

Mississippi: Target remedial goal [3]
Groundwater: 6.09 µg/L
Soil (restricted): 520 mg/kg
Soil (unrestricted): 0.581 mg/kg

Missouri: Target concentrations [2]
Groundwater: 3 µg/L
Soil (direct exposure): 150–590 mg/kg
Soil (leaching to groundwater): 0.01 mg/kg

Montana: Screening level (EPA Region 9 PRGs)
Tap water: 6.1 µg/L
Soil (industrial): 160 mg/kg
Soil (residential): 44 mg/kg

New Hampshire: Proposed risk-based remediation value [10]
Groundwater: 3 µg/L

New Mexico: Informal health advisory level from USEPA [4]
Groundwater: 568 µg/L for 10 days' exposure

New York
Brownfield program soil cleanup objectives [5]
Soil (unrestricted): 7.3 mg/kg
Soil (residential): 9.8 mg/kg
Soil (restricted residential): 13 mg/kg
Soil (commercial): 130 mg/kg
Soil (industrial): 250 mg/kg
Department of Health drinking water standard (immediate action level) [15]
Drinking water: 600 µg/L

North Carolina: Groundwater quality standard [6]
Groundwater: 7 µg/L

Ohio: EPA Region 9 PRGs for soil and groundwater [7]

Oklahoma: (EPA Region 6 screening levels) [8]
Soil (residential): 44 mg/kg
Soil (industrial—indoor worker): 520 mg/kg
Soil (industrial—outdoor worker): 170 mg/kg
Tap water: 6.1 µg/L

Oregon [20]
Risk-based concentrations:
Soil (residential—soil ingestion, dermal contact): 53 mg/kg
Soil (urban residential—soil ingestion, dermal contact): 140 mg/kg
Soil (residential—leaching to groundwater): 0.23 mg/kg

TABLE 6.1   (continued)
State, Federal (EPA), and International Regulatory Guidelines, Action Levels, and Remediation Targets for 1,4-Dioxane as of 2007

Soil (urban residential—leaching to groundwater): 0.41 mg/kg

Screening level II for assessment of ecological risk to mammals:

Soil: 63 mg/kg

Surface water: 4 mg/L

Ingestion and inhalation from tap water:

Groundwater (residential, domestic well): 5.2 ppb

Groundwater (urban residential): 8.9 ppb

Groundwater (occupational): 37 ppb

Pennsylvania: media-specific concentrations for organic-regulated substances [2]

Groundwater (residential—used aquifers): 5.6 µg/L

Groundwater (nonresidential—used aquifers): 24 µg/L

Groundwater (residential—nonuse aquifers): 56 µg/L

Groundwater (nonresidential—nonuse aquifers): 240 µg/L

Soil (nonresidential—direct contact): 210–240 mg/kg

Soil (residential—direct contact): 41 mg/kg

Soil (nonresidential—groundwater protection): 0.31–240 mg/kg

Soil (residential—groundwater protection): 0.0073–56 mg/kg

South Carolina: Drinking water health advisory [2]

Drinking water (monthly average): 70 µg/L

South Dakota: USEPA 2006 drinking water health advisories [9]

Drinking water (one-day health advisory[a]): 4000 µg/L

Drinking water (10-day health advisory): 400 µg/L

Drinking water (1:10,000 cancer risk): 300 µg/L

Tennessee: Remediation guidance levels [21]

Groundwater: 3 µg/L

Soil: 60 mg/kg

Texas: Protected concentration levels [2]

Groundwater (commercial or industrial): 18.6 µg/L

Groundwater (residential): 8.3 µg/L

Soil (total combined pathways—industrial): 2600 mg/kg

Soil (total combined pathways—residential): 552 mg/kg

Soil (groundwater protection—industrial): 0.36 mg/kg

Soil (groundwater protection—residential): 0.083 mg/kg

Virginia: Risk-based concentration [16]

Groundwater: 6.1 µg/L

Washington: Groundwater quality criteria [17]

Groundwater: 7 µg/L

West Virginia: Risk-based concentrations [2]

Groundwater: 6.1 µg/L

Soil (industrial): 5200 mg/kg

Soil (residential): 58 mg/kg

Wyoming [2]

Soil (residential): 44 mg/kg

continued

TABLE 6.1   (continued)
State, Federal (EPA), and International Regulatory Guidelines, Action Levels, and Remediation Targets for 1,4-Dioxane as of 2007

EPA National Center for Environmental Assessment [2]

Health-based advisory level (tap water): 3 µg/L

CSF: 0.011 (mg/kg d)$^{-1}$

EPA Region 3: Risk-based concentrations [2]

Tap water: 6.1 µg/L

Soil (industrial): 260 mg/kg

Soil (residential): 58 mg/kg

Soil (groundwater protection): 0.0013–0.0026 mg/kg

EPA Region 6 screening levels [2]

Tap water: 6.1 µg/L

Soil (industrial): 170–520 mg/kg

Soil (residential): 44 mg/kg

EPA Region 9 PRG [2]

Tap water: 6.1 µg/L

Soil (industrial): 160 mg/kg

Soil (residential): 44 mg/kg

Canada: Guidance value [11]

Drinking water: 30 µg/L

Japan: Drinking water quality standard [18]

Drinking water: 50 µg/L

WHO: Guidelines for drinking-water quality [23]

Drinking water: 50 µg/L

Sources: [1] ADEC (2007); [2] AFCEE (2007); [3] Mississippi-DEQ (2002); [4] NMED (2006); [5] NYSDEC (2006); [6] NCDENR (2005); [7] OEPA (2004); [8] Oklahoma DEQ (2003); [9] SDENR (2007); [10] NHDES (2004); [11] Health Canada (2005); [12] Montana-DEQ (2002); [13] OEHHA (2000); [14] ATSDR (1992); [15] ATSDR (2005a); [16] VDEQ (2007); [17] Washington Administrative Code (2003); [18] Japanese Ministry of Health, Labour, and Welfare (2003); [19] MADEP (2007); [20] ODEQ (2007); [21] TELL (1996); [22] HDOH (2005); and [23] WHO (2005).

Notes: States with no standards (not listed in Table 6.1) will typically evaluate groundwater or surface water on a case-by-case basis after 1,4-dioxane is detected. However, these states typically do not require testing, and when testing is performed, detection limits are often very high (50–100 ppb). Therefore, 1,4-dioxane has not manifested as a significant problem in these states even where chlorinated ethanes are present.

[a] One-day health advisory: The concentration of a chemical in drinking water that is not expected to cause any adverse noncarcinogenic effects for up to one day of exposure. The one-day health advisory is normally designed to protect a 10 kg child consuming 1 L of water per day.

Regulations governing acceptable levels of 1,4-dioxane in drinking water and cleanup levels for 1,4-dioxane in soil and groundwater vary widely. A compilation of state, federal, and international regulatory guidelines, action levels, and remediation targets as of 2007 is presented in Table 6.1.

6.1.2   BASIS FOR STATE REGULATIONS OR GUIDELINES: COLORADO, CALIFORNIA, CONNECTICUT, AND MICHIGAN

Chapter 5 provides a comprehensive review of the 1,4-dioxane toxicology knowledge base. The primary exposure routes leading to 1,4-dioxane toxicity are ingestion of contaminated drinking water and inhalation, which is likely to occur only in occupational settings. Liver and kidney toxicity are the primary noncancer health effects associated with exposure to 1,4-dioxane in humans and laboratory animals. The RfD value is used to demarcate safe levels of exposure. The RfD is an estimate of a daily human oral exposure that is not likely to cause an appreciable risk of deleterious noncancer health effects during a lifetime. Uncertainty associated with the RfD may span an order of magnitude. This section reviews the toxicological basis for promulgating legal standards or Advisory Action Levels by the states leading the nation in regulatory response to 1,4-dioxane contamination of groundwater. Not all states use a toxicological basis for setting 1,4-dioxane guidance; Massachusetts set its guidance value at 50 ppb because of the practical quantitation limit for 1,4-dioxane analysis.

6.1.2.1   Colorado's 1,4-Dioxane Standard

The Colorado Department of Public Health and Environment (CDPHE) finalized the first enforceable American water quality standard for 1,4-dioxane in groundwater and surface water in September 2004 with enforcement beginning in March 2005. In March 2005, the CDPHE Water Quality Control Commission adopted Regulation 41, *Basic Standards for Groundwater*, in which the standard for 1,4-dioxane is set at 6.1 µg/L for 5 years. The commission proceeded with adopting this 5-year standard for 1,4-dioxane in order to address ongoing water quality issues while USEPA reviews risk levels for 1,4-dioxane in the IRIS database. The 6.1 µg/L level was chosen as an interim standard because it was the value typically used as a cleanup level in Colorado, based on USEPA's PRG, and provides a basic level of human health protection.

The CDPHE Water Quality Control Commission's methodology for setting health-based standards yields a 3.2 µg/L standard for 1,4-dioxane. Unless USEPA provides further clarification on risk levels for 1,4-dioxane, CDPHE will change the 1,4-dioxane standard for groundwater and surface water from 6.1 to 3.2 µg/L on March 21, 2010. The CDPHE Statement of Basis and Purpose for Regulation 41 lists criteria considered in deriving the 3.2 µg/L standard as follows:

*   1,4-Dioxane is classified as a group B2 chemical, that is, a probable human carcinogen.
*   1,4-Dioxane is a groundwater contaminant in Colorado, and treated groundwater contaminated with 1,4-dioxane is discharged to Colorado surface waters.
*   1,4-Dioxane is readily treated with advanced oxidation processes (AOP) in combination with ultraviolet (UV) light.
*   Because the adopted standard for 1,4-dioxane is aligned with existing cleanup standards, it will not have a major impact on treatment costs during the initial 5-year period.
*   1,4-Dioxane is primarily used as a solvent stabilizer, and it will most likely be found in areas with known chlorinated solvent contamination.
*   1,4-Dioxane is characterized by a high solubility, moderate vapor pressure, and low Henry's law constant; therefore, it will be persistent within the aquatic environment. Available data indicate that 1,4-dioxane will not readily degrade in the environment.
*   1,4-Dioxane has been found at nine Colorado sites and is suspected at 19 others.
*   1,4-Dioxane contamination does not arise from natural sources.
*   1,4-Dioxane standards are adopted to protect domestic water supply uses.
*   The commission stated that although conflicting scientific interpretations warrant further review of 1,4-dioxane toxicity, there is no sufficient evidence as yet to invalidate the current USEPA IRIS value. The commission decided to err in the direction of protection of public health by approving the 6.1 and the 3.2 µg/L standards for 1,4-dioxane, starting in 2005 and 2010, respectively (CDPHE, 2005).

In hearings held on CDPHE's 1,4-dioxane regulation, industry groups argued that Colorado should hold off on setting a 1,4-dioxane standard until USEPA completes its review of risks for its IRIS database. Industry representatives were concerned that USEPA may update the acceptable exposure levels for 1,4-dioxane within 2 years (Inside EPA, 2004). Some parties argued that a nonlinear toxicity model should be used to characterize 1,4-dioxane toxicity, rather than the linear model used to develop the current IRIS value. The same parties also argued that a 1,4-dioxane standard should be established based on the laboratory practical quantitation limit for 1,4-dioxane.[*]

CDPHE toxicologist Raj Goyal (CDPHE, 2004) rebutted testimony by opponents to CDPHE's 1,4-dioxane regulation. Goyal's arguments included the following points:

- USEPA's new draft cancer risk-assessment guidelines[†] on the use of linear and nonlinear modeling methods support CDPHE's position. Method selection (linear and/or nonlinear) for low-dose extrapolation should depend upon the availability of information for the operative mode of action to anticipate the shape of the dose–response relationship. Toxicity assessments may use both linear and nonlinear approaches if linearity is not plausible and nonlinearity has support, but a mode of action is not well defined. Goyal's brief explains that an assessment may use both linear and nonlinear approaches if responses appear to be very different, that is, linear at low dose and nonlinear at high dose, as may occur if there are separate modes of action at high or low doses.

- USEPA's cancer risk guidelines explain that the linear approach assumes that any exposure dose has some associated risk of causing cancer to develop. Linear dose–response extrapolation implies that risk decreases proportionately with dose below the experimental dose; linearly extrapolating low doses for cancer risk assessments is generally considered to be protective of public health (USEPA, 1999).

- The nonlinear extrapolation approach assumes that cancer effects are only likely to occur if exposure dose is high enough to overwhelm the protective capacity of the receptor. Cancer effects may be a secondary effect of toxicity or of an induced physiological change that is itself a threshold phenomenon. Nonlinear dose–response relationships therefore imply that responses will decrease much more quickly than linearly with dose (USEPA, 1999). When the mode of action is understood well enough to support a nonlinear extrapolation, a dose–response curve is not estimated below the point of departure. Instead, an analysis is used to describe the margin of exposure (or distance) from the point of departure to a dose where there would be a little concern for cancer.

- The CDPHE rebuttal brief concludes that the lack of understanding on the mode of action of 1,4-dioxane justifies using both linear and nonlinear extrapolation approaches. Therefore, the USEPA IRIS toxicity value [the cancer slope factor (CSF)][‡] is not invalidated. The exact mode of action of 1,4-dioxane cancer and noncancer effects has not been established because 1,4-dioxane toxicokinetics and metabolism, as well as the toxic potential of its metabolite(s), are not yet delineated (CDPHE, 2004). Possible metabolites of 1,4-dioxane include the compounds 1,4-dioxane-2-one, β-hydroxyethoxy acetaldehyde, β-hydroxyethoxyacetic acid (HEAA), diglycolic acid, diethylene glycol, and oxalic acid. CDPHE notes that it is not known whether 1,4-dioxane is directly metabolized to HEAA or whether 1,4-dioxane-2-one is the principal metabolite, which then undergoes hydrolysis to HEAA. The two metabolites, HEAA and 1,4-dioxane-2-one, are readily interconvertible depending on the pH of the solution; HEAA is produced at alkaline pH.

- Genetic toxicity studies indicate that 1,4-dioxane is a weak genotoxin (causing direct damage of genetic material) and a strong promoter of tumors. CDPHE concludes that the mechanism or mechanisms by which 1,4-dioxane exerts its carcinogenic effects remain

---

[*] In 2003, when the Michigan MCL for 1,4-dioxane was under consideration, the example cited for using the PQL as the basis for setting a standard was Massachusetts, which at that time had a 50 ppb advisory level based on a 50 ppb PQL.

[†] USEPA (1999, 2005).

[‡] The cancer slope factor (CSF) estimates the carcinogenic potency of a chemical, characterized as a plausible upper bound on the increased human cancer risk from lifetime exposure to an average daily dose of 1 mg/kg. Therefore, the slope factor is expressed as average dose in units of $(mg/kg\ d)^{-1}$. Lower values of CSFs reflect lower carcinogenic potency of a chemical and result in higher drinking water standards.

in question because data from mechanistic studies are largely inconsistent. Similarly, information regarding tumor susceptibility following early-life exposures and exposures later in life is not yet available for 1,4-dioxane. Therefore, CDPHE concludes that the possibility of having different modes of action during different periods of life cannot be ruled out for 1,4-dioxane. This uncertainty for 1,4-dioxane also justifies using the linear low-dose extrapolation method, because a nonlinear model could underestimate risk by disproportionately reducing risk with dose reductions.

Finally, CDPHE supports its standard by noting that available evidence does not support the assumption that tumors will only develop if the exposure is high enough to saturate metabolic detoxification (CDPHE, 2004).

6.1.2.2 California's Notification Level

The California Environmental Protection Agency's Office of Environmental Health Hazard Assessment (OEHHA) publishes notification levels (NLs) to provide water utility operators with a threshold for announcing detections of contaminants in drinking water for which MCLs have not yet been established. Prior to 2004, California NLs were called action levels. When chemicals are found at concentrations greater than these levels, water utilities are encouraged to notify consumers.

The NL is usually set at the theoretical lifetime risk of up to one excess case of cancer in a population of 1,000,000 people—the 10^{-6} risk level. If the chemical is not detectable as low as the NL by available laboratory analytical methods, detectability prevails.

The California Department of Public Health (CDPH) regulates drinking water utilities. It recommends that wells in which unregulated contaminants have been detected be taken off-line when the chemical's concentration is 10 times the NL, if based on noncancer endpoints, or 100 times the NL, if based on cancer risk established at the 10^{-6} risk level (CDPH, 2007a). A level 100 times the NL corresponds to a theoretical lifetime risk of up to one excess case of cancer in 10,000 people.

A California water utility may continue to operate a well or other water source following detection of an unregulated contaminant. To continue using a well with detections of 1,4-dioxane or other unregulated contaminants, the utility must notify the local governing body (i.e., the city council) to advise that water exceeding the chemical's NL is being served to customers. CDPH also recommends that water system operators directly notify all the water system's customers and issue a press release advising that the contaminant is present in drinking water at a concentration greater than its NL. Following notification, CDPH recommends that water system operators proceed with (1) monthly sampling and analysis of the drinking water supply for as long as the contaminant exceeds its NL and (2) quarterly sampling for a year after the concentration drops below the NL (CDPH, 2007a). Although the guidance for notification of water utility customers following detection of an unregulated contaminant is in the form of recommendations, in practice, most California water utilities are inclined to place wells with detections on standby to protect their customers.

California NLs are calculated by using standard risk-assessment methods for noncancer and cancer endpoints and typical exposure assumptions, including a 2 L/d ingestion rate, a 70 kg adult body weight, and a 70-year lifetime. For noncarcinogenic chemicals, the NL is derived from the No Observed Adverse Effect Level (NOAEL—the highest dose at which none of the animals in a test population displays a toxic effect), adjusted by uncertainty factors (UFs). An estimate of drinking water's contribution to total exposure to the contaminant, the RSC, is also included. The commonly used RSC is 20% of the total exposure originating from the ingestion of contaminated drinking water.

The California Action Level or NL for carcinogens is calculated as

$$NL = \frac{BW \times R}{q_1^* \times V_{day}}, \tag{6.1}$$

where NL is the California Notification Level (in mg/L); BW is the adult body weight (in kg); R is the *de minimis* level for lifetime excess individual cancer risk; $q_1^*{}_{\text{human}}$ is the CSF (in $[\text{mg/kg d}]^{-1}$), the upper 95% confidence limit on the cancer potency slope calculated by the linearized multistage (LMS) model and the potency estimate, which is converted from animal to human equivalent by using (body weight ratio)$^{3/4}$ scaling; and V_{day} (in L/d) is the daily volume of water consumed by an adult.

The California NL for the carcinogenic effects of 1,4-dioxane is calculated on the basis of a carcinogenic potency of 1.4×10^{-2} $(\text{mg/kg d})^{-1}$ as described below (Cal EPA, 1998).

California OEHHA selected the cancer potency $[q_1^*{}_{\text{human}}]$ of 0.027 $(\text{mg/kg d})^{-1}$ calculated for the combined incidence of hepatocellular adenomas and carcinomas in female mice in the 1978 National Cancer Institute study for 1,4-dioxane (NCI, 1978; described in Chapter 5). The study used an interspecies scaling power of 2/3 as recommended by USEPA's 1986 Cancer Risk Assessment guidelines and shown in Equation 6.2a. The currently recommended interspecies scaling power is 3/4, as shown in Equation 6.2b (USEPA, 1986, 1992b, 1996a).

$$q_1^*{}_{\text{human}} = q_1^*{}_{\text{animal}} \times \left(\frac{L_{\text{animal}}}{D}\right)^2 \times \left(\frac{\text{BW}_{\text{human}}}{\text{BW}_{\text{animal}}}\right)^{1/3}, \tag{6.2a}$$

$$q_1^*{}_{\text{human}} = q_1^*{}_{\text{animal}} \times \left(\frac{L_{\text{animal}}}{D}\right)^3 \times \left(\frac{\text{BW}_{\text{human}}}{\text{BW}_{\text{animal}}}\right)^{1/4}, \tag{6.2b}$$

where L is the life span (in weeks) and D is the experimental duration (in weeks). In 1989, OEHHA derived a cancer potency $[q_1^*{}_{\text{human}}]$ of 0.027 $(\text{mg/kg d})^{-1}$ by using Equation 6.2a, as shown in Equation 6.3 (Cal EPA, 2007c):

$$q_1^*{}_{\text{human}} = 0.0014 \times \left(\frac{104}{90}\right)^2 \times \left(\frac{70}{0.035}\right)^{1/3} = 0.027. \tag{6.3}$$

OEHHA derived a new cancer potency value $[q_1^*{}_{\text{human}}]$ of 0.014 $(\text{mg/kg d})^{-1}$ by using the scaling power of 3/4, as shown in Equation 6.4:

$$q_1^*{}_{\text{human}} = 0.0014 \times \left(\frac{104}{90}\right)^3 \times \left(\frac{70}{0.035}\right)^{1/4} = 0.014. \tag{6.4}$$

The IRIS database uses data from the drinking water study by NCI (1978) and derives a cancer potency of 0.011 $(\text{mg/kg d})^{-1}$, on the basis of the incidence of squamous cell carcinoma of the nasal turbinates in male rats.[*]

To calculate the NL,

$$\text{NL} = \frac{70 \times 10^{-6}}{1.4 \times 10^{-2} \times 2} = 2.5\ \mu\text{g/L} \approx 3\ \mu\text{g/L}, \tag{6.5}$$

where BW = 70 kg is taken as the default adult male human body weight), $R = 10^{-6}$ is the default *de minimis* lifetime excess individual cancer risk, the CSF used is $q_1^*{}_{\text{human}} = 1.4 \times 10^{-2}$ $(\text{mg/kg d})^{-1}$, and the default daily water consumption = 2 L/d.

---

[*] Hoch-Ligeti and Argus (1970) and Argus et al. (1973); see Chapter 5.

By using the cancer potency of 0.011 (mg/kg d)$^{-1}$ from the same study (see footnote), the NL is calculated as follows:

$$NL = \frac{70 \times 10^{-6}}{1.1 \times 10^{-2} \times 2} = 3.2 \ \mu g/L \approx 3 \ \mu g/L. \tag{6.6}$$

6.1.2.3 Connecticut's Comparison Value for 1,4-Dioxane

The Connecticut Department of Public Health (CT DPH) derived an advisory comparison value of 20 ppb for 1,4-dioxane in drinking water. CT DPH notes that 1,4-dioxane can cause liver cancer at and above 1000 mg per kilogram body weight per day (mg/[kg d]); 1,4-dioxane is somewhat toxic to the liver and possibly carcinogenic at 100 mg/(kg d); and there is a slight suggestion of carcinogenic response at 25 mg/(kg d) in rats. At 10 mg/(kg d), research studies note no tumors or toxicity in the liver in rats. The chronic cancer NOAEL is therefore taken to be 10 mg/(kg d). CT DPH also considers this NOAEL to be protective for other toxic effects, as it is 17,500 times greater than the daily dose derived from drinking 2 L of water with a 1,4-dioxane concentration of 20 µg/L (Chute and Ginsberg, 2004).

The Connecticut Comparison Value was derived by applying the NOAEL (10 mg/(kg d)) and a UF of 3000 to obtain an RfD of 3.3 µg/(kg d). This dose is then calculated for a 70 kg person drinking 2 L/day, which results in a threshold of 115 µg/L if 100% exposure is from drinking water. The RSC of 20% exposure from drinking water is then applied, resulting in a comparison value of 23.1 ppb, which is rounded down to 20 ppb for simplicity (Chute and Ginsberg, 2004, 2005).

A 1,4-dioxane contamination case in Durham, Connecticut, led to a Health Consultation prepared jointly by the Agency for Toxic Substances and Disease Registry (ATSDR) and CT DPH. The Health Consultation focused on consequences of 1,4-dioxane in drinking water at concentrations of up to 27 ppb by up to 80 people obtaining drinking water from 20 private wells that were contaminated by chlorinated solvents from nearby manufacturing businesses. The businesses produced metal boxes and used methyl chloroform for vapor degreasing for 23 years; however, the duration of exposure to 1,4-dioxane in well water is unknown. The majority of 1,4-dioxane detections was below Connecticut's 20 ppb advisory comparison value. Because 1,4-dioxane concentrations were only slightly above the 20 ppb comparison value, and because the great majority of detections were below 20 ppb, CT DPH does not think that these exposures are likely to cause adverse health effects (ATSDR, 2005c).

The 20 ppb comparison value provides a factor of safety ranging from 8000 to 40,000 depending on assumptions of total exposure. Although past exposures in the Durham case could have been higher than those measured recently, regulatory toxicologists estimate that the toxicity from exposure to TCE, which is also present in well water and is now removed by carbon filters, would likely have posed a much greater health threat than exposure to 1,4-dioxane (ATSDR, 2005c). CT DPH and ATSDR concluded that the 1,4-dioxane in well water does not pose a public health hazard because exposure doses and risks are well below the threshold where adverse health effects caused by exposure could reasonably be expected to occur. Nevertheless, as a precaution, CT DPH recommended that residents use bottled water instead of treated well water, because the carbon treatment did not remove 1,4-dioxane (ATSDR, 2005c). Table 6.2 summarizes the risk factors incorporated into establishing regulatory guidance and statutes in California, Connecticut, and Colorado.

6.1.2.4 Michigan's Proposed Maximum Contaminant Level

Michigan pioneered the regulatory response to 1,4-dioxane contamination because of the large release of 1,4-dioxane from the Pall Life Sciences/Gelman site near Ann Arbor (PLS/Gelman; see case history of this site in Chapter 8). When the contamination near Ann Arbor was first discovered in 1985, Michigan's advisory generic residential cleanup criterion for 1,4-dioxane was 3 ppb for groundwater and 60 ppb for soils. In June 1995, the state legislature amended Part 201 of the state code (the section addressing environmental remediation), which resulted in an increase of the generic residential cleanup

TABLE 6.2
1,4-Dioxane Risk Calculation Parameters for State Drinking Water Thresholds

Parameter	California	Colorado	Connecticut
Cancer slope factor (mg/kg d)$^{-1}$	0.014	0.011	0.011
Uncertainty factor	See text	3000[a]	3000[b]
NOAEL (mg/(kg d))	See text	9.6	10
Reference dose (μg/(kg d))	See text	9.6	3.3
Body weight (kg)	70	70	70
Ingested volume (L/d)	2	2	2
Relative source contribution	See text	0.2	0.2
Drinking water concentration (μg/L)	2.5	6.7	23.1
Advisory level (μg/L)	3	6.1	20

Sources: Cal EPA, 1998, 1,4-Dioxane action level. California Environmental Protection Agency, Office of Environmental Health Hazard Assessment; CDPHE, 2004, Rebuttal testimony of Raj Goyal, Ph.D. Re: Proposed revisions to the basic standards and methodologies for surface water, Regulation #31 (5CCR 1002-31), the basic standards for groundwater, Regulation #41 (5CCR 1002-41), and the site-specific water quality classifications and standards for groundwater, Regulation #42 (5CCR 1002-42). Colorado Department of Public Health and Environment; and Chute, S.K. and Ginsberg, G., 2004, Comparison value determination for 1,4-dioxane in drinking water. Connecticut Department of Public Health, http://www.ct.gov/dph/ (accessed November 11, 2006).

[a] 10× for LOAEL to NOAEL; 10× for animal to human; 10× for human variability; 3× for database deficiency.

[b] 1000× for animal to human; sensitive individuals; promoter; 3× for database deficiencies.

criteria to 77 ppb for groundwater and 1500 ppb for soils. In June 2000, the Michigan Department of Environmental Quality (MDEQ) again modified its risk-based cleanup criteria to 85 ppb for groundwater and 1700 ppb for soils. The concentration in surface water considered safe by MDEQ for human contact and the environment is 2800 ppb. However, if that surface water is used as a source of drinking water, the concentration considered safe is 34 ppb, assuming drinking water consumption of 2 L/d and fish consumption of 15 g/d (MDEQ, 2005a). The 34 ppb surface-water safety level was based on the incidence of liver tumors in female mice rather than on the increase in nasal turbinate tumors in rats used by USEPA, because MDEQ judged that the nasal tumors were caused by topical exposure of 1,4-dioxane to the rats' nasal cavities during drinking water consumption (MDEQ, 2003). Advisory cleanup criteria are not legally binding standards for drinking water, leaving state regulators some discretion to consider site-specific conditions in their implementation.

In 2004, MDEQ proposed a 35 μg/L Michigan MCL for 1,4-dioxane by using a 1 in 100,000 (10^{-5}) cancer risk factor rather than a 1 in 1,000,000 (10^{-6}) cancer risk factor. Michigan has no other MCLs independent of USEPA's MCLs. MDEQ staff was motivated to propose an MCL because of the large-scale impacts of 1,4-dioxane to domestic and municipal drinking water supplies serving a large population in Washtenaw County.

MDEQ solicited comments and held a public hearing on the proposed addition of a 1,4-dioxane MCL to the Michigan SDWA. MDEQ received comments overwhelmingly opposed to the proposed MCL. Stakeholders objected to the (10^{-5}) cancer risk factor as the basis for the MCL; they noted that using a 10^{-6} cancer risk factor would be more protective and result in an MCL closer to 3 μg/L. Industry and responsible party representatives favored a higher MCL or none at all.

The Michigan Manufacturers Association (MMA) issued comments objecting to the MCL and asserting that MDEQ had not followed a proper methodology for setting a drinking water standard. MMA objected to (1) the lack of a comprehensive effort to prioritize compounds for regulation and (2) the lack of consideration of cost to treat 1,4-dioxane, which can be two to three times the cost of treating more volatile solvents. MMA also opined that MDEQ had not incorporated all of the available toxicological data in its determination and these omissions led to a proposed MCL that was unnecessarily low. MMA pointed out that USEPA has reopened its risk assessment because studies showed that the IRIS analysis may dramatically overstate the cancer risk posed by 1,4-dioxane (MMA, 2004).

MDEQ's survey of water supplies affected by 1,4-dioxane was also cited by MMA as a reason not to establish a state MCL. The survey showed that about 1% of public water supplies in Michigan were affected by 1,4-dioxane, and no public water supplies contained concentrations approaching the proposed MCL of 35 µg/L (MMA, 2004). Eight sites out of about 3200 in MDEQ's state code Part 201 site list include 1,4-dioxane as a listed contaminant; more than 100 sites include 1,1,1-trichloroethane (methyl chloroform). It is unlikely that all sites with methyl chloroform have been tested for 1,4-dioxane.

The city of Ann Arbor issued comments to MDEQ on the proposed MCL objecting to the use of a 10^{-5} cancer risk factor and argued that using 10^{-6} to obtain an MCL of 3.5 of 3 ppb is more protective. Ann Arbor's population was 114,000 in 2000; water supply is obtained from municipal wells and the Huron River, and several of Ann Arbor's wells are in the path of 1,4-dioxane migration from the PLS/Gelman plumes.

The city of Ann Arbor now asks developers faced with contamination cleanup to use a 10^{-6} residual cancer risk for carcinogens, including 1,4-dioxane. Both the city of Ann Arbor and Washtenaw County passed resolutions for a Best Available Control Technology (BACT) standard of 10 ppb daily maximum and 3 ppb monthly average for the remediation of 1,4 dioxane. Existing cleanup operations at PLS/Gelman routinely reduce 1-4 dioxane concentrations to 3 ppb or less (City of Ann Arbor, 2004).

Washtenaw County also issued comments objecting to MDEQ's proposed 35 ppb MCL. The county contended that the effect of 1,4-dioxane exposure on susceptible subpopulations had not been adequately addressed and that exposure from inhalation was also not given sufficient consideration. Washtenaw County officials also objected because adopting a high MCL would require the county to issue permits for new wells in areas where 1,4-dioxane is expected to be present below the MCL.

MDEQ withdrew the proposed 1,4-dioxane 35 ppb standard based on the objections from stakeholders and regulated parties (TOSC, 2007). Section 8.3 presents a detailed review of the PLS/Gelman site.

6.1.3 NEED FOR REGULATION: OCCURRENCE OF 1,4-DIOXANE IN DRINKING WATER

The extent of 1,4-dioxane occurrence in drinking water is unknown because most water utilities do not analyze samples for 1,4-dioxane. Nevertheless, 1,4-dioxane has certainly been found as a drinking water contaminant in many water systems. Given the small percentage of water systems that have undergone testing for 1,4-dioxane, the number of affected systems is likely to increase as more systems are tested. A survey of 5500 drinking water analyses in the mid-1970s revealed 15 detections of 1,4-dioxane at concentrations greater than 10^{-7} g/L (i.e., greater than 100 ppb) (Donaldson, 1977). More recent detections of 1,4-dioxane in the United States, Canada, Japan, South Korea, and the Netherlands are listed in Table 6.3.

Table 6.3 presents known detections of 1,4-dioxane in drinking water; however, this list is by no means inclusive of all detections, and no effort was made to determine the rate of detection by tallying the frequency of nondetect results for drinking water analyses that include 1,4-dioxane. A highly correlated association with methyl chloroform occurrence was shown in a survey of 1,4-dioxane detections in drinking water sources in Japan (Abe, 1999). In the United States, a survey of analytical results in the U.S. Geological Survey's (USGS) National Water-Quality Assessment (NAWQA) Program, including more than 5000 wells[*] between 1985 and 2002, indicated a 7% frequency of detection for methyl chloroform (Moran et al., 2007). Although broad generalizations

---

[*] The wells in the NAWQA survey include drinking water sources, shallow urban groundwater, and groundwater underlying agricultural areas.

TABLE 6.3
1,4-Dioxane Detections in Drinking Water

Location	Maximum Detected Concentration	Year	References
Kitchener, Ontario, Canada	285 ppb in a single well[a]; 31 ppb in finished water	2005	1
Nakdong River, South Korea	119 ppb in raw river water; 92 ppb in treated river water	2003	2
Tama District, Tokyo, Japan	113 ppb in a well; detected in >70% of 338 wells tested; average 4.5 ppb in deep wells	2005	3
Kanagawa Prefecture, Japan	95 ppb in a production well[a]	1997	4
Bally, Pennsylvania	24–77 ppb in a single well (30–50 ppb more frequent range)	2004	5
Ann Arbor, Michigan	3–38 ppb in four single-family domestic wells; earlier results up to 71 ppb in a single domestic well	2007	6
Banning, California	35 ppb in a single well[a]	2005	7
Durham Meadows, Connecticut	27 ppb in a single well, unfiltered; 27 ppb in a GAC-filtered well[b]	2004	8
Santa Monica, California	22 ppb in a single well[a]	2002	9
Costa Mesa, California	16 ppb maximum; 3.3 ppb average	2005	10
Spokane, Washington, DC	13.8 ppb in a single well[c]	2006	11
Fountain Valley, California	7.7 ppb	2006	12
Downey, California	5.6 ppb in a single well; average 4 ppb in four wells with detections of eight total in system[a,d]	2005	13
Japan[e]	5.52 ppb maximum detected in groundwater; average of 22 detections in groundwater was 1.0 ppb	2006	14
Irvine, California	5 ppb[a]	2007	15
Tucson, Arizona	2.3 ppb	2003	16
Suffolk County, New York	0.7–2.3 ppb	2007	17
City of Commerce, California	2.2 ppb maximum; detections in four wells[a]	2004	18
Ann Arbor, Michigan	2 ppb in a single well[f]	2001	19
Bell Gardens, California	1.92 ppb in a single well; average 1.7 ppb[a]	2003	20
Kanagawa Prefecture, Japan	0.2–1.5 ppb in tap water	1995	21
City of Clare, Michigan	1.0 ppb	2006	22
the Netherlands	0.5 ppb	1999	23
Niigata Prefecture, Japan	0.39 ppb in river water used to supply drinking water	2002	24

Sources: [1] Waterloo (2005); [2] Park et al. (2005); [3] Toshinari et al. (2005); [4] Abe (1999); [5] USEPA (2004a); [6] MDEQ (2007); [7] City of Banning (2005); [8] USEPA (2004b); [9] City of Santa Monica (2002); [10] MCWD (2005); [11] WSDOH (2006); [12] City of Fountain Valley (2006); [13] Bellflower/Norwalk/Park Water Company (2005); [14] Simazaki et al. (2006); [15] IRWD (2007); [16] Tucson Water (2004); [17] SCWA (2007); [18] CWSC (2004); [19] City of Ann Arbor (2001); [20] SCWC (2004); [21] Abe (1997); [22] City of Clare (2007); [23] VROM (1999); and [24] Kawata et al. (2003).

[a] Water utilities often manage wells by blending water from contaminated wells with clean water from uncontaminated wells, so that tap water in homes is below regulatory thresholds or advisory action levels, if not laboratory detection levels.

[b] GAC = Granular Activated Carbon; Chapter 7 presents information on GAC limitations to 1,4-dioxane removal with GAC.

[c] WSDOH conducted a health consultation jointly with ATSDR for the maximum 1,4-dioxane occurrence of 13.8 ppb in a well supplying 34 homes near the Colbert Landfill in Spokane, Washington. The review concluded that no apparent public health hazard exists for all routes of exposure to all exposed populations at this concentration (WSDOH, 2006).

[d] Only 15% of the Bellflower/Norwalk/Park Water Company System water supply comes from groundwater; its main source is imported surface water from the Metropolitan Water District of Southern California. The surface water/groundwater mix for other water systems on this list was not evaluated.

[e] Detected in 39 of 91 raw water samples from drinking water treatment plants; 22 out of 29 groundwater samples contained 1,4-dioxane.

[f] Ann Arbor Montgomery Street well was not in service at time of detection; it remains off-line.

must be tested on the scale of the particular circumstances of drinking water source contamination, the findings of Abe and Moran taken together suggest that it is likely that there are many more drinking water sources affected by 1,4-dioxane that have not yet been discovered because 1,4-dioxane is not commonly included in drinking water analysis. Where production wells have a history of detections of methyl chloroform and its breakdown products, 1,1-dichloroethylene and 1,2-dichloroethane, screening level testing for 1,4-dioxane is advisable.

6.2 WATER QUALITY REGULATIONS: PROTECTION OF ECOLOGICAL RECEPTORS

Remedial project managers from the regulatory and regulated communities must determine the appropriate course of action when industrial effluent or treated groundwater from which solvents have been removed contains low concentrations of 1,4-dioxane. Aquatic toxicity of contaminants discharged to streams usually drives the determination of discharge limits for surface-water discharges; however, it may also be necessary to review the potential for drinking water impacts. Human exposure becomes the driver for regulating 1,4-dioxane discharges where the surface-water body is either used directly for drinking water or recharges an aquifer from which groundwater is pumped to supply drinking water. Human exposure may also be the driver for discharge of 1,4-dioxane to sanitary sewers. The removal efficiency of 1,4-dioxane in conventional wastewater treatment plants is low, and where treated wastewater is recycled for irrigation of landscape or indirect potable reuse to recharge groundwater, drinking water impacts may result. This section examines the toxicity of 1,4-dioxane to ecological receptors (i.e., algae, aquatic insects, fish, and terrestrial animals) and reviews examples of how 1,4-dioxane has been regulated in surface-water discharges.

6.2.1 NATIONAL POLLUTION DISCHARGE ELIMINATION SYSTEM REGULATIONS

NPDES permits are issued to dischargers of treated and untreated industrial wastewater, stormwater, groundwater remediation effluent, and other effluents. The goal of the NPDES is to restrict or eliminate impacts to aquatic life in surface water from direct discharges to streams or indirect discharges to publicly owned treatment works (POTWs), including discharges with high biochemical oxygen demand, toxic metals and organic compounds, high temperatures, foaming agents, and other potential impacts. Preserving aquatic habitat, recreational uses, and the aesthetic value of natural waters has grown in importance since the NPDES program was first introduced.

Chlorinated solvent release sites are often remediated with pump-and-treat systems using air stripping or granular activated carbon (GAC). On numerous occasions, dischargers and regulators alike were surprised to learn that the discharge contained additional, untreated compounds including 1,4-dioxane and perchlorate (USEPA, 2007a). The consequences of discharging 1,4-dioxane to streams, rivers, lakes, estuaries, and bays depend on local conditions. Aquatic toxicity is unlikely to result from the 1,4-dioxane concentrations commonly encountered in effluents from groundwater treatment systems designed to treat chlorinated solvents. The beneficial uses of the surface-water body and the interaction of surface water with groundwater may therefore be the determining factors in regulating 1,4-dioxane discharges. Where surface water is used directly for drinking water or where it makes up a substantial proportion of the recharge to a drinking water aquifer, treatment of 1,4-dioxane may be warranted to protect drinking water. The antidegradation policies in federal and state water regulations may also prohibit discharge of contaminants to surface waters. To date, the regulation of 1,4-dioxane discharges to surface waters has varied considerably by regulatory agency and the context of the discharge.

6.2.2 TOXICITY OF 1,4-DIOXANE TO AQUATIC ORGANISMS

1,4-Dioxane is not highly toxic to aquatic organisms. The aquatic bioassay results tabulated in this section show that 1,4-dioxane is acutely toxic to bacteria, algae, invertebrates, and fish at

TABLE 6.4
Abbreviations of Common Toxicity Thresholds for Aquatic Organisms

EC_0	Concentration causing an effect in 0% of the test organisms
EC_{50}	Median concentration causing an effect in 50% of the test organisms
LC_{50}	Median concentration lethal to 50% of the test organisms
LC_{100}	Concentration lethal to 100% of the test organisms
LD_{50}	Median lethal dose for 50% of the test population
LOEC	Lowest observed effect concentration
NOEC	No observed effect concentration
LOEL	Lowest observed effect level
NOEL	No observed effect level
PNEC	Predicted no effect concentration
TRV	Toxicity reference value

concentrations ranging from hundreds to tens of thousands of milligrams per liter (parts per million). 1,4-Dioxane is not bioconcentrated in aquatic organisms, animals, or plants; however, it may accumulate in plants by transpiration (ATSDR, 2004). The European Chemicals Bureau Predicted No Effect Concentration for aquatic organisms is set at 2700 mg/L (ECB, 2002). There are a variety of toxicity thresholds used to describe toxic effects of contaminants measured by different tests, with some expressed as aqueous or mass concentration and others expressed in terms of the dose per day as a ratio to the body weight of the organism (toxic threshold given in mg/[kg d]). Table 6.4 lists the aquatic toxicity thresholds discussed in this chapter.

Toxicity reference values (TRVs) provide a benchmark for the toxicity of a contaminant to all the ecological receptors in an ecosystem, such as freshwater, marine, or sediment ecosystems. TRVs have not been developed for soil invertebrates and terrestrial plants as toxicity data are not available (USEPA, 1999). Table 6.5 summarizes the TRVs derived by USEPA for 1,4-dioxane. Similarly, predicted no effects concentrations (PNEC) are used in Europe to establish the concentration at which organisms of an ecological compartment are not affected by a contaminant. Table 6.6 presents PNEC values for the toxicity of 1,4-dioxane to different indicator species.

1,4-Dioxane toxicity data are available for algae, bacteria and protozoa, aquatic invertebrates, and fish, as presented in Tables 6.7 through 6.10. Aquatic toxicity data tabulated here were obtained primarily from secondary sources (BUA, 1991; NICNAS, 1998; Verschueren, 2001; ECB, 2002; ATSDR, 2004). Citations of original studies are provided for the reader's convenience but were not researched for this book.

TABLE 6.5
Aquatic TRVs for 1,4-Dioxane

Ecosystem	TRV [mg/(kg d)]	Indicator Species	References
Freshwater	62,100	*Daphnia magna* (water flea)	Bringmann and Kühn (1982)
Freshwater bed sediment	2176	Calculated from freshwater TRV[a]	USEPA (1999)
Marine–Estuarine surface water	67,000	*Menidia beryllina* (inland silverside)	Dawson and Jennings (1977)
Marine bed sediment	2348	Calculated from marine TRV[a]	USEPA (1999)

Source: U.S. Environmental Protection Agency (USEPA), 1999, Screening level ecological risk assessment protocol. U.S. Environmental Protection Agency, Office of Solid Waste.

Notes: Sediment TRVs calculated by using equilibrium partitioning assumptions, an assumed fraction of organic carbon in sediment of 4%, and a K_{oc} of 0.876 for 1,4-dioxane (USEPA, 1999).

[a] $TRV_{sed} = K_{oc} \times f_{oc} \times TRV_{sw}$.

TABLE 6.6
European PNECs Calculated for 1,4-Dioxane in the Aquatic Environment

Compartment or Organisms	PNECs	Indicator Species	References
Microorganisms	2700 mg/L	*Pseudomonas putida*	Bringmann and Kühn (1977b); BASF (1979)
Sediment compartment	43.3 mg/kg (wet weight)	Calculated with EUSES	ECB (2002)
Aquatic compartment (algae, invertebrates, fish)	57.5 mg/L	*Mycrocystis aeruginosa* (blue green algae)	Bringmann and Kühn (1976)
Terrestrial compartment	14 mg/kg (wet weight)	Calculated with EUSES	ECB (2002)

Source: European Chemicals Bureau (ECB), 2002, European Union risk assessment report: 1,4-Dioxane. Institute for Health and Consumer Protection, European Union.

Note: EUSES is the European Union System for the Evaluation of Substances model.

6.2.3 Toxicity of 1,4-Dioxane to Aquatic and Terrestrial Organisms

In soil, surface water, and sediment ecosystems, 1,4-dioxane is not persistent because of its low affinity for adsorption to organic carbon, which also causes a low potential for 1,4-dioxane to be bioconcentrated in aquatic receptors. Wildlife can be exposed to 1,4-dioxane through ingestion, inhalation, and dermal contact; however, 1,4-dioxane does not bioaccumulate in food chains (USEPA, 1999).

USEPA derived a 1,4-dioxane TRV for mammals from a 23-month chronic study of lung tumors in guinea pigs to determine the Lowest Observed Adverse Effects Level (LOAEL; Hoch-Ligeti and Argus, 1970, see Chapter 5). The LOAEL dose, 1070 mg/d, was converted to an equivalent dose based on body weight and intake rate for the guinea pig. Applying a UF of 0.1, the mammalian TRV is 107 mg/(kg d) (USEPA, 1999).

Exposures to terrestrial wildlife through the food chain are not significant with respect to the levels of 1,4-dioxane typically observed in contaminated soils (Huntley et al., 2004). Bioconcentration factors were estimated for wild birds and rodents, including the American robin, canvas-back duck, deer mouse, least shrew, mallard duck, marsh rice rat, marsh wren, mourning dove, muskrat, northern bobwhite, salt-marsh harvest mouse, short-tailed shrew, western meadow lark, and white-footed mouse. Estimated bioconcentration factors are on the order of 10^{-9}–10^{-8}; therefore, bioconcentration in these and in similar species is not of a significant concern (USEPA, 1999; U.S. Army, 2000). The Ontario Ministry of the Environment estimated TRVs for several wildlife species by using a 1.0 mg/(kg d) LOAEL for the rat, as tabulated in Table 6.11 (OMOE, 2007).

6.2.4 Regulating 1,4-Dioxane Discharges from Groundwater Cleanup Sites

The discovery of 1,4-dioxane in effluent from a groundwater treatment system designed to remove chlorinated solvents can present a vexing dilemma to remedial project managers. As described in

TABLE 6.7
Long-Term Toxicity of 1,4-Dioxane to Aquatic Algae

Species	Limit	Result (mg/L)	Method	References
Scenedesmus quadricauda (green algae)	8-day cell multiplication inhibition	5600	Static	Bringmann and Kühn (1977b)
Microcystis aeruginosa (cyanobacteria; blue-green algae)	8-day cell multiplication inhibition	575	Static	Bringmann and Kühn (1976)

TABLE 6.8
Toxicity of 1,4-Dioxane to Microorganisms and Protozoa

Species	Limit	Result (mg/L)	Method	References
Chilomonas paramecium (a cryptomonad used to measure growth)	48 h NOEC, cell multiplication inhibition	>10,000[a]	Static	Bringmann et al. (1980)
Entosiphon sulcatum (a flagellate bacterium used to measure population growth rate)	72 h NOEC, cell multiplication inhibition	5340[a]	Static	Bringmann (1978)
Pseudomonas fluorescens (a common gram-negative, rod-shaped bacterium)	16 h NOEC, glucose assimilation inhibition	2700[a]	Static	Bringmann (1973)
Pseudomonas putida (a gram-negative rod-shaped saprophytic soil bacterium)	16 h NOEC, cell multiplication inhibition	2700[a]	Static	Bringmann and Kühn (1977b)
Uronema parduczi (a ciliate protist)	20 h NOEC (EC$_0$), cell multiplication inhibition	5620[a]	Static	Bringmann and Kühn (1980)
Photobacterium phosphoreum (a bioluminescent gram-negative bacteria)	EC$_{50}$	6000	Other direct assay— luminescence reduction	Thomulka and McGee (1992)
Vibrio harveyi (a bioluminescent marine bacteria)	EC$_{50}$	16,000	Other direct assay— luminescence reduction	Thomulka and McGee (1992)
Vibrio harveyi	EC$_{50}$	6500	Other growth assay— luminescence reduction	Thomulka and McGee (1992)

[a] Nominal concentration.

TABLE 6.9
Short-Term Toxicity of 1,4-Dioxane to Aquatic Invertebrates

Species	Limit (mg/L)	Test Endpoint or Result (mg/L)	References
Daphnia magna Straus (water flea)	24 EC$_0$	6210	Bringmann and Kühn (1982)
Daphnia magna Straus	24 EC$_{50}$	8450 (nominal concentration) range: 8040–8880	Bringmann and Kühn (1982)
Daphnia magna Straus	24 EC$_{100}$	>10,000	Bringmann and Kühn (1982)
Daphnia magna (water flea)	24 EC$_0$	2070 (nominal concentration)	Bringmann and Kühn (1977a)
Daphnia magna	24 EC$_{50}$	4700 (nominal concentration)	Bringmann and Kühn (1977a)
Daphnia magna	24 EC$_{100}$	10,000	Bringmann and Kühn (1977a)
Ceriodaphnia dubia (water flea)	24 EC$_{50}$	299	Dow Chemical Company (1989b)
	48 EC$_{50}$	163	
Ceriodaphnia dubia	168 EC$_{50}$	625	Springborn Laboratories Inc. (1989)
Gammarus pseudolimnaeus (amphipod)	48 EC$_{50}$	2274	Dow Chemical Company (1989b)
Aedes agyptii (midge larvae)	4 LC$_{50}$	40,000	Kramer et al. (1983)

TABLE 6.10
Short-Term Toxicity of 1,4-Dioxane to Fish Species

Species	Limit (mg/L)	Test Result	Method	References
Pimephales promelas (fathead minnow)	96 LC_{50} 95% C.I.	9850[a]	Flow through	Geiger et al. (1990)
Pimephales promelas	96 LC_{50}	10,800[a]	Flow through	Geiger et al. (1990)
Pimephales promelas	96 LC_{50}	13,000 (10,000– 17,000)[b]	Static	Dow Chemical Company (1987, 1989a)
Pimephales promelas	72 LC_{50}	>100	Static	Dow Chemical Company (cited in BUA, 1991)
Pimephales promelas	96 LC_{50}	>100	Static	Dow Chemical Company (cited in BUA, 1991)
Menidia beryllina (inland silverside or sand smelt)	96 LC_{50}	6700[b]	Semistatic— synthetic seawater	Dawson and Jennings (1977)
Lepomis macrochirus (bluegill sunfish)	96 LC_{50}	>10,000[b]	Semistatic— synthetic seawater	Dawson and Jennings (1977)
Oryzias latipes (medaka, high-eyes, or Japanese killifish)	48 LC_{50}	10,500[b]	Semistatic	CITI-Japan (1992)
Leuciscus idus (ide or golden orfe)	48 LC_{0}	6180; 6700	Static	Juhnke and Ludemann (1978)
Leuciscus idus	48 LC_{50}	8450	Static	Juhnke and Ludemann (1978)
Leuciscus idus	48 LC_{50}	9630	Static	Juhnke and Ludemann (1978)
Leuciscus idus	48 LC_{100}	9760	Static	Juhnke and Ludemann (1978)
Cyprinus carpio (European carp)	48 LC_{50}	12,000	Other	Nishiushi (1981)
Oncorhyncus mykiss (rainbow trout)	96 LC_{50}	7961	Flow through	USEPA (1996b)
Ictalurus punctatus (channel catfish)	96 LC_{50}	6155	Flow through	USEPA (1996b)

[a] Actual concentration (measured).
[b] Nominal concentration (calculated).

TABLE 6.11
1,4-Dioxane TRVs for Wildlife and Livestock, Ontario Ministry of the Environment

Species	TRV Used in Model [mg/(kg d)]	Soil Cleanup Value (mg/kg)
Short-tailed shrew	2.2	176
Meadow vole	1.68	1.82
Red fox	0.53	625
Sheep	0.28	0.174

Source: OMOE, 2007, Rationale for the development of generic soil and groundwater standards for use at contaminated sites in Ontario. Ontario Ministry of the Environment, Standards Development Branch.

Chapter 7, the treatment technology selected for removing solvents often is not suitable for removing 1,4-dioxane. For example, the 1,4-dioxane concentration in air-stripper effluent is often only slightly less than in air-stripper influent. Discharge of concentrations of 1,4-dioxane well below aquatic toxicity thresholds may nevertheless be restricted to uphold antidegradation policies. State and federal regulators have generally arrived at a pragmatic framework for resolving the question of appropriate effluent limits for 1,4-dioxane in a manner that considers local conditions.

In California, the San Francisco Bay Regional Water Quality Control Board (SFB Water Board) uses a concentration-based trigger of 5.0 µg/L for discharges of 1,4-dioxane from groundwater treatment systems; concentration triggers are also set for 146 other compounds, including another solvent stabilizer, nitromethane. The SFB Water Board requires a one-time sample for 1,4-dioxane in treated groundwater effluent upon system start-up and twice per year thereafter where chlorinated solvents are present in groundwater. If 1,4-dioxane is not detected above the SFB Water Board's 5.0 µg/L trigger for 1,4-dioxane, dischargers may reduce the monitoring schedule to once every 3 years (Cal EPA, 2004a). If the trigger is exceeded, three additional samples are required within the next quarter; in case of additional detections, the discharger must report the median concentration and mass load discharged or provide a technical rationale for not monitoring 1,4-dioxane in the effluent. If the calculated median and mass discharge of 1,4-dioxane in the effluent exceed the trigger, the discharger must evaluate the technical feasibility and cost of treatment alternatives for 1,4-dioxane. If treatment is deemed not feasible, then the SFB Water Board requires an evaluation of impacts to beneficial uses of the receiving water. If that evaluation indicates that beneficial uses are threatened, the discharger then has the option of proposing alternatives, such as discharging to the sewer instead of surface waters, using in situ treatment instead of groundwater extraction, or other means of eliminating or reducing the impact.

In practice, the SFB Water Board has restricted 1,4-dioxane discharge to streams where there is a potential to impact drinking water sources. The groundwater and surface-water cleanup standards at a large solvent cleanup site were set to 3 µg/L, equivalent to California's NL (formerly, action level), because groundwater at the site discharges to streams, and the streams drain to a large reservoir for drinking water (Cal EPA, 2004b). The Los Angeles Water Board applies 3 µg/L as a screening level for 1,4-dioxane in effluents regulated by general NPDES permits (Cal EPA, 2002a). Table 6.12 summarizes discharge limits set for 1,4-dioxane at several groundwater cleanup sites and industrial facilities in the United States and Canada.

The volume of groundwater effluent discharged to surface-water bodies is usually small compared to the volume of receiving water flow, resulting in large dilution ratios. 1,4-Dioxane in groundwater treatment system effluent may quickly drop below detection thresholds because of dilution in receiving waters; consequently, detection of 1,4-dioxane in surface-water bodies is usually limited to large-volume industrial discharge scenarios or to watersheds in which treated municipal wastewater effluent makes up a substantial proportion of the flow. Table 6.13 lists a few examples of 1,4-dioxane detections in surface water.

6.2.5   REGULATING 1,4-DIOXANE IN INDUSTRIAL WASTEWATER DISCHARGES AND RECYCLED WATER

1,4-Dioxane is not easily removed through aerobic wastewater treatment, the primary means used to treat domestic sewage (Abe, 1999). Discharge of 1,4-dioxane to sewers should be restricted where the discharge would cause a detectable concentration in the plant influent or effluent. In addition to water quality considerations, discharge of 1,4-dioxane to sanitary sewers is potentially hazardous because 1,4-dioxane is flammable.

1,4-Dioxane is not routinely analyzed in industrial pretreatment programs.[*] Title 33 of the Code of Federal Regulations governs navigable waters and includes Chapter 26, which addresses water

---

[*] Mohr et al. (2007) reviewed two decades of organic compound analyses at five POTWs in the San Francisco Bay area as part of a study on historical dry cleaner discharges.

TABLE 6.12

Discharge Limits for 1,4-Dioxane at Groundwater Cleanup Sites and Industrial Facilities in the United States and Canada

Site	1,4-Dioxane Discharge Limit	Regulatory Agency[a]	References
Rocket Motor Facility, San Jose, California, Upstream of Drinking Water Reservoir	3 µg/L	SFBRWQCB	Cal EPA (2004b)
Rocket Motor Facility, Rancho Cordova, California	10 µg/L—daily maximum 3 µg/L—monthly average	CVRWQCB	Cal EPA (2002b)
Ontario, Canada—Interim Provincial Water Quality Objectives	20 µg/L	Ministry of the Environment	OMOE (1994)
PLS/Gelman Site, Ann Arbor, Michigan	220 µg/L—daily maximum 7 µg/L—monthly average	MDEQ	MDEQ (2005b)
Florida	120 µg/L—surface water cleanup target level	FDEP	AFCEE (2007)
Pennsylvania Manufacturer	280 µg/L—daily maximum 112 µg/L—monthly average	PADEP	PADEP (2004)
PLS/Gelman Site, Ann Arbor, Michigan	2800 µg/L—groundwater/surface water interface limit	MDEQ	MDEQ (2007)
Solvents Refinery, Circleville, Ohio	3600 µg/L—river concentration standard protective of fish consumption (recent discharge is ~35 µg/L; between 1990 and 1993, average discharge was 15 lb/day)	Ohio EPA	OEPA (2004); EWG (1996)
Oregon	4.000 µg/L—level II screening level for ecological risk assessment for mammals	DEQ	ODEQ (2007)

[a] SFBRWQCB, San Francisco Bay Regional Water Quality Control Board; CVRWQCB, Central Valley Regional Water Quality Control Board; MDEQ, Michigan Department of Environmental Quality; FDEP, Florida Department of Environmental Protection; PADEP, Pennsylvania Department of Environmental Protection; and DEQ, Department of Environmental Quality.

pollution prevention and control. The National Pretreatment Program for wastewater is derived from these and other regulations. The program is a cooperative effort of federal, state, and local regulatory environmental agencies established to protect water quality by reducing the level of pollutants discharged by industry and other nondomestic wastewater sources into municipal sewer systems. Pretreatment of industrial wastewater prior to sewer discharge is the primary mechanism used to limit the amount of pollutants released to the sewer. The National Pretreatment Program

TABLE 6.13

Documented 1,4-Dioxane Detections in Surface Waters

Location	Year	Detection (ppb)	References
Quinnipiac River, Connecticut	1987	1700	ATSDR (2007)
Tuscarawas River, Ohio	1993	>300	OEPA (1994)
Nakdong River, Korea	2003	119	Park et al. (2005)
Shinano River, Japan	2002	0.39	Kawata et al. (2003)
Agano River, Japan	2005	0.26	Tanabe et al. (2006)
Shinano River, Japan	2005	0.10	Tanabe et al. (2006)

protects POTWs from pollutants that could disrupt the chemical and biological balance necessary to operate the POTW and prevent introduction of pollutants to the POTW that may pass through the treatment works untreated. Limiting the chemical load arriving at the POTW improves opportunities to reuse wastewater and sludges generated by the POTW. Concentration limits for discharge to sewers are established by USEPA, state, or regional and local sewer districts and municipalities.

6.2.5.1 Industrial Wastewater Discharge Regulations

Pharmaceutical production is among the industries generating wastewater containing 1,4-dioxane. In 1995, revisions to the Clean Water Act introduced guidelines for effluent limitations and pretreatment standards for pollutants discharged to POTWs from pharmaceutical manufacturing plants (USEPA, 1997b). The guidelines addressed 1,4-dioxane discharges by establishing Pretreatment Standards for Existing Sources, and Best Available Technology Effluent Limitations, as listed in Table 6.14.

USEPA's review of chemicals in pharmaceutical manufacturing effluent noted the occurrence of 1,4-dioxane. Although all six plants surveyed discharged 1,4-dioxane, USEPA determined that the average loading (0.061% of total loading, about 24,000 pounds annually) did not warrant regulation. In making this determination, USEPA acknowledged that 1,4-dioxane might pass untreated through POTWs and therefore solicited data from dischargers regarding the fate of 1,4-dioxane in biological treatment processes. As a placeholder, USEPA assumed that 1,4-dioxane should be at least as biodegradable as tetrahydrofuran on the basis of structural similarities (USEPA, 1998a). Subsequent research has shown that the first-order aerobic biodegradation decay constant for tetrahydrofuran is approximately twice that of 1,4-dioxane; therefore, 1,4-dioxane is unlikely to degrade as quickly as tetrahydrofuran (Zenker et al., 2000).

USEPA groups 1,4-dioxane in pharmaceutical wastewater with "alcohols and related pollutants," including methanol, ethanol, *n*-propanol, isopropanol, *n*-butyl alcohol, *tert*-butyl alcohol, amyl alcohol, formamide, N,N-dimethylaniline, pyridine, aniline, and petroleum naphtha. The economic feasibility of removing this entire group of compounds was evaluated by USEPA, but the cost, $40 million/year, was deemed excessive compared to the benefit, because USEPA assumed all compounds in the group were easily biodegradable.

Substantial efforts have been made to eliminate 1,4-dioxane discharge in treated wastewater effluent from the production of polyester fibers. In Spartanburg, North Carolina, the Hoechst Celanese polyester plant invested millions of dollars in a large-scale treatment system including construction of a 54,000-gallon holding tank, an azeotropic distillation column for separation of dioxane from the process wastewater, and a gas-fired thermal oxidizer for the destruction of 1,4-dioxane (O'Neal Inc., 2007).

1,4-Dioxane is required for effluent monitoring from leachate treatment at several landfills, for example, Laraway Waste Management facility in Elwood, Illinois (IEPA, 2007). At the Lowry Landfill Superfund Site in Aurora, Colorado, wastewater containing 1,4-dioxane is permitted for discharge to the City of Aurora's wastewater treatment plant.

TABLE 6.14
1,4-Dioxane Pretreatment Standards for Pharmaceutical Manufacturing Effluent

Standard	Long-Term Mean Concentration (mg/L)	Maximum for Any One Day (mg/L)	Monthly Average (mg/L)
Best available technology	0.8	8.4	2.6
Pretreatment standards for existing sources	1240	3160	1760

Source: U.S. Environmental Protection Agency (USEPA), 1997b, *Federal Register* 62(153): 42,720–42,732.

Permissible discharges of 1,4-dioxane to the sewer decrease with increasing flow rates. Discharges of 10, 20, and 30 gallons/min are allowed to contain 1,4-dioxane at 5930, 3950, and 2970 µg/L, respectively (MWRD, 1999). As discussed in Chapter 7, aerobic wastewater treatment is not expected to effectively remove 1,4-dioxane, but it can dilute concentrations to below detection limits. In a 1988 chemical analysis survey of sewage sludge from 208 wastewater treatment plants, 1,4-dioxane was detected in sludges from three plants (USEPA, 1988).

6.2.5.2 Recycled Water Regulations

Groundwater recharge projects in California using recycled water from POTWs in which 1,4-dioxane is present in the sewer influent must use reverse osmosis and advanced oxidation to achieve at least 0.5 log reduction in concentration. Post-treatment concentrations can be no greater than California's NL for 1,4-dioxane, 3 µg/L (CDPH, 2007b).

In Orange County, California, 1,4-dioxane was detected in wastewater treated by advanced methods and subsequently in nine municipal wells. The wells draw from a groundwater basin in which the treated wastewater had been reinjected for more than a decade to provide a hydraulic barrier to seawater intrusion. The advanced treatment, including reverse osmosis and UV light with hydrogen peroxide, was insufficient to remove 1,4-dioxane.[*] The nine wells in three cities were temporarily shut down, requiring replacement of 34 million gallons/day of groundwater with imported surface water. 1,4-Dioxane concentrations in the influent to the advanced wastewater treatment plant ranged from nondetect to episodic peaks as high as 200 ppb, but mostly less than 100 ppb. The presence of 1,4-dioxane in the influent was traced to a manufacturing facility that used 1,4-dioxane to produce cellulose acetate membrane filters, whose discharger voluntarily ceased the discharge, and the 1,4-dioxane declined to 1 ppb, the level generally associated with domestic wastewater (Woodside and Wehner, 2002).

A solvent remediation site in Tampa, Florida, found 1,4-dioxane concentrations in extracted groundwater at less than 20 µg/L; however, treated effluent in Florida may not exceed 5 µg/L, the Florida drinking water standard. Because discharge from the treatment system is to the sanitary sewer, water agency officials required reduction or elimination of the 1,4-dioxane discharge, which could end up in reclaimed water being used to recharge groundwater (J.C. Alonso, personal communication, 2001).

6.3 AIR QUALITY REGULATIONS: OCCUPATIONAL HEALTH AND SAFETY AND AMBIENT AIR QUALITY

As described in Section 2.7, ambient air sampling at 45 locations in 12 cities during the early 1980s confirmed the presence of 1,4-dioxane concentrations ranging from less than 1–30 ppb. Because 90% of the 1,4-dioxane produced for the United States was for stabilizing methyl chloroform, it is likely that the source of 1,4-dioxane in ambient air was from vapor degreasing with methyl chloroform, as well as the other direct industrial uses of 1,4-dioxane and the industries that produce 1,4-dioxane as a by-product. Section 6.5.3 summarizes occupations in which a potential existed for inhalation of 1,4-dioxane when methyl chloroform was in widespread use.

The explosion hazard of 1,4-dioxane should not be underestimated. 1,4-Dioxane is a hazardous material that presents fire and explosion dangers from its direct use as a solvent. 1,4-Dioxane is a highly volatile and highly flammable liquid with an explosive/flammability limit of 2% in air at standard temperature and pressure. In addition, the vapor density of 1,4-dioxane is about three times denser than air, causing its vapors to sink. The combination of 1,4-dioxane's volatility, flammability,

---

[*] The combination of UV light and peroxide is effective at removing 1,4-dioxane when used in a configuration optimized for 1,4-dioxane, as described in Chapter 7. The level of peroxide at this plant, 5 mg/L, was too low; 10–15 mg/L peroxide is more effective at removing 1,4-dioxane (Woodside and Wehner, 2002).

and vapor density account for its potential to travel considerable distances to an ignition source with a consequent risk of "flash back." The accident that killed the welder working on the space shuttle solid rocket booster casings (described in Section 1.2.2.1) serves as a grim reminder of the hazard that 1,4-dioxane, nitromethane, and other flammable stabilizers can pose to workers who handle these chemicals.

As described in Chapter 4, 1,4-dioxane can form explosive peroxide crystals on the threads of screw cap bottles. 1,4-Dioxane may also react vigorously with oxidizing agents. Risks from products containing 1,4-dioxane as an ingredient depend on the physicochemical properties of other ingredients present in the product and should be determined for the mixture as a whole (NICNAS, 1998).

Regulation of 1,4-dioxane as a workplace inhalation hazard and as a toxic air contaminant is reviewed in the remainder of this section.

6.3.1 OCCUPATIONAL HEALTH AND SAFETY REGULATIONS FOR 1,4-DIOXANE

As described in Section 6.5.3, occupational use of 1,4-dioxane or work at facilities that produce 1,4-dioxane as a by-product creates the greatest potential for inhalation exposure to 1,4-dioxane. Beginning in the late 1970s, major advances were made in worker health and safety regulations and practices in the United States and other developed nations. Today, the likelihood of exposure is greatly diminished, if not eliminated, because methyl chloroform was banned in the 1990 Montreal Protocol.

A variety of regulations have been promulgated by health and labor agencies to ensure that exposure to airborne contaminants in the workplace is controlled to prevent adverse health effects. Regulations establish safe exposure levels. The wide variety in established exposure levels reflects differences in the time period for the measurement, the underlying toxicity values upon which the exposure limits are based, and the safety factor applied. Regulatory bodies throughout the world use exposure standards including the Threshold Limit Value (TLV), Occupational Exposure Standard (OES), Maximum Exposure Limit (MEL), maximum allowable concentration (MAK), and Short-Term Exposure Limit (STEL). The common convention for an exposure period is either an 8-hour workday for chronic toxins or a 15-minute exposure period for acute toxins. Table 6.15 provides definitions for the common exposure standards, and Table 6.16 lists inhalation exposure limits developed for 1,4-dioxane. Figure 6.1 presents a graphical comparison of health limits and regulatory advisory thresholds for inhalation of 1,4-dioxane.

6.3.2 EMISSIONS STANDARDS AND AIR QUALITY REGULATIONS FOR 1,4-DIOXANE

1,4-Dioxane is designated as a hazardous air pollutant in Title 42, Public Health and Welfare, Chapter 85, on Air Pollution Prevention and Control. As early as 1978, 1,4-dioxane was recognized in New Jersey as among the top eight air pollutants for which emission reductions through regulation were necessary, together with benzene, carbon tetrachloride, chloroform, ethylene amine, ethylene dibromide, tetrachloroethylene, and TCE (*Chemical Week*, 1978).

Most ambient air quality monitoring programs do not monitor for 1,4-dioxane; however, there are many indications that it was likely to be present in solvent-polluted air in air basins with significant industrial activity. In 1999, 1,4-dioxane was ranked as 14th among 40 priority pollutants by the California Air Resources Board (Kyle et al., 2001; Cal EPA, 1999).

Figure 2.5a and b in Chapter 2 present annual atmospheric emissions of 1,4-dioxane from industrial operations reported in USEPA's Toxic Release Inventory. The average emissions of 1,4-dioxane across the United States from 1988 through 2004 was more than 300,000 pounds/year; in 1999, annual 1,4-dioxane emissions in California were estimated to be about 150,000 pounds (CARB, 2008).

TABLE 6.15
Common Occupational Exposure Standards (OESs) in the United States and Europe

Standard	Acronym	Definition
Threshold Limit Value (United States, EU)	TLV	The time-weighted average concentration of airborne substances to which nearly all workers may be repeatedly exposed, without adverse effect
Time-Weighted Average (United States, EU)	TWA	The time-weighted average concentration for a conventional 8-hour day and 40-hour week (ACGIH and EU)
Short-Term Exposure Limit (United States, EU)	STEL	The concentration to which workers can be exposed for short periods of time, usually 15 min, 4 times per day, without suffering adverse effects (ACGIH and EU)
Occupational Exposure Standard (UK)	OES	The concentration of an airborne substance averaged over a reference period, at which, according to current knowledge, there is no evidence that it is likely to be injurious to employees, if they are exposed, by inhalation, day after day, to that concentration (EH/40 Occupational Exposure Limits, Health & Safety Executive, UK)
Maximum Exposure Limit (UK)	MEL	The *maximum* concentration of an airborne substance, averaged over a reference period, to which employees may be exposed by inhalation under any circumstances. Unlike the OES, there may be a residual risk to health at this level of exposure.
Maximum Workplace Concentration (Germany)	MAK	The *maximum* allowable concentration of a working substance in the workplace atmosphere as a gas, steam, or aerosol that according to current knowledge does not impair the health of employees exposed during 8-hour working days over the long term (German Commission for the Investigation of Health Hazards of Chemical Compounds in the Work Area)
Maximum Workplace Concentration (Netherlands)	MAC	The maximum concentration of a gas, steam, or aerosol of a substance in workplace air, which when inhaled during a workday has no adverse impact on the health of the employees (Dutch Expert Committee for Occupational Standards, Netherlands)
Level Limit Value and Ceiling Limit Value	LLV and CLV	LLV: An occupational exposure limit value for exposure during one working day; CLV: An occupational exposure limit value for exposure during a reference period of 15 min (Swedish Work Environment Authority)
Permissible Exposure Limits	PEL	The legal limit for exposure of an employee to a substance, usually given as a time-weighted average (TWA) (U.S. Occupational Safety and Health Administration)
Occupational Exposure Limit	OEL	The maximum concentration of an airborne substance, averaged over a reference period, to which employees may be exposed by inhalation under any circumstances (South African Variation of UK OES)
Immediately Dangerous to Life and Health	IDLH	The limit beyond which an individual will not be capable of escaping death or permanent injury without help in less than 30 min (U.S. NIOSH)
Minimal Risk Level	MRL	An estimate of the daily human exposure to a hazardous substance that is likely to be without appreciable risk of adverse noncancer health effects over a specified duration of exposure (U.S. ATSDR)

Notes: ACGIH, American Conference of Governmental Industrial Hygienists; EU, European Union; and UK, United Kingdom.

Regulation of 1,4-dioxane emissions varied by industry and by state. For example, vapor degreasing operations, a major source of 1,4-dioxane emissions, were regulated by the National Emission Standards for Halogenated Solvent Cleaning (40 CFR, 63.460 through 63.469). Local air pollution control districts inspect and enforce standards on degreasing operations to control solvent emissions. After 1998, wood furniture-manufacturing facilities releasing more than 1200 pounds of 1,4-dioxane per year must report emissions and track annual usage (USEPA, 1998b). California's Toxic Hot

TABLE 6.16

Recommended or Regulated Inhalation Exposure Limits for 1,4-Dioxane

Agency	Level (ppm)/(mg/m³)	Context	References
NIOSH	500/1800	IDLH	NIOSH (2005)
OSHA	100/360	PEL; 8-h TWA	OSHA (2007)
United Kingdom	25/91	Long-term exposure limit: 8-h TWA	UKHSE (2007)
	100/366	STEL: 15-min reference period	
South Africa	25/90	OEL: 8 h	SAIOH (2007)
	100/360	OEL: 15-min peak	
Finland	25/91	HTP: 8-h exposure level	STM (2005)
	40/150	HTP: 15-min exposure level	
Germany	20/72	MAK: 8 h	Greim (2000) (2005)[a]
	40/144	MAK: 15 min—based on eye irritation	
the Netherlands	12/43	MAC: 8 h	ECB (1999)
	24/66	MAC: 15 min	
Argentina	20/72	CMP	ILO (2007)
Hungary	10/36	Occupational exposure level: 8 h	ILO (2007)
National Authority for Occupational Safety and Health (Ireland)	20/72	OEL value: 8 h (reference period)	NAOSH (2002)
France	10/35	VME: 8-h exposure level	INRS (2006)
	40/150	VCLT: Level 15-min exposure level	
Japan	10/36	Occupational exposure level: 8 h	JSOH (2007)
Swedish Work Environment Authority	10/35	LLV: 8 h	SWEA (2005) (1996)[a]
	25/90	CLV: 15 min	
United States	2/7.2	MRL: 14-day inhalation	ATSDR (2004)
	1/3.5	MRL: 365-day inhalation	

Note: CMP, concentración Máxima Permisible; HTP, Haitallisiksi Tunnetut Pitoisuudet (concentrations known to be hazardous); NIOSH, U.S. National Institute for Occupational Safety and Health; OSHA, U.S. Occupational Safety and Health Administration; VCLT, Valeurs Limites Court Term (short-term exposure limits, i.e., STELs); VME, Valeur Limite de Moyenne d'Exposition (average exposure limit values).

[a] Dates in parentheses indicate the year in which the exposure limit was first adopted.

Spots rule requires all facilities emitting more than 85 pounds of 1,4-dioxane per year to prepare emission inventory plans and reports (CARB, 2008). Connecticut's emission regulation stipulates that 1,4-dioxane emissions must not exceed an 8-hour hazard limiting value of 450 μg/m³ or a half-hour maximum of 2250 μg/m³ (CDEP, 2002).

Verschueren (2001) offered the following equation to convert the mass concentration of an airborne contaminant (W/V) to its volume concentration (C_{vol}):

$$C_{vol} = \frac{W}{V} \frac{RT}{Pm}, \tag{6.7}$$

where C_{vol} is the volumetric concentration of the contaminant (in ppb), W is the mass of the compound present in volume V (in grams), V is the volume of contaminated air (in m³; $V = 1$ m³),

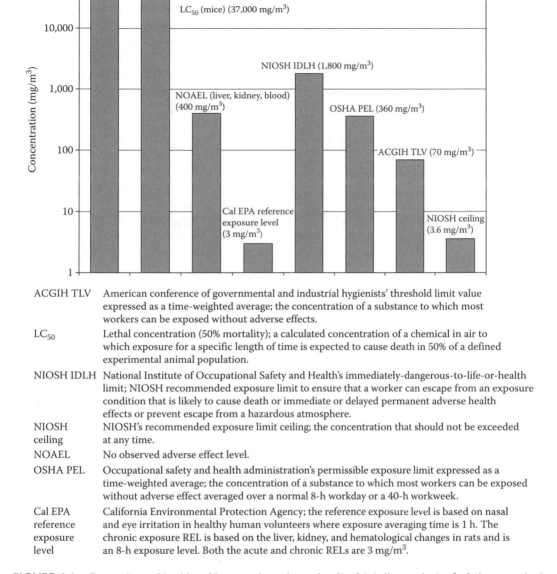

ACGIH TLV   American conference of governmental and industrial hygienists' threshold limit value expressed as a time-weighted average; the concentration of a substance to which most workers can be exposed without adverse effects.

LC_{50} Lethal concentration (50% mortality); a calculated concentration of a chemical in air to which exposure for a specific length of time is expected to cause death in 50% of a defined experimental animal population.

NIOSH IDLH   National Institute of Occupational Safety and Health's immediately-dangerous-to-life-or-health limit; NIOSH recommended exposure limit to ensure that a worker can escape from an exposure condition that is likely to cause death or immediate or delayed permanent adverse health effects or prevent escape from a hazardous atmosphere.

NIOSH ceiling   NIOSH's recommended exposure limit ceiling; the concentration that should not be exceeded at any time.

NOAEL   No observed adverse effect level.

OSHA PEL   Occupational safety and health administration's permissible exposure limit expressed as a time-weighted average; the concentration of a substance to which most workers can be exposed without adverse effect averaged over a normal 8-h workday or a 40-h workweek.

Cal EPA reference exposure level   California Environmental Protection Agency; the reference exposure level is based on nasal and eye irritation in healthy human volunteers where exposure averaging time is 1 h. The chronic exposure REL is based on the liver, kidney, and hematological changes in rats and is an 8-h exposure level. Both the acute and chronic RELs are 3 mg/m³.

FIGURE 6.1   Comparison of health guidance and regulatory levels of 1,4-dioxane in 1 m³ of air at standard temperature and pressure for inhalation of 1,4-dioxane. [After U.S. Environmental Protection Agency (USEPA), 2007 g, 1,4-Dioxane (1,4-diethyleneoxide) hazard summary. U.S. Environmental Protection Agency, Technology Transfer Network, Office of Air and Radiation. http://www.epa.gov/ttn/atw/hlthef/dioxane.html (accessed November 11, 2007)].

T is the temperature (in Kelvin; $T = 298.16$ K [25°C]), P is the pressure (in atmospheres; $P = 1$ atm = 101.325 kPa), m is the molecular weight (in g/mol), $R = 8.205 \times 10^{-5}$ (m³ atm)/(mol K), and $W/m = n$, the number of moles.

For 1,4-dioxane at 25°C (molecular weight, 88.20), 1 µg/m³ is equivalent to 0.278 ppb, and 1 ppb is equivalent to 3.6 µg/m³.

6.4   REGULATING 1,4-DIOXANE IN FOOD, PHARMACEUTICALS, COSMETICS, AND SUNDRIES

Chapter 2 describes the inclusion of 1,4-dioxane in food, personal care products, and pharmaceuticals as a by-product of the formulation of ethoxylated surfactants and polyethylene glycols (PEGs). 1,4-Dioxane has been measured in food, pharmaceuticals, and personal care products including cosmetics, shampoos, and lotions, as well as dish-washing detergents. Seven papers from the food chemistry literature identified 1,4-dioxane in chicken, shrimp, tomatoes, coffee, and fish sauce (see Table 2.13); however, 1,4-dioxane was not quantified and the analytical reliability in the timeframe in which these studies were conducted was not as good as current methods. A Japanese study conducted in 2003 identified 1,4-dioxane in many food products in the low ppb range, possibly associated with contaminated irrigation or food-processing water (Nishimura et al., 2004; see Table 2.14).

The United Nations Food and Agriculture Organization created the *Codex Alimentarius Commission* (also called the Food Codex) in 1963 to develop food standards and guidelines to protect consumer health. The U.S. Food and Drug Administration (FDA) and European Union agencies operate similar regulatory programs for contaminants of foods, cosmetics, veterinary medicines, and pharmaceuticals; all these regulatory programs promulgate standards for 1,4-dioxane.

6.4.1   FOOD REGULATIONS

1,4-Dioxane may occur as a residual in some food additives, including glycerides and polyglycides of hydrogenated vegetable oils (such as found in vitamin supplements) and polysorbates (present in ice cream, gluten-free bread, and other foods). Chlorinated solvents and their stabilizers may previously have been incorporated in food products from cleaning food-processing and canning equipment. Where food rich in fats is stored or processed in locations where concentrations of airborne solvents are high, solvents may partition from vapors into the food. For example, butter and lard purchased in shops near dry cleaners contained up to 763 µg/kg perchloroethylene. Methyl chloroform has also been measured in butter and lard, but at lower concentrations (Harrison, 2001). Presumably, 1,4-dioxane and other stabilizers prone to vapor partitioning could also enter foods stored near vapor degreasing facilities that use 1,4-dioxane. Plastic wraps, coated paper, and other food-packaging materials manufactured by processes using solvents may have also contributed solvent residuals including 1,4-dioxane in the past, before methyl chloroform was banned by the 1996 Montreal Protocol.

6.4.1.1   1,4-Dioxane in Dietary Supplements

Glycerides and polyglycides are used as an excipient in dietary supplement tablets, capsules, and liquid formulations (FDA, 2006). Glycerides and polyglycides are manufactured by heating a mixture of hydrogenated vegetable oils and PEG with an alkaline catalyst, followed by neutralization with an acid (FDA, 2006). As discussed in Chapter 2, PEG compounds often contain 1,4-dioxane as a by-product at levels up to 5 ppm, but more commonly less than 1 ppm. Title 21 of the Code of Federal Regulations permits up to 10 mg 1,4-dioxane per kilogram of glycerides or polyglycides as food additives (FDA, 2006).

The FDA estimates that a person's daily exposure to 1,4-dioxane from use of glycerides and polyglycides as an excipient in dietary supplement tablets, capsules, and liquid formulations is 800 ng. The dietary supplements are intended for ingestion in daily quantities measured in drops; for comparison, a drop of water is about 0.025 mL or 0.025 g.

On the basis of the estimated daily per person exposure to 1,4-dioxane of less than 800 ng, the FDA estimates that the upper-bound limit of lifetime human risk is 2.8×10^{-8} or 28 in 1 billion. The actual lifetime-averaged individual exposure to 1,4-dioxane is likely to be substantially less because the FDA used numerous conservative assumptions in calculating the exposure estimate. The FDA has concluded that no harm from exposure to 1,4-dioxane would result from consumption of glycerides and polyglycides in dietary supplements.

6.4.1.2   1,4-Dioxane in Ice Cream

1,4-Dioxane can also be present in polysorbate 60,[*] an emulsifier in ice cream, frozen custard, fruit sherbet, and other frozen desserts (FDA, 1999a). Polysorbate 60, in combination with polysorbate 65 and/or polysorbate 80, is added to ice cream in concentrations of 0.2–0.5% to stabilize the mixture of water, milk fat, milk protein (serum solids), sucrose or glucose sweeteners, and flavorings. Polysorbates 60, 65, and 80 are made by reacting fatty acids with sorbitol. Polysorbates prevent flavor oils such as vanilla, strawberry, and lemon from separating out of ice creams, beverages, and candies and extend the shelf life of doughnuts and other pastries (Klockenbrink, 1987). Sorbates are used to inhibit yeast growth in low-pH food products, such as wine, fruit juice, dried fruit, cottage cheese, meat, fish products, salad dressings, tomato products, and carbonated beverages (deMan, 1999). Polysorbates are also used in preparation of chewing gum, emulsified sauces, soups, and candies (ECSCF, 2002b). All polysorbates are likely to contain trace amounts of 1,4-dioxane.

The FDA estimated that exposure to 1,4-dioxane from using polysorbate 60 to produce frozen dairy desserts does not exceed 19 ng/person/day. The FDA estimates that the upper-bound limit of lifetime human risk from using polysorbate 60 in frozen dairy desserts is 6.7×10^{-10} or 6.7 in 10 billion. The European Union's Health & Consumer Protection Directorate-General published an opinion on the impurities of ethylene oxide in food additives. The polysorbates[†] are all required to meet a purity criterion of less than 5 mg of 1,4-dioxane per kilogram of polysorbate (ECSCF, 2002a).

6.4.1.3   1,4-Dioxane in Gluten-Free Bread

Preparation of gluten-free bread uses modified cellulose, including ethyl hydroxyethyl cellulose (EHEC). EHEC contains trace amounts of 1,4-dioxane as an impurity. The European Commission's Scientific Committee on Food recommends that EHEC can be permitted for use in gluten-free bread at mass fractions of less than 0.5%. If it is assumed that an individual consumes 300 g of bread or other gluten-free baked goods per day, the corresponding daily exposure to 1,4-dioxane is estimated to be 0.5 μg of 1,4-dioxane (ECSCF, 2002b). The daily dose from this consumption would be 0.008 μg 1,4-dioxane per kilogram body weight, which is considerably smaller than the overall NOAEL for the prevention of liver damage, 10 mg/(kg d). Gluten-free bread also includes polysorbates, which can contribute to 1,4-dioxane consumption. The potential maximum daily exposure of 1,4-dioxane from the permitted use of EHEC and polysorbates in bread is 0.008 and 0.05 μg/kg body weight, that is, about 170,000 times lower than the daily 10 mg/kg NOAEL, and is therefore of no toxicological concern (ECSCF, 2002b).

6.4.1.4   1,4-Dioxane in Food-Packaging Materials

1,4-Dioxane occurs as an impurity in 10 food-packaging materials and one sanitizing solution used on food-contact surfaces (U.S. Congress, 1987). Food-packaging materials generally include 1,4-dioxane as an impurity of ethoxylated surfactants; however, 1,4-dioxane may also be introduced through methyl chloroform used in adhesives, coatings, and labeling systems. For example, adhesives in microwave popcorn bags contained methyl chloroform and 1,4-dioxane and produced measurable worker exposure at a plant that manufactured the popcorn packages (Crandall and Wilcox, 1987). Polymer films used to wrap food for refrigeration or retail packaging are classified as Indirect Food Additives in Title 21 of the Code of Federal Regulations. Phosphate compounds used in cellophane coating, including mono- and bis-(octadecyldiethylene oxide)phosphates may contain 1,4-dioxane as an impurity of manufacturing.

---

[*] Also called Tween 60 or Polyoxyethylene (20) sorbitan monostearate.

[†] Polyoxyethylene (40) stearate; polyoxyethylene sorbitan monolaurate (polysorbate 20); polyoxyethylene sorbitan monooleate (polysorbate 80); polyoxyethylene sorbitan monopalmitate (polysorbate 40); polyoxyethylene sorbitan monostearate (polysorbate 60); polyoxyethylene sorbitan tristearate (polysorbate 65); polyethylene glycol 6000.

Daily exposure to 1,4-dioxane from use of phosphate compounds in cellophane coatings is estimated to be 0.22 part per trillion of a 3 kg diet, or 0.66 ng 1,4-dioxane per person. The FDA estimates that the upper-bound limit of lifetime human risk from residual 1,4-dioxane in the phosphate-bearing release agents is 2.3×10^{-11}, or 2.3 excess cancers in a hypothetical population of 100 billion (more than 43,000 times lower risk than the standard 1:1 million risk applied for regulatory standards) (FDA, 1999b).

6.4.1.5 The Delaney Clause

The Federal Food, Drug, and Cosmetics Act[*] establishes general safety standards for food additives. Food additives must be evaluated by the FDA to establish that the additive is safe for the intended use. The FDA's regulations for food additives define safe as "a reasonable certainty in the minds of competent scientists that the substance is not harmful under the intended conditions of use."[†] The Delaney clause of the Food, Drug and Cosmetics Act, also called the food additives anticancer clause, prohibits the use of carcinogenic substances as food additives. The Delaney clause declares that no food additive shall be deemed safe if it is found to induce cancer when ingested by man or animal.[‡] However, the Delaney clause applies only to the additive itself and not to impurities in the additive. Case law[§] has shown that where an additive has not itself been shown to cause cancer, but contains a carcinogenic impurity such as 1,4-dioxane or ethylene oxide, the additive is evaluated properly by using risk-assessment procedures, that is, 1,4-dioxane need not be considered.

6.4.2 Regulation of 1,4-Dioxane in Pharmaceutical Products

The synthesis of pharmaceuticals often uses solvents to enhance the yield or to determine characteristics such as crystal form, purity, and solubility. 1,4-Dioxane is among the solvents used in biological or natural extraction operations (USEPA, 1998a). The finished product may retain residual solvent from the manufacture of drugs or excipients when solvents are not completely removed in the manufacturing process. The FDA advises pharmaceutical companies to remove all residual solvents to the extent practical to ensure that no unsafe levels remain in the product. Guidelines for safe levels of solvent residuals in pharmaceutical products are established by an international committee called ICH.[¶] ICH groups solvent residuals in pharmaceutical products into three classes:

- Class 1: Solvents that should be excluded from the manufacture of drugs, excipients, and drug products because of their unacceptable toxicity or their deleterious environmental effects, that is, benzene, with a concentration limit of 2 ppm, and methyl chloroform, with a concentration limit of 1500 ppm
- Class 2: Solvents that should be limited in pharmaceutical products because of their inherent toxicity, that is, 1,4-dioxane, with a concentration limit of 380 ppm, and nitromethane, with a concentration limit of 50 ppm
- Class 3: Solvents that are regarded as less toxic and that therefore pose a lower risk to human health at levels normally accepted in pharmaceuticals (FDA, 1997, 2003)

The allowable limit for 1,4-dioxane present as a residual solvent in veterinary and human medicines in the United States and Europe is 3.8 mg/d (the "permitted daily exposure," or PDE), and the permissible concentration limit in veterinary and human pharmaceuticals is 380 ppm (FDA, 2001, 2003; EMEA, 2006). The allowable concentration limit for methyl chloroform, 1500 ppm, could

---

[*] FDA (2006) describes 21 U.S.C. 348.
[†] FDA (1999a) describes 21 CFR 170.3(i).
[‡] FDA (2006) describes role of section 409 of the Federal Food, Drug, and Cosmetic Act (21 U.S.C. 348).
[§] U.S. Court of Appeals for the 6th Circuit, 1984: Scott v. FDA.
[¶] International Conference on Harmonisation of Technical Requirements for Registration of Pharmaceuticals for Human Use (ICH).

result in a 1,4-dioxane residual concentration of up to 45 ppm if the methyl chloroform is stabilized with 1,4-dioxane at a mass ratio of 3%.

Exposure to 1,4-dioxane from ingestion of pharmaceuticals is likely to be substantially less than the permitted daily exposure, 3.8 ppm. No surveys of solvent residuals identifying 1,4-dioxane in finished pharmaceutical products have been published, possibly because (1) analysis of 1,4-dioxane in pharmaceuticals presents difficult analytical challenges and (2) the small mass and infrequent consumption of pharmaceutical products are inherently unlikely to pose a risk of exposure.

An indirect indicator of the extent to which 1,4-dioxane may be present in current or previous pharmaceutical products is the regulation of pharmaceutical manufacturers for 1,4-dioxane in plant emissions and effluent. USEPA (1998a) reported that 1,4-dioxane comprises 0.06% of overall pharmaceutical manufacturing effluent loading, or 24,422 pounds per year for all six plants included in the review. To model the performance of pharmaceutical plant effluent treatment works, USEPA assumed a 1,4-dioxane concentration of 180 mg/L of effluent. 1,4-Dioxane is certainly used in and associated with the pharmaceutical industry; however, its presence in retail pharmaceutical products has not been documented and is indeterminate.

6.4.3   REGULATION OF 1,4-DIOXANE IN COSMETICS AND SHAMPOOS

1,4-Dioxane may be present at low levels in some cosmetics, shampoos, and other personal care products because it occurs as a by-product during the production of ethoxylated surfactants used as ingredients in such products. These ingredients include certain detergents, foaming agents, emulsifiers, and solvents identifiable by the prefix, word, or syllables "PEG," "polyethylene," "polyethylene glycol," "polyoxyethylene," "-eth-," or "-oxynol-" (FDA, 2007). 1,4-Dioxane is not used directly as a cosmetic ingredient; it only occurs as a by-product. In shampoos, hair-care foaming agents such as fatty alcohols and ether sulfates may contain 1,4-dioxane as an impurity (Wilkinson, 1989). The FDA determines the levels at which a chemical is considered harmful; the levels depend on the conditions of use. The FDA (2007) has concluded that the levels of 1,4-dioxane observed in monitoring of cosmetics do not present a hazard to consumers.

To examine the potential toxicity of cosmetic impurities, the FDA performed skin absorption studies, which showed that 1,4-dioxane can penetrate animal and human skin when applied in certain preparations, such as lotions. However, studies determined that 1,4-dioxane evaporates readily from the skin, diminishing the amount available for skin absorption, even in products that remain on the skin for hours (Bronaugh, 1982). The FDA did not at first establish or recommend a specific limit on the level of 1,4-dioxane in cosmetics, but did provide guidance to manufacturers of ethoxylated surfactant ingredients, alerting them to the health concerns over 1,4-dioxane and opportunities to eliminate 1,4-dioxane by using vacuum stripping at the end of the polymerization process. The guidance was included in the *Cosmetic Handbook for Industry* (FDA, 1992). The Cosmetic Handbook includes 1,4-dioxane in the list of "explicitly prohibited ingredients." Retail cosmetics must list ingredients; however, 1,4-dioxane is a by-product and not an ingredient and has therefore not been included in the list of ingredients in cosmetics (Title 21, Code of Federal Regulations, Section 701.3). As reported by Pearl (2007), the FDA issued guidance in 2000 recommending that cosmetic products should not contain 1,4-dioxane at concentrations greater than 10 ppm.

As profiled in Chapter 2, contamination of ethoxylated surfactants with dioxane was first reported by the FDA in 1978 (FDA, 1981). Many of the raw materials analyzed since then have been found to contain 1,4-dioxane; some contained as much as 100 ppm or more. The FDA reports that the changes made in the manufacturing process have resulted in a significant decline in 1,4-dioxane levels. A recent survey of 1,4-dioxane in personal care products reported that 1,4-dioxane remains present in a number of shampoos and children's bubble bath products, some at levels in excess of the FDA's 10 ppm guidance (Campaign for Safe Cosmetics, 2007). The analysis of cosmetics was originally performed by David Steinman and published in the book *Safe Trip to Eden* (Steinman, 2007). Results from Steinman's analysis of sundries are listed in Table 6.17.

TABLE 6.17
Analyses of Consumer Products for 1,4-Dioxane

Product and References[a,b]	1,4-Dioxane Concentration (ppm)
Clairol Herbal Essence Rainforest Flowers Shampoo[a]	23
Olay Complete Body Wash with Vitamins (normal skin)[a]	23
Hello Kitty Bubble Bath (Kid Care)[a]	12
Johnson's Kids Shampoo Watermelon Explosion (Johnson & Johnson)[a]	10
Disney Clean as Can Bee Hair & Body Wash (Water Jel Technologies)[a]	8.8
Gerber Grins & Giggles Gentle & Mild Aloe Vera Baby Shampoo[a]	8.4
Johnson's Kids Tigger Bath Bubbles (Johnson & Johnson)[a]	5.6–7.9
Sesame Street Wet Wild Watermelon Bubble Bath (The Village Company)[a]	7.4
Johnson's Head-to-Toe Baby Wash (Johnson & Johnson)[a]	5.3–6.1
Rite-Aid Tearless Baby Shampoo[a]	4.3
Huggies Natural Care Baby Wash Extra Gentle and Tear Free[a]	4.2
Huggies Baby Wash Shea Butter[a]	4.0
Lil' Bratz Mild Bubble Bath (Kid Care)[a]	3.7
Scooby-Doo Mild Bubble Bath (Kid Care)[a]	3.0
Disney/Pixar Cars Piston Cup Bubble Bath (MZB Personal Care)[a]	2.2
Suave Naturals Passion Flower[a]	2.0
L'Oreal Kids Orange Mango Smoothie Shampoo[a]	2.0
Mr. Bubble Bubble Bath Gentle Formula with Aloe[a]	1.5
Citrus Magic 100% Natural Dish Liquid[b]	97.1
NutriBiotic Super Shower Gel Shampoo with GSE (fresh fruit)[b]	32.2
Method Dish Naturally Derived Ultra Concentrate[b]	27.5
Planet Ultra Dishwashing Liquid[b]	20.3
365° Everyday Value Shower Gel[b]	20.1
Earth Friendly Products Ultra Dishmate[b]	19.0
Alba Passion Fruit Body Wash[b]	18.2
Aura Cacia Natural Aromatherapy Bubble Bath[b]	14.9
Earth Friendly Products Ultra Dishmate (Natural Almond)[b]	13.6
Jason Fragrance Free Satin Soap[b]	11.9
Giovanni Cleanse Body Wash[b]	10.6
Jason Apricot Satin Soap[b]	9.2
Healthy Times Baby's Herbal Garden Pansy Flower Shampoo[b]	8.2
Jason Tea Tree Scalp Normalizing Shampoo[b]	7.9
Grandpa's Pine Tar Shampoo[b]	7.8
Sea-Chi Organics Tasmanian Lavender Shampoo[b]	7.5
Method Creamy Hand Wash[b]	7.0
Kiss My Face Moisture Soap[b]	6.3
Kiss My Face Early to Bed Shower Gel & Foaming Bath[b]	6.2
Healthy Times Baby's Herbal Garden Honeysuckle Baby Bath[b]	5.3
Giovanni Golden Wheat Shampoo[b]	4.1
Nature's Gate Awapuhi Volumizing Shampoo[b]	3.5
Origins for Men Skin Diver Active Charcoal Body Wash[b]	3.1
ShenMin Vitalize Shampoo for Thin, Fine Hair[b]	2.7
Shikai Moisturizing Shower Gel[b]	2.5
Ecover Dish-washing Liquid[b]	2.4
Nature's Gate Moisturizing Liquid Soap[b]	2.4
365° Everyday Value Shampoo[b]	2.2

continued

TABLE 6.17 (continued)
Analyses of Consumer Products for 1,4-Dioxane

Product and References[a,b]	1,4-Dioxane Concentration (ppm)
Circle of Friends No Tearski Shampoo[b]	2.1
Giovanni Smooth as Silk Deep Moisture Shampoo[b]	2.1
Origins Ginger Burst Savory Body Wash[b]	2.0
Seventh Generation Natural Dish Liquid[b]	1.9
Emerald Forest with Sapayul Botanical Shampoo[b]	1.7
365° Everyday Value Dish Liquid[b]	1.6
Nature's Gate Baby Soothing Shampoo[b]	1.6
Seventh Generation Lemongrass & Clementine Zest Natural Dish Liquid[b]	1.5
Rainbow Gentle Antibacterial Soap[b]	1.3
Ultra-Hair Conditioning Shampoo[b]	1.2
Rainbow Soap for Kids[b]	1.1
Earth Therapeutics Loofah Exfoliating Scrub Oatmeal & Honey[b]	0.9
Alba Replenishing Shampoo[b]	0.7
Life Tree Citrus Fresh Dish Soap[b]	0.7
Hugo Vanilla & Sweet Orange Conditioner[b]	0.6
Kiss My Face Obsessively Organic Lavender & Chamomile Self Foaming Liquid Soap[b]	0.3
Nature's Gate Organics Fruit Blend Liquid Soap[b]	0.3
Ecco Bella Moisturizing Shampoo Vanilla[b]	0.2

Source: From Steinman, D., 2007, *Safe Trip to Eden: 10 Steps to Save Planet Earth from the Global Warming Meltdown*. NY: Thunder's Mouth Press. With permission.

Note: The Campaign for Safe Cosmetics and the Environmental Working Group have continued to track levels of 1,4-dioxane in consumer products (see www.safecosmetics.org/ and www.cosmeticsdatabase.com/).

[a] Analyses in 2006 of children's shampoo and bath products were published in David Steinman's (2007) Safe Trip to Eden (used with permission); CSC (2007) lists some of these products and some for adults as well.

[b] Analyses in 2007 and 2008 of sundries and dish-washing products branded as natural and organic (Organic Consumers Association, 2008). The 54 products branded natural or organic but not listed here were nondetect for 1,4-dioxane, including all USDA-certified brands such as Dr. Bronner's, Sensibility Soaps (Nourish brand), Terressentials, German Natural "BDIH"-certified brands including Aubrey Organics, and Dr. Hauschka. Method: An aliquot of sample was accurately weighed into a vial with 5 mL of water and 1 g of sodium sulfate. Internal standard (5 µg 1,4-dioxane-d_8) was added. The vial was capped and heated at 95°C for 60 min. A 1 mL aliquot of the headspace over the sample was analyzed by direct injection. Conditions: Instrument: Agilent 5973N; column: 25 m × 0.20 mm HP-624, 1.12 µm film; column temperature: 40°C (hold 3 min) to 100°C at 10°C/min, then to 180°C at 25°C/min (hold 5 min); injector temperature: 220°C; mass range (selected-ion monitoring): masses 43, 58, 88 (dioxane): 64 and 96 (dioxane-d_8); and 1.72 cycles per second.

6.4.4 CALIFORNIA PROPOSITION 65

In 1986, California voters approved an initiative that requires the state to publish a list of toxic chemicals known to cause cancer, birth defects, or other reproductive harm. The initiative, called the Safe Drinking Water and Toxic Enforcement Act of 1986, came to be known as Proposition 65. The list of chemicals must be updated at least annually; it now includes approximately 775 chemicals (Cal EPA, 2007a). California businesses are required to notify the public about significant amounts of chemicals in the products for sale or used in their homes or workplaces, so that they may make informed decisions about protecting themselves from exposure to the listed chemicals. Proposition 65 also prohibits discharge of the listed chemicals to sources of drinking water. Proposition 65 is enforced by the California Attorney General's Office.

The California Environmental Protection Agency (Cal EPA) lists chemicals if a panel of independent scientists and health professionals finds that the chemical causes cancer, birth defects, or other reproductive harm. Chemicals may also be listed if other organizations, including USEPA, FDA, National Institute for Occupational Safety and Health (NIOSH), National Toxicology Program (NTP), and International Agency for Research on Cancer (IARC), list a compound in any of these categories. If a chemical on the list is present in a product, businesses must post a warning unless exposure is low enough to pose no significant risk of cancer or is significantly below levels observed to cause birth defects or other reproductive harm.

For carcinogens, the No Significant Risk Level (NSRL) is the level of exposure that would result in not more than one excess case of cancer in 100,000 individuals exposed to the chemical over a 70-year lifetime. For chemicals that are listed as causing birth defects or reproductive harm, the No Observable Effect Level (NOEL) for the chemical is divided by 1000 to provide a large margin of safety. If a product or workplace would cause exposures to chemicals listed as causing birth defects or reproductive harm that exceed 1/1000th of the NOEL, businesses are required to provide a warning to consumers or employees.

To help businesses determine when they are subject to the rules of Proposition 65, Cal EPA's OEHHA publishes "safe harbor numbers," below which businesses are exempt from the Proposition 65 warning requirements. California's safe harbor numbers consist of NSRLs for chemicals listed as causing cancer and Maximum Allowable Dose Levels (MADLs) for chemicals listed as causing birth defects or other reproductive harm.

1,4-Dioxane first appeared on the California Proposition 65 list of chemicals "known to the state to cause cancer" on January 1, 1988. The NSRL for 1,4-dioxane, that is, the 1:100,000 excess cancer risk level, is 30 µg/d (Cal EPA, 2007b). Because 1,4-dioxane causes neither birth defects nor reproductive toxicity, an MADL has not been established.

6.5 RISK ASSESSMENTS OF 1,4-DIOXANE

Risk assessments determine the likelihood and severity of harm to human health and environmental receptors that may result from exposure to a chemical substance. Risk assessments include four components: source or release assessment, exposure assessment, dose–response assessment, and risk characterization. Source assessment includes monitoring and modeling to estimate the magnitude of the release and the potential for chemical exposure resulting from the release. Exposure assessments evaluate the chemical concentrations to which populations are exposed by identifying toxic contaminants present in the release, quantifying the magnitude of exposure, determining the routes of exposure, and assessing the known or potential duration of exposure. Exposure assessment may include biomonitoring, that is, the sampling of fluid or tissue samples, as well as modeling, to estimate exposure levels.

6.5.1 OVERVIEW OF RISK-ASSESSMENT PROCESS AND DATA REQUIREMENTS

Dose–response assessment is the quantitative evaluation of toxicity as a function of exposure. The relationship between the dose of a chemical and the incidence of adverse health effects forms the basis for the quantitative dose response. Toxicity values (i.e., RfDs and slope factors) are derived from these relationships and can be used to estimate the incidence or potential for adverse effects in an exposed population. For carcinogens, the process of dose–response assessment involves the use of modeling to determine a point of departure followed by low-dose extrapolation. The method of low-dose extrapolation is dependent on the mode of action for the chemical; linear methods are used for chemicals that cause genotoxicity, and nonlinear approaches are suggested for chemicals that produce cancer via a nongenotoxic mode of action. For noncarcinogens, the point of departure may be the NOAEL or a benchmark dose level that is determined by modeling. UFs are applied to these values to derive RfDs and acceptable or tolerable daily intakes (TDIs).

Risk characterization integrates source assessment, exposure assessment, and dose–response assessment into a statement of the types and magnitudes of adverse effects that the contaminant may cause, qualified by probability of occurrence and uncertainty associated with the estimates (Cohrssen and Covello, 1989).

Table 6.18 summarizes the information needed to estimate the total dose by all routes of exposure to 1,4-dioxane in residential water.

6.5.2   EXPOSURE PATHWAYS AND ROUTES

Risk assessments typically profile 1,4-dioxane toxicity data by exposure route. The primary exposure routes for 1,4-dioxane are oral, inhalation, and dermal. European risk assessments generally rely upon the NOAEL or UF approach, whereas current U.S. risk assessments use a slope factor based on linear low-dose extrapolation. In the United States, inhalation toxicity thresholds are developed by the ATSDR, not by USEPA.

6.5.3   1,4-DIOXANE TOXICITY DATA USED IN RISK ASSESSMENTS

A key decision in preparing a risk assessment is the selection of suitable toxicity data for a given route of exposure. Risk assessments review available toxicological studies and evaluate the relevance of the derived toxicity values to the exposure framework under examination. For example, exposure to 1,4-dioxane vapors can cause toxicity to the liver through inhalation exposure. ATSDR (2004) calculated an acute-duration inhalation minimal risk level (MRL) of 2.0 ppm and a chronic-duration inhalation MRL of 1.0 ppm for 1,4-dioxane. The chronic-duration inhalation MRL was also adopted as the intermediate-duration inhalation MRL. ATSDR used the LOAEL of 50 ppm from the study by

---

TABLE 6.18
Data Requirements for Estimating Total Dose by Different Routes of Exposure to 1,4-Dioxane in Water

Direct ingestion of 1,4-dioxane through drinking:

    Quantity water consumed (2 L for adults; 1 L for a 10 kg child)
    Fraction of ingested 1,4-dioxane absorbed through gastrointestinal tract

Inhalation of 1,4-dioxane:

    Air concentrations from showering, bathing, swimming, etc.
    Duration of activities and amount of contaminated air breathed
    Fraction of inhaled 1,4-dioxane absorbed through lungs

Skin absorption of 1,4-dioxane in water:

    Period of time spent bathing or swimming
    1,4-Dioxane content of cosmetics, shampoos, lotions
    Quantity of residual shampoo, lotion applied, or cosmetic applied
    Area of skin to which dioxane-containing sundries and cosmetics applied
    Fraction of 1,4-dioxane absorbed through skin during bathing or swimming

Ingestion of 1,4-dioxane in food and drugs:

    Concentration of 1,4-dioxane in food or drugs
    Quantity of food and drugs consumed daily
    Fraction of 1,4-dioxane absorbed through gastrointestinal tract

Source: Cohrssen, J.J. and Covello, V.T., 1989, Risk analysis: A guide to principles and methods for analyzing health and environmental risks. Washington, DC: Council on Environmental Quality.
Note: Average body weight is needed for all routes of exposure.

Young et al. (1977) to derive the 2 ppm acute inhalation MRL. In a 2-year inhalation bioassay in rats exposed intermittently to 111 ppm 1,4-dioxane, no adverse effects were observed (Torkelson et al., 1974). ATSDR used the 111 ppm NOAEL to derive the chronic-duration inhalation MRL of 1 ppm for 1,4-dioxane.

Risk estimates for ingested 1,4-dioxane are usually based upon oral rat or oral mouse studies for which an NOAEL was established. For cancer risks, the CSF is the key index for evaluating exposure risk. The noncancer toxic effects of 1,4-dioxane drinking water exposure in a 2-year rat study included severe effects on the nasal cavity, lungs, liver, and kidneys; the NOAEL was equivalent to 10 mg/(kg d) (ECB, 2002).

ATSDR's draft Toxicological Profile for 1,4-dioxane derives MRLs for different durations of exposure to 1,4-dioxane. The MRL is defined as the daily human exposure to a substance that is likely to be without an appreciable risk of adverse noncarcinogenic effects over a specified duration of exposure. MRLs are based only on noncancerous health effects and do not consider carcinogenic effects. The MRL does not account for delayed health effects or those resulting from repeated high-dosage exposures; such effects include hypersensitivity reactions, asthma, or chronic bronchitis from acute inhalation exposure (ATSDR, 2004).

ATSDR determined an acute-duration (14 days or less) MRL of 4 mg/(kg d) oral exposure to 1,4-dioxane, based on an NOAEL of 370 mg/(kg d) for nasal effects in the 1998 Japan Bioassay Research Center (JBRC) rat study summarized in Chapter 5 (JBRC, 1998; Yamazaki et al., 1994; ATSDR, 2004). For intermediate-duration oral exposure (15–364 days), ATSDR developed an MRL of 0.6 mg/(kg d), based on an NOAEL of 60 mg 1,4-dioxane per kilogram of body weight per day for liver effects in rats (JBRC, 1998). ATSDR's determination of the chronic-duration oral-exposure (365 days or more) MRL for 1,4-dioxane, 0.1 mg/(kg d), used the NOAEL of 9.6 mg 1,4-dioxane per kilogram of body weight per day for liver effects in male rats from the study by Kociba et al. (1974).

A number of comprehensive risk assessments have been published by various government agencies. The key risk summaries reviewed for this book are listed in Table 6.19.

Risks associated with 1,4-dioxane exposure may occur through the occupational and domestic use of products containing varying amounts of 1,4-dioxane. Inhalation is the most common route of exposure to chemicals used in the workplace, but occupational exposure can also lead to dermal absorption of 1,4-dioxane, as described in the next section. Chapter 2 provides a detailed description of 1,4-dioxane occurrence in various manufacturing settings, and Table 2.7 lists the industrial processes in which 1,4-dioxane emissions may occur. Table 6.20 lists the leading occupations with potential

TABLE 6.19
Key Review Reports on 1,4-Dioxane Exposure by Ingestion, Inhalation, and Dermal Exposure

Year	Title	Publishing Agency
2005	"1,4-Dioxane in Drinking-Water"	WHO
2004	"Draft Toxicological Profile for 1,4-Dioxane"	ATSDR, USA
2002	"European Union Risk Assessment Report: 1,4-Dioxane"	European Union Institute for Health and Consumer Protection
1999	"Risk Assessment: 1,4-Dioxane"	Netherlands (VROM and several other Dutch government agencies)
1998	"Priority Existing Chemical No. 7—1,4-Dioxane"	National Industrial Chemicals Notification and Assessment Scheme (NICNAS)—Australian National Occupational Health and Safety Commission, and Environment Australia
1991	"1,4-Dioxane BUA Report 80"	Committee on Existing Chemicals of Environmental Relevance, German Chemical Society

TABLE 6.20
Occupational Settings with *Potential* Exposure to 1,4-Dioxane

Occupational Setting	1,4-Dioxane Occurrence	Type of Exposure	Exposure Level	References
Research and analytical laboratories	Solvent and reagent for research, development, and analysis	Inhalation, dermal	TWA—1.8 ppm	1–3
Celluloid film processing	Producing, splicing, repairing, or cementing movie film	Inhalation, dermal	TWA—1 ppm 7–14 ppm	4
Optical lens manufacture	Solvent fumes from coating ovens	Inhalation	Modeled: 10–50 ppm (high estimate)	1
Production of ethoxylated surfactants[a]	Closed-loop process; potential for exposure is during drumming off of finished product	Inhalation	Alkyl ether sulfates (100–200 ppm before vacuum stripping); current potential for worker exposure, <1 ppm	1
Production of microwave popcorn bags	In heat-resistant water-based epoxy adhesives	Dermal, ocular, inhalation	Less than hazardous levels but irritation observed	6
Automotive parts manufacturing	Vapor degreasing and parts washing	Inhalation, dermal	1.5–13.3 ppm in 15-min PBZ samples; 2.5–51 ppm at breathing level near sink and degreaser	7
Assembling and cleaning shirts from finished textiles	Dry cleaning	Inhalation	<0.1–0.5 ppm	8
Painting steel	Paint solvents (methyl chloroform)	Inhalation, dermal	up to 1.7 mg/m^3	9
Food packaging	Solvent printing and coating materials for individual serving packages of condiments	Inhalation	Not investigated	10
Resin production	Unsaturated polyester resins	Inhalation	Modeled: Maximum annual average, 0.121 μg/m^3; maximum one-hour average, 3.26 μg/m^3	11
Histology using electron microscopy	Direct use of 1,4-dioxane to dehydrate samples	Inhalation, dermal	Not investigated	12
Pharmaceutical industry	Used in chemical synthesis of pharmaceuticals	Inhalation	<3.6 mg/m^3 6.5 mg/m^3	13
Medical radioisotope production	Used in ion-exchange process to recover radioisotopes	Inhalation	Not investigated	14
Wet textile processing	Wool scouring and finishing, woven and knit fabric finishing, carpet finishing, yarn finishing, felted fabric processing	Inhalation	Total emissions can be up to 0.65 lb/h; exposure not investigated	15

continued

TABLE 6.20   (continued)
Occupational Settings with *Potential* Exposure to 1,4-Dioxane

Occupational Setting	1,4-Dioxane Occurrence	Type of Exposure	Exposure Level	References
Production of dyes	Solvent in colorants and adhesives in felt tip pens and fabric dyes	Inhalation, dermal (also oral in the case of children)	Felt tip pen dye, 11 mg/kg; fabric dye, 4.7 mg/kg; worker exposure not measured; modeled child exposure, 0.303 µg/(kg d) for felt tip pens used 0.5 h/day	16
Production of cellulose membranes	Direct use in solution deposition of cellulose acetate and cellulose triacetate membranes for production of filters used in water treatment and kidney dialysis	Inhalation, oral if used in reverse osmosis of drinking water, renal if used in kidney dialysis	Risk is absent unless filters are not adequately flushed in final production stage; worker exposure not investigated	17
Production of magnetic tape	Solvent for magnetic tape pigment, up to 40% 1,4-dioxane	Inhalation	37–75 mg/m^3	18
Production of 1,4-Dioxane at chemical plants[b]	Synthesis (closed loop)	Inhalation	0.18 mg/m^3 (59)/0.9 mg/m^3 0.08 mg/m^3 (18)/1.1 mg/m^3	18
	Storage/drumming	Inhalation	0.07 mg/m^3 (37)/40 mg/m^3 0.10 mg/m^3 (8)/10 mg/m^3	
	Waste disposal	Inhalation	0.07 mg/m^3 (35)/0.4 mg/m^3 0.15–4.7 mg/m^3 (2)	
	Laboratory	Inhalation	0.11 mg/m^3 (305)/0.6 mg/m^3	
	Production and drumming	Inhalation	43 mg/m^3 (30)	
	Around storage tanks	Inhalation	360–2880 mg/m^3 (2)	

Sources: [1] NICNAS (1998); [2] Hertlein (1980, cited in NICNAS, 1998); [3] Rimatori et al. (1994, cited in NICNAS, 1998); [4] Okawa and Coye (1982, cited in NICNAS, 1998); [5] Salisbury and Arnold (1984); [6] Crandall and Wilcox (1987); [7] Krake and Herrera-Moreno (1995); [8] Kullman (1993); [9] Hills et al. (1989); [10] MDH (2000); [11] ATSDR (2005b); [12] Shearer and Hunsicker (1980); [13] USEPA (1997a); [14] Taylor et al. (1995); [15] Smith (1986); [16] Hansen (2005); [17] Sirkar et al. (1978); and [18] VROM (1999).

Notes: TWA, Time-weighted average; PBZ, Personal breathing zone.

[a] Ethoxylated surfactants comprise alkyl and alkyl phenol ether sulfates, PEGs, and ethoxylates of alcohols, alkyl phenols, sorbitan esters, amides, and amines.

[b] First value is median, numbers in parentheses are the number of samples; value following slash "/" is 90th percentile; multiple rows are for data from different companies or different periods of record.

exposure to 1,4-dioxane; actual exposure is increasingly unlikely in the developed countries because of improvements in worker health and safety training, use of personal protective equipment, and regulation and monitoring of worker exposure. Occupational exposure to chemicals is a continuing issue in manufacturing centers located in countries with less developed health and safety regulations.

6.5.4   Risk from Dermal Exposure to 1,4-Dioxane

Toxicity from dermal exposure to 1,4-dioxane is generally considered to be relatively low (ECB, 2002; ATSDR, 2004; VROM, 1999; NICNAS, 1998). 1,4-Dioxane was negative in the complete skin carcinogenicity test using dermal exposure (King et al., 1973). As noted in Chapter 5, rhesus monkeys showed little dermal penetration from skin lotion. The primary effects of dermal exposure to 1,4-dioxane are irritation, dermatitis, eczema, and dried skin. As noted in Section 2.3.3, 1,4-dioxane is preferred in histology labs as a tissue preservative because it dehydrates tissues; therefore, avoiding direct skin contact with 1,4-dioxane in pure form is advisable. 1,4-Dioxane readily penetrates animal and human skin (Gosselin et al., 1984); however, most of the dioxane applied to the skin evaporates and thus may not be available for skin absorption (FDA, 2007).

Direct dermal exposure to 1,4-dioxane is primarily caused by a risk of incidental contact by workers handling 1,4-dioxane in laboratories, manufacturing facilities, and chemical production facilities. Significant dermal exposure to 1,4-dioxane probably occurred in earlier decades, when dioxane-stabilized methyl chloroform was widely used in vapor degreasing, and worker health and safety protective measures were less comprehensive than current standards. Occupational exposure regulations and enforcement, together with worker training and use of personal protective equipment, now minimize the potential for dermal exposure, and the number of industries in which 1,4-dioxane is routinely used is relatively limited. The general population may have skin contact with low concentrations of 1,4-dioxane while bathing in water contaminated by 1,4-dioxane (see Table 6.3). Dermal exposure may also occur from use or application of shampoos, lotions, and cosmetics containing 1,4-dioxane as a residual in the ethoxylated surfactant ingredients, as described in Section 2.4 and Table 6.17.

Incidental contact with solutions or cosmetic products containing 1,4-dioxane may result in absorption by diffusion through the outermost layer of skin (the dead stratum corneum) and traversal of the living epidermis, where metabolic processes may be involved. Unlike the stratum corneum, the cutaneous tissue is primarily an aqueous medium into which hydrophilic compounds like 1,4-dioxane will partition. Some of the chemical in contact with the skin will be absorbed, while some may be subject to evaporation. Many of the measured values of dermal permeability for chemicals are obtained by percutaneous absorption studies conducted by covering the site of application with an occlusive wrap to prevent loss of the compound. Values obtained by this approach may overestimate skin permeability for volatile compounds. Up to 3.2% of applied 1,4-dioxane (dissolved in lotion) was absorbed under occlusion for 3.5 h, whereas 0.3% absorption occurred under nonoccluded conditions (Bronaugh, 1982). The difference is due to the relative volatility of dioxane (USEPA, 2002).

Skin absorption is dominated by passive diffusion; the rate of absorption is inversely proportional to the thickness of the barrier layer (USEPA, 1992a). Skin permeation by diffusion follows Fick's first law and assumes steady-state conditions, that is, the volumes of solutions on either side of the skin membrane must be much greater than the effective volume of the membrane, solutions are well mixed, skin composition and properties are homogeneous, and concentration of the compound at the membrane surface is constant (USEPA, 1992a).

Skin penetration may also occur via skin appendages such as hair follicles, sebaceous glands, and sweat glands, which may allow dioxane to diffuse through otherwise rate-limiting barriers. Less than 1% of the human skin surface is composed of these appendages; therefore, this route of dermal exposure is negligible for lipophilic compounds such as the chlorinated solvents. 1,4-Dioxane is expected to diffuse slowly through the stratum corneum; therefore, skin appendages may play a more important role than diffusion for 1,4-dioxane passage through the skin. Polar compounds such as 1,4-dioxane are generally poorly absorbed through the skin, whereas nonpolar compounds are more readily absorbed (USEPA, 1992a). A different view is presented by the Danish EPA, which notes that 1,4-dioxane is a "fat solvent," and 1,4-dioxane can cause eczema upon prolonged or repeated contact (Danish EPA, 2005).

Rates of chemical permeation of the skin vary in humans by gender, skin type, skin condition, skin hydration, skin temperature, circulation of blood to the skin, capacity of the compound to bind to keratin, and the anatomic region exposed. For example, absorption of hydrocortisone through skin on the forehead is six times greater than through skin on the forearm, and the dermis of the scrotum absorbs hydrocortisone six times more than on the forehead, because of substantial variation of the stratum corneum thickness by anatomic region (Marzulli, 1962, cited in USEPA, 1992a). Dermal permeability in humans is relatively invariant by skin age (Wester et al., 1985). Much of the experimental data for dermal exposure factors are obtained from tests on laboratory animals such as rabbits and guinea pigs. Intraspecies variability therefore becomes an important factor for estimating human dermal exposure. Toxicologists and risk assessors have developed a weight-of-evidence scoring system to normalize the myriad sources of variability and uncertainty in dermal toxicological data. The absence of available dose–response relationships for dermal contact makes for considerable uncertainty in estimates of risk from dermal exposure (USEPA, 1992a).

The experimentally measured permeability coefficient[*] for 1,4-dioxane in aqueous media is estimated to be 4×10^{-4} cm/h ($\pm 0.36 \times 10^{-4}$), measured *in vitro* on human skin. Permeability coefficients were measured by using diffusion cells that (1) did not fully mimic the capacity of the circulatory system to remove penetrated chemicals and maintain a negligible contaminant concentration under the skin and (2) did not estimate incidental metabolism (USEPA, 1992a). The *in vitro* value agrees well with an empirical estimation technique, which uses molecular weight, octanol-water partitioning coefficient, and several statistically derived empirical factors to estimate the permeability coefficient value. The empirically estimated permeability coefficient for 1,4-dioxane is 3.6×10^{-4} (USEPA, 1992a). For comparison, the average dermal permeability coefficient of water is 1.55×10^{-3} cm/h (Bronaugh and Stewart, 1986; USEPA, 1992a).

No publications provide measured 1,4-dioxane exposure data from the use of shampoos, baby lotions, hand dish-washing liquids, or other sundries known to contain 1,4-dioxane as a by-product of manufacturing (ECB, 2002). Vacuum stripping used in the production of soaps and shampoos now effectively reduces 1,4-dioxane in shampoos, dish-washing liquids, and baby lotions to less than the regulated 10 ppm level. The European Chemicals Bureau nevertheless modeled exposure to 1,4-dioxane from use of these products, assuming elevated concentrations at levels measured in shampoos, soaps, and lotions in the 1980s. Table 6.21 summarizes the assumptions used to model dermal exposure to 1,4-dioxane in sundries.

The 1,4-dioxane concentrations in products listed in Table 6.21 correspond to measured concentrations from the 1980s and early 1990s (see Table 2.9). Recent surveys of 1,4-dioxane concentrations in shampoos and other sundries document a substantial reduction in most products (Fuh et al., 2005; Campaign for Safe Cosmetics, 2007). Both current and past dermal exposure studies for contact with 1,4-dioxane from shampoos, baby lotions, and dish-washing soaps conclude a margin of safety ranging from 6000 to 1,000,000 (ECB, 2002). The margin of safety is obtained by dividing the European Chemicals Bureau value for the calculated 1,4-dioxane dermal NOAEL [2000 mg/(kg d)] by the calculated cumulative worst-case uptake, that is, 0.0009 mg of 1,4-dioxane per kilogram body weight per day. Dermal NOAELs for 1,4-dioxane have been determined for male Wistar rats [8300 mg/(kg d)], guinea pigs [143 mg/(kg d)], and rabbits [57 mg/(kg d)] (ATSDR, 2004).

If incidental workplace exposure to 1,4-dioxane is assumed, the dermal uptake [D_{sk}, in mg/(kg d)] can be calculated as a relationship of the following dermal exposure parameters (NICNAS, 1998):

$$D_{sk} = \frac{W \times S \times A \times E \times F}{BW}, \tag{6.8}$$

where W is the weight fraction of 1,4-dioxane in the substance applied to the skin (i.e., 0.01 = 1%), S is the skin absorption rate [i.e., 0.3 mg/(cm^2 h)], A is the skin surface area exposed (i.e., hands

---

[*] Permeability coefficient ($K_{p,s}$) is a flux value, normalized for concentration, that represents the rate at which the chemical penetrates the skin (cm/h) (USEPA, 1992a).

TABLE 6.21
Exposure Estimates for Dermal Exposure to 1,4-Dioxane from Sundries

1. Shampooing Scenario: Adult Dermal Exposure

Weight: 65 kg (143 lb)	Duration: 10 min; frequency: 2–7 per week	Volume: 2 cm³ shampoo (12 g)	Concentration: 50 mg 1,4-dioxane per 1 kg shampoo (50 ppm)
Dilution: 10-fold	Absorbed fraction: 100%	Cumulative worst-case exposure: 21.4 mg/year	Cumulative worst-case daily exposure: 0.0009 mg/kg

2a. Infant Dermal Exposure Scenario for 1,4-Dioxane in Baby Lotion

Weight: 8 kg (~18 lb)	Frequency: 1 application per day	Volume: 2 cm³ lotion (2.4 g)	1,4-Dioxane weight fraction: 0.00001 (10 ppm)
Concentration of 1,4-dioxane in lotion: 0.012 mg/cm³	Absorbed fraction: 100%	Cumulative worst-case exposure (estimate): 8.8 mg/year	Cumulative worst-case daily exposure (estimate):0.003 mg/kg

2b. Adult Dermal Exposure Scenario for 1,4-Dioxane in Baby Lotion

Weight: 65 kg (143 lb)	Frequency: 1–2 applications per day	Volume: 7.5 cm³ lotion (7.5 g)	1,4-Dioxane weight fraction: 0.00001 (10 ppm)
Concentration of 1,4-dioxane in lotion: 0.01 mg/kg	Absorbed fraction: 100%	Cumulative worst-case uptake (estimate): 54 mg/year	Daily cumulative worst-case uptake (estimate): 0.00023 mg/kg

3. Adult Dermal Exposure Scenario for 1,4-Dioxane in Dish-Washing Liquid

Weight: 70 kg (154 lb)	Duration: 40 min/day	Volume: 2–10 cm³ liquid (2–10 g)	Concentration: 30 mg 1,4-dioxane per 1 kg liquid (30 ppm)
Dilution: 2000-fold	Absorbed fraction: 100%	Cumulative worst-case uptake (estimate): 0.053 mg/year	Daily cumulative worst-case uptake (estimate): 2.08×10^{-6} mg/kg

Margins of Safety: Calculated Exposures versus Daily Dermal NOAEL of 20 mg/kg

Scenario	Cumulative Worst-Case Daily Uptake (mg/kg)	Margin of Safety Compared to NOAEL[a]
Adult shampooing	0.0009	22,222
Infant baby lotion	0.003	6666
Adult baby lotion	0.00023	89,000
Adult dish-washing soap	2.08×10^{-6}	>1,000,000

Sources: European Chemicals Bureau (ECB), 2002, European Union risk assessment report: 1,4-Dioxane. Institute for Health and Consumer Protection, European Union; VROM, 1999, *Risk assessment: 1,4-Dioxane*. Bilthoven, Netherlands: S. P. Netherlands Ministry of Housing and the Environment (VROM), Chemical Substances Bureau.

Note: The European Chemicals Bureau calculates a dermal NOAEL for 1,4-dioxane of 20 mg/(kg d); other risk assessors note that a reliable NOAEL for chronic dermal effects has not been determined (NICNAS, 1998).

[a] The margin of safety is obtained by dividing the European Chemicals Bureau value for the calculated 1,4-dioxane dermal NOAEL (20 mg/(kg d)), by the calculated cumulative worst-case daily uptake.

only = 840 cm²), E is the exposure duration (i.e., 8 h/day), F is the skin contact time as a daily percentage (i.e., 4.16% = 0.042 day or 1 h), and BW is the average body weight of worker (70 kg).

In summary, the dermal exposure route can be significant for unprotected occupational exposure. The Netherlands risk assessment for 1,4-dioxane notes that repeated-dose toxicity and carcinogenicity from dermal exposure in the workplace cannot be ruled out for occupations using cleaning agents that contain 1,4-dioxane (VROM, 1999). The relatively high NOAELs and the absence of dose–response values for dermal exposure make it difficult to estimate the health effects from repeated exposure to low levels of 1,4-dioxane in contaminated shower water or in shampoos and soaps. The majority of published 1,4-dioxane risk summaries conclude that the risk of adverse health effects from dermal exposure to low concentrations in water or sundries is not significant.

6.5.5 RISK FROM INHALATION EXPOSURE TO 1,4-DIOXANE

Inhalation exposure to 1,4-dioxane may result from a variety of occupational and domestic activities involving use of products that contain 1,4-dioxane as an additive or impurity. Chapter 2 profiles the many products that contain 1,4-dioxane; use of these products indoors increases the chance of 1,4-dioxane being inhaled. 1,4-Dioxane may also be present in indoor air from adhesives and coatings used in construction and interior finishes. For example, wood parquet floors may use adhesives containing 1,4-dioxane up to 0.5%, whereas wood glues may contain 1,4-dioxane from 1% to 5% (DAP Inc., 1994). Coatings and varnishes in office cubicles as well as draperies also emit measurable levels of 1,4-dioxane (see Section 2.3.8). A NASA survey of indoor air sources determined that a minor percentage of commonly encountered home construction practices can introduce 1,4-ioxane. By testing materials in closed containers at elevated temperatures not ordinarily encountered in the home, researchers were able to measure 1,4-dioxane present in 16 construction adhesives, coatings, and rubber materials. Adhesives were the major source of 1,4-dioxane; their median emission rate was 4.3 µg/g of adhesive, and their maximum emission rate was 21 µg/g of adhesive (McDonnell Douglas Corporation, 1986). Draperies and drapery linings emitted 1.3 µg/m^2/h in one study (Bayer and Papanicolopoulos, 1990). A German study identified 1,4-dioxane at less than 50 µg/m^3 (14 ppbv) among volatile organic compounds emitted in subcarpet construction coatings (Fischer, 1998). Polyurethane foam insulation may emit 1,4-dioxane at elevated temperatures during installation or incineration. At 80°C, the average 1,4-dioxane concentration of eight air samples of emissions from one freshly sprayed and three cured polyurethane foams was 2400 ± 1500 ppbv; at 40°C (104°F), 1,4-dioxane was not detected (Krzymien, 1989).

A sampling survey of 12 homes in Woodland, California, found a geometric mean concentration of 0.03 ppb 1,4-dioxane in indoor air, an arithmetic mean of 0.39 ppb, and a maximum value of 39 ppb (Sheldon et al., 1991; Hodgson and Levin, 2003). An earlier review of surveys estimated that the geometric mean 1,4-dioxane concentration in indoor air was 0.02 ppb (Pellizzari et al., 1986). A National VOC Indoor Air Database, with 585 measurements for 1,4-dioxane, lists the average 1,4-dioxane concentration at 1.03 ppbv (Shah and Heyerdahl, 1988; Shah and Singh, 1988). A survey of indoor air in Los Angeles and Contra Costa County, California, found 1,4-dioxane present in 12–55% of the homes sampled. In Los Angeles, 112 homes sampled in the summer of 1984 had an average 1,4-dioxane concentration of 0.15 ppbv; the maximum detected value was 1.2 ppbv. A winter 1984 survey of 51 homes in Los Angeles detected 1,4-dioxane at an average concentration of 0.044 ppbv. The sampling of indoor air in Contra Costa homes occurred in the summer of 1984 and found an average concentration of 0.042 ppbv and a maximum of 0.81 ppbv (Wallace, 1986).

Exposure to 1,4-dioxane emitted from architectural coatings and other sources mentioned in this section is unlikely to pose a significant inhalation risk. The greatest inhalation risk from 1,4-dioxane exposure is with its direct use without proper personal protective equipment and ventilation measures. Unfortunately, 1,4-dioxane does not provide workers with warning properties such as a distinct and low-threshold odor or an effect such as watery eyes or respiratory irritation. The odor threshold for 1,4-dioxane in air is 24 ppm (Amoore and Hautala, 1983).

The greatest occupational exposure to 1,4-dioxane by inhalation is likely to have occurred during degreasing operations in earlier decades. Degreasing with methyl chloroform containing up to 3% 1,4-dioxane will cause inhalation exposure from 3 to 15 ppm (12–60 mg/m^3). The scenario under which such exposure would occur is an open-top vapor degreaser without supplied ventilation (VROM, 1999). Since use of methyl chloroform was banned by the Montreal Protocol, this route of exposure is no longer significant. Some occupational exposure to 1,4-dioxane may remain where it is used in laboratories, as a solvent for producing cellulose acetate membranes, or where it occurs as a by-product from the production of ethoxylated surfactants. Nevertheless, many more people may experience nonoccupational exposure to low concentrations in shower air containing 1,4-dioxane from contaminated water or from shampoos, soaps, lotions, cosmetics, and related products. Even so, the low concentrations of 1,4-dioxane likely to result from showering and using soaps and shampoos are well below the inhalation slope factor used in California, 0.027 (mg/kg d)$^{-1}$, and the California chronic inhalation Reference Exposure Level for 1,4-dioxane, 3000 µg/m^3 (Cal EPA, 1999, 2007c).

Inhalation exposure from 1,4-dioxane in contaminated shower water is profiled in the following section; this section reviews incidental low-level exposure that could result from the presence of residual 1,4-dioxane in soaps and shampoos. If it is assumed that shampoo contains the 1980s levels of 1,4-dioxane (50 ppm), an average inhalation exposure per event is calculated at 13 µg/m^3, and the yearly average is then projected to be 0.0576 µg/m^3. These values yield an internal dose of 0.02 µg 1,4-dioxane per kilogram body weight per day. Assumed parameter values used to calculate estimates of inhalation exposure to 1,4-dioxane at levels measured in the 1980s in shampoos, baby lotions, and dish-washing soaps are profiled in Table 6.22 (ECB, 2002).

Japanese domestic wastewater was evaluated for 1,4-dioxane content derived from shampoos and soaps; the potential exposure was determined to be approximately 0.25 mg/person/day (Abe, 1999). The highest cumulative worst-case uptake in the ECB 2002 risk assessment would amount to only 4% of the total mass of 1,4-dioxane ascribed to soaps and sundries in Japan. The majority of 1,4-dioxane that may have been, or continues to be, present in shampoos and sundries ends up in the sewer. Inhalation exposure to 1,4-dioxane from shampoos, baby lotions, and dish-washing soaps is four to five orders of magnitude lower than the NOAEL. The concentrations one might breathe while shampooing, coupled with dermal exposure, are inconsequential and do not constitute a health risk (ECB, 2002).

6.5.5.1 Lactational Transfer Following Occupational Exposure

Infants nursed by mothers breathing elevated levels of 1,4-dioxane in the workplace may be exposed to 1,4-dioxane through breast milk. A PBPK[*] modeling study simulated maternal exposure to 1,4-dioxane at the 25 ppm TLV established for workplace safety regulation. The PBPK model used estimated blood/air and milk/air partitioning coefficients to calculate blood/milk partitioning. The model predicted that perchloroethylene, bromochloroethane, and 1,4-dioxane would be present in breast milk at concentrations exceeding the USEPA noncancer ingestion rates for children. The PBPK simulation of lactational transfer is not well developed for volatile organic compounds; the study necessarily applies assumptions regarding the kinetics and properties of lactational transfer, but nonetheless suggests that this route of exposure could be significant for infants nursed by mothers exposed to 1,4-dioxane by breathing vapors in the workplace (Fisher et al., 1997). Analysis of breast milk has shown that the inhalation-lactational transfer route of exposure was completed in a woman who visited her husband at his dry cleaning establishment for the lunch hour each day. Her breast milk contained elevated levels of perchloroethylene because daily exposure saturated her body's ability to eliminate the chemical (Bagnell and Ellenberger, 1977). Methyl chloroform, carbon tetrachloride, dichloromethane, TCE, and perchloroethylene have all been detected in breast milk (Pellizzari et al., 1982; Labrèche and Goldberg, 1997; Solomon and Weiss, 2002).

6.5.5.2 Inhalation Risk from Showering in 1,4-Dioxane Contaminated Water

Showering in water contaminated with 1,4-dioxane will cause inhalation exposure. Although 1,4-dioxane is hydrophilic and less susceptible to partitioning from the liquid to vapor phase, the rate of volatilization is increased when water contaminated with 1,4-dioxane is heated. The total surface area available for mass transfer is a key determinant of the rate of volatilization. Shower nozzles break up the water stream into fine droplets, maximizing the opportunity for volatilization and

---

[*] PBPK = physiologically based pharmacokinetic model; see Chapter 5.

TABLE 6.22

Exposure Estimates for Inhalation Exposure to 1,4-Dioxane from Sundries

1. Shampooing Scenario: Adult Inhalation Exposure

Weight: 65 kg (143 lb)	Duration: 10 min; frequency: 2–7 per week	Volume: 2 cm³ shampoo (12 g)	Concentration: 50 mg 1,4-dioxane per 1 kg shampoo (50 ppm)
Surface area: 1200 cm²	Temperature: 38°C	Ventilation rate: 4 m³/h	Room volume: 1.6 m³
Exposure: 12.75 µg/m³ (mean event concentration)	Exposure: 0.06 µg/m³ (annual average concentration)	Uptake: 330 µg/year	Dose: 1.391×10^{-5} mg/(kg d)

2a. Infant Inhalation Exposure Scenario for 1,4-Dioxane in Baby Lotion

Weight: 8 kg (~18 lb)	1 application per day	Volume: 2 cm³ lotion (2.4 g)	Concentration: 0.012 mg of 1.4-dioxane per 1 cm³ lotion; 1,4-dioxane weight fraction: 0.00001 (10 ppm)
Release area: 2000–4300 cm²	Temperature: 25°C	Ventilation rate: 33.8 m³/h	Room volume: 22.5 m³
Exposure: 0.029 µg/m³ (mean event concentration)	Exposure: 0.029 µg/m³ (annual average concentration)	Uptake: 34 µg/year	Dose: 1.16×10^{-5} mg/(kg d)

2b. Adult Inhalation Exposure Scenario for 1,4-Dioxane in Baby Lotion

Weight: 65 kg (143 lb)	1–2 applications per day	Volume applied: 7.5 cm³ lotion (7.5 g)	Concentration: 0.01 mg per 1 kg lotion; 1,4-dioxane weight fraction: 0.00001 (10 ppm)
Release area: 1200 cm²	Temperature: 21°C	Ventilation rate: 111 m³/h	Room volume: 74 m³
Exposure: 0.035 µg/m³ (mean event concentration)	Exposure: 0.052 µg/m³ (annual average concentration)	Uptake: 300 µg/year	Dose: 1.26×10^{-5} mg/(kg d)

3. Adult Inhalation Exposure Scenario for 1,4-Dioxane in Dish-Washing Liquid

Weight: 70 kg (154 lb)	Duration: 40 min per day	Volume: 2–10 cm³ liquid (2–10 g)	Concentration: 30 mg 1,4-dioxane per 1 kg liquid (30 ppm); dilution: 2000-fold
Release area: 0.25 m²	Temperature: 50°C	Ventilation rate: 40 m³/h	Room volume: 20 m³
Exposure: 19 µg/m³ (mean event concentration)	Exposure: 0.55 µg/m³ (annual average concentration)	Uptake: 3.33 mg/year	Dose: 1.3×10^{-4} mg/(kg d)

Margins of Safety: Calculated Exposures versus Daily Inhalation NOAEL of 10 mg/(kg d)

Scenario	Cumulative Worst-Case Daily Uptake [mg/(kg d)]	Margin of Safety Compared to NOAEL[a]
Adult shampooing	2.09×10^{-5}	~480,000
Infant baby lotion	1.16×10^{-5}	~860,000
Adult baby lotion	1.6×10^{-5}	~625,000
Adult dish-washing soap	1.3×10^{-4}	~80,000

Source: European Chemicals Bureau (ECB), 2002, European Union risk assessment report: 1,4-Dioxane. Institute for Health and Consumer Protection, European Union.

[a] NOAEL = 10 mg/(kg d).

production of aerosols. The relatively confined space of a shower in a bathroom will exacerbate the potential for exposure to chemical contaminants in shower water.

Numerous risk assessments have been developed to estimate shower exposure. Shower exposure models account for the factors that contribute to volatilization, the exposure duration, and attributes of the exposed population. The groundwater concentration of 1,4-dioxane exceeding the risk of one excess cancer in 1 million exposed was calculated as 3 µg/L by staff at the Massachusetts Department of Environmental Protection (MADEP spreadsheet model for inhalation exposures from showering—based on Foster & Chrostowski shower model by Paul W. Locke and Rafael McDonald, Massachusetts Department of Environmental Protection, personal communication, 2006). The MADEP calculation used a chronic-inhalation RfC of 0.12 mg/m^3 (USEPA, 1992b) and the IRIS oral CSF of 0.011 (mg/kg d)$^{-1}$ The MADEP model calculates an internal dose that is compared to the oral slope factor. Table 6.23 lists the model parameters considered in the MADEP shower model. The assumption of a single value for many of these parameters is a simplification to avoid

---

TABLE 6.23
Model Parameters Used for Shower Model of 1,4-Dioxane Inhalation Exposure

	Value
Physical Considerations	
Receptor: Young child, 1–8 years old	16.8 kg
Receptor: Older child, 8–15 years old	39.7 kg
Receptor: Adult, 15–30 years old	54.2 kg
Shower water temperature	318 K
Ventilation rate	15 L/min
Shower droplet diameter	1 mm
Shower droplet time	2 s
Universal gas constant	8.205×10^{-5}(m^3 atm)/(mol K)
Gas-film mass transfer coefficient, water	3000 cm/h
Liquid-film mass transfer coefficient	20 cm/h
Calibration water temperature for air mass transfer	293 K
Water viscosity at mass transfer temperature	1.002 centipoise
Water viscosity at shower temperature	0.596 centipoise
Shower duration	15 min
Shower room volume	6 m^3
1,4-Dioxane Attributes	
Henry's law constant	4.91×10^{-6} (atm m^3)/mol
Molecular weight	88 g/mol
Gas-film mass transfer coefficient	1356.8 (kg cm)/h
Liquid-film mass transfer coefficient	14.14 (kg cm)/h
Overall mass transfer coefficient	14.14 cm/h
Adjusted mass transfer coefficient	19.1 cm/h
Concentration leaving water droplet	0.47 µg/L
Indoor-air generation rate	0.79 µg/(m^3 min)
Exposure factor—noncancer	0.0965 (µg/m^3)/(µg/L)
Exposure factor—cancer risk	0.0414 (µg/m^3)/(µg/L)
Noncancer risk	200 µg/L
Cancer risk	3 µg/L

Source: European Chemicals Bureau (ECB), 2002, European Union risk assessment report: 1,4-Dioxane. Institute for Health and Consumer Protection, European Union.

developing a more complex, probabilistic approach that accounts for the range of values encountered; for example, the distribution of droplet sizes produced by a shower nozzle, starting and final air temperature, etc. (Giardino et al., 1992).

The MADEP shower model results suggest that inhalation cancer risk can be limited if the 1,4-dioxane level in shower water is maintained below 3 µg/L. As discussed in Chapter 5, 1,4-dioxane exposure studies by Torkelson et al. (1974) on inhalation and by Kociba et al. (1974) on ingestion find that cancer potency of 1,4-dioxane is greater by the oral route of exposure. This finding suggests that 1,4-dioxane toxicity is related to metabolism, which is not involved with inhalation exposure; therefore, people should be less vulnerable to 1,4-dioxane toxicity from the inhalation route than from the ingestion route. However, ATSDR's 2004 draft Toxicological Profile of 1,4-dioxane states that exposure to 1,4-dioxane in tap water through inhalation during showering or other indoor activities can result in higher exposures to 1,4-dioxane compared to ingestion of drinking water (ATSDR, 2004). Health risk assessments should carefully consider shower exposure to address inhalation exposure.

6.5.6 Risk of 1,4-Dioxane Oral Ingestion in Drinking Water, Food, and Drugs

USEPA's IRIS previously listed the oral CSF value of 0.011 $(mg/kg\ d)^{-1}$ based on linear extrapolation to low doses from the 1978 National Cancer Institute study of nasal tumors in rats. As of February 9, 2004, 1,4-dioxane is being reassessed under the IRIS Program. IRIS lists the 1 in 1,000,000 risk level for 1,4-dioxane in drinking water as 3 µg/L, on the basis of the previously listed slope factor of 0.011 $(mg/kg\ d)^{-1}$ (USEPA, 2007 h). The previous IRIS risk level used the LMS extrapolation procedure, which has been questioned because a nonlinear dose–response curve is observed for 1,4-dioxane. Alternative derivations of safe drinking water concentrations for 1,4-dioxane using PBPK modeling range from 2 to 120 mg/L (Stickney et al., 2003), as presented in Table 5.5 in Chapter 5.

The World Health Organization (WHO) established a 50 µg/L advisory level for 1,4-dioxane. The WHO selected the LMS model to estimate cancer risk because studies they reviewed indicated that 1,4-dioxane induces multiple tumors including hepatic and nasal cavity tumors in rodents in most long-term oral studies conducted and tumors in peritoneum,[*] skin, and mammary glands observed in rats given a high dose (WHO, 2005). By using the 1 in 100,000 cancer risk for hepatic tumors observed in a 2-year study of rats and mice by Yamazaki et al. (1994) for the JBRC, the WHO determined the drinking water equivalency level at 54 µg/L. For the noncarcinogenic endpoint, the WHO used the TDI approach, which, like the MCLG, is an estimate of the amount of a substance that can be ingested or absorbed over a specified period of time without appreciable health risk. The long-term study of 1,4-dioxane in drinking water supplied to rats by Kociba et al. (1974) was used for the noncarcinogenic NOAEL of 16 mg/(kg d), coupled with a 1000-fold UF for inter- and intraspecies variation, to obtain a TDI of 16 µg/(kg d). The equivalent concentration in drinking water is calculated to be 48 µg/L. Taken together with a 54 µg/L TDI calculated from the Kociba study, the WHO set the Drinking Water Quality Guideline at 50 ppb (WHO, 2005).

The drinking water standards development process discussed in Section 6.1 provides the basic framework used by several American states to determine the risk threshold for ingestion of 1,4-dioxane in drinking water. Exposure to 1,4-dioxane from ingesting food or drugs containing residual 1,4-dioxane is likely to be very low and possibly inconsequential, as discussed in Section 6.4.2.

6.5.7 Relative Risks of Solvent Stabilizers

The relative risk of exposure to stabilizer compounds can be discerned by comparing the key acute and subchronic toxicity indicators for the major routes of exposure. The key indicators are the

---

[*] The peritoneum is the abdominal cavity lining.

statistically derived median single lethal dose for 50% of animals tested (LD_{50}) by the oral/ingestion, inhalation/breathing, and dermal/skin exposure routes. The lethal dose varies by species, strain, age, and sex of the laboratory animal tested. For example, rat pups, rat dams (females), and male rats may each have their own LD_{50}, whereas mice, guinea pigs, and rabbits may have considerably larger or smaller LD_{50} values. The species exhibiting the greatest sensitivity to the toxin is usually selected as the human surrogate from which estimates of toxic effect levels for humans can be extrapolated.

Subacute assays assess the toxic effects in an animal population from daily exposure to the toxicant over about 10% of the animal's lifetime. Subacute toxicity assays identify many different toxic endpoints, such as dysfunction of different organs (Kamrin, 1988). The highest dose at which none of the animals in a test population displays a toxic effect (NOAEL) is typically divided by 100 to reflect greater variation among human populations.

Carcinogenic chemicals are distinguished by their capacity to produce *neoplasms* ("new growths," or tumors) in the receptor. Any one of the following four types of evidence is generally used to classify a chemical as carcinogenic:

1. The exposed population develops tumors not seen among animals in the control population
2. There is an increased incidence of tumors in the exposed population of the same type that also commonly occurs at a low incidence rate among the control population
3. Tumors develop earlier in the exposed population than in the control population
4. Tumors in the exposed population multiply more quickly than in the control population (Williams and Weisburger, 1986)

The carcinogenic agent that actually converts a normal cell to a neoplastic (tumor) cell and finally into an overt neoplasm may be either (1) the original chemical that enters the body by ingestion, inhalation, dermal exposure, or other means or (2) a metabolic by-product that results in organ-specific exposure, for example, in the liver. The route of exposure may control the carcinogenicity of a compound. For example, a compound may not be carcinogenic when absorbed into the bloodstream through the lungs, but could be carcinogenic when metabolized following ingestion. Carcinogens are classified by the weight of evidence for their likelihood to initiate or promote cancerous tumors.

Table 6.24 lists the oral and inhalation rat LD_{50} values for selected solvent-stabilizer compounds. Although it is preferable to compare RfDs or RfCs, these are generally not available for the list of stabilizer compounds included in Table 6.24. *The reader is cautioned that NOAELs for different species, routes of exposure, toxicity endpoints, exposure durations, and study designs cannot be directly compared; the values are provided to list available information for these compounds.*

TABLE 6.24
Relative Risk Thresholds of Solvent Stabilizers Compounds

Stabilizer	Oral Rat LD_{50}	Inhalation LC_{50}	NOAEL	Carcinogenicity, Mutagenicity, and Genotoxicity
Epichlorohydrin	40 mg/kg [1] 90 mg/kg [2]	LC_{50} rat inhalation 500 ppm for 4 h [1]	3.4 mg/m³ 6 h/day; 5 day/week [3]	IRIS B2—probable human carcinogen [4]; confirmed animal carcinogen [5]; IARC group 2A probably carcinogenic to humans [6]
Acetonitrile	175 mg/kg [7]	LC_{50} rat inhalation [1] 330 ppm for 90 days	100 ppm maternal rat [8]	A4: Not classifiable as a human carcinogen [9]

continued

TABLE 6.24   (continued)
Relative Risk Thresholds of Solvent Stabilizers Compounds

Stabilizer	Oral Rat LD$_{50}$	Inhalation LC$_{50}$	NOAEL	Carcinogenicity, Mutagenicity, and Genotoxicity
Phenol	317 mg/kg [2] 530 mg/kg [10]	LC$_{50}$ rat inhalation 316 mg/m^3 [11]	1000 ppm = NOAEL for reproductive toxicity, in drinking water supplied to rats [12]	IRIS D: Not classifiable as to human carcinogenicity [10]; IARC group 3: Not classifiable as to its carcinogenicity to humans [6]
Triethylamine	460 mg/kg [13]	1000 ppm for 4 h [14]	247 ppm = NOAEL for rat inhalation [15]	A4: Not classifiable as a human carcinogen [5]
1,2-Butylene oxide	500 mg/kg [16] 1170 mg/kg [17]	398–6550 ppm for 4 h [18] 200 ppm: Rat inhalation for 6 h/day, 5 day/week, 65 exposures in total [18]	None observed [19] <50 ppm [17]	Male rats = clear evidence; female rats = equivocal evidence; male and female mice = no evidence [17]; nasal carcinogen in male rats [20]
Diisopropylamine	770 mg/kg [2]	4800 mg/m^3 for 2 h [2]	41 ppm = rat inhalation, 6 h/day, 5 day/week, 4 weeks [21]	Not carcinogenic; negative mutagenicity and genotoxicity [22]
Nitromethane	940 mg/kg [14]	Central nervous system symptoms with exposures at 3% or 5% vapors for >1 h or 1% nitromethane for 5 h in rabbits and guinea pigs [23]	200 ppm = rat inhalation, 6 h/day, 5 day/week, 4 weeks [24]	A3: Confirmed animal carcinogen with unknown relevance to humans [5]; clear evidence of carcinogenic activity of nitromethane in rats and mice [25]; cancer potency 0.18 (mg/(kg d))$^{-1}$; No Significant risk level = 39 μg/d [26]
tert-Amyl alcohol	1000—2000 mg/kg [27]	14,000 mg/m^3 for 6 h in rat [28]	225 ppm in rats [29]	—
Nitroethane	1100 mg/kg [2]	—	—	Not carcinogenic [20]
2-Methyl-3-butyn-2-ol	1420 mg/kg [30]	>21,300 mg/m^3 [30]	50 mg/kg of body weight [30]	No animal data or information on potential human carcinogenicity [30]
1,3-Dioxolane	2000 mg/kg [31] 3000 mg/kg [32] 5200 mg/kg [33]	4-h acute LC$_{50}$ is 68.4 mg/L [34]	500 ppm rat inhalation [33]	Not likely to be carcinogenic [33]; dioxolane did not induce chromosomal aberrations; not mutagenic [35]
Butoxymethyl oxirane (*n*-butyl glycidyl ether)	"2000 mg/kg" [36] 2050 mg/kg [37]	670 ppm [38]	25 ppm rat inhalation; 6 h/day, 5 day/week; 13 weeks [39]	*n*-Butyl glycidyl ether was mutagenic *in vitro* in various bacterial and mammalian cells and produced chromosomal aberrations in an *in vitro* assay [40]

continued

TABLE 6.24   (continued)
Relative Risk Thresholds of Solvent Stabilizers Compounds

Stabilizer	Oral Rat LD_{50}	Inhalation LC_{50}	NOAEL	Carcinogenicity, Mutagenicity, and Genotoxicity
1,4-Dioxane	2500 mg/kg [41]	1000 ppm for 4 h [42]	20 ppm human inhalation, 2 h [3]	IARC group 2B: Possibly carcinogenic to humans; IRIS group B2: Probable human carcinogen [43]
Isopropyl acetate	3000 mg/kg [2]	200 ppm TCLo human [44]	—	—
tert-Butyl alcohol	3500 mg/kg [45]	LD_{50} 14,100 ppm for 4 h [46]	Rats (oral, drinking water); none established [46]	A4: Not classifiable as a human carcinogen—data on which to classify the agent in terms of its carcinogenicity in humans are inadequate [16]. Two carcinogenicity studies provide some evidence for carcinogenicity of *tert*-butyl alcohol in rats and mice. Not mutagenic or genotoxic [46]
Ethyl acetate	5600 mg/kg [47]	4000 ppm for 4 h [48] 1600 ppm for 8 h [16]	—	—
sec-Butyl alcohol	6480 mg/kg [49]	16,000 ppm rat LCLo [50]	—	—

Sources: References (shown in table by numbers in brackets) are as follows: [1] Verschueren (1983); [2] Lewis (1996); [3] Quast et al. (1979); [4] USEPA (2006c); [5] ACGIH (2005); [6] IARC (1976); [7] Rom (1992); [8] NIEHS (2002); [9] ACGIH (1994); [10] O'Neil (2001); [11] ECB (2000); [12] Ryan et al. (2001); [13] Waters Corporation (2007); [14] ITII (1988); [15] Lynch et al. (1990); [16] NIOSH (1987); [17] National Toxicology Program (1988); [18] ECB (2001); [19] OEHHA (2000); [20] UCB (2006); [21] Health Council of the Netherlands (2004); [22] Health Council of the Netherlands (2004); [23] Machle et al. (1940); [24] NRC (1996); [25] NIH (1997); [26] CalEPA (2007d); [27] Clayton and Clayton (1982); [28] Scala and Burtis (1973); [29] Dow Chemical Company (1992); [30] ECB (1998); [31] Hoechst AG (1987); [32] Smyth et al. (1949); [33] Dioxolane Manufacturers Consortium (2000a); [34] Dioxolane Manufacturers Consortium (2000b, 2000c); [35] USEPA (2007f); [36] Dow Chemical Company (1972); [37] Smyth et al. (1962); [38] Hine et al. (1956); [39] Miller and Quast (1984); [40] USEPA (2007d); [41] JBRC (1998); [42] Drew et al. (1978); [43] Ernstgård et al. (2006); [44] von Oettingen (1960); [45] ACGIH (1991); [46] USEPA (2007e); [47] Clayton and Clayton (1993); [48] Snyder (1992); [49] Windholz et al. (1983); and [50] NIOSH (1992).

Notes: IARC, International Agency for Research on Cancer; IRIS, USEPA's Integrated Risk Information System; LCLo, lowest published lethal concentration. See Table 5.2 for a complete listing of toxicity thresholds for 1,4-dioxane; LC_{50}, median lethal concentration, that is, the concentration required to kill half the members of a tested population within the specified time period; LD_{50}, median lethal dose, that is, the dose required to kill half the members of a tested population; NOAEL, No Observable Adverse Effect Level, that is, the level of exposure at which an organism in the exposed population experiences no increase in adverse effects; TCLo, lowest published toxic concentration.

ACKNOWLEDGMENT

The critical review, questions, corrections, and comments from Julie Stickney, Syracuse Research Corporation, are greatly appreciated. The author retains responsibility for the interpretation and accuracy of the information provided in this chapter.

BIBLIOGRAPHY

Abe, A., 1997, Determination method for 1,4-dioxane in water samples by solid phase extraction GC/MS. *Journal of Environmental Chemistry* 7: 96–100.

Abe, A., 1999, Distribution of 1,4-dioxane in relation to possible sources in the water environment. *The Science of the Total Environment* 227: 41–48.

ACGIH, 1991, *Documentation of the Threshold Limit Values and Biological Exposure Indices*, 6th Edition, Vol. I, II, III. Cincinnati: American Conference of Governmental Industrial Hygienists.

ACGIH, 1994, *Threshold Limit Values for Chemical Substances and Physical Agents and Biological Exposure Indices for 1994–1995*. Cincinnati: American Conference of Governmental Industrial Hygienists.

ACGIH, 2005, *Threshold Limit Values for Chemical Substances and Physical Agents and Biological Exposure Indices for 2005*. Cincinnati: American Conference of Governmental Industrial Hygienists.

ADEC, 2007, Public review: Draft cleanup levels guidance. Alaska Department of Environmental Conservation, Division of Spill Prevention and Response Contaminated Sites Program.

AFCEE, 2007, 1,4-Dioxane—a primer for air force remedial program managers and risk assessors. Air Force Center for Environmental Excellence.

Amoore, J.E. and Hautala, E., 1983, Odor as an aid to chemical safety: Odor thresholds compared with threshold limit values and volatilities for 214 industrial chemicals in air and water dilution. *Journal of Applied Toxicology* 3(6): 272–290.

Argus, M.F., Sohal, R.S., Bryant, G.M., Hoch-Ligeti, C., and Arcos, J.C., 1973, Dose–response and ultrastructural alterations in dioxane carcinogenesis: Influence of methylcholanthrene on acute toxicity. *European Journal of Cancer* 9: 237–243.

ATSDR, 1992, Public health assessment: Oak Grove Sanitary Landfill, Oak Grove Township, Anoka County, Minnesota [Recommended Allowable Limits (RALs) for drinking water contaminants]. Centers for Disease Control and Prevention (CDC), Agency for Toxic Substances and Disease Registry.

ATSDR, 2004, Draft toxicological profile for 1,4-dioxane. Centers for Disease Control and Prevention (CDC), Agency for Toxic Substance Disease Registry.

ATSDR, 2005a, Public health assessment for Mohonk Road Industrial Plant, Marbletown, Ulster County, New York. Centers for Disease Control and Prevention (CDC), Agency for Toxic Substances and Disease Registry.

ATSDR, 2005b, Health consultation—Interplastic Corporation: Review of emission testing data, City of Minneapolis, Hennepin County, Minnesota. Centers for Disease Control and Prevention (CDC), Agency for Toxic Substances and Disease Registry.

ATSDR, 2005c, Health consultation—public health implications of 1,4-dioxane-contaminated groundwater at the Durham Meadows site, Durham, Connecticut, EPA Facility ID: CTD001452093. Centers for Disease Control and Prevention (CDC), Agency for Toxic Substances and Disease Registry.

ATSDR, 2007, Public health assessment: Solvents Recovery Services of New England, Southington, Hartford County, Connecticut. Centers for Disease Control and Prevention (CDC), Agency for Toxic Substances and Disease Registry.

Bagnell, P.C. and Ellenberger, H.A., 1977, Obstructive jaundice due to a chlorinated hydrocarbon in breast milk. *Canadian Medical Association Journal* 117(9): 1047–1048.

BASF, 1979, Bericht über die Prüfung von 1,4-Dioxan (stabilisiert mit 25 ppm 2,6-Di-*tert*-butyl-*p*-kresol) im Ames-Test [Report of the testing of 1,4-dioxane stabilized with 25 ppm 2,6-di-*tert*-butyl-*p*-cresol in the Ames Test]. Unpublished report (79/135) of BASF, Gewerbehygiene und Toxikologie, dated February 4, 1979 [cited in European Union risk assessment report (European Chemicals Bureau, 2002)].

Bayer, C.W. and Papanicolopoulos, C.D., 1990, Exposure assessments to volatile organic compound emissions from textile products. In *Proceedings of the 5th International Conference on Indoor Air Quality and Climate*, Toronto, Canada, July 29 to August 3, Vol. 3, pp. 725–730.

Bellflower/Norwalk/Park Water Company, 2005, Annual water quality report, 2004/2005.

Bringmann, G., 1973, Determination of the adverse biological effects of water contaminants from the inhibition of glucose assimilation in *Pseudomonas fluorescens* [in German]. *Gesundheits-Ingenieur* [*Sanitation Engineer*] 94(12): 366–369.

Bringmann, G., 1978, Determination of the harmful biological action of water contaminants on protozoa: I. Bacteria fed flagellates. *Zeitschrift für Wasser- und Abwasser-Forschung* [*Journal of Water and Wastewater Research*] 11: 210–215.

Bringmann, G. and Kühn, R., 1976, Vergleichende Befunde der Schadwirkung wassergefahrdender Stoffe gegen Bakterien (*Pseudomonas putida*) und Blaualgen (*Microcystis aeruginosa*) [Comparative results of the adverse effects of aquatic toxins on bacteria (*Pseudomonas putida*) and blue algae (*Microcystis aeruginosa*)]. *GWF-Wasser/Abwasser* 117: 410–413.

Bringmann, G. and Kühn, R., 1977a, Findings of the adverse effects of water contaminants on *Daphnia magna* [in German]. *Zeitschrift für Wasser- und Abwasser-Forschung* [*Journal of Water and Wastewater Research*] 10: 161–166.

Bringmann, G. and Kühn, R., 1977b, Limiting values of the harmful action of water contaminants on bacteria (*Pseudomonas putida*) and green algae (*Scenedesmus quadricauda*) in the cell multiplication inhibition test. *Zeitschrift für Wasser- und Abwasser-Forschung* [*Journal of Water and Wastewater Research*] 10: 87–98.

Bringmann, G. and Kühn, R., 1980, Determination of the harmful biological action of water contaminants on protozoa: II bacteria fed ciliates. *Zeitschrift für Wasser- und Abwasser-Forschung* [*Journal of Water and Wastewater Research*] 13: 26–31.

Bringmann, G. and Kühn, R., 1982, Results of the adverse effects of water contaminants against *Daphnia magna* in an improved standardized test procedure [in German]. *Zeitschrift für Wasser- und Abwasser-Forschung* [*Journal of Water and Wastewater Research*] 15: 1–6.

Bringmann, G., Kühn, R., and Winter, A., 1980, Determination of the harmful biological action of water contaminants on protozoa: III Saprozoic flagellates. *Zeitschrift für Wasser- und Abwasser-Forschung* [*Journal of Water and Wastewater Research*] 13: 170–173.

Bronaugh, R.L., 1982, Percutaneous absorption of cosmetic ingredients. In P. Frost (Ed.), *Principles of Cosmetics for the Dermatologist*, pp. 277–284. St. Louis: The C.V. Mosby Company.

Bronaugh, R.L. and Stewart, R.F., 1986, Methods for *in vitro* percutaneous absorption studies. VI: Preparation of the barrier layer. *Journal of Pharmaceutical Sciences* 75(5): 487–491.

BUA, 1991, 1,4-Dioxane BUA report 80. German Chemical Society, Committee on Existing Chemicals of Environmental Relevance, *Beratergremium für Umweltrelevante Altstoffe* (BUA) [Advisory Committee for Environmentally Relevant Materials] S. Hirzel Wissenschaftliche Verlagsgesellschaft.

Cal EPA, 1998, 1,4-Dioxane action level. California Environmental Protection Agency, Office of Environmental Health Hazard Assessment.

Cal EPA, 1999, Air toxics hot spots program risk assessment guidelines, part I: The determination of acute reference exposure levels for airborne toxicants. California Environmental Protection Agency, Office of Environmental Health Hazard Assessment.

Cal EPA, 2002a, Los Angeles Regional Water Quality Control Board, Order No. R4-2002-0125: Waste discharge requirements for treated groundwater and other wastewaters from investigation and/or cleanup of petroleum fuel–contaminated sites to surface waters in coastal watersheds of Los Angeles and Ventura Counties (General NPDES Permit No. CAG834001).

Cal EPA, 2002b, California Regional Water Quality Control Board, Central Valley Region, Order No. R5-2002-0128 and NPDES No. Ca0083861: Waste discharge requirements for Aerojet-General Corporation interim groundwater extraction and treatment system, American River Study Area and GET E/F, Sacramento County.

Cal EPA, 2004a, San Francisco Bay Regional Water Quality Control Board, Response to comments: Reissuance of general waste discharge requirements for discharge or reuse of extracted and treated groundwater resulting from the cleanup of groundwater polluted by volatile organic compounds, NPDES Permit No. CAG912003.

Cal EPA, 2004b, San Francisco Bay Regional Water Quality Control Board Order R2-2004-0032 for final site cleanup requirements for United Technologies Corporation, San Jose, Santa Clara County (California).

Cal EPA, 2007a, Safe Drinking Water and Toxic Enforcement Act of 1986. California Proposition 65. California Environmental Protection Agency, Office of Environmental Health Hazard Assessment. http://www.oehha.ca.gov/Prop65/background/p65plain.html (accessed December 1, 2007).

Cal EPA, 2007b, Proposition 65 safe harbor levels. California Environmental Protection Agency, Office of Environmental Health Hazard Assessment.

Cal EPA, 2007c, Toxicity criteria database. California Environmental Protection Agency, Office of Environmental Health Hazard Assessment.

Cal EPA, 2007d, No Significant Risk Level (NSRL) for the Proposition 65 carcinogen nitromethane. California Environmental Protection Agency, Office of Environmental Health Hazard Assessment.

CARB, 2008, California toxics inventory database. California Air Resources Board. http://www.arb.ca.gov/ toxics/cti/cti.htm (accessed January 11, 2008).

CDEP, 2002, Hazardous air pollutants, Regulation 22a-174-29. Enfield, CT: Connecticut Department of Environmental Protection.

CDPH, 2007a, Drinking water notification levels and response levels: An overview. California Department of Public Health, Drinking Water Program.

CDPH, 2007b, Groundwater recharge reuse draft regulation. Title 22, California Code of Regulations, Division 4, Environmental Health, Chapter 3, Recycling Criteria. California Department of Public Health.

CDPHE, 2004, Rebuttal testimony of Raj Goyal, Ph.D. Re: Proposed revisions to the basic standards and methodologies for surface water, Regulation #31 (5CCR 1002-31), the basic standards for groundwater, Regulation #41 (5CCR 1002-41), and the site-specific water quality classifications and standards for groundwater, Regulation #42 (5CCR 1002-42). Colorado Department of Public Health and Environment.

CDPHE, 2005, The basic standards for ground water. Colorado Code of Regulations, 5 CCR 1002-41. Colorado Department of Public Health and Environment.

Chemical Week, 1978, Jersey fights smog. *Chemical Week* 1978(15): 24.

Chute, S.K. and Ginsberg, G., 2004, Comparison value determination for 1,4-dioxane in drinking water. Connecticut Department of Public Health. http://www.ct.gov/dph/ (accessed November 11, 2006).

Chute, S.K. and Ginsberg, G., 2005, A comparison value for 1,4-dioxane in drinking water. Paper presented at New England Private Drinking Water Well Symposium. Portsmouth, NH, November 14, 2005. Cooperative State Research Education and Extension Service. http://www.usawaterquality.org/ (accessed November 11, 2006).

CITI-Japan, 1992, Biodegradation and bioaccumulation data of existing chemicals based on CSCL Japan. Chemicals Inspection and Testing Institute [cited in European Chemicals Bureau, 2002].

City of Ann Arbor, 2001, Regular city testing locates small amount of 1,4-dioxane: Use of a city well discontinued pending investigation. City of Ann Arbor, MI: Water Utilities Department.

City of Ann Arbor, 2004, Frequently asked questions: 1,4-Dioxane and Pall Life Sciences (Gelman). Ann Arbor, Michigan. http://www.a2gov.org/government/publicservices/systems_planning/environment/pls/ Pages/faq.aspx (accessed December 26, 2007).

City of Banning, 2005, 2005 Annual water quality report. Banning, CA: Public Utilities Department, Water Division.

City of Clare, 2007, Annual drinking water quality report. Clare, MI: Municipal Water System.

City of Fountain Valley, 2006, The 2006 water quality report. Fountain Valley, CA: Field Services and Water Department.

City of Santa Monica, 2002, Discussion of groundwater contamination at the city's Olympic Well Field. Santa Monica, CA: City Council Information Item, Utility Division.

Clayton, G.D. and Clayton, F.E. (Eds), 1982, *Patty's Industrial Hygiene and Toxicology*, 3rd Edition, Vol. 2A, 2B, 2C. *Toxicology*, New York: John Wiley.

Clayton, G.D. and Clayton, F.E. (Eds), 1993, *Patty's Industrial Hygiene and Toxicology*, 4th Edition, Vol. 2A, 2B, 2C, 2D, 2E, 2F. *Toxicology*, New York: Wiley.

Cohrssen, J.J. and Covello, V.T., 1989, Risk analysis: A guide to principles and methods for analyzing health and environmental risks. Washington, DC: Council on Environmental Quality.

Crandall, M.S. and Wilcox, T.G., 1987, *Graphic Packaging Corporation, Paoli, Pennsylvania*. Atlanta, GA: National Institute for Occupational Safety and Health Report No. HETA-87-181-0000.

CSC (Campaign for Safe Cosmetics), 2007, Cancer-causing chemical found in children's bath products; women's shampoos and body wash also contaminated. Campaign for Safe Cosmetics. http://www.safecosmetics.org/newsroom/press.cfm?pressReleaseID=21 (accessed February 8, 2007).

CWSC, 2004, East Los Angeles District 2004 water quality report for the City of Commerce water system. California Water Service Company.

Danish EPA, 2005, Screening for health effects from chemical substances in textile colorants. Survey of Chemical Substances in Consumer Products no. 57, 2005. English translation on Miljøministeriet website: http://www.mim.dk/ (accessed January 2, 2007).

DAP Inc., 1994, Material Safety Data Sheet for Weldwood Contact Cement, issued January 4, 1994, Dayton, Ohio. FSC: 8040 LIIN: 00N056641.

Dawson, G.W. and Jennings, A.L., 1977, The acute toxicity of 47 industrial chemicals to fresh and saltwater fishes. *Journal of Hazardous Materials* 1: 303–318.

deMan, J.M., 1999, *Principles of Food Chemistry*, 3rd Edition. Gaithersburg, MD: Aspen Publishers Inc.

Dioxolane Manufacturers Consortium, 2000a, 1,3-Dioxolane. USEPA High Production Volume Challenge Program Submission.

Dioxolane Manufacturers Consortium, 2000b, An acute inhalation toxicity study of C-121 [1,3-Dioxolane] in the rat. Project No. 79-7304, Bio/dynamics Inc., July 11, 1980 [cited in Dioxolane Manufacturers Consortium, 2000a].

Dioxolane Manufacturers Consortium, 2000c, 1,3-Dioxolane: A two-week vapor inhalation study in Fischer 344 rats. The Dow Chemical Company, Toxicology Research Laboratory, Health and Environmental Sciences, Laboratory Project Study ID K-010634-005, August 28, 1989 [cited in Dioxolane Manufacturers Consortium, 2000a].

Donaldson, W.T., 1977, Trace organics: Refined methods are needed to document their occurrence and concentrations. *Environmental Science and Technology* 11(4): 348–351.

Dow Chemical Company, 1972, Toxicological properties and industrial handling hazards of 1-tertiary-butoxy-2,3-epoxy propane. Study T36.1-59376-2.

Dow Chemical Company, 1989a, Toxic Substances Control Act test submission, October 17th, 1989. OTS0000719, Dow Chemical Company, D004057, 0158-0179. (cited in *Environment Canada*, 2008, Substance profile for the challenge: 1,4-Dioxane: http://www.ec.gc.ca/substances/ese/eng/challenge/batch7/batch7_123-91-1.cfm [accessed November 11, 2008]).

Dow Chemical Company, 1989b, Summaries of environmental data for phenol, 1,4-dioxane and acrylonitrile with attached reports and cover letter. U.S. Environmental Protection Agency, Office of Science and Technology.

Dow Chemical Company, 1992, Initial submission: Tertiary amyl alcohol: Subchronic toxicity and pharmacokinetics in CD-1 mice, Fischer 344 rats and male beagle dogs with cover letter dated 04/30/92; 04/15/81. EPA No. 88-920002262; Fiche No. OTS0539279.

Drew, R.T., Patel, J.M., and Lin, F.N., 1978, Changes in serum enzymes in rats after inhalation of organic solvents singly and in combination. *Toxicology and Applied Pharmacology* 45(3): 809–819.

ECSCF, 2002a, Opinion of the scientific committee on food on impurities of ethylene oxide in food additives. European Commission, Scientific Committee on Food.

ECSCF, 2002b, Impurities of 1,4-dioxane, 2-chloroethanol and mono- and diethylene glycol in currently permitted food additives and in proposed use of ethyl hydroxyethyl cellulose in gluten-free bread. European Commission, Scientific Committee on Food, Health and Consumer Protection Directorate.

EMEA, 2006, Note for guidance on impurities: Residual solvents. *International Conference on Harmonisation (ICH) of Technical Requirements for Registration of Pharmaceuticals for Human Use.* 24. London, UK: European Medicines Agency.

Ernstgård, L., Iregren, A., Sjögren, B., and Johanson, G., 2006, Acute effects of exposure to vapours of dioxane in humans. *Human & Experimental Toxicology* 25: 723–729.

European Chemicals Bureau (ECB), 1998, 2-Methylbut-3-yn-2-ol screening information dataset (SIDS) initial assessment report for the 8th SIDS Initial Assessment Meeting (Paris, 28–30 October 1998). United Nations Environment Programme.

European Chemicals Bureau (ECB), 1999, Risk assessment report. 1,4-Dioxane. Joint Research Center, European Chemicals Bureau.

European Chemicals Bureau (ECB), 2001, International uniform chemical information database data set for 1,2-epoxybutane. European Organisation for Economic Co-operation and Deve High Production Volume Chemicals Programme, SIDS Dossier, approved at SIAM 11, January 23–26.

European Chemicals Bureau (ECB), 2002, European Union risk assessment report: 1,4-Dioxane. Institute for Health and Consumer Protection, European Union.

EWG, 1996, Dishonorable discharge: Toxic pollution of America's waters. Environmental Working Group. http://www.ewg.org/book/export/html/7623 (accessed November 11, 2006).

Fischer, M., 1998, Teppichböden müssen nicht schuld an Innenraumbelastungen sein [Carpeted floors are not always the reason for indoor air contamination]. *Umwelt & Gesundheit* [*Environment & Health*], 1998(1): 6–13. http://www.iug-umwelt-gesundheit.de/uundg.php (accessed October 5, 2005).

Fisher, J., Mahle, D., Bankston, L., Greene, R., and Gearhart, J., 1997, Lactational transfer of volatile chemicals in breast milk. *American Industrial Hygiene Association Journal* 58(6): 425–431.

Food and Drug Administration (FDA), 1981, Progress report on the analysis of cosmetics raw materials and finished cosmetics products for 1,4-dioxane. Division of Cosmetics Technology, 15.

Food and Drug Administration (FDA), 1992, *Cosmetic Handbook for Industry.* Center for Food Safety and Applied Nutrition, College Park, MD.

Food and Drug Administration (FDA), 1997, *Guidance for Industry—Q3C Impurities: Residual Solvents.* Center for Drug Evaluation and Research, College Park, MD.

Food and Drug Administration (FDA), 1999a, Food additives permitted for direct addition to food for human consumption: Polysorbate 60. Department of Health and Human Services. 21 CFR Part 172 57,974–57,976.

Food and Drug Administration (FDA), 1999b, Indirect food additives: Polymers. *Federal Register* 64(208): 57,976–57,978.

Food and Drug Administration (FDA), 2001, Guidance for industry: Impurities: Residual solvents in new veterinary medicinal products, active substances and excipients VICH GL18. U.S. Department of Health and Human Services, Center for Veterinary Medicine.

Food and Drug Administration (FDA), 2003, Guidance for industry Q3C—tables and list. U.S. Department of Health and Human Services, Center for Drug Evaluation and Research, Center for Biologics Evaluation and Research.

Food and Drug Administration (FDA), 2006, Food additives permitted for direct addition to food for human consumption: Glycerides and polyglycides. 21 CFR Part 172. *Federal Register* 71: 12618–12621.

Food and Drug Administration (FDA), 2007, 1,4-Dioxane. http://www.cfsan.fda.gov/~dms/cos-diox.html (accessed July 26, 2007).

Fuh, C.B., Lai, M., Tsai, H.Y., and Chang, C.M., 2005, Impurity analysis of 1,4-dioxane in nonionic surfactants and cosmetics using headspace solid-phase microextraction. *Journal of Chromatography A* 1071(1): 141–145.

Geiger, D.L., Brooke, L.T., and Call, D.J. (Eds), 1990, *Acute Toxicities of Organic Chemicals to Fathead Minnows (Pimephales Promelas)*, Vol. 5. Superior, WI: University of Wisconsin, Center for Lake Superior Environmental Studies.

Giardino, N.J., Esmen, N.A., and Andelman, J.B., 1992, Modeling volatilization of trichloroethylene from a domestic shower spray: The role of drop-size distribution. *Environmental Science and Technology* 26(8): 1602–1606.

Gosselin, R., Smith, R., and Hodge, H., 1984, *Clinical Toxicology of Commercial Products.* Baltimore, MD: Williams and Wilkins.

Greim, H., 2000, 1,4-Dioxan. *Gesundheitsschädliche Arbeitsstoffe, Toxikologisch-arbeitsmedizinische Begründungen von MAK-Werten [1,4-Dioxane. Workplace Health Hazards from Chemicals: Toxicological and Occupational Health Rationale for Maximum Workplace Concentration Values].* Weinheim, Germany: VCH Verlag.

Hansen, O.C., 2005, Screening for health effects from chemical substances in textile colorants. In *Survey of Chemical Substances in Consumer Products.* Danish Ministry of the Environment, Danish Technological Institute, Report No. 57. http://miljoestyrelsen.dk/homepage/ (accessed October 9, 2006).

Harrison, N., 2001, *Food Chemical Safety*, Vol. 1: Contaminants. Boca Raton: CRC Press.

HDOH, 2005, Screening for environmental concerns at sites with contaminated soil and groundwater. Hawai'i Department of Health, Environmental Management Division.

Health Canada, 2005, Advisory Council on Drinking Water Quality and Testing Standards, Minutes from October 21, 2005 Meeting. www.odwac.gov.on.ca/minutes/2005/102105%20ODWAC%20Minutes.pdf (accessed November 11, 2007).

Health Council of the Netherlands, 2004, *Isopropylamine: Health-Based Reassessment of Administrative Occupational Exposure Limits.* The Hague, Netherlands: Committee on Updating of Occupational Exposure Limits, No. 2000/15OSH/122.

Hertlein, F., 1980, Monitoring airborne contaminants in chemical laboratories. Analytical Techniques in Occupational Health Chemistry, ACS Symposium Series No. 120: 215–230.

Hills, B., Klincewicz, S., Blade, L.M., and Sack, D., 1989, *Health Hazard Evaluation (HHE) report for BMY Corporation, a Division of Harsco Corporation, York, Pennsylvania.* Atlanta, GA: National Institute for Occupational Safety and Health Report No. HETA 87-367-1987.

Hine, C.H., Kodama, J.K., Wellington, J.S., Dunlap, M.K., and Anderson, H.H., 1956, The toxicology of glycidol and some glycidyl ethers. *American Medical Association Archives of Industrial Health* 1956(14): 250–264.

Hoch-Ligeti, C. and Argus, M.F., 1970, *Effects of Carcinogens on the Lung of Guinea Pigs: Morphology of Experimental Respiratory Carcinogenesis.* Oak Ridge National Laboratory, TN: U.S. Atomic Energy Commission Symposium Series.

Hodgson, A.T. and Levin, H., 2003, *Volatile Organic Compounds in Indoor Air: A Review of Concentrations Measured in North America Since 1990.* Berkeley, CA: Lawrence Berkeley National Laboratory Report LBNL-51715.

Hoechst, A.G., 1987, 1,3-Dioxolan—Profung der akuten oralen Toxizitatan der Wistar Ratte [1,3-Dioxolane acute oral toxicity testing of the Wistar rat]. Hoechst AG Pharma Research Toxicology Report No. 87.1441, September 1987 [cited in Dioxolane Manufacturers Consortium, 2000a].

Huntley, S., Amaral, M., and Schell, J., 2004, Toxicokinetics of 1,4-dioxane and implications for assessing hazards to terrestrial wildlife. Paper presented at the Society of Environmental Toxicology and Chemistry, 25th Annual Meeting, Portland, OR, November 14–18.

IARC, 1976, IARC *Monographs On The Evaluation Of The Carcinogenic Risk Of Chemicals To Man: Cadmium, Nickel, Some Epoxides, Miscellaneous Industrial Chemicals And General Considerations On Volatile Anaesthetics*, Vol. 11. Lyon, France: World Health Organization, International Agency for Research on Cancer.

IEPA, 2007, NPDES permit no. IL0063479 draft reissued NPDES permit to discharge into waters of the state. Illinois Environmental Protection Agency.

ILO, 2007, Chemical exposure limits. International Occupational Safety and Health Information Centre. http://www.ilo.org/public/english/protection/safework/cis/products/explim.htm#swe (accessed November 11, 2007).

INRS, 2006, Valeurs limites d'exposition professionnelle aux agents chimiques en France, Institut National de Recherche et de Securite [Exposure limit values for chemical agents in the workplace (in French), National Institute of Research and Safety]. http://www.inrs.fr/htm/valeurs_limites_exposition_professionnelle_agents.html (accessed March 11, 2007).

Inside EPA, 2004, 1,4-Dioxane: New state standard for Colorado: Colorado sets first-time water standard for widespread chemical. *Inside EPA* 25(39): 1–2.

IRWD, 2007, The 2007 water quality report: Irvine Ranch Water District, California.

ITII, 1988, *Toxic and Hazardous Industrial Chemicals Safety Manual.* Tokyo: The International Technical Information Institute.

Japanese Ministry of Health, Labour, and Welfare, 2003, Revision of drinking water quality standards and QA/QC for drinking water quality analysis in Japan. Office of Drinking Water Quality Management.

JBRC, 1998, *Thirteen-Week Studies of 1,4-Dioxane in F344 Rats and BDF1 Mice (Drinking Water Studies).* Kanagawa, Japan: Japan Bioassay Research Center.

JSOH, 2007, Recommendation of occupational exposure limits. *Journal of Occupational Health* 49: 328–344.

Juhnke, I. and Ludemann, D., 1978, Ergebnisse der Untersuchung von 200 chemischen Verbindungen auf akute Fischtoxizitat mit dem Goldorfentest [Results of the investigation of 200 chemical compounds on acute fish toxicity with the Golden Orfe Test]. *Zeitschrift für Wasser- und Abwasser-Forschung* [*Journal of Water and Wastewater Research*] 11: 161–164.

Kamrin, M.A., 1988, *Toxicology: A Primer on Toxicology Principles and Applications.* Chelsea, MI: Lewis Publishers.

Kawata, K., Ibaraki, T., Tanabe, A., and Yasuhara, A., 2003, Distribution of 1,4-dioxane and *N,N*-dimethylformamide in river water from Niigata, Japan. *Bulletin of Environmental Contamination and Toxicology* 70(5): 876–882.

King, M.E., Shefner, A.M., and Bates, R.R., 1973, Carcinogenesis bioassay of chlorinated dibenzodioxins and related chemicals. *Environmental Health Perspectives* 5: 163–170.

Klockenbrink, M., 1987, How to read a label. New York: *New York Times* (September 27, 1987).

Kociba, R.J., McCollister, S.B., Park, C., Torkelson, T.R., and Gehring, P.J., 1974, 1,4-Dioxane: I. Results of a 2-year ingestion study in rats. *Toxicology and Applied Pharmacology* 30: 275–286.

Krake, A.M. and Herrera-Moreno, V., 1995, *Automotive Controls Corporation, Independence, Kansas.* Atlanta, GA: National Institute for Occupational Safety and Health Report No. HETA-95-0296-2547.

Kramer, V.C., Schnell, D.J., and Nickerson, K.W., 1983, Relative toxicity of organic solvents to *Aedes aegypti* larvae. *Journal of Invertebrate Pathology* 42: 285–287.

Krzymien, M.E., 1989, GC-MS analysis of organic vapors emitted from polyurethane foam insulation. *International Journal of Environmental Analytical Chemistry* 36(4): 193–207.

Kullman, G.J., 1993, *Meyersdale Manufacturing Co., Meyersdale, Pennsylvania.* Atlanta, GA: National Institute for Occupational Safety and Health Report No. MHETA-88-269-1993.

Kyle, A.D., Wright, C.C., Caldwell, J.C., Buffler, P.A., and Woodruff, T.J., 2001, Evaluating the health significance of hazardous air pollutants using monitoring data. *Public Health Reports* 116: 32–44.

Labrèche, F.P. and Goldberg, M.S., 1997, Exposure to organic solvents and breast cancer in women: A hypothesis. *American Journal of Industrial Medicine* 32(1): 1–14.

Lewis, R.J., 1996, *Sax's Dangerous Properties of Industrial Materials*, 9th Edition, Vols 1–3. New York: Van Nostrand Reinhold.

Lynch, D.W., Moorman, W.J., Lewis, T.R., Stober, P., Hamlin, R., and Schueler, R.L., 1990, Subchronic inhalation of triethylamine vapor in Fisher-344 rats: Organ system toxicity. *Toxicology and Industrial Health* 6(3/4): 403–414.

Machle, W., Scott, E.W., and Treon, J., 1940, The physiological response of animals to some simple mononi-troparaffins and to certain derivatives of these compounds. *Journal of Industrial Hygiene and Toxicology* 22: 315–332.

Macler, B., 2006, Regulating emerging contaminants . . . . What can you expect? Paper presented at Groundwater Resources Association of California symposium Emerging Contaminants in Groundwater—A Continually Moving Target, Concord, CA, June 7–8. http://www.grac.org (accessed August 30, 2006).

MADEP, 2007, Standards and guidelines for contaminants in Massachusetts drinking waters. Massachusetts Department of Environmental Protection, Department of Environmental Services Drinking Water Program.

Marzulli, F.N., 1962, Barriers to skin penetration. *Journal of Investigative Dermatology* 39: 397–393.

McDonnell Douglas Corporation, 1986, Materials Testing Data Base. Software Technology Development Laboratory, Houston, Texas. [Data reported in A.T. Hodgson and J.D. Wooley, 1991, Assessment of indoor concentrations, indoor sources and source emissions of selected volatile organic compounds. Prepared for Research Division, California Air Resources Board by Indoor Environmental Program, Lawrence Berkeley Laboratory, University of California, Berkeley.]

MCWD, 2005, *The 2005 Water Quality Report.* Costa Mesa, CA: Mesa Consolidated Water District.

MDEQ, 2003, Briefing paper: Drinking water standard for 1,4-dioxane: Memo to Director Steven E. Chester by Dennis Bush, toxicologist. Michigan Department of Environmental Quality, Surface Water Quality Assessment Section.

MDEQ, 2005a, Information bulletin: Gelman Sciences Inc. site. Michigan Department of Environmental Quality. http://www.deq.state.mi.us/documents/deq-rrd-GS-InformationBulletinMarch2004.pdf (accessed May 1, 2007).

MDEQ, 2005b, Permit no. MI0048453 issued to Pall Life Sciences Inc.: Authorization to discharge under the National Pollutant Discharge Elimination System. Michigan Department of Environmental Quality.

MDEQ, 2007, Gelman Sciences, Inc. site of contamination information page. Michigan Department of Environmental Quality. http://www.michigan.gov/deq (accessed May 1, 2007).

MDH, 2000, Health consultation: Pechiney Plastic Packaging Inc., Minneapolis, Hennepin County, Minnesota. Minnesota Department of Health. www.health.state.mn.us/divs/eh/hazardous/sites/hennepin/pchney1100.pdf (accessed May 1, 2007).

Miller, R.R. and Quast, J.F., 1984, *t-Butyl Glycidyl Ether (TBGE): 90-Day Inhalation Study in Rats, Mice and Rabbits.* Midland, MI: Dow Chemical USA, Health & Environmental Sciences, Toxicology Research Laboratory, Study K-59376-(7).

Mississippi-DEQ, 2002, Final brownfields regulations: SubPart II. Mississippi Department of Environmental Quality.

MMA, 2004, *Michigan Manufacturing Association: Comments on Proposed Rule 2004-011 EQ1.* Lansing, MI: Michigan Department of Environmental Quality.

Mohr, T.K.G., Tulloch, C., Chan, S., Cook, G., Giri, S., Broughton, P., and Crowley, J., 2007, *Study of Potential for Groundwater Contamination from Past Dry Cleaner Operations in Santa Clara County.* San Jose: Santa Clara Valley Water District. http://www.valleywater.org/Water/Water_Quality/Protecting_your_water/_Solvents/_PDFs/DC_Study_Final__09_20_07_2.pdf

Montana-DEQ, 2002, Voluntary cleanup and redevelopment act application guide. Montana Department of Environmental Quality.

Moran, M.J., Zogorski, J.S., and Squillace, P.J., 2007, Chlorinated solvents in groundwater of the United States. *Environmental Science and Technology* 41(1): 74–81.

MWRD, 1999, Industrial Wastewater Discharge Permit 1-118. City of Aurora Utilities Department, Metro Wastewater Reclamation District.

NAOSH, 2002, Code of Practice for the Safety, Health and Welfare at Work. [Irish] National Authority for Occupational Safety and Health.

National Toxicology Program, 1988, Toxicology and carcinogenesis studies of trichloroethylene (CAS No. 79-01-6) in four strains of rats (ACI, August, Marshall, Osborne-Mendel) (gavage studies). Research Triangle Park, NC: National Institutes of Health Technical Report Series No. 273, Publication No. 86-2529.

NCDENR, 2005, Subchapter 2L—groundwater classifications and standards. North Carolina Administrative Code Title 15a. Department of Environment and Natural Resources.

NCI, 1978, Bioassay of 1,4-dioxane for possible carcinogenicity. Bethesda, MD: National Cancer Institute. National Institutes of Health Publication No. 78-1330 NCICGTR-80.

NDWAC, 2006, Report on the CCL Classification Process to the U.S. Environmental Protection Agency. National Drinking Water Advisory Council.

NHDES, 2004, Proposed risk-based remediation value for 1,4-dioxane (CAS #123-91-1) in groundwater. New Hampshire Department of Environmental Services.

NICNAS, 1998, Priority existing chemical assessment no. 7—1,4-dioxane. Department of Health and Aging, Australian National Industrial Chemicals Notification and Assessment Scheme (NICNAS).

NIEHS, 2002, Inhalation development toxicology studies: Acetonitrile (CAS No. 75-05-8) in rats. National Institute of Environmental Health Sciences, National Toxicology Program, NTP Study No. TER91039 (February 1994). http://ntp-server.niehs.nih.gov/htdocs/pub-TT0.html (accessed May 28, 2006).

NIH, 1997, *NIH Toxicology & Carcinogenesis Studies of Nitromethane in F344/N Rats and B6C3F1 Mice (Inhalation Studies)*. National Institutes of Health, Technical Report Series No. 461, Publication No. 97-3377, North Carolina: Research Triangle Park.

NIOSH, 1987, *Registry of Toxic Effects of Chemical Substances (RTECS)*, Vol. 2. U.S. Department of Health & Human Services, National Institute for Occupational Safety and Health Publication No. 87–114.

NIOSH, 1992, Occupational Safety and Health Guideline for *sec*-butyl alcohol. Division of Standards Development and Technology Transfer. www.cdc.gov/NIOSH/pdfs/0077-rev.pdf (accessed March 29, 2004).

NIOSH, 2005, NIOSH pocket guide to chemical hazards. U.S. Department of Health and Human Services, Centers for Disease Control and Prevention and National Institute for Occupational Safety and Health.

Nishimura, T., Iizuka, S., Kibune, N., and Ando, M., 2004, Study of 1,4-dioxane intake in the total diet using the market-basket method. *Journal of Health Science* 50(1): 101–107.

Nishiushi, Y., 1981, Evaluation of the effect of pesticides on some aquatic organisms. *Setai Kagaku* 4(3): 45–47 [cited in European Chemicals Bureau, 2002].

NMED, 2006, New Mexico and U.S. EPA's groundwater standards. New Mexico Environment Department, New Mexico Groundwater Quality Bureau. http://www.nmenv.state.nm.us/gwb/New_Pages/gw_standards.htm (accessed November 11, 2007).

NRC, 1996, *Spacecraft Maximum Allowable Concentrations for Selected Airborne Contaminants*, Vol. 2. National Research Council, Board on Environmental Studies and Toxicology. Washington, DC: National Academy Press.

NYSDEC, 2006, New York State Brownfield cleanup program: Development of soil cleanup objectives. New York Department of Environmental Conservation.

ODEQ, 2007, Risk-based concentrations. Oregon Department of Environmental Quality, Land Quality Division.

OEHHA, 2000, *Chronic Toxicity Summary 1,2-Epoxybutane [1,2-Butylene Oxide]: Determination of Noncancer Chronic Reference Exposure Levels*. Oakland, CA: California Office of Environmental Health Hazard Assessment.

OEPA, 1994, Biological, sediment, and water quality study of the Tuscarawas River, Wolf Creek, and Hudson Run, Summit and Stark Counties. State of Ohio Environmental Protection Agency, Division of Emergency and Remedial Response.

OEPA, 2004, Use of U.S. EPA Region 9 PRGs as screening values in human health risk assessments. State of Ohio Environmental Protection Agency, Division of Emergency and Remedial Response.

Okawa, M. and Coye, M.J., 1982, *HETA 80-144-1109, Film Processing Industry, Hollywood, California*. Atlanta, GA: National Institute for Occupational Safety and Health.

Oklahoma DEQ, 2003, Region 6 human health medium—specific screening levels. Oklahoma Department of Environmental Quality. http://www.deq.state.ok.us/WQDNew/ (accessed November 12, 2007).

OMOE, 1994, Provincial water quality objectives of the Ministry of Environment and Energy. Ontario Ministry of the Environment. http://www.ene.gov.on.ca/ (accessed December 31, 2007).

OMOE, 2007, Rationale for the development of generic soil and groundwater standards for use at contaminated sites in Ontario. Ontario Ministry of the Environment, Standards Development Branch.

O'Neal Inc., 2007, Dioxane reduction system. Hoechst Celanese. http://www.onealinc.com/process.cfm (accessed October 31, 2007).

O'Neil, M.J. (Ed.) 2001, *The Merck index—an Encyclopedia of Chemicals, Drugs, and Biologicals*. Whitehouse Station, NJ: Merck and Co., Inc.

Organic Consumers Association, 2008, Carcinogenic 1,4-dioxane found in leading "organic" brand personal care products: USDA certified products test dioxane-free, March 14th 2008 press release. http://www.organicconsumers.org/bodycare/DioxaneRelease08.cfm (accessed November 16, 2008).

OSHA, 2007, Occupational safety and health standards. U.S. Department of Labor, Occupational Safety and Health Administration, 1910.000, Table Z-1 (Limits for Air Contaminants).

PADEP, 2004, Applications for new or expanded facility permits. Pennsylvania Department of Environmental Protection. *The Pennsylvania Bulletin—Notices* No. 26:3985, Appendix A, Table 1 (Medium-Specific Concentrations (MSCs) for groundwater in aquifers).

Park, Y.-M., Pyoa, H., Park, S.-J., and Park, S.-K., 2005, Development of the analytical method for 1,4-dioxane in water by liquid-liquid extraction. *Analytica Chimica Acta* 548: 109–115.

Pearl, L., 2007, Report warns of investor risks from toxic chemicals in cosmetics. *Pesticide & Toxic Chemical News* 35(18): 4–17.

Pellizzari, E.D., Hartwell, T.D., Harris, B.S.H., Waddell, R.D., Whitaker, D.A., and Erickson, M.D., 1982, Purgeable organic compounds in mother's milk. *Bulletin of Environmental Contamination and Toxicology* 28: 322–328.

Pellizzari, E.D., Hartwell, T.D., Perritt, R.L., Sparacino, C.M., Sheldon, L.S., Zelon, H.S., Whitmore, R.W., Breen, J.J., and Wallace, L., 1986, Comparison of indoor and outdoor residential levels of volatile organic chemicals in five U.S. geographical areas. *Environment International* 12(6): 619–623.

Quast, J.F., Henck, J.W., Postma, B.J., Schuetz, D.J., and McKenna, M.J., 1979, Epichlorohydrin—subchronic studies: I. A 90-day inhalation study in laboratory rodents. http://www.epa.gov/IRIS/subst/0050.htm (accessed March 10, 2003).

Rimatori, V., Fronduto, M., Staiti, D., and Niu, Q., 1994, Controllo e valutazione dell'esposizione lavorativa di addetti di laboratorio (Monitoring and evaluation of laboratory worker exposure [in Italian]). *Archivio di Scienze del Lavoro* 10(3–4): 481–484 [cited in NICNAS, 1998].

Rom, W.N. (Ed.), 1992, *Environmental and Occupational Medicine*, 2nd Edition. Boston, MA: Little, Brown and Company.

Ryan, B.M., Selby, R., Gingell, R., Waechter, J.M., Jr., Butala, J.H., Dimond, S.S., Dunn, B.J., House, R., and Morrissey, R., 2001, Two-generation reproduction study and immunotoxicity screen in rats dosed with phenol via the drinking water. *International Journal of Toxicology* 20(3): 121–142.

SAIOH, 2007, *Occupational Exposure Levels*. Parktown, South Africa: South African Institute for Occupational Hygiene.

Salisbury, S.A. and Arnold, S., 1984, Niemand Industries, Inc., Statesville, North Carolina. Atlanta, GA: National Institute for Occupational Safety and Health Report No. HETA-84-108-1821.

Scala, R.A. and Burtis, E.G., 1973, Acute toxicity of a homologous series of branched-chain primary alcohols. *American Industrial Hygiene Association Journal* 34(11): 493–499.

SCWA, 2007, 2007 *Annual Drinking Water Quality Report*. NY: Suffolk County Water Authority.

SCWC, 2004, Water quality report, 2004. Bell-Bell Gardens Water System, Southern California Water Company.

SDENR, 2007, New water system planning manual. Department of Environmental Services Drinking Water Program, South Dakota.

Shah, J. and Heyerdahl, E.K., 1988, National ambient volatile organic compounds (VOCs) database update. Research Triangle Park, NC: U.S. Environmental Protection Agency, EPA 600/3-88/010.

Shah, J. and Singh, H.B., 1988, Distribution of volatile organic chemicals in outdoor and indoor air: A national VOCs data base. *Environmental Science and Technology* 22(12): 1381–1388.

Shearer, T.P. and Hunsicker, L.G., 1980, Rapid method for embedding tissues for electron microscopy using 1,4-dioxane and polybed 812. *Journal of Histochemistry and Cytochemistry* 28(5): 465–467.

Sheldon, L.S., Clayton, A., and Jones, B., 1991, Indoor pollutant concentrations and exposures. Final report. Sacramento: California Air Resources Board.

Simazaki, D., Asami, M., Nishimura, T., Kunikane, S., Aizawa, T., and Magara, Y., 2006, Occurrence of 1,4-dioxane and MtBE in drinking water sources in Japan. *Water Science & Technology: Water Supply* 6(2): 47–53.

Sirkar, K.K., Agarwal, N.K., and Rangaiah, G.P., 1978, The effect of short air exposure periods on the performance of cellulose acetate membranes from casting solutions with high cellulose acetate content. *Journal of Applied Polymer Science* 22(7): 1919–1944.

Smith, B., 1986, Identification and reduction of pollution sources in textile wet processing. Raleigh, NC, North Carolina State University, School of Textiles.

Smyth, H.F., Carpenter, C.P., and Weil, C.S., 1949, Range-finding toxicity data, list III. *Journal of Industrial Hygiene and Toxicology* 1949(31): 60–62.

Smyth, H.F., Carpenter, C.P., Weil, C.S., Pozzani, U.C., and Striegel, J.A., 1962, Range-finding toxicity data: List VI. *American Industrial Hygiene Association Journal* 23(2): 95–107.

Snyder, R. (Ed.), 1992, *Ethel Browning's Toxicity and Metabolism of Industrial Solvents*, 2nd Edition, Vol. 3. Alcohols and esters. NY: Elsevier.

Solomon, G. and Weiss, P., 2002, Chemical contaminants in breast milk: Time trends and regional variability. *Environmental Health Perspectives* 110(6): A339–A348.

Springborn Laboratories Inc., 1989, *1,4-Dioxane—Chronic Toxicity to Ceriodaphnia dubia Under Static Renewal Conditions*. Wareham, MA: Springborn Laboratories Inc., SLI Report 89-9-3089.

Steinman, D., 2007, *Safe Trip to Eden: 10 Steps to Save Planet Earth from the Global Warming Meltdown*. NY: Thunder's Mouth Press.

Stickney, J.A., Sager, S.L., Clarkson, J.R., Smith, L.A., Locey, B.J., Bock, M.J., Hartung, R., and Olp, S.F., 2003, An updated evaluation of the carcinogenic potential of 1,4-dioxane. *Regulatory Toxicology and Pharmacology* 38(2003): 183–195.

STM, 2005, *HTP-arvot 2005 (Finnish Occupational Exposure Levels)*. Helsinki, Finland: Ministry of Social Affairs and Health.

SWEA, 2005, Occupational exposure limit values and measures against air contaminants. Swedish Work Environment Authority, Solna, Sweden. www.av.se/dokument/inenglish/legislations/eng0517.pdf (accessed March 18, 2007).

Tanabe, A., Tsuchida, Y., Ibaraki, T., and Kawata, K., 2006, Impact of 1,4-dioxane from domestic effluent on the Agano and Shinano Rivers, Japan. *Bulletin of Environmental Contamination and Toxicology* 76: 44–51.

Taylor, W.A., Jamriska, D.J., Hamilton, V.T., et al., 1995, Waste minimization in the Los Alamos medical radio-isotope program. *Journal of Radioanalytical and Nuclear Chemistry* 195(2): 287–295.

TELL, 1996, Division of Solid Waste Management revises remediation guidance levels. *Tennessee Environmental Law Letter* 8(2): 15–21.

Thomulka, K.W. and McGee, D.J., 1992, Evaluation of organic compounds in water using *Photobacterium phosphoreum* and *Vibrio harveyi* bioassays. *Fresenius Environmental Bulletin* 1: 815–820.

Torkelson, R., Leong, B.K.J., Kociba, R.J., Richter, W.A., and Gehring, P.J., 1974, 1,4 Dioxane. II. Results of a 2-year inhalation study in rats. *Toxicology and Applied Pharmacology* 30: 287–298.

TOSC, 2007, Michigan's Department of Environmental Quality (DEQ) decision concerning 1,4-dioxane. Technical Outreach Services for Communities Program. http://www.egr.msu.edu/tosc/gelman/1,4-dioxane_contaminant_level.doc (accessed December 28, 2007).

Toshinari, S., Tsuyoshi, I., Mihoko, U., Kazuo, Y., and Kumiko, Y., 2005, Monitoring of 1,4-dioxane in groundwater and river water at the Tama District in Tokyo. *Journal of Japan Society on Water Environment* 28(2): 139–143.

Tucson Water, 2004, 2003 Annual water quality report. Water Quality Management Division.

UCB, 2006, The carcinogenic potency database [Lois Swirsky Gold, Ph.D., Director, University of California, Berkeley]. http://potency.berkeley.edu/ (accessed November 6, 2006).

UKHSE, 2007, EH/40 Occupational Exposure Limits, United Kingdom Health & Safety Executive. http://www.hse.gov.uk/coshh/table1.pdf (accessed November 17, 2007).

U.S. Army, 2000, Standard practice for wildlife toxicity reference values. U.S. Army Center for Health Promotion and Preventive Medicine.

U.S. Congress, 1987, Identifying and regulating carcinogens. Washington, DC: Office of Technology Assessment, U.S. Government Printing Office.

U.S. Court of Appeals for the Sixth Circuit, 1984, Scott v. Food and Drug Administration, 728 F. 2d 322 (6th Cir. 1984), Dept. of Justice, Antitrust Division, Washington, D.C. http://www.altlaw.org/v1/cases/495895 (accessed November 15, 2008).

U.S. Environmental Protection Agency (USEPA), 1986, Guidelines for carcinogen risk assessment. U.S. Environmental Protection Agency 51: 38,992–34,003.

U.S. Environmental Protection Agency (USEPA), 1988, National Sewage Sludge Survey. NTIS: PB93-500403.

U.S. Environmental Protection Agency (USEPA), 1992a, Dermal exposure assessment: Principles and applications. U.S. Environmental Protection Agency, Office of Health and Environmental Assessment.

U.S. Environmental Protection Agency (USEPA), 1992b, Draft report: A cross-species scaling factor for carcinogen risk assessment based on equivalence of mg/kg$^{3/4}$/Day. *Federal Register* 57(409): 24,152–24,173.

U.S. Environmental Protection Agency (USEPA), 1996a, 40 CFR Part 131, proposed guidelines for carcinogen risk assessment. *Federal Register* 61: 17,959–18,011.

U.S. Environmental Protection Agency (USEPA), 1996b, ASTER database. U.S. Environmental Protection Agency, National Health and Environmental Effects Research Laboratory.

U.S. Environmental Protection Agency (USEPA), 1997a, Profile of the pharmaceutical manufacturing industry. U.S. Environmental Protection Agency, Office of Enforcement and Compliance Assurance.

U.S. Environmental Protection Agency (USEPA), 1997b, Notice of availability; effluent limitations guidelines, pretreatment standards, and new source performance standards: Pharmaceutical manufacturing category—40 CFR part 439. *Federal Register* 62(153): 42,720–42,732.

U.S. Environmental Protection Agency (USEPA), 1998a, Development document for final effluent limitations guidelines and standards for the pharmaceutical manufacturing point source category. U.S. Environmental Protection Agency, Office of Science and Technology, Engineering and Analysis Division.

U.S. Environmental Protection Agency (USEPA), 1998b, National emission standards for hazardous air pollutants: Wood furniture manufacturing operations. *Federal Register* 63(248): 71,376–71,385.

U.S. Environmental Protection Agency (USEPA), 1999, Screening level ecological risk assessment protocol. U.S. Environmental Protection Agency, Office of Solid Waste.

U.S. Environmental Protection Agency (USEPA), 2002, 1,4-Dioxane (CAS Reg. No. 123-91-1) proposed Acute Exposure Guidelines Levels (AEGLs). U.S. Environmental Protection Agency, Office of Pollution Prevention and Toxics, *Federal Register* 67(154): 51,849–51,850.

U.S. Environmental Protection Agency (USEPA), 2004a, Engineering forum teleconference minutes, March 3, 2004. U.S. Environmental Protection Agency, Technical Support Program Engineering Forum Telecon.

U.S. Environmental Protection Agency (USEPA), 2004b, Community update—Durham Meadows Superfund Site: 1,4-Dioxane. U.S. Environmental Protection Agency.

U.S. Environmental Protection Agency (USEPA), 2004c, Integrated Risk Information System (IRIS); Announcement of 2004 program; Request for information. *Federal Register* 69(26): 5971–5976, February 9, 2004. http://www.wais.access.gpo.gov (accessed June 17, 2006).

U.S. Environmental Protection Agency (USEPA), 2005, *Guidelines for Carcinogen Risk Assessment*. EPA/630/P-03/001F.

U.S. Environmental Protection Agency (USEPA), 2006a, Water quality standards. http://www.epa.gov/waterscience/standards/ (accessed November 11, 2007).

U.S. Environmental Protection Agency (USEPA), 2006b, US EPA request for nominations of drinking water contaminants for the contaminant candidate list. *Federal Register* 71(199): 60,704–60,708.

U.S. Environmental Protection Agency (USEPA), 2006c, Integrated Risk Information System (IRIS): Epichlorohydrin (106-89-8). http://cfpub.epa.gov/ncea/iris/index.cfm (accessed June 17, 2006).

U.S. Environmental Protection Agency (USEPA), 2007a, Options for discharging treated water from pump and treat systems. U.S. Environmental Protection Agency, Office of Solid Waste and Emergency Response.

U.S. Environmental Protection Agency (USEPA), 2007b, Drinking water contaminants. http://www.epa.gov/safewater/contaminants/index.html#9 (accessed November 11, 2007).

U.S. Environmental Protection Agency (USEPA), 2007d, Screening-level hazard characterization of high production volume chemicals: Sponsored chemical *n*-butyl glycidyl ether (CAS No. 2426-08-6) [9th CI Name: (butoxymethyl)-oxirane]. U.S. Environmental Protection Agency, Office of Pollution Prevention and Toxics, Risk Assessment Division, High Production Volume Chemicals Branch.

U.S. Environmental Protection Agency (USEPA), 2007e, Screening-level hazard characterization of high production volume chemicals: Sponsored chemical *t*-butyl alcohol (CAS No. 75-65-0) [9th CI name: 2-propanol, 2-methyl]. U.S. Environmental Protection Agency, Office of Pollution Prevention and Toxics, Risk Assessment Division, High Production Volume Chemicals Branch.

U.S. Environmental Protection Agency (USEPA), 2007f, Screening-level hazard characterization of high production volume chemicals: 1,3-Dioxolane (CAS No. 646-06-0). U.S. Environmental Protection Agency, Office of Pollution Prevention and Toxics, Risk Assessment Division, High Production Volume Chemicals Branch.

U.S. Environmental Protection Agency (USEPA), 2007 g, 1,4-Dioxane (1,4-diethyleneoxide) hazard summary. U.S. Environmental Protection Agency, Technology Transfer Network, Office of Air and Radiation. http://www.epa.gov/ttn/atw/hlthef/dioxane.html (accessed November 11, 2007).

U.S. Environmental Protection Agency (USEPA), 2007 h, IRIS toxicological review and summary documents for 1,4-dioxane—peer review plan. http://oaspub.epa.gov/eims/eimsapi.dispdetail?deid=81310 (accessed November 11, 2007).

U.S. Environmental Protection Agency (USEPA), 2008, Drinking water contaminant candidate list 3—draft. *Federal Register* 73(5): 9627–9654, February 21, 2008. wais.access.gpo.gov (accessed February 21, 2008).

VDEQ, 2007, Voluntary remediation program risk assessment guidance. Virginia Department of Environmental Quality. http://www.deq.state.va.us/vrprisk/ (accessed November 16, 2007).

Verschueren, K., 1983, *Handbook of Environmental Data on Organic Chemicals*, 1st Edition. NY: Van Nostrand Reinhold.

Verschueren, K., 2001, *Handbook of Environmental Data on Organic Chemicals*, 4th Edition. NY: John Wiley.

von Oettingen, W.F., 1960, The aliphatic acids and their esters: Toxicity and potential dangers. *A.M.A. Archives of Industrial Health*, 1960(21): 28–65.

VROM, 1999, *Risk assessment: 1,4-Dioxane*. Bilthoven, Netherlands: S. P. Netherlands Ministry of Housing and the Environment (VROM), Chemical Substances Bureau.

Wallace, L.A., 1986, Personal exposure, indoor and outdoor air concentrations, and exhaled breath concentrations of selected volatile organic compounds measured for 600 residents of New Jersey, North Dakota, North Carolina and California. *Toxicological and Environmental Chemistry* 12(3): 215–236.

Washington Administrative Code, 2003, 173-200-040: Groundwater criteria. http://apps.leg.wa.gov (accessed November 11, 2007).

Waterloo, 2005, Greenbrook water supply system upgrade class environmental assessment. Waterloo, Ontario, Canada: Regional Municipality of Waterloo, Water Supply Division.

Waters Corporation, 2007, Material Safety Data Sheet for triethylamine. http://www.waters.nl/WatersDivision/pdfs/triethyl.pdf (accessed June 14, 2007).

Wester, R.C., Maibach, H.I., Surinchak, J., and Bucks, D.A.W., 1985, Predictability of *in vitro* diffusion systems: Effect of skin types and ages on percutaneous absorption of triclocarban. *Dermatology* 1985(6): 223–226.

WHO, 2005, 1,4-Dioxane in drinking-water: Background document for development of WHO guidelines for drinking-water quality. World Health Organization, WHO/SDE/WSH/05.08/120.

Wilkinson, S., 1989, Raw materials are more than skin deep. *Chemical Week*, November 15, 1989.

Williams, G.M. and Weisburger, J.H., 1986, Chemical carcinogens. In C.D. Klaasen, M.O. Amdur, and J. Doull (Eds), *Casarett and Doull's Toxicology: The Basic Science of Poisons*, pp. 99–173. NY: Macmillan Publishing Company.

Windholz, M., Budavari, S., Stroumtsos, L.Y., and Fertig, M.N. (Eds), 1983, *The Merck Index*, 10th Edition. Rahway, NJ: Merck Co., Inc.

Woodside, G.D. and Wehner, M.P., 2002, Lessons learned from the occurrence of 1,4-dioxane at Water Factory 21 in Orange County, California. In *Proceedings of the 2002 Water Reuse Annual Symposium*. Alexandria, VA: WateReuse Association.

WSDOH, 2006, 1,4-Dioxane contamination in North Glen Water Association well near Colbert Landfill NPL site, Colbert, Spokane County, Washington EPA Facility ID: WAD980514541. Washington State Department of Health.

Yamazaki, K., Ohno, H., Asakura, M., et al., 1994, Two-year toxicological and carcinogenesis studies of 1,4-dioxane in F344 rats and BDF1 mice. In *Proceedings of the Second Asia-Pacific Symposium on Environmental and Occupational Health*. Kobe, Japan, Kobe University, pp. 193–198.

Young, J.D., Braun, W.H., Rampy, L.W., Chenowith, M.B., and Blau, G.E., 1977, Pharmacokinetics of 1,4 dioxane in humans. *Journal of Toxicology and Environmental Health* 3: 507–520.

Zenker, M.J., Borden, R.C., and Barlaz, M.A., 2000, Mineralization of 1,4-dioxane in the presence of a structural analog. *Biodegradation* 11: 239–246.

7 Remediation Technologies

William H. DiGuiseppi

As discussed in Chapter 3, 1,4-dioxane has a low K_{oc} (soil adsorption coefficient for soil organic matter) and a high solubility and is therefore not highly sorbed to soils. Because of these characteristics, 1,4-dioxane is dominantly found in groundwater. Therefore, the following discussion of remedial technologies is focused solely on groundwater remediation methods, both *ex situ* and *in situ*, and does not address soil remediation methods. In this chapter, we first describe each technology, and then we present literature studies to demonstrate the effectiveness of the technology at 1,4-dioxane removal or destruction. The literature studies include laboratory bench-scale studies and field-scale tests, documenting the removal of 1,4-dioxane from groundwater or effluent. Finally, we assess the technology's likelihood of success in the remediation of actual sites contaminated with 1,4-dioxane.

Remediation technologies presented are

a. Vapor-phase transfer (e.g., air stripping)
b. Sorption
c. Natural attenuation
d. Ultraviolet photolysis
e. Phytoremediation
f. Bioremediation
g. Chemical oxidation

The assessment of treatment technologies for 1,4-dioxane presented here is comprehensive; however, it is not intended to capture each and every study of 1,4-dioxane treatment. The selected citations provide an assessment of the viability of a given technology to remediate groundwater contaminated with 1,4-dioxane. Preference is given to citations that provide insight into the fate of 1,4-dioxane or the treatment mechanism.

7.1 VAPOR-PHASE TRANSFER

7.1.1 AIR STRIPPING

Air stripping is a common method of removing volatile organic compounds (VOCs) from groundwater because the relationship between solubility and volatility is optimal in this group of contaminants. This method is typically conducted in a packed tower or a low-profile aeration system. Water is introduced to the top of the stripper and flows down through the system under gravity. Air is pumped into the bottom of the stripper under pressure and exits the top. This arrangement, called countercurrent flow, optimizes the exchange of contaminants from water to air. Air stripping is a nondestructive transfer of contaminants from the liquid phase to the vapor phase. The resulting gaseous phase contaminants often require treatment. Granular activated carbon (GAC) adsorption and thermal destruction are commonly used to treat contaminated air before it is discharged to the atmosphere. Although 1,4-dioxane is highly volatile when in pure form, its infinite solubility in water is generally considered to prevent efficient air-stripping removal from groundwater to concentrations suitable for discharge. Nevertheless, several studies have been performed to assess the ability of air stripping to remove 1,4-dioxane from groundwater.

An air-stripping optimization test was performed at the U.S. Air Force Plant 44 (AFP 44), located in Tucson, Arizona, to determine whether adjustments to the operating parameters for the large-scale treatment system could effectively reduce the influent 1,4-dioxane concentrations (10–15 µg/L) to meet a target level of 6.1 µg/L (Earth Tech, 2004). The AFP 44 groundwater treatment plant utilizes three parallel trains of two-stage air-stripping towers (primary and secondary stripping towers) with a design capacity of ~5000 gallons/min. At an air:water ratio of seven, the 25-feet-tall primary towers, in combination with GAC treatment of the off-gas, are capable of removing as much as 85% of the trichloroethene (TCE) and 1,1-dichloroethylene. At an air:water ratio of 25, the 42-feet-tall secondary towers are designed to remove the remaining TCE and 1,1-dichloroethylene from the water; the stripped vapors are then vented to the atmosphere. This system had no measurable effect on 1,4-dioxane removal. Optimization of the existing treatment system to remove 1,4-dioxane was accomplished by modifying the system to allow higher air-flow rates, thereby increasing the air:water ratios. The air:water ratio in the primary tower was set at 69, the maximum air:water ratio possible, and the secondary tower's air:water ratio was varied from 183 to as high as 291. The study indicated that the maximum removal rate for 1,4-dioxane was ~10%, even with air:water ratios as high as 291, which is more than 10 times the designed air:water ratio for the removal of TCE and dichloroethene. Chlorinated VOC removal was complete regardless of the air:water ratio. The results of the optimization test indicated that utilizing the air-stripping system as the primary means to remove 1,4-dioxane would not achieve the target level of 6.1 µg/L.

7.1.2 Accelerated Remediation Technologies, LLC In-Well System

A more focused and comprehensive mass-transfer approach—combining air stripping, air sparging, soil-vapor extraction, enhanced bioremediation, and underground circulation—has been demonstrated to be effective for 1,4-dioxane removal under specific circumstances (Odah et al., 2005). The Accelerated Remediation Technologies, LLC (ART) In-Well System is designed to operate in a 4-inch or larger well, screened both above and below the water table. The system uses several concurrent physical, chemical, and biological processes to reduce contaminant levels in groundwater (Figure 7.1). Air is forced into the lower part of the well to remove some levels of contamination in the groundwater through sparging. Additionally, the aeration of the groundwater reduces its density, which causes a drop in hydraulic head in the well. The lower head causes cleaner, low-density groundwater to be pushed out from the upper part of the well and contaminated, high-density groundwater to be drawn into the lower part of the well to create what the inventor refers to as dynamic subsurface circulation. The circulated water effectively flushes soluble contaminants from the lower vadose zone. Water in the well bore is also pumped to the top of the well and sprayed downward through the rising air column, creating an air-stripping effect without using a packing medium to increase the surface area available for phase transfer. The In-Well stripping requires multiple stripping passes, to compensate for 1,4-dioxane's general resistance to stripping from groundwater. Vapors and air are extracted from the wellhead, creating a soil-vapor extraction effect for the screened section of the well above the water table. Lastly, the increased movement of air from all this sparging and extraction leads to higher levels of dissolved oxygen in the groundwater and more oxygenation of the vadose zone, potentially stimulating aerobic biologic processes that may otherwise be absent under natural conditions. The radius of influence of the ART In-Well System has been documented at ~10 times the water-column height, both up gradient and down gradient from the well (Odah et al., 2005). The ART In-Well technology was applied at a major aerospace manufacturing facility in North Carolina, where the subsurface was saprolitic soil over fractured bedrock. A single ART well was installed, and monitor wells were positioned 10 and 20 ft down gradient (MW-1 and MW-2, respectively) to gauge the system's effectiveness. 1,4-Dioxane concentrations in that part of the aquifer were as high as 43,000 µg/L, and high levels of chlorinated VOCs were also present. Within 90 days, levels in MW-1 declined from 25,000 to below 7400 µg/L, and those in MW-2 declined from 28,000 to 2400 µg/L. These reductions represented more than

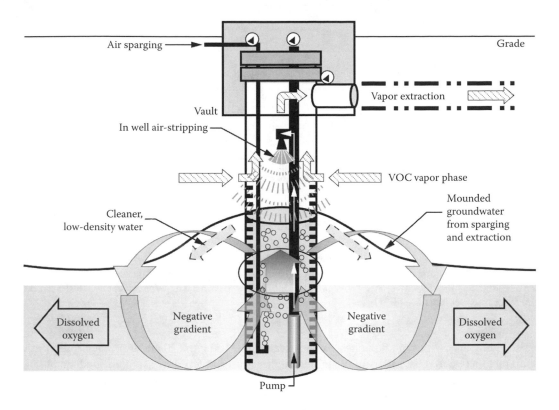

FIGURE 7.1 Various removal processes simultaneously applied in the ART In-Well System. (From Odah, M.M., Powell, R., and Riddle, D.J., 2005, *Remediation Journal* 15(3): 51–64.)

90% removal. The fact that 1,4-dioxane levels in extracted vapors were as high as μg/m³ demonstrated that removal in the vapor phase was a significant mechanism. Chlorinated VOC removal rates were predictably high as well. The results of this pilot study with a single well led the state to approve the ART In-Well technology as the primary remediation method for the site.

7.1.3 PERVAPORATION

An additional vaporization technique for the separation of contaminants from water, generally in industrial applications, is referred to as pervaporation. This process is a combination of evaporation and membrane permeation, which uses relatively low temperatures and pressures for cost-effective *ex situ* separation of liquids with similar boiling points. In the pervaporation process, the membrane is gas permeable, which allows VOCs in the gas phase to pass through the membrane, while water is prevented from passing (Figure 7.2). Chemicals in a mixture can be separated from each other because their partial pressures and membrane permeabilities differ. Pervaporation effectively separates azeotropic mixtures, which have similar boiling points and evaporation rates.

Specific studies related to separation of 1,4-dioxane from water have been performed by using a variety of innovative pervaporation membranes to increase flux rates, separation efficiency, or resistance to chemical and thermal degradation. Pintauro and Jian (1995) assessed the capability of separation of 1,4-dioxane from water with a polyvinylidene fluoride resin membrane. Their results indicated that pervaporation separation of 1,4-dioxane from an aqueous solution did occur, but at significantly lower rates than for other contaminants, such as benzene, toluene, or chloroform. Flux rates through the membrane, at percent-level concentrations of 1,4-dioxane, were on the order of 5 g/(m² h).

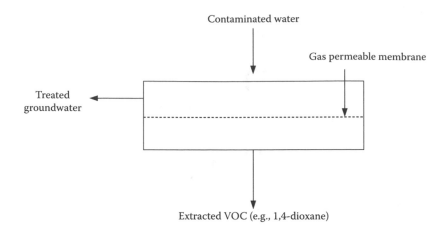

Contaminated water

Gas permeable membrane

Treated
groundwater

Extracted VOC (e.g., 1,4-dioxane)

FIGURE 7.2   Overview of pervaporation process. [After Pintauro, P.N. and Jian, K., 1995, United States Patent 5,387,378: Integral asymmetric fluoropolymer pervaporation membranes and method of making the same. Assignee: Tulane University, New Orleans. http://www.patentstorm.us/patents/5387378-description. html (accessed October 12, 2007).]

Much higher 1,4-dioxane flux rates, on the order of 30–100 g/(m² h) and efficiencies, with separation factors two to three orders of magnitude higher, were achieved by using blended membranes composed of chitosan and nylon 66 (Smitha et al., 2006). Proportions of these two components were varied, and their tensile, mechanical, thermal, and molecular characteristics were evaluated with a variety of 1,4-dioxane feed mixtures. Because this testing was performed at percent-level concentrations and the pervaporation method is generally used for separation of liquids in industrial manufacturing scenarios, it is not clear whether these materials would be effective at the part per billion concentration levels and gallons per minute flow rates typically being evaluated in groundwater remediation programs.

7.1.4   SUMMARY

Some level of 1,4-dioxane removal from conventional *ex situ* air-stripping systems is achievable under ideal circumstances, but in most cases, it is unlikely that this technology will achieve the removal levels required. The ART In-Well technology—which maximizes air stripping—along with several other physical, chemical, and biological processes has been shown to reduce 1,4-dioxane levels from high concentrations, but the exact mechanism for the removal is not known. Pervaporation has been demonstrated to be capable of separating 1,4-dioxane from water at high percent levels at rates that may be applicable in some *ex situ* situations. It is not clear how effective these latter two technologies would be at the lower concentration levels typically found at most 1,4-dioxane-contaminated sites.

7.2   SORPTION

The "sorption" process refers to a combination of chemical adsorption and physical absorption of a contaminant to a "sorptive" medium. Some materials, like engineered resins, are specifically used for one process or the other (i.e., they are either "absorptive" or "adsorptive"), whereas other materials, such as clays and activated carbons, can be used for both processes. Different chemicals often have an affinity for one process over another. Many types of sorptive media are used to remove contaminants from groundwater, most commonly in tandem with a "pump-and-treat" system, although some technologies can be applied *in situ*. For the typical *ex situ* system, the sorptive medium is contained in a series of vessels connected to the extracted groundwater stream. The groundwater passes through the vessels containing the medium, and contaminants are transferred to the medium before the treated groundwater exits the vessel. The contaminants removed are trapped within or form a coating on the medium's sorptive surfaces until they reach the point of saturation, after which no additional contaminants can be removed and breakthrough occurs. The saturation point typically varies in proportion to the applied concentration and flow rates. As with air stripping, sorption technologies constitute a transference of contamination from one medium to another, and the contaminated sorptive media must be disposed of properly and/or treated. Typical treatment and disposal options include stream regeneration for GAC, regeneration of resins, or disposal in landfills. Newer sorbents may have the ability to treat contaminants after capture [e.g., Trap and Treat® (Remediation Products, Inc., Golden, Colorado) (R. Elliott, personal communication, 2007)], which extends the life of the sorptive medium until both the treatment and sorptive capacities are exceeded. Because of the low partitioning coefficient of 1,4-dioxane, it is generally considered to be resistant to sorbing to GAC (a common adsorptive medium). As discussed in the following sections, research into the efficiency of several types of sorptive media has been performed to assess their ability to remove 1,4-dioxane from groundwater.

7.2.1 ADSORPTIVE MEDIA

A bench-scale treatability test was performed for AFP 44 to assess the removal efficiency of several adsorptive media that were identified as potentially applicable for 1,4-dioxane treatment (Earth Tech, 2004). These included a surfactant-modified zeolite (SMZ) (referred to as 14-40 SMZ) and an SMZ with zero-valent iron (ZVI), both provided by Robert Bowman, New Mexico Institute of Mining and Technology, and activated Tri-Base Pelletized Carbon™ (Tri-Base Carbon™) (Hiatt Distributors Limited, Long Beach, California). Zeolites are natural or synthetic aluminum silicates that form a regular crystal lattice and release water at high temperature. Zeolites are polar in nature and have been used for the filtration of a number of organic and inorganic contaminants (Bowman et al., 2000). The results of the partitioning phase of the study indicate that the SMZ with ZVI media adsorbed measurable concentrations of 1,4-dioxane at media:water ratios of 1:20 through 1:2, indicating that sorption is possible at low 1,4-dioxane concentrations and low media:water ratios. No measurable sorption of the 1,4-dioxane to the standard SMZ medium took place. Tri-Base Carbon adsorbed 1,4-dioxane measurably. However, a high media:water ratio was required in order to significantly reduce 1,4-dioxane concentrations. An additional potential filtration technology using macroporous polymer (MPP) extraction was also tested by the manufacturer (Akzo Nobel, Incorporated), but was ineffective at removing 1,4-dioxane from the tested water. In conclusion, the Tri-Base Carbon exhibited a measurable sorption rate for groundwater samples from the site. However, results of the testing and calculations to scale-up the test led to the estimate that ~1500 tons of Tri-Base Carbon would be required to treat AFP 44 regional groundwater per year. This volume of material is prohibitive from a logistical and cost standpoint.

7.2.2 ACTIVATED CARBON

Johns et al. (1998) performed a series of laboratory studies to determine the sorption characteristics of activated carbons derived from a variety of waste agricultural by-products. Two groups of materials were evaluated: soft materials (made from rice straw, soybean hulls, sugarcane waste, and peanut shells) and hard materials (made from pecan and walnut shells). Materials were commingled with molasses as a binder and then pressed into briquettes or pellets. All materials were carbonized and then crushed and sieved for a consistent size fraction, prior to activating with carbon dioxide or steam. The study specifically utilized lower carbonization temperatures (750°C) than normally used in the industry (1000°C) to demonstrate potential reductions in energy demands and overall manufacturing costs. The sorption effectiveness of the agricultural products was evaluated for a variety of common organic contaminants, including benzene, toluene, 1,4-dioxane, acetonitrile, acetone, and methanol. Several types of commercially available GAC were also tested for comparison. Of the chemicals evaluated, 1,4-dioxane had the lowest overall adsorption rates, as well as the most variability in sorption rates during the tests. The only GAC capable of effective removal of 1,4-dioxane was made from hard nutshells, which achieved a little more than 50% removal. Johns et al. (1998) concluded that the low removal rates could have been due to competition for sorption sites with other adsorbates, steric hindrance, reduced intraparticle diffusion, or lower GAC binding affinity for 1,4-dioxane.

GAC has been reported to effectively remove 1,4-dioxane from extracted groundwater at the Stanford Linear Accelerator Center (SLAC) in California (USEPA, 2006; Sabba and Witebsky, 2003). On the basis of the evaluation of several different kinds of GAC by Johns et al. (1998), it is unlikely that the reductions noted at SLAC are due solely to the sorption to the GAC. The GAC unit at the SLAC site has moderate influent concentrations of 1,4-dioxane (average 725 µg/L) and very low flow rates (~0.5 gpm). The observed reduction in 1,4-dioxane concentration (<10 µg/L in effluent) may result from cometabolic biodegradation, in which a cocontaminant, in this case tetrahydrofuran (THF), supports biological activity while a by-product of the biological activity (i.e., an enzyme) effectively destroys the target contaminant. Cometabolic degradation of 1,4-dioxane in the presence of THF is well documented in the literature (Zenker et al., 2000; Parales et al., 1994) and is discussed further in Section 7.6. Further discussion of the GAC unit at SLAC is provided in Section 8.5.4.

7.2.3 ORGANOCLAYS

Organoclays have been used as adsorbents at wood-treating sites for polynuclear aromatic hydrocarbon and creosote removal. They are typically bentonite clays (primarily montmorillonite), chemically modified to make them hydrophobic (water repelling) and organophilic (oil attracting). One application uses quaternary amines to produce an ion exchange capacity of ~70–90 milliequivalents/g (Alther, 2006). Amine chains bind to the clay particles and attach to organic compounds in the influent water, binding contaminants to the organoclay material. Organoclays are sometimes mixed with GAC, or a secondary carbon unit may be used on the effluent to achieve higher removal efficiencies. Alther (2006) performed a number of laboratory tests to evaluate the removal efficiency of organoclays on a variety of contaminants, including 1,4-dioxane. Laboratory batch tests were run by combining water containing 1,4-dioxane with nonionic organoclay in glass vials and by shaking for one day. Aqueous concentration of 1,4-dioxane was reduced by only 2.8% from levels as high as 958,000 µg/L, which are higher than those that are typically found on contaminated sites.

A proprietary modification technology used by Aqua Technologies of Wyoming, Inc. (A.B. Brown, personal communication, July 2 and 5, 2007), resulted in an organoclay product (ET-1) capable of removing up to 83% of 1,4-dioxane, from 726,500 to 126,000 µg/L. Test waters were provided by a plastics manufacturer in South Carolina from a contaminated site. The results were verified by an independent laboratory using Environmental Protection Agency Method 8260 with selected-ion monitoring (SIM). Although this removal rate is impressive, the very high concentrations in the extracted groundwater are not typical of most hazardous waste sites contaminated with 1,4-dioxane, and the treatment end points are well above most 1,4-dioxane standards.

7.2.4 PALLADIUM-111

Azad et al. (2000) evaluated the sorption characteristics of 1,4-dioxane on palladium-111 (^{111}Pd). Their research indicated that sorption occurred at very low temperatures (–193°C) in a multilayer that desorbs to leave an overlayer at a somewhat higher temperature (–113°C). The 1,4-dioxane overlayer desorbs and thermally decomposes at ~27°C. The temperatures involved in this process are not likely to be achieved cost-effectively in environmental remediation projects.

7.2.5 SUMMARY

The following sorptive media have been tested to evaluate their capacity to remove 1,4-dioxane from water: SMZs, SMZ with ZVI, Tri-Base Carbon, MPPs, GACs made from agricultural by-products, organoclays, and palladium-111. Although some of the media tested demonstrated removal rates of up to 50%, the test conditions required high media:water ratios that would not be feasible in many field applications. Specific sorbents may be a consideration in situations with low levels of 1,4-dioxane and low groundwater extraction rates.

7.3 NATURAL ATTENUATION

Natural attenuation, or monitored natural attenuation (MNA), has been used to demonstrate passive remediation at sites contaminated with a variety of organic and inorganic contaminants. MNA is most commonly applied to sites contaminated with chlorinated ethenes and petroleum compounds, primarily because these are among the most common contaminants at hazardous waste sites and there are published protocols for leveraging natural attenuation of these compounds for site remediation. The U.S. Environmental Protection Agency's (USEPA's) policy directive on natural attenuation describes MNA processes as those that "act without human intervention to reduce the mass, toxicity, volume or concentrations of contaminants in soil or groundwater" (USEPA, 1999). Natural attenuation of contaminant mass is attributable to a variety of physical, chemical, or biological processes that can include biodegradation, dilution, dispersion, advection, sorption, volatilization, radioactive decay, and chemical and biological stabilization, transformation, and destruction of contaminants; destructive processes are preferred over nondestructive processes (USEPA, 1999).

7.3.1 EVIDENCE OF THE EFFECTIVENESS OF NATURAL ATTENUATION

Demonstrating that MNA will reliably achieve the desired passive remediation typically requires identifying the active attenuation process and then establishing its effectiveness through several different lines of evidence. USEPA (1999) has established the following evaluation elements, or lines of evidence, that should be assessed:

1. Establish historical trends
2. Indirectly verify the attenuation process by using hydrogeologic and chemical data
3. Directly confirm the attenuation process; for instance, microbial activity can be proven through microcosm studies

Implementing an MNA solution typically involves demonstrating any or all available lines of evidence, the availability of MNA protocols for the contaminant of interest, or other factors. Because MNA has not been commonly used for 1,4-dioxane, implementing MNA for 1,4-dioxane will probably require a more rigorous weight of evidence to satisfy regulatory authorities. Furthermore, many of the MNA processes commonly invoked for chlorinated solvent sites may be less effective with 1,4-dioxane because of its chemical characteristics, especially its miscibility, low soil partitioning coefficient, and resistance to biodegradation under natural conditions. Demonstrating a natural attenuation mechanism for 1,4-dioxane is expected to be difficult.

7.3.2 DIFFUSION INTO IMMOBILE POROSITY

One potentially applicable attenuation mechanism relates to the hydraulic characteristics of fine-grained materials in the subsurface. Finer-grained, less permeable aquifer materials such as clays and silts in contact with more transmissive, coarser-grained aquifer materials may trap contaminants migrating through the aquifer. Fine-grained sediments reduce the effective porosity of the aquifer to the point that molecules of the contaminant can diffuse from the transmissive zones into dead-end pore spaces within the finer-grained aquifer material. Diffusion into dead-end pores will cause a reduction in monitored aqueous concentrations similar to the reduction caused by chemical sorption of typical organic contaminants to organic materials in the aquifer. Fine-grained clay and silt are present in most unconsolidated aquifers. Clays, though highly porous, are relatively impermeable and hence restrict the free flow of groundwater. Dead-end pores exist that are connected to the more permeable materials of the aquifer, but are not subject to through-flow. Because water is not actually flowing into or through these dead-end pore spaces, the contaminant migration process is governed by molecular diffusion. The transfer of 1,4-dioxane into dead-end pores is a slow process driven by the concentration gradient between the contaminated free-flowing groundwater and the relatively clean water in the dead-end pore spaces. Unfortunately, unlike chemical adsorption, where the contaminant may be permanently bound to the organic material in the aquifer, this attenuation process involving diffusion into dead-end pores is a reversible process. Back-diffusion from the dead-end pore spaces into the flowing groundwater occurs after the concentration gradient reverses to cleaner water in the aquifer and more contaminated water in the dead-end pore spaces.

7.3.3 NORTH CAROLINA EXAMPLE

Statistical trend analysis and a numerical flow model were used to gain the approval of a natural attenuation remedial approach for an industrial site in North Carolina (Chiang et al., 2006, 2007b, 2008). Active sources of 1,4-dioxane had been removed, and dissolved-phase plumes were present in four distinct hydrogeologic layers beneath the site. Data collected from 60 wells at the site since the early 1990s were evaluated yearly from 2003 through 2005 by using Mann-Kendall trend analysis. The evaluation in 2005 noted that 12 wells (20%) exhibited a downward trend and 21 wells (35%) exhibited a stable trend. Twenty-six wells (43%) exhibited no discernable trend. An upward trend was noted in only one well (2%). To demonstrate a continued attenuation trend into the future, a Modular Three-Dimensional Groundwater Flow and Transport Model (MODFLOWT) was constructed by using four hydrogeologic layers representative of the varying hydraulic conductivities in subsurface sand versus clay or silt units. The 1,4-dioxane plume was assumed to be more sensitive to advection than to hydrodynamic dispersion. (Advection refers to the process of contaminant migration due to movement of groundwater. Hydrodynamic dispersion is the sum of mechanical dispersion and diffusion.) The modeling effort simulated advective–dispersive transport, but considered diffusion to be relatively insignificant and therefore omitted this term from the model.

The flow model was calibrated by using measured heads from monitoring and production wells at the site. Particle tracking and transport modeling were performed to estimate future concentrations of 1,4-dioxane at a local surface-water body to the east. The modeling under an assumption of no degradation underestimated the observed decline in 1,4-dioxane concentrations. To assess the degradation rate, historical field measurements were evaluated at several monitoring wells within different hydrogeologic layers. Time-series plots were prepared under the assumption of no degradation versus degradation half-lives of three, five, and seven years. The optimal correlation between observed and predicted values occurred with a seven-year degradation half-life. For each of the four layers, Figure 7.3 presents the ten-year predictions, which show significant reductions in the distribution of 1,4-dioxane as a result of using a 7-year half-life. The modeling results predict that 1,4-dioxane migration at levels above North Carolina's surface-water standards would not reach the surface-water body via any of the modeled hydrogeologic layers. Among the natural attenuation mechanisms, Chiang et al. surmised that 1,4-dioxane is more highly influenced by advection because it is not readily sorbed to the aquifer matrix. Because no active sources of contamination existed at the site and the combination of modeling and field observations demonstrated that the plume was stable or shrinking, the proposed MNA approach was approved by the regulatory agency for implementation as the final remedy (Chiang et al., 2008).

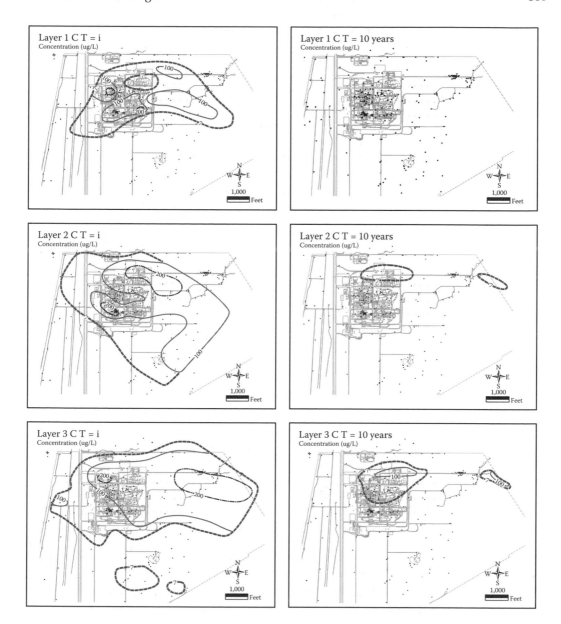

FIGURE 7.3 Ten-year predictions of 1,4-dioxane simulations in three hydrogeologic layers under pump-off scenario with 1,4-dioxane half-life of seven years. (After Chiang, D., Zhang, Y., Glover, E., Harrigan, J., and Woodward, D., 2006, In: B.M. Sass (Ed.), *Proceedings of the Fifth International Conference on Remediation of Chlorinated and Recalcitrant Compounds*, Paper G-70. Columbus, OH: Battelle Press.)

Possible biological mechanisms for natural attenuation of 1,4-dioxane are being evaluated at Air Force Plant 44, in Tucson, Arizona, using a parallel approach to that typically undertaken for the evaluation of natural TCE biodegradation. Baseline geochemical conditions were defined, including analysis of cations, anions, dissolved gases, total organic carbon, and alkalinity. BioTrap™ sampling for quantitative polymerase chain reaction (qPCR) was performed to verify the presence and abundance of bacteria and enzymes capable of aerobically degrading TCE and 1,4-dioxane. The targeted microorganisms and enzymes include methanotrophs (MOB), soluble methane monooxygenase (sMMO), phenol hydroxylase (PHE), ring hydroxylating toluene monooxygenase (RMO), and toluene dioxygenase (TOD). This phase of study revealed that MOB and

TOD are ubiquitous with the density of MOB ranging from 10^3 to 10^7 cell/mL and TOD ranging from 10^4 to 10^8 cell/mL. PHE and sMMO are also present at significant levels. Based on the presence of key bacteria and enzymes in all samples collected, further testing for naturally occurring biodegradation will be performed. Stable isotope probing (SIP) will be conducted using BioTraps™ baited with 1,4-dioxane tagged with the ^{13}C isotope. The ^{13}C-1,4-dioxane-baited BioTraps™ will be evaluated for ^{13}C fractioning (indicating biodegradation of 1,4-dioxane), incorporation of ^{13}C into cell phospholipid fatty acids (i.e., PLFA), and generation of ^{13}C carbon dioxide. The SIP results will directly indicate whether or not 1,4-dioxane is being biodegraded in this aerobic aquifer. The last step of the study will be to assess enzyme activity directly, which should indicate which active enzymes have the capacity for degrading 1,4-dioxane. This natural attenuation study will assess the intrinsic biodegradation capability of the aquifer, and will help define whether bioaugmentation or amendments could enhance this capability.

7.3.4 SUMMARY

Natural attenuation processes may reduce concentrations of 1,4-dioxane in the groundwater under specific site conditions. The most likely processes are advection, diffusion (especially into dead-end pores), and dispersion. Intrinsic biodegradation may also be a viable mechanism, but additional field study is necessary to verify occurrence and effectiveness. Volatilization and sorption to conventional soil materials are not likely to be significant attenuation mechanisms for 1,4-dioxane. Further discussion of MNA is provided in Section 10.4.1.

7.4 ULTRAVIOLET PHOTOLYSIS

The energy from an ultraviolet lamp can often directly break apart the chemical bonds of an organic contaminant, ultimately resulting in harmless constituent compounds, such as carbon dioxide and water. When photons are absorbed by a compound, the chemical bonds holding the elements together may be broken if their energy is less than the energy contained in the photon. On the basis of historical research into quantum physics by Albert Einstein and Max Planck, the energy E of a "light quanta" is described by the equation $E = hc/x$, where h is Planck's constant (4.14×10^{-15} eV s), c is the speed of light (3.0×10^8 m/s), and x is the wavelength (Tailored Lighting, 2007). Thus, shorter-wavelength light, including ultraviolet light, contains more energy than longer-wavelength light.

Hentz and Parrish (1971) demonstrated that the decomposition of 1,4-dioxane occurred at a "far" (ISO, 2005) ultraviolet wavelength of 147 nm, which carries 6.2–12.4 eV of energy per photon. Ninety-eight percent of the total decomposition yielded ethane and formaldehyde. In general, this application is restricted to ultrapure, laboratory-grade water, and restricted low wavelengths of ultraviolet light.

A. Festger (personal communication, 2007) of Trojan Technologies has stated that manufacturers of ultraviolet-peroxide water treatment systems use 1,4-dioxane as a reagent in bench-scale tests quantifying the effects of ultraviolet energy on oxidation systems because 1,4-dioxane does not react by strict ultraviolet photolysis at middle ultraviolet wavelengths commonly applied to water treatment (230–300 nm), without the addition of oxidation chemicals. The middle ultraviolet wavelengths contain less energy (4.13–6.20 eV), which likely limits the destructive effect on 1,4-dioxane. The use of 1,4-dioxane in bench-scale tests allows Trojan Technologies to demonstrate the treatment efficacy of ultraviolet plus oxidation, as opposed to limited ultraviolet photolysis, by quantifying the degradation of 1,4-dioxane.

Although there is some evidence of direct ultraviolet photolysis of 1,4-dioxane, the demonstrated applications are at wavelengths not typically applied to groundwater treatment and were tested on extremely pure water. Therefore, direct ultraviolet photolysis is not considered as a viable remediation technology for 1,4-dioxane.

7.5 PHYTOREMEDIATION

Phytoremediation refers to contaminant removal from the soil and groundwater through one or more of several plant-mediated processes. Transpiration is a major phytoremediation process. Transpiration involves the transfer of soluble groundwater contaminants through the plant's root, stem, and leaf systems to the atmosphere. Additional documented phytoremediation mechanisms include stimulation of microbial activity and biological degradation in the root zone through enzymatic action, incorporation of contaminants into the plant material, and transformation of contaminants into less toxic forms through plant metabolism. Obvious limitations of phytoremediation include depth to groundwater, growing season, and soil conditions.

7.5.1 BENCH-SCALE TESTS

Imperial Carolina hybrid poplar tree cuttings were rooted hydroponically in a series of experiments by Aitchison et al. (2000). Cuttings were tested in individual flasks containing an aqueous solution of 20,000 µg/L 1,4-dioxane. Several variables were evaluated in this study, including plant uptake, transpiration, incorporation into the stem material, sorption to roots, and degradation by microorganisms associated with the root zone. The researchers used a carbon isotope (^{14}C) to label 1,4-dioxane and assess its distribution among the plant roots, stems, and leaves. Between 30% and 79% (average 54%) of the 1,4-dioxane was removed from the solution after eight days. However, the fact that only 10% was removed from the solution in experiments in which the leaves had been detached from the stem suggests that transpiration was an important process. Eight percent removal occurred in the control flask that did not contain a poplar cutting, confirming that the 1,4-dioxane was not degrading or volatilizing from the flasks by some other mechanism. The data indicate that transpiration was the dominant removal mechanism and that volatilization of the 1,4-dioxane from the plant's leaves accounted for 77% of the 1,4-dioxane removed by the plant. Degradation within the root zone was considered to be minimal.

Researchers at the University of Iowa experimented with adding *Amycolata sp.* CB1190 to planted and unplanted soil (Kelley et al., 2001). CB1190 has been demonstrated to effectively degrade 1,4-dioxane in water (Parales et al., 1994). The effectiveness of bioaugmentation of poplar (*Populus sp.*) root zones for the destruction of 1,4-dioxane was evaluated by using a known 1,4-dioxane-degrading bacteria to enhance phytoremediation. Soil and basal salt medium (BSM) microcosms were prepared and spiked with 100,000 µg/L of 1,4-dioxane prior to the introduction of CB1190 bacteria, which were grown on 1,000,000 µg/L of THF. Several controls were also used, including one in an aqueous solution with CB1190 and no soil, to demonstrate the viability of the bacteria to degrade 1,4-dioxane. Additional controls demonstrated that neither sterile conditions nor the native soil bacteria were effective in 1,4-dioxane degradation. Complete removal of the 1,4-dioxane occurred in all CB1190-amended microcosms within 30 days in the BSM and within 45 days in a soil amended with poplar root extract.

Adding the THF apparently led to enhanced enzyme creation. Poplar root extract and 1-butanol stimulated additional growth of CB1190. Planting hybrid poplar trees in 1,4-dioxane-contaminated soils caused some of the dioxane to be evapotranspired through the trees. Adding CB1190 to planted soils was less effective because evapotranspiration removed the dioxane, making it less available for cometabolic biodegradation. 1,4-Dioxane was not directly degraded by poplar root exudates; the root exudates only enhanced the growth of CB1190. Only THF produced a dioxane-specific activity.

Unplanted soils amended with CB1190 at 10^7 cells/g of soil completely destroyed ~25% ± 5% of the ^{14}C-labeled 1,4-dioxane added to the soil. CB1190 can survive in the soil substrate and compete with indigenous soil and rhizosphere microflora. CB1190 could colonize the poplar rhizosphere, which suggests that bioaugmenting poplar roots with CB1190 might enhance the remediation process. However, cells grow slowly and the cell yield is relatively low. To bioaugment 1 acre of contaminated soil over a 1-foot thickness at 10 mg of dry cell mass/kg of soil (i.e., ~10^7 cells/g), ~15,000 gallons of cell suspension would be needed (Kelley et al., 2001).

7.5.2 MATHEMATICAL MODELING

Ying (2002) modified an existing one-dimensional mathematical model [CTSPAC—Coupled Transport of Water, Solutes, and Heat in the Soil–Plant–Atmosphere Continuum (Cawfield et al., 1990)] to predict the removal of 1,4-dioxane from groundwater and soil. Calibration of the model was performed by using the hydroponic and soil experimental results of Aitchison et al. (2000). Model simulations indicated that uptake from soil would be relatively rapid and that 30% of the 1,4-dioxane mass would be removed within seven days. The modeled distribution of 1,4-dioxane in the various parts of the plant correlated well with research by others. The findings, shown in Figure 7.4, indicate that 1,4-dioxane was concentrated early into the leaves, and, after seven days, the leaves still contained the greatest mass, relative to the proportional mass of the various plant parts. The stem contained some 1,4-dioxane, but the root system held little, if any, at the conclusion of the simulations. Ying

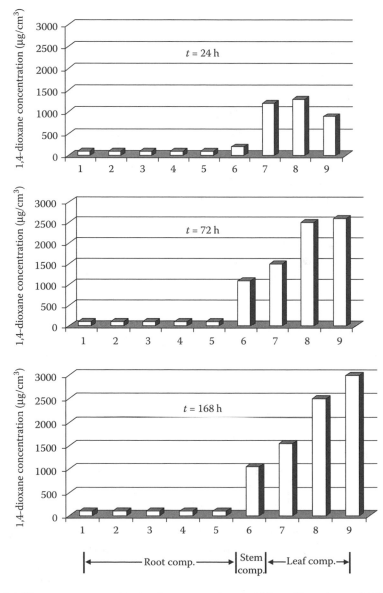

FIGURE 7.4 1,4-Dioxane concentration in plant compartments ("Comp.") at times of 24, 72, and 168 h. [From Ying, O.Y., 2002, *Journal of Hydrology* 266(1–2): 66–82.]

(2002) concluded that the model effectively simulated the removal rates and distribution of 1,4-dioxane in the plant material that have been documented by others in laboratory experiments. The CTSPAC model could therefore be used to estimate removal rates from contaminated sites.

In a second modeling study, Ying (2008) developed a dynamic model for the uptake and translocation of contaminants from a soil–plant ecosystem (UTCSP) using the STELLA® modeling tool (isee systems, inc., Lebanon, New Hampshire) UTCSP assesses simultaneous transport, accumulation, and transpiration of water and contaminants in plant systems. The model was calibrated with experimental data and then used to predict 1,4-dioxane phytoremediation from a sandy soil by a poplar tree. Results indicated ~20% 1,4-dioxane removal from the soil in 90 days. Ying (2008) identified a steadily increasing mass of 1,4-dioxane in the plant material over time, as well as a typical diurnal distribution pattern in the uptake and translocation in the plant system, resulting from daily variations of plant leaf transpiration. The study concluded that the UTCSP model could be useful for estimating 1,4-dioxane phytoremediation in soil–plant systems.

7.5.3 GREENHOUSE AND FIELD STUDIES

Ferro et al. (2005) performed greenhouse and field pilot studies toward the design of a full-scale 1,4-dioxane phytoremediation project at an industrial site in North Carolina. Greenhouse studies were utilized to identify candidate species of both deciduous and coniferous trees to employ in the full-scale design. The study concluded that the full-scale project should go forward with both types of trees in use for continuous groundwater phytoremediation. In this scenario, the coniferous trees address the contaminants during the winter when the deciduous trees, which are the major water-extraction plants, are dormant.

Documented field application of phytoremediation of 1,4-dioxane has been limited to a single study by Chiang et al. (2007a). A phytoremediation field pilot study was performed in an area of approximately 8000 ft^2 to address a groundwater seep with the potential to impact surface water. Over 100 poplar trees were planted in 12 rows perpendicular to the groundwater flow direction. Later, an additional 100 hybrid poplar cuttings were planted between the trees to increase the immediate water uptake capacity. Seep sample locations were not able to be sampled the following summer because they were dry, which was interpreted to be a direct result of the dewatering (i.e., water uptake) capacity of the trees. Thus, although the specific removal of 1,4-dioxane could not be documented, the removal of impacted groundwater was deemed a successful outcome.

7.5.4 SUMMARY

Phytoremediation to remove 1,4-dioxane from soil and groundwater has been repeatedly demonstrated in bench-scale studies, predominantly through the use of hybrid poplars. The primary removal mechanism includes uptake into the plant and transpiration of the 1,4-dioxane out of the leaves to the atmosphere. Because of the short half-life of 1,4-dioxane in the atmosphere, this process could be considered an effective long-term method for the removal and destruction of 1,4-dioxane. Climatic (i.e., growing season) and hydrogeologic (i.e., depth to groundwater) limitations are important factors when considering phytoremediation of 1,4-dioxane.

7.6 BIOREMEDIATION

Bioremediation of common organic contaminants such as chlorinated solvents and petroleum hydrocarbons has been demonstrated to be effective at many project sites, such that it is now considered a presumptive remedy for certain contaminants. In general, bioremediation is implemented by providing amendments to the existing biocommunity in the form of nutrients, food, or oxygen. In some cases, the native bacterial community is not robust enough to effectively address the contaminants present. In these cases, bioaugmentation with specialized bacteria is often performed.

1,4-Dioxane is a relatively stable molecule with strong internal chemical bonding. Therefore, it has been commonly considered to be nonbiodegradable. However, a substantial amount of research has been performed to identify specific organisms that can degrade 1,4-dioxane, the mechanisms of degradation, and the geochemical conditions necessary for these organisms to thrive.

Information presented in this chapter will focus on studies related to furthering the possibility of field application of bioremediation as a technology to carry sites toward closure. Additional information has been presented in Chapter 3 that addresses the identification of organisms capable of degrading 1,4-dioxane and the definition of degradation pathways. Although there is some overlap between these two chapters, the theoretical approaches are primarily addressed in Chapter 3; Chapter 7 deals with potential applied approaches.

7.6.1 Mixed Bacterial Cultures

7.6.1.1 Actinomycete Culture

Adamus et al. (1995) patented a culture comprising an actinomycete, genus *Amycolata*, identified as ATCC 55486, which they claimed was capable of utilizing dioxane as a sole carbon growth source under growth or resting aerobic conditions. The actinomycete strain was tested between 15°C and 35°C on 1,4-dioxane concentrations as high as 10,000,000 µg/L. Adamus et al. (1995) noted in their patent application that, among the cyclic ethers, 1,4-dioxane was considered to be remarkably refractory to biological degradation and generally considered to be not readily biodegradable.

The *Amycolata* strain can be stored for at least 1 month at 4°C with little loss of activity. Its activity is maintained indefinitely at −70°C without cryoprotectants such as skim milk or glycerol, provided that cultures are stored in BSM.

Adamus et al. incubated bacteria in the presence of 100,000–300,000 µg/L of THF in a BSM containing 0.1% weight per volume yeast extract. Once THF degradation is confirmed, the culture is supplied with both THF and 100,000–500,000 µg/L of 1,4-dioxane. Transfer cultures degraded the THF and dioxane to less than 1000 µg/L within 78 h.

The culture—designated as "CB1184"—represents a mixed population including irregularly branched, filamentous organisms with clumps or mats of organisms formed within these filaments. The specific rate of 1,4-dioxane degradation, normalized to 1 mg of the protein of these organisms, was ~0.2–1.2 µg/min when the 1,4-dioxane concentration was maintained at 300,000 µg/L (i.e., 2–3 nmol/min per 1 mg of the organisms' protein).

Because the mixed population includes both eukaryotic microorganisms (protozoa, fungi, and yeast) and prokaryotic organisms (bacteria), tests were conducted to determine which degraded 1,4-dioxane. Antibiotics, which terminate activity in specific bacteria, were added to the culture; treatment with 50 µg/mL tetracycline, streptomycin, and penicillin inhibited the degradation of 1,4-dioxane, whereas treatment with cycloheximide (inhibitor of eukaryotic protein synthesis) did not inhibit the degradation of 1,4-dioxane. Therefore, the degradation of dioxane was traceable to the prokaryotic organisms.

Adamus et al. built a large-scale (40 L capacity) packed-bed reactor to treat groundwater contaminated with 1,4-dioxane, biphenyl, biphenyl ether, ethylene glycol, chlorobenzene, acetone, chloroform, 1,1-dichloroethane, 1,1-dichloroethylene, 1,2-dichloroethene, methylene chloride, methyl chloroform, and toluene. At an effluent temperature of ~15°C, influent and effluent 1,4-dioxane concentrations were 255,000 and 40,000 µg/L, respectively, and the hourly volumetric degradation rate was calculated to be 0.016 g/L. When the bioreactor temperature was 27.4°C, influent and effluent concentrations were 250,000 and less than 1000 µg/L, respectively, and the rate was calculated to be 0.019 g/L.

7.6.1.2 Indigenous Soil Microbes

Taylor et al. (1997) investigated the ability of indigenous soil microbes to degrade a series of contaminants related to the production of plastic films in Rochester, New York. Aerobic and anaerobic tests were conducted on representative groundwater samples. Quantification and isolation of the aerobic and anaerobic bacterial populations and fungi were conducted for growth on the site contaminants acetone, isopropanol, methanol, toluene, dichloroethane, dichloromethane, cyclohexane, and 1,4-dioxane as the sole source of carbon and energy. All the major site contaminants, except for 1,4-dioxane, were utilized by the aerobic and anaerobic bacteria and fungi.

Zenker et al. (1999) sought to determine whether cyclic ethers and alkyl ether oxygenates such as MTBE biodegrade under ambient conditions in soil microcosms. The microcosms used 1,4-dioxane concentrations of 1000 and 200,000 μg/L as well as mixtures of 1,4-dioxane and THF at 200,000 μg/L (Zenker et al., 1999). The researchers noted the complete biodegradation of 1,4-dioxane and THF at 35°C and high concentrations (200,000 μg/L) with the addition of nitrogen, phosphorous, and trace minerals. Biodegradation of 1,4-dioxane was dependent upon the presence of THF. Microcosms using activated sludge and incubated under ambient conditions exhibited no biodegradation of 1,4-dioxane or THF, which suggests that breakdown of 1,4-dioxane is unlikely to succeed by intrinsic bioremediation. The latter involves a demonstration that indigenous subsurface microflora do not have the capability to biodegrade the target contaminants without artificial augmentation.

7.6.1.3 Degradation of Tetrahydrofuran and 1,4-Dioxane Together

In the second important paper on the biodegradation of 1,4-dioxane from North Carolina State University, "Mineralization of 1,4-Dioxane in the Presence of a Structural Analog," a mixed bacterial culture (a "consortium") capable of aerobically biodegrading 1,4-dioxane in the obligate presence of THF was enriched from a 1,4-dioxane-contaminated aquifer sediment. Zenker et al. (2000) used a continuous-flow rotating biological contactor (RBC) to inoculate samples from each soil microcosm bottle at 35°C. The RBC was supplied with 6 L/day of a 25% BSM containing 20 mg/L of THF and 30 mg/L uniformly ^{14}C-labeled 1,4-dioxane. Radio-labeled 1,4-dioxane was utilized to monitor destruction and assimilation (labeling molecules allows the tracing and quantitation of by-products).

THF and 1,4-dioxane disappeared after three months in one microcosm and after 10 months in the remaining two. Microcosms incubated without THF did not exhibit 1,4-dioxane biodegradation, even when respiked with additional 1,4-dioxane. When more THF was added, both compounds were degraded within 75 days. Zenker explained that these data show that the enrichment culture can completely oxidize 1,4-dioxane. A small amount (16.1%) of ^{14}C-labeled 1,4-dioxane was observed in the particulate fraction, which the researchers considered to be cell mass. The majority of the 1,4-dioxane (78.1%) was converted to CO_2, whereas 5.8% remained in the liquid.

Zenker et al. noted that the biodegradation of 1,4-dioxane was dependent on the presence of THF. Repeated addition of 1,4-dioxane in the absence of THF failed to stimulate biodegradation. During the 400 days in which the RBC was operated, samples of the bioreactor culture were periodically placed in BSM containing only 1,4-dioxane. No disappearance of 1,4-dioxane or microbial growth was observed during the course of these experiments.

The toxicity to by-products is due to the formation of HEAA that can denature enzymes or other cellular components. Therefore, the continued biodegradation of the nongrowth substrate is limited by the effects of product toxicity. In contrast, 1,4-dioxane biodegradation did not decrease the observed THF biodegradation rate.

The study showed that 1,4-dioxane does not begin to disappear until THF reaches relatively low levels. This result suggests competitive inhibition, a commonly observed phenomenon in cometabolic processes. Even though Zenker et al.'s study shows that 1,4-dioxane is completely destroyed, produces no toxic by-products, and is partially assimilated into biomass, the ability of the bacterial culture to biodegrade 1,4-dioxane is lost in the absence of THF.

The North Carolina State University research suggests that the loss of biodegradation activity may be due to the depletion of enzymatic cofactors, or coenzymes. The biodegradation of 1,4-dioxane may exert an abnormally high demand on the resources of the consortium. For example, the consortium may invest more energy and/or coenzymes in the biodegradation of 1,4-dioxane than is ultimately yielded. This greater required bacterial effort is consistent with the high energy demand required for scission of the ether bond, in contrast to the small yield of assimilable carbon. The loss of biodegrading activity could also be due to a rapid turnover of the active enzyme once the THF is depleted.

7.6.1.4 Kinetics of Cometabolism

In Zenker et al.'s (2002) third noteworthy contribution to the body of 1,4-dioxane biodegradation knowledge, the kinetics of cometabolism of cyclic ethers by a mixed culture was modeled by using data from their 2000 study to estimate the maximum specific utilization rate of 1,4-dioxane. Note that in biodegradation studies, total suspended solids (TSS) is often used as a surrogate for biomass. The maximum specific utilization rates were determined to be 1.09 mg of THF per milligram of TSS per day and 0.45 mg of 1,4-dioxane per milligram of TSS per day. The half-saturation coefficients (K_S) were measured as 10.8 and 12.6 mg/L for THF and 1,4-dioxane, respectively. No evidence of toxic by-products was observed, and Zenker et al. noted that THF biodegradation was not inhibited by the presence of 1,4-dioxane. However, 1,4-dioxane biodegradation may be competitively inhibited by the presence of THF. The model was capable of predicting the biodegradation of 1,4-dioxane and THF at 1,4-dioxane:THF molar ratios of 0.9:3.3.

The presence of the nongrowth substrate, 1,4-dioxane, did not significantly affect the biodegradation of the growth substrate, THF. If the same enzyme is responsible for the biotransformation of both ethers, then THF apparently binds much more aggressively to the enzyme than does 1,4-dioxane. Zenker et al. also noted that in systems treating dilute industrial effluents, 1,4-dioxane concentrations may be present below the threshold necessary to maintain growth.

7.6.2 BACTERIA GENUS: *RHODOCOCCUS*

Additional work by Cowan et al. (1994) identified a bacterium from the genus *Rhodococcus* that could degrade 1,4-dioxane as the sole carbon and energy source. They found that the bacterial growth rate and the associated biodegradation rate were optimal at 35°C and decreased dramatically above and below that temperature. Cowan et al. evaluated the performance of a fluidized bed reactor inoculated with the *Rhodococcus*, which effectively reduced an initial 1,4-dioxane concentration of 100,000 to less than 1000 µg/L. However, the growth rates were very slow; adequate removal was only achievable with long residence times.

7.6.3 BACTERIA GENUS: *PSEUDONOCARDIA*

7.6.3.1 Degradation of Multiple Ether Pollutants

Vainberg et al. (2006) published a study of the biodegradation of THF, 1,4-dioxane, 1,3-dioxolane, bis-2-chloroethylether (BCEE), and methyl *tert*-butyl ether (MTBE) by *Pseudonocardia* sp. strain ENV478. Strain ENV478 degraded 1,4-dioxane after growth on sucrose, sodium lactate, yeast extract, 2-propanol, and propane, which showed that there was some level of constitutive degradation activity. The highest rates of 1,4-dioxane degradation occurred after growth on THF. Degradation of 1,4-dioxane caused the accumulation of 2-hydroxyethoxyacetic acid (2HEAA). Although ENV478 does not grow directly on 1,4-dioxane, ENV478 degraded 1,4-dioxane for more than 80 days in aquifer microcosms in the presence of THF. The inability of strain ENV478 to grow solely on 1,4-dioxane is related to its inability to efficiently metabolize the 1,4-dioxane degradation product 2HEAA. Vainberg et al. proposed that strain ENV478 may nonetheless be useful as a biocatalyst for remediating 1,4-dioxane-contaminated aquifers.

The Vainberg et al. paper discusses the identification of urinary metabolites of 1,4-dioxane from toxicology studies using rats as indicators of possible bacterial 1,4-dioxane biodegradation pathways. Two urinary metabolites of 1,4-dioxane are 2HEAA and 1,4-dioxane-2-one (PDX). δ-Hydroxy acids are generally unstable in aqueous solutions except as salts; they more commonly

Strain ENV478 degradation rates for bacteria grown on THF were as follows (TSS is a surrogate for biomass):

Organic Contaminant	Rate per Gram of TSS (mg/h)
1,4-Dioxane	21
1,3-Dioxolane	19
THF	63
BCEE	12
MTBE	9.1

The BCEE degradation rates were about threefold higher after ENV478 growth on propane (32 mg/h/g of TSS) than after ENV478 growth on THF, and MTBE degradation resulted in the accumulation of *tert*-butyl alcohol.

occur in the lactone form. The lactonization of 2HEAA produces PDX, as identified in the study of Woo et al. (1977). The initial monooxidation of 1,4-dioxane likely results in the production of 1,4-dioxane-2-ol, the hemiacetal of 2-hydroxyethoxy-2-acetaldehyde. A proposed degradation pathway for the biodegradation of 1,4-dioxane by Strain ENV478 is shown in Figure 7.5.

Strain ENV478, when first grown on THF, degraded 1,4-dioxane at an initial rate of ~21 mg/h/g of TSS, which was approximately one-third the rate at which ENV478 degraded THF. When the THF and 1,4-dioxane were added together in equal amounts, 1,4-dioxane degradation did not occur until all the THF in the medium had been degraded (Figure 7.6). When Strain ENV478 cells grown on THF were incubated with both 1,4-dioxane and 1,3-dioxolane, they degraded both compounds simultaneously at approximately the same rate (Vainberg et al., 2006).

Much slower cell growth and lower 1,4-dioxane degradation rates (per gram of TSS) were observed when Strain ENV478 was grown on sucrose (0.71 mg/h), yeast extract (1.1 mg/h), sodium lactate (0.6 mg/h), 2-propanol (1.5 mg/h), and propane (3.2 mg/h). After growing on lactate, the strain also degraded 1,3-dioxolane (1.0 mg/h). The team concluded that Strain ENV478 has a low level of constitutive activity with these ethers. The team also found that Strain ENV478 does not

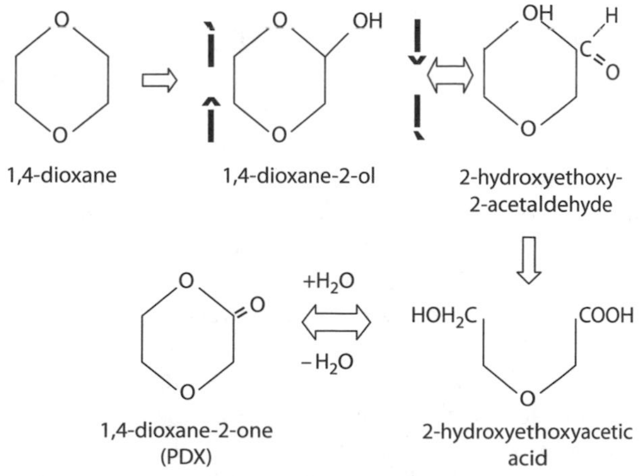

FIGURE 7.5 Proposed partial pathway for biodegradation of 1,4-dioxane by strain ENV478. (From Vainberg, S., McClay, K., Masuda, H., et al., 2006, *Applied and Environmental Microbiology* 72(8): 5218–5224. With permission.)

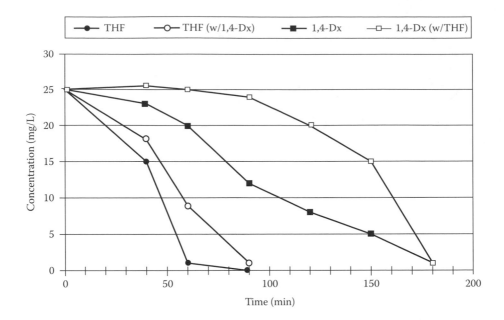

FIGURE 7.6 Biodegradation of 1,4-dioxane and THF by THF-grown strain ENV478 when the compounds were added alone or as a 50:50 mixture. The concentration of volatile suspended solids was 0.38 mg/L. [From Vainberg, S., McClay, K., Masuda, H., et al., 2006, *Applied and Environmental Microbiology* 72(8): 5218–5224.]

grow on 2HEAA as a sole carbon source. To identify the biodegradation products, radio-labeled 1,4-dioxane was used. Analysis of the culture liquor showed that the [^{14}C] dioxane converted to a product that coeluted with authentic 2HEAA. The apparent 2HEAA product was not degraded further, even after 402 h of incubation at 25°C. Addition of THF to cultures that had depleted the 1,4-dioxane and accumulated 2HEAA also did not result in the further degradation of 2HEAA. At the end of the experiment, the radioactivity in the supernatant was not less than the radioactivity in the control. This observation indicated that no carbon dioxide had been produced and no 1,4-dioxane was converted to the Strain ENV478 biomass.

Vainberg et al. incubated aerobic aquifer microcosms for 80 days at 15°C to simulate ambient aquifer temperatures. 1,4-Dioxane at ~700 µg/L was rapidly degraded in the augmented microcosms before the first sampling time (day 20); when repeatedly added at ~1000 µg/L, the 1,4-dioxane was also rapidly degraded. Fresh groundwater was added to the microcosms after 53 days of incubation to replace the volume that had been removed for sampling. This addition created an ~30% dilution of the resident microbial population in the microcosms, with a commensurate decrease in 1,4-dioxane degradation. In the study by Zenker et al. (2000), the system required augmentation with relatively large amounts of THF on a continuous basis to maintain 1,4-dioxane degradation activity. The ability of Strain ENV478 to maintain 1,4-dioxane degradation activity in microcosms for 80 days without THF was, therefore, surprising and encouraging; however, additional work is needed to confirm whether inoculating contaminated aquifers with Strain ENV478 will be effective for economical remediation.

Strain ENV478 is phylogenetically related to Strains CB1190 and K1, but is not able to grow on 1,4-dioxane as a sole carbon source even though it appears that the same enzyme or enzymes in the strain are required to oxidize both substrates. The inability of Strain ENV478 and other THF-degrading bacteria to grow on 1,4-dioxane alone could be related to a lack of induction of the requisite enzyme(s) by 1,4-dioxane or to an inability to efficiently process metabolites for energy production. The research team focused on evaluating why this strain cannot grow on 1,4-dioxane and on whether this strain can be used to degrade 1,4-dioxane in contaminated environments.

Degradation of 1,4-dioxane by Strain ENV478 was greatest after its growth on THF; the presence of THF inhibited the degradation of 1,4-dioxane by this strain. During 1,4-dioxane degradation, Strain ENV478 appeared to accumulate 2HEAA, an expected product of a monooxygenation reaction with 1,4-dioxane. The research team therefore concluded that the degradation of 1,4-dioxane and THF by Strain ENV478 involves an initial oxidation of the cyclic ether by a monooxygenase that presumably is homologous to the THF monooxygenase of Strain K1 (Vainberg et al., 2006).

7.6.3.2 Degradation of 1,4-Dioxane

Parales et al. (1994) isolated CB1190, a pure culture of a nocardioform actinomycete, likely of the family *Pseudonocardia*, in mixed cultures of bacteria from industrial sludge generated at a 1,4-dioxane-contaminated site in Darlington, South Carolina. Bacteria were then grown aerobically on BSM, yeast extract, and THF. CB1190 was demonstrated to have a specific activity for dioxane degradation at a rate of 0.33 mg dioxane/min/mg of protein. Initial concentrations of 5.5 mM 1,4-dioxane (48.4 μg/L) were degraded to less than 0.55 mM (4.84 μg/L).

Parales et al. also confirmed the complete destruction of 1,4-dioxane by CB1190, when 4000 μM 1,4-dioxane and CB1190 were incubated at 30°C for 18 h and degraded to less than 10 μM. About 60% of the carbon in the dioxane molecule was converted to carbon dioxide. CB1190 was also found to degrade the ethers THF, tetrahydropyran, 1,3-dioxane, diethyl ether, and 2-methyl-1,3-dioxolane, as well as the alcohols ethanol, isopropanol, 1-butanol, and 2-butanol. CB1190 would not degrade 1,4-dioxane at temperatures higher than 42°C. Moderately increased salinity did not affect the growth of CB1190 on 1,4-dioxane; 2% NaCl did not slow the growth at all, and some growth was observed at up to 4% NaCl. No growth was observed at 8% NaCl.

Attempts to grow the organism directly on 1,4-dioxane were unsuccessful. Early experiments apparently cometabolized 1,4-dioxane in the presence of THF. Long-term attempts eventually yielded a culture capable of growth on 1,4-dioxane alone, but THF remains the preferred substrate for growth of CB1190. Once the THF was exhausted and only 1,4-dioxane was provided, the cultures were capable of degrading 10 mM 1,4-dioxane daily (Parales et al., 1994).

7.6.4 GASOTROPHIC BACTERIA: COMETABOLIC BIODEGRADATION

Propane-utilizing aerobic bacteria (propanotrophs) capable of cometabolic 1,4-dioxane degradation in laboratory studies have been cultured from contaminated groundwater at several sites (Fam et al., 2005). Natural enzyme systems within these cultures, which are designed to oxidize propane, are capable of cometabolically converting 1,4-dioxane to carbon dioxide. Propane degradation occurs initially, and the destruction of 1,4-dioxane occurs only after most of the propane has been destroyed. The bacterial cultures are able to degrade up to 10,000 μg/L of 1,4-dioxane, but higher concentrations inhibit degradation. The aerobic propanotrophs compete for oxygen with other aerobically degraded contaminants such as acetone, Freon, and chlorinated solvents; therefore field applications with competing contaminants may inhibit propanotroph growth on 1,4-dioxane.

Innovative Engineering Solutions, Inc. (IESI) (2008) performed microcosm studies using groundwater from a 1,4-dioxane-contaminated industrial facility in Michigan and designed a field pilot study as a follow-on. Baseline conditions indicated that the site groundwater was only slightly anaerobic, in the range of nitrate and iron reduction, and therefore was conducive to bio-enhancement to stimulate aerobic propanotrophs. Aerobic microcosms were constructed and filled with site groundwater and nutrients. The presence of propane-utilizing bacteria was confirmed through the addition of propane and observation of species growth. Subsequently, 1,4-dioxane was added to the reactors at the rate of approximately 3 mg/L, and 1–4% propane was sparged through the test solution. Two of the well samples were confirmed to have the appropriate propanotrophic bacteria in sufficient quantities and were observed to reduce 1,4-dioxane concentrations. Bioaugmentation with SL-D, IESI's proprietary propanotrophic bacteria consortium, resulted in greater microbial growth and more rapid 1,4-dioxane degradation. IESI did not observe any evidence that the 1,4-dioxane was directly degraded as the sole carbon and energy source during the 88-day test period.

Global BioSciences, Inc. (2007) evaluated the capability of butane biostimulation to enhance the degradation of 1,4-dioxane in groundwater from a brownfield site in Massachusetts. Samples were enriched by using standard microbiological subculturing techniques and incubated at ~10°C in a BSM for four weeks. Microcosm studies were performed by using static headspace methods over a five-day period to evaluate butane consumption and 1,4-dioxane degradation. Duplicate serum bottles of the biostimulated and control microcosms were sacrificed each day for analysis. Samples demonstrated 1,4-dioxane losses from 6.73 to 7.66 mg/L over the 5-day period, whereas the control bottles exhibited losses from 1.35 to 2.07 mg/L, which were attributed to abiotic losses, specifically leaks through a serum bottle septum. The difference between the total loss and abiotic loss was attributed to undefined microbial activity, either direct metabolism or cometabolism. After 2 days, the 1,4-dioxane concentrations in the serum bottles were below the laboratory detection limit of 0.1 mg/L.

A pilot project is being initiated involving the use of cometabolic propane biosparging at an industrial manufacturing site in the southeast with 1,4-dioxane levels ranging from 18,000 to 580,000 mg/L (T. Renn, AECOM, personal communication, 2009). As is discussed in Section 7.6.4, gasotrophic bacteria have been demonstrated in the laboratory to cometabolically degrade 1,4-dioxane to low levels; therefore, this approach has the potential to achieve the 1,4-dioxane groundwater protection standard through bioremediation.

7.6.5 BIOREACTORS AND BIOBARRIERS

Building on the laboratory work related to the bacterial degradation of 1,4-dioxane, several authors have constructed and tested bioreactors or have proposed the use of bacteria in a biobarrier concept.

Zenker et al. (2004) demonstrated the effectiveness of a trickling filter inoculated with bacteria derived from a 1,4-dioxane-contaminated site. The bacteria were grown in an RBC in the presence of THF and 1,4-dioxane at concentrations of 20,000 and 30,000 μg/L respectively. The bacteria were then inserted into the trickling filter system, which was maintained at 35°C. Various THF and 1,4-dioxane concentration regimes were evaluated; 1,4-dioxane levels were tested at 1000 and 200 μg/L. Removal efficiencies for 1,4-dioxane were 93% at both concentrations. Zenker et al. (2004) confirmed the obligate presence of THF because biodegradation of 1,4-dioxane slowed and then disappeared once THF was depleted.

A bioreactor for *ex situ* cometabolic biotreatment of 1,4-dioxane was designed and implemented at the Lowry Landfill, in suburban Denver, Colorado, following a series of bench-scale studies described in Stanfill et al. (2004). An existing multiprocess treatment plant, including ultraviolet oxidation and GAC, was effective at removing other organic contaminants, but did not effectively remove 1,4-dioxane because of low light transmittance of the influent groundwater and potential interference by inorganic compounds. Groundwater at the landfill contained high levels of 1,4-dioxane (up to 60,000 μg/L) and other organic contaminants. The extracted groundwater was fortuitously contaminated with THF (up to 200,000 μg/L), which had been shown by Zenker et al. (2000, 2004) and others to foster cometabolic degradation of 1,4-dioxane. The bioreactors were constructed to operate in a batch processing mode, with an anoxic initial stage, which created activated sludge. The sludge flowed to a multichamber aerobic reactor, the first stage of which was aerated and agitated; the second stage was designed as a clarifier. The reactors were seeded with biomass derived from a polyester manufacturing facility that treats 1,4-dioxane. Two distinctly different feed waters were used: One was from the toe of the contamination plume and contained high levels of 1,4-dioxane and THF, as well as other organic compounds, and the other was a combination of the high-organic-content water and water from an extraction trench with much lower organic contamination levels. This combination was used to evaluate 1,4-dioxane degradation in the potential combined influent that would be used in a full-scale operating system. A significant proportion of the organic content in both tested waters was attributable to the high concentrations of 1,4-dioxane and THF, which were demonstrated to be degraded in the bioreactor tests. In the highly organic water, the removal rates were lower than in the blended water, suggesting that toxic chemicals in the more contaminated water may have had an inhibitory effect on the biodegradation of both 1,4-dioxane and THF. The removal rate was as high as 97.5% for 1,4-dioxane and 99.3% for THF in the blended water tests, but only 87.3% and 97.9%, respectively, in the highly organic water. Although the field pilot-testing indicated successful reduction of 1,4-dioxane, below the detection limit in some cases, the detection limit was 360 μg/L, well above health-based and regulatory cleanup criteria in many states and EPA regions.

Evans et al. (2006) measured the specific activity of 1,4-dioxane metabolized by CB1190 at 0.33 μg/min per milligram of protein. A variety of bioreactor configurations was tested to measure maximum achievable biodegradation rates under optimized conditions. The bioreactors used the microbe strain CB1185, a mixed culture that includes CB1190, grown in batch on BSM with initial 1,4-dioxane concentrations of 5000 mg/L and fed additional 1,4-dioxane as needed to maintain the required biomass and growth rate. Reactors supported by silica sand were inoculated with 100 L cultures containing 2500 mg of protein per liter. The dimensions of the packed-bed reactor were 0.5 ft in diameter and 10 ft in height; the groundwater flow rate was 0.5–0.8 L/min, and the air-flow rate was 2 L/min. Evans et al. measured the best performance in the sequence batch reactor with biomass carriers. Biodegradation rates were slow, and reactor efficiency declined over time, possibly because of bacteria washout, low temperatures, toxic contaminants or by-products, and imbalance between the rates of bacteria growth and die-off (Evans et al., 2006). Table 7.1 presents a summary of the laboratory testing results of various bioreactor designs and 1,4-dioxane concentrations for influent and effluent.

Microvi Biotech LLC has developed a process that removes 1,4-dioxane from groundwater in an *in situ* biobarrier or an *ex situ* bioreactor (F. Shirazi, written communication, 2007). This treatment uses *Rhodococcus* sp., which under aerobic conditions can degrade 1,4-dioxane as the sole carbon or energy source, without the addition of a secondary substrate (Cowan et al., 1994, as referenced in Zenker et al., 2000). The bacteria are isolated, grown on glucose, and encapsulated in a porous polymeric matrix to increase survivability, reduce bacterial washout, and provide an improved environment for the stability and longevity of the bacteria. Figure 7.7 shows a scanning electron microscope photograph of the encapsulated microorganisms within the matrix. Preliminary research and bench testing demonstrated that

TABLE 7.1
Laboratory Bioreactor Results Summary

Bioreactor	Run Time (h)	1,4-Dioxane In (mg/L)	1,4-Dioxane Out (mg/L)	Biodegradation Rate[a]
Activated sludge with recycle	60	100	0.42	1.8
Activated sludge with recycle	43	1200	270	24
RBC	55	91	1.0	1.8
RBC	55	5000	1800	54
Packed bed reactor	<9	100	<0.04	12
Packed bed reactor	<9	2	<0.05	0.24
Sequencing batch reactor	24	100	<1.0	1.8
Sequencing batch reactor with biomass carriers	12	1200	<1.0	250
Fluidized bed reactor	15	10	<1.0	117

Source: Evans, P.J., Parales, R.E., and Parales, J.V., 2006, *18th Symposium in Groundwater Resources Association Series on Groundwater Contaminants: Emerging Contaminants in Groundwater: A Continually Moving Target, June 7–8, 2006.* Concord, CA. Housed at the University of California Water Resources Archives, Berkeley, CA.

[a] Units are milligrams of protein per liter per hour.

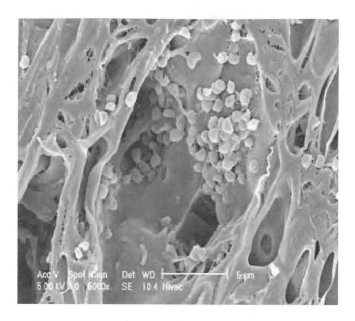

FIGURE 7.7   Encapsulated microorganisms shown inside MicroBead. (From F. Shirazi, personal communication, 2007.)

the isolated *Rhodococcus* sp. was capable of degrading 1,4-dioxane in concentrations of greater than 10,000 µg/L in less than 50 h (Figure 7.8); however, the effectiveness at lower 1,4-dioxane concentrations has not been demonstrated. The rate of 1,4-dioxane degradation per day (46,000–56,000 µg/L) was maintained for more than 300 days in the laboratory reactor without preconditioning the bacteria to 1,4-dioxane or adding any substrate to support cometabolic biodegradation.

7.6.6   ANAEROBIC BACTERIA

The anaerobic degradation of 1,4-dioxane was evaluated under iron-reducing conditions involving anaerobic sludges by Pan and Chen (2006) of Zhejiang University in Hangzhou, China. They determined that the iron-reducing bacteria present in the anaerobic sludge could intrinsically degrade 40% of 1,4-dioxane in the microcosm without the need for amendments. Adding humic substances, which may have acted as a carbon source, stimulated the anaerobic degradation of up to 70% of the 1,4-dioxane present. The degradation was caused by iron-reducing bacteria in which Fe(III) was available as an electron acceptor. 1,4-Dioxane biodegradation was greatest in the presence of added humic substances and Fe(III) under all the redox conditions. Complete degradation of 1,4-dioxane was observed (Pan and Chen, 2006).

Shen et al. (2008) presented additional information related to their studies of iron-reducing bacteria for 1,4-dioxane degradation. The degradation observed was determined to be entirely from biological activity, because sterile conditions resulted in no statistically significant declines in 40 days. Degradation of 1,4-dioxane in reactors incubated with iron-reducing bacteria derived from the anaerobic sludge resulted in a decline of 25% (from 13 mg/L) over 40 days with no amendments. Introduction of Fe(III) oxide caused a decline of 37% and adding Fe(III) oxides and humic acids led to a reduction of 62% over the 40-day period. The humic acids are thought to act as a catalyst, promoting biodegradation by the iron-reducing bacteria, rather than creating reducing conditions, as is common in biodegradation of other contaminants, such as chlorinated VOCs. A second set of experiments conducted after the initial 40-day period, using the same inoculated bottles but replenishing 1,4-dioxane up to 50 mg/L, resulted in even higher daily

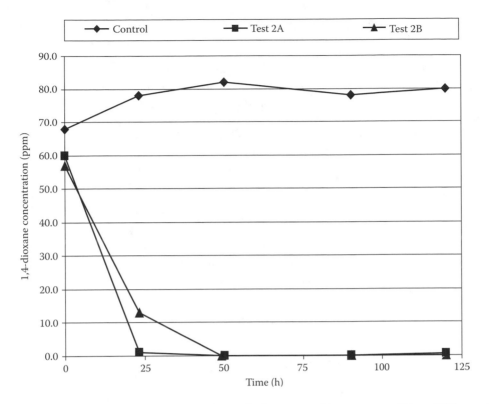

FIGURE 7.8   Reduction in 1,4-dioxane in duplicate *Rhodococcus* sp. bioreactor tests (2A and 2B), as well as a sterile control. (From F. Shirazi, personal communication, 2007.)

degradation rates: 0.7 mg/L versus the initial rate of 0.2 mg/L. Shen et al. attributed this higher rate to the presence of an established and robust bacterial population. They concluded that the iron-reducing bacteria were using the 1,4-dioxane as the sole source of carbon; however, depending on the source of humic material added, other organic compounds may fovored cometabolization of 1,4-dioxane.

7.6.7 FUNGI

A patented remediation process to reduce concentrations of 1,4-dioxane in groundwater has been developed by Patt and Abebe (1995). The remediation requires the addition of a pure culture of a fungal microorganism that degrades cyclic ethers, a cometabolite (ideally THF), and oxygen. The process involves fermentation of the chemical pollutant as the substrate—in this case 1,4-dioxane—in the presence of THF and oxygen by a nonbacterial fungus (*Aureobasidium pullulans*), identified by the Northern Regional Research Laboratories as NRRL 21064. Compared to bacterial cultures, the fungus has a broader range of pH tolerance and therefore would be suited to a wider range of site conditions. Additionally, sufficient dissolved oxygen, as is normally found in natural groundwater, is a prerequisite for fungal survival. The fungus is also not as sensitive to temperature, being functional between 5°C and 37°C, although lower temperatures result in slower metabolism and higher temperatures kill the fungus. The process is proposed as being applicable in batch mode (lakes, ponds, and reactors) or in continuous mode (reactors, ponds with inflow and outflow, and *in situ* groundwater). The information provided in the patent does not elaborate on the relationship between the THF and the 1,4-dioxane except to clearly define the THF as an obligate cometabolite.

Patt and Abebe's patent cites two applied examples of 1,4-dioxane degradation using *Aureobasidium pullulans*. In the first application, a 2-million-gallon aeration lagoon containing 1,4-dioxane at a concentration of 8000 μg/L and THF at concentrations between 5000 and 8000 μg/L was treated with a 300-gallon volume of *Aureobasidium pullulans* inoculum containing about a million cells per milliliter; hence, ~1.1 trillion cells were added to the 2-million-gallon lagoon. 1,4-Dioxane concentrations decreased to below an unspecified detection limit within 2 months.

In the second example, a fluidized-bed reactor (FBR) was inoculated with 10 billion cells of *Aureobasidium pullulans* and supplied with groundwater contaminated with 1,4-dioxane at a concentration of 8000 μg/L and supplemented with THF. The FBR retention time was 1 h, and the pH was between 6 and 7. Effluent contained 1,4-dioxane in concentrations from just a few micrograms per liter to 2000 μg/L and less THF than the 500 μg/L detection limit (Patt and Abebe, 1995).

Kim et al. (2006) made a comparison between the destruction of 1,4-dioxane by advanced oxidation technology (AOT) and biodegradation using BAC-TERRA, a filamentous fungi. AOT was demonstrated with ultraviolet + hydrogen peroxide and a modified Fenton's chemistry, both of which demonstrated pseudo first-order kinetics, with a rate coefficient of 5×10^{-4} s^{-1}. The fungal degradation also exhibited pseudo first-order kinetics, but with a rate coefficient that was two orders of magnitude lower (2.38×10^{-6} s^{-1}) than that shown for AOT.

7.6.8 ENHANCING BIODEGRADABILITY WITH *IN SITU* CHEMICAL OXIDATION

Several authors have performed various studies to determine the effect of *in situ* chemical oxidation (ISCO) on the biodegradation of 1,4-dioxane.

Adams et al. (1994) evaluated the use of hydrogen peroxide and ozone to enhance the biodegradation of 1,4-dioxane using a mixed microbial culture obtained from a municipal wastewater treatment plant. Upon subjecting the test cells to hydrogen peroxide/ozone ratios of 0.0 (ozone only), 0.5, and 1.0, the biological oxygen demand (BOD$_5$) increased, and, following a lag period, the chemical oxygen demand (COD) decreased. This rise in BOD$_5$ and the limited drop in COD suggests that biodegradability of the initial oxidation by-products occurred more readily than the biodegradation of the original 1,4-dioxane. Biodegradability was also assessed by using shaker-table bioassays to measure COD removal rates at different hydrogen peroxide/ozone molar ratios. Minimal biodegradation was observed in the samples with no oxidant, but up to 85% reduction in COD was observed in 57 h in oxidized samples. Higher residual COD (up to 37%) was present in the samples with lower oxidant input because of residual untreated 1,4-dioxane. The ratio of oxidant to 1,4-dioxane in these experiments was lower than the ratio required for direct and complete oxidation of the 1,4-dioxane. Adams et al. (1994) concluded that, for the treatment of 1,4-dioxane in wastewater, significant reduction in oxidant usage could be obtained by optimizing an oxidation system to enhance biodegradability of the 1,4-dioxane, rather than full oxidation of the contaminant.

Suh and Mohseni (2004) looked at the kinetics of 1,4-dioxane oxidation with ozone and peroxide and the relationship of these oxidants to enhanced biodegradability. They also looked at the effects of 1,4-dioxane concentration, pH, and hydrogen peroxide concentration on 1,4-dioxane biodegradability. The fact that initial seeded solutions contained minimal BOD$_5$ indicated that 1,4-dioxane was not biodegradable if conventional bacterial methods were used. After treatment with hydrogen peroxide and ozone, 1,4-dioxane and COD decreased and BOD$_5$ increased, indicating oxidation of 1,4-dioxane and creation of biodegradable intermediates. Oxidation of 1,4-dioxane was found to follow a Langmuir–Hinshelwood–type model following first-order kinetics and transitioning to zero-order kinetics at higher concentrations (>60,000 μg/L). The overall kinetics, as well as the transition from first order to zero order, is affected by pH, hydrogen peroxide, and presence of other intermediate species. Increases in pH, up to ~9, increase biodegradability. Increasing hydrogen peroxide levels enhances biodegradability up to a point (0.4–0.45 mol:mol hydrogen peroxide:ozone), after which the hydrogen peroxide has a negative effect, reduces 1,4-dioxane destruction, and thereby decreases biodegradability.

7.6.9 BACTERIA SUMMARY

Laboratory and field pilot studies have identified several different types of bacteria that are effective at either utilizing 1,4-dioxane as an energy source or cometabolically degrading 1,4-dioxane while consuming some other energy source. Chapter 3 summarized the laboratory studies aimed at identifying the requisite bacteria and optimal conditions, defining degradation pathways, and determining biodegradation rates. The available literature related to the potential field application for soil or groundwater remediation of 1,4-dioxane indicates that the bacteria are dominantly aerobic and, although they may be naturally occurring, the natural conditions generally encountered at hazardous waste sites are not conducive to high degradation rates without some form of biostimulation. Biodegradation of 1,4-dioxane has been applied *ex situ* in bioreactors at the field scale, but to date, there have been no field-scale bioaugmentation or biostimulation projects utilizing *in situ* aerobic bioremediation for 1,4-dioxane for an entire groundwater plume.

7.7 CHEMICAL OXIDATION

Oxidation is a contaminant-destruction technology in which oxidation–reduction (redox) chemical reactions convert hazardous contaminants to nonhazardous or less toxic compounds. Redox reactions involve the transfer of electrons from one compound to another. During chemical oxidation, the redox reaction oxidizes one reactant (the contaminant, which loses electrons), whereas the other reactant (the oxidizer, which gains electrons) is reduced. Oxidizing agents that may be used in the treatment of drinking water, wastewater, and groundwater include ozone, hydrogen peroxide, persulfate, hypochlorite, chlorine, potassium permanganate, and Fenton's reagent (hydrogen peroxide and iron). Applications of chlorine-based methods for the treatment of groundwater are not common. The strength of the oxidant—measured as oxidation potential—varies, and oxidation potentials of more than 2.0 V are generally considered to effectively address 1,4-dioxane. The strongest oxidant is fluoride, which is impractical for water treatment. The hydroxyl radical is nearly as strong as fluoride and has been effectively employed for both *in situ* and *ex situ* remediation. Table 7.2 presents a compilation of common oxidants, their oxidation potential, and their strength relative to chlorine. Chemical oxidation for the destruction of 1,4-dioxane *ex situ* is a proven technology demonstrated by numerous full-scale applications.

Because 1,4-dioxane is fully miscible in water, it is an ideal candidate for extraction and *ex situ* treatment (e.g., "pump and treat"). Figure 7.9 presents a comparison of laboratory simulations of

---

TABLE 7.2
Oxidation Potentials for Some Common Oxidizers

Oxidant	Oxidation Potential (V)	Relative Strength (Chlorine = 1)
Hydroxyl radical (OH$^\bullet$)	2.7	2
Sulfate radical (SO$_4^-$)	2.6	1.8
Ozone (O$_3$)	2.2	1.5
Persulfate anion (S$_2$O$_8^{2-}$)	2.1	1.5
Hydrogen peroxide (H$_2$O$_2$)	1.8	1.3
Permanganate ion (MnO$_4^-$)	1.7	1.2
Chlorine (Cl$^-$)	1.4	1
Oxygen (O$_2$)	1.2	0.9

Source: DiGuiseppi, W.H. and Whitesides, C., 2007, *Environmental Engineer* 43(2): 36–41.

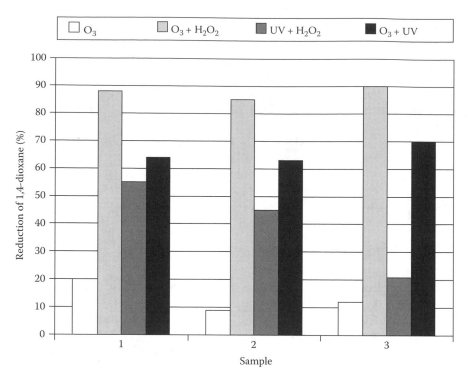

FIGURE 7.9 Reduction in 1,4-dioxane in three samples of drinking water (start value: 230 μg/L) through the use of four different remediation treatments. (H. Davis, personal communication, 2007; Courtesy of ITT—WEDECO.)

various *ex situ* chemical oxidation treatment technologies for 1,4-dioxane removal, provided by ITT-WEDECO (H. Davis, personal communication, 2007). Tap water samples from Herford, Germany, were spiked with 1,4-dioxane at a concentration of 230 μg/L and treated with ozone (O_3), ozone plus hydrogen peroxide (H_2O_2), hydrogen peroxide plus ultraviolet light, and ozone plus ultraviolet light. These specific technologies are discussed in further detail in this section, but with 1,4-dioxane removal rates of up to 90%, these are clearly demonstrated methods to be considered for 1,4-dioxane treatment. Site heterogeneity and undiscovered source areas often hinder successful applications of pump-and-treat technologies. When applying oxidation technologies, the effect of oxidizing conditions on inorganic elements present in extracted groundwater or *in situ* materials must be evaluated and addressed. Most notably, bromate formation from naturally occurring bromide may occur in chemical oxidation applications involving hydroxyl radicals. Additionally, hexavalent chromium, which is often a cocontaminant at solvent sites, is not addressed by chemical oxidation, and its concentration may be increased by creating strong oxidizing conditions in the subsurface. Arsenic, selenium, and other metals may also exhibit increased solubility, causing elevated groundwater concentrations under oxidizing conditions.

7.7.1 Hydrogen Peroxide and Ozone—*Ex Situ*

The combination of ozone (O_3) and hydrogen peroxide (H_2O_2) is known in chemistry as peroxone and has been used for specific contaminant remediation applications, such as TCE and explosives (Fleming et al., 1997). At equilibrium in water, hydrogen peroxide is present as $HO_2^- + H^+$. The hydroperoxide ion (HO_2^-) reacts with ozone to form hydroxyl radicals (OH•). *Ex situ* peroxone applications have been demonstrated in the treatment of drinking water and in the remediation of contaminated groundwater.

Bowman et al. (2007) demonstrated the effectiveness of injecting hydrogen peroxide and ozone under high pressure in a continuous in-line process to treat feed water containing an aqueous mixture with 1,4-dioxane and chlorinated compounds. In the HiPOx™ (Applied Process Technology, Inc., Pleasant Hill, CA) system, hydrogen peroxide is injected into the water stream first, followed by high-pressure ozone. Ozone is produced at 40–50 psig on site from a solid-state ozone generator supplied with liquid oxygen or from oxygen from the atmosphere, generated with a pressure swing absorption system. The ozone is injected at specific locations along the in-line reactor flow path to avoid the creation of undesirable by-products, such as bromate (Figure 7.10).

The HiPOx system was tested with water from three groundwater extraction and treatment systems operating in California with elevated chlorinated VOC and 1,4-dioxane levels. A system at South El Monte, California, was pilot-tested at 10 gpm for perchloroethylene levels of 150 µg/L and 1,4-dioxane of 20.2 µg/L. Effluent levels in initial testing achieved perchloroethylene reductions to 2.1 µg/L and 1,4-dioxane reductions to the detection limit at 2 µg/L. Because the existing water treatment for the contaminated drinking-water production well included liquid-phase GAC for effective treatment of chlorinated VOCs, the scaled-up design for the permanent production well solution was optimized for the removal of 1,4-dioxane. The final 490-gpm system was installed pre-GAC and effectively lowered the 1,4-dioxane concentration from 4.6 µg/L to below the detection limit at 2 µg/L (Bowman et al., 2007), but allowed low levels of chlorinated VOCs to exit the system. This approach addressed the 1,4-dioxane issue, while keeping the inexpensive GAC system operational and extending the life of the existing GAC media by lowering influent chlorinated VOC levels.

A second system in the City of Industry, California, was tested as pre- and post-treatments to an existing air stripper that was effectively treating chlorinated ethenes and ethanes as high as 730 µg/L (for methyl chloroform) to below California's regulatory threshold. 1,4-Dioxane levels as high as 610 µg/L decreased slightly in the air stripper to 430 µg/L, although the exact mechanism for its removal was not determined or documented. Pilot-testing was optimized for 1,4-dioxane removal and not for the removal of chlorinated ethanes (e.g., methyl chloroform, 1,1-dichloroethane) because these compounds are approximately one-tenth as susceptible to hydroxyl radical oxidation as the

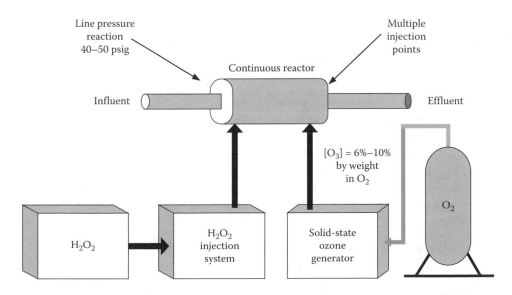

FIGURE 7.10 Schematic of HiPOx™ advanced oxidation system. [From Bowman, R.H., Miller, P., Purchase, M., and Schoellerman, R., 2007, Ozone-peroxide advanced oxidation water treatment system for treatment of chlorinated solvents and 1,4-dioxane. http://www.aptwater.com/assets/tech_papers/Paper-1.4Dioxane.pdf (accessed August 1, 2007).]

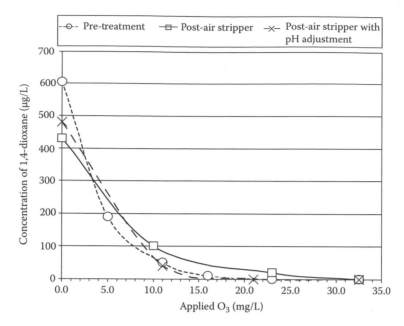

FIGURE 7.11 Plot of 1,4-dioxane concentration as a function of applied ozone at the City of Industry AOP Pilot Study. (From Bowman, R.H., Miller, P., Purchase, M., and Schoellerman, R., 2007, Ozone-peroxide advanced oxidation water treatment system for treatment of chlorinated solvents and 1,4-dioxane. http://www. aptwater.com/assets/tech_papers/Paper-1.4Dioxane.pdf [accessed August 1, 2007].)

corresponding chlorinated ethene (Bowman et al., 2007). Increasing hydrogen peroxide and ozone injections to eliminate the methyl chloroform would have increased costs dramatically for the air stripper system that was already fully capable of removing chlorinated ethanes. Pilot-testing of the HiPOx advanced oxidation process (AOP) system as a post-treatment option revealed that the elevated pH of the stripper effluent, caused by the addition of an antifouling agent, reduced the efficiency of the AOP. Several different injection rates for ozone were tested to maximize 1,4-dioxane removal and assess the optimal order of treatment for the system (i.e., pre- or post-air stripper) (Figure 7.11). On the basis of the results, the system was designed for use before the existing air stripper and was optimized to achieve the removal of 1,4-dioxane to levels below the detection limit of 2 μg/L. Chlorinated VOCs leaving the AOP system were effectively treated in the air stripper.

The third test site, in Mountain View, California, was treating chlorinated VOCs (as high as 8800 μg/L for TCE). 1,4-Dioxane levels as high as 16 μg/L were untreated by the existing system (Boarer and Milne, 2004). Testing confirmed that treatment goals for all the contaminants of concern, including 1,4-dioxane, could be achieved with the HiPOx as a replacement for the air stripper while retaining the GAC system as a polishing unit to address compounds resistant to complete removal by oxidation (e.g., 1,1-dichloroethane) (Bowman et al., 2007). Overall, the HiPOx technology has been demonstrated in a variety of applications and conditions and is capable of complete destruction of 1,4-dioxane and associated chlorinated ethenes. The HiPOx system can also be used to destroy 1,4-dioxane and decrease the chlorinated VOCs to sufficiently low levels for enhancing or prolonging the effectiveness of existing treatment systems.

7.7.2 HYDROGEN PEROXIDE AND OZONE—*IN SITU*

As of the writing of this book, there are no completed and documented full-scale remediation projects involving ISCO in the literature. Several pilot studies have been performed confirming the efficacy of the technology.

Schreier et al. (2006) performed a series of bench-scale studies assessing the viability of using ozone alone for ISCO, as opposed to mixing ozone and hydrogen peroxide *in situ*, which can be difficult in heterogeneous aquifer materials. The work was performed in support of remedial action at the Cooper Drum Superfund site in South Gate, Los Angeles County, California, where 1,4-dioxane concentrations targeted for treatment by ISCO were as high as 750 µg/L (Yunker, 2007). Peroxone, the mixture of ozone and hydrogen peroxide, is a demonstrated oxidizer, as already described, but ozone alone had been commonly regarded as not being of sufficient oxidation potential to destroy 1,4-dioxane. Ozone sparging of site groundwater resulted in the complete removal of 1,4-dioxane. To assess whether the removal was due to volatilization or destruction, vapor samples were collected from the off-gas and a sparge with inert gas (nitrogen) was evaluated. The data indicate that 1,4-dioxane removal was due to oxidation. Schreier et al. postulated that naturally occurring inorganic materials in the site soil or groundwater were enhancing the capability of ozone, which could eliminate or reduce the need for hydrogen peroxide. Ferrous iron, chelated iron, bicarbonate, and cocontaminants such as TCE were evaluated as potential oxidation enhancers. A test of spiked deionized water indicated that ozone alone was capable of reducing concentrations from 400 to 98 µg/L, but not to the detection limit of 3 µg/L. Additional tests run with ferrous iron, chelated iron, and bicarbonate resulted in reductions down to the detection limit. 1,4-Dioxane removal in the presence of TCE did not significantly differ from the removal observed with ozone alone. Schreier et al. (2006) concluded that natural waters may contain elements or compounds that allow ozone to succeed at complete 1,4-dioxane removal.

A follow-on field pilot study at the same site (Sadeghi et al., 2006; Sadeghi and Gruber, 2007; USEPA, 2007) involved the use of specially constructed wells with an ozone diffuser in the bottom of the well and a hydrogen peroxide diffuser higher in the well, such that the two amendments interact in the subsurface to form the hydroxyl radical. This system is patented by Applied Process Technology as the Pulse-OX™ (Applied Process Technology, Inc., Pleasant Hill, CA) system. Initial injections were of ozone alone, which resulted in significant reductions in 1,4-dioxane (and TCE) concentrations (Figure 7.12). Five months later, hydrogen peroxide injections were initiated, and it was determined that optimal results were achieved with the combination of 16% hydrogen peroxide and 2 pounds per day per injection well or 1 pound per day per injection interval of ozone. The

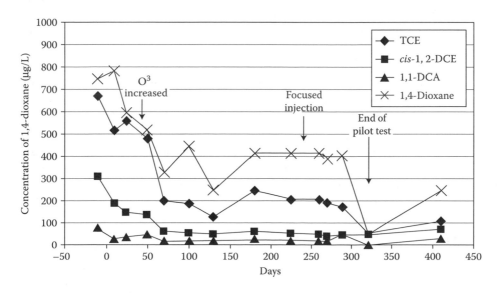

FIGURE 7.12 ISCO field pilot study results. (From Sadeghi, V.M. and Gruber, D.J., 2007, *In situ* oxidation of 1,4-dioxane with ozone and hydrogen peroxide. Poster presentation at URS's Environmental Technology and Management Seminar, Oakland, CA.)

injection interval was best situated not more than 10 ft below the target interval for maximum removal. Reductions as high as 88% were achieved in some of the monitoring wells (Table 7.3), and all the target contaminant concentrations decreased during the pilot test period. However, none of the concentrations were reduced to the detection limits during the field pilot test. The recommendation of this study was for the design and implementation of a full-scale system with more intense oxidation in the source zone and lesser efforts in the dissolved portion of the plume. The project is ongoing at the time of publication.

Kerfoot and Brolowski (2006) and Kerfoot (2008) described bench- and field-scale pilot-testing for 1,4-dioxane oxidation using ozone alone and ozone with hydrogen peroxide. Ozone reactivity exhibited a 1,4-dioxane removal rate of 0.15% per minute and total removal rates of 69–78% in soil slurries contaminated with 1,4-dioxane and chloroethene. Overall, the reaction rates were computed to be in the range of values previously defined for TCE and perchloroethylene, indicating that ISCO application of this technology could be as effective for 1,4-dioxane removal as for TCE and perchloroethylene removal. A field application was performed with the Perozone® system (Kerfoot Technologies, Inc., Mashpee, MA), which uses hydrogen peroxide-coated ozone microbubbles (< 50 μm) delivered with a proprietary sparging system, referred to as the C-Sparger® (Kerfoot Technologies, Inc., Mashpee, MA) (Kessel et al., 2005). The two-month-long study evaluated the technology application in a barrier configuration designed to protect an open-water channel affected by groundwater discharges. Preliminary results indicated decreases in 1,4-dioxane concentrations within the ~25-feet radius of influence of the injection wells. Kerfoot and Brolowski (2006) noted that the removal rates depended upon contaminant concentrations and competitive destruction, because of the presence of other organic contaminants. Later bench-scale testing of nanoscale (250 nm) ozone sparge bubbles, produced by a miniature Laminar Spargepoint® (Kerfoot Technologies, Inc., Mashpee, MA), indicated equivalent effectiveness between ozone alone and ozone plus hydrogen peroxide. The nanobubbles exhibited a greater resistance to collapse and a longer half-life (more than 20 h) than microbubbles (Kerfoot, 2008). This technology was applied as a full-scale remedy for the site, but results were not available at the time of publication.

A field pilot study performed by Pall Corporation (2004) involved injecting hydrogen peroxide alone (to be combined with naturally occurring ferrous iron as a Fenton's reagent reaction), ozone alone, and ozone plus hydrogen peroxide. Results indicated that all three methods were effective to varying degrees. Ozone injection caused significant decreases in 1,4-dioxane concentrations, but also caused the formation of bromate above the maximum contaminant level of 10 ppb. The

TABLE 7.3
ISCO Field Pilot Study Results

Well	COC	Initial Concentration in July 2005 (μg/L)	Concentration in June 2006 (μg/L)	Change Since Start of Pilot Study (%)	Concentration in August 2006 (μg/L)	Change Since Start of Pilot Study (%)
EW-1 (63 ft)	TCE	660	65	−90	120	−82
EW-1 (63 ft)	1,4-Dioxane	750	47	−94	250	−67
MW-33A	TCE	940	180	−81	130	−86
MW-33A	1,4-Dioxane	630	99	−84	74	−88
MW-20	TCE	520	110	−79	140	−73
MW-20	1,4-Dioxane	140	79	−44	71	−49

Source: Sadeghi, V.M. and Gruber, D.J., 2007, *In situ* oxidation of 1,4-dioxane with ozone and hydrogen peroxide. Poster presentation at URS's Environmental Technology and Management Seminar, Oakland, CA.

contaminated aquifer exhibited high levels of bromide, which exacerbated the bromate formation issue. Above-ground treatment using ozone and peroxide can be adjusted to control the reactions, whereas subsurface reactions cannot be adequately controlled to ensure that bromate levels do not become a problem.

7.7.3 HYDROGEN PEROXIDE AND ULTRAVIOLET LIGHT

Ultraviolet + peroxide oxidation systems cleave the hydrogen peroxide molecule (H_2O_2) into two hydroxyl radicals (OH·) by direct photolysis caused by the energy absorbed from an ultraviolet lamp.

Experiments performed by Stefan and Bolton (1998) defined the degradation pathways for 1,4-dioxane by hydroxyl radicals generated by direct ultraviolet photolysis of hydrogen peroxide. They determined that the destruction of 1,4-dioxane by ultraviolet + peroxide follows pseudo first-order kinetics and was more rapid than direct photolytic destruction of 1,4-dioxane using ultraviolet light. Within the first 5 min of irradiation, almost 90% of the initial concentration of 1,4-dioxane was depleted.

Ultraviolet + peroxide is a reliable, demonstrated *ex situ* technology, that is commercially available from a number of vendors. A case study presented in *Pollution Engineering* in July 1999 described a system installed to remove 1,4-dioxane following an air stripper that removes chlorinated compounds. The system demonstrated 1,4-dioxane removal from an initial concentration of 600 μg/L to below unspecified detection limits at a rate of 100 gpm.

An improvement to the typical ultraviolet + peroxide system was proposed and tested by Safarzadeh-Amiri et al. (1996), who used an ultraviolet + ferrioxalate + hydrogen peroxide system for treatment of highly contaminated waste streams. The ferrioxalate complex [$Fe(C_2O_4)_3^{3-}$] is widely studied and is used to measure light intensity. The photolysis of ferrioxalate yields ferrous iron, which then reacts with the hydrogen peroxide to generate the hydroxyl radicals and provide a constant source of Fenton's reagent (discussed further in the following section). Safarzadeh-Amiri et al. (1996) demonstrated removal of 1,4-dioxane, among other contaminants, to below unspecified detection limits at substantially higher efficiency than that of ultraviolet + peroxide systems. This greater efficiency would result in fewer ultraviolet bulbs being needed and therefore lower maintenance and energy costs.

In a German application, bench testing for 5000 gallons per day of process wastewater containing 15,000,000–20,000,000 μg/L 1,4-dioxane achieved a 99.95% removal efficiency by using ultraviolet light and peroxide (Sörensen and Weckenmann, 2006).

7.7.4 HYDROGEN PEROXIDE AND FERROUS IRON

Fenton's chemistry involves catalyzing hydrogen peroxide, typically with a transition metal such as ferrous iron, to produce hydroxyl radicals, which are strong, nonspecific oxidizers. For the reaction to work properly, water must be acidified to a pH between 3 and 5. The iron catalyst is typically added as iron sulfate ($FeSO_4$), and then hydrogen peroxide is slowly added to control generation of heat from the reaction.

Pall Corporation (2004) performed multiple phases of field testing to evaluate Fenton's chemistry for the treatment of 1,4-dioxane *in situ*. Initial testing of hydrogen peroxide injections, which would have relied on naturally occurring ferrous iron, indicated that the 1,4-dioxane was completely broken down in the test area, without the production of bromate. However, a follow-up field study concluded that the 1,4-dioxane could not be destroyed *in situ* to a significant extent by peroxide injections alone. Potential injections of ferrous iron and peroxide were evaluated but are not documented in the literature.

Rushing et al. (2006) performed large-scale batch experiments using contaminated soil and groundwater from a coastal-plain pharmaceutical site. Naturally occurring ferrous iron was sufficient to catalyze the hydrogen peroxide injected at a concentration of 10 g/L into the reactors that were maintained at a pH of 4. Samples were collected before adding peroxide and after 24 h of reaction time. The concentration of 1,4-dioxane, which was not the focus of the study, was reduced 99% from 15,000 to 110 µg/L. Rushing et al. (2006) also noted a relatively small decline in COD and total organic carbon (TOC) compared to the large reductions in the targeted contaminants of concern. They concluded that the parent compounds were degraded, but complete destruction of the organic compounds was not achieved.

An additional Fenton's reagent pilot study was performed to address 1,4-dioxane in saprolitic soils and groundwater at an active printing facility in North Carolina (Masten, 2004). Analytical data, which were collected 45 days postinjection, demonstrated that 1,4-dioxane was decreased by 98% in the pilot-testing area and similar concentration reductions were confirmed down gradient.

Laboratory bench-scale tests using modified Fenton's chemistry (Isotec, 2008) resulted in greater than 99% reduction in 1,4-dioxane levels in groundwater from 26,500 µg/L to ND at 55.1 µg/L. Although this detection level is well above many states' promulgated groundwater standards, this level of reduction, with one application, of a demonstrated aggressive oxidant suggests that achievement of lower concentration standards is likely. Soil remediation testing was also performed with comparably positive results, with 1,4-dioxane declining from 330 to less than the 1.1 µg/kg detection limit. This removal from soil took two doses of the modified Fenton's oxidant to achieve.

7.7.5 Persulfate

Persulfate is a strong oxidizer, with a standard oxidation potential of 2.1 V, which is higher than that of hydrogen peroxide (1.8 V) and the permanganate anion (1.7 V) (Block et al., 2004). Persulfate can be "activated" in a number of ways to create sulfate radicals (SO_4^-) that have an oxidation potential (2.6 V) almost as high as that of the hydroxyl radical (2.7 V). Activation methods for persulfate include heat, alkali, chelated iron, and hydrogen peroxide. Reactions involving the sulfate radical are kinetically fast, and the sulfate radical is more stable, and therefore longer lived, than the hydroxyl radical. The hydroxyl radical has a half-life of minutes to hours, whereas the sulfate radical has a half-life of hours to weeks, depending on aquifer geochemistry (Huling and Pivetz, 2006), allowing greater migration distances in the subsurface when injected. Persulfate also has a lower affinity for natural soil organic matter than some other oxidizers, most notably permanganate, and would therefore be more effective in highly organic soils. Note that the sulfate formed through activation has a secondary drinking water standard of 250,000 µg/L for taste.

Evaluation of the effectiveness of alkaline activation of sodium persulfate ($Na_2S_2O_8$) for 1,4-dioxane degradation was performed by Block et al. (2004); in this study, potassium hydroxide (KOH) was used as the base to increase pH to >10. Several contaminants were tested, including chlorinated VOCs and oxygenated compounds (e.g., MTBE, TBA, and 1,4-dioxane), at a range of molar ratios of KOH:persulfate. The laboratory-scale studies demonstrated that alkaline activation of persulfate, in addition to the buffering effect of the KOH, resulted in significant declines in target compound concentrations (Figure 7.13). Block et al. (2004) additionally reviewed activation with inorganic compounds and found that Fe^{2+} was an effective activator; however, natural conditions would not favor using ferrous iron for activation because of solubility issues. Evaluation of various chelating agents and inorganic compounds identified a combination of iron and ethylenediaminetetraacetic acid (EDTA) to be an effective activator at typical pH ranges found in natural waters.

Félix-Navarro et al. (2007) studied the sensitivity of sodium persulfate oxidation of 1,4-dioxane at different persulfate concentrations, temperatures, and pH levels. At temperatures ranging from 25°C to 50°C, a 25 mM persulfate solution at pH 7 oxidized 1.13 mM 1,4-dioxane at half-lives ranging from 5 to 64 min; higher temperatures yielded more rapid destruction. Persulfate solutions ranging from 12.5 to 100 mM, at pH 7 and 25°C, oxidized 1.13 mM 1,4-dioxane within half-lives of 38–122 min; the higher-concentration solution oxidized the fastest. Dependency on pH was also

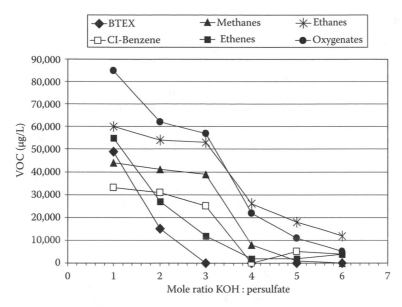

FIGURE 7.13 Effect of KOH ratio on persulfate reactivity. (From Block, P.A., Brown, R.A., and Robinson, D., 2004, In: A.R. Gavaskar and A.S.C. Chen [Eds], *Proceedings of the Fourth International Conference on Remediation of Chlorinated and Recalcitrant Compounds*, Paper 2A-05. Columbus, OH: Battelle Press.)

evaluated at pH values of 3, 5, 7, 9, and 11, for a 25 mM persulfate solution at 25°C containing 1.13 mM 1,4-dioxane. The reaction rate was found to decrease with increasing pH values. Apparently, sulfate and hydroxyl radicals decay rapidly because of reactions with hydroxyl ions in high-pH solutions. Additionally, carbon dioxide from oxidation leads to the formation of bicarbonate and carbonate ions, which may hinder the oxidation of organic compounds.

FMC (2007) discussed an industrial site with a methyl chloroform rail car loading/unloading operation where contaminant concentrations were elevated as follows: methyl chloroform, 203,000 μg/L; 1,1-dichloroethylene, 82,000 μg/L; and 1,4-dioxane, 50,000 μg/L. Initial testing at the site using injected calcium hydroxide (hydrated lime) and steam encountered difficulties with steam daylighting—that is, steam escape through to the ground surface—resulting in an inability to provide the required heat to the treatment area. Sodium hydroxide was then used to catalyze the persulfate with high alkalinity. Postinjection concentrations of 1,4-dioxane were below the local cleanup standard and detection limit of 5 μg/L. Chlorinated VOCs were reduced dramatically within the first year and continued to decline during the following nine-month monitoring period.

Removal of 1,4-dioxane with persulfate has been successfully demonstrated in numerous field applications with a variety of activation methods. Following a successful bench-scale study in which 100% of 1,4-dioxane and 85–99% of chlorinated VOCs were removed, Cronk (2006) executed a source-area removal at a site in Huntington Beach, California, with 1,4-dioxane levels as high as 260,000 μg/L, as well as TCE and dichloromethane as contaminants of concern. Approximately 5000 pounds of persulfate activated with catalyzed hydrogen peroxide was applied to the tight clayey soil through eight application wells with a radius of influence of 10–12 ft each. 1,4-Dioxane levels were reduced by 88–95% in the source zone, down to 21,000 μg/L, during the source-removal project. In La Mirada, California, another field pilot study used chelated iron (EDTA) activation of sodium persulfate. 1,4-Dioxane concentrations in three wells located down gradient of the injection area declined from between 100 and 225 μg/L to nondetectable levels over the nine-month test period. Levels have remained below detection limits for several months, and the groundwater will be monitored periodically to ensure that rebound does not occur (Cronk, 2008).

7.7.6 Permanganate

Potassium permanganate ($KMnO_4$) is a common oxidizer used in water treatment. A recent survey of surface-water treatment systems serving more than 10,000 people reveals that 36.8% of the treatment systems use potassium permanganate for the removal of organic compounds (Waldemer and Tratnyek, 2004). Although ISCO with potassium permanganate is an accepted and widely used remedial method for chlorinated VOCs, it has been generally thought that the oxidation potential (1.7 V) of potassium permanganate is too low to have an impact on 1,4-dioxane. However, Waldemer and Tratnyek (2006) determined the kinetics of permanganate oxidation for a number of environmental contaminants and found that the secondary rate constant for permanganate oxidation of 1,4-dioxane was four orders of magnitude lower than that found for TCE or perchloroethylene. Although at a very slow rate, oxidation of 1,4-dioxane was nonetheless observed. Given the very rapid destruction of chlorinated VOCs (on the order of minutes), an oxidation rate for 1,4-dioxane four orders of magnitude slower would be on the order of months, which may be suitable for some applications.

Evidence from an ongoing ISCO pilot study at Air Force Plant 44 (AFP 44), in Tucson, Arizona—a site contaminated with chlorinated VOCs and 1,4-dioxane—indicates long-term decline in 1,4-dioxane concentrations following injection of potassium permanganate (Earth Tech, 2008). 1,4-Dioxane levels in two monitoring wells located near the center of the $KMnO_4$ injection area (M-96 and M-98) declined from 250 to 100 mg/L and from 650 to 170 mg/L, respectively, between May 2004 and May 2007. Although the permanganate oxidation kinetics for 1,4-dioxane are slow, the reaction is likely occurring and, given enough time, could reduce concentrations sufficiently to meet regulatory standards. $KMnO_4$ oxidation of 1,4-dioxane was further studied at AFP 44 through a bench-scale test using site groundwater contaminated with 1,4-dioxane in the range of 40 μg/L. Triplicate samples analyzed quarterly, where one sample was an untreated control, one sample was acidified to sterilize and eliminate possible bioactivity, and one sample was treated with approximately 4 mg/L of $KMnO_4$, intended to approximate the $KMnO_4$ concentrations maintained in site groundwater since 2004. Results from the first quarter indicate 1,4-dioxane reductions of ~20% in the control and the acidified control, likely because of interactions with dissolved minerals in the site groundwater. The $KMnO_4$-treated replicate indicated ~30% reduction over the same time period, which suggested that oxidation was occurring. Results from subsequent quarters were ambiguous, indicating that 1,4-dioxane destruction was not unequivocal. Therefore, the test was terminated.

7.7.7 Photocatalytic Oxidation

In aqueous solutions, ultraviolet light activates a catalyst, such as titanium dioxide (TiO_2), to produce hydroxyl radicals on its surface from oxygen in air or dissolved oxygen in water.

Hill et al. (1996) determined that 1,4-dioxane in water can be completely broken down by exposure to oxygen and ultraviolet light, in the presence of titanium dioxide as a catalyst. No reaction occurred in any experiment conducted that omitted any one of the three requisite elements. The study focused on degradation intermediates, particularly ethylene diformate, which appears to be created more readily than it is destroyed. Formate esters, which are likely present even in "pure" 1,4-dioxane, should therefore be considered in toxicological studies of the effects of 1,4-dioxane.

Mehrvar et al. (2000) further evaluated the photocatalytic destruction of 1,4-dioxane, alone and in an aqueous solution combined with THF, which is associated with 1,4-dioxane at many sites. They defined the hypothetical reaction mechanism pathways for both THF and 1,4-dioxane and developed kinetic models to predict degradation behavior. They concluded that 1,4-dioxane is converted to hydroxylated 1,4-dioxane; this process leads to ring opening and formation of organic acids, which ultimately convert to CO_2 and water.

Yanagida et al. (2006) evaluated the effect on an applied voltage to a titanium dioxide–impregnated screen exposed to ultraviolet light for the enhanced treatment of 1,4-dioxane. They determined that the photocatalytic decomposition rate of 1,4-dioxane was increased when a voltage swing of ±0.4 V was applied, and formation of toxic intermediates was suppressed.

Ex situ photocatalytic treatment of 1,4-dioxane in extracted groundwater is a demonstrated technology, as described in the trade literature (www.purifics.com). Several case studies are presented, with influent 1,4-dioxane levels as high as 1500 μg/L and post-treatment levels of below 3 μg/L.

An *ex situ* advanced oxidation system evaluation was performed by Wannamaker (2005), where cost and anticipated performance were compared between systems utilizing hydrogen peroxide + ultraviolet, hydrogen peroxide + ozone, and ultraviolet + titanium dioxide (photocatalytic oxidation). Additionally, bromate production was evaluated for each of the technologies. Field tests of the ultraviolet + titanium dioxide system were performed that yielded a 1,4-dioxane decrease from 150 to less than 1.9 μg/L (the detection limit). Bromate conversion, from naturally occurring bromide at 600 μg/L, was minimal in the effluent from the ultraviolet + titanium dioxide system.

7.7.8 OZONATION

Kishimoto et al. (2005) studied ozonation combined with electrolysis for COD removal from 1,4-dioxane solution. They defined the destruction pathway as starting with hydroxyl radicals produced through ozonation near the cathode, in a high-pH environment. Additional cathodic ozone reduction occurred, which helps destroy the 1,4-dioxane. Ozone was noted to then destroy the 1,4-dioxane degradation products produced in the initial stages. Carbon dioxide produced by the oxidation of organics typically forms bicarbonate, which inhibits 1,4-dioxane oxidation; however, this process was mitigated by using a two-compartment flow cell to strip off the carbon dioxide in the anodic cell. Chlorine, produced from the oxidation of chloride at the anode, was found to enhance oxidation of the degradation by-products in a one-compartment flow-cell test.

Suh et al. (2005) used a palladium (Pd) catalyst deposited on activated carbon to evaluate 1,4-dioxane oxidation using ozone, ozone + peroxide, and ozone + Pd catalyst. No significant concentration decreases were observed with ozone alone (Figure 7.14), and there were no significant

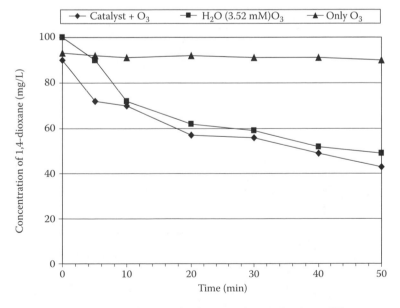

FIGURE 7.14 Destruction of dioxane during oxidation experiments for three different sources of oxidant; pH = 10; ozone dosage = 10 mg/min. (From Suh, J.H., Kang, D.J., Park, J.D., and Lee, H.S., 2005, *Proceedings from the 9th Russian-Korean International Symposium on Science and Technology* [KORUS].)

differences between the 1,4-dioxane concentration decrease from ozone + peroxide or ozone + Pd catalyst. Suh et al. noted that oxidation of 1,4-dioxane to more easily biodegraded intermediates would support the integration of advanced oxidation and more conventional biological treatment methods for wastewater.

7.7.9   SONOCHEMICAL OXIDATION

Sonochemical oxidation of organic contaminants is caused by the production of hydroxyl radicals and other oxidizing species during the collapse of cavitation bubbles formed during high-frequency acoustic stimulation of the aqueous media. High temperatures (up to 4000°C) and pressures (1000 atmospheres) can be present in the cavitation bubbles when they collapse. VOCs are degraded through combustion, high-temperature chemical reactions, supercritical water oxidation, and the production of free radicals. Hydroxyl and other free radicals formed during this process are capable of oxidizing organic compounds, such as 1,4-dioxane.

Beckett and Hua (2000) and Hua (2000) performed an evaluation of the viability of using sono-chemical destruction for 1,4-dioxane and defined the decomposition products and pathways for destruction of 1,4-dioxane during sonolysis at discrete ultrasonic frequencies. Ninety-six percent of the 1,4-dioxane was completely broken down in the first 120 min, and Beckett and Hua (2000) and Hua (2000) identified five major reaction intermediates: ethylene glycol diformate (EGD), meth-oxyacetic acid, formic acid, glycolic acid, and formaldehyde (Figure 7.15). In general, these break-down products are less toxic and less recalcitrant than 1,4-dioxane. The frequency proved to be an important factor because the bubble size, and resultant collapse energy, is inversely correlated to the sonic frequency.

Beckett and Hua (2003) assessed how the addition of ferrous iron affected 1,4-dioxane sono-chemical degradation rates. Ferrous iron addition increased the degradation rate and efficiency of 1,4-dioxane decomposition in all tested ultrasonic frequencies. The combination of ferrous iron and ultrasound yielded a higher concentration of the hydroxyl radical, thereby causing more rapid destruction than either sonication or Fenton's reagent alone. Nakajima et al. (2007) further assessed the effect on sonochemical degradation rates by addition of reduced titanium dioxide powder, as

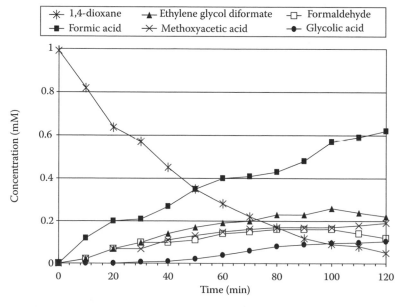

FIGURE 7.15   Formation of sonolytic by-products of 1,4-dioxane over time. (From Beckett, M.A. and Hua, I., 2000, *Environmental Science and Technology* 34(19): 3944–3953.)

compared to an inert powder (SiO_2). They found that the titanium dioxide increased both the temperature and cavitation in the aqueous solutions. These changes led to an increase in the ultrasonication energy consumed in the water and the destruction efficiency of 1,4-dioxane. Nakajima et al. (2004) coupled photocatalytic treatment of 1,4-dioxane using ultraviolet and titanium dioxide with sonication in the presence of titanium dioxide. They also assessed hydrofluoric acid treatment of titanium dioxide, which had been shown elsewhere to enhance sorption of organic contaminants to the titanium dioxide surface. The research findings indicate that the synergistic effect of these combined methods improves the decomposition rate of 1,4-dioxane. The reaction constant for the decomposition of 1,4-dioxane using the combination of ultrasound, ultraviolet, and a TiO_2 catalyst is three times higher than that derived by using only ultraviolet and the TiO_2 catalyst.

7.7.10 CHEMICAL OXIDATION SUMMARY

Chemical oxidation effectively destroys 1,4-dioxane, and the degradation mechanisms and kinetics are well understood. *Ex situ* treatment systems utilizing several of the technologies described in this chapter are in place and functioning as designed, decreasing 1,4-dioxane concentrations to below clean-up standards at several sites. *In situ* applications of chemical oxidation technologies have thus far been limited to pilot studies and other smaller-scale implementations. The effectiveness of ISCO application is limited by the oxidation demand of natural organic materials and inorganic elements in the aquifer, as well as aquifer hydraulic characteristics, which are limitations with all *in situ* technologies. Full-scale ISCO projects have not been implemented and operated to complete the remediation of 1,4-dioxane as of this writing (2009).

BIBLIOGRAPHY

Adams, C.D., Scanlan, P.A., and Secrist, N.D., 1994, Oxidation and biodegradability enhancement of 1,4-dioxane using hydrogen peroxide and ozone. *Environmental Science and Technology* 28: 1812–1818.

Adamus, J.E., May, H.D., Paone, D.A., Evans, P.J., and Parales, R.E., 1995, United States Patent 5,474,934: Biodegradation of ethers. Assignee: Celgene Corporation, Warren, NJ.

Aitchison, E.W., Kelley, S.L., Alvarez, P.J.J., and Schnoor, J.L., 2000, Phytoremediation of 1,4-dioxane by hybrid poplar trees. *Water Environment Research* 72: 313–321.

Alther, G.R., 2006, Organoclays trap PNAHs and creosote in permeable barriers. In: B.M. Sass (Ed.), *Proceedings of the Fifth International Conference on Remediation of Chlorinated and Recalcitrant Compounds*, Paper L-71. Columbus, OH: Battelle Press.

Azad, S., Bennett, D.W., and Tysoe, W.T., 2000, Investigation of the surface chemistry of crown ethers: The adsorption and reaction of 1,4-dioxane on palladium (111). *Surface Science* 464: 183–192.

Beckett, M.A. and Hua, I., 2000, Elucidation of the 1,4-dioxane decomposition pathway at discrete ultrasonic frequencies. *Environmental Science and Technology* 34(19): 3944–3953.

Beckett, M.A. and Hua, I., 2003, Enhanced sonochemical decomposition of 1,4-dioxane by ferrous iron. *Water Research* 37(10): 2372–2376.

Block, P.A., Brown, R.A., and Robinson, D., 2004, Novel activation technologies for sodium persulfate *in situ* chemical oxidation. In: A.R. Gavaskar and A.S.C. Chen (Eds), *Proceedings of the Fourth International Conference on Remediation of Chlorinated and Recalcitrant Compounds*, Paper 2A-05. Columbus, OH: Battelle Press.

Boarer, J.E. and Milne, R., 2004, 1,4-Dioxane treatment in Mountain View, California. *Pollution Engineering* 36(7): 68–69.

Bowman, R.H., Miller, P., Purchase, M., and Schoellerman, R., 2007, Ozone-peroxide advanced oxidation water treatment system for treatment of chlorinated solvents and 1,4-dioxane. http://www.aptwater.com/assets/tech_papers/Paper-1.4Dioxane.pdf (accessed August 1, 2007).

Bowman, R.S., Sullivan, E.J., and Li, Z., 2000, Mechanisms of cationic, anionic, and nonpolar organic solute sorption by surfactant-modified zeolite. In: C. Colella and F.A. Mumpton (Eds), *Natural Zeolites '97: Occurrence, Properties, and Use*, pp. 287–297. Brockport, NY: International Committee on Natural Zeolites.

Cawfield, D.E., Lindstrom, F.T., and Boersma, L., 1990, *User's Guide to CTSPAC: Mathematical Model for Coupled Transport of Water, Solutes, and Heat in the Soil–Plant–Atmosphere Continuum*. Corvallis, OR: Oregon State University, Agricultural Experiment Station.

Chiang, D., Glover, E., Dahlgren, B., Peterman, J., Harrigan, J., and Woodward, D., 2007a, Approach, assessment and implementation of monitored natural attenuation at a 1,4-dioxane-contaminated site. In: A.R. Gavaskar and C.F. Silver (Eds), *Proceedings of the Ninth International In Situ and On-Site Bioremediation Symposium*, Paper M-03. Columbus, OH: Battelle Press.

Chiang, D., Glover, E., Dahlgren, B., Woodward, D., and Aitchison, E.W., 2007b, Phytoremediation of 1,4-dioxane for groundwater seep remediation. In: A.R. Gavaskar and C.F. Silver (Eds), *Proceedings of the Ninth International In Situ and On-Site Bioremediation Symposium*, Paper M-04. Columbus, OH: Battelle Press.

Chiang, D., Glover, E., Peterman, J., Harrigan, J., DiGuiseppi, W., and Woodward, D., 2008, Evaluation of natural attenuation at a 1,4-dioxane contaminated site. *The Remediation Journal* 19(1): 19–37.

Chiang, D., Zhang, Y., Glover, E., Harrigan, J., and Woodward, D., 2006, 1,4-Dioxane solute transport modeling in support of natural attenuation determination. In: B.M. Sass (Ed.), *Proceedings of the Fifth International Conference on Remediation of Chlorinated and Recalcitrant Compounds*, Paper G-70. Columbus, OH: Battelle Press.

Cowan, R.M., Morin, M.D., Sock, S.M., Grady, C.P.L., Jr., and Hughes, T.A., 1994, Isolation and identification of microorganisms responsible for 1,4-dioxane mineralization. In: *Abstracts of the 94th General Meeting of the American Society of Microbiology*, Abstract Q-444. Washington, D.C.: American Society of Microbiology.

Cronk, G., 2006, Effective treatment of recalcitrant compounds using activated sodium persulfate. In: B.M. Sass (Ed.), *Proceedings of the Fifth International Conference on Remediation of Chlorinated and Recalcitrant Compounds*, Paper D-81. Columbus, OH: Battelle Press.

Cronk, G., 2008, Case study: Comparison of multiple activation methods for sodium persulfate ISCO treatment. In: B.M. Sass (Ed.), *Proceedings of the Sixth International Conference on Remediation of Chlorinated and Recalcitrant Compounds*. Columbus, OH: Battelle Press.

Davis, H., 2007. Personal communication.

DiGuiseppi, W.H. and Whitesides, C., 2007, Treatment options for remediation of 1,4-dioxane in groundwater. *Environmental Engineer* 43(2): 36–41.

Earth Tech, Inc., 2004, Technology evaluation for treatment of 1,4-dioxane in groundwater. Prepared for the Air Force Center for Environmental Excellence, December 2004.

Earth Tech, Inc., 2008, Air Force Plant 44 semi-annual report. Prepared for the Air Force Center for Engineering and the Environment, May 2008.

Evans, P.J., Parales, R.E., and Parales, J.V., 2006, Bioreactors for treatment of 1,4-dioxane. In: *18th Symposium in Groundwater Resources Association Series on Groundwater Contaminants: Emerging Contaminants in Groundwater: A Continually Moving Target, June 7–8, 2006*. Concord, CA. Housed at the University of California Water Resources Archives, Berkeley, CA.

Fam, S.A., Fogel, S., and Findlay, M., 2005, Rapid degradation of 1,4-dioxane using a cultured propanotroph. In: B.C. Alleman and M.E. Kelley (Eds), *Proceedings of the Eighth International In Situ and On-Site Bioremediation Symposium*, Paper I-10. Columbus, OH: Battelle Press.

Félix-Navarro, R.M., Lin-Ho, S.W., Barrera-Díaz, N., and Pérez-Sicairos, S., 2007, Kinetics of the degradation of 1,4-dioxane using persulfate. *Journal of the Mexican Chemical Society* 1(2): 67–71.

Ferro, A., Tammi, C.E., Hodgen, J., and LaRue, J., 2005, Treatment of recovered groundwater containing 1,4-dioxane: Year-round phytovolatilization by irrigating stands of deciduous and coniferous trees. In: *The Annual Conference on Soils, Sediments and Water*, University of Massachusetts, Amherst, Massachusetts. http://www.umasssoils.com/abstracts2005/Tuesday/ecological%20restoration.htm (accessed September 15, 2007).

Fleming, E.C., Zappi, M.E., Toro, E., Hernandez, R., and Myers, K., 1997, *Laboratory Assessment of Advanced Oxidation Processes for Treatment of Explosives and Chlorinated Solvents in Groundwater from the Former Nebraska Ordnance Plant*. Vicksburg, MS: U.S. Army Engineer Waterways Experiment Station.

FMC, 2007, In-situ chemical oxidation with Klozur™ activated persulfate: Co-mingled plume of chlorinated solvents and 1,4-dioxane. http://www.envsolutions.fmc.com/Klozur/ResourceCenter/tabid/356/Default.aspx (accessed August 14, 2007).

Global BioSciences, Inc., 2007, Enhancing natural attenuation: 1,4-Dioxane degradation using butane biostimulation. http://www.globalbiosciences.com/downloads/Natural_Attenuation.pdf (accessed October 26, 2007).

Hentz, R.R. and Parrish, C.F., 1971, *Photolysis of Gaseous 1,4-Dioxane at 1470 Å. Journal of Physical Chemistry* 75(25): 3899–3901.

Hill, R.R., Jeffs, G.E., and Roberts, D.R., 1996, *Photocatalytic Degradation of 1,4-Dioxane in Aqueous Solution*. Milton Keynes, UK: The Open University, Department of Chemistry, Walton Hall.

Hua, I., 2000, An investigation of homogeneous and heterogeneous sonochemistry for destruction of hazardous waste. Final Report 09/16/1996–09/14/2000. Department of Energy, Office of Scientific and Technical

Information. http://www.osti.gov/bridge/servlets/purl/13487-CVTUXk/webviewable/13487.PDF (accessed September 4, 2007).

Huling, S.G. and Pivetz, B.G., 2006, In-situ chemical oxidation. EPA Engineering Issue Paper EPA/600/R-06/072.

Innovative Engineering Solutions, Inc. (IESI), 2008, Microcosm testing and design of a bioenhancement remediation system, Michigan (1,4 dioxane). http://www.iesionline.com/projects/Michigan.pdf (accessed August 24, 2008).

International Organization for Standardization (ISO), 2005, Space environment (natural and artificial)—process for determining solar irradiances. ISO TC 20/SC 14, December 1, 2005. http://www.spacewx.com/pdf/ISO_DIS_21348_E_revB.pdf (accessed September 21, 2008).

Isotec, 2008, ISOTEC technology effectively treats 1,4-dioxane. http://www.insituoxidation.com/images/Isotec%201-4%20dioxane.pdf (accessed December 14, 2008).

Johns, M.M., Marshall, W.E., and Toles, C.A., 1998, Agricultural by-products as granular activated carbons for absorbing dissolved metals and organics. *Journal of Chemical Technology and Biotechnology* 71: 131–140.

Kelley, S.L., Aitchison, E.W., Deshpande, M., Schnoor, J.L., and Alvarez, P.J.J., 2001, Biodegradation of 1,4-dioxane in planted and unplanted soil: Effect of bioaugmentation with *Amycolata* sp. CB1190. *Water Research* 35(16): 3791–3800.

Kerfoot, W.B., 2008, *In situ* 1,4-dioxane remediation at HVOC sites. In: B.M. Sass (Ed.), *Proceedings of the Sixth International Conference on Remediation of Chlorinated and Recalcitrant Compounds*. Columbus, OH: Battelle Press.

Kerfoot, W.B. and Brolowski, A., 2006, *In-situ* 1,4 dioxane remediation in HVOC sites. Paper presented at The Annual Conference on Soils, Sediments and Water, University of Massachusetts, Amherst. http://www.umasssoils.com/abstracts2005/Tuesday/ozone.htm (accessed August 15, 2007).

Kessel, L., Hajali, P., Nielsen, B., and Olivo, K., 2005, Competitive contaminant destruction during ozone-peroxide chemical oxidation: Implications for in-situ remediation of 1,4 dioxane. In: *15th Annual West Coast Conference on Soil, Sediment and Water of the Association for Environmental Health and Sciences (AEHS)*, San Diego, CA.

Kim, C.G., Seo, H.J., and Lee, B.R., 2006, Decomposition of 1,4-dioxane by advanced oxidation and biochemical process. *Journal of Environmental Science and Health, Part A* 41: 599–611.

Kishimoto, N., Yasuda, Y., Oomura, T., and Ono, Y., 2005, Ozonation combined with electrolysis as an alternative technology for removing 1,4-dioxane from landfill leachates. In: R. Cossu and R. Stegmann (Eds) *Sardinia 2005, 10th International Waste Management and Landfill Symposium*, Cagliari, Italy. CISA, Environmental Sanitary Engineering Centre, Italy.

Masten, S.J., 2004, Chemical oxidation treatment technologies: Gelman Sciences Dioxane Project, Ann Arbor, MI. Technical Outreach Services for Communities (TOSC). http://www.egr.msu.edu/tosc/gelman/Masten_Community_Meeting.ppt (accessed October 3, 2007).

Mehrvar, M., Anderson, W.A., and Moo-Young, M., 2000, Photocatalytic degradation of aqueous tetrahydrofuran, 1,4-dioxane, and their mixture with TiO_2. *International Journal of Photoenergy* 2: 67–80.

Nakajima, A., Sasaki, H., Kameshima, Y., Okada, K., and Harada, H., 2007, Effect of TiO_2 powder addition on sonochemical destruction of 1,4-dioxane in aqueous systems. *Ultrasonics Sonochemistry* 14(2): 197–200.

Nakajima, A., Tanaka, M., Kameshima, Y., and Okada, K., 2004, Sonophotocatalytic destruction of 1,4-dioxane in aqueous systems by HF-treated TiO_2 powder. *Journal of Photochemistry and Photobiology A: Chemistry* 167: 75–79.

Odah, M.M., Powell, R., and Riddle, D.J., 2005, ART in-well technology proves effective in treating 1,4-dioxane contamination. *Remediation Journal* 15(3): 51–64.

Pall Corporation, 2004, *Final Feasibility Study & Proposed Interim Response Plan for the Unit E Plume*. Ann Arbor, MI: Pall Life Sciences Corporation.

Pan, S. and Chen, H., 2006, Anaerobic degradation of 1,4-dioxane under humic-reducing or Fe(III)-reducing conditions. The 231st American Chemical Society National Meeting, Abstract ENVR 144.

Parales, R.E., Adamus, J.E., White, N., and May, H.D., 1994, Degradation of 1,4-dioxane by an *actinomycete* in pure culture. *Applied and Environmental Microbiology* 60(12): 4527–4530.

Patt, T.E. and Abebe, H.M., 1995, United States Patent 5,399,495: Microbial degradation of chemical pollutants. Assignee: The Upjohn Company, Kalamazoo, MI. http://www.patentstorm.us/patents/5399495-fulltext.html (accessed September 12, 2007).

Pintauro, P.N. and Jian, K., 1995, United States Patent 5,387,378: Integral asymmetric fluoropolymer pervaporation membranes and method of making the same. Assignee: Tulane University, New Orleans. http://www.patentstorm.us/patents/5387378-description.html (accessed October 12, 2007).

Pollution Engineering, 1999, UV oxidation provides the missing link. *Pollution Engineering* 31(7): 2.

Renn, T., 2009. AECOM, personal communication.

Rushing, J., Vikesland, R., Love, N., Mutuc, M., Chan, K., Casselberry, R., and Cichy, P., 2006, Evaluating in-situ chemical and biological treatment approaches for two chlorinated aliphatic ethers, BCEE and BCEM. In: B.M. Sass (Ed.), *Proceedings of the Fifth International Conference on Remediation of Chlorinated and Recalcitrant Compounds*, Paper D-62. Columbus, OH: Battelle Press.

Sabba, D. and Witebsky, S., 2003, Documenting the nature and extent of 1,4-dioxane at a solvent plume site. In: *Symposium Resource Binder: 1,4-Dioxane and Other Solvent Stabilizer Compounds in the Environment*, December 10, 2003. San Jose, CA, Groundwater Resources Association of California (housed at the University of California Water Resources Archives, Berkeley, CA).

Sadeghi, V.M. and Gruber, D.J., 2007, *In situ* oxidation of 1,4-dioxane with ozone and hydrogen peroxide. Poster presentation at URS's Environmental Technology and Management Seminar, Oakland, CA.

Sadeghi, V.M., Gruber, D.J., Yunker, E., Simon, M., and Gustafson, D., 2006, *In situ* oxidation of 1,4-dioxane with ozone and hydrogen peroxide. In: B.M. Sass (Ed.), *Proceedings of the Fifth International Conference on Remediation of Chlorinated and Recalcitrant Compounds*, Paper D-31. Columbus, OH: Battelle Press..

Safarzadeh-Amiri, A., Bolton, J.R., and Cater, S.R., 1996, *Calgon Carbon Oxidation Technologies*. Ontario, Canada: Markham.

Schreier, C.G., Sadeghi, V.M., Gruber, D.J., Brackin, J., Simon, M., and Yunker, E., 2006, In-situ oxidation of 1,4-dioxane. In: B.M. Sass (Ed.), *Proceedings of the Fifth International Conference on Remediation of Chlorinated and Recalcitrant Compounds*, Paper D-21. Columbus, OH: Battelle Press.

Shen, W., Chen, H., and Pan, S., 2008, Anaerobic biodegradation of 1,4-dioxane by sludge enriched with iron-reducing microorganisms. *BioResources Technology* 99(7): 2483–2487.

Shirazi, F., 2007. Personal communication.

Smitha, B., Dhanuja, G., and Sridha, S., 2006, Dehydration of 1,4-dioxane by pervaporation using modified blend membranes of chitosan and nylon 66. *Carbohydrate Polymers* 66(4): 463–472.

Sörensen, M. and Weckenmann, J., 2006, 1,4-Dioxan: Was tun? [1,4-Dioxane: What to do?]. *Process*. http://www.process.vogel.de/index.cfm?pid=2511&pk=58794 (accessed October 10, 2007).

Stanfill, J.C., Koon, J., Shangraw, T., and Bollmann, D., 2004, 1,4-Dioxane bio-degradation bench study at the Lowry Landfill Superfund Site. In: *Proceedings of the Water Environment Federation, 10th Annual Industrial Wastes Technical and Regulatory Conference*, WEF/A&WMA Industrial Wastes, pp. 420–433 (14).

Stefan, M.I., and Bolton, J.R., 1998, Mechanism of the degradation of 1,4-dioxane in dilute aqueous solutions using the UV/hydrogen peroxide process. *Environmental Science and Technology* 32: 1588–1595.

Suh, J.H., Kang, D.J., Park, J.D., and Lee, H.S., 2005, A study on the catalytic ozonation of 1,4-dioxane. In: *Proceedings from the 9th Russian-Korean International Symposium on Science and Technology* (KORUS), Novosibirsk, Russia.

Suh, J.H., and Mohseni, M., 2004, A study on the relationship between biodegradability enhancement and oxidation of 1,4-dioxane using ozone and hydrogen peroxide. *Water Research* 38(10): 2596–2604.

Tailored Lighting, Inc., 2007, Solux ultraviolet radiation primer. http://spectralfidelity.com/cgi-bin/tlistore/infopages/uv.html (accessed September 20, 2008).

Taylor, S.W., Lange, C.R., and Lesold, E.A., 1997, Biofouling of contaminated ground-water recovery wells: Characterization of microorganisms. *Ground Water* 35(6): 973–980.

U.S. Environmental Protection Agency (USEPA), 1999, *Use of Monitored Natural Attenuation at Superfund, RCRA Corrective Action, and Underground Storage Tank Sites*. Washington, DC: U.S. Environmental Protection Agency Office of Solid Waste and Emergency Response Directive 9200.4-17P.

U.S. Environmental Protection Agency (USEPA), 2006, *Treatment Technologies for 1,4-Dioxane: Fundamentals and Field Applications*. Washington, DC: U.S. Environmental Protection Agency Office of Solid Waste and Emergency Response, EPA-542-R-06-009.

U.S. Environmental Protection Agency (USEPA), 2007, *Superfund and Technology Liaison Region 9 Newsletter*, 39th Edition. Spring 2007. San Francisco, CA: Office of Research and Development. http://www.epa.gov/osp/hstl/sn_spring2007.pdf (accessed October 23, 2007).

Vainberg, S., McClay, K., Masuda, H., Root, D., Condee, C., Zylstra, G.J., and Steffan, R.J., 2006, Biodegradation of ether pollutants by *Pseudonocardia* sp. strain ENV478. *Applied and Environmental Microbiology* 72(8): 5218–5224.

Waldemer, R.H., and Tratnyek, P.G., 2004, The efficient determination of rate constants for oxidations by permanganate. In: A.R. Gavaskar and A.S.C. Chen (Eds), *Proceedings of the Fourth International Conference on Remediation of Chlorinated and Recalcitrant Compounds*, Paper 2A-09. Columbus, OH: Battelle Press.

Waldemer, R.H. and Tratnyek, P.G., 2006, Kinetics of contaminant degradation by permanganate. *Environmental Science and Technology* 40: 1055–1061.

Wannamaker, E., 2005, Treatment of 1,4-dioxane in groundwater. Presented at Colorado Hazardous Waste Management Society (CHWMS) luncheon, Denver, CO.

Woo, Y.T., Argus, M.F., and Arcos, J.C., 1977, Metabolism *in vivo* of dioxane: Effect of inducers and inhibitors of hepatic mixed function oxidases. *Biochemical Pharmacology* 26: 1539–1542.

Yanagida, S., Nakajima, A., Kameshina, Y., and Okada, K., 2006, Effect of applying voltage on photocatalytic destruction of 1,4-dioxane in aqueous system. *Catalysis Communications* 7: 1042–1046.

Ying, O.Y., 2002, Phytoremediation: Modeling plant uptake and contaminant transport in the soil-plant-atmosphere continuum. *Journal of Hydrology* 266(1–2): 66–82.

Ying, O.Y., 2008, Modeling the mechanisms for uptake and translocation of dioxane in a soil-plant ecosystem with STELLA. *Journal of Contaminant Hydrology* 95(1–2): 17–29.

Yunker, E., 2007, *In Situ Chemical Oxidation of 1,4-Dioxane and VOCs with Ozone and Hydrogen Peroxide.* Pittsburgh, PA: The National Association of Remedial Project Managers (NARPM) 17th Annual Training Conference.

Zenker, M.J., Borden, R.C., and Barlaz, M.A., 1999, Investigation of the intrinsic biodegradation of alkyl and cyclic ethers. In: B.C. Alleman and A. Leeson (Eds), *Natural Attenuation of Chlorinated Solvents, Petroleum Hydrocarbons, and Other Organic Compounds, The Fifth International In Situ and On-Site Bioremediation Symposium,* pp. 165–170. San Diego, CA: Battelle Press.

Zenker, M.J., Borden, R.C., and Barlaz, M.A., 2000, Mineralization of 1,4-dioxane in the presence of a structural analog. *Biodegradation* 11: 239–246.

Zenker, M.J., Borden, R.C., and Barlaz, M.A., 2002, Modeling cometabolism of cyclic ethers. *Environmental Engineering Science* 19(4): 14.

Zenker, M.J., Borden, R.C., and Barlaz, M.A., 2004, Biodegradation of 1,4-dioxane using trickling filter. *Journal of Environmental Engineering* 130(9): 926–931.

8 Case Studies of 1,4-Dioxane Releases, Treatment, and Drinking Water Contamination

Thomas K.G. Mohr

Groundwater contamination case studies provide an appreciation for the range of release modes and contaminant behavior one might expect to encounter at different contamination sites. The rate of migration of 1,4-dioxane relative to that of chlorinated solvents is among the more interesting attributes of this contaminant. Migration rates reflect both the higher mobility of 1,4-dioxane and its refractory nature relative to methyl chloroform because 1,4-dioxane is miscible and resistant to biodegradation and abiotic degradation. Will 1,4-dioxane be flushed out of the aquifer by flowing groundwater and diluted to inconsequential concentrations? Should one expect to find a substantial mass of residual 1,4-dioxane near the point of release, withheld in the fine-grained pores of silts and clays? What conditions support persistence of 1,4-dioxane such that the quality of drinking water is degraded? The answers to these questions are helpful for planning site characterization at 1,4-dioxane release sites.

Case studies provide useful benchmarks for the success or failure and cost of remedies selected to treat water contaminated with 1,4-dioxane. The experience of the water utilities and the drinking water regulators in places where 1,4-dioxane has contaminated the drinking water is helpful to other utilities that unfortunately discover 1,4-dioxane in their drinking water supplies.

This chapter provides several case studies profiling migration patterns, treatment solutions, and responses to drinking water contamination.[*] Benchmarking previously studied sites to determine solutions for another site is useful to bracket the range of possible conditions one might encounter, but there is no substitute for field data from the site in question to make informed site-characterization decisions.

8.1 SUPERFUND SITE AT SEYMOUR, INDIANA

In Jackson County, Indiana, Seymour Recycling Corporation (SRC) operated a waste chemical storage and processing facility on a 14-acre property next to the former Freeman Army Air Field about 2 miles southwest of the town of Seymour, at the junction of U.S. Highway 50 and State Route 11. The town of Seymour's population in the mid-1980s was approximately 15,300; about 100 homes were located within a 1-mile radius of the SRC site (USEPA, 1986). SRC processed, stored, and incinerated chemical wastes at the site from about 1970 to early 1980. Most of the area immediately surrounding the site is used for agriculture. Approximately 100 residences and businesses relied on private wells that used groundwater from the shallow aquifer near the site before these users were connected to a municipal water system. The neighboring Freeman Municipal Airport obtains its water supply from a well completed in the deep aquifer (USEPA, 1987).

---

[*] The sites in these case studies were all characterized by professional geologists and engineers who have a great deal more familiarity with site conditions than the author; any unintended misrepresentation of site conditions is the responsibility of the author alone.

FIGURE 8.1   Photograph of the Seymour Superfund Site in 1980, before the drums were removed.

The Seymour community experienced problems with the SRC site beginning in 1976. Residents complained about air and surface-water discharges migrating from the site, which they blamed for health problems. In March 1980, about 100 homes were evacuated when a reaction among incompatible chemicals at the site released toxic fumes that drifted into a residential neighborhood. Following that incident, USEPA began enforcement actions at the site (USEPA, 1987). SRC had accumulated about 60,000 55-gallon drums and 98 bulk tanks that contained waste solvents, metal finishing wastes, phenols, cyanides, acids, and C-56 (a pesticide by-product), as well as hundreds of small containers of hazardous materials, primarily from laboratory operations. Some wastes were highly explosive (USEPA, 1983). These wastes leaked and spilled from their containers, leading to fire and odor problems, as well as substantial soil, groundwater, surface water, and sediment contamination (USEPA, 1986). The site was ordered to be closed in 1978 after violating a State of Indiana order to stop receiving wastes; the property was placed in receivership by court order in 1980, and USEPA entered into a consent decree with PRPs (potentially responsible parties) in 1982. USEPA began removal and emergency response actions in 1980. These actions included fencing the site and constructing runoff control dikes. Two PRPs removed several thousand drums in 1980, and in 1981, USEPA removed chemicals from the bulk tanks and transferred those wastes to authorized disposal sites (USEPA, 1987). By 1984, all of the drums and bulk storage tanks were removed, and the top foot of contaminated soil was removed from about 75% of the site and also transported to authorized disposal sites; clean fill replaced the contaminated soil (USEPA, 1987). Figure 8.1 shows a photograph of the site in 1980, before the drums were removed.

8.1.1 CONTAMINANT DISTRIBUTION AND HYDROGEOLOGY

Chemicals from the SRC site contaminated a shallow aquifer, 6–8 ft below ground, and a "deep" aquifer, about 55–70 ft below ground, beneath a leaky silt–clay aquitard. Groundwater in the shallow aquifer discharges into a local creek and a drainage ditch during the wet season; in the dry season, the creek is dry and groundwater flows beneath the creek bed toward the nearby private wells. Although flows in the shallow aquifer are to the northwest, flows in the deep aquifer are primarily to the south, toward supply wells at the airport east of the site.

Sampling via monitoring wells showed that the shallow aquifer beneath the 14-acre site was highly contaminated with more than 35 different hazardous organic chemicals. In 1985, the VOC plume maps of the shallow aquifer showed high-concentration contamination extending about 400 ft beyond the site boundaries and trace amounts of organic chemicals as far as 1100 ft downgradient of the site boundary. Shallow groundwater flow was estimated to be 400 ft/year (USEPA, 1987; Feldman, 2000).

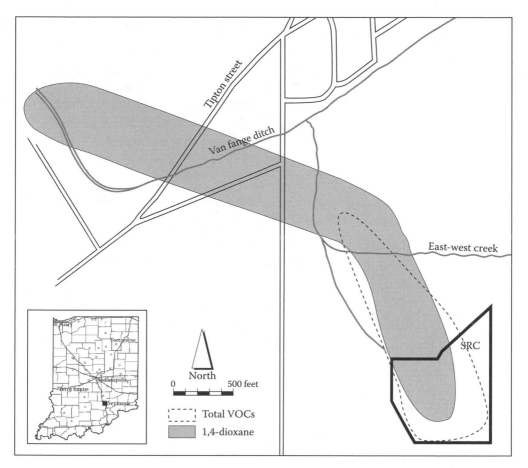

FIGURE 8.2 July 1999 distribution of 1,4-dioxane (10 ppb contour) relative to total VOCs at Seymour Indiana Superfund site, where the SRC operated (Feldman, S.M., 2000).

1,4-Dioxane is associated with soil, groundwater, and sediment at the SRC site (USEPA, 2007a). The soils were contaminated by 54 organic chemicals, including high concentrations of 1,1,2-trichloroethane, carbon tetrachloride, 1,1,2,2-tetrachloroethane, and trichloroethylene. The presence of 1,4-dioxane in the groundwater was discovered by placement and sampling of monitoring wells during the remedial design phase of the project (USEPA, 1997). Subsequent sampling revealed that 1,4-dioxane had migrated about 3900 ft north-northwest of the site boundary, as shown in Figure 8.2. Although diverse chemicals, including methyl chloroform and THF (which extends northwest of the site boundary by about 1300 ft), have contaminated the groundwater near the SRC site, the plume itself is defined by the 1,4-dioxane. In the space of six years, between 1984 and 1990, the 1,4-dioxane plume advanced approximately 2000 ft (Feldman, 2000).

In mid-1990, at a monitoring well located on the plume centerline about 1500 ft downgradient of the site boundary (where the plume turns to the left in Figure 8.2), the 1,4-dioxane concentrations had decreased from about 2300 ppb to less than 100 ppb. Other shallow monitoring wells along the plume core generally decreased from about 500 ppb to less than 100 ppb (Feldman, 2000).

8.1.2 Risk Management

The foremost risk at the SRC site included offsite migration of highly mobile organic contaminants that could arrive at the nearest private well, one-quarter mile northwest of the site, at concentrations greater than those equivalent to the 1 in 1,000,000 excess lifetime cancer risk. 1,4-Dioxane was specifically identified as potentially exceeding the recommended risk levels at the nearest private well in less than five years (USEPA, 1987). USEPA analyzed the cumulative excess cancer risk from all carcinogenic compounds emanating from the site, including benzene, methylene chloride, chloroform, perchloroethylene, 1,2-dichloroethane, 1,1,2-trichloroethane, 1,1-dichloroethylene (DCE), trichloroethylene, 1,4-dioxane, and vinyl chloride. To prevent exposure, the airport well and the nearest operational downgradient private supply well were monitored for volatile organic contaminants (USEPA, 1987).

8.1.3 TREATMENT

A remedial investigation and feasibility study concluded that pump and treat prior to discharge to the local POTWs would achieve City of Seymour pretreatment requirements (total toxic organics <2.13 ppm).[*] Pretreatment consisted of mixed media filtration (a sand filter to remove iron to prevent formation of iron oxide in the air stripper), followed by an air stripper and carbon adsorption. During remedial investigations, USEPA discovered that bacteria were naturally aiding in soil remediation, which allowed construction of a smaller groundwater treatment facility than originally planned, leading to substantial savings (USEPA, 2007c). USEPA capitalized on contaminant biodegradability by applying nutrients to contaminated soil to stimulate biodegradation (U.S. Congress, 1988). Soil–vapor extraction (SVE) and a multimedia cap over the most contaminated soils were also included in the remedy. SVE was preferred over soil excavation because of the dangers of worker exposure associated with excavating soil with high concentrations of volatile compounds. The selected remedial alternative was projected to cost $18 million (U.S. Congress, 1988).

The discharge of treated groundwater to the POTW was projected at 150 gpm, which was approximately 20% of the excess capacity of the Seymour POTW and approximately 15% of the average daily wastewater flow. An evaluation showed that there would be no detectable levels of contaminants in the discharge of the POTW to the South Fork of the White River (USEPA, 1987).

The SVE system removed the bulk of the contaminant mass during its first year of operation; more than 20,000 pounds of VOCs were vented. The system reached asymptotic levels after approximately 34,000 pounds were removed, and operations were discontinued. About 85% of the total mass of VOCs extracted by the SVE system over a four-year period consisted of methyl chloroform (31.8%), trichloroethylene (23.2%), perchloroethylene (9.7%), cis-1,2-DCE (8.7%), 1,1,2-trichloro-1,2,2-trifluoroethane (Freon 113) (7.0%), and toluene (4.8%).

After 12 years of operating the groundwater pump and treatment system, more than 90% of the contamination had been removed from the groundwater (USEPA, 2002). USEPA concluded that the plume was stable and shrinking, because the areal extent of the 1-4 dioxane and THF plumes had decreased when compared to data from four years earlier. The pump and treat system was shut down in October 2001 because it was deemed no longer efficient in removing the lower levels of groundwater contaminants that were still above cleanup standards. The approval for the shutdown required (1) increased groundwater monitoring and (2) a contingency to restart the groundwater treatment system if the groundwater plume expanded from its position at system shutdown (USEPA, 2002). The groundwater treatment was originally estimated to take 28–42 years to complete groundwater remediation, but system shutdown was approved within less than 20 years (U.S. Congress, 1988).

The five-year review report of the groundwater monitoring network found that the plume remained stable, but two additional monitoring wells were needed to sample for 1,4-dioxane (USEPA, 2007b). The overall extent and concentration distribution of 1,4-dioxane in the shallow aquifer has not changed significantly since 2001. 1,4-Dioxane concentrations in one deep-aquifer monitoring well near the capped area showed a significant increase to 690 μg/L as of the five-year review; however, other monitoring wells continued to show low or nondetectable levels of 1,4-dioxane.

---

[*] Most total toxic organics analyses are not performed in a manner that is likely to detect 1,4-dioxane.

8.1.4 Discussion

Figure 8.2 shows that by 1999, conditions at the SRC site caused 1,4-dioxane to migrate about 2.3 times farther than total VOCs. Contaminant transport modeling accounting only for retardation and dispersion, but discounting the biotransformation of contaminants, accurately predicted 1,4-dioxane migration. The model overestimated the migration of chloroethane and THF because these contaminants are biodegradable (Nyer et al., 1991).

The remedies selected for this site—soil bioremediation, SVE, air stripping, and carbon adsorption—are effective at removing the bulk of the chlorinated VOCs at the site, but are relatively ineffective at removing 1,4-dioxane. The areal extent of 1,4-dioxane distribution changed little between 1999 and 2008, which suggests that natural attenuation is also not actively removing 1,4-dioxane at this site. The primary means of attenuating 1,4-dioxane have apparently been by pumping 1,4-dioxane out of the shallow aquifer and discharging to the POTW, where it is substantially diluted prior to discharge to the receiving water, where it is presumably diluted further. The part of the plume outside the influence of pumping wells will eventually attenuate through dispersion and dilution from recharge due to rainfall, irrigation, and losing streams. Risk of exposure at this site was eliminated by connecting private well owners to the municipal water supply, which was not contaminated by this site, and by frequent monitoring of the nearest actively used supply wells. Maximum concentrations of 1,4-dioxane measured along the plume centerline in 1990, ~2300 ppb, were relatively low compared to the much higher concentrations seen at other solvent-waste handling and recycling facilities. The 2300 ppb maximum was measured 10 years after the cessation of active discharges, raising the possibility that the bulk of the 1,4-dioxane mass has dispersed much farther downgradient.

8.2 SOLVENT SERVICES SITE, SAN JOSE, CALIFORNIA

In Silicon Valley, Solvent Service Inc. (SSI) opened a facility in 1973 to reclaim spent solvents and other wastes from printed circuit board manufacturers, wafer fabs (semiconductor manufacturers), and other electronics and industrial manufacturing facilities. SSI treated solvents under a treatment, storage, and disposal facility (TSDF) permit at a facility located on a 3-acre parcel in an industrial area of San Jose, California. Subsurface contamination was investigated beginning in 1983 when leaking tanks were discovered (Cal EPA, 2000). SSI employed a wide range of chemical engineering processes to refine solvents, metal plating wastes, and other waste streams. Solvents were refined through the use of distillation and gravity phase separation. As discussed in Chapter 1, distillation of vapor-degreaser still bottoms will further concentrate the "high boiler" stabilizer compounds such as 1,4-dioxane.

VOCs were first detected in shallow groundwater beneath underground solvent-storage tanks, wash-down sumps, a drum-storage area, and an unloading area for solvent tank trucks (Cameron-Cole, 2001). Initially, more than 20 contaminants were discovered, including VOCs associated with production of printed circuit boards, semiconductors, and other electronic parts. The VOCs were detected at extremely high levels, including occurrences of free-phase solvent. The chemicals detected included methyl ethyl ketone, acetone, xylene isomers, perchloroethylene, trichloroethylene, methyl chloroform, 1,1-dichloroethane, *cis*-1,2-dichloroethylene, and others. Soil concentrations of total VOCs in some locations were in excess of 10,000 ppm (i.e., >1%). Ethenes (perchloroethylene and trichloroethylene) were quantified in soil at more than 3000 ppm, and ethanes (primarily dichloromethane and methyl chloroform) were present at more than 1700 ppm. In groundwater, ketones (methyl ethyl ketone and acetone) were found as high as 21,000 ppm; ethenes were at 153 ppm, and ethanes were at 468 ppm. In addition to having some very high concentrations of contaminants, the site also gained notoriety for a 1992 acid leak that caused a toxic cloud and injured 11 people, as well as for shutting down a 3-mile stretch of the U.S. 101 freeway and causing evacuation of a nearby neighborhood (Chui, 1995). The SSI site was briefly added to the Superfund List on June 21, 1988, before it was returned to state oversight (Benson, 1988). Eleven public water supply wells were located within 1 mile of the site; the closest supply well is one-quarter-mile cross-gradient. Contamination at the SSI site occurs primarily in the shallow aquifer and has not affected deeper aquifers used for drinking water (Todd Engineers, 1987).

Early analyses (in 1989) included a number of TICs, for which the laboratory could suggest a chemical identity based on the chromatographic peak, but could not provide a reliable quantitation because methods for these analytes either had not been developed or the laboratory was not certified for the appropriate methods. TICs at the SSI site included 1,4-dioxane, THF, and 1,3-dioxolane, as well as several others. Although these compounds frequently appeared as TICs, they were not identified as chemicals of concern (COCs) and were not considered during the remedial designs because they were not reliably quantified (Todd Engineers, 1998). After the COCs were established for the site, TICs were no longer included in the laboratory analyses for ongoing monitoring.

In 1998, the analytical laboratory inadvertently analyzed groundwater samples for an extended list of chemicals that were not historically part of the analytical program. The extended list included 1,4-dioxane and THF, which were detected frequently and at elevated concentrations (up to 56,000 and 850 µg/L, respectively) (McCraven, 2006). As discussed in Chapter 4, methods for reliably detecting 1,4-dioxane below 150 ppb were not generally available in commercial laboratories before 1997, and it was necessary to request that the laboratory analyze specifically for 1,4-dioxane.

8.2.1 TREATMENT

Three groundwater extraction trenches and a steam injection and vapor extraction system were constructed to contain the major source area. The steam injection and vapor extraction system was converted to a dual-phase groundwater and SVE system in 1996. Subsequently, five additional dual-phase groundwater and SVE wells were installed off-site. A cap was installed over the areas with the highest levels of soil contamination. The retention of a large mass of VOCs in soil and groundwater is attributed to lenses of low-permeability soils in a highly heterogeneous hydrostratigraphy at the SSI site. A paleochannel* was discovered to cause pronounced anisotropy, leading to contaminant migration in a direction as much as 30° away from the prevailing groundwater-flow direction inferred from groundwater-elevation contour maps. Regional flow is generally southwest, whereas contaminant flow is to the west-southwest (Cameron-Cole, 2006; McCraven, 2006).

The SSI site has continued as an operating hazardous waste treatment facility under different owners throughout the site cleanup history. On-site facilities are available for the treatment of highly contaminated groundwater. VOCs in groundwater are removed by biologic treatment, air stripping, and carbon adsorption. Both the treated groundwater and low-level contaminated groundwater, for which pretreatment is not required, are used as cooling water for the hazardous waste treatment operations. Contact evaporation during cooling acts to passively strip the VOCs from low-level contaminated water. Both types of cooling water are then discharged to the sanitary sewer.

The SSI facility holds an Industrial Wastewater Discharge Permit allowing discharge of 10,000 gallons/day of cooling-tower water containing up to 2.13 ppm total toxic organic compounds. The discharge limit is set by the Bay Area Air Quality Management District (BAAQMD) to regulate the partitioning of VOCs to the atmosphere from the cooling tower.

Currently, highly contaminated groundwater extracted from the source area and high-concentration liquid condensate captured in the SVE system are hauled to an off-site disposal facility. Vapors collected from the SVE are condensed to remove water vapor and then directed through two activated-carbon adsorption vessels in parallel. Treated vapors are then discharged to the atmosphere under a permit from the BAAQMD. By year 2000, 128,000 pounds of VOCs had been removed by the combined extraction of free product, soil vapor, and groundwater. Between

---

* A paleochannel is a natural, semicontinuous deposit of sand and gravel laid down by an ancient stream and subsequently buried by younger deposits.

3 and 6 pounds of 1,4-dioxane per month are recovered from the on-site and off-site SVE systems (Cole-Cameron, 2006).

Prior to discovery and characterization of 1,4-dioxane at the SSI site, it was believed that the extent of groundwater contamination had been defined. Discovery of 1,4-dioxane and THF led to an investigation to characterize their downgradient extent, which was significantly greater than that of the chlorinated solvents, as shown in Figure 8.3. The additional characterization effort defined a 1,4-dioxane plume extending beyond the capture zone of the existing remediation systems. These findings were met with regulatory directives to expand the monitoring network by installing additional monitoring wells and expanding the groundwater extraction system. The operational schedule of the existing remedial systems was increased to 24 h/day, seven days/week (24/7) for on-site SVE. An additional, 87-foot-long off-site groundwater extraction trench was installed to address 1,4-dioxane presence beyond the capture zone of the existing system (Cameron-Cole, 2006). Although the presence of fine-grained deposits limited the VOC migration, 1,4-dioxane and THF were estimated to migrate 41 and 35 ft/year, respectively.

The off-site occurrence of 1,4-dioxane beyond the groundwater extraction trench is being addressed by MNA. Consultants for the SSI site acknowledge that 1,4-dioxane is not prone to natural *in situ* degradation. Therefore, the MNA approach relies only on dispersion, diffusion, and dilution from incidental recharge. Because the sources of 1,4-dioxane contamination are cut off by extraction trenches, the consultants have submitted that concentrations should continue to decline, and perimeter wells are not expected to produce samples with detectable concentrations of 1,4-dioxane. Declining concentrations of 1,4-dioxane in monitoring wells between the off-site extraction trench and the perimeter monitoring wells support this assertion.

8.2.2 1,4-Dioxane Detection and Laboratory Challenges

In the first quarter of 1998, the laboratory inadvertently analyzed eight samples for an extended list of VOCs that included 1,4-dioxane and THF. Detections of 1,4-dioxane and THF were at first uncertain because 1,4-dioxane also appeared in trip blanks and in upgradient wells; nevertheless, the consultant decided to reanalyze the full complement of 48 quarterly groundwater samples for 1,4-dioxane. 1,4-Dioxane was detected in samples from all but one monitoring well; the concentrations ranged from 500 to 56,000 ppb. Subsequent sampling revealed a maximum 1,4-dioxane concentration of 340,000 ppb (Todd Engineers, 1998; McCraven, 2006).

The distribution of 1,4-dioxane is shown in Figure 8.3. Delineation of the extent of 1,4-dioxane required installation of at least a dozen additional monitoring wells. At the SSI site, 1,4-dioxane was present at higher concentrations than most of the other VOCs at the time when 1,4-dioxane was discovered, nearly a decade after remediation began and 15 years after contamination was first reported. By 2006, 1,4-dioxane concentrations in three wells had increased as much as sixfold; the cause is under investigation. Many of the shallow monitoring wells at the SSI site display large swings in 1,4-dioxane concentration; the variation exceeds 100%.

1,4-Dioxane analysis was carried out with EPA Method 8260 SIM selected-ion mode. Vapor samples are analyzed for standard-list VOCs by using Method TO14, which includes 1,4-dioxane. Many samples with elevated 1,4-dioxane concentrations required a high dilution level to obtain results within the calibration range of the analytical laboratory equipment. All analytical results for duplicate samples were within acceptable limits established in the Quality Assurance Project Plan except those listed in Table 8.1 (Cameron-Cole, 2006). The few duplicate results that failed are included here as a reminder of the importance of running the full complement of duplicate samples. Apparently, the accuracy of 1,4-dioxane analysis by EPA Method 8260 SIM can be compromised by a challenging matrix with elevated concentrations of other organic compounds, including ketones, aromatics, ethenes, and ethanes.

The Water Board requests that any analytical results exceeding the historical range of results at a particular well be called out in a table of exceedances. Many recent results exceed the historical

Trichloroethylene

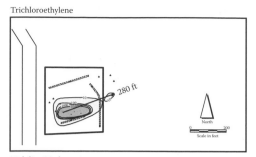

Mobility: Moderate
Solubility: 1100 mg/L
Highest concentration: 180,000 ug/L

Vinyl chloride

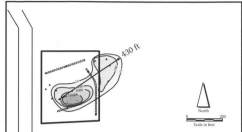

Mobility: Very high
Solubility: 2670 mg/L
Highest concentration: 37,000 ug/L

1,1-dichloroethylene

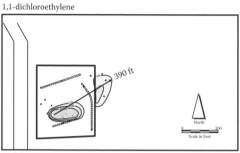

Mobility: Moderate
Solubility: 400 mg/L
Highest concentration: 9500 ug/L

Tetrahydrofuran

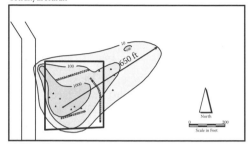

Mobility: Very high
Solubility: Miscible
Highest concentration: 6170 ug/L

Acetone

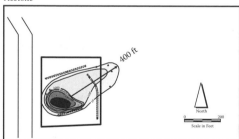

Mobility: Very high
Solubility: Miscible
Highest Concentration: 7,800,000 ug/L

1,4-Dioxane

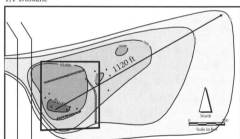

Mobility: Extremely high
Solubility: Miscible
Highest concentration: 34,000

cis-1,2-Dichloroethylene

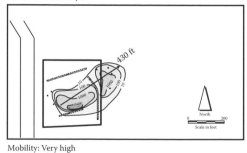

Mobility: Very high
Solubility: 2700 mg/L
Highest concentration: 190,000 ug/L

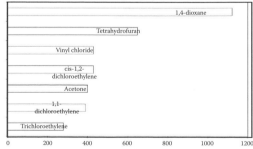

FIGURE 8.3 Distribution of 1,4-dioxane relative to VOCs and THF at the San Jose solvent recycling facility. (After McCraven, S., 2006, Occurrence, fate, and transport of 1,4-dioxane in groundwater—A case study. Presented at Emerging Contaminants in Groundwater: A Continually Moving Target, Groundwater Resources Association of California, Concord, CA, June 7–8. With permission.)

TABLE 8.1
Exceptional Results for Duplicate 1,4-Dioxane Analyses by USEPA 8260 SIM

Well	Analyte	Primary Concentration (ppb)	Duplicate Concentration (ppb)	Relative Percent Difference	Status
88A	1,4-Dioxane	1100	800	32	Accepted/flag
90A	1,4-Dioxane	14,000	20,000	35	Accepted/flag
RW-1A	1,4-Dioxane	1600	4700	98	Rejected
RW-2A	1,4-Dioxane	210	<50	157[a]	Rejected

[a] Half the reporting limit, 25 ppb, was used to calculate the relative percent difference.

range because historical samples were often reported as nondetect at elevated reporting limits. In the fourth-quarter 2006 monitoring report, 1,4-dioxane remained elevated at 47,000 ppb in one extraction trench; that detection exceeded historical maximum values. In some wells, a definite decreasing trend from 3000–1000 ppb is interrupted by a single detection at 6000 ppb; the fact that the values then return to the previously defined trend suggests an analytical anomaly.

8.2.3 REGULATION

The regulated party in the SSI case recommended that the Water Board delay imposing a cleanup level for 1,4-dioxane until a health-based standard is promulgated. The basis for the request was a citation of the paper on PBPK (physiologically based pharmaco-kinetic modeling) modeling of 1,4-dioxane toxicity and dose–response by Reitz et al. (1990), in which a 20,000 ppb cleanup level is suggested. The Water Board denied this request and required continuous treatment to achieve containment and attainment of California's action level of 3 ppb for 1,4-dioxane and the USEPA Region 9 PRG of 8.8 ppb for THF. In addition, the board required a report documenting complete investigation of the extent of 1,4-dioxane in groundwater, delineated to 3 ppb. The board also required a feasibility study for 1,4-dioxane and THF to evaluate remedial measures that will effectively address these chemicals. The primary fate of 1,4-dioxane at this site has been sewer discharge; the treatment train was not modified to add treatment modules capable of 1,4-dioxane removal.

A risk assessment was developed for the site for future hypothetical exposure scenarios involving residential land use and construction-worker inhalation. Drinking water was not among the 1,4-dioxane exposure scenarios evaluated because only the shallow groundwater is affected and installation of drinking water wells is restricted by institutional controls. Site-specific risk-based cleanup levels (RBCLs) were based on a cumulative cancer risk of 1×10^{-5} and a noncancer total hazard index of 1.0. The RBCLs for 1,4-dioxane ranged from 472 to 4512 ppm, and for THF, from 253 to 86,544 ppm. On-site and off-site concentrations of 1,4-dioxane are well below the calculated RBCLs.

Risk from potential vapor intrusion to indoor air was evaluated by using the Johnson and Ettinger model (Johnson and Ettinger, 1991). The estimated residential indoor air concentrations that could result from underlying groundwater 1,4-dioxane concentrations averaging 32 ppm was 0.23 µg/m^3 (0.83 ppbv), under the assumption that future homes will be constructed atop the plume core. These concentrations were determined to present no significant risk by the risk-assessment authors.

8.2.4 DISCUSSION

The SSI case illustrates the repercussions of discovering a new contaminant after more than a decade of cleanup and investigation. 1,4-Dioxane was not a listed COC for the case until its accidental discovery. The laboratory reported the discovery of 1,4-dioxane to the consultant, which probably led to an interesting conversation with the client, because 1,4-dioxane was not on the specified analyte list. The cleanup order for the case did not include either of the newly discovered contaminants, 1,4-dioxane or THF; however, there is a standard clause in Water Board cleanup orders requiring that any investigations and findings conducted or obtained beyond the requirements of the order be reported. The consultant and the facility operator fulfilled that requirement by reporting the results, which completely changed the course of the cleanup.

The consequences of 1,4-dioxane discovery at this site included

- Increasing treatment operations
- Expanding the monitoring network by adding soil borings and 12 monitoring wells
- Adding additional analyses to the monitoring regimen
- Constructing a new off-site extraction trench
- Conducting a feasibility analysis for additional treatment capable of addressing 1,4-dioxane

The selected remedy was extraction of low-concentration 1,4-dioxane from an extraction trench in the off-site plume and discharge under permit to the POTW. The fate of 1,4-dioxane in the POTW and recycled water produced at the POTW was not evaluated, most likely because the POTW treats in excess of 100 MGD (million gallons per day), whereas the discharge volume containing 1,4-dioxane is very small in comparison. The highest concentrations of 1,4-dioxane are handled together with high-concentration VOCs by off-site disposal.

8.3 PALL/GELMAN SCIENCES INC., ANN ARBOR, MICHIGAN

Gelman Sciences Inc., acquired by Pall Life Sciences in 1997 (hereafter P/GSI), produced cellulose triacetate filters used in medical and other applications. As described in Chapter 2, 1,4-dioxane is used for producing microporous cellulose filters for scientific and medical applications. In 1965, company scientists began experimenting with a polymer called cellulose triacetate that had higher thermal stability and better chemical compatibility than any other membrane on the market. Because this polymer was more stable, traditional solvents, like acetone, could not dissolve it to create pores. In 1966, the company began using 1,4-dioxane to create a stronger solvent system (Fotouhi et al., 2006). In 1986, the company discontinued the use of 1,4-dioxane and phased out the line of filters that required its use in production. The line was eventually replaced by an improved filter product. At the time that P/GSI stopped using 1,4-dioxane, it was discovered that 30 private wells in the Scio Township area were contaminated with 1,4-dioxane. Approximately 850,000 pounds of 1,4-dioxane was used to form triacetate filters during the 20 years before contamination was discovered (Fotouhi et al., 2006).

The P/GSI site has a long and complicated history that includes protracted legal disputes, complex hydrogeology, a vociferous citizenry, and an evolution in regulatory requirements. Pall Corporation (Pall) acquired GSI in 1997, well after the discovery and initial litigation of 1,4-dioxane contamination. P/GSI acquired GSI's environmental liability and has been engaged in an epic effort to contain a multidirectional set of plumes that have migrated into developed neighborhoods.

The social equation inherent to cases of drinking water pollution often includes the obligatory public flogging of the responsible party in the local press, public meetings, and courts. In this case, the currently responsible party is a company from New York that never handled 1,4-dioxane in Michigan but acquired cleanup responsibility after the fact. The engineers at P/GSI, and the regulators with the State of Michigan and Washtenaw County, together with scientists at the University of Michigan, were pioneers who developed 1,4-dioxane analysis, treatment technologies, toxicological profiles, and regulatory solutions to a massive 1,4-dioxane contamination problem. The P/GSI site is the oldest and largest 1,4-dioxane contamination case in the United States.

8.3.1 WASTEWATER DISPOSAL

When Gelman Instruments, an early predecessor to P/GSI, moved to Scio Township, there were no sanitary sewers available on-site. P/GSI explains that the Michigan Department of Natural

Resources (DNR) recommended discharge of wastewater into infiltration lagoons and removal of natural geologic barriers between the pond bottom and the water table 15–40 ft below to facilitate slow infiltration to groundwater (Kellogg, 2005). The state issued a permit in 1964 to install a three-million-gallon wastewater lagoon to hold and infiltrate process wastes, which may have contained 1,4-dioxane in concentrations as high as 25 mg/L (City of Ann Arbor, 2006; SRSW, 2006); other sources familiar with the case state that the 1,4-dioxane concentration in wastewater may have been higher, in the range of 200–300 mg/L.[*] After installing two additional lagoons in 1966 and 1967, the company still required additional wastewater capacity. In 1976, DNR granted P/GSI temporary permission to try managing treated process wastes by using spray irrigation. Seven months later, in May 1977, the Michigan Water Resources Commission issued a groundwater discharge permit authorizing effluent spraying onto open fields on the company's 40-acre property. Later, USEPA recommended that Gelman drill an injection well, which they did in 1981. The injection well was drilled at a cost of approximately $1,000,000. The well was drilled more than 1 mile into sandstone (the Mt. Simon Formation), which is geologically isolated from shallow production aquifers (Fotouhi et al., 2006).

In 1984, a University of Michigan graduate student from the School of Public Health, Dan Bicknell, discovered 1,4-dioxane in Third Sister Lake near the P/GSI property in a section of the Saginaw Forest owned by the university. The following year, 1,4-dioxane contamination was discovered in nearby private wells by an investigation conducted by the Washtenaw County Health Board. After learning of the discovery of 1,4-dioxane contamination in late 1985, P/GSI used a water supply well to remove more than 25,000 pounds of 1,4-dioxane from groundwater near the plant between 1987 and 1994. The contaminated water was not treated; instead, it was discharged into the injection well under a USEPA permit. In 1994, P/GSI decided not to renew the injection well permit because of the cost of upgrading and operating the well, and the well was sealed (Kellogg, 2005). P/GSI had permits for its wastewater disposal practices for the entire period in which 1,4-dioxane-laden wastewater was discharged; however, citizen groups allege that permit limits were at times exceeded and other unpermitted discharges may have occurred. In any case, the permitted practice of treating 1,4-dioxane-laden wastewater in unlined lagoons and spray irrigating wastewater on lawns and fields clearly contributed to the widespread contamination of groundwater with 1,4-dioxane.

8.3.2 GEOLOGIC SETTING

The advance and retreat of glaciers defines the highly heterogeneous hydrostratigraphy of the Ann Arbor/Scio Township area. The subsurface consists of unconsolidated glacial outwash sediments, tills, and lacustrine deposits.[†] Multiple glacial outwash channels provide high-permeability aquifers facilitating rapid contaminant migration in different directions. The glacial deposits are up to 300 ft thick and overlie the Coldwater Formation of Mississippian age (i.e., somewhat older than 300 million years), which is predominantly shale (Brode et al., 2005; Fotouhi et al., 2006). Groundwater is generally shallow, averaging 15 ft below the ground surface. Three major aquifers were initially identified; they are designated as the Unit C3 (Core Area), Unit D0 (Western System), and Unit D2 (Evergreen System) aquifers. A Unit E aquifer that is deeper was later determined to be present.

A confining layer within the glacial sediments was thought to isolate deeper aquifers from contamination. Subsurface characterization performed in 2001 revealed a window in the confining layer that allowed contamination to migrate into deeper aquifers. The complex geology in the vicinity of the P/GSI property contributes to the widespread nature of the contamination (MDEQ, 2005).

---

[*] Jim Brode (of Fishbeck, Thompson, Carr and Huber, Kalamazoo, Michigan, February 2008, personal communication).
[†] Lacustrine deposits are lake bed deposits.

8.3.3 Delineating 1,4-Dioxane Contamination

Beginning in 1986, investigations by the company identified soil contamination on the P/GSI property and groundwater contamination extending off the property. Groundwater concentrations of 1,4-dioxane were as high as 221,000 µg/L. The plumes collectively encompass an area of approximately 0.6 square miles, as defined by the Michigan drinking water action level of 85 µg/L (Brode et al., 2005). Citizen groups define the plume to a 1 µg/L contour and describe the plume of deeper 1,4-dioxane contamination as "18 million square feet and growing, three miles long and one mile wide" (Kellogg, 2005). 1,4-Dioxane method detection levels have ranged from 100 to 1 ppb during the course of the project (Fotouhi et al., 2006). As of 2005, groundwater was monitored routinely at 50–100 locations (Brode et al., 2005). Figure 8.4 displays the extent of the shallow and deep plumes delineated to 85 ppb, the Michigan 1,4-Dioxane Generic Residential Cleanup Criteria [Part 201 of Natural Resources and Environmental Protection Act (NREPA)].

Deeper contamination was discovered during an investigation in spring 2001, following a detection of 1,4-dioxane at 2 ppb in the City of Ann Arbor's Northwest Supply well. The city turned off the well immediately following the detection, and investigations of deep aquifer contamination began. After an intensive effort to characterize deep contamination, concentrations as high as approximately 5000 ppb were found in the deep Unit E aquifer. A clay layer separating the two shallower aquifers from the deeper Unit E aquifer in areas to the east was found to be absent in areas to the west of the site. The Unit E aquifer generally flows to the east toward the Huron River, a major regional hydraulic feature (MDEQ, 2005). State regulators have required vigilant monitoring to ensure that spreading plumes of 1,4-dioxane do not pose a threat to private wells outside of the city. The plume is headed in the direction of the Huron River, which is the source of about 80% of the City of Ann Arbor's water supply; however, Ann Arbor's drinking

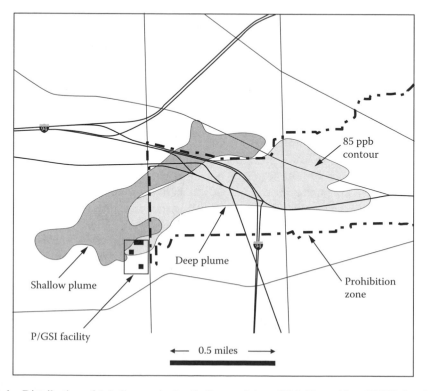

FIGURE 8.4   Distribution of 1,4-dioxane in the shallow and deep (Unit E) aquifers, P/GSI site, Ann Arbor, Michigan (Fotouhi, F., 2006).

water supply is not threatened by 1,4-dioxane because the intake location is upstream of the groundwater discharge area.

The first private well contamination was discovered in 1986, but it was not until the late 1990s before full cleanup of off-site groundwater began. The delay was partly due to legal actions filed by the State of Michigan and the city and to countersuits by P/GSI, as described in Section 8.3.6. For more than 10 years, off-site cleanup was limited to some extraction of contaminated groundwater from a former supply well, and 1,4-dioxane continued to migrate away from the site. The plumes have expanded at rates very close to the linear groundwater velocities (Fotouhi et al., 2006). 1,4-Dioxane has migrated 3000 ft to the west of the site and as far as 9000 ft east of the site (according to the MDEQ 85 ppb threshold) or farther when mapping is based on the limit of detection.

8.3.4 REGULATION

Upon discovery of private well contamination by 1,4-dioxane, P/GSI provided bottled water for residents and businesses and agreed to contribute toward the cost of extending municipal water lines to affected subdivisions (City of Ann Arbor, 2006). Because there were concerns over inhalation and dermal exposure in addition to drinking water exposure, P/GSI arranged for families whose wells had high levels of 1,4-dioxane to shower at a nearby hotel (Fotouhi et al., 2006). By 2000, more than 120 homes with contaminated private water supply wells had been connected to the municipal water supply system (MDEQ, 2000). Washtenaw County staff collects regular samples from an additional 50 operating private wells near the edge of the plume. In 13 of these 50 wells, 1,4-dioxane levels range from 1 to 24 ppb; the remainder are nondetect. P/GSI pays the costs for sampling and connecting to the municipal supply (MDEQ, 2000).

Michigan does not have a state Maximum Contaminant Level (MCL) for 1,4-dioxane. Interested elected officials sought to adopt a Michigan MCL at 35 ppb. If adopted, the 1,4-dioxane MCL would have been the first time that the State of Michigan developed a state drinking water MCL that was not a federal MCL. The proposed MCL was opposed by the City of Ann Arbor, Washtenaw County, and citizen groups as not being protective enough, while the Michigan Manufacturers Association, P/GSI, and others objected that it was too restrictive. The basis for the proposed MCL is explained in Chapter 6. In 1986, the State of Michigan recommended 3 ppb as a safe standard for 1,4-dioxane in drinking water (City of Ann Arbor, 2006). In June 1995, the state legislature amended the NREPA, increasing the generic residential cleanup criteria for 1,4-dioxane to 77 ppb for groundwater. In June 2000, the state adopted USEPA's methodology for calculating risk-based cleanup criteria, which resulted in the generic residential cleanup criteria and the current drinking water criteria of 85 ppb. The 1,4-dioxane concentration in groundwater that Michigan considers to be safe for venting to surface water used for human contact is 2800 ppb (MDEQ, 2004).

Among the more interesting regulatory actions in the P/GSI case is the prohibition zone (PZ) (depicted in Figure 8.4). In order to protect the public, the Washtenaw County Circuit Court issued an order on May 17, 2005, to restrict the use of groundwater in portions of the City of Ann Arbor and in Ann Arbor and Scio Townships. The order prohibits the use of groundwater that is or may become contaminated with unacceptable levels of 1,4-dioxane in a mapped area called the PZ. A few of the properties in this area receive their drinking water from private wells; however, most areas are already connected to the city water supply (MDEQ, 2007). In the PZ area, the order prohibits the consumption or use of well water, the installation of new water supply wells, and requires the proper destruction of all private wells at P/GSI's expense. P/GSI must also fund connecting properties within the PZ that are on private wells to the municipal water supply. Exceptions are available for ground-source heat-pump systems that operate with closed circulation loops and for dewatering wells used in construction and maintenance. Property owners must disclose the order restrictions upon selling or leasing a property within the PZ, as well as disclosing the presence of 1,4-dioxane contamination beneath the property. Because the depth to groundwater in most areas of the PZ is about 40 ft, typical construction activities are unlikely to encounter contaminated groundwater. The order requires that P/GSI continue monitoring the deep plume beneath the PZ and install additional monitoring wells if concentrations change in a manner suggesting that the plume is heading in a different direction than anticipated in the delineated PZ. The restriction on new wells is enforced by the Washtenaw County Department of Planning and Environment, which issues well permits in the PZ.

The PZ order is an institutional control mechanism that is effective to the extent property owners within the designated area are aware of it and adhere to its requirements. During 2006, a small lake in west Ann Arbor showed detections of 1,4-dioxane at 6 ppb. The lake had not previously shown any 1,4-dioxane detections. Investigation of the contamination revealed that the owners of a local inn and restaurant were using a well to cool the buildings and water the lawns. The owners did not realize that they were in the PZ or that there was a 1,4-dioxane contamination problem in the area. 1,4-Dioxane levels in their cooling system water, which discharges into a storm drain that drained to the lake, was several times higher than the 85 ppb action level (Davis, 2006b). In another example of the limitations of institutional controls, the notification requirements for landlords and property sellers published as official notices in the local newspapers apparently were not uniformly heeded in the PZ. At an April 2007 public meeting on the P/GSI case, a resident complained that she had rented a property within the PZ for a year before buying a house, also in the PZ, and neither her landlord nor the house seller disclosed that the properties were in the PZ. The property owner complained that she now owns a house in the path of or overlying a 1,4-dioxane plume that was not disclosed to her, but she is required to disclose it when she sells. At the meeting, state officials announced a small expansion of the PZ in light of new monitoring data and resolved to reissue notifications to property owners (Davis, 2007).

8.3.5 TREATMENT

P/GSI embarked on a series of treatability studies in the decades following the discovery of 1,4-dioxane contamination in private wells. Early treatability studies focused on separation methods using gas-phase separation, such as air stripping and steam stripping, and on solid-phase separation using GAC. P/GSI's treatability studies also tested destructive treatment methods including chemical destruction using hydrogen peroxide, ozone, and chlorine; physical destruction using UV light; and combinations of physical and chemical destruction methods. Biological treatability studies included microbial methods such as activated sludge and bioactivated carbon, enzymes, and inoculation with engineered organisms (Fotouhi et al., 2006).

Site remediation began with contaminated soil removal. The most highly contaminated soils were removed in 1979 and 1987 (MDEQ, 2000). Groundwater remediation near the source zone began in 1987 with the reinjection of contaminated water from a nearby supply well. This remediation continued until 1994, when P/GSI found it too expensive and impractical to renew their injection well permit. The injection well also had limited capacity; therefore, it could not accommodate the purge volumes produced at the site. Between 1994 and 1997, there was a hiatus in off-site groundwater remediation while P/GSI responded to a lawsuit filed by the Michigan attorney general on behalf of the Department of Natural Resources to force P/GSI to clean up groundwater (City of Ann Arbor, 2006).

Remediation of the two shallow aquifers has been under way since 1997. Treated groundwater is discharged to Honey Creek, a small ephemeral tributary to the Huron River. This remediation has significantly decreased the concentration and mass of 1,4-dioxane contamination in the shallow aquifers.

The selected remedy for full-scale groundwater treatment was UV light + hydrogen peroxide (H_2O_2). P/GSI operated a full-scale UV + H_2O_2 system at its facility to remediate groundwater containing high concentrations of 1,4-dioxane. As of 2004, more than 60,000 pounds of 1,4-dioxane had been extracted from the groundwater and destroyed by UV + H_2O_2 since treatment began in 1997 (Brode et al., 2005).

Installation of a centralized treatment system with high-cost equipment and hazardous chemicals creates a logistical challenge. It is not practical or safe to install many treatment centers throughout the wide area of the plumes, yet it is difficult to pipe extracted groundwater back to a centralized treatment center. Residents in developed neighborhoods overwhelmingly oppose putting extraction wells and pipelines conveying contaminated water in their residential streets. In order to overcome property owner and local government objections, P/GSI was forced to install a 4479-ft-long horizontal infiltration well to extract and divert groundwater to the treatment system (Brode et al., 2005). This well may be the longest trenchless horizontal well yet installed for environmental remediation purposes anywhere in the world.

The UV + H_2O_2 remediation treated 1,4-dioxane-contaminated water produced at up to 1300 gpm from 18 extraction wells. Influent 1,4-dioxane concentrations were 5000–10,000 ppb, and effluent concentration was 5 ppb or less (Fotouhi et al., 2006). Treated water was discharged to Honey Creek, a tributary to the Huron River that enters the river upstream of the City of Ann Arbor's water supply intake. Prior to treatment, extracted water entered a lined pretreatment pond where it was mixed with a 93% solution of sulfuric acid to lower the pH to 3.8, which P/GSI determined to be optimal for UV + H_2O_2 reactions (Brode et al., 2005). Following pH reduction, a 50% H_2O_2 solution was injected into the treatment line and mixed into groundwater by a static mixer. The mixed solution was then exposed to UV when it passed through multiple chambers containing 22 UV lamps. Following UV exposure, the pH was raised back to 6.9 with a 40% solution of sodium hydroxide. Sodium bisulfite was added in order to comply with surface discharge requirements and to remove excess H_2O_2 (Brode et al., 2005). Although the areal extent of the plume did not decrease significantly as a result of this treatment, 1,4-dioxane concentrations in the plume core decreased 100-fold to below 10,000 ppb.

P/GSI is permitted to discharge up to 1300 gpm to Honey Creek. The system was the largest known 1,4-dioxane treatment operation anywhere in the world, operating 24/7/365, excluding brief maintenance periods (Fotouhi et al., 2006). P/GSI's UV + H_2O_2 treatment system operated from October 1997 through March 2005, prior to changing the treatment technology to ozone-peroxide.

The successful operation of the UV + H_2O_2 treatment system required large volumes of reactive chemicals (hydrogen peroxide, sulfuric acid, sodium hydroxide, and sodium bisulfite), and the UV bulbs consumed a lot of electricity, averaging 600 kW h/day. The overall cost of the UV + H_2O_2 treatment system was $3.50/1000 gallons. Following careful engineering analysis, P/GSI determined that an ozone-peroxide system would use considerably less energy and would result in a significant reduction in chemical use, making the technology safer, more environmentally friendly, and less expensive to operate (Brode et al., 2005).

The ozone-peroxide system requires 50% less H_2O_2 consumption and eliminates the need for sulfuric acid and sodium hydroxide (Brode et al., 2005). The new system transfers water from extraction wells to a pretreatment pond where insoluble iron settles out. Peroxide is introduced via injector quills and then ozone is separately administered through a series of Venturi injectors. Following the addition of post-treatment chemicals, the water again enters a settling pond prior to surface discharge. The flow rate for the ozone-peroxide system is 1200 gpm. Influent 1,4-dioxane concentrations range from 2000 to 7000 ppb, whereas effluent 1,4-dioxane concentrations are below 5 ppb. Treatment costs using the ozone-peroxide technology are approximately $1.50/1000 gallons, yielding a savings of approximately $450K compared to the UV + H_2O_2 costs of approximately $1.4 M/year for a 1200 gpm flow (Brode et al., 2005; Fotouhi et al., 2006). In addition to the reduction in chemical use, the electrical demand for the ozone-peroxide treatment system is approximately one-third the demand of the UV + H_2O_2 treatment system. P/GSI also deployed a remotely operated mobile ozone-peroxide treatment unit to treat the Unit E plume (Fotouhi et al., 2006).

An important consideration for operating ozone-peroxide treatment is the formation of bromate, as described in Chapter 7. Bromate, a probable human carcinogen, forms as a by-product of ozone treatment of groundwater containing bromide. The MCL for bromate is 10 ppb; a 5 ppb MCL was considered but not adopted. Michigan allows discharge to surface water of bromate in concentrations of up to 50 ppb; P/GSI's treatment system attains bromate concentrations of less than 10 ppb in the effluent of its ozone-peroxide treatment system for 1,4-dioxane (Kellogg, 2005). Residual 1,4-dioxane and bromate in the effluent are a concern because Honey Creek runs past several private wells that could be affected, and Honey Creek discharges to the Huron River upstream of the City of Ann Arbor's drinking water intake.

In 2003, P/GSI submitted a work plan to evaluate the feasibility of an *in situ* oxidation system using hydrogen peroxide and ozone to degrade 1,4-dioxane in the aquifers without removing the groundwater. ISCO was pilot-tested by injecting H_2O_2 as a catalyst and Fenton's reagent (see Chapter 7) into one of the confined aquifers, but only a minor decrease in 1,4-dioxane concentrations was observed. Bromate formation exceeded the 10 µg/L MCL. The *in situ* oxidation approach was therefore considered infeasible for large-scale subsurface remediation.

Altogether, P/GSI has pumped about 4 billion gallons of water and removed about 100,000 pounds of 1,4-dioxane (Fotouhi et al., 2006). Since 1997, the company has been spending $5.5 million/year to clean up 1,4-dioxane pollution (Kellogg, 2005). The value of the affected groundwater resources supplying residents of Ann Arbor, Scio Township, and Ann Arbor Township has not been established.

8.3.6 LEGAL ACTIONS

In addition to major technical and logistical challenges, the P/GSI case has been punctuated by many lawsuits. Litigants against P/GSI have included the State of Michigan, the City of Ann Arbor, several citizen groups, local businesses, a class-action suit, and an insurance company. In addition, P/GSI filed a product liability suit against 1,4-dioxane manufacturers (Fotouhi et al., 2006). In 1988, the Michigan attorney general, on behalf of the Department of Natural Resources, sued GSI[*] to force a cleanup of the groundwater (City of Ann Arbor, 2006). The attorney general sought penalties of more than $400,000 as reimbursement for costs associated with providing clean water to 50 homes and businesses whose wells were contaminated with 1,4-dioxane as well as additional fines and a court order requiring clean up (*Wall Street Journal*, 1988). GSI prevailed against the cost-recovery part of the attorney general's suit and did not have to pay the state's $400,000 bill for hooking up homes to municipal water supplies (Bodwin, 1989).

GSI filed a counter-claim against the state's cleanup suit, claiming that Michigan had never issued criteria for cleaning up 1,4-dioxane. P/GSI again prevailed, but the state won an appeal to a federal court in Detroit, which led to a consent decree. In 2000, Judge Donald Shelton ordered the state to issue cleanup criteria, and P/GSI was ordered to conduct a study of cleanup alternatives and implement a major remediation plan (Kellogg, 2005). Judge Shelton ordered P/GSI to clean up groundwater to within Michigan's 85 ppb action level within five years. That deadline has passed; the goal was not achieved.

GSI's product liability suit against 1,4-dioxane producers and distributors claimed that they failed to alert GSI that 1,4-dioxane is not biodegradable. GSI named eight companies in its suit and claimed that they failed to adequately test 1,4-dioxane to determine its persistence in the environment. The companies named included Dow Chemical Company of Midland, Michigan; PVS-Nolwood Chemicals Inc. of Detroit, Michigan; Ecclestone Industrial Chemical Co. of Warren, Michigan; Chemcentral Detroit Corp. of Romulus, Michigan; Union Carbide Corp. of Danbury, Connecticut; Ashland Chemical Co. of Ashland, Kentucky; McKesson Corp. of San Francisco; and Van Waters & Rogers Inc. of San Mateo, California (Raphael, 1988). The parties settled out of court. Dow Chemical's 1,4-dioxane containers now include a label that advises it is not biodegradable (Kellogg, 2005).

The City of Ann Arbor sued P/GSI over detections of trace levels of 1,4-dioxane in a well that is used by the city as a winter water source (the Montgomery Well). The well was on standby at the

---

[*] GSI refers to Gelman Sciences Inc., the name of the firm before it was acquired by Pall Life Sciences in 1997.

time 1,4-dioxane was detected and has not been used since. This well provided less than 5% of the City of Ann Arbor's water. P/GSI noted that the city voluntarily stopped using the well, which had 1,4-dioxane concentrations of 2–4 ppb, whereas the drinking water action level was 85 ppb. Other sources of 1,4-dioxane could have caused the contamination, according to P/GSI, and the well had other water quality problems unrelated to P/GSI's release (Shelton, 2004; Kellogg, 2005). Among the other possible sources, P/GSI noted that three landfills, including the city's closed landfill and the University of Michigan landfill, had detections of 1,4-dioxane in leachate. The city allows untreated groundwater from those landfills to be pumped into the sanitary sewer and discharged into the Huron River because the small contaminant mass is diluted in the voluminous flow of the receiving water (Kellogg, 2005). Leachate from the city's closed landfill contains 1,4-dioxane in excess of 100 ppb (Davis, 2002).

Prior to the litigation over the detection of 1,4-dioxane concentrations of 2–4 ppb in the Montgomery well, the city appealed a decision by MDEQ to allow an increase in the 1,4-dioxane concentration limit to a daily maximum of 60 ppb and a monthly average of 10 ppb for discharge of P/GSI's treated groundwater into Honey Creek, a tributary to the Huron River upstream of the city's water intake. The city also prohibited P/GSI's discharge of treated groundwater into its storm sewer system owing to volume limitations. P/GSI contrasted the city's practice of allowing the discharge of leachate containing 1,4-dioxane at concentrations up to 100 ppb into its sewers, while P/GSI discharges treated groundwater with less than 5 ppb into the same receiving water at a high cost (Davis, 2002). The founder of GSI, Charles Gelman, argued vehemently for consistent application of regulatory standards, and regulation of all sources of 1,4-dioxane, including antifreeze, aircraft deicing fluid, and other sources (Bodwin, 1989).

The delineation of the PZ was also the subject of litigation between the city and P/GSI. P/GSI objected to the city well being included in the PZ. The court agreed to exclude the well, pending resolution of the separate lawsuit the city filed against P/GSI regarding the contamination of the city's Montgomery well (MDEQ, 2005).

The City of Ann Arbor and P/GSI crafted an agreement that settled three related state and federal lawsuits regarding well contamination and the PZ. Under the terms of the agreement, the city was to receive a $285,000 payment, and P/GSI was required to conduct more groundwater monitoring to safeguard the city water supply against 1,4-dioxane contamination. The agreement provides for P/GSI to pay the city more than $4 million if 1,4-dioxane concentrations exceed the state's 85 ppb drinking water limit (Davis, 2006a).

The longest running lawsuit—lasting 16 years—related to the P/GSI case was between P/GSI and the State of Michigan over cleanup requirements. In addition to the existing 1300 gpm ozone-peroxide treatment system that P/GSI installed at its main facility, the MDEQ required that P/GSI install a treatment system large enough to accommodate 1150 gpm in a commercial area. P/GSI argued that such a facility would not be safe because of the large volumes of hazardous chemicals required to treat that volume of contaminated water. The court questioned the feasibility and safety of MDEQ's proposal, because of the challenges of acquiring and rezoning enough land to site both the treatment facility and the required treatment ponds in a congested commercial area (Shelton, 2004). The court also noted that MDEQ's requirements would involve piping millions of gallons of water contaminated with 1,4-dioxane beneath the city to the proposed treatment facility and would require miles of pipelines including at least 1.5 miles of pipelines in residential Ann Arbor neighborhoods. The judge noted that the City of Ann Arbor concurred with MDEQ's requirements, but did not offer a commitment to facilitate the installation of pipelines for contaminated groundwater through Ann Arbor's residential streets (Shelton, 2004).

Judge Shelton determined that MDEQ's requirement that P/GSI comprehensively treat and eliminate 1,4-dioxane was not feasible, was unwarranted, and was not supported by competent, material, and substantial evidence. However, P/GSI designed and positioned a 200-gpm mobile ozone-peroxide system to intercept 1,4-dioxane contamination to the State of Michigan Groundwater Surface Water Interface criteria of 2800 µg/L. The court did not offer a different remedial solution to remove contaminants from the plume that spread eastward into the deep Unit E aquifer. Judge Shelton noted, "It will never be possible to extract all of the 1,4-dioxane from this deep aquifer and the geology is such that it will ultimately end up in the Huron River and be diluted far below currently acceptable standards. But the goal must be to remove as much of the contaminant as possible, as quickly as possible, so that the ultimate dilution will take place with minimal impact on the water resource" (Shelton, 2004).

The court's unusual solution was to allow the leading edge of the 1,4-dioxane plume to migrate through Ann Arbor unabated. The court ruled that this option is allowed by Michigan law, provided unacceptable exposures to human health and the environment are prevented, including an institutional control to prevent use of the groundwater in the expected path of the Unit E plume (Shelton, 2004). This ruling led to the creation of the PZ, which is the antithesis of the antidegradation policy used for groundwater resources protection elsewhere in the United States. Nevertheless, MDEQ has effectively eliminated exposure and ensured water supply reliability, while P/GSI has continued its substantial progress with remediating this massive 1,4-dioxane plume and monitoring its migration.

In July 2008, Charles and Rita Gelman donated $5 million to the University of Michigan's Risk Science Center, which promotes the application of multidisciplinary skills to the study of the diverse health hazards that people face. The university expects the gift to position its Risk Science Center to become the nation's premier center for assessing, quantifying, and communicating risks to public health. The gift also supports graduate student fellowships and the hiring of two additional faculty members in the field and establishes a named professorship for a senior researcher. The Gelmans announced that their vision is to "help inform industry, government and the public about how to properly assess the benefits and hazards posed by technology (and chemicals in particular) in our society." Mr Gelman believes that settlement of the P/GSI 1,4-dioxane cases would have come far earlier if a neutral entity such as the U-M Risk Science Center had existed (University of Michigan, 2008).

8.4 FORMER AMERICAN BERYLLIUM COMPANY, TALLEVAST, FLORIDA

The former American Beryllium Company (ABC) site in Tallevast, Florida, is an example of the impact that the late discovery of 1,4-dioxane in a cleanup can have on-site investigation, remediation, and community relations. The bulk of the information presented herein was derived from the very comprehensive website for the project (http://www.tallevast.info/) and from the Florida Department of Environmental Protection (FDEP) website (http://www.dep.state.fl.us/).

Located in Florida's Manatee County, Tallevast traces its roots back to the 1890s, when shacks were built there for a community of African-American laborers. The laborers who first lived in Tallevast worked tapping sap from the local long-leaf "slash" pine forests and boiled it to make turpentine for use in the nation's shipyards and harbors. Residents of this "turp camp" also grew sugarcane, celery, and strawberries in area fields or worked in orange groves or on dairy farms (Lerner, 2008). At present, Tallevast is a blend of single-family residential homes and light commercial and industrial development (Cilek et al., 2004). Tallevast is located between Bradenton and Sarasota and less than 2 miles from Florida's west coast.

8.4.1 GEOLOGIC SETTING

Tallevast is situated in north-central Manatee County, in the terraced coastal lowlands physiographic province, a featureless plain extending to the Gulf of Mexico. The stratigraphy in the area consists of sandy sediments overlying a sequence of limestones, clays, and sandy limestones. The near-surface geology to a depth of 18 ft in the vicinity of the American Beryllium site consists of an upper unit of undifferentiated sandy and silty sediments of Holocene to Pliocene age and a lower undifferentiated carbonate unit of the Miocene Arcadia Formation extending to a depth of 300 ft below ground surface (bgs). The carbonate unit is predominantly dolostone with thin beds of sand and clay and phosphate deposits (Cilek et al., 2004).

The percentage of silt in the shallow Holocene sediments increases with depth, and the sand grain size decreases from fine to very fine. The transition to the undifferentiated Arcadia Formation occurs at approximately 15–18 ft bgs, where phosphatic sands and pebble-size quartz grains appear. With increasing depth, there is a gradual increase of phosphatic sand and percentage of clay (nearly 20%). At about 30 ft bgs is a thick clay unit, the Venice Clay, described as a mix of gray to olive-green clayey sand, tan lime mud, and limestone or dolostone fragments with quartz and phosphatic sands (Cilek et al., 2004). At approximately 300 ft bgs, the 3000-ft-thick carbonate Floridan aquifer begins (Campbell and Scott, 1993; Cilek et al., 2004).

The production well nearest the site is an agricultural well located one-quarter mile to the east. The agricultural water supply well is 805 ft deep and cased to 368 ft bgs (TetraTech, 1997). Most of the irrigation and potable wells in the Tallevast community are less than 100 ft deep; they are cased from land surface to between 20 and 30 ft bgs with an open-hole interval to total depth. The majority of irrigation and potable wells at the site produce from the regional intermediate aquifer system.

Karst conditions are suggested by topographic features in the area, including low-lying marshy areas and closed depressions; however, no sinkholes are reportedly present within 2 miles of the site. Groundwater is located approximately 2 ft bgs (TetraTech, 1997). The groundwater hydraulic high within the surficial aquifer is near the south-central part of the ABC site, which also represents the area's topographic high point. Proceeding away from this topographic high, the surficial aquifer's hydraulic head parallels the slope of decreasing ground surface elevation. Consequently, groundwater flows outward from the site, predominantly toward the northwest, north, northeast, east, and southeast (Cilek et al., 2004).

In a stunning example of an unfortunate coincidence, the hydraulic mounding was later discovered to be caused by a municipal waterline leak at the facility under a concrete slab adjacent to the former sump location; the sump was the probable source of the contamination described here. The broken pipe was discharging as much as 70,000 gallons/day, possibly for several years, and was probably the principal cause of groundwater and contaminants flowing radially outward and away from the former ABC facility (Arcadis-BBL, 2007a).

8.4.2   Site History

ABC operated on a 5-acre parcel located near the northeast corner of the Sarasota–Bradenton airport. In 1957, a small machine shop called Visioneering opened its factory in Tallevast. In May 1961, Visioneering became the ABC, a subsidiary of Loral Metals Technology, which operated as a machining and metalworking plant until 1996. ABC was one of the largest precision machiners of beryllium, a lightweight, toxic metal that can withstand harsh environments. ABC became a contract manufacturer of components for aerospace guidance systems and nuclear reactors. Because beryllium has a low density and is stronger than steel, it is used by aerospace industry companies. At ABC, beryllium machine parts were milled, lathed, and drilled into various components. Components were finished by electroplating, anodizing, and vapor degreasing and ultrasonic solvent cleaning (Cilek et al., 2004; FDOH, 2007).

Solvent use and wastewater handling are inferred from aerial photographs and from company records. A 1000-gallon solvent tank was used from 1960 until it was removed in 1986, when it was replaced with drum storage for solvents. Air permit records show that three portable vapor degreasers with 10- to 15-gallon methyl chloroform solvent tanks were used at the site. The first degreaser was acquired in the 1960s; the other two in the 1970s. The methyl chloroform was supplied by Ashland Chemical Company and contained 90–95% methyl chloroform and a proprietary inhibitor package that included 1,4-dioxane (TetraTech, 1997; Ashland Chemical, 1986). All three methyl chloroform degreasers were replaced in 1988 with ultrasonic cleaners that used Freon (TetraTech, 1997).

A small wastewater treatment system pumped all the facility's wastewater streams to an evaporation pond before the facility was connected to the public sewer system in 1986. The wastewater treatment system was subsequently removed and the pond backfilled and paved (www.tallevast.info). An unspecified quantity of soil was removed from the evaporation pond at the time of closure (TetraTech, 1997). A 400-ft-deep supply well provided water to the site; it was located near the plating shop where the wastewater sumps were located. This well was reported to have been in use through the 1970s; it was abandoned when the facility was connected to the municipal water supply.

A series of sumps were reportedly located near the wastewater treatment facility (TetraTech, 1997). One sump, dug near the plating shop, was described as an unlined pit filled with gravel and used between 1968 and 1984 for disposal of acid baths. The pit was paved with concrete in 1984 (TetraTech, 1997). The sumps were determined to be the predominant source of groundwater impacts at the site (Cilek et al., 2004).

The first indication of the presence of solvents in groundwater was noted in 1987, when 1,1-dichloroethane was detected at 65 µg/L in a single groundwater sample (TetraTech, 1997). A 1993 investigation of a fuel leak at a nearby fiberglass boat manufacturer revealed TCE in groundwater at concentrations of less than 25 µg/L (Cilek et al., 2004).

In 1996, a major aerospace contractor, referred to hereafter as "the responsible party," acquired the ABC and closed the Tallevast facility in the same year. In 2000, the responsible party sold the facility to Wiring Pro International (WPI) but retained the groundwater and soil remediation responsibility (FDOH, 2007). In 2000, the responsible party identified VOCs during a due diligence investigation to support the land sale to WPI. Manatee County Environmental Management and the FDEP were notified voluntarily of the VOC contamination caused by American Beryllium's past operations in July 2000 by the responsible party (www.tallevast.info). Between 2000 and 2003, the responsible party performed voluntary assessment and soil removal at the former ABC site, with oversight by the FDEP.

During the course of the voluntary investigations in 2002 and 2003, VOCs, chromium, and beryllium were identified in soil and groundwater. In 2001, 538 tons of impacted soil were removed and landfilled offsite as an interim cleanup measure. In July 2003, FDEP approved the Contamination Assessment Report that delineated impacts from former solvent use, which were largely contained within the site except for a low-concentration plume that extended a short distance off site to the north and east of the facility. The conclusion was based on data available from monitoring wells installed for the investigation, which were later determined by the FDEP Site Investigation Section to underrepresent the extent of groundwater impacts.

In 2002, FDEP asked the Florida Department of Health (FDOH) to assess potential impacts to off-site private wells. FDOH visited residential areas surrounding the site and tentatively identified several private wells north of the site and outside of the VOC plume as delineated by groundwater samples taken from existing monitoring wells (FDEP, 2004). The responsible party conducted a survey of water supply wells in the adjacent community and identified several domestic drinking water wells beyond the delineated plume. Because data collected by their consultants suggested that the plume was stable or shrinking, and no domestic wells were identified within or immediately adjacent to the plume, the domestic wells outside of the delineated plume were not sampled (FDEP, 2004).

In the fall of 2003, Tallevast residents living near the site learned of the off-site activities after observing a consultant conducting field work in the area. Although the responsible party notified the regulatory agencies in 2000, those agencies apparently did not notify the community about the site investigation activities over that three-year period (there is no regulatory requirement for a responsible party conducting voluntary investigation and remediation activities to notify the local community). In the fall of 2003, the responsible party provided information to the community about voluntary site assessment and remediation activities that were being performed. At that time, local residents voiced concerns about the potential for impacts to private wells. In 2004, community leaders provided the responsible party with additional information regarding the location of domestic wells, and FDEP, FDOH, and Manatee County surveyed the area and located and sampled 17 wells (FDEP, 2004).

FDEP and Manatee convened a community meeting, at which the responsible party asked residents for help identifying private well locations and agreed to test wells and pay for connection to public water if impacts were detected (FDEP, 2004). Several iterations of sampling detected solvents above MCLs in some domestic wells and irrigation wells, and affected residents were provided with bottled water. Temporary county water hookups were provided to all 17 homes that were not previously connected to the public supply. Some of the affected private wells were located a significant distance from the delineated plume; hence, further investigation of the plume was conducted (FDEP, 2004).

Comparisons of contaminant levels between private and monitoring well data sets are difficult to evaluate owing to the multiple aquifers in this geologic setting and the unknown construction details for most of the private wells that have been tested. Because of these limitations, FDEP generally does not rely on private well data for contamination delineation or remediation monitoring purposes (FDEP, 2006).

Because the impacted private wells were located outside the established VOC plume boundaries, as delineated by the responsible party's consultant, FDEP conducted an investigation to identify the source(s) of contamination and to resample all monitoring wells (Cilek et al., 2004). FDEP collected 39 groundwater samples from the surficial aquifer; most of the samples were collected near its base. Direct-push groundwater grab samples were collected at locations adjacent to existing monitoring wells, selected irrigation wells, and potential source areas (Cilek et al., 2004). The highest TCE concentration reported in the initial site groundwater survey, 2050 µg/L, was a direct-push grab sample retrieved near the eastern boundary of the facility (Cilek et al., 2004; FDEP, 2004).

FDEP's results indicated a larger chlorinated solvent groundwater plume with higher concentrations of chlorinated solvents than that delineated by the responsible party's monitoring wells. For example, a TCE concentration of 550 µg/L was reported in one well, whereas analysis of the nearby grab groundwater sample collected nearby reported a TCE concentration of 29,000 µg/L (35,000 µg/L in a duplicate sample). FDEP interprets this result to be indicative of DNAPL (dense, nonaqueous-phase liquid) because the concentration is greater than 1% of TCE's solubility in water (Cilek et al., 2004). No analyses for 1,4-dioxane were performed in FDEP's or the responsible party's 2004 investigations.

In May 2004, some Tallevast neighborhood residents reported that they had received soil excavated from the former American Beryllium site to fill in low spots on their properties. Others, concerned that they might be breathing vapors from contaminated groundwater underneath their homes, requested indoor air testing of their homes (FDOH, 2007). During public meetings held in June and July 2004, residents expressed a variety of health concerns they believed American Beryllium caused, including high rates of cancer, fertility problems, and illnesses that they attributed to exposures to dust and surface soils transferred from American Beryllium to resident yards (FDOH, 2007). In June 2004, the FDEP and the responsible party's consultants tested neighborhood surface soil for metals, VOCs, semi-VOCs, and hydrocarbons (FDOH, 2007). Although several of the soil samples were found to contain elevated levels of arsenic, lead, and petroleum-related organic compounds[*] (Cilek et al., 2004), concentrations were generally only slightly above standards, and no pattern of pervasive soil contamination was identified.

In the fall of 2004, FDOH and the Manatee County Health Department tested for 61 VOCs in the indoor air of three homes and in the community center located adjacent to the American Beryllium site. The indoor air-testing sites are located over the highest concentrations of TCE in groundwater. Summa canisters were used to collect three 8-h samples at each location. FDOH mailed letters to each indoor air-testing participant explaining chemical results obtained from their indoor air, including copies of the laboratory results and information on each chemical found. At all of the four locations tested, most of the 61 chemicals analyzed in indoor air were nondetect. FDOH enlisted the

---

[*] 1,4-Dioxane was not tested on soil samples, but because of its high volatility and low sorption, is commonly not found in soil.

assistance of the Agency for Toxic Substances Disease Registry (ATSDR) to prepare a health consultation and exposure investigation report. For those with known comparison guidance concentrations, all levels detected were below ATSDR's guidance concentrations (ATSDR, 2005; FDOH, 2007). 1,4-Dioxane was not detected in any of the four indoor air samples (ATSDR, 2005).* Subsequent vapor-intrusion investigations also tested for 1,4-dioxane, including 23 soil–vapor samples and two ambient-air samples in June 2006; however, analyses did not detect 1,4-dioxane above its 18 µg/m³ reporting limit (5 ppbv) (Arcadis-BBL, 2007b).

Five years into the site investigation and remediation activities, in February 2005, a Site Assessment Report Addendum presented additional groundwater findings, including the detection of 1,4-dioxane in 24 upper-aquifer monitoring wells and in 16 lower-aquifer monitoring wells. Of these results, 21 upper-aquifer monitoring wells had 1,4-dioxane greater than Florida's Groundwater Cleanup Target Level of 3.2 µg/L, and 13 lower-aquifer monitoring wells exceeded the 3.2 µg/L target level (TetraTech, 2005).

The discovery of 1,4-dioxane substantially increased the plume size because 1,4-dioxane had spread much farther than the methyl chloroform† from which it originated and much farther than the TCE and PCE that was also released from the site. Prior to the discovery of 1,4-dioxane, the plume of chlorinated ethenes was mapped at 50 acres; following the discovery of 1,4-dioxane, the plume was mapped at 131 acres and eventually was delineated at about 200 acres.

An independent consultant, Environmental Science & Technology (ES&T), was retained by the local community group (but funded by the responsible party) to complete an independent sampling study of private wells. ES&T suggested that their findings indicated the plume was moving more rapidly than was first thought. They attributed the plume's movement to changes in hydraulic pressure that occurred when Tallevast residents were connected to county water in the summer of 2004 and stopped pumping their wells. Cessation of pumping near the center of the plume, coupled with continued pumping by surrounding residents and businesses on the periphery of the plume caused an apparent shift in the directions and rates of plume migration (ES&T, 2005).

The responsible party's February 2005 1,4-dioxane results did not correlate with ES&T's December 2005 reported levels. ES&T's 1,4-dioxane results were obtained by EPA Method 8260 using a reporting limit of 20 µg/L, whereas the responsible party's consultant used FDEP-approved EPA Method 8270, which is generally regarded in the industry as a more reliable method. ES&T's EPA 8260 results suffered from quality control problems. For example, one duplicate sample was nondetect (<20 µg/L) in the first analysis and 150 µg/L in the second analysis. Overall, ES&T's EPA 8260 results showed higher concentrations of 1,4-dioxane. ES&T suggested that the differences between their sample results and those obtained by the responsible party's consultant may have been related to how deep samples were retrieved. ES&T noticed that samples collected from the deeper, open-cased bore hole had higher concentrations than those from the upper, cased section of the well (ES&T, 2005). The large differences in 1,4-dioxane results from the two sampling surveys, while collected at different times and with different methods, led to public confusion and suspicion of the lower concentration results.

The responsible party funded a field study in 2006 to collect whole-volume split samples for analysis of 1,4-dioxane by three methods at three laboratories. A new consultant, Arcadis-BBL, collected field samples, including some samples spiked with standards at three different concentrations. Analysis was performed by EPA Methods 8270C, 8260B, and 8270 with isotope dilution ("8270-ID") at three different commercial laboratories. 1,4-Dioxane concentrations in the duplicate samples analyzed by Method 8260B were highest among the three methods; some results were more than an order of magnitude higher than the Method 8270C results. Detailed information regarding this laboratory study is presented in Chapter 4.

---

* A common laboratory method for analysis of soil vapor and indoor air samples is EPA Method TO-15, which includes 1,4-dioxane; however, 1,4-dioxane was not the focus of the indoor air sampling.
† The methyl chloroform is virtually absent from the Tallevast plume; its abiotic and microbial degradation products, 1,1-dichloroethane and 1,1-dichloroethylene, are all that remain.

8.4.3   REGULATORY RESPONSE: REMEDIATION AND WATER SUPPLY REPLACEMENT

The record for the Tallevast site reflects significant effort by the responsible party to delineate the contamination, report findings to the regulatory agencies, and initiate voluntary remedial actions. Although not required under environmental regulations to communicate directly with the community, the responsible party made significant attempts to keep them informed, address their concerns, and provide financial recompense to the affected community. Substantial additional effort by FDEP, FDOH, and Manatee County Environmental Management was required to oversee the investigation, engage an active and concerned community, address health issues, and respond to a heavy involvement by local, state, and federal elected officials, as well as intensive coverage by the local media. FDEP worked with the responsible party to aggressively investigate the extent of contamination, which was accomplished by installation and sampling of 250 monitoring wells in the space of only a few years. A soil removal action and construction and operation of an interim remedial action were also used to remove contaminants from the source area. 1,4-Dioxane turned out to be the most widespread contaminant. This finding greatly expanded the scope of the investigation and presented a challenge to both analysis and remediation.

The primary remedial strategy is groundwater extraction and treatment for removal of trichloroethylene, perchloroethylene, *cis*-1,2-dichloroethylene, 1,1-dichloroethylene, 1,1-dichloroethane, and 1,4-dioxane. The groundwater extraction plans include on-site extraction wells, six extraction trenches in the upper aquifer totaling more than 1 mile of linear trenching to a depth of approximately 25 ft bgs, and 50 off-site extraction wells in several lower-aquifer units (Arcadis-BBL, 2007c). Figure 8.5 shows the configuration of extraction trenches at the former American Beryllium site.

Pilot testing was performed to evaluate treatment technologies—including photo-oxidation, ozone-peroxide oxidation, and biostimulation—to remove 1,4-dioxane and chlorinated solvents. Pilot-test outcomes are described in the following sections. The treatment technology selected at that time for the on-site Interim Remedial Action was the Photo-Cat™ advanced oxidation system coupled with an iron oxidation pretreatment system consisting of a 20,000-gallon influent tank to facilitate contact time with a coagulant and precipitation of iron. A series of bag and cartridge filters provides secondary filtration for removal of iron hydroxide precipitate. Effluent from the Photo-Cat system is further treated by two larger GAC canisters installed in series in a lead/lag fashion that remove any remaining 1,1-DCA and polish the treated groundwater. Following GAC treatment, the groundwater is discharged to the Manatee County wastewater treatment plant (Arcadis-BBL, 2007c).

The Photo-Cat water treatment system destroys VOCs and 1,4-dioxane by mixing titanium dioxide (TiO_2) catalyst with the groundwater and exposing it to UV light in a reactor; the contaminants are converted to carbon dioxide, water, and salts. Pilot testing (described further in the next section) confirmed that the Photo-Cat system reliably treats trichloroethylene (TCE) and 1,4-dioxane to less than 3 µg/L, which is at or below the anticipated discharge limits to be established by Manatee County (Arcadis-BBL, 2007c).

8.4.3.1   UV Oxidation Pilot Treatability Study

Purifics ES, Inc. conducted an on-site pilot test using the Photo-Cat UVOx treatment process, a patented closed-loop TiO_2 slurry-based photocatalytic process to remove 1,4-dioxane and VOCs from groundwater at the former ABC facility.[*] Bicarbonate ions, abundant in the carbonate aquifers at the Tallevast site, scavenge the hydroxyl radicals generated by the Photo-Cat process and can limit the effectiveness of the photocatalytically generated hydroxyl radicals. Bicarbonate is controlled by lowering pH with sulfuric acid, nominally to 4.6, where bicarbonate is converted to CO_2; however, to control formation of iron hydroxide precipitates from dissolved iron present in the upper aquifer, pH must be further lowered to 3.0 (Powell, 2005).

---

[*] In United States, Photo-CAT is sold under licenses from Purifics ES by Basin Water; the technology is NSF-61 certified for drinking water systems.

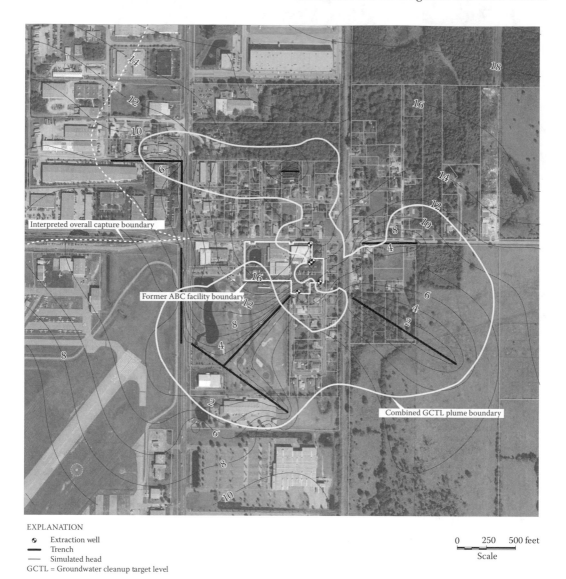

EXPLANATION

⚬ Extraction well
━ Trench
── Simulated head
GCTL = Groundwater cleanup target level

0   250   500 feet
Scale

FIGURE 8.5 Configuration of extraction trenches and extraction wells proposed for remediating 1,4-dioxane and VOCs at the former American Beryllium site (Arcadis-BBL, 2007c).

To normalize operating parameters between pilot tests, a first-order rate relationship was used to establish a rate constant for the Photo-Cat reaction with the site contaminants. The simple first-order equation for the rate constant k is

$$k = \frac{1}{t} \cdot \ln\left(\frac{C_0}{C}\right) \propto \left(\frac{v_f}{P}\right) \cdot \ln\left(\frac{C_0}{C}\right), \tag{8.1}$$

where C_0 is the initial concentration, C is the final concentration, t is the time (in s), k is the first-order rate constant, v_f is the flow rate (in L/min), and P is the power (in kW). Because time is directly proportional to the power used to operate the Photo-Cat system, flow in liters per minute is divided by power in kilowatts, giving units of (L/min)/kW for the first-order rate constant in Equation 8.1.

Influent concentrations in various pilot-test configurations ranged from 250 to 900 μg/L 1,4-dioxane and 2800–10,000 μg/L trichloroethylene. The corresponding rate constant for TCE, averaged from both the upper- and lower-aquifer pilot tests, was 36.8 (L/min)/kW and for 1,4-dioxane, 24.9 (L/min)/kW. The lower aquifer has higher chloride concentration, which also scavenges hydroxides and affects TCE destruction, but has less impact on the destruction of 1,4-dioxane (Powell, 2005).

By modeling different extraction and contaminant migration scenarios, the timeframe to achieve 90% overall mass removal in approximately 30 years was estimated; however, the time to achieve FDEP's groundwater target level, 3.2 ppb for 1,4-dioxane, may be greater than 100 years. A screening-level analysis was performed for ISCO and enhanced biological degradation to eliminate 1,4-dioxane in the source zone; this analysis determined that these technologies would have limited effectiveness at the Tallevast site. As of April 2008, nearly 6 million gallons of groundwater have been pumped and treated. The expanded extraction system with trenches, extraction wells, and large-capacity treatment is expected to be fully operational by March 2009 (Arcadis-BBL, 2007c).

8.4.3.2   Ozone and Peroxide Oxidation Pilot Study

Samples of contaminated site groundwater were sent to Applied Process Technology, Inc. (APT) in California to evaluate treatability of 1,4-dioxane and chlorinated solvents using APT's HiPOx™ technology. HiPOx, described further in Chapter 7, uses a continuous, in-line, at-pressure AOP to destroy VOCs and 1,4-dioxane. The HiPOx process employs ozone (O_3) and hydrogen peroxide (H_2O_2) chemistry in a uniquely designed oxidation reactor. These reactants are injected under high pressure directly into the treatment water stream at specified ratios and locations, generating hydroxyl radicals that attack the bonds in organic molecules, progressively oxidizing contaminants and their by-products into carbon dioxide, water, and salts. Table 8.2 summarizes influent and effluent concentrations for Tallevast groundwater treated with the HiPOx process and the corresponding ozone demand (Herlihy and Bowman, 2005).

The reasons for selecting Photo-Cat instead of HiPOx are not stated in the consultant reports, but the HiPOx system generally has a higher capital cost, but lower long-term operating and maintenance costs, and the higher capital cost may have influenced the decision. Both appear to be effective at removing 1,4-dioxane and chlorinated solvents, and both require the use of hazardous chemicals (sulfuric acid for Photo-Cat and hydrogen peroxide for HiPOx). Systems dependent on UV light may be impeded where nitrate, colloids, or other light-blockers are present, whereas systems using ozone may form undesirable bromates where bromide is present.

8.4.3.3   *In Situ* Biostimulation and Bioaugmentation Treatability Study

Primary and secondary evidence was established for reductive dechlorination of chlorinated solvents, but it was concluded that biodegradation of 1,4-dioxane through biostimulation or bioaugmentation was unlikely to be effective. The primary positive evidence includes the presence of reductive dechlorination by-products, including *cis*-1,2-DCE, 1,1-DCE, 1,1-DCA, and ethene in some groundwater samples; secondary evidence includes reducing geochemical conditions,

TABLE 8.2
Ozone Demand for HiPOx Treatment of 1,4-Dioxane and Trichloroethylene

Influent TCE (μg/L)	Effluent TCE (μg/L)	Influent 1,4-Dioxane (μg/L)	Effluent 1,4-Dioxane (μg/L)	Applied Ozone (mg/L)
2700	2	500	3.2	15.2
2700	4.2	300	3.2	13.6
2700	3	300	2.5	14.3

Source: Herlihy, P. and Bowman, R., 2005, HiPOX-TM Technology Lab Testing, Former American Beryllium site, Tallevast, Florida. Pleasant Hill, CA: Applied Process Technologies Inc.

specifically iron-reducing, sulfate-reducing, and methanogenic conditions (Arcadis-BBL, 2007c). The presence of reducing geochemical conditions in all monitored zones is supported by low dissolved oxygen concentrations (less than 1.0 mg/L), negative ORP (oxidation–reduction potential) values, abundant naturally occurring organic carbon in site groundwater, elevated sulfide concentrations (above 500 mg/L), and elevated methane concentrations (above 50 ppb in some groundwater samples). The lack of further degradation products beyond *cis*-1,2-DCE (i.e., vinyl chloride, ethene) in most monitoring wells suggests that natural biodegradation of chlorinated ethenes may be stalled, possibly owing to the presence of excessive alternative electron acceptors such as sulfate and chloride (Arcadis-BBL, 2007c).

SiREM laboratories used polymerase chain reaction (PCR) tests to determine whether indigenous *Dehalococcoides*, an ethene-degrading microbe, was present. SiREM also performed biostimulation tests on-site groundwater augmented with EOS™ as an electron donor, with and without bioaugmentation using a proprietary bacterial culture, KB-1™.* The fact that the *Dehalococcoides* assay on two tested samples was negative indicated that bioaugmentation would be necessary; however, site data are indicative of active reductive dechlorination. The widespread and persistent occurrence of 1,4-dioxane, in conjunction with evidence of chemically reducing conditions, which are the opposite of the aerobic conditions that 1,4-dioxane-degrading bacteria generally require, suggested that geochemical conditions do not favor natural biodegradation of 1,4-dioxane.

8.4.4 COMMUNITY RESPONSE

Communication with the Tallevast community has been a major challenge for the regulators and the responsible party due in large part to the dynamic nature of characterizing a plume and the late discovery of 1,4-dioxane. Because some community members felt that they should have been informed of the site investigation earlier, these residents were not inclined to trust the assurances they received that their concerns would be addressed. Local newspaper articles suggest that residents viewed much of the information they received through a lens of distrust. The events that followed the initial discovery of contamination in private wells apparently strengthened the residents' assumptions that they were not being told the full story of the contamination. These events include the chlorinated solvent plume's turning out to be much larger than originally determined, the discovery of 1,4-dioxane in 2006, inconsistent laboratory results, and a health study by FDOH that mistakenly used incorrect data.

Tallevast residents formed a citizen action group called FOCUS—Family Oriented Community United Strong. FOCUS was successful in obtaining substantial funding from the responsible party to retain an independent consultant to review and advise community members on the content of technical documents in lay terms. The independent consultant also performed additional soil and groundwater sampling and a supplemental health risk assessment and participated in technical strategy discussions with the responsible party and regulators. The Vice President of FOCUS, Ms Wanda Washington, provided the local newspapers with a series of insightful quotations that underscore the challenges that regulators and responsible parties face when plumes affect private wells. These quotes are included here not to malign the FDEP, FDOH, the responsible party, or their consultants; rather, the quotes are provided to emphasize the dilemma faced by regulators, responsible parties, consultants, and the community alike when 1,4-dioxane contamination is discovered late in the cycle of site investigation and remediation.

Regarding the expanding plume, Ms Washington provided this comment:

"These experts get up and talk about my safety and welfare and say everybody is OK," said Wanda Washington, vice president of Tallevast's community group FOCUS. "Then they draw this 50-acre plot,

---

* EOS™ is an emulsified edible oil used to stimulate anaerobic biodegradation of chlorinated solvents and other contaminants; KB-1 Dechlorinator (KB-1™) is a natural microbial culture used to introduce *Dehalococcoides* organisms to sites where they are absent to promote the complete dechlorination of PCE, TCE, DCE, and vinyl chloride in groundwater.

and now it is 131 acres? My confidence is just shrinking by the minute. Just when you think it can't get any worse, it gets worse." (*Bradenton-Herald*, Radway and Wright, 2005)

Regarding the discovery of 1,4-dioxane late in the investigative process, an editorial in the *Bradenton-Herald* expressed empathy for Tallevast residents:

> One wonders why it took until now to test for [1,4-] dioxane—or other possible chemicals. Wouldn't it be cheaper to check all potential pollutants at once rather than going back to previous drilling sites? This drip-drip-drip process of discovering and releasing bad news leaves affected residents uneasy. The very existence of a pollution threat was withheld from residents for almost four years. It's no wonder that [Ms.] Washington and members of the community group FOCUS are suspicious of almost everything they're being told. (Fishbein, 2005)

Some members of the Tallevast community are convinced that residents' health was impaired by ABC's historical operations and the resulting groundwater contamination that affected their wells. Some Tallevast residents point out that dozens of their neighbors have taken ill with cancer and other health issues that they attribute to the presence of industrial contaminants in their well water. The FDOH drafted a report on the potential health impacts to the community but found only four cases of cancer, a finding in stark contrast to an informal survey by residents that showed about 90 cases of cancer or other diseases in the Tallevast community (Fishbein, 2008).

When FDOH officials met with FOCUS in March 2008, they acknowledged that their numbers, based on a state database and figures from a local hospital, were "wildly off the mark." FDOH also admitted that they had studied the wrong ZIP code: although Tallevast has a post office, most Tallevast residents live in a Sarasota ZIP code (Fishbein, 2008). By using a statewide database to review health records, FDOH administrators apparently missed important demographic features of the Tallevast community.

Despite the adverse perceptions by the community, newspaper articles and regulatory records clearly show that once the private well contamination was discovered, state and local regulators and the responsible party provided a comprehensive response to ensure a safe supply of drinking water. An account of the Tallevast case by Steve Lerner posted on the Collaborative on Health and the Environment's website,[*] describes the response as follows:

> Once news reports of the contamination began to appear, county officials were galvanized into action. They appeared in Tallevast at 8 p.m. one evening handing out five-gallon plastic bottles of water and warning residents not to drink water if it came from a well. Subsequently, all wells in town were capped and above-ground blue plastic pipes were installed as a "temporary hook-up" to county water lines. The plastic pipes remain in place today, four years later. The increase in pressure from the county water hookup caused numerous leaks in faucets and hot water heaters in Tallevast homes and the telephones of Laura Ward and Wanda Washington began ringing with requests for help with plumbing problems. (Lerner, 2008)

The responsible party initiated a financial incentive program to abandon private water supply wells in the area, and the Manatee County government restricts the construction of new wells in the affected area.[†] The goals of both programs are to reduce the potential for exposure to affected groundwater and to limit potential cross-connection between vertically distinct groundwater aquifer zones (Arcadis-BBL, 2007c). More than 50 private wells were closed under this program (Arcadis-BBL, 2007a).

The community nevertheless remained dissatisfied, and construction work to install monitoring wells and pilot trenches associated with the groundwater remedial action apparently alarmed residents even more. Residents demanded that Tallevast remediation work stop until the community is moved out of harm's way. Their cause was taken up by the Tallevast representative to the state

---

[*] Collaborative on Health and the Environment (www.healthandenvironment.org).
[†] In this context, "abandon" means to properly seal the well according to a technical standard that will prevent cross-contamination.

legislature, Republican Bill Galvano. Representative Galvano proposed that the whole town be moved to a new site and sent the responsible party a proposal suggesting that they pay for most of the cost of relocating residents to a new site of their selection in exchange for dropping their 2005 lawsuit. Tallevast residents apparently asked for assistance after some of them were declined refinancing of their mortgages because of the plume of contamination from the former American Beryllium plant (Marsteller, 2005).

To address residents' concerns over their property values, the responsible party established the Tallevast Property Value Protection Program for current homeowners interested in selling or refinancing their residence. The program offers a payment for the reasonable difference (up to 40%), if any, between the actual sale price negotiated and the appraised fair market value if there were no associated groundwater impact. The program also provides assistance with residential seller support services such as appraisals, home inspections, brokerage services, and home marketing assistance (www.tallevast.info).

The Tallevast site provides an excellent example of how to manage dynamic community relations through multiple channels to ensure that relevant information is distributed to community stakeholders in a timely manner. The responsible party hosted frequent and widely announced public meetings to explain the ongoing work and staffed these meetings with the consultant team that was performing the work. In addition, the responsible party maintained a website with a complete report repository and prepared dozens of informative newsletters that were mailed to the affected residents. Communications have at times been contentious, but the responsible party ensured that residents had ample opportunity to register their concerns and inquire directly with the parties performing the work.

8.4.5 Discussion

The Tallevast saga continues. Looking back on all that has transpired, it might be tempting to second-guess how the site was addressed. But several features of this case would probably have caused most environmental professionals to follow the same course and arrive at a similar conceptual model, only to discover later that it does not quite fit. By following conventional wisdom and orienting an investigation around biodegradable hydrophobic chlorinated solvents—as is typically done at these types of sites—the persistent and extremely mobile nature of 1,4-dioxane was not effectively assessed. The bias in the sampling and analytical methods for 1,4-dioxane in open-cased irrigation and domestic wells made a significant difference as well. The peculiar mounding seen in the site potentiometric surface map was generally consistent with the site location on a relative topographic high; yet, the magnitude of the mounding appeared greater than the local topographic relief. Nevertheless, most hydrogeologists would probably not start out with a conceptual model that included an unknown 50 gallons/min water main leak next to the source area! This feature alone was probably responsible for a much larger mass fraction leaving the source area in multiple directions than could have been anticipated to occur under a natural gradient.

The Tallevast case is a reminder that a conceptual model is always a work in progress, to be revisited and updated frequently as more data become available, and whose assumptions should be periodically questioned. On the basis of the documents reviewed by the author, it appears that the responsible party, consultants, and regulators involved have been quite effective at adapting to changing conditions and have made substantial progress to address a complex and vexing 1,4-dioxane plume.

8.5 STANFORD LINEAR ACCELERATOR CENTER GROUNDWATER CLEANUP SITES

The Stanford Linear Accelerator Center (SLAC National Accelerator Laboratory, hereafter SLAC) site provides examples of 1,4-dioxane migration behavior in low conductivity bedrock aquifer materials. The SLAC data also show the apparently serendipitous removal of 1,4-dioxane on GAC, possibly by passive biodegradation or chlorine oxidation.

8.5.1 SLAC Site History

SLAC is located in Menlo Park, neighboring the Stanford University Campus. SLAC is a high-energy physics research facility owned and operated by Stanford University under contract to the U.S. Department of Energy. The site covers 426 acres. The facility is best known for the 2-mile-long Stanford Linear Collider, the longest linear accelerator in the world, which is claimed to be "the world's straightest object." The linear accelerator can accelerate electrons and positrons up to 50 GeV (giga-electron volts) and has been operational since 1966. Other research projects located at SLAC include the Positron-Electron Project storage ring, the Stanford Positron-Electron Asymmetric Ring, and the Stanford Synchrotron Radiation Laboratory. Among numerous other superlative distinctions, SLAC hosted the world's first web site.

Past research and support operations at SLAC included the storage and use of solvents for cleaning parts in machining and plating shops and for paint work, as well as other common solvent uses. A 1984 site investigation discovered contaminated soil and groundwater, and in 1985, the California Regional Water Quality Control Board issued a cleanup order for a former solvent underground storage tank (FSUST) area at the SLAC facility. Since 1985, SLAC has conducted numerous soil and groundwater investigations that included the installation of more than 100 groundwater monitoring wells. Site cleanup activities have included (1) several efforts to remove soil contaminated with VOCs and (2) extraction and treatment of contaminated groundwater within several small plumes (Cal EPA, 2005).

8.5.2 Geologic and Hydrologic Setting

The SLAC facility is located in the rolling foothills of northern California's Santa Cruz Mountains, above an alluvial plain that borders the western margin of the south end of San Francisco Bay. The Santa Cruz Mountains were created in an active tectonic province characterized by ongoing strike-slip and compressional movement of the nearby San Andreas Fault (SLAC, 2006a). The San Andreas Fault is located 2 miles west of the SLAC site.

The SLAC facility is situated within the San Francisquito Creek Watershed, which encompasses approximately 40 square miles and extends from the ridge of the Santa Cruz Mountains to the San Francisco Bay (Cal EPA, 2005). Contaminated groundwater at the eastern end of SLAC flows primarily within the Ladera Sandstone, a thick sequence of silty marine sandstone that dominates SLAC's geology. The Ladera Sandstone typically contains about 60% sand and 40% silt and is part of a 2000-ft-thick sequence of consolidated marine sedimentary rocks of Eocene to Miocene age (i.e., 55–5 million years old) (SLAC, 2006a). The Ladera is a sandy siltstone with discontinuous fractures and very low hydraulic conductivity ranging from 10^{-4} to 10^{-7} cm/s (Sabba and Witebsky, 2003). Nonmarine silts, sands, and gravels of the upper Pliocene to Pleistocene Santa Clara Formation (about 2 million to 100,000 years old) rest unconformably atop the bedrock (SLAC, 2006a). The bedrock typically has a weak to friable weathered zone that extends to 30 ft bgs.

Groundwater well yields and natural water quality have been evaluated to determine potential beneficial uses of groundwater at SLAC. Results of the assessment indicate that groundwater in most areas beneath the facility would not be suitable as a drinking water source on the basis of well production rates lower than 200 gallons/day and concentrations of total dissolved solids above 3000 ppm (Cal EPA, 2005). Groundwater in the area has naturally high concentrations of total dissolved solids ranging from 890 to 8300 mg/L, and well yields measured in extraction wells screened within the Ladera Sandstone are low yielding. Two SLAC extraction wells completed in the Ladera Sandstone yielded 50 and 170 gallons/day (SLAC, 2004).

8.5.3 CONTAMINANT RELEASES

There are four small VOC plumes at SLAC, three of which involve release of methyl chloroform and 1,4-dioxane. The plumes emanate from a variety of minor releases that occurred through chemical handling beginning in the early 1960s. Because the Santa Clara Formation and Ladera Sandstone have low hydraulic conductivities, the plumes migrate very slowly (<10 ft/year) and are all less than 200 ft long. The 1,4-dioxane plumes occur at facilities within SLAC known as the Former Solvent Underground Storage Tank (FSUST), the Plating Shop Area, and the Former Hazardous Waste Storage Area (FHWSA) (Sabba and Witebsky, 2003). SLAC staff has not found evidence of direct disposal of solvent wastes, and DNAPLs (dense, nonaqueous-phase liquids) or large masses of solvent have not been encountered at the site. Aerial photographs show that the FHWSA was used as a drum storage area during the 1970s (SLAC, 2004). In addition to possible drum leaks, other release mechanisms at SLAC may have included steam-cleaning wastewater, vapor degreasing operations, floor sumps, leakage through the floor of the plating shop, spills and leaks in chemical storage areas, and spills and leaks at a rinse-water treatment operation. Some combination of these sources contributed to solvent and 1,4-dioxane releases at the Plating Shop, FHWSA, and FSUST.

SLAC reports that vapor degreasing operations switched from trichloroethylene to methyl chloroform before 1980; this change provides a time marker for solvent releases bearing trichloroethylene. Releases including trichloroethylene and its breakdown products are considered to have started between 1963 and 1968 and stopped prior to 1980 (SLAC, 2003). At the plating shop area, 1,4-dioxane is present in plumes having no remaining methyl chloroform, suggesting complete abiotic degradation of methyl chloroform to acetic acid and 1,1-dichloroethylene and biologic degradation to 1,1-dichloroethane and chloroethane.

SLAC staff and consultants cite the DNAPL literature and note that VOC migration in the Ladera Sandstone depends on the mode of release. A sudden, large-volume spill would be expected to rapidly migrate downward and laterally, leaving a relatively large volume of residual VOCs in the unsaturated zone, whereas a slow leak occurring over long periods of time is expected to migrate downward along permeable pathways and be less prone to lateral migration. Therefore, daily de minimis losses from spills, leaks, splashes, and drips are likely to deliver more VOCs to the water table beneath a thick unsaturated zone (SLAC, 2003). Conditions observed in the Plating Shop Area at SLAC are consistent with long-term releases of small volumes of solvent.

The local saturated hydraulic conductivity of the Ladera Sandstone at the FHWSA is estimated to range from 3.3×10^{-5} to 1.4×10^{-4} cm/s. Yields from three pumping wells screened in the Ladera Sandstone at the FHWSA have ranged from approximately 53–170 gallons/day, and groundwater velocity ranges from 5 to 10 ft/year (SLAC, 2004).

Seasonal water level fluctuations beneath the FHWSA are from 3 to 5 ft. Since 1997, depth to groundwater has ranged from 10 to 32 ft bgs. The linear accelerator subdrain system drains groundwater beneath SLAC at 35 ft bgs and exerts an influence over local groundwater flow, reversing the natural southward and eastward gradients. Groundwater collected by the subdrain system is discharged to San Francisquito Creek at about 5 gpm (SLAC, 2004). Two other large subterranean structures used for high-energy particle research at SLAC are equipped with subdrains. Two curving tunnels were bored through the Ladera Sandstone at depths ranging from 60 to 90 ft bgs, each several thousand feet long. The tunnels housing the experimental equipment act as groundwater drains, but discharge only 1.5 and 6.2 gpm over the entire tunnel lengths. The low rate of groundwater discharge in these two tunnels and in the linear accelerator subdrain support porous media flow as the primary mode of groundwater and contaminant transport, although fractures and other conduits may locally influence hydraulic gradients (SLAC, 2004). The complex hydraulics influenced by drains, topography, fracture orientation, and the strikes and dips of bedded subunits in the sandstone combine to cause a large range of groundwater flow directions. Three of the small groundwater plumes at SLAC are migrating toward the linear accelerator subdrain system, which is monitored for contaminants prior to discharge to San Francisquito Creek (SLAC, 2006b).

Analyses of groundwater samples from FHWSA monitoring wells show presence of perchloro-ethylene, trichloroethylene, and *cis*-1,2-dichloroethylene, suggesting a historical release of perchloroethylene and reductive dechlorination occurring in the subsurface (SLAC, 2004). In addition, detections of methyl chloroform, 1,1-dichloroethylene, and 1,1-dichloroethane in samples further suggest abiotic and microbially mediated reductive dechlorination of methyl chloroform (SLAC, 2004).

There are two apparent primary release areas at the FHWSA for methyl chloroform and 1,4-dioxane. SLAC staff attributes the detection of 1,4-dioxane to its use as a stabilizer for methyl chloroform, as there are no other known significant uses of 1,4-dioxane at the SLAC facility (Cal EPA, 2005). 1,4-Dioxane was first analyzed in groundwater extraction wells in 1997 by using EPA Method 8270 in selected-ion mode (SIM) with a 25 µg/L reporting limit; in 1998, the reporting limit was lowered to 10 µg/L (Sabba and Witebsky, 2003). Further 1,4-dioxane characterization in groundwater began in 2000 using EPA 8270-SIM and a reporting limit of 1.0 µg/L (SLAC, 2006c; Cal EPA, 2008). Although detected in soil vapor and in 30% of groundwater samples, 1,4-dioxane has only been detected in one soil sample collected at the FHWSA, at 0.74 mg/kg (SLAC, 2004). The distributions of 1,4-dioxane and the breakdown products of methyl chloroform at the FHWSA are mapped in Figure 8.6.

A fate and transport study of VOCs and 1,4-dioxane in groundwater at the FHWSA was performed by Erler & Kalinowski, Inc. (EKI), a consultant to SLAC (SLAC, 2006a). EKI modeled the migration of 1,4-dioxane, perchloroethylene, and DCE in two separate chemical plumes using a one-dimensional model and site-specific data. One of the plumes is slowly migrating north toward the accelerator subdrain, and the other plume is slowly migrating south toward San Francisquito Creek. EKI interprets the model results to predict that a maximum concentration of 0.2 µg/L of

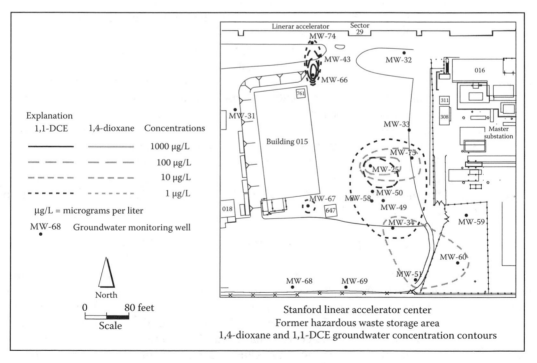

FIGURE 8.6 Distribution of 1,4-dioxane and 1,1-dichloroethylene at the SLAC Former Hazardous Waste Storage Area. (Adapted from SLAC, 2004, *Site Characterization Report for the Former Hazardous Waste Storage Area*, Vol. 1 of 4, submitted to the Regional Water Quality Control Board, San Francisco Bay Region (RWQCB) SLAC-I-750-3A33H-015, SLAC National Accelerator Laboratory Environmental Health and Safety Division.)

1,4-dioxane would reach the linear accelerator subdrain about 80 years from 2001, whereas a maximum concentration of 1 μg/L of DCE would arrive at the subdrain about 150 years from 2001. The model suggests that perchloroethylene would never arrive at the San Francisquito creek (685 years from 2001 if the maximum concentration is 1.6×10^{-6} μg/L). The regulatory agency's review of VOC and 1,4-dioxane migration at the FHWSA concluded that these compounds are not likely to impact downgradient offsite groundwater or surface water at detectable concentrations (Cal EPA, 2005).

8.5.3.1   Remediation at Former Hazardous Waste Storage Area

Interim remediation at FHWSA was a dual-phase soil–vapor and groundwater extraction system that operated from December 2003 to March 2006 and removed an estimated 35.6 pounds[*] of VOCs before it was integrated into the full-scale system. The full-scale remediation system at FHWSA consists of 23 dual-phase extraction (DPE) wells, which pump groundwater and soil vapor intermittently and discharge to an air stripper. Since start-up in March 2006, the FHWSA DPE system removed about 15 lbs of VOCs. Extracted groundwater is discharged into the sanitary sewer after treatment by an air stripper under permit from the local sewer authorities. Extracted soil vapor is vented without treatment to the atmosphere under a permit from the Bay Area Air Quality Management District (BAAQMD) (SLAC, 2008). Influent to the air stripper from the DPE system included 1,4-dioxane ranging from 1.8 to 22 μg/L in 2006 and 2007 (Cal EPA, 2008). SLAC's location in a hilly, windy area about a half-mile from the nearest residential area is compatible with release of low-concentration VOCs in air-stripper emissions.

8.5.3.2   Former Solvent Underground Storage Tank Area

A 2400-gallon underground storage tank was used to store paint shop wastes from 1967 to 1978 at the location now known as the FSUST. The tank and contaminated soil were removed in 1983, and investigations were conducted in 1984 and 1985. Despite two major excavations performed to remove contaminated soil, solvents at concentrations greater than 1000 mg/kg in soils were detected in samples collected in 1996. Concentrations in soil decrease laterally over a short distance, and highest soil concentrations generally occur in the saturated zone between 8 and 18 ft bgs but extend as deep as 30 feet below the FSUST. In addition to chlorinated solvents and 1,4-dioxane, other detected contaminants include bis-2-ethylhexyl phthalate, acetone, toluene, and methyl ethyl ketone (Cal EPA, 2005; SLAC, 2008). The chemicals are migrating slowly in the groundwater, at less than 8 ft/year, and have not yet arrived at the linear accelerator subdrain located 350 ft south of the FSUST. In 2001, SLAC installed a groundwater extraction system to hydraulically control contaminant migration, supplemented by five additional extraction wells in 2006. The extraction wells discharge to a GAC treatment system (Cal EPA, 2005; SLAC, 2006b, 2008). Figure 8.7 presents a map comparing distribution of 1,4-dioxane and methyl chloroform degradation products at the SLAC FSUST site.

As of late 2007, the FSUST system utilizes six dual-phase (groundwater and soil vapor) and two groundwater extraction wells, an SVE system, and two 2000-pound-capacity GAC vessels to provide hydraulic control of groundwater contamination and treatment of extracted groundwater (SLAC, 2008). The influent is chlorinated prior to entering the first vessel to prevent the formation of hydrogen sulfide. Chlorination is expected to cause indiscriminate oxidation of organics. Chlorine oxidation of organic compounds in the influent probably limits the potential for subsequent biodegradation of organics in the GAC; however, the GAC units were not intended to perform as bioreactors and have not been tested for their ability to sustain microbial degradation of organic contaminants.

---

[*] This seemingly small quantity underscores how small a part per billion is. If the regulatory threshold for cleaning up the VOC(s) is 5 ppb, then these 35.6 lbs (16.2 kg) would be enough to render $3\frac{1}{4}$ billion liters of water contaminated to a VOC level greater than 5 ppb; that is enough water to irrigate a square-mile more than 4 ft deep.

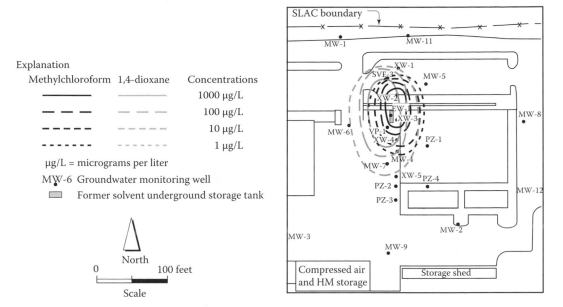

FIGURE 8.7   1,4-Dioxane distribution compared to methyl chloroform plume at the SLAC Former Solvent Underground Storage Tank facility. [Adapted form SLAC, 2008, *Semi-Annual Self-Monitoring Report*, Winter 2008, Part 1, submitted to the Regional Water Quality Control Board, San Francisco Bay Region (RWQCB), SLAC-I-750-2A15H-023, SLAC National Accelerator Laboratory Environmental Health and Safety Division.]

The system extracted an average of 243 gallons of groundwater per day, or 0.169 gpm, between startup in 2001 and March 2008. 1,4-Dioxane concentrations in a single extraction well have been as high as 3100 μg/L, where the maximum detected concentrations are as follows: methyl chloroform—13,000 μg/L, 1,1-dichloroethane—13,000 μg/L, and 1,1-dichloroethylene—3500 μg/L (SLAC, 2008). Combined extraction-well influent concentrations of 1,4-dioxane in 2006 and 2007 have ranged from 75 to 480 μg/L; the midpoint sampling location had one detection at 1.3 μg/L out of eight quarterly measurements. Effluent from the GAC has been consistently nondetect for 1,4-dioxane (Cal EPA, 2008). Figure 8.8 plots 1,4-dioxane concentrations in GAC influent and effluent from 2001 through 2008.

Individual SVE wells contain 1,4-dioxane at concentrations ranging from 3 to 267 ppbv (measured by EPA Method TO-15) and a reporting limit of 1 ppbv. The combined influent of the SVE system to the GAC soil–vapor treatment unit is 0.0111 ppmv, and the soil–vapor effluent at the GAC treatment unit is nondetect for 1,4-dioxane at a 5 ppbv reporting limit (SLAC, 2006b, 2008). Since 2001, a total of about 2.5 pounds of 1,4-dioxane has been extracted through pump and treat; this amount comprises less than 1% of the total mass of VOCs removed from groundwater at the FSUST site (SLAC, 2008).

8.5.4   POSSIBLE REASONS FOR 1,4-DIOXANE REMOVAL BY GRANULAR ACTIVATED CARBON

The complete removal of 1,4-dioxane by GAC treatment vessels is not expected because of 1,4-dioxane's low K_{oc}. Among the several possible mechanisms by which 1,4-dioxane may be removed by GAC, three are considered here.

The first possible mechanism is that 1,4-dioxane is undergoing cometabolism or another form of biodegradation. The slow rate of flow, large surface areas, and possible warm temperatures during much of the year may create an environment in which passive biodegradation of 1,4-dioxane could occur. The practice of running two carbon vessels in series can sustain a consortium of viable

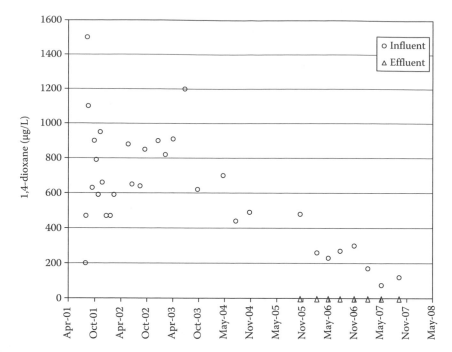

FIGURE 8.8   1,4-Dioxane concentrations in SLAC GAC influent and effluent.

bacteria or actinomycetes capable of 1,4-dioxane metabolism, and the lag vessel acquires the microbial community of the lead vessel over time.

Speculating on the serendipitous cometabolism of 1,4-dioxane under passive conditions by the fortuitous and accidental culturing of just the right microbes may be viewed as wishful thinking; however, the discovery of PM-1, a mixed-culture microbial consortium capable of biodegrading MTBE, was made in a very similar setting. The PM-1 consortium was recovered by the University of California, Davis graduate students from a compost-based biofilter used for gas-phase treatment of vapors at a Water Pollution Control Plant operated by the Los Angeles Sanitation District (Hanson et al., 1998). The GAC at SLAC has not yet been cultured to identify the possible presence of 1,4-dioxane-degrading microbes.

The second speculation for a possible mechanism by which the GAC at SLAC may remove 1,4-dioxane is absorption into the GAC pores, as opposed to adsorption onto GAC surfaces. The very slow rate of flow, 0.2 gpm, coupled with the 106-cubic-foot capacity in each GAC vessel (634 gallons in both vessels, accounting for a 40% void volume) leads to a slow bed-volume exchange (0.45 bed volumes per day) and a long contact duration (53 h of transit time through the two GAC vessels, i.e. 2.2 days). Under these conditions, diffusion of 1,4-dioxane into the pore structures of activated carbon may play a much greater role than would be the case at higher flow rates. Although 1,4-dioxane may be easily displaced by methyl ethyl ketone or other compounds from the surface of the GAC by competitive sorption (see Section 3.3.2.7), it may remain diffused in pore spaces unavailable to other compounds, depending on the polar state of the predominant 1,4-dioxane conformation and the electrostatic charge properties of the GAC. 1,4-Dioxane's polarity in its dominant conformation may favor its diffusion into pores spaces over nonpolar compounds that may be more likely to adsorb to the surfaces. The interaction of 1,4-dioxane and GAC pores by diffusion was not studied to support this speculation, but the opportunity for diffusion to play a more dominant role under very low flow conditions is nevertheless a possible explanation for 1,4-dioxane removal by GAC.

The grade of carbon selected for removal of VOCs at SLAC is Westates® coconut shell-based GAC—AquaCarb® 1230C, which is designed to remove poorly adsorbable organics from water such as MTBE and other trace-level organics. AquaCarb 1230C is an acid-washed pH-neutral GAC with

---

TABLE 8.3
Properties of GAC Used at SLAC

Parameter	AquaCarb® 1230C
Carbon type	Coconut shell
Mesh size, U.S. Sieve	12×30
Effective size (mm)	0.6–0.85
Uniformity coefficient	2.0
Iodine number (mg I_2/g)	1100
Hardness no. (wt%)	95
Abrasion no. (wt%)	85
Apparent density (g/cm³)	0.46–0.52
Water soluble ash (weight percent)	2
Contact pH	9–10

Source: Siemens, 2008, Westates® coconut shell based granular activated carbon. www.siemens.com/water (accessed September 15, 2008).

---

a low ash content designed for use in potable water systems and in high-purity water systems for the microelectronics and other industries. This grade of carbon passes the ANSI/NSF Standard 61 for use in potable water applications; its specifications are summarized in Table 8.3 (Siemens, 2008).

A third plausible theory for consideration is a combination of chemical, physical, and biological mechanisms that may have developed within the GAC system operating conditions at SLAC. In this theory, part of the 1,4-dioxane in the influent is chemically oxidized by the addition of chlorine, and the remaining 1,4-dioxane is biodegraded by the appropriate microbial consortia in the GAC. Microbial consortia may thrive in the GAC if organic compounds and/or toxic inhibitory degradation by-products that would otherwise minimize or negate microbial degradation are removed through physical adsorption. In such a case, the enhanced environmental conditions possibly created within the GAC vessel might support growth of a microbial population of 1,4-dioxane-degrading microorganisms. For example, cometabolites such as toluene may facilitate biodegradation by sustaining the growth of monooxygenase-expressing strains of bacteria. Research conducted at University of California, Berkeley has shown that monooxygenase-expressing bacteria are able to degrade 1,4-dioxane, including those induced by toluene (Mahendra and Alvarez-Cohen, 2006). Toluene is consistently present in the GAC influent at concentrations exceeding 1 mg/L (SLAC, 2008). Similar physical–biological interactions have been observed in fixed-film GAC biological reactors as well as in wastewater treatment processes employing powdered activated carbon. At SLAC, the potential occurrence of physical–chemical–biological removal mechanisms is facilitated by the unusually long system retention times (i.e., >2 days) and the in-series operation of the GAC vessels (SLAC, personal communication, 2008).[*]

8.6　1,4-DIOXANE AND ORANGE COUNTY WATER DISTRICT OCWD GROUNDWATER REPLENISHMENT SYSTEM

The unexpected discovery of 1,4-dioxane in advance-treated wastewater used in OCWD's Groundwater Replenishment System underscores the challenge of using recycled water for groundwater basin management. Fortunately, most recycled water is not tainted by the large-volume industrial discharge of 1,4-dioxane that caused problems in Orange County; however, as this study shows, 1,4-dioxane is nevertheless ubiquitous at low concentrations in treated wastewater because of the

---

[*] Personal communication with SLAC staff on possible mechanisms for 1,4-dioxane removal by GAC, 2008; SLAC National Accelerator Laboratory Environmental Health and Safety Division, Menlo Park, CA.

presence of 1,4-dioxane as an impurity of surfactants in household sundries such as shampoos and liquid detergents. This study provides a cautionary tale for planners of recycled water projects and shows why it is important to test for and consider 1,4-dioxane when designing advanced wastewater treatment projects. This case study is also a success story: OCWD is the gold-standard bearer for recycled water projects and is the destination for engineers from across the globe to learn about successfully transforming wastewater into a valuable water resource.

8.6.1 HISTORY OF THE ORANGE COUNTY WATER DISTRICT GROUNDWATER REPLENISHMENT SYSTEM

OCWD was formed in 1933 to manage water supply and groundwater in a 360-square-mile area in the lower Santa Ana River watershed, covering much of Orange County on the southern California coast. OCWD is the water provider to about 2.3 million people in more than 20 cities, including Newport Beach, Huntington Beach, and Seal Beach on the coast, Cypress and Fullerton on the northern end, Anaheim and Yorba Linda to the east, and Santa Ana and Irvine to the south. The groundwater basin is a 3000-ft-thick wedge of alluvium overlying bedrock formations; water supply wells extend as deep as 1000 ft. Under normal operating conditions, groundwater is the resource for about 70% of the water provided by OCWD (Deshmukh, 2007). The alluvial basins are vulnerable to seawater intrusion at several "gaps" where bedrock has been eroded by rivers to leave a direct alluvial aquifer interconnection between the main groundwater basin and the Pacific Ocean. The 2.5-mile-wide Talbert Gap between the Newport and Huntington Mesas presents the greatest seawater intrusion challenge to groundwater-basin operations in Orange County (OCWD, 2004).

To sustain the groundwater basin and prevent overdraft while managing seawater intrusion, OCWD actively recharges the groundwater basin, primarily with water from the Santa Ana River and to a lesser extent with imported water. Percolation ponds introduce water into the groundwater basin in the upland cities of Anaheim and Orange. OCWD operates one of the oldest and most sophisticated conjunctive use and groundwater protection programs in the world, sampling more than 700 wells to collect more than 13,000 water samples and conduct more than 300,000 groundwater analyses every year. The OCWD monitoring program analyzes more than 300 constituents, including emerging contaminants beyond regulatory requirements (Yamachika, 2007).

Between 1945 and 1969, the basin went into overdraft because of increased pumping and prolonged drought conditions with only two intervening wet years during these three decades. An annual overdraft of 100,000 acre-feet brought the water table to 15 ft below sea level, leading to seawater intrusion into some aquifers; coastal wells soon began producing brackish water and had to be abandoned. OCWD increased its artificial recharge using Santa Ana River water and, beginning in 1948, water imported from the Colorado River. To pay for the imported water, OCWD assessed a pump tax on all producers of groundwater and began the long-term process of replenishing the overdrafted groundwater basin. By 1956, Orange County's population was booming, and the water level dropped further; the cumulative overdraft grew to 700,000 acre-feet, and salt water invaded aquifers as much as 5 miles inland. By 1964, the replenishment program caught up with extraction and began refilling the basin. During this timeframe, the Santa Ana River became increasingly unreliable as a source of replenishment water. In the face of increasingly expensive imported water and less reliable local surface water, OCWD turned to reclaimed wastewater, that is, recycled water (OCWD, 2008).

In 1965, OCWD began a pilot project to inject treated wastewater to create a hydraulic barrier to seawater intrusion along the Talbert Gap in Huntington Beach. In 1971, officials of the California Department of Health Services approved a full-scale project after reviewing data from the pilot project. OCWD drilled a series of 23 multipoint injection wells distributed across the 2.5-mile-wide Talbert Gap. Construction on an advanced wastewater treatment system called Water Factory 21 began in 1972. Full-scale operation of the project started in 1976 (OCWD, 1996).

Water Factory 21 was dedicated in a 1977 ceremony that included serving punch made with reclaimed water. The plant was very successful and won international acclaim. By 1991, OCWD received a permit to inject 100% recycled wastewater in the Talbert Gap wells without blending, accompanied by intensive monitoring (Sovich, 2001; OCWD, 2008). Over 95% of the injected water flows inland and contributes to basin replenishment. As of 2004, 12 MGD was injected into four aquifers in the Talbert Gap; the injected water consisted of one-third Water Factory 21 recycled water, one-third deep well water, and one-third made up of a blend of imported water and ground-water (OCWD, 2004).

The advanced treatment system consisted of chemical clarification, air stripping (for removal of ammonia), recarbonation, filtration, GAC absorption, and chlorination; the system treated 15 MGD. OCWD treated as much as 5 MGD by blending advanced treated wastewater with water purified and demineralized by reverse osmosis. The injection effort was augmented by drilling and operating seven extraction wells between the injection wells and the shoreline to intercept salt water and return it to the ocean (OCWD, 1996).

8.6.2   1,4-Dioxane Detected at Water Factory 21

In December 2001, 1,4-dioxane was discovered in Water Factory 21 influent and effluent at concentrations ranging from 1 to 75 µg/L, as well as two detections at 150 and 200 µg/L. Because the Talbert Gap injection barrier had been receiving highly treated water from Water Factory 21 for decades, in January 2002 the OCWD staff decided to sample 19 water supply wells that drew water from the affected aquifers. 1,4-Dioxane was detected in nine of these supply wells at concentrations ranging from 4 to 20 µg/L—that is, above California's 3 µg/L action level, but below the threshold for shutting down the water supply wells (Mehta, 2002a). OCWD notified the three affected water agencies—the cities of Newport Beach and Fountain Valley, and the Mesa Consolidated Water District—which made the decisions to close their wells. The nine wells served more than 100,000 people. As discussed in Chapter 6, California's 3 µg/L action level, now called a Notification Level, requires notification of consumers when 1,4-dioxane is detected at higher concentrations, but does not require shutting down a water supply unless the concentration exceeds the Notification Level by 100-fold. Local water utility officials nevertheless decided to shut down the nine wells while researching the newly discovered 1,4-dioxane problem, forfeiting approximately 34 MGD of groundwater supply (Woodside and Wehner, 2002). The City of Newport Beach staff decision to shut down all its wells, which supply all of the city's winter drinking water, led it to purchase costly imported water to replace the supply. One-third of the supply to two other cities was similarly shut off after the discovery of 1,4-dioxane (Mehta, 2002a).

OCWD was among the first in California to test for 1,4-dioxane after encountering it while investigating contaminated groundwater in an industrial section of northern Orange County. Although the industrial groundwater contamination by 1,4-dioxane did not threaten drinking water, it led OCWD's analytical chemists (1) to adapt and refine analytical methods to detect 1,4-dioxane and (2) to decide to test for 1,4-dioxane at Water Factory 21 (Mehta, 2002a). OCWD's laboratory analytical experience with 1,4-dioxane is described in a laboratory case study in Section 4.3. Within a month of detecting 1,4-dioxane in 9 out of 19 wells, OCWD proceeded to test 51 more supply wells, revealing one more detection in a Santa Ana supply well that was not in use (Mehta, 2002b).

Water Factory 21 operators and OCWD water quality professionals intensified their monitoring for 1,4-dioxane in the influent and the effluent. The fact that influent concentrations of 1,4-dioxane were highly variable suggested intermittent industrial discharges as the likely source of the observed 1,4-dioxane levels. The Orange County Sanitation District quickly reviewed potential industrial discharges of 1,4-dioxane and identified one significant source, a manufacturer of cellulose acetate membranes, ironically a producer of reverse osmosis filters similar to those used at Water Factory 21. The Orange County Sanitation District worked cooperatively with the discharger, who voluntarily ceased the 1,4-dioxane discharge. Concentrations of 1,4-dioxane subsequently declined to levels associated with domestic wastewater (near 1 µg/L, attributed to shampoos and detergents), as shown

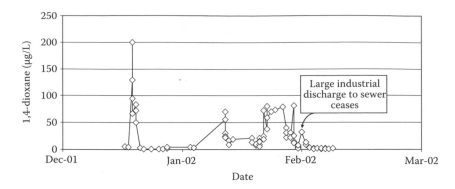

FIGURE 8.9   Influent 1,4-dioxane concentrations at Water Factory 21. (From Woodside, G.D. and Wehner, M.P., 2002, Lessons learned from the occurrence of 1,4-dioxane at Water Factory 21 in Orange County, California. Proceedings of the 2002 Water Reuse Annual Symposium. Alexandria, VA: WateReuse Association. With permission. Courtesy of the WateReuse Association.)

in Figure 8.9 (Woodside and Wehner, 2002). According to USEPA's *Toxic Release Inventory*, from 1993 to 1996, the membrane manufacturer discharged 1,4-dioxane to the sewer at rates ranging from 48,000 to 87,000 lbs/year (USEPA, 2008a). This mass of 1,4-dioxane was apparently sufficient to overcome the dilution occurring over more than 5 miles of transit in sewers to Water Factory 21.

8.6.3   Wastewater Treatment and 1,4-Dioxane Removal

The passage of 1,4-dioxane through Water Factory 21 shows that the series of advanced wastewater treatment steps were not configured for 1,4-dioxane removal. The treatment technologies included GAC, air stripping for ammonium removal, and reverse osmosis, all of which are ineffective at removing 1,4-dioxane. In addition, treatment at Water Factory 21 in 2002 included UV light with hydrogen peroxide, a technology that is widely used for 1,4-dioxane removal. The UV-peroxide treatment step uses a 70-MGD Trojan UV-Phox low-pressure, high-output lamp system. The Trojan UV-Phox was added in 2001 to remove nitroso-dimethyl amine (NDMA), another potentially carcinogenic contaminant discovered in 2000 (Deshmukh, 2007). Hydrogen peroxide was added at less than 5 mg/L, which was apparently insufficient to remove 1,4-dioxane to below the action level. UV light with hydrogen peroxide concentrations in the range of about 10–15 mg/L would have achieved more significant removal of 1,4-dioxane; however, the Orange County Sanitation District's successful effort to locate and eliminate the discharge made increasing the hydrogen peroxide feed unnecessary (Woodside and Wehner, 2002).

8.6.4   Communicating with Water Utilities and the Public about 1,4-Dioxane in Drinking Water Wells

Removing the source of 1,4-dioxane in Water Factory 21 did not put the local water utilities and the public at ease, as the nine wells remained contaminated with 1,4-dioxane. CDPH officials urged the public not to worry and noted that the local water agencies acted out of an abundance of caution by testing for 1,4-dioxane and closing wells despite relatively low levels of contamination. CDPH staff reminded city officials that the three parts per billion level is the amount expected to cause one extra case of cancer per million people who drink 2 L of contaminated water every day for 70 years (Mehta, 2002a). CDPH evaluated the relatively low concentrations of 1,4-dioxane detected in the nine supply wells and assured water utility officials that the concentration of 1,4-dioxane did not represent a significant risk to health. On the basis of CPDH's recommendation, the production wells were returned to service (Woodside and Wehner, 2002).

To put the perceived threat of 1,4-dioxane in drinking water in perspective, CDPH staff prepared a "Q&A" sheet to assist water utilities with public outreach. The discovery of 1,4-dioxane in Water Factory 21 effluent was a blow to the campaign to promote the reuse of recycled water in the Talbert Gap intrusion barrier and other recharge projects, coming only two years after the discovery of NDMA. To deter the inevitable mentality of paralysis in which recycled water projects are opposed because they *might* introduce some unknown contaminant, CDPH adopted the strategy of acknowledging that recycled water projects indeed *will* introduce contaminants for which tests have not yet been developed. An excerpt from OCWD's Q&A on 1,4-dioxane follows:

Q—With the discovery of NDMA in 2000, and now 1,4-dioxane this year—both cancer-causing compounds put into the groundwater by reusing sewer water—shouldn't we not reuse sewer water in the future as part of our water supply?
A—No, we will always be finding new contaminants in water, no matter what the source—for example, perchlorate in water from the Colorado River. Finding new possible contaminants, regardless of where the water comes from, is part of keeping the water safe. What is important is the action that is taken when the new compounds are found and determined to be a possible threat. The water industry has a superior record in this area. At OCWD we routinely test our water with state-of-the-art methods to ensure water quality. Remember, we are living longer today [owing] in large part to the new technology that allows us to find these new substances.

The nine contaminated wells remained shut off during the month it took to identify and eliminate the source of the 1,4-dioxane; water utilities' costs for purchasing imported water mounted. The mayor of the City of Newport Beach reported that annual costs to purchase imported water would be close to $4 million, an expense for which the city requested reimbursement from OCWD, because it injected 1,4-dioxane into the groundwater basin. Similarly, the City of Costa Mesa projected an imported water cost of $850,000 annually, for which they too sought reimbursement from OCWD (Mehta, 2002c). In late February, CDPH's Chief of Drinking Water, David Spath, met with water utility officials at OCWD's offices to further explain the role of Notification Levels and why 1,4-dioxane does not pose a risk to water utility operations in the 4–20 µg/L concentration range. Dr. Spath advised the agencies that, consistent with CDPH guidance, removing the wells from service is not necessary, nor is treatment necessary to remove 1,4-dioxane at the low levels detected. However, continued and frequent monitoring was recommended because the level of 1,4-dioxane exceeded the Notification Level (OCWD, 2002). To reassure the public, Dr. Spath told reporters, "We feel there isn't any significant risk to the public with the use of these wells" (Mehta, 2002b).

CDPH and OCWD officials were confident that well operations could resume in large part because of OCWD's advanced understanding of the groundwater-basin hydrogeology and water balance. OCWD's groundwater flow model showed that the approximate residence time of Water Factory 21 effluent injected in the Talbert Gap injection barrier was 20 years. Because 1,4-dioxane is miscible and resistant to sorption, OCWD's hydrogeologists estimated that the remaining 1,4-dioxane in the aquifer would be pumped out within 20 years. Therefore, the 70-year exposure time upon which the Notification Level for carcinogens is predicated would not be met, and the risk from exposure would diminish correspondingly.* This estimate of 1,4-dioxane's fate may be neglecting its tendency to be stored in fine-grained sediments, as discussed in Section 3.5. Because 1,4-dioxane migrates rapidly and resists sorption and abiotic or biologic degradation, the unintended injection of 1,4-dioxane creates a time marker for tracing the long-term movement of injected water in the groundwater basin.

As of July 2008, two of the four affected Mesa Consolidated Water District wells had decreasing concentrations of 1,4-dioxane, whereas two wells had increasing concentrations of 1,4-dioxane

---

* Personal communication with Dr. David Spath, Chief of Drinking Water, California Department of Health Services (now CDPH), April 2003. Discussion of CDPH role in Water Factory 21 1,4-dioxane episode.

(MCWD, 2008a). The range of detections in Mesa's wells in 2007 was nondetect to 16 µg/L (MCWD, 2008b). The City of Fountain Valley's wells continue to be tested for 1,4-dioxane; in 2007, results ranged from nondetect to 6.6 µg/L (City of Fountain Valley, 2008). Blending with water from uncontaminated wells lowers 1,4-dioxane concentrations such that levels in water delivered to homes may be below the 3 µg/L California Notification Level.

8.6.5 INSTITUTIONAL MEASURES TO PREVENT FUTURE INDUSTRIAL DISCHARGES ADVERSE TO RECYCLING WASTEWATER

In order to ensure that Water Factory 21 is not again subjected to unexpected high influent concentrations of 1,4-dioxane, the Orange County Sanitation District entered into an agreement with OCWD to establish a 5 µg/L target concentration for 1,4-dioxane in the wastewater treatment plant effluent (i.e., the influent to the Water Factory 21 advanced treatment system) (Mowbray, 2008). OCWD continues to pioneer new advances in the reuse of treated wastewater, the identification of emerging contaminants, and their treatment. USEPA recently awarded OCWD a $300,000 grant to research the fate of emerging contaminants including 1,4-dioxane in treatments using a variety of reverse osmosis membranes (Yamachika, 2007).

8.7 AIR FORCE PLANT 44, TUCSON, ARIZONA

William H. DiGuiseppi, AECOM

The Air Force Plant 44 case study reflects work funded by the United States Air Force (USAF), through the Air Force Center for Engineering and the Environment (AFCEE). However, the following is solely the opinion and assessment of AECOM Environment (formerly Earth Tech, Inc.) and does not reflect the opinion or approval of the USAF.

8.7.1 BACKGROUND

Air Force Plant (AFP) 44 is located 15 miles southwest of downtown Tucson, Arizona. It is bounded on the east by Tucson International Airport and on the west by Nogales Highway (Route 89), the Union Pacific Tucson-Nogales railroad spur, and the San Xavier Papago Indian Reservation, of the Tohono O'odham nation. To the south is the Hughes Access Road, vacant land, and light commercial property. AFP 44 covers approximately 1266 acres and has industrial facilities occupying a total building area in excess of 1.2 million square feet. The government-owned, contractor-operated defense industrial plant was constructed by Hughes Aircraft Company (Hughes) in 1951 for the purpose of manufacturing Falcon air-to-air missiles. Hughes sold the plant to the U.S. government in 1951. Hughes then operated AFP 44 under a series of facility contracts from 1951 to 1997. In December 1997, Hughes merged with Raytheon Company (Raytheon), which has operated the plant since that time. AFP 44 has been used for the production of weapons systems since its inception in 1951 (Earth Tech, 1994).

Since 1981, the USAF, in conjunction with Hughes, has investigated soil and groundwater impacts at AFP 44 under the DOD (U.S. Department of Defense) Installation Restoration Program (IRP). The Tucson International Airport Area (TIAA) was placed on the preliminary USEPA National Priorities List (NPL) in July 1982. It was declared a Superfund Site and placed on the final NPL list in September 1992 (Earth Tech, 1994). AFP 44 is one of six project areas within the TIAA Superfund Site. The USAF is the lead federal agency at AFP 44 and is responsible for environmental response actions that are being conducted in accordance with CERCLA, as amended. In 1983 and 1984, additional USAF/Hughes studies defined an off-site groundwater TCE plume substantially attributable to AFP 44. The plume encompassed an area containing City of Tucson water supply wells, several of which were closed following verification of solvent impacts through additional sampling.

In 1984, the USAF accepted responsibility for remediation activities south of Los Reales Road. An informal agreement was reached among the USAF, Hughes, the USEPA, and the Arizona Department of Health Services (ADHS) to begin groundwater remediation at AFP 44 first, with subsequent historical disposal site/source area investigation and remediation. The USAF was also party to an agreement and provided funding for design, installation, and operation of an extraction well field and treatment plant to replace the lost water supply capacity for the City of Tucson.

8.7.2   Geologic and Hydrogeologic Setting

Groundwater beneath AFP 44 occurs within Tucson Basin alluvial sediments, starting at depths of 90–140 ft bgs. Figure 8.10 presents a conceptual site model intended to schematically portray the various water-bearing units and potential migration pathways for 1,4-dioxane at the site. The unsaturated zone above the regional aquifer is made up of discontinuous sand and gravel lenses interbedded with layers of sandy clay, clayey sand, and clay. The material is unconsolidated except where cemented with caliche. A continuous stratum of reddish brown sandy clay and clay occurs between about 80 and 130 ft deep and is locally referred to as the upper clay unit. In the northwestern portion of AFP 44, perched groundwater is present above and within the upper clay unit in an area referred to as the "Shallow Groundwater Zone (SGZ)."

The regional aquifer beneath the site consists of an upper and a lower-aquifer zone separated by a thick, laterally extensive, clay aquitard. The upper zone extends down to the contact with the clay

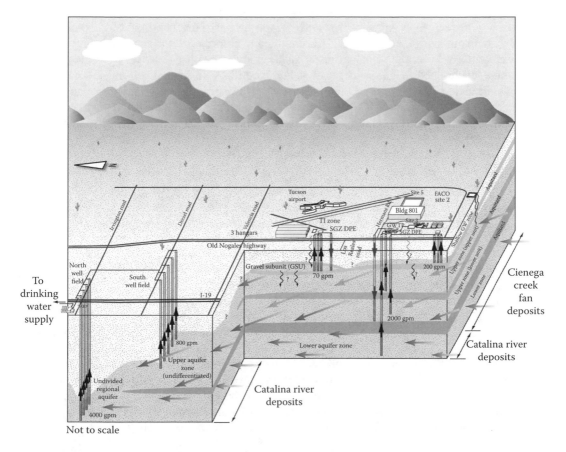

FIGURE 8.10   Schematic conceptual site model for Air Force Plant 44.

aquitard, at depths of 180–225 ft. The average thickness throughout the remediation area is 70 ft. The upper zone consists of unconsolidated, highly permeable sand and gravel layers interbedded with layers of clayey, impermeable sediments. The stratigraphy is similar to that of the unsaturated zone. Water within the individual sand and gravel beds occurs under unconfined to semiconfined conditions. In some parts of the area, the individual sand and gravel beds appear to be hydraulically isolated from beds above and below at the same location.

Locally, the upper zone of the regional aquifer is the most productive zone. The upper zone will support sustainable pumping rates of 50–150 gpm in the southeast part of the remediation area and as much as 350 gpm in the northwest (where more coarse-grained, permeable beds predominate). The upper zone also contains the bulk of the solvent contamination. An elongated contaminant plume formed as a result of a continuing migration from source areas (below former disposal sites) combined with relatively high groundwater flow rates in the upper zone. Regional groundwater flow in the upper zone is northward to northwestward across the remediation area.

The lower zone of the regional aquifer begins at about 300–350 ft bgs at the base of the confining clay and continues to an unknown depth. The lower zone contains minor levels of solvent contamination. To the northwest, the upper zone and lower zone combine to become the undivided regional aquifer, which is the principal aquifer for large volumes of drinking water for the City of Tucson.

8.7.3 CHLORINATED SOLVENT USE, RELEASE MECHANISMS, AND DISCOVERY

Historical industrial processes conducted at AFP 44, including the production, maintenance, and modification of weapons systems, have resulted in wastewater and general industrial waste including solvents, paint sludge, and thinners. Chromium and chlorinated solvents, including TCE and methyl chloroform, were the principal materials used at the facility. Releases occurred primarily from sludge-drying beds, lagoons, degreasers, and landfills. Chlorinated solvents associated with AFP 44 were identified in off-site groundwater to the northwest, as were similar compounds released within other project areas at the TIAA Superfund Site (Figure 8.11). Chlorinated compounds were present in both shallow and deep soil; shallow soils were contaminated primarily by chromium and other inorganic constituents.

TCE was used at AFP 44 from the 1950s until it was replaced by methyl chloroform in 1974. Methyl chloroform was used from 1974 until the mid-1990s although the last decade of usage was at a much lower rate owing to process changes and efforts to minimize waste. The chlorinated solvents identified in the groundwater consist of TCE with lesser amounts of DCE present. 1,1-DCE was not used as a solvent on-site, but resulted from the breakdown of methyl chloroform. Other chlorinated daughter products are present on the site at insignificant levels. The primary sources of solvents that contaminated the groundwater, in the order of greatest probable contribution, are Site 2—FACO Landfill, Site 3—Inactive Drainage Channel Disposal Pits, Site 5—Sludge Drying Beds/Building 801, and Site 14—Shallow Groundwater Zone.

The chromium plume was much smaller than the solvent plume. At the start of remediation in 1987, it covered 190 acres and was almost entirely on AFP 44 property. The sources of chromium in the groundwater were plating wastes placed in unlined sludge-drying beds at Site 5, plating wastes placed in disposal pits at Sites 3 and 4, and wastes accumulated in associated drainage channels at Site 6. The chromium plume extent was reduced dramatically once these source areas were excavated in the early to mid-1990s. Chromium is only present above standards in a few wells at the site, which are in areas of known disposal or release. Hexavalent chromium is the more soluble and more toxic oxidation state for chromium; however, at AFP 44, natural aquifer conditions are conducive to the reduction of hexavalent chromium to trivalent chromium, which is the less soluble and less toxic oxidation state.

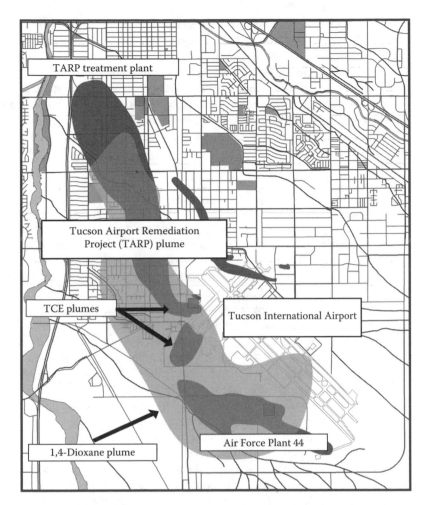

FIGURE 8.11   Groundwater contamination occurrence in the TIAA Superfund Site Project areas.

8.7.4 COMMUNITY INVOLVEMENT

In the face of rising community concerns about the TCE discovered in the groundwater on the south side of Tucson, the USAF in the early 1990s developed a facility-wide community involvement program. This included holding meetings of a group that evolved to become the Unified Community Advisory Board (UCAB), which involves all the parties at the Superfund Site. The USAF participates in the UCAB because it is the lead agent for the cleanup of AFP 44. Negative publicity for the site (measured by media coverage, letters to elected officials, and other activities) decreased by 89% following the onset of this program. In addition to support provided to the UCAB, the USAF has coordinated and documented public meetings, responded to community comments, and provided information about the ongoing cleanup at AFP 44 to community residents. Through this proactive public participation program, the USAF has been able to improve public opinion about the overall cleanup effort.

In addition to its support for the UCAB, the USAF has cohosted an ongoing community outreach program that includes tours of AFP 44 and special events such as ribbon cuttings and environmental fairs for middle schools located near the Superfund Site. The Air Force and the UCAB (with as many as 20 other area organizations) have provided opportunities for more than 1600 students from the south Tucson area to learn about environmental science issues and technologies. Additionally, the USAF works with three different area middle schools to highlight their environmental science projects at the Tucson Earth Day Festival. The festival has become the region's largest Earth Day event. Most recently, the UCAB sponsored and supported the development of a middle and high school curriculum that introduces local students to the history of the site and examines the social, scientific, and political ramifications of the groundwater contamination.

Despite the progress the USAF has made to build trust among community members, the USAF continues to be challenged by shifting community attitudes about those responsible for the Superfund cleanup. In addition to filing lawsuits over health effects, some community members continue to distrust (or have become distrustful of) the USAF, other government agencies, or others responsible for the cleanup.

8.7.5 REMEDIAL ACTIONS

In 1986, the USAF released a Remedial Action Plan (RAP) that proposed groundwater extraction, treatment, and reinjection to address the contamination at the south of Los Reales Road (USAF, 1986). Construction of the groundwater treatment plant (GWTP) began that same year (Figure 8.12). This "Groundwater Reclamation System" was intended to reduce or eliminate off-site migration of TCE and other chlorinated VOCs from AFP 44. The well field is designed to contain and remediate the southern half of the TIAA Superfund Site contaminant plume, which formed in the upper zone of the regional aquifer. A separate remediation system, referred to as the Tucson Area Remediation Project (TARP), handles the northern half of the plume, which includes solvent contamination emanating from the Tucson Airport property and 1,4-dioxane that may ultimately be determined to be coming from the airport as well as AFP 44. The regional aquifer in the Tucson Basin is a heavily used drinking water source, so restoring the contaminated part of the aquifer is a very high priority for the USAF, local governments, regulatory agencies, and the community.

The AFP 44 well field utilizes extraction and injection wells across a 2-square-mile area to achieve hydraulic containment of the contamination plume by extracting groundwater from the center of the plume and reinjecting it along the outside perimeter of the plume. This reinjection was mandated because the site is within the Tucson Active [groundwater] Management Area, which requires that groundwater extraction does not result in an overall decline in available water resources in the sole drinking water aquifer for the City of Tucson. In April 1987, the GWTP began to extract and treat groundwater at a capacity of approximately 5000 gpm. This flow rate was reduced over time, based on MODFLOW groundwater modeling results, to optimize capture while minimizing the quantity of

FIGURE 8.12   Aerial view of Air Force Plant 44 GWTP.

groundwater removed. The plant operated between 2000 and 2700 gpm throughout most of the 1990s. AFP 44 currently has the capacity to operate approximately 25 active upper-zone and four lower-zone extraction wells, and 10 active upper-zone recharge wells. In addition to the extraction wells, there are 65 upper-zone and 10 lower-zone wells in which water levels and water quality are monitored on a regular basis. The air-stripper system continues to operate today in accordance with the 1986 RAP; however, as will be discussed subsequently, the system is being upgraded with an advanced oxidation treatment (AOT) system specifically to address 1,4-dioxane contamination.

Once the AFP 44 remediation system was installed and demonstrated to be operating as designed, subsequent remedial actions focused on potential groundwater contamination sources, especially those associated with VOCs and chromium waste. SVE systems were installed between 1994 and 1996 at solvent contamination sites 1, 2, and 3. Soil excavation and off-site disposal was conducted at shallow metals-contaminated sites 4, 5, and 6. Additionally, DPE was conducted at site 5 to address deep vadose zone VOC contamination. Most of these remediation systems successfully reduced contaminant levels to below agreed-upon standards, and the sites were documented as having remedial action complete.

An additional set of remedial actions involved *in situ* treatment methods including chemical oxidation (ISCO) and bioremediation/chemical reduction. The ISCO was focused on TCE and 1,1-DCE and utilized potassium permanganate ($KMnO_4$). This oxidizer has been effectively demonstrated at numerous solvent sites and was pilot-tested in a series of large-scale injection programs in the regional groundwater beneath soil sites 2 and 3. Over 1 million gallons of water containing more than 40,000 pounds of $KMnO_4$ at concentrations ranging from 0.5% to 2% were injected at these two sites over the course of the multiyear program. VOC destruction has been encouraging: less than 10% of the maximum-measured VOC concentrations in groundwater remained after this massive $KMnO_4$ injection. Monitoring and evaluation of the program is ongoing. An *in situ* bioremediation pilot study was performed in an area with chlorinated solvent and hexavalent chromium contamination in the deep vadose zone and groundwater. Because hexavalent chromium was present in this area, oxidation technologies could not be used. An emulsified mixture of vegetable oil and sodium lactate was injected to create chemically reducing conditions, which would lead to chemical reduction of hexavalent chromium to the less mobile and less toxic form, trivalent chromium. Additionally, the injection would create favorable conditions for the growth of indigenous bacteria capable of consuming the chlorinated contaminants, either directly or cometabolically. This 2006–2008 study yielded mixed results and will need to be further evaluated to determine whether the technology warrants application on a larger scale.

The TARP plant was constructed near Irvington Road and Interstate 19, northwest of AFP 44, and placed in operation in September 1994 to contain the leading edge of the TCE plume. The system has been operating continuously since that time, with the exception of brief maintenance-related shutdowns. The system includes two separate well fields, the south well field, extracting approximately 1000 gpm from five wells in the upper zone of the regional aquifer, and the north well field, extracting 4000 gpm from four wells in the undivided regional aquifer. The combined extraction of approximately 4900 gpm is treated by using air strippers, with GAC for off-gas treatment, and then blended with water from other sources and distributed to the community. An estimated 3500 pounds of chlorinated VOCs was removed from the drinking water aquifer between 1994 and 2008 (ADEQ, 2008). Because this water is provided to the community, a significant amount of public scrutiny has been directed at the pumping, blending, and distribution, especially with respect to concentrations of 1,4-dioxane in the groundwater.

8.7.6 1,4-Dioxane Discovery and Regulation

In 2002, 1,4-dioxane was discovered in influent and effluent groundwater at the AFP 44 GWTP at concentrations of approximately 10 μg/L. Ultimately, the 1,4-dioxane plume was determined to extend over 5 miles and to have a slightly wider footprint than the chlorinated VOC plume. The identification of 1,4-dioxane at the site led to the supposition, and later empirical demonstration, that the existing treatment system was not adequate to address this new-found recalcitrant contaminant. In response to this finding, the extraction well-field operation was reconfigured to containment-only at minimal extraction rates, resulting in a reduction in 1,4-dioxane levels in the effluent/recharge water from approximately 10 µg/L to below 4 µg/L. Additionally, the USAF conducted a Technology Evaluation (Earth Tech, 2004) that determined that an AOT system could treat both the VOCs and the 1,4-dioxane for about the same annual cost as the present system (when taking into account costs for chemicals, GAC, natural gas, evaporative water loss, electricity, and ongoing maintenance of the 20-year-old stripping towers).

After the USAF decided to install an AOT system, the EPA issued an Administrative Order to the USAF in July 2007 under Section 1431 of the Safe Drinking Water Act (SDWA), stating that the 1,4-dioxane in the regional aquifer posed an "imminent and substantial endangerment" to the public (USEPA, 2007d). The Administrative Order required installation of an AOT system at the AFP 44 GWTP to address 1,4-dioxane identified in the groundwater above a Region 9 PRG of 6.1 µg/L and to ensure that recharge water contained less than 3 µg/L. A proposed Settlement Agreement separate from the order requested that the USAF conduct a focused remedial investigation (FRI) of 1,4-dioxane in the entire plume area. The USAF believes that the SDWA Administrative Order is invalid, because the remediation is occurring under CERCLA, and therefore other regulatory enforcement mechanisms are not applicable, among other reasons. In spite of this disagreement, the USAF expedited the design and installation of the AOT system, which was programmed for installation in the fiscal year 2009. The startup of the system is currently scheduled for May 2009. The USAF also initiated a Phase I FRI, which also would address the 1,4-dioxane concerns expressed by the USEPA in the proposed Settlement Agreement submitted in April 2007. The scope of the Phase I FRI is (1) to evaluate the geology, hydrogeology, and water quality with data gathered from all the TIAA Superfund Site parties, (2) to assess the groundwater model that had been used by the Environmental Protection Agency, and (3) to provide recommendations to fill any data gaps identified. The Phase I FRI was completed in November 2008. In July 2008, while the Phase I FRI was being conducted, the EPA issued a Special Notice Letter to the TARP parties, which required a full remedial investigation to be performed (to include drilling, well installation, well abandonment, and other tasks) (USEPA, 2008b). This scope includes tasks from Phase I and Phase II, which was funded by the USAF in March 2009 and is anticipated to be completed in the spring of 2010. The requirements of the Special Notice Letter did not materially change the USAF activities that were funded and ongoing.

Most recently, in August 2008, the USEPA issued a request to the TIAA that a specific area near their Three-Hangars building be investigated as a potential source of 1,4-dioxane. This request was stimulated by sampling conducted by the USGS on the Tucson Airport property that identified elevated 1,4-dioxane in a location and geologic unit that could not have been related to AFP 44 contaminant migration. Ongoing and future investigations by several parties will focus on whether additional potential sources, besides AFP 44, exist for the 1,4-dioxane in the regional aquifer.

8.7.7 Advanced Oxidation Treatment System Upgrade

The USAF accelerated the design and installation of an AOT system to address both chlorinated VOCs and 1,4-dioxane. The system selected (following an extensive review of available technologies) was APT's HiPOx™ system. The AOT system utilizes ozone and hydrogen peroxide to form hydroxyl radicals. This application of peroxone chemistry is well documented and is further detailed in Chapter 7. Hydroxyl radicals are second only to fluorine in their oxidizing potential. Hydroxyl radicals react very rapidly to oxidize organic compounds into nonhazardous compounds including carbon dioxide and water. This technology has been demonstrated to be effective on recalcitrant compounds, including 1,4-dioxane. The oxidation process occurs in the aqueous phase and does not increase the temperature or pressure of the water because it usually occurs at very low concentrations (<1%). The fundamental difference between other peroxone technologies and HiPOx is in the associated mass transfer. In "traditional" commercial processes, air containing 2%–3% ozone by weight is injected through a diffuser and bubbles upward through a contactor at or near atmospheric pressure. Mass transfer of ozone from the gas phase to the liquid (and diffusion/mixing within the liquid) is a very slow and inefficient process. As a result, destruction requires contact or residence times as long as 20 min. HiPOx incorporates high-pressure, high-precision distribution of ozone and reagents and high-efficiency mixing in order to maximize mass transfer, manage localized reagent and ozone concentrations, and reduce the mass-transfer limitation on the overall oxidation rate by orders of magnitude. Figure 7.10 illustrates how higher ozone concentrations (8%–10% by weight in air), higher operating pressure (25–30 pounds per square inch gauge [psig] versus ambient pressure), and efficient, in-line mixing are combined to maximize mass transfer and reaction efficiency. The dosed fluid flows immediately through a static mixer followed by a reaction zone specifically designed for the required residence time. A critical design feature of the distributed-injection approach is the low, local concentrations of ozone and hydrogen peroxide.

While the AOT system was being designed and fabricated, the existing treatment system piping was modified to accommodate the system upgrade. The piping is configured with valves to allow bypass of the existing air-stripper systems or bypass of the HiPOx system, or operation of both systems in series, as a safety precaution. Some of the existing piping and pipe racks were utilized in the final installation configuration.

The AOT system was delivered to AFP 44 in June 2008 on a flatbed truck trailer. The AOT unit, which is installed in a steel CONEX box, was placed on an existing, unused concrete slab with a crane (Figure 8.13). The unit is connected to the power supplies utilizing existing transformers.

FIGURE 8.13   Advanced oxidation unit delivery and placement.

Fiber-optic cables are connected to allow automated operation and control from the integrated control systems in place for the GWTP operations. These systems allow remote monitoring and operation of the system. Influent and effluent water samples will be collected and analyzed to provide assessment of the system's operational effectiveness during the 30-day start-up phase, anticipated to be in May 2009. Optimization of the system will occur during the initial week of operations, while both treatment systems are operated in series. This approach ensures that TCE treatment will be performed as before, regardless of the performance of the HiPOx system. Once effective removal is verified, the air strippers will be bypassed, and the system will be operated for an additional test period of 30 days, before turning the system over to the Operation and Maintenance Contractor at the facility for long-term operations.

8.7.8   Cost Analysis

Operating the new AOT system instead of using the existing air strippers will result in an annual electrical cost reduction of $25,000. Additionally, the elimination of the air-stripper systems will save $25,000 in sulfuric acid neutralization costs, $10,000 in GAC cost, and $30,000 in evaporative water loss per year. However, operating the HiPOx system will add $60,000 in peroxide cost and $5000 in the generation of oxygen for ozone production. The AOT upgrade represents a net overall savings of approximately $25,000/year in materials and electrical costs for water treatment. In addition, the new treatment plant replaces the aging air strippers with new equipment and has fewer moving parts; hence, future maintenance costs will be reduced. The existing air strippers are over 20 years old, and maintenance costs are increasing over time as the system has exceeded its 15-year designed life expectancy. The new system is also modular and therefore capable of being relocated and reused at another location in the future.

8.7.9   Future Plans

Long-term operation and maintenance of the HiPOx system will begin in 2009 and may continue for decades, given the scale and levels of remaining TCE, DCE, and 1,4-dioxane contamination in the regional groundwater. The Phase II FRI includes the following:

- Drilling and installing wells to better define the nature and extent of 1,4-dioxane contamination in three dimensions in the regional aquifer
- Groundwater modeling to understand the migration potential and patterns, as well as the potential for human exposure to 1,4-dioxane

All steps in the Phase II FRI are expected to be completed by the spring of 2010. Ongoing monitoring of the $KMnO_4$ pilot-test areas will focus on identifying the distribution of 1,4-dioxane (in low-permeability units present in the saturated zone) and will specifically look for 1,4-dioxane trapped in pores as well as for evidence of chemical oxidation of 1,4-dioxane. A bench-scale $KMnO_4$ oxidation test for 1,4-dioxane begun in spring 2008 involved extraction of contaminated groundwater from the site, treatment of one set of bottles with $KMnO_4$, acidification of another set of bottles to eliminate possible biodegradation, and use of an untreated control. These triplicate sets will be sacrificed quarterly for several years to assess the impact of $KMnO_4$ on 1,4-dioxane concentrations. Initial results from the first quarter indicate 13% greater reduction in 1,4-dioxane concentration in the $KMnO_4$-spiked bottles. Planned future studies include assessment of the biological communities native in the groundwater to determine if any bacteria capable of degrading 1,4-dioxane are present in the aquifer and to identify any applicable amendments with the potential to stimulate the biological communities for more effective degradation.

ACKNOWLEDGMENTS

The critical reviews, comments, corrections, and questions from the following people are greatly appreciated: Jeff Lorenzo, Lorenzo Law Firm in Seymour, Indiana; Sally McCraven, Todd Engineers in Alameda, California; Farsad Fotouhi, Pall Life Sciences in Ann Arbor, Michigan; Jim Brode of Fishbeck, Tompson and Carr in Kalamazoo, Michigan; Staff of the SLAC, Environmental Health and Safety Division, in Menlo Park, California; Roy Herndon and Greg Woodside of the OCWD in Fountain Valley, California; Jim Hatton and Ed Meyers of AECOM (formerly Earth Tech) in Greenwood Village, Colorado, and Orlando, Florida (respectively); and Sarah Raker, Mactec Engineering and Infrastructure in Petaluma, California. The authors are solely responsible for the accuracy and interpretation of the foregoing case studies.

BIBLIOGRAPHY

ADEQ, 2008, Tucson Airport Remediation Project (TARP). http://www.azdEquationEquationgov/environ/waste/sps/download/tucson/tucsona.pdf (accessed September 24, 2008).

Arcadis-BBL, 2007a, *Vapor Intrusion Assessment Report: Former American Beryllium Company Site, Tallevast, Florida*. Cranbury, NJ: Arcadis-Blasland Bouck and Lee, Inc.

Arcadis-BBL, 2007b, *Interim Data Report*. Cranbury, NJ: Arcadis-Blasland Bouck and Lee, Inc.

Arcadis-BBL, 2007c, *Remedial Action Plan for the Former American Beryllium Company (ABC) Facility and Site, Tallevast, Florida*. Tampa, FL: Arcadis-Blasland Bouck and Lee, Inc.

Ashland Chemical Company, 1986, MSDS for 1,1,1-trichloroethane degrease cold vapors, Columbus, Ohio. MSDS No. 6810-00-476-5613.

ATSDR, 2005, *Health Consultation: Exposure Investigation Report—Indoor Air Testing*. Former American Beryllium site, Tallevast, Manatee County, FL: Agency for Toxic Substances and Disease Registry.

Benson, M., 1988, 13 New Bay Area Superfund sites—9 South Bay plants are added to list. *San Jose Mercury News*, San Jose, CA, June 22.

Bodwin, A., 1989, Lonesome cowboy: Gelman vs. DNR. *Crain's Detroit Business*, November 20, p. 8.

Brode, J., Fotouhi, F., and Kolon, S., 2005, Ultraviolet and hydrogen peroxide treatment removes 1,4-dioxane from multiple aquifers. *US EPA's Technology News and Trends*, January.

Cal EPA, 2000, *Board Order No. 00-010, Site Cleanup Requirements for [Solvent Recycling Facility], San Jose, California*. San Francisco Bay Region, Oakland, CA: California Environmental Protection Agency, Regional Water Quality Control Board.

Cal EPA, 2005, *Board Order No. R2-2005-0022, Site Cleanup Requirements for Stanford University and the United States Department of Energy*. San Francisco Bay Region, Oakland, CA: California

Cal EPA, 2008, GeoTracker. California Environmental Protection Agency, State Water Resources Control Board. http://geotracker.swrcb.ca.gov/ (accessed 2005, 2006 Environmental Protection Agency, Regional Water Quality Control Board., 2007, 2008).

Cameron-Cole, LLC, 2001, *Off-Site Groundwater Extraction System Design Report for [Solvent Recycling] Facility, San Jose, California*, Oakland, CA: Report to San Francisco Bay Regional Water Quality Control Board.

Cameron-Cole, LLC, 2006, *Annual groundwater monitoring report [solvent recycling] facility, San Jose, California*, Oakland, CA: Report to San Francisco Bay Regional Water Quality Control Board.

Campbell, K.M. and Scott, T.M., 1993, *Geologic Map of Manatee County, Florida*. Tallahassee, FL: Florida Geological Survey.

Chui, G., 1995, San Jose firm agrees to pay $850,000 for falsifying toxic waste test results. *San Jose Mercury News*, San Jose, CA, January 31.

Cilek, R., Jensen, D.A., Meyers, D., Martin, W.A., McClain, K., Newton J.L., and Winters, J. 2004, Tallevast Community; Tallevast, Manatee County Florida: Preliminary contamination assessment report. Florida Department of Environmental Protection, Site Investigation Section.

City of Ann Arbor, 2006, A brief history of the Ann Arbor groundwater contamination problem. http://www.a2gov.org/government/publicservices/systems_planning/Environment/pls/Pagestimeline.aspx (accessed June 9, 2006).

City of Fountain Valley, 2008, *2008 Water Quality Report*. City of Fountain Valley, CA: City of Fountain Valley Water Department.

Davis, T., 2002, Delay, denial part of battle over dioxane; First well contamination found in 1986. *Ann Arbor News*, Ann Arbor, MI, April 7, 2002.

Davis, T., 2006a, City, Pall reach deal: water quality still being monitored. *Ann Arbor News*, Ann Arbor, MI, November 21, 2006.

Davis, T., 2006b, Contaminant traced to [Ann Arbor] Inn well. *Ann Arbor News*, Ann Arbor, MI, June 29, 2006.

Davis, T., 2007, New wells track dioxane. *Ann Arbor News*, Ann Arbor, MI, May 1.

Deshmukh, S., 2007, Emerging contaminants and the groundwater replenishment system. Presented at April Water Issues Meeting, Southern California Alliance of Publicly Owned Treatment Works.

Earth Tech, 1994, Remedial investigation report, U.S. Air Force Plant 44, Tucson, Arizona, January 1994.

Earth Tech, 2004, Final technology evaluation for treatment of 1,4-dioxane in groundwater. December.

ES&T, 2005, Sampling and analysis report, Tallevast Community. Prepared for FOCUS Inc. Lakeland, FL: Environmental Sciences & Technologies, Inc.

FDEP, 2004, Letter from FDEP Waste Management Division Director Michael W. Sole to U.S. EPA Waste Management Division Director Winston Smith. Florida Department of Environmental Protection.

FDEP, 2006, SARA #3, response to FDEP comments, former American Beryllium site, Tallevast, Manatee County, Florida, Florida Department of Environmental Protection.

FDOH, 2007, DOH timeline of activities around former American Beryllium site, Tallevast Community. http://www.doh.state.fl.us/Environment/community/SUPERFUND/hazwastework/currenttimeline.htm (accessed January 21, 2008).

Feldman, S.M., 2000, *Principal Scientist/Hydrogeologist*, Melville, NY: ARCADIS Geraghty & Miller, Inc.

Fishbein, N., 2005, Level with Tallevast pollution victims. *Bradenton Herald*, Bradenton, FL, April 20.

Fishbein, N., 2008, State admits Tallevast pollution study way off mark. *Sarasota Herald-Tribune*, Sarasota, FL, March 31.

Fotouhi, F., Tousi, S., and Brode, J.W., 2006, Managing a significant release of 1,4-dioxane into a complex glacial depositional environment: The integration of hydrogeology, remedial engineering and politics. Presented at Emerging Contaminants in Groundwater: A Continually Moving Target, Groundwater Resources Association of California, Concord, CA, June 7–8.

Hanson, J.R., Scow, K.M., Eweis, J.B., Schroeder, E.D., and Chang, D.P.Y., 1998, Isolation and characterization of MTBE-degrading bacterial cultures. Presented at First International Conference on Remediation of Chlorinated and Recalcitrant Compounds, Battelle International, Monterey, CA, May 18–21.

Herlihy, P. and Bowman, R., 2005, *HiPOX-TM Technology Lab Testing, Former American Beryllium site, Tallevast, Florida*. Pleasant Hill, CA: Applied Process Technologies Inc.

Johnson, P.C. and Ettinger, R.A., 1991, Heuristic model for predicting the intrusion of contaminant vapors into buildings. *Environmental Science and Technology* 25(8):1445–1452.

Kellogg, K., 2005, On-going cleanup by Pall Life Sciences: A continuing controversy. *Ann Arbor Business News*, July.

Lerner, S., 2008, Tallevast, Florida: Rural residents live atop groundwater contaminated by high-tech weapons company. http://www.healthandenvironment.org/articles/homepage/3860 (accessed August 16, 2008).

Mahendra, S., and Alvarez-Cohen, L., 2006, Kinetics of 1,4-dioxane biodegradation by monooxygenase-expressing bacteria. *Environmental Science and Technology* 40(17): 5435–5442.

Marsteller, D., 2005, Tallevast home values still in limbo. *Bradenton Herald*, Bradenton, FL, July 24.

McCraven, S., 2006, Occurrence, fate, and transport of 1,4-dioxane in groundwater—A case study. Presented at Emerging Contaminants in Groundwater: A Continually Moving Target, Groundwater Resources Association of California, Concord, CA, June 7–8.

MCWD, 2008a, *Water Quality Program Update*. Costa Mesa, CA: Mesa Consolidated Water District.

MCWD, 2008b, *The 2008 Water Quality Report*. Costa Mesa, CA: Mesa Consolidated Water District.

MDEQ, 2000, *Information bulletin: Gelman Sciences, Inc. site, Scio Township, Washtenaw County, Michigan, July 31*. Lansing, MA: Michigan Department of Environmental Quality.

MDEQ, 2004, *Operational Memorandum no. 5—Groundwater–Surface Water Interface Criteria*. Lansing, MA: Michigan Department of Environmental Quality, Remediation and Redevelopment Division.

MDEQ, 2005, *Fact sheet—Prohibition Zone to Restrict Use of Groundwater, Gelman Sciences, Inc., Unit E Aquifer, Groundwater Contamination, Washtenaw County*. Lansing, MA: Michigan Department of Environmental Quality, Remediation and Redevelopment Division.

MDEQ, 2007, *Fact Sheet—Prohibition Zone, Groundwater Use Restrictions, Gelman Sciences, Inc., Unit E Aquifer, 1,4-Dioxane Groundwater Contamination, Washtenaw County, March*. Lansing, MA: Michigan Department of Environmental Quality, Remediation and Redevelopment Division.

Mehta, S., 2002a, 9 O.C. drinking wells closed. *Los Angeles Times*, January 30.

Mehta, S., 2002b, No health threat from dioxane-tainted wells, officials say. *Los Angeles Times*, March 7.

Mehta, S., 2002c, Cities seek restitution for tainted well water. *Los Angeles Times*, February 22.

Mowbray, S.L., 2008, Removal of NDMA and 1,4-dioxane during secondary wastewater treatment at OCSD. Orange County—American Chemical Society Environmental Program, Chapter Meeting, Kennedy/ Jenks Consultants, Irvine, CA: American Chemical Society.

Nyer, E.K., Kramer, V., and Valkenburg, N., 1991, Biochemical effects on contaminant fate and transport. *Ground Water Monitoring Review* 11(Spring): 80–82.

OCWD, 1996, *Water Factory 21*. Fountain Valley, CA: Orange County Water District.

OCWD, 2002, *Press Release: OCWD Receives Advice from Department of Health Services on Wells with Low Levels of 1,4-Dioxane*. Fountain Valley, CA: Orange County Water District.

OCWD, 2004, *Orange County Water District Groundwater Management Plan*. Fountain Valley, CA: Orange County Water District.

OCWD, 2008, Orange County Water District web site. http://www.ocwd.com (accessed March 22, 2008).

Powell, T., 2005, *Test Report: Photo-Cat Treatment of Groundwater at the Former ABC Site, Sarasota*. London, Ontario: Purifics ES Inc.

Radway, S., and Wright, D., 2005, Tallevast plume reaches 131 acres. *Bradenton Herald*, Bradenton, FL, April 16.

Raphael, S., 1988, Gelman sues chemical giants. *Crain's Detroit Business* 4(8): 2.

Reitz, R.H., McCroskey, P.S., Park, C.N., Andersen, M.E., and Gargas, M.L., 1990, Development of a physiologically based pharmacokinetic model for risk assessment with 1,4-dioxane. *Toxicology and Applied Pharmacology* 105(1): 37–54.

Sabba, D. and Witebsky, S., 2003, Documenting the nature and extent of 1,4-dioxane at a solvents plume site. Presented at 1,4-Dioxane and Other Solvent Stabilizers in the Environment, Groundwater Resources Association of California, San Jose, CA, December 10, 2003.

Shelton, D.E., 2004, Opinion and order regarding remediation of the contamination of the Unit E aquifer. Circuit Court for the County of Washtenaw, MI, Honorable Donald E. Shelton, State of Michigan Case No. 88-34734-CE.

Siemens, 2008, Westates® coconut shell based granular activated carbon. www.siemens.com/water (accessed September 15, 2008).

SLAC, 2003, *Site Characterization for the Plating Shop Area*. Menlo Park, CA: SLAC National Accelerator Laboratory Environment Safety & Health Division.

SLAC, 2004, *Site Characterization Report for the Former Hazardous Waste Storage Area*, Vol. 1 of 4, submitted to the Regional Water Quality Control Board, San Francisco Bay Region (RWQCB) SLAC-I-750-3A33H-015, SLAC National Accelerator Laboratory Environmental Health and Safety Division.

SLAC, 2006a, *The Geology of Stanford Linear Accelerator Center*. Menlo Park, CA: SLAC National Accelerator Laboratory Environment Safety & Health Division.

SLAC, 2006b, *Semi-annual Self-Monitoring Report, Summer 2006*. Menlo Park, CA: SLAC National Accelerator Laboratory Environment Safety & Health Division.

SLAC, 2006c, *Quality Assurance Project Plan for the Environmental Restoration Program*, Revision 003. Menlo Park, CA: SLAC National Accelerator Laboratory Environment Safety & Health Division.

SLAC, 2008, *Semi-Annual Self-Monitoring Report*, Winter 2008, Part 1, submitted to the Regional Water Quality Control Board, San Francisco Bay Region (RWQCB), SLAC-I-750-2A15H-023, SLAC National Accelerator Laboratory Environmental Health and Safety Division.

Sovich, T., 2001, Operation and maintenance of the OCWD Talbert barrier injection well system. Presented at Seawater Intrusion Barrier Workshop—Operation, Maintenance & Repairs: Lessons Learned in Southern California, American Water Resources Association, Loyola Marymount University, Los Angeles, October 13, 2001.

SRSW, 2006, Scio residents for safe water web site: History of the Pall/Gelman site. www.http://srsw.org/ (accessed June 9, 2006).

Todd Engineers, 1987, Feasibility study of remedial action plans of soil and groundwater at Solvent Service Inc., San Jose. Oakland, California: Report to California Environmental Protection Agency, Regional Water Quality Control Board—San Francisco Bay Region.

Todd Engineers, 1998, Quarterly report for Solvent Service Inc. San Jose, California, to San Francisco Bay Regional Water Quality Control Board. Report to California Environmental Protection Agency, Regional Water Quality Control Board—San Francisco Bay Region, Oakland, CA.

TetraTech, 1997, *Phase I Environmental Assessment—Former American Beryllium Company, 1600 Tallevast Road, Tallevast, Florida*. Pasadena, CA: Tetra Tech Inc.

TetraTech, 2005, *February 2005 Site Assessment Report Addendum, Former American Beryllium Company, 1600 Tallevast Road, Tallevast, Florida*. San Diego, CA: Tetra Tech Inc.

USAF, 1986, Remedial Action Plan, Air Force Plant 44, Tucson, AR.

U.S. Congress, 1988, *Are We Cleaning Up? 10 Superfund Case Studies—Special Report*. Washington, DC: Office of Technology Assessment.

USEPA, 1983, NPL site narrative for Seymour Recycling Corp. Seymour, Indiana. *Federal Register* Notice, September 8.

USEPA, 1986, EPA Superfund record of decision: Seymour Recycling Corp. EPA ID: IND040313017 Operable Unit 1 Seymour, Indiana, 9/30/1986. U.S. Environmental Protection Agency ROD/R05-86/046.

USEPA, 1987, EPA Superfund record of decision: Seymour Recycling Corp. EPA ID: IND040313017 Operable Unit 02 Seymour, Indiana, 9/30/1987. U.S. Environmental Protection Agency ROD/R05-87/050 1987: 31.

USEPA, 1997, *Five-year Review Report Seymour Superfund Site, Seymour, Indiana*. Region 5, Chicago, IL: U.S. Environmental Protection Agency.

USEPA, 2002, *Five Year Review, Seymour Recycling Corp Superfund Site, Seymour, Indiana*. Region 5, Chicago, IL: U.S. Environmental Protection Agency.

USEPA, 2007a, Superfund information systems. http://www.epa.govR5Super/npl/indiana/IND040313017.htm (accessed February 12, 2007).

USEPA, 2007b, *Third Five Year Review Report—Seymour Recycling Superfund Site, Seymour, Jackson County, Indiana*. Region 5, Chicago, IL: U.S. Environmental Protection Agency.

USEPA, 2007c, September 19th, 2007, Superfund 20th anniversary report—Making the program faster, fairer, and more efficient. http://www.epa.gov/superfund/20years/ch4pg9.htm (accessed November 11, 2007).

USEPA, 2007d, Administrative settlement agreement and order on consent for focused remedial investigation to address 1,4-dioxane. Region 9, San Francisco, CA: U.S. Environmental Protection Agency.

USEPA, 2008a, Toxic release inventory database. http://www.epa.gov/triexplorer/ (accessed January 17, 2008).

USEPA, 2008b, Special notice letter for the upcoming RI/FS activities, Tucson International Airport Area Superfund Site, Pima County, AR, July.

University of Michigan, 2008, Gift positions U-M as national leader in studying risks to public health. University of Michigan News Service Press Release, July 1, 2008. http://www.ns.umich.edu (accessed December 6, 2008).

Wall Street Journal, 1988, Gelman Sciences Inc. accused by Michigan of water pollution. *Wall Street Journal*, March 1.

Woodside, G.D. and Wehner, M.P., 2002, Lessons learned from the occurrence of 1,4-dioxane at Water Factory 21 in Orange County, California. *Proceedings of the 2002 Water Reuse Annual Symposium*. Alexandria, VA: WateReuse Association.

Yamachika, N., 2007, Orange County Water District. Presented at Workshop on Trace Organics: Mapping a Collaborative Research Roadmap, San Francisco, CA, May 17–18, 2007.

9 Forensic Applications for 1,4-Dioxane and Solvent Stabilizers

Thomas K.G. Mohr

In the town of Solventville, a printing shop, a former dry cleaner, an electrical utility substation with several large transformers, and a printed circuit board manufacturer are all located within a short distance of each other. The neighboring gas station investigates a fuel tank leak and discovers perchloroethylene (PCE) in groundwater samples. The regulator suspects that the PCE originated from the printed circuit board manufacturer, which is located up-gradient from the gas station. The electronics firm retains a consultant, namely you, to advise it of its potential liability in the PCE spill. Your initial review shows that PCE was used in varying amounts at different times at all the four facilities, and even though two facilities are cross-gradient of the current known occurrence at the gas station's monitoring wells, sewer lines and storm drains may have provided a horizontal conduit through which PCE could have migrated. How will you proceed? It is time to open your environmental forensics toolbox and see what you have got to work with. Solvent stabilizers are a potentially useful forensic tool; however, very little work has been published documenting their application in completed studies.

An all-encompassing definition for the widely used term "environmental forensics" is found in the definitive primer from two leading authors on the subject:

> Environmental forensics is the systematic and scientific evaluation of physical, chemical, and historical information for the purpose of developing defensible scientific and legal conclusions regarding the source or age of a contaminant release into the environment.[*]

Forensic investigations into environmental contamination are usually oriented toward attributing responsibility to parties who used and released a contaminant and must therefore share in the burden of investigation and remediation (Murphy and Morrison, 2007). Environmental forensic investigations often focus on cost allocation; however, contaminant hydrogeologists can derive many other dividends by applying environmental forensic techniques to verify and enhance their site conceptual model.

Chlorinated solvents have been the subject of many forensic investigations. Methods to distinguish sources of solvents include the following:

- Mapping ratios of the solvent to its breakdown products
- Correlating historical operations to the general pattern of solvent usage constrained by regulations, equipment developments, and markets
- Back-calculating the timing of a release from the length of a plume and rates of groundwater flow

---

[*] From the foreword to *Environmental Forensics—Contaminant Specific Guide* by Robert D. Morrison and Brian L. Murphy (Morrison and Murphy, 2006).

- Compound-specific isotopic analysis
- Use of impurities and solvent stabilizers as marker chemicals to distinguish two sources of the same solvent (Morrison et al., 2006)

This chapter focuses on the potential for solvent stabilizers to persist and be detected in solvent plumes in a manner useful for various methods of analysis that can be applied toward deconvoluting commingled plumes of a common solvent.

9.1 LEVERAGING 1,4-DIOXANE DATA FOR PLUME ANALYSIS AND SOURCE APPORTIONMENT

Forensic analysis of environmental data is hampered by a high degree of uncertainty regarding many of the parameters that can be leveraged to interpret the source and timing of a release. Uncertainty can be reduced by collecting more data from more locations, using more accurate measurement techniques, and adding additional types of data to the analysis. When multiple lines of evidence supported by a variety of data types point predominantly in the same direction, the forensic scientist begins to get a warm and fuzzy feeling, and with a little bit of luck, she or he may actually attain the much sought after "Aha!" moment and reach a conclusion.

Perhaps the most important and least available feature of solvent contamination sites is the mass and timing of the release. It is the rare exception in which both are known with certainty, for example, when an accident such as a truck overturn, train derailment, or above-ground tank failure is documented. The potential utility of solvent stabilizers to assist in establishing the mass released and timing (or rate and duration of the release) depends on whether the mass fraction of the stabilizer is sufficient to be detected in environmental samples and whether the stabilizer compounds persist in the subsurface. The ability to reliably identify and quantify solvent stabilizers through laboratory analysis is of course essential to their use in forensic investigations. Detailed site operations records and first-hand accounts of operating practices are also very useful in reconstructing the manner in which solvents were used at the site. Finally, the study of solvent stabilizers at a release site cannot by itself answer source apportionment questions. The incorporation of solvent stabilizers should be viewed as an additional tool in the forensic toolbox whose usefulness will depend on the unique circumstances of the site's history and hydrogeology. This section examines some of the factors that may contribute to or detract from the utility of solvent stabilizers in forensic investigations.

9.1.1 ESTIMATING THE MASS OF THE RELEASE

Methods for estimating the release mass generally utilize a combination of the site's operating history—for example, the timeframe during which a leaking underground solvent-waste tank was used and the portion of that timeframe that the tank is assumed to have leaked—and the interpolation of subsurface data points. The site history analysis is based on a study of tank construction, photographs, and field notes that can be helpful in determining the mode of failure. Common modes of failure include

- Corrosion from an external galvanic cell in the soil
- Acidic corrosion from within the tank due to mixtures of acid wastes or solvent breakdown to hydrochloric acid
- Physical weakening of the tank due to the practice of dropping a tank dipstick used to measure product volume

Subsurface data points encompass soil-core and groundwater data to delineate the mass concentration at different points within the source area and various interpolation routines to allow the investigator to infer the amount of pore space in the heterogeneous media occupied by the solvent.

The interpolation analysis is made difficult by the heterogeneity of the subsurface. The percentages and interconnected porosity of gravels, sands, silts, and clays can be estimated from soil borings; however, the solution to this inverse problem lies more in the domain of probabilistic conditional simulation than in the more available "back of the envelope" approach using spreadsheet software.

Once an estimate of percentages of different geologic materials and their respective porosities and permeabilities is made, the investigator must also infer the propensity for the solvent to migrate into these media. The ability of dissolved solvent to migrate into the pore space of the analyzed soil core depends not only on advective flow and diffusion, but also on sorption of the solvent to soil organic matter and mineral surfaces and its abiotic and biologic breakdown.

Where 1,4-dioxane is present, the uncertainty due to sorption and abiotic and biologic transformation is reduced because 1,4-dioxane is presumed to be resistant to these fates. If the starting concentration of 1,4-dioxane in the release is known or estimated, the 1,4-dioxane concentration measured in down-gradient wells can be useful for inferring the mass released and the mass remaining in the source area. For example, 1,4-dioxane could be used to establish an order-of-magnitude estimate to the upper limit of the volume of the subsurface contaminated by the release, as 1,4-dioxane is expected to migrate farthest among the contaminants released at facilities using methyl chloroform for vapor degreasing.

The range of 1,4-dioxane concentrations present in the solvent wastes released at a vapor degreasing site can be estimated from the vapor degreaser's operating history or from measurements made at other facilities that used methyl chloroform for vapor degreasing in a similar fashion. As discussed in Section 1.2.7.1, the boiling-point difference between 1,4-dioxane and methyl chloroform (101°C versus 78°C) causes 1,4-dioxane to become concentrated in vapor degreaser still bottoms. Experiments have measured the 1,4-dioxane liquid-vapor partitioning factor at methyl chloroform's boiling point. In the operating vapor degreaser, 27% of the 1,4-dioxane will partition to the vapor phase, while 73% will remain in the liquid phase (Spencer and Archer, 1981). Through continued use, a vapor degreaser iteratively partitions 1,4-dioxane such that the proportion of 1,4-dioxane in the sump will increase over time. In several weeks of daily use, the composition of the liquid solvent can evolve to contain as much as 10–20% of 1,4-dioxane. For example, the measured 1,4-dioxane concentration accumulated in the solvent sump in a laboratory vapor degreaser used for 24 days was 7.5%, whereas the originally supplied methyl chloroform had been stabilized with only 2.8% 1,4-dioxane (Spencer and Archer, 1981). Laboratory analysis of new and spent methyl chloroform sampled from a vapor degreaser at the Hayes International Corporation showed a 68% increase in 1,4-dioxane concentration, from 1.7% to 2.9% (by mass) (Tarrer et al., 1989).

It is possible to estimate the probable concentration of 1,4-dioxane in vapor degreaser still bottoms if the operating practices are known from records or personnel interviews. For example, a Model MLW-120 stainless steel Baron-Blakeslee vapor-immersion degreaser unit has a 45.4 L solvent-sump capacity. In this estimation, 45.4 L of methyl chloroform solvent with 3% 1,4-dioxane fills the sump, and 73% of the 1,4-dioxane remains in the liquid phase. Moreover, once in each five-day work week, 6 L of makeup solvent was added to the boiling sump. Solvent vapor carryout, 1,4-dioxane removal in the water trap (see Section 1.1.1.3), and accumulation of dirt and oils in the sump are ignored. The calculations using these assumptions indicate that the 1,4-dioxane content of the solvent waste could build up to 10% by mass within 20 weeks.

Calculating the 1,4-dioxane content of still bottoms in this manner produces lower estimates than the measured data, most likely because of the significant role that vapor drag-out and liquid solvent carryout plays in removing about three times more solvent from the degreaser than 1,4-dioxane. As discussed in Section 1.1.1.3, solvent drag-out rates have been estimated at 1–2 gallons of solvent per ton of small parts cleaned when a covered, in-line conveyor system is used, whereas small parts cleaned at a similar rate in manually operated open-top degreasers lost 20% more of the solvent through vapor drag-out (ASTM, 1962). The average annual solvent losses of three methyl chloroform vapor degreasing operations at southeastern U.S. military installations (summarized in Table 1.8) was 65%.

A broad estimate of the rate and concentration of 1,4-dioxane accumulation in solvent waste can be inferred from solvent-waste recovery studies. The vapor degreasing operations using methyl chloroform at the southeastern air force bases listed in Table 1.8 are profiled more extensively in Table 9.1 for their potential to accumulate 1,4-dioxane in liquid solvent waste.

Estimates of the total mass of 1,4-dioxane at release sites are not widely available. Approximately 16,000 gallons of methyl chloroform from the Rocky Flats Closure Project were buried at the Idaho National Engineering Laboratory (INEL). Engineers at INEL estimated the 1,4-dioxane content of the methyl chloroform from the industrial literature because the methyl chloroform in question was not used for vapor degreasing where 1,4-dioxane would become progressively concentrated; rather, the methyl chloroform was used in uranium processing (Shuckrow et al., 1982). On the basis of an average 1,4-dioxane content in methyl chloroform of 2.75 vol% (2.1 wt%) and a maximum 1,4-dioxane content of 3.5 vol% (2.67 wt%), the corresponding average and maximum 1,4-dioxane masses in the buried methyl chloroform were 1720 kg and 5900 kg, respectively. The only documented value for 1,4-dioxane in methyl chloroform at the site of origin was 1.9% (INEL, 2006).

A major Midwest manufacturer studied the stabilizer content of solvents to improve the quality of solvents recovered by distillation for reuse. The firm's industrial engineers sought to identify the solvent stabilizers in methyl chloroform, TCE, and PCE to restore the original composition following on-site distillation. The stabilizers in TCE were identified as 1,2-butylene oxide, cyclohexene oxide, p-tert-butyl phenol, and 1-propanol. Stabilizers in new methyl chloroform supplied by two solvent producers were identified. The first, presumably supplied by Dow Chemical, contained the familiar formula of 1,4-dioxane (3.17 wt%), nitromethane (0.35%), and 1,2-butylene oxide (0.45%), whereas the second, presumably supplied by Vulcan Chemical, contained 1,3-dioxolane (2.45%), nitromethane (0.3%), and 1,2-butylene oxide (1.45%) (Bohnert and Carey, 1991).[*] The concentration of stabilizers in the solvent that was removed from vapor degreasers and stored in drums was not measured, as the study objective concerned refining distillation techniques. Following a filtering, desiccation, and distillation process, the distilled methyl chloroform solvent waste had a lower weight percent of 1,4-dioxane, as shown in Table 9.2.

This finding of less 1,4-dioxane in the distilled solvent waste is consistent with the expectation of increased 1,4-dioxane concentrations in vapor degreasing waste due to boiling-point differences. Distillation retains only the vapor fraction of 1,4-dioxane (~27%). The Tarrer et al. study of southeastern air force base vapor degreasers also observed a decrease in 1,4-dioxane concentration following distillation of methyl chloroform solvent waste. The waste had 2.9% 1,4-dioxane initially and 1.96% after distillation, a decrease of 32% (Tarrer et al., 1989). A second study by the same Midwest manufacturer analyzed the amount of 1,4-dioxane in distilled methyl chloroform and found a 65% decrease in 1,4-dioxane concentration (Holt, 1990) (see Table 9.7).

The solvent distillation process profiled in the Bohnert and Carey report used a preparatory drying step to return the solvent to required moisture conditions (less than 100 ppm water in methyl chloroform and TCE). Drying was achieved with a recirculating pump and a 3 Å (angstrom) (0.3 nm) molecular sieve desiccant bed. Solvent waste was recirculated through the sieve for 24–48 h or as needed to decrease the moisture level to the target of 100 ppm water (Bohnert and Carey, 1991).

While removing the water that accumulated in the solvent, water-soluble stabilizers are also likely to be removed, particularly 1,4-dioxane. The fate of the removed water then becomes important to the forensic investigation. If the desiccant was regenerated by using a centrifuge or gravimetric technique, then the handling of the wastewater should be investigated to confirm whether it was discharged on site or hauled away as hazardous waste. If the desiccant was regenerated through an evaporation step, then it is likely that the 1,4-dioxane was emitted to the atmosphere, where it is

---

[*] The report did not identify the solvent producers; Dow held the first patent for use of 1,4-dioxane in methyl chloroform, and Vulcan held an early patent for use of 1,3-dioxolane to stabilize methyl chloroform (Bursack et al., 1963). Both Dow and Vulcan produced methyl chloroform stabilized by 1,4-dioxane or 1,3-dioxolane. The third major solvent producer, PPG, often used dimethoxymethane and tert-butyl alcohol instead of either 1,4-dioxane or 1,3-dioxolane (PPG, 1998).

TABLE 9.1

Annual Operating Parameters for Methyl Chloroform Vapor Degreasers at Southeastern U.S. Air Force Bases and Potential Accumulation of 1,4-Dioxane in Solvent Waste

Vapor Degreaser Number	Vapor Degreaser Dimensions (ft)	Annual Number of Batch Changes	Reservoir Volume (gallons)	Annual Solvent Usage (gallons)	Annual Solvent Waste Generation (gallons)	Annual Emissions Loss (%)	Annual 1,4-Dioxane Consumption (lb/kg)	Annual 1,4-Dioxane in Solvent Waste (lb/kg)	Annual 1,4-Dioxane in Solvent Waste (wt%)	Annual 1,4-Dioxane Lost to Vapor (lb/kg)
1	4.5 × 3 × 5.6	10	50	654	275	58	216/98	158/72	6.9	34/15
2	8 × 4 × 8	6	110	3850	990	74.3	1271/578	927/422	11.2	255/116
3	10 × 3 × 10	8	110	4180	1320	68.4	1379/627	1007/458	9.1	255/116

Source: Data from Tarrer, A.R., et al., 1989, *Reclamation and Reprocessing of Spent Solvent*. Park Ridge, NJ: Noyes Data Corporation.

Notes: 1,4-Dioxane consumption was based on an assumed initial content of 3 wt%. 1,4-Dioxane content in liquid waste applies the 73% 1,4-dioxane partitioning factor for 1,4-dioxane at methyl chloroform's 78°C boiling point (Spencer and Archer, 1981). Weight percent 1,4-dioxane in liquid waste was obtained as the ratio of calculated kilograms of 1,4-dioxane in the waste to the volume of the waste in liters. This screening-level calculation neglects 1,4-dioxane losses due to removal in the water trap and does not account for any addition of make-up solvent to the vapor degreaser between solvent changes. See Section 1.1.1.3 for more on vapor degreasing.

TABLE 9.2
Stabilizer Content in Distilled Methyl Chloroform Solvent Waste at a Major Midwest Manufacturer

Stabilizer	Amount in New Methyl Chloroform (wt%)	Amount in Distilled Methyl Chloroform Solvent Waste (wt%)	Difference
1,4-Dioxane	3.17	0.92–2.79	−12% to −71%
1,2-Butylene oxide	0.45	0.26–1.59	−42% to +253%
Nitromethane	0.35	0.26–0.54	−26% to +54%

Source: Bohnert, G.W. and Carey D.A., 1991, Scale-up of recovery process for waste solvents. Technical Communications. Kansas City, Missouri: Allied Signal Aerospace Company.

expected to photo-oxidize, disperse, or return to the surface through precipitation or condensation at inconsequential concentrations (see Section 3.1.4).

9.1.2 IMPURITIES[*]

The impurities in chlorinated solvents mentioned in Section 1.2.7 can also be important to forensic investigations and can be used together with stabilizers for determining whether iterative concentration in vapor degreasing cycles due to boiling-point differences could produce enough impurity mass to be used as a marker chemical to trace the plume. Studies of the occurrence and fate of solvent impurities provide a model for the analysis of "high-boiler" stabilizers that boil at temperatures higher than the solvent. The presence of impurities in solvents is difficult to document as most producers list their products in terms of their purity, attributing the fraction that is not the solvent to unspecified proprietary stabilizers. Nevertheless, some documentation of impurities in solvents is available, as tabulated in Table 9.3.

Two published studies describe the potential for PCE to be present at TCE release sites where PCE was never used, through the presence of PCE as an impurity of TCE production and the concentration of PCE during vapor degreasing because of boiling-point differences. The key distinction between the two profiles is the assumption of the starting PCE concentration that may be present as an impurity of TCE.

The first profile of PCE as an impurity of TCE is provided in *Environmental Forensics—Contaminant Specific Guide* (Morrison and Murphy, 2006). Morrison, Murphy, and Doherty have noted that these solvents are produced by the same process and then separated by fractional distillation (Morrison et al., 2006), which is facilitated by the large boiling-point difference between TCE and PCE, that is, 87.2°C and 121.2°C, respectively (Shepherd, 1962). Morrison et al. (2006) cited a recent chemical supplier's specifications website for impurity content in degreasing grade TCE of up to 3–4% chlorinated compounds.[†] As described subsequently in this section, this estimate is probably too high for most grades of TCE.

The second profile of PCE as an impurity of TCE is provided by Lane and Smith (2006),[‡] who have cited Dow Chemical's records showing that Dow's TCE had no detectable "perchloroethylene as a manufacturing impurity" ("PCEMI") of TCE since about 1990. The reporting limit for analyses performed by Dow Chemical was 1 ppm PCE in TCE. Dow has been one of the leading providers of TCE in the United States since the 1960s. PCEMI may have been present in TCE produced by Dow before the 1990s, or in TCE from other solvent producers, as suggested by older MSDS and other citations.

[*] From Gauthier, T.D. and Murphy, B.L., 2003, Environmental Forensics 2003(4): 205–213. With permission.
[†] Unfortunately, the cited website is no longer operational.
[‡] See Lane and Smith (2006).

TABLE 9.3
Documented Impurities in Chlorinated Solvents and Their Potential for Enrichment During Vapor Degreasing

References	Impurities in Specific Chlorinated Solvents	Impurity Amounts (Total)	Potential Concentration after Vapor Degreasing Cycle
	Perchloroethylene (b.p. = 121°C)		
[1]	Carbon tetrachloride, methyl chloroform, dichloromethane, trichloroethylene, and water	<1%	All would evaporate
[2]	Trichloroethylene	1.6%	Evaporates
[2]	"Dichloromethane or 1,1-dichloroethylene"	0.13%	Evaporates
[2]	Vinyl chloride	0.07%	Evaporates
[3]	1,2-Dichloroethane	<0.3%	Evaporates
	Methyl chloroform		Evaporates
	1,1,2-Trichloroethane		Evaporates
	Tetrachloroethane		*Concentrates*
	Pentachloroethane		*Concentrates*
	Hexachloroethane		*Concentrates*
[5]	Hexachlorobenzene may be present as a by-product	NA	Not calculated
[4]	Methyl chloroform, dichloromethane	See note[a]	
	Trichloroethylene (b.p. = 87°C)		
[4]	Methyl chloroform, dichloromethane	See note[a]	
[5]	Hexachlorobenzene may be present as a by-product	NA	Not calculated
[6]	Perchloroethylene, 1,1,2-trichloroethane, methyl dichloroacetate, 1,1-dichloroethylene, and chloroform[b]	<2%	Perchloroethylene and 1,1,2-trichloroethane will concentrate—see text
[7]	Carbon tetrachloride[c]	0.05%	All would evaporate
	Methyl chloroform[c]	0.035%	
	Chloroform[c]	0.01%	
[8]	Tetrachloroethylene[d]	<0.03%	*Concentrates*
	1,1-Dichloroethylene[d]	<0.01%	Evaporates
	Methyl chloroform[d]	<0.01%	Evaporates
	Chloroform[d]	<0.01%	Evaporates
	Carbon tetrachloride[d]	<0.005%	Evaporates
	Dichloromethane[d]	<0.001%	Evaporates
	Bromodichloromethane[d] (b.p. = 90°C)	<0.1%	*Concentrates*
	Methyl chloroform (b.p. = 73.8°C)		
[9]	1,1-Dichloroethylene (in 18 of 22 samples)[e]	30–900 mg/L; avg 300 mg/L	Evaporates
[9]	1,1-Dichloroethane[e] (in 11 of 22 samples)	680 mg/L	Evaporates
	Trichloroethylene[e] (in 12 of 22 samples)	1885 mg/L	*Concentrates*
	1,1,2-Trichloroethane[e] (in 9 of 22 samples)	1340 mg/L	*Concentrates*
[10]	1,2-Dichloroethane, 1,1-dichloroethane, chloroform, carbon tetrachloride, trichloroethylene, 1,1,2-trichloroethane, and 1,1-dichloroethylene		
[11]	Chloroform[f]	100 ppm	Evaporates
	Carbon tetrachloride[f] (b.p. = 76.8°C)	250 ppm	*Concentrates*
	1,1-Dichloroethane[f]	426 ppm	Evaporates
	1,2-Dichloroethane[f] (b.p. = 83.5°C)	2300 ppm	*Concentrates*

continued

TABLE 9.3   (continued)
Documented Impurities in Chlorinated Solvents and Their Potential for Enrichment During Vapor Degreasing

References	Impurities in Specific Chlorinated Solvents	Impurity Amounts (Total)	Potential Concentration after Vapor Degreasing Cycle
	1,1-Dichloroethylene[f]	398 ppm	Evaporates
	Trans-1,2-dichloroethylene[f]	50 ppm	Evaporates
	Trichloroethylene[f] (b.p. = 87°C)	200 ppm	Concentrates
	Tetrachloroethylene[f] (b.p. = 121°C)	475 ppm	Concentrates
	Dichloromethane technical grade (b.p. = 39.7°C)		
[12]	May contain methyl chloride		Evaporates
	Chloroform (b.p. = 61.2°C)		Concentrates
	1,1-Dichloroethane (b.p. = 57.2°C)		Concentrates
	1,2-Dichloroethane (b.p. = 83.5°C)		Concentrates

Sources: [1] European Chemicals Bureau (2004); [2] Parsons et al. (1984); [3] Borror and Rowe (1981); [4] ASTM (2006); [5] UNEP (2005); [6] Totonidis (2005); [7] Henschler et al. (1977); [8] European Chemicals Bureau (2004); [9] Henschler et al. (1980); [10] Stewart et al. (1969); [11] Maltoni et al. (1986); and [12] IARC (1986).

NA = Not available.

[a] Both would evaporate in PCE & TCE; ASTM test method to determine MC and DCM in TCE & PCE suggests these impurities were present.

[b] Industrial grade TCE.

[c] Industrial grade TCE (1977).

[d] Trichloroethylene (Europe, pre-1995).

[e] German technical grade.

[f] Technical grade.

One of the attributes that producers of solvents marketed for a competitive edge is the quality and reliability of their products. There was an inherent business interest in producing high-quality, high-purity TCE; hence, it is likely that the purity improved over time and as end-use demands for TCE became more specialized and sensitive to impurities, for example, in the electronics industry.

PCE may occur as a coproduct when TCE is produced by the chlorination of 1,2-dichloroethane or the oxychlorination of ethylene (Doherty, 2000a). As noted in Section 1.2.3, typical technical or industrial grade TCE has been known to include about 0.15% carbon tetrachloride as the major chlorinated hydrocarbon impurity since the 1950s; however, PCEMI was not mentioned (Willis and Christian, 1957). PCEMI was not reported in analyses of pre-1977 American technical grade TCE (Henschler et al., 1977) or in pre-1990 American technical grade TCE (Holt, 1990). The Japanese analysis of pre-1983 industrial grade TCE detected several impurities, but PCE was not among them (Tsuruta et al., 1983). TCE produced in Europe before 1995 had up to 0.03% PCE as an impurity (European Chemicals Bureau, 2004). Indian TCE and British reagent grade TCE both contained detectable PCEMI in more recent analyses (Totonidis, 2005). In America, a widely used grade of vapor degreasing TCE, Dow Chemical's Neu-Tri™, had a purity of 99.4% in 1981, as indicated in Table 1.14, with stabilizers composing the balance. However, a technical grade of TCE, meeting military specification O-T-634 and distributed by Ashland Chemical in 1986, indicated a TCE content of 95% or more. Dow's technical grade TCE for military specification O-T-634 in 1981 was the same as its Neu-Tri™ reported as 99.4% TCE and 0.6% 1,2-butylene oxide with identical composition listed in 1991, 1994, and 1999 (Lane and Smith, 2006). Table 9.4 lists some of the MSDS descriptions of TCE in the 1980s and 1990s.

The purity listing on an MSDS does not necessarily imply that the balance of the solvent may include impurities; solvent stabilizers often composed the balance. It is only through the laboratory analysis of solvents that PCEMI and other impurities are definitively revealed.

TABLE 9.4
MSDS Descriptions of Trichloroethylene Purity in the 1980s and 1990s

Company	Date	Id Number	Name/Info	Trichloroethylene (%)	1,2-Butylene Oxide (%)
Dow Chemical USA	12/14/81	FSC: 6810 NIIN: 00-924-7107	Trichloroethylene, Technical O-T-634	99.4	0.6
Dow Chemical USA	12/14/81	FSC: 6810 NIIN: 00-924-7107	Neu-Tri Trichloroethylene #56530	99.4	0.6
Dow Chemical USA	1/1/85	FSC: 6810 NIIN: 00-754-2813	Trichloroethylene-E	>95	0.6
Allied Chemical Corp.	1/1/85	FSC: 6810 NIIN: 01-031-5512	Trichloroethylene Low Mobile Ion	>97	—
Ashland Chemical Company	3/1/86	FSC: 6810 NIIN: 00-678-4418	Trichloroethylene Deg Cold/ Vap; Trichloroethylene, Technical O-T-634; ASTM D 4080	95	—
Baxter Healthcare Corp/Burdick and Jackson Div	12/1/89	FSC: 6505 NIIN: Liin: 00f038115	Trichloroethylene	99	1
Chem Central Kansas City	12/1/94	FSC: 6810 NIIN: 00-184-4800	Trichloroethylene, Technical	99.4	0.5

   Whether or not PCEMI occurs in the part per million range or in the percent range, if present, it is likely to become concentrated in TCE through use. In vapor degreasing operations, TCE volatilizes more rapidly than PCE, and PCE is concentrated in the remaining liquid TCE. Morrison et al. (2006) explained that even at sites where PCE was allegedly never used, it may be found where spent TCE vapor degreasing waste was discharged because of the concentration of PCE as a high-boiling impurity in TCE.

   Morrison et al. (2006) estimated the accumulation of PCEMI in TCE by assuming a starting ratio of 1% PCEMI, calculating the total volume of PCEMI in the volume of TCE present in the vapor degreaser solvent reservoir, and then estimating how much PCEMI volatilizes during operations. They also accounted for the addition of PCEMI in new TCE added to the degreaser between batch changes. In their example evaluation of the potential for PCEMI to become concentrated, they assumed a degreaser with a 350-gallon solvent capacity and the addition of 55 gallons of new TCE per week to make up the volume lost to volatilization, drag-out, and carryout. Morrison et al. further assumed that the degreaser is cleaned out four times per year and that a total of 100 gallons of TCE and accumulated oils and other waste are removed per year.

   To estimate the rate of volatilization of PCEMI in TCE, the vapor pressure of PCE at TCE's boiling point is determined by using the Clausius–Clapeyron equation, shown here in two forms:

$$\ln P_{\mathrm{vap}} = k\left(\frac{1}{T_{\mathrm{B}}} - \frac{1}{T}\right),$$ (9.1a)

$$\ln\left(\frac{P_1}{P_2}\right) = \frac{\Delta H_{\mathrm{vap}}}{R}\left(\frac{1}{T_2} - \frac{1}{T_1}\right),$$ (9.1b)

where P_{vap} is the vapor pressure of PCE at the TCE boiling point in atmospheres, T_{B} is the boiling point of PCE (394.35 K), T is the TCE boiling point (360.35 K), and k is a constant relating the ideal gas constant (R) and the enthalpy of vaporization of PCE (ΔH_{vap}). The relative rate of solvent loss is proportional to the vapor pressures, under the assumption that PCE is infinitely soluble in TCE (Morrison et al., 2006).

Lane and Smith began their analysis of PCEMI in TCE with a starting concentration between 1 and 10 ppm; they cited communication with Dow officials to support that assumption. PCEMI will accumulate owing to boiling-point differences in the still bottoms. Lane and Smith cited a typical still bottom composition of 20% TCE and 80% oils and other waste. As discussed in Section 1.1.1.3 and Table 1.10, operators were motivated to remove solvent from service when oil and grease reached 25% to avoid solvent breakdown due to overheating; hence, the inverse of this ratio is more likely to have been the composition of solvent wastes removed from vapor degreasers. The American Society for Testing Materials (ASTM) manual for vapor degreasing recommends that solvent be removed from service at 25% oil (ASTM, 1962).

Starting with an assumed PCEMI content of 5 ppm, Lane and Smith assumed a 500-fold increase in PCEMI content through iterative concentration in vapor degreasing cycles and so ended up with 2500 ppm (or 0.25%). Their result is a substantially higher degree of concentration than the seven-fold increase obtained in the analysis by Morrison et al. to obtain 7% PCEMI from their starting value of 1% PCEMI. Even with the conservative assumption of tremendous concentration, the ending ratio in the Lane and Smith evaluation gives 400 times more TCE than PCEMI. Given that the solubility of TCE in water is substantially higher than that of PCE (1100 versus 150 mg/L, i.e., a solubility ratio of 7.33),[*] TCE will preferentially dissolve out of the solvent waste once it has migrated into the saturated zone. With a 400-fold greater concentration and a 5.5 times greater solubility, the concentration of TCE in groundwater from the solvent waste under discussion is likely to be 2200 times greater than the concentration of PCEMI (not accounting for other fate and transport properties such as sorption).

Lane and Smith pointed out that the effective solubility (see Section 3.3.1.1) must be considered when estimating the likely concentrations in groundwater for TCE and PCEMI. By way of review, $S_{eff} = X_i S_{water}$, where S_{eff} is the effective solubility of a compound in a mixture (lower than its solubility when measured alone in water), X_i is the mole fraction of a compound in the mixture, and S_{water} is the compound's single-component aqueous solubility. Table 9.5 provides the terms for calculating and comparing the effective solubility of TCE and PCEMI.

A quantity of solvent in still bottoms released to groundwater may completely dissolve over time. As the Lane and Smith analysis shows, TCE will initially be a great deal more soluble in water than PCEMI owing to its much higher effective solubility. As the residual mass of solvent waste shrinks, the fraction that is PCEMI will increase, and it will eventually dissolve into water as well. The mass of PCEMI is probably small to begin with; therefore, detection of PCE at sites where TCE was used exclusively is more likely to reflect migration from other sources of PCE such as dry cleaners, brake shops, and printing operations, or such detection may reflect incomplete site operations records, obscuring some period during which PCE was used at the facility.

The analysis of PCEMI suggests that other solvent-stabilizer compounds present in solvents in small quantities would have low potential to concentrate to detectable levels even when boiling-point differences favor their concentration. Table 9.6 summarizes the potential for the high-boiling stabilizers listed in Tables 1.27 through 1.30 to become concentrated, as calculated by using the Clausius–Clapeyron equation to determine vapor pressures and the resulting mass fraction and Raoult's law to determine effective solubilities.

Calculating the iterative enrichment due to boiling-point and vapor-pressure differences and the resulting effective solubilities focuses attention on the probable composition of the still bottoms and

---

[*] These solubility values differ from those used by Lane and Smith and from those cited in Table 3.9. Solubility of the chlorinated solvents has a range of values in the literature. Solubility values used here were obtained from Dow Chemical Company (2006).

TABLE 9.5
Comparison of Effective Solubility and Corresponding Groundwater Concentrations of Trichloroethylene and Perchloroethylene as a Manufacturing Impurity in Trichloroethylene Vapor Degreasing Waste

Parameter	Trichloroethylene	Perchloroethylene	Ratio
Molecular weight (Da)	131.4	165.8	—
Aqueous solubility (mg/L)	1100	150	7.33
Mass fraction in still bottoms[a]	99.75%	0.25%	400
Mole fraction in still bottoms[a]	0.998	0.001982	503.5
Effective solubility (mg/L)	1097.8	0.2974	3692
Probable groundwater concentration (μg/L)	15,000	4	—

Source: Lane, V. and Smith, J., 2006, Fact or fiction? The source of perchloroethylene contamination in groundwater is a manufacturing impurity in chlorinated solvents. Presented to the National Groundwater Association Ground Water and Environmental Law Conference, July 7, 2006. Chicago, Illinois: National Groundwater Association. http://www.ngwa.org (accessed August 2, 2006).

Notes: This analysis ignores volatilization that may occur at the surface of the spill site or in the unsaturated zone, differential sorption in transit to the saturated zone, diffusion, biodegradation, and other fate processes that would alter the probable groundwater concentration. The 15 ppm concentration of trichloroethylene in groundwater is an assumed value found at more highly contaminated trichloroethylene release sites; the 4 ppb concentration of perchloroethylene is calculated from the trichloroethylene value.

[a] After assumed 500-fold concentration.

other solvent wastes that are the source of the many chlorinated solvent plumes and the subject of ongoing cleanup. The resulting high effective solubility of 1,4-dioxane explains the elevated concentrations at which it is sometimes encountered at methyl chloroform release sites. Table 9.6 also profiles the potential for other solvent stabilizers to become concentrated in waste and preferentially dissolve into groundwater. For example, 1,3-dioxolane is rarely analyzed in groundwater samples; yet it may be present at some dichloromethane release sites. However, where 1,3-dioxolane was a stabilizer of methyl chloroform, the small boiling-point difference (only 4°C) does not favor enrichment: a starting formulation of 2 wt% 1,3-dioxolane in methyl chloroform will have an ending 1,3-dioxolane concentration of 4.5 wt% after seven weeks and an effective solubility only seven times higher than methyl chloroform. This small increase could still be sufficient to produce detections useful as a marker chemical, but it is unlikely that a 1,3-dioxolane plume would develop at a methyl chloroform release site.

The numerous assumptions necessary to estimate partitioning and calculate stabilizer concentrations will not provide a technically rigorous or legally defensible analysis of the likely origin of solvents, as indicated by detections of stabilizers in groundwater. The most reliable evidence for stabilizer enrichment in vapor degreasing is laboratory analyses done in the context of industrial engineering studies. The data from Bohnert and Carey (1991) and Tarrer et al. (1989, 1993) are based on such studies, as are the data from Holt (1990), presented in Table 9.7. The Holt study measured a 65% increase in 1,4-dioxane concentration in a vapor degreasing sump, which corroborates the analyses by Tarrer et al. in which a 68% increase in 1,4-dioxane in the sump was measured. The decrease in 1,4-dioxane concentration in solvent recovered by distillation (the last column in Table 9.7) is due once again to the boiling-point difference; 1,4-dioxane remains behind in the distillation process, as discussed in Section 9.1.1.

TABLE 9.6
Summary of the Potential for High-Boiling Stabilizers to Become Concentrated

Operating Parameters for Vapor Degreasers Used in Stabilizer-Enrichment Calculations[a]

Vapor Degreaser Type	Dimensions (ft)	Solvent	Batch and Reservoir Volume (gallons)	Solvent Added per Week (gallons)	Batch Changes per Year	Annual Solvent Usage (gallons)	Annual Solvent Waste (gallons)	Emissions Losses (%)
Open top	5 × 2.5 × 6.5	Perchloroethylene	70/50	27	17	1824	420	77
Open top	8 × 4 × 8	Methyl chloroform	165/110	55	6	3850	990	74.3
Open top	12 × 4.5 × 4.5	Trichloroethylene	280/180	35	12	7200	1800	75
Open top	5 × 3 × 3	Dichloromethane	75/50	30	12	2200	280	85

Vapor Pressures of "High-Boiling" Stabilizers at Solvent Boiling Point

	Boiling Point (K)	Concentration (wt%)	Enthalpy of Vaporization[b] (ΔH_{vap}) (J/mol)	Calculated Stabilizer Vapor Pressure at Solvent Boiling Point (atm)
Stabilizers of Methyl Chloroform (b.p. = 346.8 K)				
1,4-Dioxane	374.1	3	36,500	0.40
Nitromethane	374.2	0.5	34,000	0.42
2-Methyl-3-butyn-2-ol	376.5	2.5	42,000	0.32
Nitroethane	387	0.5	38,270	0.25
Stabilizers of Trichloroethylene (b.p. = 360.2 K)				
Pyridine	388.2	0.5	37,300	0.41
Epichlorohydrin	390.9	0.15	42,900	0.32
p-tert-Amyl phenol	535.5	0.01	58,200	0.0017
Stabilizers of Perchloroethylene (b.p. = 394.2 K)				
Butoxymethyl oxirane	437	0.0043	NA	*Not calculated*
Cyclohexene oxide	402.5	0.0011	NA	*Not calculated*
Stabilizers of Dichloromethane (b.p. = 346.8 K)				
Nitromethane	374.2	0.43	34,000	0.12
1,4-Dioxane	374.1	0.8	36,500	0.10

Cyclohexane	353.7	0.01		32,300		0.24	
1,3-Dioxolane	351	2		34,600		0.23	

Iterative Enrichment of High-Boiling Stabilizers through Emission Losses and Solvent Addition

	Initial Mass Fraction (wt%)	Vapor Pressure Ratio	Mass Fraction of Stabilizer in Liquid Solvent (wt%) Weeks of Operation with Additions before Clean out						
			1	2	3	4	5	6	7...
Stabilizers of Methyl Chloroform (b.p. = 346.8 K; total solvent volume = 120 gallons; 45 gallons added per week; 55 gallons of emission losses per week; nine weeks per batch)									
1,4-Dioxane	3.00	0.40	3.6	4.3	5.4	6.8	8.8	11.9	16.6
Nitromethane	0.50	0.42	0.6	0.7	0.9	1.1	1.4	1.9	2.7
2-Methyl-3-butyn-2-ol	2.50	0.32	3.1	3.8	4.8	6.1	8.0	10.9	15.3
Nitroethane	0.50	0.25	0.6	0.8	1.0	1.3	1.7	2.3	3.3
Stabilizers of trichloroethylene (b.p. = 360.2 K; total solvent volume = 275 gallons; 90 gallons added per week; 104 gallons of emission losses per week; four weeks per batch)									
Pyridine	0.50	0.41	0.59	0.69	0.80	0.94	—	—	—
Epichlorohydrin	0.15	0.32	0.18	0.22	0.26	0.30	—	—	—
p-tert-Amyl phenol	0.01	0.00	0.013	0.017	0.021	0.026	—	—	—
Stabilizers of Perchloroethylene (b.p. = 394.2 K; total solvent volume = 460 gallons; 25 gallons added per week; 28 gallons emission losses per week; three weeks per batch)									
Butoxymethyl oxirane (CF: 1.75)[c]	0.0043	0.0075	0.013	0.023	—	—	—	—	—
Cyclohexene oxide (CF: 0.94)[c]	0.0011	0.0010	0.00097	0.0009	—	—	—	—	—
Stabilizers of Dichloromethane (b.p. = 346.8 K; total solvent volume = 125 gallons; 30 gallons added per week; 35 gallons emission losses per week; four weeks per batch)									
Nitromethane	0.43	0.12	0.52	0.63	0.79	0.99	—	—	—
1,4-Dioxane	0.80	0.10	0.97	1.19	1.47	1.87	—	—	—
Cyclohexane	0.10	0.24	0.12	0.14	0.17	0.21	—	—	—
1,3-Dioxolane	2.00	0.23	2.35	2.81	3.43	4.30	—	—	—

continued

TABLE 9.6 (continued)
Summary of the Potential for High-Boiling Stabilizers to Become Concentrated

	Mass Fraction in Still Bottoms (wt%)	Effective Solubilities of Stabilizers Concentrated in Still Bottoms				
		Mole Fraction	Aqueous Solubility (mg/L)	Effective Solubility (mg/L)	Solubility Ratio (Solvent:Stabilizer)	Stabilizer Concentration[d] (μg/L)
Stabilizers of Methyl Chloroform						
1,4-Dioxane	16.6	0.20	10,000,000	2,045,261	0.00017	>100,000
Nitromethane	2.7	0.048	11,100	530	0.67	>10,000
2-Methyl-3-butyn-2-ol	15.3	0.20	10,000,000	1,963,840	0.00018	>100,000
Nitroethane	3.3	0.047	45,000	2127	0.17	>10,000
Methyl chloroform	62.1	0.50	700	353	—	—
Stabilizers of Trichloroethylene						
Pyridine	0.94	0.015	10,000,000	154,850	0.007	>100,000
Epichlorohydrin	0.30	0.0043	65,900	281	3.83	>1000
p-tert-Amyl phenol	0.026	0.00022	168	0.037	28,911	<1
Trichloroethylene	98.6	0.98	1100	1076	—	—

Stabilizers of Perchloroethylene					
Butoxymethyl oxirane	0.023	0.00029	20,000	5.87	25.549	<100
Cyclohexene oxide	0.0009	0.000015	—	—	—	<1
Perchloroethylene	99.98	0.9997	150	150	—	—
Stabilizers of Dichloromethane					
Nitromethane	1.0	0.0137	11,100	152	103	<100
1,4-Dioxane	1.9	0.0178	10,000,000	178,312	0.087	>10,000
Cyclohexane	0.21	0.0021	55	0.12	132,346	<1
1,3-Dioxolane	4.3	0.0488	10,000,000	487,945	0.032	>100,000
Dichloromethane	92.6	0.9175	17,000	15,598	—	—

Notes: The values given here are rough approximations; the calculations that yielded these estimates do not account for the ~25% of oily wastes that accumulate in vapor degreaser still bottoms.

[a] Operating parameters for perchloroethylene and methyl chloroform vapor degreasers are from Tarrer et al. (1993); parameters for trichloroethylene and dichloromethane vapor degreasers are inferred from data in USEPA (1977).

[b] Enthalpy data are from Chickos and Acree (2003). Enthalpies of vaporization were not located for butoxymethyl oxirane or cyclohexene oxide. Where enthalpy data are not available, enthalpy of vaporization can be calculated by using Antoine coefficients with the method presented in the Handbook of Chemical Property Estimation Methods (Lyman et al., 1990); however, Antoine coefficients were also not located for these two compounds.

[c] CF = Concentration factor from Table 1.26.

[d] Where solvent concentration is 10 mg/L. Final stabilizer concentration in groundwater resulting from a spill of still bottoms is estimated to an order of magnitude without considering fine-grained storage or other fates of stabilizers in the subsurface.

TABLE 9.7

Gas Chromatographic Measurements of Solvents, Stabilizers, and Impurities

Solvents with Stabilizers (S) and Impurities (I)	Fraction in Solution (wt%)		
	Before Use	In Sump	Distilled
Trichloroethylene	96.24	98.89	95.41
1,2-Butylene oxide (S)	0.5	0.22	0.78
Methyl chloroform (I)	0.022	0.015	0.089
1,1,2-Trichloroethane (I)	0.099	0.15	0.046
Unknowns	3.14	0.72	3.67
Methyl chloroform	96.25	96.34	96.21
sec-Butanol (S)	0.97	0.80	1.12
1,4-Dioxane (S)	0.91	1.50	0.80
1,3-Dioxolane (S)	0.84	0.75	0.95
Nitromethane (S)	0.57	0.34	0.77
1,2-Butylene oxide (S)	0.32	0.25	0.38
1,2-Dichloroethane (I)	0.12	0.13	0.11
Trichloroethylene (I)	0.0087	0.013	0.0074
Cyclohexane (S)	0.0049	0.0046	0.0088
Nitroethane (S)	0.0047	0.0063	0.0027

Source: After Holt, R.D., 1990, Physical properties of contaminated TCE and 1,1,1-trichloroethane. Technical Communications. Kansas City, Missouri: Allied Signal Aerospace Company.

9.2   RATIO ANALYSIS

At many large-scale solvent release sites, passive remediation approaches such as monitored natural attenuation (MNA) are favored where risk can be managed and where processes that degrade or destroy contaminants can be demonstrated. The MNA approach is often a cost-effective and preferred remedial strategy where the contaminated property has a single owner or where long-term passive remediation will not interfere with beneficial land and groundwater uses.

To demonstrate MNA, remediation consultants must prove that natural processes, whether abiotic or microbially mediated, are transforming the solvents and associated chemicals into breakdown products that are less harmful, thereby decreasing risk and reducing contaminant mass. This demonstration is often attempted by plotting changes in ratios of the molar concentrations of the contaminant and its breakdown products with distance from the source. The ratio can serve as an indicator of the degree to which the breakdown reaction or microbial process has progressed. However, the original contaminant and its breakdown product may behave differently in the subsurface, leading to different rates of migration. For example, the molar ratio in a down-gradient monitoring well may reflect differences in the rate of adsorption due to the organic carbon partition coefficients for these compounds (TCE has an average K_{oc} value of 101, while cis-1,2-dichloroethylene has an estimated K_{oc} value of 240). Consequently, the molar ratios of these two compounds may reflect both the rate of biotransformation and differences in rates of migration. Because the more mobile compound is likely to migrate farther down-gradient, the ratio of the more mobile compound to the less mobile compound will decrease in monitoring wells located near the source zone and increase farther from the source zone. Similarly, the ratio of the compound more resistant to degradation to the compound more prone to degradation will increase with time, especially when the compound more prone to degradation produces the more resistant compound as a daughter product

(McNab and Narasimhan, 1994). Determining rates of transformation from groundwater samples is especially challenging—because the groundwater is moving, the solutes are subject to dispersion, diffusion, and sometimes multiple pathways of biodegradation or abiotic degradation. Contaminant migration is often retarded owing to sorption and matrix diffusion.

To discern the rates of biodegradation that are independent of variable migration rates among parent and breakdown products, a conservative tracer can be useful to confirm the rate of groundwater flow. Where 1,4-dioxane was released together with methyl chloroform, the 1,4-dioxane can serve as a tracer. Molar concentration ratios of 1,4-dioxane to methyl chloroform and its elimination reaction product, 1,1-dichloroethylene, can reveal patterns in the abiotic breakdown of methyl chloroform. Similarly, ratios of 1,4-dioxane to methyl chloroform and its reductive dechlorination products, 1,1-dichloroethane and chloroethane, can profile the rate of biologic transformation of methyl chloroform. As a conservative tracer, 1,4-dioxane can demonstrate the degree to which physical fates such as transverse dispersion or matrix diffusion may complicate the interpretation of molar ratios to discern degradation rates. With sufficient monitoring points, sampling for 1,4-dioxane can reveal the degree to which subsurface heterogeneity at a site introduces uncertainty into the degradation rates inferred from constituent ratios. This section examines the utility of ratio comparisons to demonstrate MNA with 1,4-dioxane as a conservative tracer.

1,4-Dioxane has not been directly observed to undergo breakdown at field sites;[*] therefore, its presence at a site may conflict with the requirements for demonstrating MNA to the satisfaction of the U.S. Environmental Protection Agency's (USEPA's) (1999) Directive on MNA. The USEPA Directive notes that the fuel-oxygenate methyl *tert*-butyl ether (MTBE) migrates farther and threatens down-gradient water supplies at the same sites where the migration of fuel constituents has either reached steady state (i.e., plume stabilization) or diminished by natural attenuation. 1,4-Dioxane released at methyl chloroform sites presents a similar dilemma. USEPA's Directive cautions that compounds such as MTBE and 1,4-dioxane that are not readily degradable in the subsurface and that may represent an actual or potential threat to the beneficial uses of groundwater should be assessed when evaluating the appropriateness of MNA remedies (USEPA, 1999).

9.2.1 Ratio Analysis of Methyl Chloroform Breakdown Using 1,4-Dioxane

When in use, methyl chloroform is usually stabilized against oxidative attack by and reaction with metals and acids; once released to the subsurface, it breaks down fairly quickly due to hydrolysis and biodegradation. Elimination reactions remove HCl from methyl chloroform to produce 1,1-dichloroethylene, and hydrolysis of methyl chloroform produces acetic acid. Reductive dechlorination transforms methyl chloroform to 1,1-dichloroethane (Vogel and McCarty, 1987a, 1987b). The hydrolysis of methyl chloroform to acetic acid occurs about five times more rapidly than the elimination reaction that produces 1,1-dichloroethylene (Vogel and McCarty, 1987b). Acetic acid (vinegar) is not a contaminant of interest and is not usually monitored because it is quickly mineralized by soil microbes. One can expect the detection of 1,1-dichloroethylene at concentrations of at least 5 µg/L where methyl chloroform has been in groundwater at concentrations greater than 120 µg/L for more than one year (Vogel and McCarty, 1987a).

The reported half-life of methyl chloroform in the elimination reaction has been variously cited as ranging from 1.7 to 3.8 years at 20°C, whereas the yields of the abiotic elimination and hydrolysis reactions of methyl chloroform range from 20% to 27% 1,1-dichloroethylene and 73% to 80% acetic acid, respectively (literature survey summarized in Wing, 1997). For each molecule of 1,1-dichloroethylene generated by the elimination of HCl from methyl chloroform, approximately three molecules of acetic acid are concurrently generated by the hydrolysis of methyl chloroform (Wing, 1997). The fact that 1,1-dichloroethylene is itself resistant to both hydrolysis and biodegradation permits its use in ratio

---

[*] As this book was going to press, Dr. Shaily Mahendra of UCLA reported evidence for biodegradation of 1,4-dioxane in arctic soils; findings will be presented at the Battelle Chlorinated Solvents Conference in Monterey, CA, in May 2010.

analysis to assess the rates of methyl chloroform breakdown. In a study that analyzed groundwater data collected over a 10-year period from a methyl chloroform release site, methyl chloroform exhibited first-order kinetics and a half-life of 2.9 years at 15°C for the abiotic transformation to 1,1-dichloroethylene and acetic acid (Wing, 1997). This reaction is described further in Section 9.2.2.

Methyl chloroform is less soluble and has a higher affinity for soil organic matter than 1,1-dichloroethylene; therefore, its migration is relatively retarded. Because 1,1-dichloroethylene is less prone to degradation, ratios of 1,1-dichloroethylene to methyl chloroform should increase with distance from the source zone. The superimposed effect of retardation may complicate the interpretation of methyl chloroform degradation. To determine the degree to which retardation hampers interpretation of degradation rates, McNab and Narasimhan (1994) modeled data from the Lawrence Livermore National Laboratory (LLNL) site by using a transport code that accounts for sequential decay chains. The modeling effort led to the conclusion that the effect of degradation of methyl chloroform to 1,1-dichloroethylene overwhelmed the effect of retardation at the LLNL site (McNab and Narasimhan, 1994). This conclusion suggests that sorption, dispersion, and matrix diffusion play only a minor role in the ratios of methyl chloroform to 1,1-dichloroethylene and so permit their use to interpret degradation rates. Their modeled methyl chloroform abiotic degradation rate (half-life of two years) agreed well with literature values. It may therefore be unnecessary to use 1,4-dioxane as a control on migration rates when using ratios to interpret abiotic degradation of methyl chloroform; however, 1,4-dioxane may still be a useful tracer for interpreting rates of reductive dechlorination of methyl chloroform to 1,1-dichloroethane.

9.2.2 USING RATIOS TO APPROXIMATE THE DATE OF A RELEASE

Ratios of 1,1-dichloroethylene to methyl chloroform can be used to infer the timing of a release when the release was episodic in nature. "Dating" a plume applies knowledge of the degradation half-life for the chemical and the source-zone concentration to estimate when the chemical was released into the subsurface (Morrison, 2006). The ratio approach to determining the timeframe of a methyl chloroform release requires knowledge of (1) the rate of chemical hydrolysis of methyl chloroform to acetic acid and (2) the rate of the elimination reaction to form 1,1-dichloroethylene based on laboratory data or field measurements. Several assumptions are also required (Gauthier and Murphy, 2003):[*]

- The rate of hydrolysis is constant over time.
- Because the hydrolysis of methyl chloroform exhibits first-order kinetics and is relatively independent of pH between 4 and 9, its rate depends only on temperature. Assuming a constant rate of hydrolysis implies an assumption of constant groundwater temperature.
- Biodegradation of methyl chloroform and 1,1-dichloroethylene is either unimportant or can be adequately modeled.
- Chromatographic separation effects such as differing volatilization rates or sorption rates do not skew the ratio of 1,1-dichloroethylene to methyl chloroform, that is, they are either negligible or can be modeled.

The rate of the elimination reaction is first order; therefore, the Arrhenius equation can be used to describe the temperature dependence of the rate of methyl chloroform degradation (Gauthier and Murphy, 2003). The general form of the Arrhenius equation is shown as follows:

$$k = Ae^{E_a RT},$$ (9.2)

---

[*] The ratio approach to establishing the timeframe of a release described here summarizes Gauthier and Murphy's (2003) original article in the *Journal of Environmental Forensics*, to which the reader is referred for a more complete treatment of the topic: "Age Dating Groundwater Plumes Based on the Ratio of 1,1-Dichloroethylene to 1,1,1-Trichloroethane: An Uncertainty Analysis," Thomas D. Gauthier and Brian L. Murphy, Exponent, Tampa, FL, USA. Used with permission.

where k is the first-order abiotic rate constant for methyl chloroform loss, A is the Arrhenius constant, E_a is the activation energy (in kJ/mol), R is the gas constant [8.3145×10^{-3} kJ/(mol K)], and T is the absolute temperature (in K). For the abiotic degradation of methyl chloroform, the rate constant k is the lumped sum of the reaction rates for the elimination reaction, k_e, and the hydrolysis or substitution reaction, k_s: $k = k_e + k_s$. The elapsed time of reaction corresponding to the ratio of the concentration of the elimination product, 1,1-dichloroethylene, to that of its parent, methyl chloroform, can be determined by applying the literature values of the reaction rates, as shown in Equation 9.3 (Gauthier and Murphy, 2003).

$$ t = \frac{1}{k} \ln \left(1 + \frac{k \left[1,1\text{-dichloroethylene} \right]}{k_e \left[\text{methyl chloroform} \right]} \right). \tag{9.3} $$

A "field-estimate" of the abiotic rate constant k can be derived by plotting time-series data as the natural logarithm of the molar concentration ratios from monitoring points. The natural logarithm of the molar concentration ratio in Equation 9.3 is plotted on the y-axis and time on the x-axis as dates or Julian days. A best-fit straight line then has a slope equal to the estimated k, and the x-intercept provides an estimate of the date at which hydrolysis began, that is, the approximate date when dissolved methyl chloroform was first subjected to abiotic degradation. This date does not account for the time since the actual spill, disposal, or tank leak, but reflects the date when methyl chloroform entered groundwater as a dissolved phase. Pure (nonaqueous phase) methyl chloroform does not participate in abiotic degradation reactions; however, condensed water vapor in vapor degreasers is likely to facilitate abiotic reactions that form 1,1-dichloroethylene. The molar ratios measured in down-gradient monitoring points within the plume can only represent a travel time from the source zone and not the date when the spill occurred (Gauthier and Murphy, 2003). In order for k to be determined by time-series data from a single well, that well must be at the plume front. If the methyl chloroform was released in the dissolved phase, for example, from wastewater discharge out of a vapor degreaser water trap or other wastewater, measurement of the ratio of 1,1-dichloroethylene to methyl chloroform at any location or time can be used to determine the date when hydrolysis began (Gauthier and Murphy, 2003).

In performing ratio analysis to discern the timeframe of a release, uncertainty due to variable rates of migration among the constituents should be considered. The primary transport parameter that may affect the molar ratio of 1,1-dichloroethylene to methyl chloroform is sorption, which is governed by the fraction of organic carbon in soil (f_{oc}) and the compound's organic carbon partition coefficient, K_{oc}. The K_{oc} for methyl chloroform is more than twice that of 1,1-dichloroethylene (152 and 65 mL/g, respectively; Pankow and Cherry, 1996). The degree to which retardation of methyl chloroform relative to 1,1-dichloroethylene affects the ratio depends on the f_{oc} and can be significant where f_{oc} is greater than 0.1%, as summarized in Table 9.8 (Gauthier and Murphy, 2003). At higher f_{oc} values, the migration rate of 1,1-dichloroethylene may substantially exceed that of methyl chloroform and so affect the molar concentration ratios. This possibility is seen by the measured and predicted range of volume concentration ratios observed with increasing distance from the methyl chloroform source, as shown in Figure 9.1.

Where biologic transformation of methyl chloroform is also evident from detection of 1,1-dichloroethane and chloroethane, Equation 9.4 can be used:

$$ t = \frac{1}{k_T} \ln \left(1 + \frac{k}{k_e} \left(\frac{\left[1,1\text{-dichloroethylene} \right]}{\left[\text{methyl chloroform} \right]} \right) + \frac{\left[1,1\text{-dichloroethane} \right]}{\left[\text{methyl chloroform} \right]} \right), \tag{9.4} $$

where the total rate constant $k_T = k + k_b$, and k_b is the biodegradation rate constant for the reductive dechlorination of methyl chloroform to 1,1-dichloroethane (Gauthier and Murphy, 2003).

TABLE 9.8
Retardation Coefficients and Relative Velocities as a Function of Fraction of Organic Carbon

Fraction of Organic Carbon, f_{oc} (%)	Retardation Coefficient,[a] R		Ratio of 1,1-Dichloroethylene Velocity to Methyl Chloroform Velocity
	Methyl Chloroform	1,1-Dichloroethylene	
0.01	1.07	1.03	1.04
0.02	1.14	1.06	1.08
0.05	1.35	1.15	1.17
0.1	1.69	1.30	1.31
0.2	2.39	1.59	1.50
0.5	4.47	2.49	1.80
1.0	7.95	3.97	2.00

Source: After Gauthier, T.D. and Murphy, B.L., 2003, *Environmental Forensics* 2003(4): 205–213.

[a] These calculations of retardation coefficient $R = 1 + (\rho_b K_d/\Theta_e)$ assume bulk soil density $\rho_b = 1.6$ g/cm³, water-filled porosity $\Theta = 0.35\%$, and partition coefficient $K_d = K_{oc} \times f_{oc}$. Because f_{oc} can vary within an aquifer and may be higher in fine-grained deposits where hydraulic conductivity is lower, its heterogeneous distribution may impart uncertainty to the analysis of ratios to estimate the timeframe of a release. Overall retardation can be estimated by comparing the distance that methyl chloroform and its breakdown products have migrated compared to that of 1,4-dioxane.

One case study provides a remarkable example of the application of ratio analysis to interpret the timeframe of a release. The estimate predicts the date of release to within one week of the actual date (Wing, 1997). On the morning of August 14, 1984, a pipe supplying a vapor degreaser at a precision medical instrument plant in Santa Clara, California, failed and released as much as 20 gallons of methyl chloroform. Three hundred cubic yards of contaminated soil were removed within

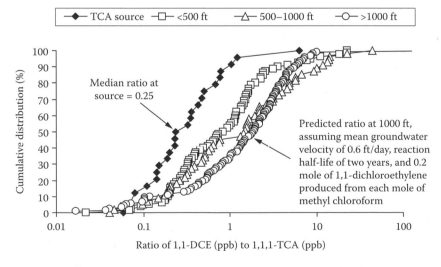

FIGURE 9.1 Field and modeled volume concentration ratios of methyl chloroform (1,1,1-TCA) and 1,1-dichloroethylene (1,1-DCE). Predicted volume concentration ratio is given at 1000-ft distance from plume source. The modeling assumes a mean velocity of 0.6 ft/day, reactant half-life of 2 years, and 0.2 mol of 1,1-dichloroethylene produced from each mole of methyl chloroform. (From McNab, W.W., et al., 1999, Historical case analysis of chlorinated volatile organic compound plumes. Lawrence Livermore National Laboratory report UCRL-AR-133361. With permission.)

a year of the spill. Methyl chloroform concentrations in groundwater were up to 11,000 µg/L; a pump-and-treat system removed 41 million gallons of contaminated water until 1994, when the system was shut down because cleanup goals had been attained at most monitoring wells. By 2003, methyl chloroform concentrations had decreased to 2.3 µg/L, with 5.1 µg/L 1,1-dichloroethylene and 4.7 µg/L 1,1-dichloroethane. The results of analyses for 1,4-dioxane at the site were nondetect at less than 2 µg/L in all monitoring wells, based on one-time sampling in 2002, 18 years after the initial release (Cal EPA, 2003).

Using data from monitoring wells placed within the 500-feet-long methyl chloroform plume at the Santa Clara site, Wing (1997) plotted time-series concentration data from two groundwater extraction trenches as the log of the molar ratios of the sum of 1,1-dichloroethylene and 1,1-dichloroethane to methyl chloroform against sample dates. Because no data for acetic acid was available, Wing applied the approximate kinetic ratio of 3:1 (molecules acetic acid to 1,1-dichloroethylene produced from methyl chloroform) and multiplied the molar concentrations of 1,1-dichloroethylene by four. The specific ratio plotted to interpolate the timeframe of release was (Wing, 1997)

$$\ln\left(\frac{4\left[1,1\text{-dichloroethylene}\right] + \left[1,1\text{-dichloroethane}\right]}{\left[\text{methyl chloroform}\right] + 1}\right).$$

Wing's analysis corroborated the finding by McNab and Narasimhan which showed that degradation effects overwhelm retardation effects. The slope of the linear regression line (correlation coefficient $r^2 = 0.81$) was 0.298 year^{-1}, from which an overall half-life of methyl chloroform of 2.3 years was obtained. To account for only the elimination reaction that transforms methyl chloroform to its field-measured degradation product, 1,1-dichloroethylene, Wing removed 1,1-dichloroethane and the factor of four that accounted for acetic acid to obtain a slope of 0.243 year^{-1}. The resulting half-life of methyl chloroform for the elimination reaction was determined to be 2.9 years (Wing, 1997).

The study of field data by Wing produced unexpectedly accurate results, providing encouragement for applying this technique where an episodic release is suspected (e.g., a plume of unknown origin that may have originated as a "midnight dumping" episode). However, caution is in order despite the high degree of accuracy obtained by Wing. Sources of error and uncertainty include hydrodynamic dispersion and molecular diffusion into fine-grained sediments, which can be expected to decline with distance from a continuous source (McNab and Dooher, 1998). The number, placement, and construction of monitoring wells may invalidate assumptions of a continuum between data points where anisotropy creates preferential pathways (e.g., paleochannels) affecting some monitoring wells and not others. Heterogeneity of hydraulic conductivity, redox conditions, and temperature may all affect the ratios used to interpret the timeframe of release. Seasonal variations in hydraulic conditions such as water level fluctuations or seasonal recharge causing variation in groundwater flow direction may also confound the application of ratios to estimating the timing of a release. To determine the extent to which these factors may be influencing the ratio data at a site, contrasting the concentrations of methyl chloroform and its degradation products to the concentrations of 1,4-dioxane where it has been established as a cocontaminant is recommended.

9.2.3 Modeling 1,4-Dioxane Migration to Approximate the Date of a Release

Two modeling studies used the 1,4-dioxane data set from the Gloucester Landfill (described in Section 3.4.2) to refine the methods of solving the inverse problem to estimate the probable timing of a 1,4-dioxane release to groundwater in a special waste unit. Laboratory wastes, presumably including scintillation wastes, were discharged to a trench and periodically burned. Wastes containing 1,4-dioxane leached from the trench to groundwater and migrated about 1600 ft down-gradient. The modeling studies sought to solve the linear inverse problem by using assumed values for contaminant transport parameters such as lateral and transverse dispersivity.

The first study by Woodbury et al. (1998) used the minimum relative entropy (MRE) method to recover the source-release history of the plume in three dimensions. The MRE method applies a probabilistic approach that treats parameters as random variables rather than as fixed deterministic values. Assessing the statistical properties of unknown model parameters is a significant challenge when using probabilistic models. Although most probabilistic studies usually assume Gaussian distributions, Woodbury et al. used the MRE approach to assign truncated exponential distributions where only the lower and upper bounds and some knowledge of the mean are given. The MRE approach supports a more robust reconstruction of the probability density function associated with the model parameters. Woodbury et al. applied hydrogeologic parameters discerned from detailed field studies conducted by Jackson et al. (1991) and inferred transport parameters from the distribution of 1,4-dioxane measured in a dense network of multilevel monitoring wells. The retardation factor assumed for 1,4-dioxane was 1.0, that is, 1,4-dioxane was assumed to move at the rate of the groundwater flow without retardation. The sample data used were from the early and late 1980s (Woodbury et al., 1998).

The recovered or modeled release history suggests a fairly narrow window of time during which 1,4-dioxane releases were likely to have occurred; two peaks were evident, coinciding with times of 8.4 and 9.7 years. The latter time corresponds to the year 1978, when approximately 1 ton of organic solvents was reported to have been disposed of within the special waste compound, whereas the former time estimate does not correspond to a known event (Woodbury et al., 1998).

The second modeling study of the Gloucester Landfill 1,4-dioxane data set used to refine techniques for back-dating the release was published by Michalak and Kitanidis in 2002. Their study used a geostatistical approach to inverse modeling, applying a Bayesian stochastic inference method to estimate the release history for 1,4-dioxane, and a novel approach to enforce nonnegative concentration values. The advantage of using a stochastic approach such as Bayesian inference is that it provides a best estimate together with the associated confidence intervals that represent the information contained in the available data (Michalak and Kitanidis, 2002). The conditional simulations performed by Michalak and Kitanidis generated 50,000 realizations, which were summarized to obtain a range of probable outcomes. The estimated timeframe of release successfully reproduced the observations to within the estimated measurement error variance, which accounted for both analytical measurement error and the limitations of the physical model (Michalak and Kitanidis, 2002). The modeling results indicated that the 1,4-dioxane release from the special waste compound continued after the end of landfill operations and identified the probable occurrence of multiple release periods spanning a 20-year timeframe.

9.2.4 COMPOUND-SPECIFIC ISOTOPE ANALYSIS

Compound-specific isotope analysis (CSIA) assesses the degree of isotopic enrichment or depletion that may occur during biotransformation, phase transitions, or chemical reactions involving a contaminant of interest. CSIA leverages variations in the naturally occurring stable isotopes of elements incorporated in the molecular structure of the contaminant under study, for example, carbon, hydrogen, and chlorine in TCE (C_2HCl_3). Carbon has six protons and six neutrons in 98.89% of all carbon atoms in nature, whereas its stable isotope, with six protons and seven neutrons, forms 1.11% of all carbon atoms. Hydrogen most often has one proton, but 0.015% of natural hydrogen atoms have one proton and one neutron (that isotope is called deuterium). Chlorine in its most abundant state has 17 protons and 18 neutrons, but 31.978% of chlorine atoms have 17 protons and 20 neutrons. Oxygen with eight protons and eight neutrons is most abundant; 0.204% of oxygen atoms have eight protons and 10 neutrons, and 0.037% of oxygen atoms have eight protons and nine neutrons.

Very sensitive isotope ratio mass spectrometers can measure the small differences in molecular mass to determine the ratios of the isotopes present, for example, the inclusion of heavier chlorine isotopes in a TCE molecule. Physical, chemical, or biological processes may favor heavier or lighter molecules and enrich or deplete the ratio of the rare to the most abundant isotope. The natural variations in the mass of the oxygen, hydrogen, and carbon atoms on the 1,4-dioxane molecule can potentially be used to interpret its fate in the subsurface, as discussed further in Section 9.2.5.

Ratios are referenced to an international standard material, such as Vienna Standard Mean Ocean Water for oxygen and hydrogen; the Peedee Belemnite (PDB), a South Carolina limestone formation, for carbon; and seawater for chlorine (Fritz and Clark, 1997). By convention, the ratio of two isotopes of the same element is described by δ notation and expressed in units of per mil (‰), as shown for carbon:

$$\delta^{13}C = \left(\frac{R_{sample}}{R_{std}} - 1 \right) \times 1000, \tag{9.5}$$

where R_{sample} is the $^{13}C/^{12}C$ ratio for a compound in a sample and R_{std} is the $^{13}C/^{12}C$ ratio for the international CO_2 standard (PDB) (Fritz and Clark, 1997). A detailed treatment of isotope hydrology and CSIA is beyond the scope of this book; the reader is directed to the leading texts and online resources on this subject.[*]

CSIA is accomplished with in-line gas chromatography-isotope ratio mass spectrometry (GC-IRMS). CSIA can be effective for differentiating solvents that have been through processes causing isotopic fractionation (i.e., enrichment or depletion, shifting the initial isotope ratio). Applications include characterizing, monitoring, and evaluating natural and engineered bioremediation (Sturchio et al., 1998). For example, the analysis of deuterium (^2H) in TCE has been used to distinguish two sources of TCE: the first due to biodegradation of PCE and the second due to a direct release of TCE. This differentiation is possible because the deuterium isotope ratio (δ^2H) in manufactured TCE is typically in the range of +466.9‰ to +681.9‰, whereas TCE that has been dechlorinated from PCE typically has a deuterium isotope ratio in the range of –351.9‰ to –320.0‰ (Shouakar-Stash et al., 2003; Morrison et al., 2006).

Depletion of heavy isotopes in the solvent can also be expected from vaporizing solvents such as will occur in a vapor degreaser or dry-cleaning operation. The enrichment of heavy isotopes in the vapor phase is due to the higher vapor pressure for heavier compounds. This "inverse isotope effect" is also attributed to a greater surface tension (i.e., intermolecular attraction) for lighter compounds. The smaller volume and heavier mass in the liquid phase confer greater motion energy for ^{13}C molecules in the liquid phase, which together with lower intermolecular cohesive forces, produces a higher volatility for the molecules with the heavy isotopes (Bouchard, 2007). Studies have also demonstrated a large enrichment of ^{13}C in the headspace above the liquid phase contained in closed vessels for chlorinated solvents (e.g., Slater et al., 1999; Hunkeler and Aravena, 2000).

CSIA has also been applied to differentiate the manufacturer of origin for chlorinated solvents. Isotopic differentiation between two grades of the same solvent produced at different manufacturers may occur because of variation in the method of solvent production, such as dehydrochlorination and dehydrogenation reactions, temperatures, and durations of the different stages of production, catalysts used, and isotopic variation in the original feedstock used to produce the solvent (Morrison and Murphy, 2006). The utility of this isotopic differentiation may be limited because carbon isotope ratios for the same solvent produced by the same manufacturer in different years have shown substantial variation and because feedstocks and methods of production may

---

[*] Texts: (1) Clark and Fritz (1997). (2) Immanuel Mazor, 1996, *Chemical and Isotopic Groundwater Hydrology*. New York: Marcel Dekker. (3) Pradeep K. Aggarwal, Joel R. Gat, and Klaus F.O. Froehlich, 2005, *Isotopes in the Water Cycle: Past, Present and Future of a Developing Science*. New York: Springer-Verlag.

Websites: (1) University of Arizona, "Sustainability of Semi-Arid Hydrology and Riparian Areas (SAHRA)": http://www.sahra.arizona.edu/programs/isotopes/index.html (2) International Atomic Energy Agency, "Environmental Isotopes in the Hydrological Cycle—Principles and Applications": http://www.iaea.org/programmes/ripc/ih/volumes/volumes.htm (3) R.J. Pirkle, 2006, "Compound Specific Isotope Analysis: The Science, Technology and Selected Examples from the Literature with Application to Fuel Oxygenates and Chlorinated Solvents": http://www.microseeps.com/pdf/csia.pdf

vary over time. A further limitation to using CSIA for solvent source identification is that once delivered to the site of use, the solvent may undergo additional isotopic depletion during vapor degreasing, dry cleaning, on-site recycling including distillation and carbon filtering, and evaporation during cold-cleaning operations. Carbon isotope ratios for solvents from different manufacturers in different years were analyzed by Shouakar-Stash et al. (2003), Van Wamerdam et al. (1995), and Beneteau et al. (1999) and compared in Morrison et al. (2006). Source identification applications of CSIA are best suited to the rare circumstance in which a sample of the original solvent waste is available. A source-zone sample may be a reasonable surrogate, particularly for studying biodegradation of solvents with CSIA.

One challenge to using CSIA has been the mass of the contaminant required on the gas chromatography column in GC-IRMS, which previously limited or even prevented CSIA from contaminant fate studies and the differentiation between sources of contaminants based on their isotopic signature. Because of the large mass required for CSIA, studies of *in situ* biotransformation of contaminants were previously restricted to source-zone samples at sites with high aqueous concentrations of contaminants to avoid collecting and processing very large volumes of samples in which the subject contaminant is present at low concentrations (Blessing et al., 2008). New methods such as SPME have been developed to improve the isotopic resolution of GC-IRMS on low-concentration samples (Zwank et al., 2003; Jochmann et al., 2006; Blessing et al., 2008). Currently, most CSIA laboratories use purge and trap as the GC-IRMS concentration method to achieve detection limits of $\delta^{13}C$ of about 1 ppb for chlorinated solvents and MTBE (Robert J. Pirkle, personal communication, Microseeps, Pittsburgh, Pennsylvania, 2008).

The limitations and uncertainty of CSIA for fingerprinting sources of chlorinated hydrocarbons may discourage its use in forensic investigations; however, it may nevertheless prove useful if approached in a stepwise fashion, beginning with a screening-level evaluation. To ascertain whether CSIA has the potential to be a useful tool for a particular investigation, the expected isotopic "end-members" of the system should be identified and sampled as a screening-level evaluation of the potential viability of CSIA. For example, where groundwater analyses show a sequence of geochemical conditions suggestive of a reducing environment and reductive dechlorination of chlorinated solvents, samples of the original solvent and its primary breakdown product(s) can be analyzed by using CSIA to confirm whether there is a sufficient shift in carbon, hydrogen, or chlorine isotope ratios to justify additional investigation. Similarly, if CSIA is to be useful for solvent-source identification for allocating cost in a commingled plume of the same solvent from two or more sources, sampling the different source areas and a few monitoring wells in the commingled plume should be the first step. If the source-area isotope ratios show a wide enough differentiation from the isotope ratios in the commingled plume, it may be possible to apply this method to establish mixing patterns. The range in values of carbon isotope ratios ($\delta^{13}C$) should be larger than the margin of analytical error, that is, typically 1.0–0.5‰ (Morrison et al., 2006).

One effort to confirm the biodegradation of TCE in a plume from a landfill in Switzerland produced the surprising result that the presence of *cis*-1,2-DCE was not the result of reductive dechlorination of TCE. Isotopic signatures showed the absence of *in situ* degradation of TCE, despite the presence of *cis*-1,2-DCE, a known metabolite of TCE (Zwank et al., 2003).

To constrain the uncertainty or ambiguity that a CSIA investigation may produce, investigators may include additional lines of evidence, such as the presence of proprietary solvent-stabilizer formulations, which together with the basic geochemical and hydrogeological framework may result in an improved conceptual model.

9.2.5 Compound-Specific Isotope Analysis of 1,4-Dioxane Biodegradation

CSIA has not yet been applied to assess the isotopic fractionation that may occur during the biodegradation of 1,4-dioxane. A grant application was filed with the National Science Foundation by Jennifer A. Field of Oregon State University to study CSIA of 1,4-dioxane biodegradation jointly with Barbara Sherwood Lollar of the University of Toronto; unfortunately, the grant was not funded.[*] The study would have pursued measurement of the degree of isotopic fractionation occurring in aerobic biodegradation of 1,4-dioxane (and anaerobic, if verifiable) following approaches used to apply CSIA to biodegradation of another ether, MTBE. The remainder of this discussion of application of CSIA to 1,4-dioxane is paraphrased from the Oregon State University CSIA study outline, in which the following approach was proposed:

1. Develop analytical methods for gas or liquid chromatography–isotope ratio mass spectrometry (GC-IRMS or LC-IRMS) for 1,4-dioxane and its expected biodegradation products.
2. Determine fractionation factors (the shifts in isotope ratio per fraction contaminant degraded) for both aerobic and anaerobic microcosms using pure and mixed cultures.
3. Apply the fractionation factors to field data to assess the extent to which the 1,4-dioxane has biodegraded and under what conditions (e.g., aerobic or anaerobic).
4. Develop protocols to ensure that analytical measurements of both concentrations and isotope values in the laboratory accurately reflect the status of field and microcosm samples.

Isotopic fractionation resulting from biodegradation follows the Rayleigh model, in which the extent of measured fractionation is proportional to the extent of biodegradation by the following relationship:

$$1000 \times \ln\left(\frac{\delta^{13}C_x + 1000}{\delta^{13}C_0 + 1000}\right) = 1000 \times (\alpha - 1) \ln F = \varepsilon \ln F, \qquad (9.6)$$

where $\delta^{13}C_x$ and $\delta^{13}C_0$ are the $^{13}C/^{12}C$ isotope ratios determined from microcosm experiments at time x and time 0, ε is the carbon isotope enrichment factor, α is the carbon fractionation factor, and F is the ratio of contaminant concentrations at time x and 0 (e.g., the fraction remaining). Carbon isotopic fractionation for nondegrading physical- or phase-transfer processes such as dissolution, vaporization, or sorption for *dissolved* TCE and PCE is not significant at equilibrium (i.e., <0.5‰). Therefore, shifts in contaminant isotopic ratios are useful as indicators of biodegradation (Slater et al., 2000).[†]

The fractionation factors (α) for each reaction can be calculated by posting the data for laboratory degradation experiments on a plot of $\ln F$ versus $1000 \times \ln(\delta^{13}C_x/\delta^{13}C_0)$. Through the use of least-squares regression, the fractionation factor, α, is determined as the slope of the plotted data. Researchers applying CSIA in biodegradation studies use fractionation enrichment factors from laboratory microcosm experiments and the kinetic isotope effect (KIE) described later in this section. Measurements of isotope ratios for contaminants in field samples, when combined with the laboratory-determined enrichment factors, permit the calculation of F, the ratio of contaminant remaining relative to the initial concentration, which is a direct measure of the extent of biodegradation of the contaminant.

Researchers have found that different microorganisms grown with different electron acceptors generated fractionation factors for chlorinated ethenes and for aromatic hydrocarbons in a relatively narrow range. This general agreement of fractionation factors for organisms of different origin or organisms using different electron acceptors suggests that different microbial populations in different environmental conditions degrade the same contaminants via similar mechanisms, which results in similar isotopic fractionation.

---

[*] Personal communication with Jennifer A. Field of the Department of Environmental and Molecular Toxicology, Oregon State University, December 2, 2005, regarding NSF Grant Proposal for 1,4-Dioxane CSIA Study.

[†] The negligible isotopic shift for dissolved chlorinated ethenes is in contrast to the expected enrichment in the vapor phase for these solvents in the vapor degreaser or dry-cleaner setting, where solvents are not in aqueous solution and a significant isotopic shift is expected.

Reaction rates for different isotopes vary because of mass-dependent differences in activation energies for the respective reactions. This variation is called the KIE, which can be used to predict the degree of both carbon and hydrogen isotopic fractionation during biodegradation. KIE describes the ratio of the reaction rates for the heavy, ^Hk, and light, ^Lk, isotopes

$$\text{KIE} = \frac{^Lk}{^Hk}. \tag{9.7}$$

Published KIE values from known reactions can be used to estimate the degree of fractionation for biodegradation of a compound whose KIE has not yet been measured (Jennifer A. Field, Oregon State University, personal communication, 2005).

KIE can be classified as primary (a large ratio, resulting from the atom being directly involved in the reaction) or secondary (a small ratio, resulting from the atom being bonded to the reactive center, but not immediately involved in the reaction). On the basis of KIE values for the hydroxylation reaction in biodegradation of other contaminants, only a moderate secondary carbon isotope effect (KIE = 1.015) is expected for 1,4-dioxane as the carbon atom is not immediately involved in the reaction. A substantial primary hydrogen isotope effect (KIE = 3 to 8) is expected as the hydrogen atom is thought to be directly involved in the reaction. Hydroxylation is proposed as the initial degradation step in one possible pathway for 1,4-dioxane degradation, in which hydrolysis to diethylene glycol occurs (see Figure 3.7) (Steffan et al., 2007). Other degradation pathways may be present and produce greater or lesser degrees of fractionation.

Given the established relationships between reaction pathway and KIE, it is possible to hypothesize reaction mechanisms from the degree of isotopic fractionation observed for a given atom. 1,4-Dioxane biodegradation may undergo an initial step of proton abstraction. If proton abstraction is involved, a substantial primary hydrogen isotope effect is expected with a theoretical KIE ranging from 3 to 8 (Elsner et al., 2005). The large difference in KIE between carbon and hydrogen results from the hydrogen atom undergoing a primary isotope effect whereas the carbon atom fractionation is a secondary isotope effect.

At sites contaminated with 1,4-dioxane from methyl chloroform releases, it is likely that 1,4-dioxane is present in the highly anoxic or anaerobic portions of the release near the source zone, at the oxygenated edges of the solvent plume, and at more distant locations where 1,4-dioxane has migrated beyond the solvent plume. It is therefore relevant to use CSIA to determine the biodegradability of 1,4-dioxane under both aerobic and anaerobic conditions (Field, 2005). Both pure and mixed cultures that possess the ability to degrade 1,4-dioxane aerobically and one study of anaerobic biodegradation are described in Chapter 7.

The rate of anaerobic biodegradation of 1,4-dioxane is expected to be slow because the anaerobic biodegradation of structurally similar compounds, such as MTBE, is slow (Kuder et al., 2005). Under anaerobic systems, the addition of proportionately larger concentrations of another biodegradable ether may be a viable strategy for promoting anaerobic cometabolism of 1,4-dioxane (Field, 2005). For example, water-soluble polyethylene glycol (PEG) is degraded by various bacteria in the absence of molecular oxygen (Yeh and Pavlostathis, 2005). The addition of PEG may stimulate "etherase" activity in the subsurface capable of cleaving the 1,4-dioxane ring (Field, 2005). Nakamiya et al. (2005) (see Chapter 7) reported that a filamentous fungus degrades 1,4-dioxane via etherase-type reactions to sequentially produce ethylene glycol, glycol-aldehyde, glycolic acid, and oxalic acid.

By combining isotope ratio measurements and biodegradation studies under aerobic and anaerobic conditions, CSIA methods could be used to assess *in situ* 1,4-dioxane biodegradation at enhanced bioremediation sites and MNA sites. This information would be highly useful for risk-based remediation decision making by facility owners, regulators, and the environmental engineering community. Numerous published studies have shown that CSIA is an excellent tool for providing the evidence of *in situ* biodegradation and to quantify the extent of *in situ* biodegradation of chlorinated solvents at field sites; it remains to be seen whether this will also be the case for 1,4-dioxane.

9.3 SOLVENT STABILIZERS AS AGE MARKERS

The date when a chemical first became commercially available is important for establishing the earliest possible date that a release could have occurred. Knowledge of the dates that chemicals were produced must be coupled with evidence that the facility used the chemical in question. An excellent compilation of dates of production and use patterns for chlorinated solvents is found in Richard Doherty's solvent history articles in the *Journal of Environmental Forensics* (Doherty, 2000a, 2000b). A brief summary of the history of solvent use is profiled in Section 1.1. A tabulation of the dates of introduction of the most widely used solvent-stabilizer packages for the major chlorinated solvents could be similarly useful. Section 1.2.6 reviews the available sources of information on solvent-stabilizer packages, and Tables 1.21 through 1.24 list the formulations. The dates listed in these tables unfortunately do not correspond to when the stabilizer was first used; rather, the dates refer to the date of publication of the citation for the inclusion of the listed stabilizer in the solvent.

In the case where there is documentation available for stabilizer use in a particular solvent, then the earliest patent claiming its use can be helpful for bracketing its introduction for commercial application. However, this information must also be integrated with the general pattern of commercial use for the major chlorinated solvents. For example, 1,4-dioxane was first patented as a stabilizer for methyl chloroform in 1957. But it was not until the early 1960s that methyl chloroform found somewhat wider use in cold-cleaning and vapor degreasing applications and not until the mid-1970s that it displaced TCE as the most widely used vapor degreasing solvent. In 1970, TCE accounted for 82% of all chlorinated solvents used in vapor degreasing (Morrison, 2006). By 1974, TCE accounted for only 46% of vapor degreasing solvent consumption (USEPA, 1977), and by 1976, TCE's share had declined to 42% (Morrison, 2006).

By the mid-1960s, the Navy and NASA were already using 1,4-dioxane-stabilized methyl chloroform. The Navy analyzed four commercially available methyl chloroform products and identified 16 compounds, including stabilizers and impurities. The report notes that most of the compounds (listed in Table 9.9) were previously found in submarine, spacecraft, and environmental test chamber atmospheres (Saunders, 1965). A chromatogram from the report (Figure 9.2) shows a distinct peak for 1,4-dioxane.

Similarly detailed composition profiles of technical grade methyl chloroform can be found in the toxicological literature in which scientists analyzed the composition of technical grade solvents to determine all of the toxic agents to which laboratory animals are exposed. For example, Table 9.10 lists the analysis of pre-1986 technical grade methyl chloroform from a study on the toxicity of methyl chloroform to male and female Sprague Dawley rat pups (Maltoni et al., 1986).

9.4 SOLVENT STABILIZERS AS MARKERS OF SOLVENT APPLICATIONS

Solvent stabilizers were tailored to prevent solvent breakdown in the demanding environments in which solvents were used. The use of certain stabilizers to prevent solvent-breakdown reactions that occur in unique applications—such as PCE used in phosphatizing baths for automotive coatings or high temperatures encountered by PCE used in transformer fluids—can provide unique associations that may allow the interpretation of solvent origins. The following circumstances occurring together would favor the use of solvent stabilizers as marker chemicals to distinguish the origins of the same solvent used in different applications:

- Stabilizer is unique to a specific application employed by only one of the potentially responsible parties assumed to have contributed to the solvent spill(s)
- Stabilizer is added in sufficient quantity to favor detection, or stabilizer is concentrated in the application so that the mass released is sufficient to favor detection
- Stabilizer that would serve as a unique marker is sufficiently mobile and persistent to favor its migration to the region in which plumes of the common solvent have commingled

TABLE 9.9

Stabilizers and Impurities in 1965 Technical Grade Methyl Chloroform from Four Suppliers

	Octagon Process Company (vol%)	Fisher Chemical Company (vol%)	Phillips–Jacobs Chemical Company (vol%)	Eastman Kodak Company (vol%)
Methyl chloroform (%)	65	94	95.5	95
Stabilizers				
1,4-Dioxane	—	3.6	3.4	4.5
1,2-Dimethoxymethane	0.24	—	—	—
tert-Butyl alcohol	3.64	—	—	—
Nitromethane	1.3	—	—	—
Diethyl ether	trace	—	—	—
sec-Butyl alcohol	—	0.2	0.2	0.3
Impurities				
Trichloroethylene	7.6	—	—	—
1,2-Dichloroethane	22	2.1	0.8	—
1,1-Dichloroethylene	0.06	Trace	Trace	Trace
1,1,2-Trichloroethane	Trace	—	—	—
1,2-Dichloroethylene	Trace	—	Trace	Trace
2-Chlorobutane	—	Trace	—	—

Source: After Saunders, R.A., 1965, *The Composition of Technical Grade Methyl Chloroform*. Washington, DC: Physical Chemistry Branch, U.S. Navy Research Laboratory.

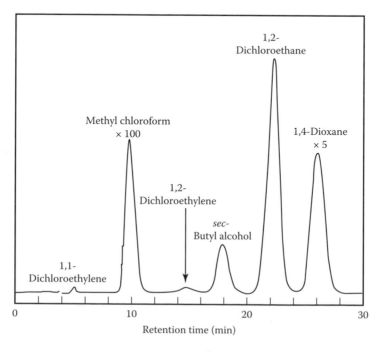

FIGURE 9.2 Chromatogram of 1965 methyl chloroform sold by Phillips–Jacobs Chemical. (From Saunders, R.A., 1965, *The Composition of Technical Grade Methyl Chloroform*. Washington, DC: Physical Chemistry Branch, U.S. Navy Research Laboratory.)

TABLE 9.10
Stabilizers and Impurities in Pre-1986 Technical Grade Methyl Chloroform

	Concentration
Stabilizers	
1,4-Dioxane	3.8 wt%
1,2-Epoxybutane	0.47 wt%
Nitromethane	0.27 wt%
n-Methyl pyrrole	<0.0001 wt%
Impurities	
Chloroform	100 ppm
Carbon tetrachloride	250 ppm
Trichloroethylene	200 ppm
1,2-Dichloroethane	2300 ppm
1,1-Dichloroethylene	398 ppm
1,2,3-Trichloroethane	41.8 ppm
trans-1,2-Dichloroethylene	50 ppm
Tetrachloroethylene	475 ppm

Source: After Maltoni, C., et al., 1986, *Acta Oncology* 7: 101–117.

- Analytical methods are available to detect the marker stabilizer
- Stabilizer compound intended to serve as a unique marker of the solvent is not commonly used in other applications that might be released in neighboring industrial operations

The combined occurrence of these circumstances is unlikely because stabilizers generally compose only a small percentage of the solvent mixture. The following examples profile the potential for solvent stabilizers to serve as marker chemicals.

9.4.1 PCE as a Dielectric Fluid

As described in Chapter 1, PCE is stable up to 150°C in the presence of air and moisture, and in the sealed environment of an electrical transformer, but in the absence of catalysts, PCE may be stable up to 500°C. The desired characteristics of a dielectric fluid include long-term resistance to decomposition at high temperatures in the presence of oxygen to prevent formation of electrically conductive or corrosive materials. PCE was used in transformers as a replacement for the more toxic class of dielectric fluids containing PCB compounds. Commercial grades of PCE may contain manufacturing impurities including any of the chlorinated ethanes at levels up to 0.3 wt% (Borror and Rowe, 1981). When PCE is used as a dielectric fluid in an electrical transformer, any chlorinated ethanes present can be expected to undergo dehydrochlorination, producing hydrochloric acid and causing corrosion. To function well as a dielectric fluid, PCE must have less than 50 ppm total chloroethanes (Borror and Rowe, 1981). In particular, dichloroethane, methyl chloroform, symmetrical tetrachloroethane, pentachloroethane, and hexachloroethane should not be present in PCE used as a dielectric fluid at masses totaling more than 10 ppm by weight, and unsymmetrical tetrachloroethane should not exceed 30 ppm by weight. Even small amounts of chloroethanes present as impurities will react with aluminum chloride or other catalysts and form acid, which will in turn begin to attack PCE. PCE destined to be used as a dielectric fluid was subjected to extra refining, including scrubbing and distillation, and was stabilized with *n*-methyl pyrrole and pentaphen (*p-tert*-amyl phenol).

Some transformers require a dielectric fluid that will vaporize at higher temperatures. PCE will react with oxygen at high temperatures, particularly in the vapor phase; therefore, the stabilizers added to PCE must be equally effective in the liquid and vapor phases. *N*-methyl pyrrole boils at 112°C (compared with 121°C for PCE) and will be carried into the vapor phase, whereas pentaphen boils at 266°C but has sufficiently high vapor pressure in the system to be partially carried into the vapor phase at the boiling point of PCE (Borror and Rowe, 1981).

N-methyl pyrrole is not unique to the stabilization of PCE in transformers; it is also used to stabilize methyl chloroform and TCE, as described in Chapter 1. Similarly, pentaphen is also used to stabilize TCE against oxidation. Neither stabilizer provides a unique marker if TCE was among the solvents released at a site that also had a PCE release from a transformer. However, if the task is to differentiate PCE from a transformer versus PCE from a dry cleaner, then there may be some potential to use these stabilizers as marker chemicals. Dry-cleaning grades of PCE apparently did not use *n*-methyl pyrrole or pentaphen. The patent literature includes citations of commercial PCE containing 0.022–0.028% *n*-methyl pyrrole as an antioxidant (Copelin, 1959); however, citations referring to pyrrole as a stabilizer of PCE are presented in the vapor degreasing/metal-cleaning context and not in the dry-cleaning context.

Dow Chemical's dry-cleaning grade of PCE, DowPer, used 4-methyl morpholine, Vulcan's PerSec used diallylamine and tripropylene compounds, and PPG's dry-cleaning grade used cyclohexene oxide and β-ethoxypropionitrile. A stabilizer concentrate for PPG's dry-cleaning grade of PCE adds cyclohexene oxide, β-ethoxypropionitrile, *n*-methyl morpholine, and 4-methoxyphenol (see Table 1.21).

The comparative fate of these stabilizers is fundamental to establishing their utility as marker chemicals. Table 9.11 contrasts the fate and transport properties of stabilizers of dry-cleaning versus transformer grades of PCE. The fate and transport properties rule out some stabilizers as marker chemicals; for example, cyclohexene oxide has low solubility and hydrolyzes easily, and both diallylamine and 4-methyl morpholine are susceptible to sorption in the ionized form they assume once dissociated in water. In the BIOWIN GCM, all of the PCE stabilizers are estimated to be easily biodegradable in a matter of weeks. As discussed in Chapter 3, BIOWIN estimates are less reliable than some of the other estimation tools in the EPIWIN suite, and field evidence suggests that at least one of the compounds in Table 9.11 is resistant to biodegradation because it has been detected years after the probable date of release.

Butoxymethyl oxirane, also called *n*-butyl glycidyl ether or *n*-BGE, was found in soils at a PCE release in Mountain View, California. The *n*-BGE was detected in soils adjacent to a chemical packaging facility that packaged paint thinners and related products, but never handled PCE. A PCE plume of unknown origin was identified adjacent to the property, and in soil samples retrieved beneath a stormwater drainage ditch, where it occurs together with *n*-BGE. The adjacent property housed a printing operation, which may have used PCE to clean printing presses and could be a source of PCE; the stormwater drainage ditch drains from the printing facility toward the chemical packaging facility. Sewer lines also traverse the area and collect wastewater from an area that included several former dry-cleaning operations. A complicating factor for interpreting the presence of *n*-BGE collocated with PCE is its possible presence in printing inks. Butoxymethyl oxirane is used as an epoxy resin diluent in some ink formulations (NIEHS, 2004). The detection of butoxymethyl oxirane did not factor into the determination of responsibility for the PCE plume.

For the hypothetical situation in which PCE was released from both a dry cleaner and an electrical transformer, the fate and transport properties suggest that it may be possible to track the movement of the transformer solvent with *n*-methyl pyrrole and the dry-cleaner solvent with *n*-butyl glycidyl ether. The scenario used for this example would be uncommon if it has occurred at all, so of course no field data are available to test the viability of leveraging these stabilizers to differentiate commingled PCE plumes. Nevertheless, there is a reasonably good potential that some combinations

TABLE 9.11
Comparison of Stabilizer Fate and Transport Properties for Electrical Transformer and Dry-Cleaning Grades of Perchloroethylene

Stabilizer	Solubility (mg/L)	log K_{oc}	Hydrolysis Potential	Acid Dissociation Constant[a] (pK_a) at pH 7	BIOWIN-Estimated Biodegradation Potential	Subsurface Mobility
Stabilizers of dielectric fluid grade perchloroethylene for transformers						
p-tert-Amyl phenol (pentaphen)	168	2.41	Low	0.00167%	Weeks to months	Minimal
n-Methyl pyrrole	71,900	2.16[b]	—	−2.9	Weeks to months	High
Stabilizers of dry-cleaning grade perchloroethylene						
4-Methyl morpholine	*Miscible*	*0.719*	—	8.49, 7.38	Weeks to months	High
Diallylamine	86,000	1.11	—	9.29	Weeks	High
Tripropylene compounds[c]	*196,000*	1	—	NA	Weeks to months	High
Cyclohexene oxide	*4530*	*1.2*	High	NA	Weeks	Minimal
3-Methoxy propionitrile	*732,000*	*0.186*	—	NA	Weeks	High
n-Butyl glycidyl ether	20,000	*0.451*	Very low	4.89	Weeks	High
Perchloroethylene[d]	150	2.32	—	—	Months	Moderate

Notes: See Chapter 3 for description of these parameters and BIOWIN estimates.

[a] A high pK_a indicates that the compound may exist in a protonated form in the environment; cations generally adsorb to organic carbon and clay more strongly, thereby reducing mobility compared to that of neutral compounds.

[b] Italicized values were estimated by using EPIWIN.

[c] "Tripropylene compounds" are assumed to be tripropylene glycol monomethyl ether, also known as DOWANOL® TPM glycol ether and ARCOSOLV® TPM; tripropylene glycol-mono-n-propyl ether, also known as DOWANOL® TPnPᵣ or tripropylene glycol mono-n-butyl ether, also known as ARCOSOLV® TPNB or DOWANOL® TPNB; listed values are for tripropylene glycol monomethyl ether.

[d] Mass fraction = 99%.

of stabilizers and solvents in addition to 1,4-dioxane and methyl chloroform will prove useful for distinguishing the origins of commingled plumes.

BIBLIOGRAPHY

ASTM, 1962, *Handbook of Vapor Degreasing*, American Society for Testing and Materials Special Technical Publication No. 310.

ASTM, 2006, Standard test methods for determination of 1,1,1-trichloroethane and methylene chloride content in stabilized TCE and tetrachloroethylene. D5320-96 (2000) (Withdrawn 2004; replaced by D6806), American Society for Testing and Materials.

Beneteau, K.M., Aravena, R., and Frape, S.K., 1999, Isotopic characterization of chlorinated solvents—laboratory and field results. *Organic Geochemistry* 30(8 Part 1): 739–753.

Blessing, M., Jochmann, M.A., Haderlein, S.B., and Schmidt, T.C., 2008, Compound-specific isotope analysis at trace concentrations: Evaluation of a large volume injection (LVI) based method. *Geophysical Research Abstracts* 10(EGU2008-A-08981).

Bohnert, G.W. and Carey, D.A., 1991, Scale-up of recovery process for waste solvents. Technical Communications. Kansas City, Missouri: Allied Signal Aerospace Company.

Borror, J.A. and Rowe, E.A.J., 1981, United States Patent 4,293,433: PCE dielectric fluid containing pyrrole and phenol. Diamond Shamrock Corporation, Dallas, TX.

Bouchard, D., 2007, Use of stable isotope analysis to assess biodegradation of volatile organic compounds in the unsaturated subsurface. PhD dissertation. Centre D'Hydrogéologie. Neuchâtel, University of Neuchâtel, Switzerland.

Bursack, K.F., Rains, J.H., and Bussard, P.A., 1966, British Patent 1,042,076: Stabilization of methyl chloroform. Assignee: Vulcan Materials Company, Birmingham, AL.

California Environmental Protection Agency (Cal EPA), 2003, Letter from San Francisco Bay Regional Water Quality Control Board to Varian Medical Systems: Request for Site Closure at 3100 Jay Street, Santa Clara, CA, San Francisco Bay Regional Water Quality Control Board, May 5th.

Chickos, J.S. and Acree, W.E., Jr., 2003, Enthalpies of vaporization of organic and organometallic compounds, 1880–2002. *Journal of Physical and Chemical Reference Data* 32(2): 519–878.

Copelin, H.B., 1959, United States Patent 2,904,600: Stabilization of chlorinated hydrocarbons. Assignee: E.I. du Pont de Nemours and Company, Wilmington, DE.

Doherty, R.E., 2000a, A history of the production and use of carbon tetrachloride, tetrachloroethylene, TCE and 1,1,1-trichloroethane in the United States: Part 1–Historical background; carbon tetrachloride and tetrachloroethylene. *Journal of Environmental Forensics* 2000(1): 69–81.

Doherty, R.E., 2000b, A history of the production and use of carbon tetrachloride, tetrachloroethylene, TCE and 1,1,1-trichloroethane in the United States: Part 2–TCE and 1,1,1-trichloroethane. *Journal of Environmental Forensics* 2000(1): 83–93.

Dow Chemical Company, 2006, Chlorinated solvents–physical properties. Form No. 100-06358. http://www.chlorinatedsolvents.co (accessed May 29, 2006).

Elsner, M., Zwank, L., Hunkeler, D., and Schwarzenbach, R.P., 2005, A new concept linking observable stable isotope fractionation to transformation pathways of organic pollutants. *Environmental Science and Technology* 39(18): 6896–6916.

European Chemicals Bureau, 2004, European Union risk assessment report: Trichloroethylene. 348. Environment Agency Chemicals Assessment Section, Ecotoxicology and Hazardous Substances National Centre, Isis House, Howbery Park Wallingford, Oxfordshire. European Commission Joint Research Centre, Institute for Health and Consumer Protection, European Chemicals Bureau.

Fritz, I.D. and Clark, P., 1997, *Environmental Isotopes in Hydrogeology*. Boca Raton, Florida: CRC Press.

Gauthier, T.D. and Murphy, B.L., 2003, Age dating groundwater plumes based on the ratio of 1,1-dichloroethylene to 1,1,1-trichloroethane: An uncertainty analysis. *Environmental Forensics* 2003(4): 205–213.

Henschler, D., Eder, E., Neudecker, T., and Metzler, M., 1977, Carcinogenicity of TCE: Fact or artifact? *Archives of Toxicology* 1977(37): 233–236.

Henschler, D., Reichert, D., and Metzler, M., 1980, Identification of potential carcinogens in technical grade 1,1,1-trichloroethane. *International Archives of Occupational and Environmental Health* 47(3): 263–268.

Holt, R.D., 1990, Physical properties of contaminated TCE and 1,1,1-trichloroethane. Technical Communications. Kansas City, Missouri: Allied Signal Aerospace Company.

Hunkeler, D. and Aravena, R., 2000, Determination of stable carbon isotope ratios of chlorinated methanes, ethanes and ethenes in aqueous samples. *Environmental Science and Technology* 34(13): 2839–2844.

INEL, 2006, Mass estimate of organic compounds in 743-series waste buried in the subsurface disposal area for operable units 7-08 and 7-13/14. U.S. Department of Energy, Idaho National Engineering and Environmental Laboratory. Idaho cleanup project report 23256.

International Agency for Research on Cancer (IARC), 1986, Monographs on the evaluation of the carcinogenic risk of chemicals to humans, Volume 41: Some halogenated hydrocarbons and pesticide exposures. International Agency for Research on Cancer, Lyon, France.

Jackson, R.E., Lesage, S., Priddle, M.W., Crowe, A.S., and Shikaze, S., 1991, *Contaminant Hydrogeology of Toxic Organic Chemicals at A Disposal Site, Gloucester, Ontario: 2. Remedial Investigation.* Canada, National Water Research Institute, Burlington, Ontario.

Jochmann, M.A., Blessing, M., Haderlein, S.B., and Schmidt, T.C., 2006, A new approach to determine method detection limits for compound-specific isotope analysis of volatile organic compounds. *Rapid Communications in Mass Spectrometry* 20(24): 3639–3648.

Kuder, T., Wilson, J.T., Kaiser, P., Kolhatkar, R., Philp, P., and Allen, J., 2005, Enrichment of stable carbon and hydrogen isotopes during anaerobic biodegradation of MTBE: Microcosm and field evidence. *Environmental Science and Technology* 39(1): 213–220.

Lane, V. and Smith, J., 2006, Fact or fiction? The source of perchloroethylene contamination in groundwater is a manufacturing impurity in chlorinated solvents. Presented to the National Groundwater Association Ground Water and Environmental Law Conference, July 7, 2006. Chicago, Illinois: National Groundwater Association. http://www.ngwa.org (accessed August 2, 2006).

Lyman, W.J., Reehl, W.F., and Rosenblatt, D.H. (Eds), 1990, *Handbook of Chemical Property Estimation Methods: Environmental Behavior of Organic Compounds.* Washington, DC: American Chemical Society.

Maltoni, C., Cotti, G., and Patella, V., 1986, Results of long-term carcinogenicity bioassays on Sprague-Dawley rats of methyl chloroform administered by ingestion. *Acta Oncology* 7: 101–117.

McNab, W.W. and Dooher, B.P., 1998, A critique of a steady-state analytical method for estimating contaminant degradation rates. *Ground Water* 36(6): 983–987.

McNab, W.W. and Narasimhan, T.S., 1994, Degradation of chlorinated hydrocarbons and groundwater geochemistry: A field study. *Environmental Science and Technology* 28(5): 769–778.

McNab, W.W., Rice, D.W., Bear, J., Ragaini, R., Tuckfield, C., and Oldenburg, C., 1999, Historical case analysis of chlorinated volatile organic compound plumes. Lawrence Livermore National Laboratory report UCRL-AR-133361.

Michalak, A.M. and Kitanidis, P.K., 2002, Application of Bayesian inference methods to inverse modeling for contaminant source identification at Gloucester Landfill, Canada. XIVth International Conference on Computational Methods in Water Resources, June 23–28, 2002, Delft, Netherlands: Elsevier.

Morrison, R.D., 2006, Forensic techniques for establishing the origin and timing of a contaminant release, Environmental Data Pages. http://www.techstuff.com/bmorrep.htm (accessed July 29, 2006).

Morrison, R.D. and Murphy, B.L., 2006, *Environmental Forensics—Contaminant Specific Guide.* San Diego: Elsevier.

Morrison, R.D., Murphy, B.L., and Doherty, R.E., 2006, Chlorinated solvents. In: R.D. Morrison and B.L. Murphy (Eds), *Environmental Forensics—Contaminant Specific Guide.* San Diego: Elsevier.

Murphy, B.L. and Morrison, R.D., 2007, *Introduction to Environmental Forensics.* San Diego: Elsevier Academic Press.

National Institute of Environmental Health Sciences (NIEHS), 2004, *n*-Butyl glycidyl ether (BGE)—[CAS No. 2426-08-6]: Review of toxicological literature. National Toxicology Program (NTP), National Institute of Environmental Health Sciences (NIEHS), Research Triangle Park, North Carolina.

Pankow, J.F. and Cherry, J.A., 1996, *Dense Chlorinated Solvents and Other DNAPLs in Groundwater: History, Behavior, and Remediation.* Portland, Oregon: Waterloo Press.

Parsons, F., Wood, P.R., and DeMarco, J., 1984, Transformations of tetrachloroethene and trichloroethene in microcosms and groundwater. *Journal of the American Water Works Association* 76(2): 56–59.

Pittsburgh Plate and Glass, 1998, Material Safety Data Sheet for tri-ethane 348, issued March 2, 1998, Pittsburgh, Pennsylvania.

Saunders, R.A., 1965, *The Composition of Technical Grade Methyl Chloroform.* Washington, DC: Physical Chemistry Branch, U.S. Navy Research Laboratory.

Shepherd, C.B., 1962, Perchloroethylene and trichloroethylene. In: S.J. Sconce (Ed.), *Chlorine: Its Manufacture, Properties and Uses*, American Chemical Society Monograph Series, No. 154, Chapter 13. American Chemical Society, Washington, DC.

Shouakar-Stash, O., Frape, S., and Drimmie, R., 2003, Stable hydrogen, carbon, and chlorine isotope measurements of selected organic solvents. *Journal of Contaminant Hydrology*, 30(3–4): 211–228.

Shuckrow, A.J., Pajak, A.P., and Touhill, C.J., 1982, *Hazardous Waste Leachate Management Manual*. Park Ridge, NJ: Noyes Data Corporation.

Slater, G.F., Ahad, J., Sherwood-Lollar, B., Allen-King, R., and Sleep, B., 2000, Carbon isotope effects resulting from equilibrium sorption of dissolved VOCs. *Analytical Chemistry* 72(22): 5669–5672.

Slater, G.F., Dempster, H.S., Sherwood-Lollar, B., and Ahad, J., 1999, Headspace analysis: A new application for isotopic characterization of dissolved organic contaminants. *Environmental Science and Technology* 33(1): 190–194.

Spencer, D.R. and Archer, W.L., 1981, British Patent 1,582,803: Stabilization of 1,1,1-trichloroethane. Assignee: Dow Chemical Company, Midland, MI.

Steffan, R.J., McClay, K.R., Masuda, H., and Zylstra, G.J., 2007, *Biodegradation of 1,4-dioxane*. Arlington, Virginia: SERDP—Strategic Environmental Research and Development Program.

Stewart, R.D., Gay, H.H., Schaffer, A.W., Erley, D.S., and Rowe, V.K., 1969, Experimental human exposure to methyl chloroform vapor. *Archives of Environmental Health* 19(4): 467–472.

Sturchio, N.C., Clausen, J.C., Heraty, L.J., Huang, L., Holt, B.D., and Abrajano, T., 1998, Stable chlorine isotope investigation of natural attenuation of trichloroethene in an aerobic aquifer. *Environmental Science and Technology* 32(20): 3037–3042.

Tarrer, A.R., Donahue, B.A., Dhamavaram, S., and Joshi, S.B., 1989, *Reclamation and Reprocessing of Spent Solvent*. Park Ridge, NJ: Noyes Data Corporation.

Tarrer, A.R., Sanjay, H.J., and Howell, G.S., 1993, Minimizing pollution in cleaning and degreasing operations—Armstrong Laboratory, Tyndall Air Force Base, Florida. U.S. Environmental Protection Agency, Cincinnati, Ohio. EPA/600/R-93/191.

Totonidis, S., 2005, A role for TCE in developing nation anaesthesia. *Kathmandu University Medical Journal* 3(2): 181–190.

Tsuruta, H., Iwasaki, K., and Fukuda, K., 1983, Analysis of trace impurities in reagent and technical grade TCE. *Industrial Health* 21(4): 293–295.

United Nations Environment Programme (UNEP), 2005, *Draft technical guidelines for the environmentally sound management of wastes consisting of, containing, or contaminated with hexachlorobenzene (HCB)*. Open-ended Working Group of the Basel Convention on the Control of Transboundary Movements of Hazardous Wastes and Their Disposal, Nairobi, Kenya. United Nations Environment Programme.

U.S. Environmental Protection Agency (USEPA), 1977, Control of volatile organic emissions from solvent metal cleaning. Research Triangle Park, North Carolina, U.S. Environmental Protection Agency, Office of Air Quality Protection Standards.

U.S. Environmental Protection Agency (USEPA), 1999, Use of monitored natural attenuation at Superfund, RCRA corrective action, and underground storage tank sites. U.S. Environmental Protection Agency, Office of Solid Waste and Emergency Response. Directive 9200.4-17P.

Van Wamerdam, E.M., Frape, S.K., Aravena, R., Drimmie, R.J., Flatt, H., and Cherry, J.A., 1995, Stable chlorine and carbon isotope measurements of selected chlorinated organic solvents. *Applied Geochemistry* 10(5): 547–552.

Vogel, T.M. and McCarty, P.L., 1987a, Abiotic and biotic transformations of 1,1,1-trichloroethane under methanogenic conditions. *Environmental Science and Technology* 21(12): 1208–1213.

Vogel, T.M. and McCarty, P.L., 1987b, Rate of abiotic formation of 1,1-dichloroethylene from 1,1,1-TCE in groundwater. *Journal of Contaminant Hydrology* 1(3): 299–308.

Willis, G.C. and Christian, C.A., 1957, United States Patent 2,803,676: TCE stabilized with propargyl alcohol and pyrrole. Assignee: Dow Chemical Company, Midland, MI.

Wing, M.R., 1997, Apparent first-order kinetics in the transformation of 1,1,1-trichloroethane in groundwater following a transient release. *Chemosphere* 34(3): 771–781.

Woodbury, A., Sudicky, E., Ulrych, T.J., and Ludwig, R., 1998, Three-dimensional plume source reconstruction using minimum relative entropy inversion. *Journal of Contaminant Hydrology* 1998(32): 131–158.

Yeh, D. and Pavlostathis, S., 2005, Anaerobic biodegradability of tween surfactants used as a carbon source for the microbial reductive dechlorination of hexachlorobenzene. *Water Science and Technology* 52: 343–349.

Zwank, L., Berg, M., Schmidt, T.C., and Haderlein, S.B., 2003, Compound-specific carbon isotope analysis of volatile organic compounds in the low-microgram per liter range. *Analytical Chemistry* 75(20): 5575–5583.

10 Regulatory Policy Implications of 1,4-Dioxane

Thomas K.G. Mohr

The growing number of groundwater contamination sites where 1,4-dioxane is present leads to several important policy questions on the regulation of 1,4-dioxane in drinking water and at cleanup sites, as the preceding chapters demonstrate. Foremost among these questions is whether other states and the federal government should follow the lead set by the CDPHE and establish regulatory standards for 1,4-dioxane. The question of setting standards depends on the ongoing debate about 1,4-dioxane's toxicity, which centers on the nonlinear dose–response for 1,4-dioxane's toxicity and carcinogenicity endpoints. Do the toxicological assays of 1,4-dioxane adequately determine carcinogenicity? In the face of toxicological uncertainty, should we apply the precautionary principle and regulate 1,4-dioxane in case it is later confirmed to be a human carcinogen? Or should we be cautious about committing scarce financial resources to remediate contamination whose potential to cause harm is still being evaluated? Among the criteria used to set drinking water standards is the estimated population exposed to 1,4-dioxane in drinking water, which requires solid data for 1,4-dioxane occurrence in public drinking water systems. This in turn leads to questions on the policies applied by federal and state agencies to administer drinking water testing requirements that best serve the public for emerging contaminants such as 1,4-dioxane. When water utilities discover 1,4-dioxane in drinking water systems, they face the difficult challenge of risk communication, that is to inform their customers while preserving their reputation for reliably serving high-quality water and protecting their legal interests.

At sites contaminated with chlorinated solvents, the late discovery of 1,4-dioxane poses a number of regulatory policy questions. Should the regulator reopen the Record of Decision or other regulatory order to add cleanup goals for 1,4-dioxane? Does the presence of 1,4-dioxane, a recalcitrant and mobile contaminant, negate the premise for previously approved chlorinated solvents cleanup plans that employ monitored natural attenuation (MNA)? Without successful methods for *in situ* remediation of 1,4-dioxane, pumping and *ex situ* treatment or in-well treatment are the primary options. Do the benefits of removing 1,4-dioxane from contaminated groundwater via high-volume pumping outweigh the environmental impacts of energy consumption and associated greenhouse gas emissions?

Finally, 1,4-dioxane is subject to some of the same regulatory policies as other chemicals put into widespread use to solve one problem while unintentionally causing another. The fuel oxygenate MTBE[*] is a prime example, perchlorate in road flares is another. When should the community of university and research scientists, consulting and industry professionals, and water resources managers have known that 1,4-dioxane would cause widespread groundwater contamination? Did we miss the signs or are we now applying 20-20 hindsight? Why do we find ourselves, once again, "blindsided"? The European Green Chemistry program, REACH (Registration, Evaluation & Authorization of Chemicals), is intended to prevent the widespread deployment of environmentally harmful chemicals. Can regulations succeed in preventing the next 1,4-dioxane-like contaminant from damaging irreplaceable groundwater resources?

---

[*] MTBE is methyl *tert*-butyl ether, the fuel oxygenate.

This chapter provides a roadmap of the regulatory landscape defining the issues raised by 1,4-dioxane contamination of water resources and a compass to anticipate the path through the wilderness of uncodified but nonetheless practiced regulatory policies.

10.1 APPLYING THE PRECAUTIONARY PRINCIPLE TO REGULATION OF 1,4-DIOXANE AND EMERGING CONTAMINANTS IN DRINKING WATER

"First, do no harm." "An ounce of prevention is worth a pound of cure." "Look before you leap." "Better safe than sorry." "Measure twice, cut once."

Was Mom right? These seemingly trite platitudes belie the wisdom they hold for averting unwanted environmental consequences, which is framed as the precautionary principle. The precautionary principle posits that when credible evidence of harm or potential harm exists and the risk is significant, authorities should take action even if some scientific uncertainty remains. The precautionary principle concept has its roots in German environmental law developed in the early 1970s, when German scientists and policy-makers grappled with "forest death." Suspecting air pollution as the cause, yet not fully convinced with scientific certainty, they developed the *Vorsorgenprinzip*, which translates to "fore caring" or "precautionary principle" and generated the German Clean Air Act of 1974 to combat acid rain and other environmental problems (McKenna and Sylvester, 2004; European Environment Agency [EEA], 2000).

Proponents of applying the precautionary principle advocate that when faced with potentially significant risk, authorities should take action even if some cause and effect relationships are not fully established. Proponents further recommend that the process of applying the precautionary principle must be open, informed, and democratic and must include parties potentially affected by the environmental consequence at issue (Global Development Research Center [GDRC], 1998). The EEA's inspired report, *Late lessons from early warnings: the precautionary principle 1896–2000* (EEA, 2000), profiles a number of examples of chemicals put into widespread service and later discovered to be harmful to human health or the environment. The examples teach important lessons, most of which are relevant to emerging contaminants such as 1,4-dioxane. The regulation of new chemicals entering the marketplace by the European Union's REACH program and California's 2008 *Green Chemistry Initiative* as well as USEPA's 1998 voluntary *High Production Volume Challenge* program, as described in Section 10.8.3, are intended to implement some of these recommendations. The EEA's recommendations regarding the application of the precautionary principle to chemicals such as ozone-depleting compounds (ODCs) and the fuel oxygenate MTBE are listed below, followed by commentary on their relevance to 1,4-dioxane:[*]

1. *Acknowledge and respond to ignorance, as well as uncertainty and risk, in technology appraisal and public policymaking.*
   Comment: In the early years of developing solvent stabilizer compounds, ignorance of the consequences of adding toxic and persistent compounds to chlorinated solvents prevailed. Indeed, the environmental consequences of the chlorinated solvents themselves were not well-appreciated until worker exposure and smog issues arose in the 1960s; however, that was not enough to motivate water quality protection with cradle-to-grave handling of solvents as hazardous materials and hazardous wastes. This protection came only with the adoption of the RCRA in 1976. The first RCRA regulations were published in 1980 and implemented in the mid-1980s. Nevertheless, there were pockets of awareness of potential health consequences of solvent stabilizers, particularly in relation to occupational exposure. For example, producers of TCE stopped adding epichlorohydrin as a stabilizer due to its toxicity. As early as 1968, chemical distributors marketed laboratory chemicals for scintillation counting as "dioxane-free" to avoid health problems associated with

---

[*] The report lists 12 recommendations, seven of which are profiled here. The other four are less relevant to 1,4-dioxane.

1,4-dioxane in Bray's solution (Beckman Instruments, 1968). The realization that both TCE and its stabilizer epichlorohydrin had the potential to cause adverse health consequences in the workplace apparently did not extend to evaluating whether solvents and their stabilizers would cause the degradation of groundwater resources. These late discoveries of health effects after chemicals were put into widespread use were not met with a comprehensive response to address the uncertainty in health effects and risks.

2. *Provide adequate long-term environmental and health monitoring and research into early warnings.*

Comment: The occupational health and exposure literature provided evidence of the lethal consequences of exposure to very high concentrations of 1,4-dioxane well before 1,4-dioxane was employed as a solvent stabilizer (Barber, 1934). By the time that 1,4-dioxane production was substantially increased to keep up with demand for stabilizing methyl chloroform in the 1970s, the toxicology literature on 1,4-dioxane had enough studies to warrant both health monitoring and further research into the potentially toxic effects of exposure. However, at that time, governmental agencies had not sufficiently developed the legal and administrative framework needed to establish long-term monitoring programs.

3. *Identify and work to reduce "blind spots" and gaps in scientific knowledge.*

Comment: A blind spot in the case of 1,4-dioxane introduction to the chemical marketplace as a solvent stabilizer was the lack of recognition of the fact that 1,4-dioxane becomes concentrated in vapor degreasing operations such that larger quantities of 1,4-dioxane are included in solvent wastes than are present in the original solvent formulation. In addition, industry and government personnel may not have fully realized that solvent waste handling practices such as directing water separator wastes to sewers or sending drummed solvent wastes to unlined landfills would cause significant groundwater contamination.

4. *Identify and reduce interdisciplinary obstacles to learning.*

Comment: As discussed in the preceding comment, industrial chemists were not generally concerned with disposal practices or downstream consequences of solvent stabilizers. On the one hand, industrial chemists worked with a body of knowledge, chemical engineering, that is ideally suited to anticipate the environmental fate and transport of solvents and their stabilizers. Industrial chemists certainly knew that the solvent stabilizers were escaping into the environment because they recommended that fresh solvent should be periodically added back to the vapor degreaser. They also developed stabilizer concentrates to replenish stabilizers lost to emissions from open-top vapor degreasers and from removal in water traps and from disposal of still bottoms. On the other hand, industrial chemists had no direct motivation or mandate to investigate the potential fate of their chemical inventions in the environment. Regulators apparently lacked an appreciation for the potential for 1,4-dioxane to become concentrated in vapor degreaser wastes. For example, a 1996 regulatory staff directive to test for 1,4-dioxane at California groundwater contamination sites was restricted to known releases from tanks containing 1,4-dioxane as a pure product. The directive advised the staff not to require 1,4-dioxane testing at sites with solvent contamination that did not have 1,4-dioxane tanks to relieve responsible parties from the burden of high analytical costs (Department of Toxic Substances and Control [DTSC], 1996a.)[*] Except for the cellulose acetate membrane filter industry, polyester manufacturing, and other chemical manufacturing operations that used 1,4-dioxane directly, 1,4-dioxane was not used in sufficiently high volumes to justify storage in tanks; therefore, the directive effectively eliminated testing at vapor degreasing sites, which are likely to have significant 1,4-dioxane contamination.

5. *Systematically scrutinize the claimed justifications and benefits alongside the potential risks.*

---

[*] The directive was penned by a political appointee and distributed over the objections of staff scientists at DTSC (Public Employees for Environmental Responsibility [PEER], 2000).

Comment: Solvent stabilizers were important for protecting worker health and safety against exposure to phosgene gas, which forms when unstabilized methyl chloroform reacts with aluminum and decomposes. The objectives solved by industrial chemists were worker protection and uninterrupted, high-quality production on vapor degreasing process lines in manufacturing operations, whose economic value was critical to the industry. From the perspective of the industry, solvent stabilization was justified and provided tremendous benefits to enable cost-effective and highly productive manufacturing operations. Nevertheless, producers of 1,4-dioxane-stabilized grades of methyl chloroform did not systematically scrutinize the risk of groundwater contamination when developing and promoting their product. Similarly, scrutiny of worker exposure to 1,4-dioxane vapors in degreasing operations lagged behind exposure assessments of the solvents used in vapor degreasers.

6. *Identify and reduce institutional obstacles to learning and action.*

   Comment: In some cases, state drinking water regulatory agencies are separate from those state agencies that deal with groundwater cleanup, leading to an organizational disconnect between the institutional knowledge of each agency. Consequently, cleanup agencies sometimes did not adequately convey their knowledge of 1,4-dioxane occurrence to drinking water regulators so that they could recommend 1,4-dioxane testing to water utilities. Regulatory caseworkers managing solvent cleanup sites may not have fully appreciated the extreme mobility of 1,4-dioxane, which could have led to directing the responsible party to test municipal or domestic wells beyond the leading edge of the solvent plume. Conversely, in some instances where drinking water testing revealed contaminants in drinking water, there has not been a corresponding action by groundwater cleanup agencies to investigate and mitigate the potential sources. These instances are probably the exception, but they also exist between agencies that regulate food and consumer products, workplace safety, air quality, pesticides, and other areas.

7. *Avoid "paralysis by analysis" by acting to reduce potential harm when there are reasonable grounds for concern.*

   Comment: The long timeframe required to investigate the potential presence of a chemical in the environment before it is considered for regulation could be described as "paralysis by analysis" when one considers that private well owners and municipal well customers have been found to be consuming 1,4-dioxane-contaminated drinking water for extended periods of time.

The absence of reliable information on the environmental impacts and health effects of 1,4-dioxane has not been the sole impediment to effective strategies to prevent further impacts or to address those impacts that have already occurred. In those instances in which information became available on the environmental consequences of emerging contaminants, there has been an apparent absence of political will to take action so as to reduce hazards in the face of conflicting costs and benefits (EEA, 2000).

The political will to act can be stymied when regulatory agency managers face substantial uncertainty about the potential environmental hazard under consideration. In the arena of emerging contaminants, those corporations with economic exposure to the ongoing or past use of potentially harmful chemicals have devised well-financed efforts to cast doubt about the hazards, including "well-crafted public relations campaigns that masquerade as independent scientific information from unimpeachable authorities" (Davis, 2007; see also Michaels, 2008). Peer-reviewed research that threatens potentially responsible parties with substantial costs has often been vilified by such campaigns as "junk science," while industry-commissioned research is sanctified as "sound science" (Michaels, 2005). Examples include the campaigns to discredit the nearly unanimous scientific consensus on climate change and the health effects of nicotine and second-hand tobacco smoke; less blatant examples include efforts to influence the outcome of the standard-setting process for perchlorate. Some efforts to influence regulatory outcomes have been undertaken both by industry groups and by environmental groups, leaving regulatory managers to face substantial uncertainty as to the validity of either argument (e.g., the Michigan MCL hearings profiled in Chapter 6) (Taverne, 2008). Nevertheless, the doubts raised by campaigns to foment uncertainty about the potential harm of emerging contaminants must not be confused with proof that emerging contaminants are harmless (Davis, 2007).

The reliability of information on emerging contaminants is sometimes eroded by the sources of the information. For example, a study of the effect of drinking naturally occurring perchlorate occurring in three Chilean towns' water supplies presented compelling evidence that neonatal thyroid hormone ratios are not adversely affected (Tellez et al., 2005), yet the study was widely disregarded by key environmental groups because it was funded by a major producer of perchlorate. The veracity of toxicology and risk assessment studies of emerging contaminants should be subjected to the same level of healthy debate and independent verification as all scientific endeavors, yet it is difficult, if not impossible, to escape industry-funded studies in the framework of limited government funding for research into emerging contaminants. Scientists employed by Dow Chemical published the definitive studies on 1,4-dioxane toxicity using physiologically based pharmacokinetic (PBPK) modeling (Reitz et al., 1990). Recent refinements to the PBPK models were funded by the Dioxane Risk Management Consortium, an industry interest group formed to address the regulatory response to 1,4-dioxane (Sweeney and Gargas, 2006; Sweeney et al., 2008).

The research-funding dilemma is well posed in a sage editorial commentary that appeared in a special issue of *Environmental Science and Technology* focused on emerging contaminants, as excerpted here (Field et al., 2006):

> At present, regulatory communities are placed in a reactive, rather than proactive, position with respect to identifying emerging contaminants and addressing public concern. This position is exacerbated in situations where federal funding is provided only on a piecemeal, short-term basis and only for specifically identified research needs, which by definition are reactionary calls to fill data gaps. This approach to funding generates short-term products for stakeholders, yet it leads to fragmentary science. In the long term, such a piecemeal, constrained, and goal-oriented approach to environmental funding does not allow for exploratory research that can be used to anticipate future environmental issues. Unfortunately, in the U.S., few, if any, competitive funding vehicles exist for the discovery of new contaminants. In addition, no cohesive plan exists to proactively screen and identify all contaminants of potential concern. On the other hand, both Canada and the EU are actively developing plans [REACH] that will place them in positions from which they can anticipate or thwart future environmental issues.[*]

The precautionary principle as a basis for the regulation of emerging contaminants is not without its detractors. Opponents to this approach are concerned that a more precautionary approach to forestalling potential environmental hazards will stifle innovation, economic productivity, and free enterprise or that it may compromise science (McKenna and Sylvester, 2004). While the costs of preventive actions are a legitimate concern for both business interests and regulatory agency budgets, they are usually tangible, clearly allocated, and often short term. In contrast, the costs of failing to act are less tangible, less clearly distributed, and usually longer term, posing problems of governance (EEA, 2000). Failure by industry and regulatory agencies to take the long-term view of environmental risks has led to long latent periods between first exposures and late health effects. As a result, a series of unstoppable consequences, decades long, were set in place before actions could have been taken to stop further environmental exposures (EEA, 2000). Following decades of releasing millions of tons of chlorofluorocarbons and other ozone-degrading compounds to the atmosphere, impacts to the lower stratospheric ozone layer were unavoidable. The fungicide hexachlorobenzene was detected in blood samples collected from 99.9% of 1800 Americans tested in 2003 and 2004 despite a 1984 ban on the use of the chemical (Englehaupt, 2009). Similarly, releases of MTBE and 1,4-dioxane to groundwater have led to the slow but inexorable migration of these contaminants toward drinking water supply wells.

---

[*] Field, J.A., Johnson, A., and Rose, J.B., 2006, What is "emerging"? *Environmental Science and Technology* 40(23): 7105. Used with permission.

Another objection to following the precautionary principle approach to regulating emerging contaminants is that it is entirely possible that initial concerns over the potential environmental harm or toxicity of a newly discovered contaminant in drinking water will not be substantiated by later studies. This objection is not without foundation. After decades of spending millions of dollars on remedial efforts to remove *cis*-1,2-dichloroethylene and *trans*-1,2-dichloroethylene[*] from groundwater to comply with California maximum contaminant levels (MCLs) of 6 and 10 micrograms per liter (µg/L), respectively, the California OEHHA drafted new Public Health Goals (PHGs) that would revise the health criteria upward to 100 and 60 µg/L, respectively (Cal EPA, 2005). Many of the groundwater cleanup sites, where millions of dollars have been expended, have concentrations of these two isomers in the range between the current MCL and the proposed PHG.

The converse has also happened: in 2001, the California OEHHA revisited the PHG for perchloroethylene, recommending a PHG of 0.06 µg/L. New findings of perchloroethylene health effects at lower concentrations came about after decades of enforcement and compliance with the California and federal perchloroethylene MCL of 5 µg/L. Similarly, USEPA Region 9 lowered the PRG for TCE after taking into account the total exposure to its metabolic byproducts, haloacetic acids, which also occur in drinking water treated by chlorination. The EEA report notes that the commonly feared risk of "false positives" occurs less often than the incidence of "false negatives"; therefore, the precautionary principle is the road to follow into the future (EEA, 2000). Accordingly, the precautionary principle is enshrined in the Treaty of the European Union.

10.1.1 EMERGING AND UNREGULATED CONTAMINANTS

Emerging contaminants include chemical and microbial constituents that have not historically been considered as contaminants but are present in the environment on a global scale. Broadly defined, an emerging contaminant is a chemical or material that is characterized by a perceived, potential, or real threat to human or ecological health but for which health standards have not yet been published. Emerging contaminants by definition are not regulated. Once the presence and health threat of an emerging contaminant such as 1,4-dioxane is recognized, it can take a decade or more before a federal or state MCL is established, and then only if warranted in the judgment of USEPA and state health agency professionals. The process for reviewing and adopting MCLs is discussed in Chapter 6. Until federal or state agencies adopt an MCL, state health agencies often promulgate interim guidelines for emerging contaminants (e.g., action levels, drinking water Guidelines, notification levels, PHGs, preliminary remediation goals, screening levels, and risk-based concentrations; see Table 6.1).[†] Until regulators make a determination to adopt a standard or conclude that regulating an emerging contaminant is not warranted, ongoing exposure of unknown proportions and unknown consequences may continue without regulatory intervention. Too often, the first indication that a drinking water supply is threatened by a potentially toxic or carcinogenic agent is its detection in the drinking water that is being served to and consumed by the public.

Under the status quo, the process for adopting regulatory standards requires a standard of proof that involves a multiyear vetting process; this means that real harm may occur before action is finally taken to prevent further harm. In practice, harm or widespread evidence of environmental contamination from a chemical used in industry seems to be a prerequisite for regulation, not unlike the stop sign that is finally erected at the intersection after a fatal accident occurs. In the case of 1,4-dioxane in groundwater, failure to test, regulate, and remediate 1,4-dioxane can mean that water from production wells from which chlorinated solvents have been removed but in which 1,4-dioxane remains is distributed to drinking water systems (e.g., the Bally, Pennsylvania, Superfund case).[‡]

---

[*] *cis*-1,2-Dichloroethylene and *trans*-1,2-dichloroethylene are products of the biodegradation of trichloroethylene.

[†] The exception is Colorado, which has already adopted an MCL for 1,4-dioxane.

[‡] See the USEPA website for the Bally, Pennsylvania Superfund site: http://www.epa.gov/reg3hwmd/super/sites/PAD061105128/index.htm

The risk of continued consumption of 1,4-dioxane in drinking water stems from long regulatory timeframes and the regulation of solvents but not solvent stabilizers. Applying the Precautionary Principle to emerging contaminants could avert such risks.

10.2   1,4-DIOXANE AND DRINKING WATER POLICY

At this time, Colorado is the only state that regulates 1,4-dioxane with a legal standard.[*] Chapter 6 provides a detailed review of the basis for Colorado's MCL for 1,4-dioxane and the basis for advisory guidelines adopted by California, Connecticut, and Michigan as well as an inventory of similar action levels in use in more than 30 other states and countries (see Table 6.1). Colorado was motivated to adopt an MCL because 1,4-dioxane contamination was documented at nine sites and suspected at 19 others, because of its classification as a probable human carcinogen, and because of its mobility, persistence, and widespread use as a solvent stabilizer for methyl chloroform. Other states such as California have many more sites with known 1,4-dioxane releases and a relatively high incidence of 1,4-dioxane in drinking water, yet California has not adopted an MCL. Whether 1,4-dioxane should be regulated in drinking water depends on both the resolution of outstanding questions regarding 1,4-dioxane's carcinogenicity and toxicity and on whether there is a sufficient threshold of contaminated drinking water supplies to justify regulation. In May 2009, USEPA published a draft toxicological review of 1,4-dioxane.[†] The report retains the original IRIS carcinogenicity classification: 1,4-dioxane is classified as likely to be carcinogenic to humans, based on evidence of liver carcinogenicity. The draft report determined an oral CSF for 1,4-dioxane of 0.19 $(mg/kg-day)^{-1}$.[‡] Because the mode of action for liver carcinogenicity of 1,4-dioxane is not known, the CSF was derived by linear low-dose extrapolation (USEPA, 2009).

Ironically, most drinking water utilities do not test for 1,4-dioxane because it is not a regulated contaminant nor there is sufficient regulatory incentive, such as a PHG or a statutory requirement to report PHG exceedances. Consequently, drinking water utilities face the dilemma of deciding whether to budget for discretionary testing beyond the list of analytes required for permit compliance and include emerging contaminants such as 1,4-dioxane, or risk being placed in a reactive position should 1,4-dioxane turn up in the water supply (see the examples of both proactive and reactive discovery of 1,4-dioxane discovered in drinking water supplies listed in Table 6.3). The decision regarding whether to test for 1,4-dioxane hinges on whether the water utility in question embraces the precautionary principle, described further in Section 10.1.1.

10.2.1   1,4-DIOXANE DETECTIONS AND DRINKING WATER TESTING PROGRAM POLICY

Monitoring water supply wells and surface water is the only available defense against unknowingly consuming or serving drinking water that contains 1,4-dioxane, as the taste and odor threshold (23 milligrams per liter [mg/L]) is substantially higher than the concentrations that are likely to appear in sources of drinking water. Most water utilities do not routinely test for 1,4-dioxane. In California, 65 public water supply wells out of 779 sampled since 1997 had detections of 1,4-dioxane above California's 3 µg/L notification level (Cal EPA, 2008). Because testing was motivated where there was a likelihood of 1,4-dioxane detection, this high rate of detection (8.3%) is unlikely to apply to the approximately 14,000 water supply wells operating in California, yet it is likely that there are more yet undetected occurrences of 1,4-dioxane in California's drinking water wells located in

---

[*] As of April, 2009.

[†] USEPA, 2009, Draft Toxicological Review of 1,4-Dioxane US Environmental Protection Agency, Washington, D.C. EPA/635/R-09/005, www.epa.gov/iris

[‡] As noted in Chapter 5, the previous oral CSF value published on USEPA's IRIS database for 1,4-dioxane was 0.011 (mg/kg-day)–1. The higher CSF published in the draft 2009 report suggests that lower drinking water thresholds for 1,4-dioxane may be appropriate if the new CSF is adopted.

urban or industrial areas. This situation occurs worldwide wherever current or past land use includes industries that used 1,4-dioxane or 1,4-dioxane-stabilized methyl chloroform, or industries that produce 1,4-dioxane as a by-product (see Chapter 2). Clearly, a coherent strategy is needed to perform screening-level testing of drinking water sources of 1,4-dioxane.

10.2.2   USEPA's CONTAMINANT CANDIDATE LIST INCLUDES 1,4-DIOXANE

As described in Section 6.1.1, the CCL includes contaminants judged by USEPA staff and a committee of experts to warrant further evaluation due to known toxicity, prevalence and magnitude of occurrence, and persistence and mobility. 1,4-Dioxane was included on the final CCL3, a list of 104 chemicals pared down from about 7500 chemicals reviewed by USEPA's staff and expert committees. USEPA's mandate is to issue a formal decision on whether to initiate a process to develop a national primary drinking water regulation for a specific contaminant. USEPA is required by law to make regulatory determinations for at least five contaminants from the most recent CCL every five years; however, the CCL list does not directly require monitoring. A related program, the Unregulated Compound Monitoring Requirement (UCMR) program, requires monitoring of selected contaminants that may be present in drinking water but do not have health-based drinking water standards. The UCMR list is revised every five years and is largely based on the CCL.

The Safe Drinking Water Act Amendments of 1996 stipulate that no more than 30 contaminants are included on the UCMR list per five-year cycle. 1,4-Dioxane is not included on the current cycle (UCMR2 from 2007 to 2010) (USEPA, 2008). It is therefore unlikely that USEPA will require monitoring for 1,4-dioxane anytime soon; however, to prepare for eventual monitoring needs for both USEPA's drinking water and Superfund programs, USEPA has developed a new analytical method for 1,4-dioxane analysis, Method 522 (see Section 4.5.7). While USEPA's drinking water programs do not yet require that water utilities monitor their supply wells for 1,4-dioxane, some water utilities have decided to monitor for 1,4-dioxane voluntarily due to known releases of 1,4-dioxane at solvent contamination sites lying within the zones of contribution to the selected wells. Voluntary monitoring has detected 1,4-dioxane in numerous public water supply wells. It is therefore important that water utilities and state drinking water regulatory agencies maintain a detailed awareness of groundwater contamination investigations near supply wells.

10.2.3   INTEGRATING AWARENESS OF GROUNDWATER CLEANUPS TO PRIORITIZE 1,4-DIOXANE MONITORING IN WATER SUPPLY WELL SAMPLING PROGRAMS

Water utilities operating municipal water supply wells are required to monitor water quality by state agencies charged with regulating drinking water. Monitoring requirements typically include a list of analytes and a schedule setting the monitoring frequency. Monitoring frequency is typically increased following detection of a contaminant. The baseline monitoring frequency for VOCs and SVOCs—the suites of analytes that would include analysis of 1,4-dioxane—is usually once every three years; hence, water utilities typically monitor one-third of their wells for VOCs and SVOCs each year. State agencies have been very effective at managing water utility compliance with required monitoring programs, and water utilities have been diligent at implementing the programs. While some water utilities have voluntarily added contaminants to the monitoring list, most water utilities are inclined to follow the requirements of the state regulatory agency and let the regulators determine whether additional contaminants should be included. Those water utilities that choose to monitor for unregulated contaminants are challenged to justify exploratory monitoring to their ratepayers, who may protest when asked to pay for discretionary monitoring or research projects. Some water utilities do not monitor for unregulated contaminants simply because it is not required. Many water utilities are reluctant to independently determine whether there is cause to test for a given contaminant as they are not in a position to take on the responsibility for determining health risks. This perspective is summarized by the question, "If you don't want to find something, why would you test for it?" Most water utilities prefer to await an official determination by regulatory agency health experts and proceed with testing at the same time and on the same basis as all the other water utilities.

To determine the nature of known and potential contamination within the zones of contribution of groundwater to water supply wells, USEPA developed a program to assess potentially contaminating activities (PCAs) that could affect sources of drinking water. The 1996 amendments to the Safe Drinking Water Act required that states develop Source Water Assessment Programs (SWAPs) to delineate wellhead protection areas within zones of contribution to wells, and to inventory PCAs that could contaminate wells. The inventory of PCAs typically focuses on known and ongoing activities that have a potential to contaminate groundwater, as inventorying past activities typically involves a greater level of effort. Because groundwater travels slowly, much of the contamination occurring in water supply wells today is the result of legacy contamination from past industrial activity. The detection of 1,4-dioxane in supply wells exemplifies legacy contamination that originated before the Montreal Protocol banned the use of methyl chloroform in 1996, when disposal of 1,4-dioxane-stabilized methyl chloroform vapor degreasing waste was the primary means by which 1,4-dioxane was released to soil and groundwater.

To anticipate 1,4-dioxane in public water supply wells, SWAPs should consider whether any of the industrial activities that use 1,4-dioxane (listed in Chapter 2) operated in proximity to water supply wells. This review of potential 1,4-dioxane releases should include industries that used methyl chloroform, industries that produced 1,4-dioxane as a by-product, and destinations for 1,4-dioxane-containing wastes such as cement kilns, landfills, and solvent recycling facilities.

10.2.4 Predicting Water Supply Well Contamination

Knowledge of a 1,4-dioxane release within the zone of contribution to a supply well does not necessarily predict that the well will become contaminated with 1,4-dioxane. Detailed analysis of plume dynamics and the hydrogeologic setting is needed to confirm that the 1,4-dioxane will arrive at the well at detectable levels. Currently, plume characterization falls under the jurisdiction of state agencies overseeing groundwater cleanups, while water supply well monitoring requirements reside with the state agencies overseeing water utility compliance with drinking water standards. In many states, these are two separate agencies that develop their own bureaucratic cultures and, while they may communicate about groundwater contamination issues, that communication often does not cause a corresponding measure of coordinated action. Drinking water agencies do not usually get involved with interpreting data from cleanup sites, and groundwater cleanup agencies are not usually focused upon the operating features of individual water supply wells. This arrangement has its strengths in that the two skill sets are different and the division of duties allows regulators in each discipline to refine their skills and apply their knowledge to the distinct challenges of groundwater cleanup and drinking water system permitting, inspection, and regulation. There are limitations to this arrangement as well: the regulatory "silos" that result can diminish the effectiveness of groundwater management strategies that must apply a holistic approach in order to succeed. Although seldom evaluated, contaminant mass discharge (or mass flux) should be the common interest of both drinking water and groundwater cleanup agencies, as described below.

Regulatory agencies overseeing groundwater cleanups often require a well search to identify all operating and inactive wells within a fixed radius of the cleanup site. This requirement to investigate whether water supply wells may be impacted from release sites should be expanded to require that the discharger estimates the date by which the contamination could be expected to arrive at active water supply wells residing within the fixed radius. A further expansion to this requirement is to use the mass discharge approach to estimate the maximum contaminant concentration that could be expected in the well.

Determining the likelihood that a well will become contaminated involves evaluating whether the mass of 1,4-dioxane at the release site is sufficient to sustain migration and arrival at the well, which requires an assessment of the capture zone of the well. Where a supply well is screened in an interval underlying a confining layer, the recharge location from which the well ultimately draws water may be located a considerable distance upgradient from the well. Delineating capture zones can be complex because the zones are three-dimensional and occupy aquifers with a range of hydraulic conductivities. Variable pumping rates, geologic controls over groundwater movement, vertical flow within the well, interference from the operation of other nearby wells, variable sources of recharge such as rainfall runoff or dam releases, and vertical conduits facilitating flow across aquitards all add to the complexity of assessing water supply well capture zones (Einarson and Mackay, 2001). In addition to conventional finite difference and finite element modeling, analytic element modeling is often used to simulate the three-dimensional water supply well capture zone to delineate wellhead protection areas (e.g., WhAEM, USEPA's Wellhead Analytic Element Model).

When the 1,4-dioxane release site lies entirely within the delineated capture zone of a water supply well, the mass discharge of 1,4-dioxane from the release site will constrain the maximum possible concentration of 1,4-dioxane that could be detected in groundwater pumped from a supply well. Contaminants arriving at the well screen are often substantially diluted in supply wells because they typically occupy a narrow radial flow tube, whereas the remaining radial flow tubes are usually occupied by clean water. Moreover, many municipal supply wells are screened in multiple intervals. Contaminants often arrive at the shallowest screened interval, while clean water is concurrently pumped from deeper screened intervals. The combined radial and vertical contributions of clean water to the well result in substantial dilution of contaminants entering the well. There are probably instances in which plumes of 1,4-dioxane have arrived at and entered water supply wells but have not yet been detected because the concentrations are below current detection limits due to in-well blending with clean water (or because 1,4-dioxane is not analyzed for insamples from the well).

Antidegradation policies prohibit discharges that degrade water quality, regardless of whether a water supply well becomes contaminated. The antidegradation approach protects all current and future beneficial uses of the groundwater resource to preserve the highest quality groundwater. Therefore, it is generally unacceptable to manage a contaminated site cleanup based on whether a supply well will become contaminated, or to use dilution in a supply well as a contamination mitigation measure. Nevertheless, determining the likelihood that a well will become contaminated from a known release within the well capture zone can be useful to help water utilities adjust monitoring priorities and operations. The analysis of mass discharge from release sites provides a simple method to estimate the maximum contaminant concentrations that can be expected to be detected in groundwater samples from the well (Einarson and Mackay, 2001).

The mass discharge of contaminant migrating from a release site located within a well capture zone should equal the contaminant mass discharge from the well if there is a continuous source that releases the contaminant at a constant rate with conservative transport, that is, contaminant retardation from sorption and diffusion are negligible and the contaminant is not transformed abiotically or through biodegradation (Einarson and Mackay, 2001). The maximum concentration of the contaminant that could occur in a downgradient supply well is obtained from the following relationship:

$$C_{SW} = \frac{M_d}{Q_{SW}}, \tag{10.1}$$

where C_{SW} is the maximum concentration of contaminant in water extracted from the supply well (mass/volume), M_d is the mass discharge from the release site (mass/time), and Q_{SW} is the pumping rate from a supply well (volume/time) (Einarson and Mackay, 2001).[*]

The total mass of contaminant leaving the release site per unit of time (e.g., grams per day) can be evaluated by performing detailed plume profiling in a transect of multilevel monitoring wells or a one-time investigation using direct-push grab-sampling methods (e.g., cone penetrometer testing

---

[*] The details for applying the mass discharge approach to estimating the maximum anticipated concentration of a contaminant in a supply well are provided in the online supplemental material to Einarson and Mackay's 2001 article, "Predicting Impacts of Groundwater Contamination," *Environmental Science and Technology* 35(3): 66A–73A.

with discrete interval groundwater sample recovery). The transect must be oriented orthogonal to the primary groundwater flow direction. The resulting geologic and contaminant concentration data can then be integrated to estimate the rate of groundwater flow and contaminant mass migrating across the grid cells in an imaginary plane defined by the contaminant concentrations at each vertical interval in the transect, yielding the estimated mass discharge from the release site. The estimate is improved where two or more orthogonal sampling transects are installed, allowing calculation of M_d at different points along the length of a dissolved plume (Einarson and Mackay, 2001).

M_d can serve as the objective function for the optimization of site remediation. Ideally, the discharger will control the source zone so that $M_d = 0$, and downgradient supply wells are not impacted. At 1,4-dioxane sites, it is often the case that the releases are old enough such that 1,4-dioxane migration has already progressed well beyond the point from which source control would be useful.

For two supply wells down-gradient of plumes, in which all attributes of the plumes, hydrogeology, and wells are identical except for the pumping rates, the well with the higher continuous pumping rate will have lower concentrations, because contaminant concentrations in continuously pumped supply wells are inversely proportional to pumping rates (Einarson and Mackay, 2001). Therefore, smaller supply wells and especially private wells are more vulnerable to contamination, because there is less dilution by blending with clean water. Wells that pump less have a smaller chance of capturing a plume of contamination, but if the plume does intersect the well capture zone, there will be less dilution and therefore higher concentrations. The mass discharge approach is therefore useful to anticipate impacts to private wells and small water supply wells.

Water utilities are very unlikely to perform mass discharge investigations by installing transects of multilevel wells due to the high cost and because the responsibility is assigned to the discharger. Because an important goal of groundwater cleanup projects is to protect against exposure to chemicals through consumption of drinking water, dischargers should conduct mass discharge investigations where other plume attributes such as length, duration, and concentrations suggest that the mass of contaminant present is sufficient to impact down-gradient supply wells. High-resolution, multilevel monitoring well transects to investigate mass discharge from contaminant release sites yield benefits to both the remediation team and to the operator of the down-gradient well. For example, mass discharge investigations can be used to establish whether natural attenuation is occurring or to estimate the rate at which contaminant transformation products are being formed (Einarson and Mackay, 2001).

10.2.5 Moving Beyond Checklist Compliance

It is not an accident that drinking water quality in the United States is among the world's best. The American drinking water regulatory framework is highly developed and nearly eliminates the risk of ingesting harmful chemicals in drinking water. State and federal regulatory agency staffs are effective at ensuring that the regulations are enforced, and local government and private water utility operators regularly deliver outstanding performance to ensure that American drinking water remains safe. In the spirit of continuous improvement, there is nevertheless an opportunity to examine features of the current regulatory framework and ask whether some adjustments could bring about additional safeguards against drinking water contamination.

As discussed in Section 4.1 (The Flawed Paradigm of Analyte Lists), when cleanup site investigations or drinking water testing programs rely on prescribed USEPA method lists of analytes without checking for unidentified peaks in the gas chromatogram, it is likely that the presence of other possibly significant contaminants will be missed. Often, the analyses are performed by staff chemists in the water utilities' laboratory. Many of these chemists hold advanced university degrees and are capable of conducting research projects and analytical method development, yet they are relegated to routine analyses using standard method lists to achieve regulatory compliance, without exploring the complete chemistry of the samples they handle. Of course there are also exceptions; for example, see the OCWD case study of method development for 1,4-dioxane analysis in Chapter 4.

Water utility managers may not be inclined to authorize analyses that extend beyond compliance requirements because the information obtained may present risk communication challenges. Informing consumers of the presence of a contaminant like 1,4-dioxane in drinking water is loaded with problems, as consumers typically do not appreciate that there can be a safe level for any synthetic or industrial chemical, particular one whose chemical name sounds very similar to the substantially more toxic group of compounds called dioxins.* A water utility may have an impeccable record of full compliance with all statutory requirements, but the utility's well-deserved image with its consumers can be damaged when they publicize the discovery of an unregulated contaminant.

In addition to facing a significant public information challenge, water utilities operated by local governments may be faced with responding to political leaders who call for action to address the perceived risk from the discovery. Consequently, substantial sums of public funds are being expended to address emerging contaminants with treatment or purchasing alternative water supplies before there is consensus on the threshold of risk for the contaminant and long before a regulatory standard is adopted. As profiled in Chapter 6, state 1,4-dioxane action levels range from 3 µg/L in California to 85 µg/L in Michigan, underscoring the lack of consensus. In the absence of well-defined federal or state regulatory guidance on emerging contaminants, local water utilities are left to decide whether they should follow the precautionary principle and test, publicize, and treat emerging contaminants. State and federal programs such as the UCMRs are intended to fill this void; however, the number of chemicals potentially present in groundwater withdrawn from wells in urban, industrial, or agricultural areas is so large that it is virtually unmanageable.

Water utility managers are seemingly faced with a lose–lose proposition when it comes to testing for emerging contaminants. The Groundwater Ambient Monitoring and Assessment (GAMA) program provides a solution by arranging for state government and federal contractors to perform testing on public water supply wells for ultra-trace levels of regulated contaminants and for emerging contaminants.† Yet some water utilities are also inclined to resist cooperation with the GAMA program for the same reason for which they are not inclined to do the testing themselves: communicating the presence of contaminants, even at ultra-trace levels, can lead to public distrust of a water system that satisfies all statutory requirements.

The solution may lie in developing risk communication programs that include stakeholder involvement before, during, and after the testing for emerging contaminants or ultra-trace levels of regulated contaminants is performed. Outreach programs to educate the interested public and local journalists are essential to ensuring that the data are understood in its proper context. The outreach programs can include public meetings with appearances by state officials with regulatory authority over the water utility to allow questions to be answered by an informed but disinterested party. If water utilities are to succeed at fulfilling their collective mission of serving healthy, safe water, it is necessary to proactively explore all aspects of the water quality and to deal forthrightly with the information thus obtained.

While water utilities may face challenges when analyzing for 1,4-dioxane and other emerging contaminants, they may also derive secondary benefits in addition to improved protection of drinking water quality. For example, using laboratory analyses to identify all detectable compounds can inform water treatment plant operators of the possible presence of dimethylamine and dithiocarbamate, which can lead to formation of nitroso-dimethyl amine upon chlorination during the treatment of drinking water. The detection of 1,4-dioxane can provide early warning that the well is capturing contamination from a solvent release site and that chlorinated solvents (most likely 1,1-dichloroethylene) are likely to follow.

---

* For example, the dioxin compound known as 2,3,7,8-tetrachlorodibenzo-p-dioxin, also called 2,3,7,8,-TCDD, has a federal MCL of 3×10^{-8} mg/L, that is, 0.00003 µg/L or 30 parts per quadrillion.
† For many VOCs, the GAMA program uses a reporting limit of 5 ng/L, i.e. five parts per trillion. For 1,4-dioxane testing in limited areas, the GAMA program used a reporting limit of 0.15 µg/L.

10.3 UPDATING THE TOXICOLOGICAL BASIS FOR 1,4-DIOXANE REGULATION

As detailed in Chapters 5 and 6, the nonlinear dose–response seen when 1,4-dioxane is administered to laboratory rats raises questions about the legitimacy of the generic assumption of linear low dose extrapolation used to determine the drinking water dose that may induce tumors in humans. A 1996 internal memorandum by the California DTSC recommended PRGs at California cleanup sites and summarized the nonlinear dose–response conundrum succinctly (DTSC, 1996b).[*] The memo explained that because 1,4-dioxane tumorgenicity in animals occurs only at intoxicating concentrations (10,000 ppm or greater when administered orally), it is thought that the pharmacokinetic profile in animals and humans may explain the disparity in interspecies cancer outcomes.[†] Rodents are less efficient at metabolizing 1,4-dioxane to β-hydroxyethoxyacetic acid than humans. When 1,4-dioxane is supplied to rodents at concentrations that overwhelm their ability to metabolize it, 1,4-dioxane accumulates, causing intoxication followed by systemic cell injury and carcinogenesis. The tissue death and proliferation mechanism in animals was not accounted for when the USEPA's CSF for 1,4-dioxane was extrapolated to humans to develop PRGs. The PBPK model by Reitz et al. (1990) accounts for this mechanism and applies a relationship between the metabolism of 1,4-dioxane in rats and humans (DTSC, 1996b). In a presentation to the Groundwater Resources of California's 2001 *Symposium on Recalcitrant and Emerging Contaminants*, DTSC staff toxicologist Calvin Willhite profiled the 1,4-dioxane nonlinear dose–response issue and USEPA's CSF extrapolation and famously declared, "These numbers are broken!"

How can environmental regulatory staff effectively manage risks from 1,4-dioxane in the face of genuine uncertainty regarding the mechanisms and degree of toxicity and possible carcinogenicity? What improvements could be made to increase confidence in the scientific basis for establishing standards to protect drinking water consumers and other populations exposed to 1,4-dioxane? The art and science of PBPK modeling has advanced considerably since the Reitz et al. (1990) paper. Sweeney et al. (2008) made refinements to estimates and *in vitro* measurements of intercompartmental partition coefficients. More reliable and independently validated PBPK models will go far to improve the consensus on a scientific basis for regulating 1,4-dioxane.

The studies finding 1,4-dioxane-induced nasal carcinomas upon which the IARC bases its Class 2B "probable human carcinogen" classification of 1,4-dioxane have been questioned as unreliable because of the way in which rats ingested drinking water containing 1,4-dioxane from sipper tubes (Stickney et al., 2003). Nasal carcinomas observed in rats may have been caused because their nasal turbinates were probably splashed repeatedly with aspirated water containing 5000–10,000 mg/L of 1,4-dioxane (Reitz et al., 1990). Rats have a more convoluted nasal turbinate system than humans, resulting in greater deposition in the upper respiratory tract. The location of the nasal tumors within the rat nasal turbinate also suggests that water containing 1,4-dioxane entered the nasal cavity during drinking (Stickney et al., 2003; see also Chapter 5). Therefore, an additional opportunity to improve the toxicological basis for regulating 1,4-dioxane in drinking water may be to repeat the 1,4-dioxane rat and mice studies of Argus et al. (1965, 1973), Hoch-Ligeti and Argus (1970), Hoch-Ligeti et al. (1970) (see Chapter 5), using drinking water administration methods that prevent the aspiration of 1,4-dioxane-laden drinking water into the rodent nasal cavity. Repeating 1,4-dioxane carcinogenicity testing could also be important because rodent carcinogenicity assays in general have been shown to be much less reproducible than might be expected. In a survey estimating the reliability of 121 replicate rodent carcinogenicity assays from the two parts of the 1997 Carcinogenic Potency Database (NCI/National Toxicology Program and Literature), researchers found a

---

[*] The DTSC's analysis of 1,4-dioxane toxicity and appropriate remediation goals was not included in Chapter 6, which focuses on the drinking water standard-setting process.

[†] The paucity of human data makes it difficult to characterize an interspecies disparity in cancer outcomes; this statement is more likely an opinion than a conclusion based on a survey of available data.

concordance of only 57% between the overall rodent carcinogenicity classifications from both sources (Gottmann et al., 2001). However, repeating 1,4-dioxane carcinogenicity testing would be extremely expensive, and the available bioassays are deemed sufficient for deriving CSFs by most regulatory toxicologists. The question then becomes whether the expense of additional or repeat carcinogenicity assays is outweighed by the cost of cleaning 1,4-dioxane up to a threshold set too low due to problems with extrapolating the nonlinear dose–response.

Even with improvements to reduce the uncertainty in PBPK models and better assays to determine 1,4-dioxane carcinogenicity and toxicity, the assessment of 1,4-dioxane's health effects may be limited because, like all other drinking water toxicity evaluations, the framework for assessing risk focuses on 1,4-dioxane in isolation from the other chemicals occurring as co-contaminants. A number of new approaches to evaluating the toxicity of chemical contaminants may impart a more comprehensive analysis of the effects of multiple contaminants through multiple exposure pathways by providing mechanistic details of events at the cellular and molecular levels (Bhogal et al., 2005).

Regulators and regulated parties are best served by clearly delineated regulatory thresholds, for which well-established toxicity values are needed. As described above, the road to human 1,4-dioxane toxicity values are constrained by the limitations of the methods used; however, new developments hold promise to remove these limitations and expand the toxicological database upon which regulations are established. Some of these new toxicological methods may improve the evaluation of 1,4-dioxane toxicity and are profiled in the following sections.

10.3.1 ADDRESSING SYNERGISTIC EFFECTS OF MULTIPLE CONTAMINANTS

Most toxicological testing is performed on single chemicals, yet human exposures are rarely limited to single chemicals as contamination at hazardous waste sites generally involve multiple contaminants. In addition to the chemical of interest, people may also be exposed to chemicals they consume in alcoholic drinks, tobacco smoke, medicines, and foods, and may involuntarily be exposed to vehicle exhaust, drinking water disinfection byproducts, and chemicals in the workplace. What is the net effect of exposure to mixtures of chemicals? Mixtures of chemicals, each of which may be present at concentrations less than their respective regulatory thresholds, may cause health effects due to additivity, interactions, or both. It is therefore important to make an exposure-based assessment of the joint toxic action of chemical mixtures (Agency for Toxic Substances Disease Registry [ATSDR], 2004).

The ATSDR published guidance on a semiquantitative screening process for the assessment of the joint toxic action of chemical mixtures (ATSDR, 2004). Practical and accessible tools are needed to facilitate assessment of this particularly complex challenge. Aspects of the complexity of the joint toxic action of chemical mixtures include the combined action of different chemicals at different concentrations, each with different dose–response relationships, different routes of exposure, different potentials to produce toxic metabolic byproducts whose interactions must also be evaluated, different target organs, and combinations of carcinogenic and noncarcinogenic effects. The composition of the mixture also changes with time and distance from the release point due to the differential fate and transport of its constituent compounds (ASTDR, 2004).

The major mechanisms that must be considered for toxicant interactions are direct chemical–chemical, pharmacokinetic, and pharmacodynamic mechanisms (ATSDR, 2004). Developing a knowledge base of these mechanisms for binary mixtures and for classes of chemicals can support the prediction of interactions for new combinations of chemicals. Chemical–chemical interactions may include potentiation or synergism[*] such as the formation of carcinogenic nitrosamines from

---

[*] *Potentiation*: when a component that does not have a toxic effect on an organ system increases the effect of a second chemical on that organ system. *Synergism*: when the effect of the mixture is greater than that estimated for additivity based on the toxicities of the components (ATSDR, 2004).

noncarcinogenic nitrites and amines in the stomach (ATSDR, 2004). Pharmacokinetic interactions include absorption, distribution, excretion, and metabolism. Pharmacodynamic interactions may include interaction at same receptor site or target molecule, interaction at different sites on the same molecule, or interaction among different receptor sites (ATSDR, 2004).

The ATSDR guidance uses variations of an old and simple approach called the hazard index to address the joint toxic action of chemical mixtures. The hazard index assumes that the noncancer health effects of the mixture can be estimated from the sum of the doses (weighted for potency) or the effects of the individual components as described in the following equation:

$$\text{HI} = \sum_{i=1}^{n} \frac{E_i}{\text{DL}_i}, \quad (10.2)$$

where E_i is the dose or level of exposure to the ith chemical, and DL is the defined level of exposure to that chemical, usually a toxicity threshold or regulatory threshold. This ratio is known as the hazard quotient. For example, the hazard quotient for a single chemical exceeds unity when the concentration exceeds the MCL; when the sum of the hazard quotients for a mixture exceeds unity, the mixture is considered to be capable of causing adverse health effects. The hazard index approach must account for each pathway and exposure duration of concern, with a separate hazard index for each (ATSDR, 2004).

The simplistic basis of the hazard index accounts for noncancer effects only and does not account for the mechanisms of carcinogenicity. Computational methods using databases of binary interactions of carcinogens with tumor initiators, tumor promoters, and tumor inhibitors can be used to develop more sophisticated models of the joint toxic action of chemical mixtures. Carcinogenesis is a multistage process and is often modeled as a synergistic response between a tumor initiator and a tumor promoter. The combination of a genotoxic contaminant that causes DNA damage combined with another chemical that enhances cell proliferation would act synergistically, and the response can be more than additive of the two chemicals' toxic effects taken independently. Integrating PBPK modeling to assess the toxic effects of chemical mixtures can further address the modes of action for mixtures of toxins. For mixtures of two chemicals, PBPK models for the individual chemical are linked at the assumed point of interaction, frequently the hepatic (liver) metabolism term. Uncertainty about the toxicological interaction of multiple chemicals is an impediment to regulation; therefore, computational toxicology methods are needed to address the complexity of the joint toxic effects of exposure to multiple contaminants (see Section 10.3.3).

10.3.2 THE PROMISE OF CELL LINES TO ACCELERATE AND IMPROVE TOXICOLOGY ASSAYS

The field of environmental toxicology is undergoing a paradigm shift from a check-list approach using animal testing (*in vivo*) to the use of new cell line methods (*in vitro*) and computational toxicological methods (*in silico*). The new methods provide mechanistic details of events at the cellular and molecular levels and are being developed primarily by pharmacologists and biotechnologists to accelerate the development and testing of pharmaceuticals; however, they also have ready application to the toxicological assessment of chemicals and chemical products. *In vitro* methods permit the observation of specific changes at the molecular level, rather than just the number of tumors, deaths, or overt clinical changes observed in test animals using *in vivo* methods. Molecular-scale changes observed *in vitro* include DNA alteration at a target organ site, and changes to proteins in cell membranes and within cells (Bhogal et al., 2005).

The European Union's new REACH policy requires the assessment of tens of thousands of chemicals. It will be unworkable to complete the task using conventional methods; therefore, new innovative tools are required to allow high-throughput screening of chemicals (Bhogal et al., 2005).

Cell lines use living animal or human tissue cells to measure responses to toxins. Cell-based *in vitro* tests permit the rapid testing or large sets of replicates to improve statistical characterization of cellular responses to environmental contaminants. Cell testing can also be considerably quicker and less expensive than *in vivo* testing; moreover, cell testing provides an indication of specific toxicological endpoints that are not usually determined from acute toxicity tests (Bhogal et al., 2005). *In vitro* systems can include tissue slices, perfused organ preparations, primary cultures, and cell lines grown either in suspension or as adherent cultures. Cells can be isolated from tissues for *ex vivo* culture by digesting tissues with enzymes that remove the extra cellular matrix, by purifying blood to isolate white blood cells, or by placing a piece of excised tissue in a growth medium. Most cell cultures are short-lived; after doubling their population multiple times, cells stop dividing but generally retain viability as an established or immortalized cell line that can sustain itself to permit testing. Cell lines can be isolated from many tissues, organs, and species, cultured over extended periods and/or cryopreserved for future use. The use of human tissues and cells has an obvious advantage because the need for interspecies extrapolation is avoided (Bhogal et al., 2005).

The liver metabolizes toxins and is central to the assessment of toxicity. A compounding factor in toxicological evaluations is that biotic and abiotic transformations of the chemical inside the target organism may lead to the formation of reactive metabolites that are toxic (USEPA, 2006). Cell line bioassay methods are now available that use human liver tissue and are sensitive enough to detect the hepatotoxicity of contaminants and their metabolites in water samples using the low density lipoprotein (LDL)-uptake activity of human hepatoblastoma cells, *Hep G2*. The LDL-update activity assay of *Hep G2* can be used to evaluate cytotoxicity for up to 48 h with high sensitivity and selectivity using a fluorescent plate reader (Shoji et al., 2000).

Studies of changes to the composition of proteins and activities of cells provide an approach to eliciting the mechanisms of toxicity using the "omics." The "omics" methods include genomics—the study of an organism's genome including mapping its DNA sequence; proteomics—the study of protein structures and functions; and metabonomics—profiling metabolic changes and metabolic products produced by exposure to toxins (Bhogal et al., 2005). The "omics" methods make it possible to develop molecular profiles to identify the key steps that trigger toxicity and cause adverse health effects to target organs and to entire living organisms (USEPA, 2006).

A number of programs are now underway to achieve these advances, including USEPA's Computational Toxicology Research Program, with the goal of shifting the field of toxicology from a descriptive to a predictive science and thereby improving USEPA's ability to assess hazards and characterize risks (USEPA, 2006).

10.3.3 APPLYING COMPUTATIONAL TOXICOLOGY TO PHYSIOLOGICALLY BASED PHARMACOKINETIC MODELING

The emerging field of computational toxicology applies mathematical and computer models and molecular biological and chemical approaches to explore both qualitative and quantitative relationships between environmental contaminants and adverse health effects. The integration of advanced computing methods with molecular biology and chemistry is enabling scientists to better prioritize data, inform decision makers on chemical risk assessments, and understand a chemical's progression from the environment to the target tissue within an organism, and ultimately, mechanisms of toxic effects. A key goal of both computational and *in vitro* toxicology is to reduce the uncertainties in the extrapolation of effects across dose, species, and chemicals (USEPA, 2006).

In silico simulation of contaminant biotransformations following exposure and descriptions of metabolic maps has shown great promise for improving the toxicological foundation of health risk assessments. Knowledge of predicted metabolites and their associated toxic effects may be useful for pollution prevention by avoiding the commercial use of those chemicals found to form toxic metabolites, again employing the precautionary principle. For example, forecasting the probable metabolites of xenobiotic chemicals and interfacing that information with toxic effect models that predict chemical binding to estrogen receptors in aquatic species can be used to anticipate the potential for endocrine disruption in aquatic organisms (USEPA, 2006).

Among the most challenging aspects in conducting risk assessments is relating the effects detected at the dose level to which the test animal was subjected to the effects that would be caused by the dose that actually reaches the target organ in humans. Toxicokinetic studies can predict the internal target organ dose, which must account for the absorption, distribution, metabolism, and excretion (ADME) of the toxin of interest. PBPK models can be used to predict ADME *in vivo* by integrating literature values and computational techniques, and by extrapolating data from *in vitro* studies between species. Improved dose–response models of *in vivo* toxicity can then be developed from the test animal data (Bhogal et al., 2005).

The work of developing a PBPK model involves tracking down lots of data to estimate the required input parameters; consequently, PBPK modeling is considered "data hungry" and "resource intensive," and thus requires a certain financial threshold before a PBPK modeling study can be launched. Efforts are underway in Britain to make PBPK modeling more efficient, less expensive, and more accessible by developing a PBPK model equation generator and a PBPK parameter database, which allow the construction of PBPK models in minutes rather than days (Health and Safety Laboratories [HSL], 2008). The PBPK parameter database contains physicochemical, biochemical, anatomical, and physiological data that have been validated for their quality. Parametric data are retained together with metadata describing the studies that generated the data and the quality of the parameter data. Developing the key PBPK model equations and populating the model parameters is then a semiautomatic task, allowing researchers to focus their skill and attention on the toxicological validity of the modeled relationships rather than on the laborious and time-consuming task of collecting and compiling data (HSL, 2008).

Another computational toxicology initiative to accelerate the prediction of chemicals' toxic effects is USEPA's ToxCast™, which was launched in 2007 to develop rapid and cost-effective tools for prioritizing the toxicity testing of large numbers of chemicals. Using data from high-throughput screening *in vitro* bioassays, the ToxCast™ project is building computational models to forecast the potential toxicity of chemicals to humans. ToxCast™ is expected to provide USEPA regulatory programs with science-based information that are helpful in prioritizing chemicals for more detailed toxicological evaluations lead to more efficient use of animal testing (USEPA, 2007).

Predicting carcinogenicity is another toxicological challenge for which computational toxicology has made substantial progress using a method called Computer Automated Structure Evaluation (CASE). The CASE system uses QSAR in the same manner as described in Section 3.1. Molecules of interest are divided into chemical fragments of two to ten heavy atoms, and a statistical distribution is performed to determine whether the fragments identified are capable of the specific biological activity necessary to induce toxic effects. The CASE system has developed predictions of carcinogenicity and mutagenicity endpoints, using a variety of database modules including Ames mutagenicity and carcinogenicity in male and female rats and mice (Bhogal et al., 2005).

Overall, the combined application of advanced biotechnology tools to develop cell lines for toxicological assays, leveraging advances in computer technology and database systems, and improvements to assembling PBPK models all signify that the quality of toxicological assessments is likely to continue to grow better. These changes will completely transform the way in which toxicological data are generated and applied. Taken together, these advances may comprise more than a paradigm shift; rather, environmental toxicology is probably undergoing a quantum leap. With sufficient funding to enable research scientists to focus on advancing the state of toxicology, the traditional laboratory animal approach to toxicity testing could soon be consigned to history. Advanced *in vitro* and *in silico* methods will become the routine approaches to assess environmental toxicology (Bhogal et al., 2005).

10.4   REGULATORY POLICY ON 1,4-DIOXANE REMEDIATION AT CLEANUP SITES

The detection of 1,4-dioxane at solvent release sites often occurs years after the solvent release has been investigated. As described in the case studies in Chapter 8, the late detection of 1,4-dioxane may cause regulators to require the installation of additional monitoring wells to characterize the extent to which 1,4-dioxane has migrated. The occurrence of actionable levels of 1,4-dioxane beyond the capture zones of pump-and-treat systems that were installed to address chlorinated solvents may require the installation of additional extraction wells or extraction trenches to capture 1,4-dioxane. The treatment technology employed to remove chlorinated solvents will most likely require modification or additional treatment technologies to remediate 1,4-dioxane. In addition, the health risk assessment may need to be revisited to evaluate risks posed by the presence of 1,4-dioxane. Finally, if 1,4-dioxane is discovered to have migrated further than the solvents and impacted private or municipal wells after assurances were given that the solvent release had been contained, the public will lose trust in both the regulatory agency and the discharger, making it more difficult to get public cooperation for future remedial actions necessitated by the late discovery of 1,4-dioxane. In short, the late detection of 1,4-dioxane at a solvent release site in the advanced stages of cleanup can reopen the case, delaying closure and adding substantially to the cost of cleanup. Regulatory agency staff members are faced with tough decisions and fierce opposition when pursuing additional requirements to address the late discovery of 1,4-dioxane at solvent release sites. This section profiles a few of the challenges faced by regulators and regulated parties.

10.4.1   Monitored Natural Attenuation and 1,4-Dioxane

Is MNA an acceptable remedy for solvent sites where 1,4-dioxane is a co-contaminant? USEPA uses the term MNA to refer to reliance on a variety of *in situ* physical, chemical, or biological processes that, under favorable conditions, act without human intervention to reduce the mass, toxicity, mobility, volume, or concentration of contaminants in soil or groundwater (USEPA, 1999). These *in situ* processes include biodegradation, dispersion, dilution, sorption, volatilization, and chemical transformation or destruction of contaminants. When relying on natural attenuation processes for site remediation, USEPA prefers those processes that degrade or destroy contaminants and indicates that it is only appropriate to adopt MNA as a remedial strategy at sites that have a low potential for contaminant migration (USEPA, 1999). MNA is typically used in conjunction with active remediation measures. For example, active remedial measures could be applied in areas with high concentrations of contaminants while MNA is used for low concentration areas, or MNA could be used as a follow-up to active remedial measures (USEPA, 1999).

USEPA's policies for site remediation direct staff to ensure that source control measures should use treatment wherever practicable and engineering controls such as containment where treatment is impracticable. In addition, the maximum beneficial uses of contaminated groundwater should be restored within a reasonable timeframe wherever practicable; when restoration is not practicable, USEPA directs staff to seek remedies that will prevent further migration of the plume, prevent exposure to the contaminated groundwater, and further reduce risk (USEPA, 1999). The USEPA directive on MNA notes "MNA will be an appropriate remediation method only where its use will be protective of human health and the environment and it will be capable of achieving site-specific remediation objectives within a timeframe that is reasonable compared to other alternatives" (USEPA, 1999).

The USEPA MNA directive includes several important points that bear on whether MNA is an acceptable remedy for 1,4-dioxane sites. The directive notes that since engineering controls are not used to control plume migration in an MNA remedy, decision makers need to ensure that MNA is appropriate to address all contaminants that represent an actual or potential threat to human health or to the environment (USEPA, 1999). The directive specifically notes that cleanup of solvent spills is complicated by the fact that a typical spill may include contaminants that *tend not to degrade* *readily in the subsurface. For example, 1,4-dioxane, which is used as a stabilizer for some chlorinated solvents, is more highly toxic, less likely to sorb to aquifer solids, and less biodegradable than some other solvent constituents under the same environmental conditions* (USEPA, 1999). The directive also calls for identifying the TICs[*] and determining whether MNA will sufficiently diminish their concentrations and eliminate risk.

The persistent and mobile nature of 1,4-dioxane in groundwater hinders its eligibility as a candidate for MNA remediation; however, because 1,4-dioxane migrates at essentially the same rate as groundwater, it will become more rapidly attenuated by physical processes such as dispersion, matrix diffusion, and dilution from recharge. Therefore, chlorinated solvent sites with 1,4-dioxane releases may nevertheless be suitable for MNA as a reasonable long-term remedy if the intent of the USEPA protocol can be upheld to the satisfaction of USEPA's staff.

1,4-Dioxane plumes can become very large due to the fate and transport properties described in Chapter 3. Where 1,4-dioxane plumes are regional in nature, the high cost of installing dozens of extraction wells and multiple advanced oxidation treatment systems may make conventional remediation impractical. The Pall Life Sciences case study in Chapter 8 profiles a regional plume where the court-ordered remedy was to establish a groundwater exclusion zone to prevent the use of 1,4-dioxane-contaminated groundwater in Ann Arbor, Michigan. 1,4-Dioxane within the exclusion zone will attenuate due to physical processes as it migrates toward the Huron River, where it is expected to be further diluted to inconsequential concentrations upon discharge to the river. In another regional plume case involving perchlorate in Santa Clara County, California, hundreds of private wells were impacted with low concentrations near California's 6 µg/L perchlorate MCL. Because perchlorate exposure from groundwater occurs primarily through drinking and cooking, providing bottled water was deemed sufficient to eliminate most exposures. This option might not be available for a regional 1,4-dioxane plume because inhalation exposure through bathing, showering, dishwashing, and other household water uses may pose health risks. To satisfy USEPA's MNA protocol requirements for 1,4-dioxane plumes, dischargers must ensure that MNA is protective of human health by verifying that there are no currently exposed receptors (especially private wells) and that beneficial uses will not become impaired over the planned timeframe of the MNA remedy.

Regulatory agency decisions on whether MNA is an acceptable remedy have typically hinged on whether there may be a more practical means of removing the groundwater contaminants more quickly than would be achieved by MNA. However, 1,4-dioxane treatment using advanced oxidation is very energy intensive. Consequently, the environmental impact of carbon dioxide (CO_2) emissions from the power generation used to supply electricity to the treatment operation may be significant. Growing concerns over the contribution of CO_2 emissions to global warming are causing the groundwater remediation community to rethink the benefits of energy-intensive pump and treat systems. "Green remediation" focuses on including CO_2 emission considerations in the remedial technology feasibility analysis. When global warming concerns are included in the evaluation of treatment options that are equally protective of human health, MNA may be favored even though the timeframe to complete cleanup and obtain closure may be longer.

10.4.2   1,4-DIOXANE AND NATIONAL POLLUTION DISCHARGE ELIMINATION SYSTEM EFFLUENT LIMITATIONS

The discharge or reuse of extracted and treated groundwater resulting from the cleanup of groundwater contamination sites is regulated by the National Pollutant Discharge Elimination System (NPDES). USEPA delegates the authority to issue NPDES permits to state regulatory agencies, which oversee the discharge of treated groundwater to surface water. The permits specify water quality conditions that must be maintained in the receiving water (e.g., the creek or river to which

---

[*] TICs are analytes that do not match the standard list of analytes reported in the laboratory method.

the treated effluent is discharged) to sustain aquatic life, and the permits limit the mass of chemical constituents that may be discharged, typically on a pounds per day basis. Many NDPES permits have been issued for chlorinated solvent cleanup sites. When 1,4-dioxane is discovered after the permit has been issued, the regulator and the regulated party are faced with evaluating appropriate discharge limits for 1,4-dioxane.

The aquatic toxicity threshold for 1,4-dioxane is high. For example, the fathead minnow LC_{50} is approximately 10,000 mg/L (see Section 6.2.2 and Tables 6.4 through 6.10 for a complete review of the aquatic toxicity of 1,4-dioxane). Therefore, the water quality objectives protected in NPDES effluent limitations are confined to preserving drinking water quality, which is often a direct beneficial use of surface waters, and may be an indirect beneficial use where the surface water body recharges aquifers from which municipal supply wells pump drinking water. Most often, the volume of the discharge from a cleanup site is several orders of magnitude smaller than the flow in the receiving water such that the discharge is substantially diluted. Nevertheless, NDPES effluent limits do not usually account for dilution, and 1,4-dioxane limits are often set to the state's drinking water action levels or USEPA PRGs, and in some states, to higher values that are less than 100 times the action level (see Section 6.2.4 and Table 6.12). California has issued NPDES permits specifying 1,4-dioxane effluent limitations equal to the drinking water NL, 3 µg/L.

Establishing low 1,4-dioxane effluent limitations in NPDES permits may be at odds with groundwater cleanup orders that leverage the dilution that is expected to occur at the groundwater–surface water interface to permit a higher groundwater cleanup level. For example, the Pall Life Sciences case study in Chapter 8 explains the reasoning that went into establishing a limit of 2800 µg/L for the groundwater–surface water interface. Another example is the industrial site in North Carolina where MNA was approved based on the modeling study profiled in Section 7.3.3 and on dilution when the plume eventually discharges to the river (Chiang et al., 2006). NPDES permits that set low limits for 1,4-dioxane are also not well-matched to USEPA wastewater effluent limits. USEPA issued a wastewater universal treatment standard (UTS) for 1,4-dioxane of 0.22 mg/L in 1996 (USEPA, 1996).

The question of what level of 1,4-dioxane is safe to discharge to a surface water body is often asked after the discharge has been ongoing for years, due to the late discovery of 1,4-dioxane in the effluent from treatment of groundwater contaminated with chlorinated solvents. Therefore, it should be possible to establish whether 1,4-dioxane is impacting beneficial uses of surface water and groundwater by sampling the receiving water and the adjacent groundwater for 1,4-dioxane. A reasonable challenge to a regulatory agency's decision to set a low NPDES effluent limitation for 1,4-dioxane would be to sample the receiving water and to install monitoring wells in the aquifers that are recharged by the receiving water body some distance downstream of the discharge point. Because 1,4-dioxane migrates very rapidly, if it was discharged at problematic concentrations, it should be detectable in groundwater. Collection of field data can lead to a more informed basis for setting 1,4-dioxane effluent limits, and may give the regulatory caseworker added confidence to waive requirements to reduce 1,4-dioxane concentrations in treated water effluent, or provide evidence to justify issuing stricter effluent limits. In addition to the regulatory agency, the water quality stakeholders (local water utilities or water districts and other downstream surface water users or downgradient groundwater pumpers) should be consulted in decisions to permit discharge of 1,4-dioxane at concentrations in excess of drinking water action levels. Water quality stakeholders will seek assurances that the cumulative impact of all discharges does not result in a detectable degradation of water quality in either surface water or groundwater.

10.4.3   Lessons Learned from 1,4-Dioxane Case Studies

The treatment examples provided in Chapter 7 and the case studies profiled in Chapter 8 summarize a number of important remediation and site investigation features that are unique to hydrophilic compounds like 1,4-dioxane. For convenience, the important lessons from these treatment examples and case studies, as well as some of the 1,4-dioxane occurrence and use information from Chapter 2 and fate and transport characteristics from Chapter 3 are summarized as follows:

1. Expect to find 1,4-dioxane at vapor degreasing sites that used 1,4-dioxane-stabilized methyl chloroform and released vapor degreasing wastes.
2. Expect to find 1,4-dioxane at solvent recycling facilities.
3. Expect to find 1,4-dioxane further downgradient than the leading edge of the solvent plume.
4. Expect to find 1,4-dioxane in groundwater within silts and clays near the source zone, but not usually in sands and gravels at release sites unless the spill is recent (it is often flushed out).
5. Expect to find 1,4-dioxane in municipal, industrial, and particularly in university landfills.
6. Expect to find 1,4-dioxane in landfill leachate and landfill gas condensate, and in groundwater contaminated by landfills.
7. Expect to find 1,4-dioxane at cement kilns that used solvent wastes for fuel.
8. Expect to find 1,4-dioxane in water bodies receiving effluent from textile manufacturing, resin production, plastics manufacturing, photographic film production, and cellulose acetate membrane production.
9. Expect to find low levels of 1,4-dioxane (1–2 µg/L) in wastewater treatment influent, in treated wastewater effluent, and in recycled water that has not been subjected to advanced oxidation.
10. Expect remedial engineering challenges for 1,4-dioxane treatment, including high costs.
11. Expect adverse public reaction when the late discovery of 1,4-dioxane at cleanup sites is publicized, especially where domestic wells have been contaminated with 1,4-dioxane.

Regulatory agency caseworkers and remedial project managers should consider these lessons when drafting site cleanup orders, Records of Decision, or other cleanup requirements. Consulting engineers and hydrogeologists performing remedial investigations and feasibility studies on behalf of dischargers also stand to benefit from taking these lessons into account when recommending solutions for remediating 1,4-dioxane and solvent releases to soil and groundwater.

10.4.4   1,4-DIOXANE TREATMENT TECHNOLOGY RESEARCH NEEDS

Remediating 1,4-dioxane is expensive and challenging, because the best options currently available are primarily limited to groundwater extraction and *ex situ* treatment using advanced oxidation. Success stories for the treatment of 1,4-dioxane *ex situ* include Applied Process Technology's HiPOx™, Basin Water's Photo-Cat™, Trojan Technologies' UVPhox™, Calgon Carbon's Rayox®, and others (see Chapter 7). New technologies successful at *ex situ* removal of 1,4-dioxane from extracted groundwater include Liquid Separation Technologies and Equipment's LSTE-10, which uses a multichambered vacuum aeration tank and a high-vacuum separator tower.

In-well treatment successes have been achieved with the ART In-Well system (see Section 7.1.2) and Applied Process Technology's Pulse-OX™ (see Section 7.7.2). ISCO technologies include FMC's persulfate solution, Klozur™ (see Section 7.7.5) and Isotec's Fenton chemistry solution (see Section 7.7.4). Research continues in most of the treatment technology areas profiled in Chapter 7.

The greatest need for research in 1,4-dioxane remediation is to overcome the challenges to achieving reliable *in situ* treatment of 1,4-dioxane using bioremediation and ISCO. The biodegradation research profiled in Chapters 3 and 7 provides a solid foundation on which the state of 1,4-dioxane bioremediation science can be advanced. Research needs include the following:

1. Isolating and sustaining the growth of 1,4-dioxane-respiring cultures that can be acclimated to the *in situ* geochemical environment
2. Developing compound-specific isotope analysis methods sensitive enough to detect isotope ratios in 1,4-dioxane and its metabolites at environmentally relevant (i.e., low) concentrations

3. Developing improvements to ISCO technology that can be safely and economically deployed to remediate 1,4-dioxane source zones
4. Developing analytical methods that can reliably detect the expected biodegradation and reaction breakdown products of 1,4-dioxane at trace levels

10.5   BLINDSIDED AGAIN?

Groundwater scientists must face the inevitable question about the appearance of 1,4-dioxane at so many solvent release sites years after adopting cleanup plans to address solvent contamination: "How could this have happened?" Have we routinely examined the full gamut of contaminants in solvent degreasing wastes and dismissed them from the list of contaminants of concern? Or have we neglected to consider what it is we are cleaning up? Site characterization must include analysis of the source waste itself wherever available for sampling, or investigations should leverage GC/MS analysis to perform open-scan searches for TICs in source-zone samples, lest we be blindsided again by surprise contaminants like 1,4-dioxane (Mohr and Crowley, 2001).

When should we have known?* Did we miss the signs? There have certainly been a few signs available to show that 1,4-dioxane was potentially harmful and that it was a very mobile contaminant. Moreover, some in the community of groundwater professionals, particularly regulatory staff in USEPA, California DTSC, water boards, and probably water quality staff in other states, recognized that 1,4-dioxane poses a significant problem and took action to investigate. At different points in time, there have been pockets of awareness, but apparently, that awareness was not sufficiently extensive to cause regulators and consultants to pursue 1,4-dioxane universally at solvent cleanup sites. When individual regulators realized the potential for 1,4-dioxane to be a significant groundwater contaminant, they were probably challenged to find the time or authorization to pursue the issue in addition to their already large caseloads. Awareness has grown substantially in the beginning of the new millennium, as measured by thousands of downloads of the *Solvent Stabilizers White Paper* (Mohr, 2001) since its presentation to the Groundwater Resources Association of California in 2001. Some signs of the potential hazard to water resources posed by 1,4-dioxane and other solvent stabilizers were available to industry and to water quality professionals through the past several decades, as summarized in Table 10.1.

A review of the toxicology literature in Chapter 5 shows that since 1,4-dioxane was patented as a metal inhibitor for Dow Chemical's grades of methyl chloroform in 1957, there have been publications on 1,4-dioxane toxicity or carcinogenicity in nearly every decade. Even so, does the above list of sources of information mean that we should have known and prioritized investigation of 1,4-dioxane releases sooner? The sources of information in Table 10.1 reside in disparate corners of the various professions whose expertise is invoked to protect groundwater quality. Moreover, although these information sources were all available in the pre-Google decades, it would have been rather difficult for individuals and institutions to assimilate all of this scattered information and draw a conclusion to prioritize 1,4-dioxane. The possible exceptions include the USEPA, which is charged with identifying environmental threats and whose libraries house much of the needed information, and Dow Chemical or other producers of 1,4-dioxane and 1,4-dioxane-stabilized methyl chloroform, whose research chemists evaluate the potential environmental hazards of their products.

A broader question is whether the community of groundwater professionals should have realized the hazards posed by the widespread use and routine disposal of chlorinated solvents to facilities we now know were ill suited to contain them. The answer is, debatably, "yes": Analytical methods to detect chlorinated solvents in groundwater at the concentrations found near source zones were well

---

* The collective "we" refers to the community of groundwater professionals and scientists, whether employed by regulatory agencies, industry, consulting firms, universities, or water districts.

TABLE 10.1
Chronology of Available Information on 1,4-Dioxane Hazards

Year	Information Source	References
1934	1,4-Dioxane animal toxicity study published	Fairley et al. (1934)
1934	Report on fatalities from massive 1,4-dioxane exposure	Barber (1934)
1965	Publication of 1,4-dioxane carcinogenicity study	Argus et al. (1965)
1965	Navy report on composition of methyl chloroform	Saunders (1965)
1968	Beckman Instruments promotes dioxane-free scintillation fluid to avoid "bad side effects associated with dioxane"	Beckman Instruments (1968)
1970	Two-year study of 1,4-dioxane inhalation in rats published	Hoch-Ligeti et al. (1970)
1970	1,4-Dioxane recognized as a by-product of PET plastic production process	Hovenkamp and Munting (1970)
1972	Beckman Instruments Advertisement: "Turn off dioxane danger"	Beckman Instruments (1972)
1976	Manufacturing Chemists' Association funds research into toxicity of epichlorohydrin, a solvent stabilizer formerly used with TCE, and an epoxy resin component[a]	MCA (1976)
1977	Drinking water analysis study detects 1,4-dioxane	Donaldson (1977)
1978	1,4-Dioxane recognized as one of the top eight air pollutants for which emission reductions are recommended in New Jersey	Chemical Week (1978)
1979	FDA announces that hundreds of cosmetics and sundries contain 1,4-dioxane	Washington Post (1979)
1979	Groundwater study detects methyl chloroform in 835 of 1071 New Jersey groundwater samples (78%) in 1977–1979	Page (1981)
1980	Publication: Identification of potential carcinogens in technical grade methyl chloroform	Henschler et al. (1980)
1981	USEPA phase 1 risk assessment for 1,4-dioxane	USEPA (1981)
1982	Controversy over 1,4-dioxane in contraceptive sponges and spermicidal lubricants	Medical Economics Publishing (1983)
1984	USEPA Groundwater Primer identifies 1,4-dioxane as a contaminant	USEPA (1984)
1986	30 private wells discovered contaminated with 1,4-dioxane near Ann Arbor, Michigan (see Chapter 8, Pall Life Sciences Case Study)	Fotouhi et al. (2006)
1988	USEPA publishes heated purge and trap method capable of 1,4-dioxane detection	Lucas et al. (1988)
1989	Publication: Reclamation and reprocessing of spent solvent	Tarrer et al. (1989)

[a] While the research by MCA did not address 1,4-dioxane toxicity, it demonstrates awareness that solvent stabilizers can impart toxicity to vapor degreasing grades of solvents.

known since at least 1950, and a method with a detection limit of 10 µg/L was published as early as 1953 (Amter and Ross, 2001). The following commentary was issued in response to an editorial titled "Once Again Blindsided!" published in the National Ground Water Association Newsletter, which raised the question of whether groundwater professionals should also have known about the threat posed by 1,4-dioxane (Mohr and Crowley, 2001): "Water quality professionals were only partially blindsided; at least some people, in some places, were concerned before others. Furthermore, past limitations in detection technology weren't as dire as is often believed, and thus can't fully explain the situation in which we now find ourselves" (Amter, S., Disposal Safety Inc., Washington, D.C., personal communication, 2002).

The blindsiding effect of 1,4-dioxane continues, extending beyond the realm of groundwater contamination. In 2007, a survey of 1,4-dioxane content in children's bath products revealed that the FDA's limits for 1,4-dioxane in ethoxylated surfactants were exceeded in numerous products, including imported products. In 2008, California's attorney general sued a major retailer of organic food and organic products for violating California's Safe Drinking Water and Toxic Enforcement Act of 1986 (Proposition 65) by failing to remove 1,4-dioxane from body washes, gels, and liquid dish soaps (Phillips, 2008). In 2009, the Saudi Food and Drug Authority pulled eight types of shampoos from store shelves in Saudi Arabia, following the sampling of 84 brands of shampoo available in the Saudi market and detection of 1,4-dioxane in eight shampoos. The action was taken over the concern that extended exposure to 1,4-dioxane can be damaging to the liver and kidney (Abdullah, 2009). Similar actions were taken for shampoos produced in India and Iran by authorities in Kuwait and the United Arab Emirates following the Saudis' analysis of shampoos.

The Saudi response seems to neglect the relevance of the route of exposure, as dermal exposure is not known to be a major pathway for 1,4-dioxane toxicity. The relatively high NOAELs and the absence of dose–response values for dermal exposure to 1,4-dioxane make it difficult to estimate the health effects from repeated exposure to low levels of 1,4-dioxane in contaminated shower water or in shampoos and soaps. The majority of published 1,4-dioxane risk summaries conclude that the risk of adverse health effects from dermal exposure to low concentrations in water or sundries is not significant (see Sections 5.6 and 6.5.4). There is a notion held by some that a small amount of a chemical carcinogen in a children's product is not dangerous because the level is very low, and low doses of cancer-causing chemicals are safe because there is a threshold for cancer induction. That notion is refuted by two prominent environmental oncologists who take issue with the presence of 1,4-dioxane in shampoos and in children's bath products and advance another notion: "The combined effects of our lifetime exposure to dioxane and other carcinogens can create synergistic effects, so what may look like low exposure levels for any one compound adds up and even multiplies" (Davis and Heberman, 2007).

The question of whether adverse synergistic effects of 1,4-dioxane with other toxins actually occur remains unproven and is an important topic for research in the arena of computational toxicology. Does the detection of 1,4-dioxane in shampoos justify removing products from store shelves? The Saudi Food and Drug Authority applied the precautionary principle and decided that removing eight shampoo products still leaves consumers with 76 safe shampoos and ensures that no exposure to 1,4-dioxane occurs. The Saudi newspaper noted, "There are processes that can remove this contaminant, but some companies forego the added expense to ensure their products do not contain the solvent" (Abdullah, 2009).

Nearly a million people in Daegu, South Korea's fourth largest city, were also blindsided by 1,4-dioxane in their drinking water supply, the Nakdong River. Daegu Metropolitan Waterworks reported in January 2009 that 1,4-dioxane levels at the Maegok water purification plant rose to 54 μg/L, exceeding the 50 μg/L World Health Organization Guideline for Drinking-Water Quality (Dong-A Ilbo, 2009). The 1,4-dioxane concentrations peaked at 88 μg/L, then stayed consistently at 65 μg/L at one monitoring point for one week. In 2003, concentrations in the Nakdong River reached 119 μg/L (see Table 6.13). The source of the 1,4-dioxane is upstream discharges from polyester manufacturing plants.

10.6   THE PROMISE OF GREEN CHEMISTRY

Can we avoid being blindsided by another 1,4-dioxane-like environmental contaminant? The existing regulatory framework for testing new chemical products for toxicological and environmental fate and transport properties has had limited effectiveness. Existing law has not prevented contaminants like MTBE, nitrosamine compounds [nitrosodimethylamine (NDMA), nitrosodiethylamine (NDEA) and others], 1,4-dioxane, polybrominated diphenyl ether (PBDE) flame retardants, poly fluorinated octanoic acid (PFOA, also called C-8), perfluorooctane sulfonic acid (PFOS) and others from being widely used and released into the environment. Some of these contaminants are particularly widespread and have attained global circulation; the insidious nature of PBDE's includes detection in polar bear fat and human breast milk (Stavelova, 2008), and the pervasive nature of PFOA includes detection in most human blood samples.

What requirements should society demand of new and existing chemical products or any other technology with potential to cause harm to human health and the environment? The question leads back to the precautionary principle: environmentally conscious decision-makers would not start with the question, "How much of this pollutant can we stand?" Instead, they would ask, "Is there an alternative to this polluting activity? Does society need this activity in the first place? What are the implications for future generations?" (McKenna and Sylvester, 2004). Chlorinated solvents were extremely useful to industry, and they could not be used effectively without solvent stabilizers like 1,4-dioxane. Yet, after the 1996 Montreal Protocol ban on using methyl chloroform, industry has adapted to find alternatives and to eliminate virtually all solvent emissions from those vapor degreasing operations that still use chlorinated solvents. When challenged, industry has proven itself remarkably adaptable and innovative, surviving and even thriving in a changed regulatory environment.

The existing regulatory framework for testing new chemicals entering the marketplace is a voluntary right-to-know program called the U.S. High Production Volume (HPV) Challenge Program.[*] The program was adopted in 1998 after the environmental nonprofit group Environmental Defense published a report called *Toxic Ignorance* (Environmental Defense, 1997). The report advised that more than 70% of industrial chemicals in active, high-volume use in the U.S. marketplace lacked sufficient data on toxicity and environmental fate to permit basic evaluation of potential environmental and human health hazards, as determined from the review of publicly available records. Follow-on surveys by both USEPA and the CMA found that the number of chemicals lacking basic hazard data was even greater than Environmental Defense reported: more than 90% of the HPV industrial chemicals in U.S. commerce lacked sufficient publicly available hazard-screening data. The findings led to a joint effort to develop a more effective means of ensuring an increased level of testing for new and existing chemicals. The effort was undertaken by Environmental Defense, USEPA and CMA (now the American Chemistry Council), and led to the adoption of the HPV Challenge Program. The HPV Challenge Program calls for chemical producers to voluntarily fill gaps in basic screening-level hazard data for HPV chemicals—those produced in the United States in amounts of one million pounds or more annually (Denison, 2004).

The HPV Challenge program originally called for the development and public release of screening-level hazard data for nearly 2800 chemicals; however, 532 of the chemicals originally included in the program were either never sponsored (by companies who will fund the research to develop screening-level hazard data) or have had initial sponsorships withdrawn. Of these 532 "unsponsored" chemicals, 156, and perhaps as many as 259, are "orphans." Until they are sponsored, gaps will persist in the public availability of important environmental data (Denison, 2004).

USEPA's National Center for Computational Toxicology performed a more extensive review of the availability of basic data for chemicals in the marketplace. The survey found that of 9912 chemicals reviewed, at least some acute hazard data were publicly available for 66% of the chemicals, while no toxicology data were available for 34%. Data were less available for specific disease endpoints: carcinogenicity data were available for 26%, developmental toxicity for 29%, reproductive toxicity for 11%, and genotoxicity for 28% of the 9912 chemicals (Judson et al., 2008). The survey underscores the ongoing need for chemical testing and evaluation, and the monumental task required to keep pace with new chemicals entering the marketplace. We need a new approach for developing and introducing new chemicals, with their environmental fate and human health impacts as a starting consideration in the product development process. That new approach is Green Chemistry.

Green Chemistry refers to design of chemical products and processes with the objective of reducing or eliminating the use and generation of hazardous substances, minimizing energy consumption, minimizing or eliminating waste generation, and incorporating renewable raw materials (Goosey, 2008). The main idea behind Green Chemistry is to carefully select and control chemicals at the

---

[*] For more information on the HPV Challenge program, see http://www.epa.gov/chemrtk/index.htm

front end of the manufacturing process rather than trying to capture, manage, or clean up waste at the back end. Some of the key principles of Green Chemistry include the following:

1. Preventing the generation of waste rather than dealing with waste disposal
2. Incorporating all of the materials used in manufacturing processes into the final product
3. Substituting less hazardous or nontoxic chemicals
4. Designing safer chemicals that achieve their performance objective with little or no toxicity
5. Using safer solvents and other auxiliary chemicals in manufacturing
6. Designing chemical processes to optimize energy efficiency; where possible, conducting chemical synthesis at ambient temperature and pressure
7. Using renewable feedstocks instead of those that deplete finite natural resources whenever technically and economically feasible
8. Designing chemical products that break down into innocuous degradation products at the end of their function and do not persist in the environment
9. Choosing safe chemical components for the chemical process to minimize the potential for chemical accidents, including releases, explosions and fires

An example of Green Chemistry is the substitution of supercritical CO_2 as a cleaning agent to replace chlorinated solvents (Strickland, 2008; Goosey, 2008).

Europe, Canada, and California now have Green Chemistry laws. In 2007, the European Union adopted the REACH regulation. REACH creates a unified regulatory framework for substances on the European market and requires that manufacturers or importers of at least 1 metric ton (1000 kg) per year of chemical substances submit a registration dossier to the European Chemicals Agency in Helsinki, Finland (Strickland, 2008; Lahl and Hawxwell, 2006). The registration dossier is required to provide information about the chemical identity of the substance in question, including its physical–chemical properties, toxicity assay profiles, and ecotoxicity properties.

The REACH approach will require consideration of water quality and air pollution factors at the registration stage of introducing a new chemical into the marketplace, including specialized hazard categories, such as endocrine disruption in aquatic organisms (Strickland, 2008). The collected data will be maintained in a publicly available database. Application to obtain authorization to use chemical substances must include an analysis of available substitutes or alternative production processes, as well as an analysis of their technical and economic feasibility (Strickland, 2008; Lahl and Hawxwell, 2006).

In California, Governor Arnold Schwarzenegger signed the DTSC's Green Chemistry Initiative into law on September 29, 2008 (Renner, 2009). The new laws (AB 1879 and SB 509) give the California DTSC two years to identify and prioritize toxic, persistent, and bioaccumulative chemicals and will create a new online Toxics Information Clearinghouse for businesses and consumers. The goal of California's new laws is to move away from a chemical-by-chemical approach and toward a more comprehensive policy towards cradle-to-cradle stewardship to prevent unintended consequences. Unlike REACH, California's program does not require companies to provide chemical data; instead, the DTSC will conduct lifecycle assessments on existing chemicals and their alternatives (Renner, 2009). California is the first American state to adopt Green Chemistry laws. However, if each state pursues its own Green Chemistry program, industries may be stymied by 50 different sets of program and registration requirements, which could encourage new production to be outsourced offshore. USEPA has begun a series of Green Chemistry initiatives, including on-line tools to help manufacturers find safer alternatives to chemicals used in their operations (see http://www.epa.gov/greenchemistry/). USEPA coined the phrase "cradle-to-cradle"* and proposed a

---

* The "cradle-to-cradle" concept posits that it is more profitable and environmentally beneficial to design and produce chemicals that may be readily recovered as raw materials once a product's useful life has ended.

"Green Pharmacy" approach to life-cycle stewardship of pharmaceuticals and personal care products (Daughton, 2003).

While California has established a tradition of pioneering environmental regulatory policies, only USEPA can implement a nationally unified regulatory framework for Green Chemistry. Therefore, USEPA should lead the way to develop a U.S. national equivalent of Europe's REACH program.

10.7   SHOULD 1,4-DIOXANE BE BANNED?

Given 1,4-dioxane's toxicity and associated groundwater contamination hazard, should all uses of 1,4-dioxane be banned? No. To do so would be to disregard the major advances in industry stewardship and regulations governing chemical handling, workplace training, enforcement, manifesting wastes, monitoring, full accounting of chemicals entering and leaving industrial operations, and more. These actions and requirements together offer orders of magnitude greater environmental protection than was in effect at the time most solvent releases occurred. Because methyl chloroform was banned as an ozone depleting compound by the Montreal Protocol in 1990, and because 90% of 1,4-dioxane produced in the 1970s and 1980s was used to stabilize methyl chloroform, the total volume of 1,4-dioxane now in use has been substantially reduced. Moreover, 1,4-dioxane is still revealing its unique properties to chemical researchers, who continue to find innovative uses for the symmetry of the dioxane ring. One example follows, although 1,4-dioxane has been an agent in many more chemical discoveries.

Researchers at the University of Edinburgh, Scotland, serendipitously created polymer doughnuts while studying potential drug-carrying microparticles. While synthesizing micro-spheres, the team added 5% of 1,4-dioxane to their usual ethanol solvent. To their surprise, the resulting microparticles were regular in size and shape, with a hole through the middle, just like a doughnut (see Figure 10.1). Their unique and uniform structure was immediately interesting to the researchers, who tested the potential usefulness of the dioxane-derived polymer doughnuts as carrier particles for drug delivery. The dioxane-derived polymer doughnuts showed a high level of liver cell-specificity (94% uptake rate), leading the team to conduct extensive *in vivo* testing in mice. Human embryonic kidney

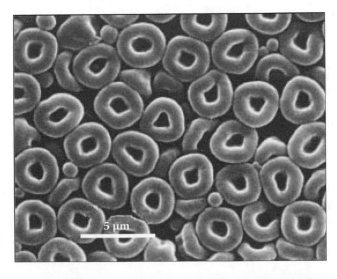

FIGURE 10.1   1,4-Dioxane-derived polymer doughnuts, with potential uses for drug delivery to liver cells, liver toxicity testing, water filtration, and other as yet undiscovered applications. (Reproduced from Royal Society of Chemistry, 2008, *Science Daily*, June 24. http://www.sciencedaily.com/releases/2008/06/080624120237. htm [accessed June 25, 2008]. With permission.)

cells demonstrated a 50% uptake of the doughnut, while human ovarian cancer cells had an 18% uptake rate. Researchers injected the doughnuts into mice tail veins and, within 4 h, the doughnuts had all migrated to the mouse livers, causing no adverse effects. The doughnuts thus appear to be ideally suited to delivering drugs or other agents to the mouse liver and may provide utility for *in vivo* organ-specific toxicity testing (Alexander et al., 2008; Royal Society of Chemistry, 2008).

10.8   REGULATORY POLICY RECOMMENDATIONS FOR ADDRESSING 1,4-DIOXANE RELEASES

The following regulatory policy recommendations are made for consideration by regulators and regulated parties with the goal of applying a consistent set of regulations and response priorities to effectively deal with 1,4-dioxane releases in a manner commensurate with the risks to human health and the environment:

1. Fund studies to further improve and validate PBPK models by performing mechanistic studies (possibly including *in vitro* and "omics" studies) to further the understanding of 1,4-dioxane mode of action for cancer in animals, focusing on the human relevance of animal cancer data.
2. Prioritize screening-level site investigations for 1,4-dioxane at vapor degreasing sites, solvent recycling sites, cement kilns that used waste solvents for fuels, and at landfills that received scintillation counting wastes (e.g., university landfills*).
3. Include 1,4-dioxane as a routine analyte for public water supply well sampling programs where facilities that used or created 1,4-dioxane as a by-product overlie the well's capture zone.
4. Test public water supply wells for 1,4-dioxane where methyl chloroform or its breakdown products (1,1-dichloroethylene and 1,1-dichloroethane) have been detected.
5. Test for 1,4-dioxane in private wells located near potential sources of 1,4-dioxane.
6. Test recycled water for 1,4-dioxane before starting indirect potable reuse projects.
7. Test for 1,4-dioxane at sites where MNA has been approved for solvent remediation.
8. Fund the 1,4-dioxane remediation research initiatives suggested in Section 10.4.4.
9. Run the gas chromatographs and MSs in open-scan mode, and report all TICs in samples from both cleanup sites and drinking water systems.
10. Fund exploratory research that can be used to anticipate future environmental issues to fulfill the promise of Green Chemistry (see Section 10.7).
11. Revisit the causes of solvent and 1,4-dioxane contamination of groundwater in the United States and Europe, and evaluate whether similar practices are now employed in offshore manufacturing operations in countries whose regulatory framework does not yet address these causes or prevent groundwater contamination by solvents and 1,4-dioxane.

The sources of funding to implement these recommendations must overcome the funding-source dilemma of implied conflict of interest when studies are funded and conducted by the producers of the subject chemicals. One way to do this is to establish a pool of funds from industry and government to perform basic, independent research administered by governmental agencies with oversight by panels of independent experts including representatives of industry, regulatory agencies, academic institutions, and potentially impacted parties such as water purveyors, citizen groups, and environmental groups.

---

* As described in Chapter 2, landfills on university campuses or landfills receiving wastes from universities and other institutions with research laboratories are likely to have scintillation counting wastes and other laboratory wastes that include 1,4-dioxane.

ACKNOWLEDGMENTS

The critical review, questions, corrections, and comments from the following individuals are greatly appreciated and have substantially improved this chapter: Tracy Hemmeter, Santa Clara Valley Water District, San Jose, California; Julie Stickney, Syracuse Research Corporation, Portland, Maine; Rula Deeb, Malcolm Pirnie, Emeryville, California; Dani Renan, Attorney and Hydrogeologist, Concord, California; Murray Einarson, AMEC Geomatrix, Oakland, California; and William DiGiuseppi, AECOM Environment, Englewood, Colorado. The opinions provided in this chapter are the author's alone and do not necessarily reflect the views of the reviewers or their employers.

BIBLIOGRAPHY

Abdullah, S., 2009, Dioxane-tainted shampoos pulled from store shelves. *Arab News*, January 8, 2009.

Alexander, L., Dhaliwal, K., Simpson, J., and Bradley, M., 2008, Dunking doughnuts into cells—selective cellular translocation and *in vivo* analysis of polymeric micro-doughnuts. *Chemical Communications* 2008(30): 3507–3509.

Amter, S. and Ross, B., 2001, Commentaries and perspectives: Was contamination of southern California groundwater by chlorinated solvents foreseen?. *Environmental Forensics* 2: 179–184.

Argus, M.F., Arcos, J.C., and Hoch-Ligeti, C., 1965. Studies on the carcinogenic activity of protein-denaturing agents: Hepatocarcinogenicity of dioxane. *Journal of the National Cancer Institute* 35: 949–958.

ATSDR, 2004, *Guidance Manual for the Assessment of Joint Toxic Action of Chemical Mixtures*. Atlanta, GA: Agency for Toxic Substances and Disease Registry.

Barber, H., 1934, Haemorrhagic nephritis and necrosis of the liver from dioxane poisoning. *Guys Hospital Report* 84: 267–280.

Bhogal, N., Grindon, C., Combes, R., and Balls, M., 2005, Toxicity testing: Creating a revolution based on new technologies. *Trends in Biotechnology* 23(6): 299–307.

Beckman Instruments, 1968, Dioxane substitute. Advertisement for Bio-Solv solubilizers for scintillation counters. *Environmental Science and Technology* 2(4): 307.

Beckman Instruments, 1972, Turn off dioxane danger. Advertisement for Bio-Solv solubilizers for scintillation counters. *Science* 152(April 14): 199.

California Environmental Protection Agency (Cal EPA), 2005, *Draft Public Health Goal for cis- and trans-1, 2-Dichloroethylene in Drinking Water*. Oakland, CA: Office of Environmental Health Hazard Assessment.

Cal EPA, 2008, Groundwater information sheet: 1,4-Dioxane. State Water Resources Control Board. http://www.swrcb.ca.gov/water_issues/programs/gama/docs/1_4_dioxane.pdf (accessed December 18, 2008).

Chemical Week, 1978, Jersey fights smog. *Chemical Week* 1978(15): 24.

Chiang, D., Zhang, Y., Glover, E., Harrigan, J., and Woodward, D., 2006, 1,4-Dioxane solute transport modeling in support of natural attenuation determination. In: B.M. Sass (Ed.), *Proceedings of the Fifth International Conference on Remediation of Chlorinated and Recalcitrant Compounds*, p. G-70. Columbus, OH: Battelle Press.

Daughton, C.G., 2003, Cradle-to-cradle stewardship of drugs for minimizing their environmental disposition while promoting human health. I. Rationale for and avenues toward a green pharmacy. *Environmental Health Perspectives* 111(5): 757–774.

Davis, D., 2007, Off-target in the war on cancer. *Washington Post*, November 4, 2007, Section B01. (See also *The Secret History of the War on Cancer*, 2007, New York: Basic Books)

Davis, D.L. and Heberman, R.B., 2007, Myths vs. facts about cancer and environmental risks. *Healthy Choices, Healthy Lives* 1(4): 1–3. Newsletter of the University of Pittsburgh Cancer Institute Center for Environmental Oncology. http://www.environmentaloncology.org/files/file/Publications/Newsletters/2007-Spring-News.pdf (accessed January 2, 2008).

Denison, R.A., 2004, Orphan chemicals in the HPV challenge: A status report. *Environmental Defense*. June. www.environmentaldefense.org (accessed March 18th, 2007).

Department of Toxic Substances and Control (DTSC), 1996a, 1,4-Dioxane. Memorandum from Barbara Coler, Chief, Statewide Cleanup Operations Division to Branch Chiefs, July 19, 1996. California Department of Toxic Substances and Control.

Department of Toxic Substances and Control (DTSC), 1996b, 1,4-Dioxane. Memorandum from Stephen M. DiZio, Senior Toxicologist, Office of Scientific Affairs to Barbara Coler, Chief, Statewide Cleanup Operations Division, July 5, 1996. California Department of Toxic Substances and Control.

Donaldson, W.T., 1977, Trace organics in water: Refined methods are needed to document their occurrence and concentrations. *Environmental Science and Technology* 11(4): 348–352.

Dong-A, I., 2009, High levels of carcinogen found in Daegu tap water http://english.donga.com/srv/service.php3?biid=2009012148818 (accessed January 21, 2009).

Einarson, M.D. and Mackay, D.M., 2001, Predicting impacts of groundwater contamination. *Environmental Science and Technology* 35(3): 66A–73A.

Engelhaupt, E., 2009, Pollutants remain in Americans' blood despite bans. *Environmental Science and Technology*, Article ASAP Publication Date (Web): January 14, 2009. http://pubs.acs.org/doi/full/10.1021/es803694d (accessed January 15th, 2009).

Environmental Defense, 1997, *Toxic Ignorance*, http://www.edf.org/documents/243_toxicignorance.pdf (accessed March 18, 2007).

European Environment Agency (EEA), 2000, In P. Harremoës et al. (Eds), *Late Lessons From Early Warnings: The Precautionary Principle 1896–2000*. Environmental Issue Report No. 22. Copenhagen, Denmark: European Environment Agency.

Fairley, A., Linton, E.C., and Ford-Moore, A.H. 1934, The toxicity to animals of 1,4 dioxane. *Journal of Hygiene* 34: 486–501.

Field, J.A., Johnson, A., and Rose, J.B., 2006, What is "emerging"? *Environmental Science and Technology* 40(23): 7105.

Fotouhi, F., Tousi, S., and Brode, J.W. 2006, Managing a significant release of 1,4-dioxane into a complex glacial depositional environment: The integration of hydrogeology, remedial engineering and politics. Presented at Emerging Contaminants in Groundwater: A Continually Moving Target, June 7–8, 2006. Concord, CA: Groundwater Resources Association of California.

Global Development Research Center (GRDC), 1998, Wingspread statement on the precautionary principle. http://www.gdrc.org/u-gov/precaution-3.html (accessed January 10, 2009).

Goosey, M., 2008, REACH and Green Chemistry. http://www.envirowise.gov.uk/ (accessed December 23, 2008).

Gottmann, E., Kramer, S., Pfahringer, B., and Helma1, C., 2001, Data quality in predictive toxicology: Reproducibility of rodent carcinogenicity experiments. *Environmental Health Perspectives* 109(5): 509–514.

Health and Safety Laboratories (HSL), 2008, British Health and Safety Laboratory Website, Joint Industry Project—the Rapid Generation of PBPK Models. http://www.hsl.gov.uk/capabilities/pbpk.htm (accessed December 26, 2008).

Henschler, D., Reichert, D., and Metzler, M., 1980, Identification of potential carcinogens in technical grade 1,1,1-trichloroethane. *International Archives of Occupational and Environmental Health* 47(3): 263–268.

Hoch-Ligeti, C. and Argus, M.F., 1970, Effect of carcinogens on the lung of guinea pigs. In P. Nettesheim, M.G. Hanna, and J.W. Deatherage (Eds), *Morphology of Experimental Respiratory Carcinogenesis*, pp. 267–279. AEC Symposium Series 21, National Cancer Institute and U.S. Atomic Energy Commission. CONF700501.

Hoch-Ligeti, C., Argus, M.F., and Arcos, J.C., 1970, Induction of carcinomas in the nasal cavity of rats by dioxane. *British Journal of Cancer* 24(1): 164–167.

Hovenkamp, S.G. and Munting, J.P., 1970, Formation of diethylene glycol as a side reaction during production of polyethylene terephthalate. *Journal of Polymer Science: Part A-1. Polymer Chemistry* 8(3): 679–682.

Judson, R., Richard, A., Dix, D.J., et al., 2008, The toxicity data landscape for environmental chemicals. *Environmental Health Perspectives* 117(5): 685–695.

Lahl, U. and Hawxwell, K.A., 2006, REACH—the New European chemicals law. *Environmental Science and Technology* 40(23): 7115–7121.

Lucas, S.V., Burkholder, H.M., and Alford-Stevens, A., 1988, Heated purge and trap method development and testing. U.S. Environmental Protection Agency, Environmental Monitoring and Support Laboratory. EPA/600/4-88/029.

Manufacturing Chemists' Association, 1976, Minutes of the 245th Meeting of the Board of Directors of the Manufacturing Chemists' Association, Inc., held at the Madison, Washington, DC, January 13.

McKenna, B. and Sylvester, T., 2004, An ounce of prevention: A precautionary principle primer. *From the Ground Up* November/December 2004. Ecology Center, Ann Arbor, MI. www.ecocenter.org (accessed January 19, 2008).

Medical Economics Publishing, 1983, Contraceptive sponge holds up under criticism. *Mark Drug Topics* 127(August 15): 23.

Michaels, D., 2005, Doubt is their product. *Scientific American* 292(6): 96–101.

Michaels, D., 2008, *Doubt is Their Product: How Industry's Assault on Science Threatens Your Health*. New York: Oxford University Press.

Mohr, T. and Crowley, J., 2001, Once again blindsided! Guest Editorial, *Association of Ground Water Scientists and Engineers Newsletter*, National Ground Water Association, October/November/December 2001, Columbus, OH.

Mohr, T.K.G., 2001, *Solvent Stabilizers White Paper*, Santa Clara Valley Water District, San Jose, CA. http://www.valleywater.org/Water/Water_Quality/Protecting_your_water/_Solvents/_PDFs/SolventStabilizers.pdf

Page, G.W., 1981, Comparison of groundwater and surface water for patterns and levels of contamination by toxic substances. *Environmental Science and Technology* 15(12): 1475–1481.

Phillips, K., 2008, California sues personal care manufacturers regarding 1,4-dioxane labeling. *Chemical Week* 170(19): 31.

Public Employees for Environmental Responsibility (PEER), 2000, Report on Department of Toxic Substances Control, California Public Employees for Environmental Responsibility, July 2000. http://www.peer.org (accessed November 12, 2004).

Reitz, R. H., McCroskey, P. S., Park, C. N., Andersen, M. E., and Gargas, M. L., 1990, Development of a physiologically based pharmacokinetic model for risk assessment with 1,4-dioxane. *Toxicology and Applied Pharmacology* 105: 37–54.

Renner, R., 2009, Environmental News: California launches nation's first Green Chemistry program. *Environmental Science and Technology* 43(1): 5.

Royal Society of Chemistry, 2008, Cells have an appetite for micro-doughnuts. *Science Daily*, June 24. http://www.sciencedaily.com /releases/2008/06/080624120237.htm (accessed June 25, 2008).

Saunders, R.A., 1965, *The Composition of Technical Grade Methyl Chloroform*. Washington DC: Physical Chemistry Branch, U.S. Navy Research Laboratory.

Shoji, R., Sakoda, A., Sakai, Y., Utsumi, H., and Suzuki, M., 2000, A new assay for evaluating hepatotoxicity and cytotoxicity using LDL-uptake activity of liver cells. *Journal of Health Science* 46(6): 493–502.

Stavelova, M., Brenner, V., and Hajslova, J., 2008, Monitoring of brominated flame retardants (BFRs) in the Czech Republic Environment. Emerging Contaminants Poster Session, Battelle's 6th International Symposium on the Remediation of Chlorinated and Recalcitrant Compounds, May 24–27, Monterey, CA.

Stickney, J.A., Sager, S.L., Clarkson, J.R., Smith, L.A., Locey, B.J., Bock, M.J., Hartung, R., and Olp, S.F., 2003, An updated evaluation of the carcinogenic potential of 1,4-dioxane. *Regulatory Toxicology and Pharmacology* 38(2): 183–195.

Strickland, K., 2008, REACH: A timely overview. *Chemical Engineering* 2008(3): 42–47.

Sweeney, L.M. and Gargas, M.L., 2006, Physiologically-based pharmacokinetic (PBPK) modeling of 1,4-dioxane in rats, mice and humans: Final report. Report to ARCADIS on behalf of the Dioxane Risk Management Consortium.

Sweeney, L.M., Thrall, K.D., Poet, T.S., Corley, R.A., Weber, T.J., Locey, B.J., Clarkson, J., Sager, S., and Gargas, M.L. 2008, Physiologically based pharmacokinetic modeling of 1,4-dioxane in rats, mice, and humans. *Toxicological Sciences* 101(1): 32–50.

Tarrer, A.R., Donahue, B.A., Dhamavaram, S., and Joshi, S.B., 1989, *Reclamation and Reprocessing of Spent Solvent*. Park Ridge, NJ: Noyes Data Corporation.

Taverne, D., 2008, Suppressing science. *Nature* 453: 857–858.

Tellez, R.T., Michaud, P., Reyes, C., Blount, B.C., Van Landingham, C.B., Crump, K.S., et al., 2005. Long-term environmental exposure to perchlorate through drinking water and thyroid function during pregnancy and the neonatal period. *Thyroid* 15(9): 963–975.

U.S. Environmental Protection Agency (USEPA), 1981, Phase I risk assessment of 1,4-dioxane. U.S. Environmental Protection Agency, EPA Contract No. 68-01-6030.

U.S. Environmental Protection Agency (USEPA), 1984, Protecting the nation's groundwater from contamination: Volume II, U.S. Congress, Office of Technology Assessment, OTA-O-276, October 1984, Washington, DC.

U.S. Environmental Protection Agency (USEPA), 1996, Land disposal restrictions phase III; Final rule and partial withdrawal and amendment of final rule: Wastewater standard for 1,4-dioxane. *Federal Register* 61(68): 15565–15660.

U.S. Environmental Protection Agency (USEPA), 1999, Use of monitored natural attenuation at superfund, RCRA corrective action, and underground storage tank sites. Office of Solid Waste and Emergency Response Directive No. 9200.4-17P, April 21. United States Environmental Protection Agency, Washington, DC.

U.S. Environmental Protection Agency (USEPA), 2004, Integrated Risk Information System (IRIS): Announcement of 2004 program. *Federal Register* 69(26): 5971–5976. wais.access.gpo.gov (accessed October 5, 2006).

U.S. Environmental Protection Agency (USEPA), 2006, ORD's computational toxicology research program implementation plan (FY 2006–2008). Office of Research and Development, April 2006.

U.S. Environmental Protection Agency (USEPA), 2007, U.S. Environmental Protection Agency National Center for Computational Toxicology website, http://www.epa.gov/ncct/toxcast/ (accessed December 26, 2008).

U.S. Environmental Protection Agency (USEPA), 2008, Ground water & drinking water web page, http://www.epa.gov/safewater/ (accessed December 22, 2008).

Washington Post, 1979, A suspect chemical is found in cosmetics. *Washington Post*, July 20, p. A16.

Appendix 1

Synopses of Selected Patents from the Art of Solvent Stabilization

1,925,602, 1933; Pitman, for E.I. Du Pont de Nemours & Co., Inc., USA

Notes that TCE, PCE, and DCM form acids upon decomposition that corrode metal storage tanks. Gasoline has been used to stabilize TCE—but large volumes are required. Notes that other stabilizers have boiling point differences causing fractionation upon distillation. Seeks a stabilizer that boils with TCE and is not malodorous. Claims that triethylamine meets these criteria, at 0.0001–0.1 vol%. If TCE is not exposed to light, only 0.0005% triethylamine is needed, and more if storage includes exposure to light. Presents data for storage stability of PCE, chloroform, and TCE. No discussion of solvent breakdown in vapor degreasing.

1,971,318, 1934; Smith and De Pree, for Dow Chemical Co., USA

Stabilizes CCl_4 and PCE with small amounts of natural resins, such as gum mastic, sandarac, and rosin. Adds 0.006–0.13% to protect solvents during storage when in contact with iron or copper.

2,108,390, 1938; Price, for Westvaco Chlorine Products Corp., USA

Stabilizes TCE with thiocyanate esters at 0.001–1.0% (preferably methyl thiocyanate). Notes that the stabilizer itself must be stable, and stand up to repeated phase transitions from liquid to vapor and vapor to liquid.

2,111,253, 1938; Stoesser and Alquist, for Dow Chemical Co., USA

Observes that all chlorinated aliphatic hydrocarbons decompose in the presence of moisture, leading to corrosion of metal surfaces. Proposes the addition of a piperazine compound such as tetramethyl piperazine or piperzine hydrate to stabilize chlorinated aliphatic hydrocarbons (PCE in particular) against breakdown from exposure to light and against formation of acid.

2,155,723, 1939; Levine and Cass, for E.I. Du Pont de Nemours & Co., Inc., USA

Notes that TCE decomposes during recovery (distillation, condensation) if not restabilized at this point in the solvent life cycle. "In order to combat this tendency to decompose during storage, use, and recovery, it has been usual to dissolve in the TCE an agent [*p-tert*-butyl phenol] which will stabilize the solvent under the conditions at which these operations are normally carried out."

2,371,644, 1945; Petering and Aitchison, for Westvaco Chlorine Products Corp., USA

Emphasizes the addition of oxygen-containing organic compounds such as organic oxides, oximes, ethers, and alcohols. The more alcohol or other oxygen compound present, the more effective the composition in restraining the "metal-induced" decomposition of solvents. Recommends using 1 mol oxygen compound per 99 mol chlorohydrocarbon. Emphasizes that alcohols boil with the solvent. A minor amount of oxygen-containing compound results in a marked decrease in metal-induced decomposition. Claims that *n*-butyl alcohol and isoamyl alcohol stabilize TCE.

2,435,312, 1948; Klabunde, for E.I. Du Pont de Nemours & Co., Inc., USA

Stabilizes TCE with di-isobutylene to inhibit the condensation decomposition reaction catalyzed by aluminum or iron. Notes that when exposed to air, light, or heat, TCE degrades by oxidation to form dichloracetyl chloride, HCl, and phosgene. At higher temperatures, TCE undergoes a condensation-type decomposition to form HCl together with a resinous material. Describes di-isobutylene as an unsaturated organic compound boiling at about 102°C. Uses di-isobutylene to stabilize TCE at 0.1–1.0%. Recommends addition of an antioxidant.

2,517,894, 1950; Larchar, for E.I. Du Pont de Nemours & Co., Inc., USA

Proposes adding cyclohexene at 0.02–0.25% to TCE to stabilize against reactions with metals. Notes that cyclohexene boils with TCE, allowing recovery on distillation. Finds that cylohexene protects against solvent breakdown in the presence of alkali metals 17 times more effectively than cyclohexane. Also finds that cyclohexene in TCE is 3 times more effective than aniline.

2,721,883, 1955; Stevens, for Columbia-Southern Chemical Corp., USA

The decomposition of perchloroethylene is more pronounced when in contact with iron or copper. Products of PCE decomposition materially injure its beneficial effect, causing metal corrosion. Notes that perchloroethylene shows poor light stability; fluorescent and ultraviolet light cause PCE to decompose, producing a measurable decrease in pH. Compounds previously used to stabilize PCE, such as benzaldehyde, cyclohexane, toluene, benzyl chloride, amylene hexylresorcinols, butyl mercaptan, ethyl acetate, guanidine, and pyridine, have been found to be unsatisfactory because they are ineffective or their odors or color are objectionable. Proposes *N*-methylmorpholine as a stabilizer to protect PCE against UV-light attack.

2,751,421, 1956; Stauffer, for E.I. Du Pont de Nemours & Co., Inc., USA

Claims that *tert*-butyl-4-hydroxyanisole (BHA, the food preservative) stabilizes TCE in storage, which otherwise oxidizes and decomposes resulting in the formation of acids. This decomposition is accelerated by heat, light, moisture, and contact with metals. The use of amines is disadvantageous because they are dissipated by reaction with the acids that the solvent may encounter during normal use. Amines have a further disadvantage: the production of amine salts accelerates the corrosion of metals.

2,775,624, 1956; Skeeters and Baldridge, for Diamond Alkali Co., USA

Claims that acetylenic alcohols, particularly the methyl butynol and methyl pentynols (e.g., 2-methyl-3-butyne-2-ol, 3-methyl-1-pentyne-3-ol, and propargyl alcohol), stabilize PCE against oxidation. Provides boiling ranges for several acetylenic alcohols that fall within or slightly below PCE's boiling range of 119–122°C. Discusses methods for purifying PCE to remove trace lower aliphatics that lead to decomposition. Asserts that acetylenic alcohols perform better than all other stabilizers for metal degreasing and wood dehydration using PCE.

2,795,623, 1957; Starks, for F.W. E.I. Du Pont de Nemours & Co., Inc., USA

Presents a stabilizer for a stabilizer: *N*-methyl pyrrole. Pyrroles added to TCE and PCE at 0.001–1 wt% are somewhat unstable and may decompose when exposed to light, heat, or oxygen. TCE containing *N*-methyl pyrrole will show discoloration on exposure to bright sunlight in as little as 2–4 h. The solvent in which *N*-methyl pyrrole has become unstable will itself destabilize and deposit black tar. This invention uses organometallic chelate compounds to stabilize *N*-methyl pyrrole added at 0.001 wt% and claims copper butylacetoacetate, copper salicylate, aluminum methylsalicylate, and aluminum ethylacetoacetate.

2,797,250, 1957; Copelin, for E.I. Du Pont de Nemours & Co., Inc., USA

Proposes the use of an organic oxide, preferably epichlorohydrin, in combination with a basic amine, such as triethylamine, to act as antacids for TCE and PCE. Basic amines are commonly used to impart alkaline conditions to chlorinated hydrocarbons. HCl formation is a problem because it catalyzes the rusting of iron or steel workpieces cleaned in the machine or in the degreasing apparatus itself. Even small amounts of HCl will catalyze the rusting of iron or steel workpieces put through the degreasing machine. If HCl formation occurs, or if metallic salts—particularly aluminum chloride—are formed, the operator must shut down the degreasing production line to neutralize the solvent.

2,811,252, 1957; Bachtel, for Dow Chemical Company, USA

Bachtel presents the first formulation for stabilizing methyl chloroform against reaction with metals using 1,4-dioxane at a preferred ratio of 4% by volume. Adding 1,4-dioxane provides stability in the presence of aluminum, even when boiling, a result that enables the use of methyl chloroform for vapor degreasing products containing aluminum and iron for the first time. Bachtel notes that over 2 months, the mixture of methyl chloroform with 1,4-dioxane stored in black iron drums will show a rusty discoloration; however, the discoloration can be prevented by including *sec*-butyl alcohol, *tert*-amyl alcohol, or 2-octanol as a stabilizer (any of these alcohols at 0.1–0.5% by volume). The patent presents experimental data showing that the 1,4-dioxane is carried over to the vapor phase, such that methyl chloroform vapors remain stable in the presence of aluminum.

2,838,458, 1958; Bachtel, for Dow Chemical Company, USA

In this patent, Bachtel improves upon his first formulation for stabilizing methyl chloroform against reaction with metals using 1,4-dioxane. He notes that 1,4-dioxane is not entirely effective at preventing corrosion of aluminum or iron exposed to vapors of methyl chloroform entrained with liquid at boiling temperatures, which is a disadvantage for vapor degreasing. Bachtel solves this problem by including a lower aliphatic monohydric acetylenic alcohol, preferably 2-methyl-3-butyn-2-ol (also called dimethyl ethynyl carbinol). The ratios tested include 1,4-dioxane at 2.5–4% by volume with 2-methyl-3-butyn-2-ol at 0.1–0.3% by volume. The effectiveness of 2-methyl-3-butyn-2-ol is not compared to the performance of the pairing of 1,4-dioxane with *sec*-butyl alcohol claimed in his first patent. Bachtel's 1958 patent seems to be an alternate embodiment of his 1957 patent, possibly to retain the rights for all combinations of effective stabilizer mixtures that include 1,4-dioxane.

2,818,446, 1957; Starks, for E.I. Du Pont de Nemours & Co., Inc., USA

Notes that many other stabilizers increase the solubility of HCl. Invention claims synergistic combination of an epoxide and an ester to stabilize TCE. The preferred combination is 0.2 wt% of ethyl acetate and 0.3 wt% of epichlorohydrin. Claims this combination provides a stabilizing effect from 7- to 20-fold greater than that of either component used singly. Notes that it may still be necessary to add a stabilizing amount of a phenolic antioxidant.

2,947,792, 1960; Skeeters, for Diamond Alkali Co., USA

Lists numerous stabilizers of PCE and suggests they have light stabilizing properties. This invention claims isoeugenol (4-hydroxy-3-methoxy-1-propenylbenzene) as a light stabilizer of PCE, and recommends it should be used in combination with methyl pentynol.

3,049,571, 1962; Brown, for Dow Chemical Co., USA

Notes that methyl chloroform (TCA) will react with zinc in the boiling vapor when stabilized with 1,4-dioxane, nitromethane, and *sec*-butyl alcohol. It is the stabilizers themselves that catalyze the reaction with zinc. This is a problem for reclaiming spent solvent from cold degreasing using TCA, because much solvent recycling equipment has galvanized iron containing zinc. Zinc-lined stills commonly used in solvent recycling would be damaged. Zinc causes degradation of the solvent vapor and evolves acid, which attacks and liberates more zinc. This patent claims the addition of vicinal monoepoxides to methyl chloroform (aliphatic epoxides containing an ethylene oxide group). 1,2-Butylene oxide is the example tested.

3,133,885, 1964; Petering and Callahan, for Detrex Chemical Industries, Inc., USA

Provides stabilizers, primarily epoxides, for amine-stabilized PCE and TCE, noting that these formulations will react with iron and copper in the absence of water. Notes that most stabilizers have high boiling points, but the wide difference in boiling points between the solvent and the stabilizers is undesirable because of the difficulty of stabilizing the vapor phase. This patent teachers that distillation of solvent may result in complete loss of stabilizer if boiling point differences are large.

990,302 (Great Britain), 1965; Tect, Inc. (USA)

Notes that the utility of methyl chloroform (TCA) as a cold cleaning agent was recognized as a preferable replacement to the more toxic carbon tetrachloride (CT) since 1950, but it was not until 1954 that addition of stabilizers enabled the use of TCA for cold cleaning without the formation of a tarry mass or releasing phosgene gas and HCl. This 1965 patent notes that until recently TCA was not considered stable for hot solvent vapor degreasing. Invention was successful in stabilizing TCA for degreasing for up to 7200 h. This patent claims a stabilizer package including 1,4-dioxane, methyl pentynol, propargyl alcohol, and *sec*-butyl alcohol.

3,326,988, 1967; Stack, for Pittsburg Plate Glass Company, USA

Alcohol-containing compositions of methyl chloroform (TCA) can be better stabilized by adding perchloroethylene to form an azeotrope. Addition of PCE at 0.5–5% weight increases the rate of TCA evaporation. Alcohol-stabilized TCA will leave spots on the work if PCE is not added. Most common TCA formulations contain 1–2% alcohol (saturated and unsaturated monohydric aliphatic alcohols, e.g., *sec*-butyl alcohol).

3,878,256, 1975; Richtzenhain and Rudolf, for Dynamit Nobel AG, Germany

Tests different proportions of a four-component system using the MIL-T-7003 metal strip method with aluminum, zinc, and iron. Finds that the best stabilizing mixture was methoxyacetonitrile, 0.5%; 1,2-epoxybutane, 0.8%; 1,4-dioxane, 2%; and nitromethane, 1%. This mixture caused no corrosion of metal strips in a 168 h test.

4,032,584, 1977; Irani, for Stauffer Chemical Co., USA

Notes that dichloromethane is the least toxic of the chloromethanes, is more stable than PCE, TCE, and MC, and has a lower boiling point. Dichloromethane is also resistant to photochemical activity and therefore does not produce smog. Dichloromethane can react with aromatic and aliphatic compounds in the presence of metals, metal halides, and combinations thereof, including aluminum, zinc, and iron, and their halides. The reaction product is generally an objectionable, high-boiling tarry substance that renders the dichloromethane unsuitable for further use. Compounds that react with methylene chloride are generally introduced from various cutting oils and lubricants used in metal fabricating operations. Stabilizers are added to dichloromethane to prevent degradation and other types of deterioration such as oxidation, hydrolysis, and pyrolysis. Stabilizing amounts of mixed amylenes, propylene oxide, butylene oxide, and tertiary butylamine are added.

Appendix 2
Fate and Transport Properties of Solvent-Stabilizer Compounds

Stabilizer Compound	Chemical Abstract Service Registry Number	Boiling Point (°C @ 760 mm Hg)	Water Solubility (mg/L @ 25°C)	Solubility Measurement Temperature (°C)	log K_{ow}	log K_{oc}	Henry's Law Constant [(atm m³)/mol]	Vapor Pressure (mm Hg)	Vapor Temperature (°C)	Vapor Density (@ 25°C) Relative to Air (unitless)
1,2-Butylene oxide	106-88-7	63.3 [1]	95,000	25 [12]	0.9 [22]	0.9 [12]	0.00018 [32]	180	25 [32]	2.2 [32]
1,3,5-Trioxane	110-88-3	114.5 [2]	212,000	25 [3]	−0.47 [23]	−0.43 [31]	— [23]	10	20 [23]	—
1,3-Dioxolane	646-06-0	78 [1]	Miscible	25 [1]	−0.37 [24]	1.17 [32]	0.000024 [23]	70	20 [6]	2.6 [51]
1,4-Dioxane	123-91-1	101.1 [3]	Miscible	25 [13]	−0.27 [24]	1.23 [22]	4.8×10^{-6} [34]	38.09	25 [38]	3.03 [6]
2,6-di-tert-Butyl-p-cresol (BHT)	128-37-0	265 [4]	0.4	20 [14]	5.1 [25]	—	0.00059 [25]	—	—	7.6 [6]
2-Butanone	78-93-3	79.6 [4]	353,000	10 [14]	0.29 [24]	1.5 [23]	0.000047 [35]	91	25 [45]	2.41 [14]
2-Methyl-3-butyn-2-ol	115-19-5	103.5 [5]	Miscible	20 [5]	0.318 [5]	1.55 [5]	9.9×10^{-7} [5]	15	20 [5]	2.49 [52]
Acetonitrile	75-05-8	81.6 [4]	Miscible	25 [15]	−0.34 [24]	1.2 [32]	0.000035 [36]	88.8	25 [46]	1.42 [46]
Butoxymethyl oxirane	2426-08-6	164 [6]	20,000	20 [16]	0.63 [24]	0.24 [33]	0.000027 [33]	3.2	25 [33]	3.78 [16]
Cyclohexane	110-82-7	80.7 [4]	0	25 [4]	3.44 [24]	2.2 [32]	0.15 [37]	97	25 [47]	2.98 [53]
Diisobutylene	107-40-4	104.9 [1]	0	25 [1]	4 [25]	2.44 [32]	0.88 [32]	35.9	25 [32]	3.8 [32]
Diisopropylamine	108-18-9	84 [7]	11,000	25 [10]	1.4 [24]	2.15 [32]	0.000096 [38]	79.4	25 [38]	3.5 [51]
Epichlorohydrin	106-89-8	117.9 [4]	65,900	25 [17]	0.45 [27]	1.6 [32]	0.00003 [32]	16.4	25 [32]	3.29 [6]
Ethanol	108-01-0	135 [4]	Miscible	25 [4]	−0.31 [24]	—	—	—	—	3.03 [6]
Ethyl acetate	141-78-6	77 [4]	64,000	25 [18]	0.73 [29]	1.77 [29]	0.00013 [29]	79.4	25 [38]	3.04 [4]
Nitroethane	79-24-3	114 [4]	45,000	20 [17]	0.18 [24]	1.48 [32]	0.000048 [32]	20.80	25 [32]	2.58 [6]
Nitromethane	75-52-5	101.2 [4]	11,100	25 [15]	−0.35 [24]	0.28 [32]	0.000026 [32]	35.8	25 [38]	2.11 [6]
N-Methylmorpholine	109-02-4	113 [8]	Miscible	25 [8]	−0.35 [26]	0.791 [26]	2.5×10^{-7} [26]	18.7	25 [26]	3.5 [8]
N-Methyl pyrrole	96-54-8	112 [9]	0	25 [9]	0.75 [24]	2.162 [26]	0.00019 [26]	19.5	25 [26]	2.8 [9]
Phenol	108-95-2	181.75 [10]	82,800	25 [19]	1.46 [24]	1.2 [32]	3.3×10^{-7} [36]	0.35	25 [32]	3.24 [54]
1,2-Propylene oxide	75-56-9	34.23 [4]	590,000	25 [12]	0.03 [24]	1.4 [32]	0.000070 [32]	538	25 [48]	2 [16]
p-tert-Amyl phenol	80-46-6	262.5 [4]	168	25 [17]	4.03 [28]	2.41 [22]	2.0×10^{-6} [39]	0.008	25 [38]	—
Pyridine	110-86-1	115.2 [4]	Miscible	20 [10]	0.65 [24]	1.7 [22]	0.00001 [40]	20.8	25 [32]	0.982 [55]
Resorcinol	108-46-3	280 [4]	71,700	25 [17]	0.8 [24]	1.81 [32]	1.1×10^{-11} [32]	0.00049	25 [49]	3.79 [16]
sec-Butyl alcohol	78-92-2	99.5 [11]	181,000	25 [10]	0.61 [24]	1.7 [32]	9.1×10^{-6} [41]	18.3	25 [38]	2.6 [51]
tert-Amyl alcohol	75-85-4	102.4 [11]	110,000	25 [20]	0.89 [24]	0.46 [32]	0.00073 [42]	16.8	25 [38]	3.0 [56]
tert-Butyl alcohol	75-65-0	82.41 [4]	Miscible	25 [15]	0.35 [24]	1.57 [32]	9.1×10^{-6} [43]	40.7	25 [38]	2.55 [32]
tert-Butyl phenol	128-37-0	265 [4]	0.6	25 [21]	6.2 [30]	2.77 [30]	0.0030 [21]	0.18825	25 [30]	7.6 [6]
Thymol	89-83-8	233 [6]	1000	25 [4]	3.3 [24]	—	—	0.095	40 [32]	—
Triethylamine	121-44-8	89.3 [1]	15,000	20 [14]	1.45 [24]	2.16 [29]	0.00015 [44]	57.1	24 [50]	3.49 [16]

Notes: References are listed by number in brackets and italics to the right of each value; see the list of references on the pages following. Dashes indicate that reliable values were not found.

REFERENCES

1. Lide, D.R. (Ed.), 2000, *CRC Handbook of Chemistry and Physics*. Boca Raton, FL: CRC Press LLC.
2. Windolz, M. (Ed.), 1983, *The Merck Index—An Encyclopedia of Chemicals, Drugs and Biologicals*. Rahway, NJ: Merck and Co., Inc.
3. Budavari, S. (Ed.), 1989, *The Merck Index—An Encyclopedia of Chemicals, Drugs and Biologicals*. Rahway, NJ: Merck and Co., Inc.
4. Budavari, S. (Ed.), 1996, *The Merck Index—An Encyclopedia of Chemicals, Drugs, and Biologicals*. Rahway, NJ: Merck and Co., Inc.
5. UNEP (United Nations Environment Programme), 1998, Screening information data set (SIDS) for 2-methyl-3-yn-2-ol. Nairobi, Kenya: United Nations Environment Programme Report 115-19-5. http://www.chem.unep.ch/irptc/sids/OECDSIDS/115195.pdf (accessed June 3, 2007).
6. Sax, N.I. and Lewis, R.J.S., 1987, *Hawley's Condensed Chemical Dictionary*. New York: Van Nostrand Reinhold Company.
7. Lide, D.R. and Kehiaian, H.V., 1994, *CRC Handbook of Thermophysical and Thermochemical Data*. Boca Raton, FL: CRC Press.
8. Huntsman, 2005, Technical bulletin: N-Methylmorpholine (NMM) CAS 109-02-4. The Woodlands, TX: Huntsman Corporation.
9. Fisher Scientific, 1998, Material Safety Data Sheet for *n*-methyl pyrrole for 96-54-8, Fair Lawn, NJ. https://fscimage.fishersci.com/msds/43291.htm (accessed on February 4, 2007).
10. Kirk, R. and Othmer, D., 1982, *Kirk-Othmer Encyclopedia of Chemical Technology*. New York: Wiley.
11. Snyder, R. (Ed.), 1992, *Ethel Browning's Toxicity and Metabolism of Industrial Solvents*, 2nd Edition, Volume 3: Alcohols and Esters. New York: Elsevier.
12. Bogyo, D.A., Lande, S.S., Meylan, W.M., Howard, P.H., and Santodonato, J., 1980, Investigation of selected potential environmental contaminants: Epoxides—final technical report. Prepared by Syracuse Research Corporation for the U.S. Environmental Protection Agency Office of Toxic Substances.
13. Hawley, G.G., 1977, *The Condensed Chemical Dictionary*. New York: Van Nostrand Reinhold Company.
14. Verschueren, K., 1996, *Handbook of Environmental Data on Organic Chemicals*, 3rd Edition. New York: Van Nostrand Reinhold Company.
15. Riddick, J.A., Bunger, W.B., and Sakano, T.K., 1985, *Techniques of Chemistry*, 4th Edition. Volume II. Organic Solvents. New York: Wiley.
16. Patty, F., 1963, *Industrial Hygiene and Toxicology: Volume II. Toxicology*. New York: Wiley Interscience Publishers.
17. Yalkowsky, S.H. and Dannenfelser, R.M., 1992, *The AQUASOL Database of Aqueous Solubility*. Tucson: University of Arizona, College of Pharmacy.
18. Wasik, S.P., Tewari, Y.B., Miller, M.M., and Martire, D.E., 1981, *Octanol/Water Partition Coefficients and Aqueous Solubilities of Organic Compounds*. Washington, DC: U.S. Department of Commerce, National Bureau of Standards. NBS TR81-2406.
19. Southworth, G.R. and Keller, J.L., 1986, Hydrophobic sorption of polar organics by low organic carbon soils. *Journal of Water, Air, and Soil Pollution* 28(3–4): 239–247.
20. Barton, A.F.M., 1984, Alcohols with water. In: A.S. Kertes (Ed.), *Solubility Data Series*, Volume 15. Oxford: Pergamon.
21. Inui, H., Akutsu, S., Itoh, K., Matsuo, M., and Miyamoto, J., 1979, Studies on biodegradation of 2,6-di-*tert*-butyl-4-methylphenol (BHT) in the environment. Part IV: The fate of ^{14}C-phenyl-BHT in aquatic model ecosystems. *Chemosphere* 8(6): 393–404.
22. Lyman, W.J., Reehl, W.F., and Rosenblatt, D.H., 1990, *Handbook of Chemical Property Estimation Methods: Environmental Behavior of Organic Compounds*. New York: McGraw-Hill.
23. U.S. Environmental Protection Agency (USEPA), 2000, High production volume challenge program submission for 1,3,5-trioxane by Trioxane Manufacturer's Consortium. Prepared by Toxicology and Regulatory Affairs, Flemington, NJ. http://www.epa.gov/hpv/pubs/summaries/triox/c12863.pdf (accessed January 4, 2009).
24. Hansch, C. and Leo, A.J., 1985, Medchem Project issue no 26. Claremont, CA, Pomona College, Medchem Project.
25. UNEP (United Nations Environment Programme), 2002, Screening information data set (SIDS) for 2,6-di-*tert*-butyl-*p*-cresol [betahydroxy toluene or BHT]: Initial assessment report for SIDS initial assessment meeting (SIAM), 14th meeting, Paris, France, March 26–28, 2002.

26. U.S. Environmental Protection Agency (USEPA), 2007, Estimations programs interface for Windows (EPI Suite), v. 3.20. http://www.epa.gov/oppt/exposure/pubs/episuite.htm (accessed November 3, 2007).

27. Deneer, J.W., Sinnige, T.L., Seinen, W., and Hermens, J.L.M., 1988, A quantitative structure-activity relationship for the acute toxicity of some epoxy compounds to the guppy. *Aquatic Toxicology* 13(3): 195–204.

28. Schultz, T.W., 1986, The use of the ionization constant (pK_a) in selecting models of toxicity in phenols. *Ecotoxicology and Environmental Safety* 14(2): 178–183.

29. Swann, R.L., Laskowski, D.A., McCall, P.J., and Vander Kuy, K., 1983, A rapid method for the estimation of the environmental parameters octanol/water partition coefficient, soil sorption constant, water to air ratio and water solubility. *Residue Reviews* 85: 17–28.

30. NICNAS (National Industrial Chemicals Notification and Assessment Scheme), 2003, Full report: 2,6-di-*t*-Butyl-4-methyl phenol. Canberra, Commonwealth of Australia: N. I. C. N. A. S. (NICNAS), National Occupational Health and Safety Commission.

31. U.S. Environmental Protection Agency (USEPA), 2007, High Production Volume Information System (HPVIS) online database. http://www.epa.gov/chemrtk/hpvis/index.html (accessed November 3, 2007).

32. NLM (National Library of Medicine), 2006, Hazardous substance data bank: MEDLARS Online Information Retrieval System. National Library of Medicine. http://toxnet.nlm.nih.gov/ (accessed April 12, 2006).

33. ERSTG (Epoxy Resin Systems Task Group), 2002, *High Production Volume (HPV) Challenge Program Test Plan and Robust Summaries for* n-*Butyl Glycidyl Ether*. U.S. Environmental Protection Agency, AR201-1400A.

34. Park, J.H., Hussam, A., Couasnon, P., Fritz, D., and Carr, P.W., 1987, Experimental reexamination of selected partition coefficients from Rohrschneider's data set. *Analytical Chemistry* 59(15): 1970–1976.

35. Bhattacharya, S.K., Qu, M., and Madura, R.L., 1996, Effects of nitrobenzene and zinc on acetate utilizing methanogens. *Water Research* 30(12): 3099–3105.

36. Gaffney, J.S., Streit, G.E., Spall, W.D., and Hall, J.H., 1987, Beyond acid rain. Do soluble oxidants and organic toxins interact with SO_2 and NO_x to increase ecosystem effects? *Environmental Science and Technology* 21(6): 519–524.

37. Bocek, K., 1976, Relationships among activity coefficients, partition coefficients and solubilities. *Experientia Supplementum* 23: 231–239.

38. Daubert, T.E. and Danner, R.P., 1985, *Data Compilation Tables of Properties of Pure Compounds*. New York: American Institute of Chemical Engineers.

39. Meylan, W.M. and Howard, P.H., 1991, Bond contribution method for estimating Henry's law constants. *Environmental Toxicology and Chemistry* 10(10): 1283–1293.

40. Hawthorne, S.B., Slevers, R.E., and Barkley, R.M., 1985, Organic emissions from shale oil wastewaters and their implications for air quality. *Environmental Science and Technology* 19(10): 992–997.

41. Snider, J.R. and Dawson, G.A., 1985, Tropospheric light alcohols, carbonyls, and acetonitrile: Concentrations in the southwestern United States and Henry's law data. *Journal of Geophysical Research* 90(D2): 3797–3805.

42. Butler, J.A.V., Ramchandani, C.N., and Thompson, D.W., 1935, The solubility of non-electrolytes: Part I. The free energy of hydration of some aliphatic alcohols. *Journal of the Chemical Society, Faraday Transactions* 1935: 280–285.

43. Altschuh, J., Berggemann, R., Santl, H., Eichinger, G., and Piringer, O.G., 1999, Henry's law constants for a diverse set of organic chemicals: Experimental determination and comparison of estimation methods. *Chemosphere* 39(11): 1871–1886.

44. Christie, A.O. and Crisp, D.J., 1967, Activity coefficients of the *n*-primary, secondary and tertiary aliphatic amines in aqueous solution. *Journal of Applied Chemistry* 17: 11–14.

45. Alarie, Y., Nielsen, G.D., Andonianhaftvan, J., and Abraham, M.H., 1995, Physicochemical properties of nonreactive volatile organic chemicals to estimate RD_{50}: Alternatives to animal studies. *Toxicology and Applied Pharmacology* 134(1): 92–99.

46. Clayton, G.D. and Clayton, F.E., 1994, *Patty's Industrial Hygiene and Toxicology*. New York: Wiley.

47. Chao, J., Lin, C.T., and Chung, T.H., 1983, Vapor pressure of coal chemicals. *Journal of Physical and Chemical Reference Data* 12(4): 1033–1063.

48. Boublik, T., Fried, V., and Hala, E., 1984, *The Vapor Pressures of Pure Substances: Selected Values of the Temperature Dependence of the Vapor Pressures of Some Pure Substances in the Normal and Low Pressure Region*. New York: Elsevier Science Publishing Company.

49. Yaws, C.L., 1994, *Handbook of Vapor Pressure: Volume 2. C$_5$ to C$_7$ Compounds*. Houston, TX: Gulf Publishing Company.

50. U.S. Coast Guard, 1984, *CHRIS–Hazardous Chemical Data*. Volume II. Washington, DC: Department of Transportation, U.S. Government Printing Office.

51. NFPA (National Fire Protection Association), 1991, *National Fire Protection Guide: Fire Protection Guide to Hazardous Materials*. Quincy, MA: National Fire Protection Association.

52. BASF, 2006, Safety Data Sheet for 2-methylbut-3-yn-2-ol, Florham Park, NJ: BASF.

53. Mackison, F.W., Stricoff, R.S., and Partridge, L.J., Jr. (Eds), 1981, *NIOSH/OSHA—Occupational Health Guidelines for Chemical Hazards*. Washington, DC: National Institute of Occupational Safety and Health (NIOSH).

54. Bingham, E., Cohrssen, B., and Powell, C.H., 2001, *Patty's Toxicology*. New York: Wiley.

55. Browning, E., 1965, *Toxicity and Metabolism of Industrial Solvents*. New York: American Elsevier.

56. Mallinckrodt Baker, 2008, Material Safety Data Sheet for *tert*-amyl alcohol—75-85-4. Phillipsburg, NJ: Mallinckrodt Baker.

Appendix 3

Compilation of Instrumental
Parameters in Literature
Studies of 1,4-Dioxane Analyses
of Various Media

Matrix	MDL	Preparation	Standard and Carrier Gas	Solvent	Drying	GC
Water	1 ppb	SPE coconut shell charcoal: SKC 226-26 Tekmar LSC-2 P&T	1,4-Difluorobenzene	Methanol + CS_2 5% v/v	3.5 M Na_2SO_4	CG-FID: HP5580
Water	GC-FID: 2.5 ppb GC-MS: 0.25 ppb	SPME on CAR-PDMS with headspace extraction	1,4-Dioxane-d_8, isopropanol, methyl chloroform Helium	None	Desorption at 310°C	GC-FID: Varian 3400 MS: Varian Saturn 2200 ion trap
Water	LLE: 0.2 ppb SPE: 6 ppb P&T: 25 ppb	LLE separatory funnel compared with J&W SPE using AC-2 on SDB and with P&T using Tekmar 3000 P&T	1,4-Dioxane-d_8, 2-bromo-1-chloropropane Helium	DCM, MTBE, NaCl, acetone for AC-2 cartridge	Centrifuge, air	GC: Agilent 6890 MS: 5973N
Water	SPE with LLE: 0.13 ppb LOQ = 0.31 ppb	Sorption on carbon disks	1,4-Dioxane-d_8 THF-d_8	Acetone	Laboratory airflow for 1 h	GC: HP 5890 Series II MS = Finnigan Mat TSQ 700
Water	SPE: 50 ppb	LLE in hexane-DCM solvent followed by SPE using C_{18} cartridges	1,4-Dioxane-d_8 Helium	Hexane: DCM 80:20 v/v	Eluted from cartridge with acetonitrile	GC: HP 5890 Series II MS: HP 5989
Water	P&T: 1 ppb	20-minute purge and 20-minute desorb Tekmar 3100 with Vocarb 4000 Trap	1,4-Dioxane-d_8 Helium	None	Air dried for 2–3 min under 2.7 kPa; centrifuged at 3000 rpm (1700 g) for 10 min	GC: Varian 3800 MS: Varian Saturn 2000 GC-MS/MS system
Water	SPE: 0.02 ppb	SPE on carbon felt, 1500 m^2/g	1,4-Dioxane-d_8 Helium	Acetone, DCM N_2 gas blowdown	Air dried for 2 min, then centrifuged at 3000 rpm for 10 min	Shimadzu GC-MS QP5050A

Injection Temperature (°C)	Column	Column Material	Final Column Temperature (°C)	Gas Flow, Pressure at Head	Monitored Ions m/z	References
50	J&W DB-5: 6 ft × 1/4 in.	Supelco: Glass 1% SP-1000 on Carbopack B	50 → 100	30 mL/min	88, 58 for 1,4-dioxane 114 for 1,4-difluorobenzene	Epstein et al. (1987)
45	Supelco PDMS: 30 m × 0.32 mm × 4 μm	Bonded polydimethyl siloxane (PDMS)	45 → 230	GC: He at 13 psi and 40 cm/s	88–98 57–67	Shirey and Linton (2006)
200	Ultra-2: 50 m × 0.2 mm id; 0.33 μm film thickness	Ultra-2: 5% diphenyl, 95% methylpolysiloxane stationary phase	40 → 200	GC: He at 0.5 mL/min	88 58, 57, 87 for 1,4-dioxane 96, 64 for 1,4-dioxane-d_8 77, 79 for 2-bromo-1-chloropropane	Park et al. (2005)
150	Supelco SPB-1: 30 m × 0.32 mm × 4 μm	Supelco SPB-1: Nonpolar methylsilicone (bonded polydimethyl-siloxane)	40 → 120	77 kPa	45 $[C_2H_4O + H]$ + for dioxane 49 $[C_2D_4O + H]$ + for dioxane-d_8 62 $[C_4D_7]$ + for THF-d_8	Isaacson et al. (2006)
130	HP-5: 25 m × 0.2 mm × 0.33 μm	Agilent HP-5: Nonpolar (5% phenyl) methylpolysiloxane	40 → 100	GC: He at 1 mL/min	88 for 1,4-dioxane 96 for 1,4-dioxane-d_8	Song and Zhang (1997)
125	Varian VRX: 60 m × 0.32 mm × 1.8 μm	J&W DB-WAX: Fused silica with polyethylene glycol (PEG)	35 → 220	GC: He at 1 mL/min	40–100	Kawata et al. (2001)
170	J&W DB-WAX: 30 m × 0.25 mm × 0.5 μm	J&W DB-WAX: Fused silica with polyethylene glycol (PEG)	35 → 190	GC: He at 1 mL/min	1,4-Dioxane: 88, 58	Song and Zhang (1997)

continued

Matrix	MDL	Preparation	Standard and Carrier Gas	Solvent	Drying	GC
Water	SPDE: 0.03 ppb	SPDE: CTC-CombiPAL autosampler by Chromtech (Idstein, Germany)	1,4-Dioxane-d_8 (99 + %)	Acetone, dichloromethane	Cartridges air dried with aspirator and centrifuged at 3000 rpm (1700 g) for 10 min	GC: TraceGC 2000 MS: TraceDSQ ThermoFinnigan
Water	P&T: 20 ppb	Heated purge and trap with sodium sulfate salt; 2 μL of 250 μg/mL 1,4-dioxane-d_8 solution added to the syringe	1,4-Dioxane-d_8 Helium	Purge at 85°C	Na_2SO_4	GC-MS Hewlett-Packard 5970 and Tekmar LSC-1000 in scan mode
Water	SPE: 5 ppb	SPE: SDB + AC	1,4-Dioxane-d_8 Helium	Acetone N_2 gas blowdown	Cartridge air dried	GC-MS: QP2010 Shimadzu MS in SIM
Synthetic leachates		Solvent extraction	1,4-Dioxane-d_8 Helium	DCM	Dry Na_2SO_4 centrifuge	JEOL Automass #50
Landfill leachates		SPE on Sep-Pak PS-2 and AC-2	1,4-Dioxane-d_8 Helium	Methanol and methyl acetate	Airflow, dry Na_2SO_4	GC: Agilent 6890A MS: Agilent 5973
Aspirin, acetominophen	1.5 ppm	Static headspace	Benzene, trichloroethylene, chloroform, dichloromethane, 1,4-dioxane Hydrogen	Water	None— aqueous headspace	GC-FID: 9001, Finnigan

Injection Temperature (°C)	Column	Column Material	Final Column Temperature (°C)	Gas Flow, Pressure at Head	Monitored Ions *m/z*	References
200	Restek Corp.: 60 m × 0.32 mm × 0.5 μm	Restek Stabilwax: Fused-silica capillary column	40 → 180	GC: He at 1.5 mL/ min	71, 55 for acrylamide 87, 44 for *N,N*-dimethylacetamide 73, 44 for *N,N*-dimethylformamide 88, 58 for 1,4-dioxane 96, 95 for furfural 98, 81 for furfuryl alcohol 102, 56 for *N*-nitrosodiethylamine 74, 42 for *N*-nitrosodimethylamine	Kawata et al. (2001)
160	ID DB-624: 75 m × 0.53 in. with a 3 μm film thickness	J&W DB-624: Medium-polarity column; equivalent to VOCOL	40 → 160at 20°C/min	GC: He at 120 mL/ min	88, 58 for 1,4-dioxane 96, 64 for 1,4-dioxane-d_8	Rudinsky et al. (1997)
200	Rtx-1701: 30 m × 0.25 mm × 1 μm	14% cyano-propylphenyl and 86% dimethyl-polysiloxane	40 → 250	GC: He at 45 cm/s	88, 58 for 1,4-dioxane 96, 64 for 1,4-dioxane-d_8	Jochmann et al. (2006)
200	DB-1: 30 m × 0.53 mm id	DB-1: 100% nonpolar dimethyl polysiloxane	30 → 100 → 200	7 psi	88, 58 for 1,4-dioxane 96, 64 for 1,4-dioxane-d_8	Yasuhara et al. (1999)
250	FFAP: 30 m × 0.25 mm id	FFAP: Nitro-terephthalic acid modified polyethylene glycol (PEG)	35 → 250	1.8 mL/ min 97.4 kPa	88, 58 for 1,4-dioxane 96, 64 for 1,4-dioxane-d_8	Yasuhara et al. (2003)
250	Restek Rtx-1701: 30 m × 0.53 mm × 1.0 μm Restek Rtx–Volatiles: 30 m × 0.53 mm × 2.0 μm	G43: 6% cyanopropyl-phenyl and 94% dimethyl polysiloxane	35 → 170	GC: He at 80 cm/s	GC-FID configuration: Precolumn and two megabore columns to analyze on three detectors	Rankin (1996)

continued

Matrix	MDL	Preparation	Standard and Carrier Gas	Solvent	Drying	GC
Tween 20 Tween 80	50 ppb 100 ppb	Thermal desorption	Nitrogen	None—thermal desorption	None—thermal desorption	GC-FID SP-6800
Cosmetics and nonionic surfactants	300 ppb 60 ppb by GC-MS	SPME: 75 µm Carboxenpoly (dimethylsiloxane) (PDMS–CAR), 85 µm poly(acrylate) (PA), and 100 µm poly (dimethylsiloxane) (PDMS) fiber assembly (Supelco)	1,4-Dioxane, THF Nitrogen	Methanol, acetonitrile, or acetone	Salting out with 5 mL of 20% NaCl	GC-FID: HP 6890 and Varian Saturn 3800 GC) coupled with a Varian 2000 ion-trap MS

Notes: Temp. = temperature;   AC2 = activated   carbon; C_{18} = 18%   activated   carbon   loading;   CAR-PDMS = 100 µm   carboxen extraction; LOQ = Limit of Quantitation; MDL = Method Detection Limit; MTBE = methyl *tert*-butyl ether; NaCl = sodium SPE = solid-phase extraction; SPME = solid-phase microextraction; THF = tetrahydrofuran.

Injection Temperature (°C)	Column	Column Material	Final Column Temperature (°C)	Gas Flow, Pressure at Head	Monitored Ions *m/z*	References
155	60 cm SS column 2 mm id packed with 3% DEGS (polydiethylene glycol succinate)	3% DEGS and 150–200 mesh SS-101 white supporter	105	0.32 MPa 42 mL/min	Not provided	Guo and Brodowsky (2000)
To GC: 40 To MS: 240	HP-1 (Agilent Technologies) column: 30 m × 0.25 mm id	HP-1 column; nonpolar, 100% dimethyl polysiloxane	160	GC: N$_2$ at 1 L/min FID: N$_2$ at 45 L/min; H$_2$ at 45 mL/min Air at 450 mL/min	88 for 1,4-dioxane 71 for THF	Fuh et al. (2005)

polydimethylsiloxane-coated wire adsorbent; DCM = dichloromethane; dry Na$_2$SO$_4$: anhydrous sodium sulfate; LLE = liquid–liquid chloride solution; P&T = purge and trap; PEG = polyethylene glycol; SDB = styrene divinyl benzene; SIM = selective ion monitoring;

Index

Note: n denotes footnote.